FOSSILS OF THE CARPATHIAN REGION

LIFE OF THE PAST

James O. Farlow, editor

FOSSILS OF THE CARPATHIAN REGION

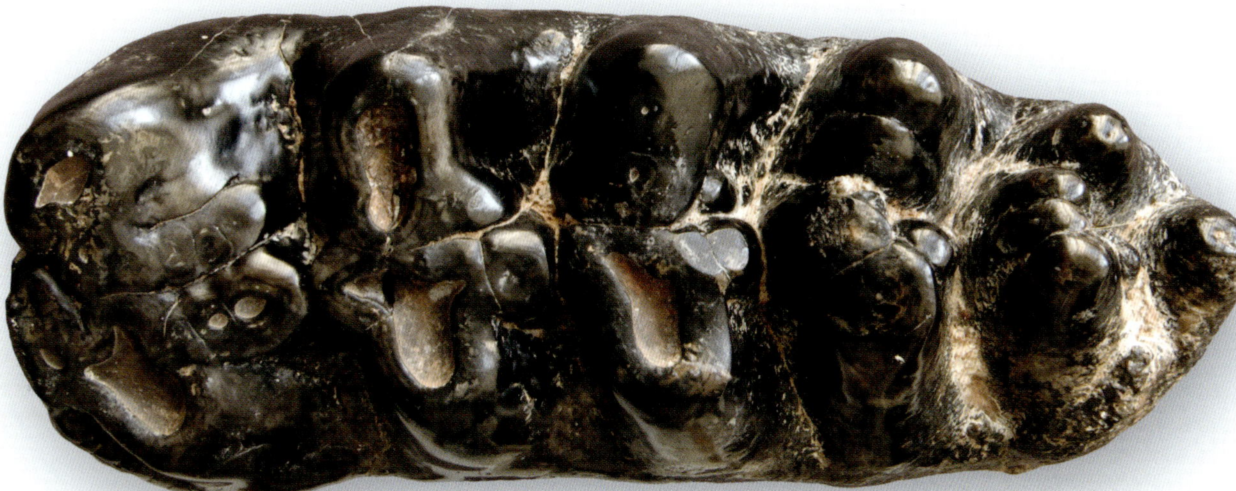

ISTVÁN FŐZY AND ISTVÁN SZENTE
GARETH DYKE, English Text Editor

Indiana University Press
Bloomington and Indianapolis

This book is a publication of

Indiana University Press
Office of Scholarly Publishing
Herman B Wells Library 350
1320 E. 10th Street
Bloomington, IN 47405-3907 USA

iupress.indiana.edu

Telephone orders 800-842-6796
Fax orders 812-855-7931

♾ The paper used in this publication meets the minimum requirements of the American National Standard for Information Sciences—Permanence of Paper for Printed Library Materials, ANSI Z39.48-1992.

Manufactured in South Korea

Library of Congress Cataloging-in-Publication Data

Főzy, István.
[A Kárpát-medence ősmaradványai. English]
Fossils of the Carpathian region / István Főzy and István Szente ; English text editor, Gareth Dyke.
pages cm. — (Life of the past)
Includes bibliographical references and index.
ISBN 978-0-253-00982-1 (cloth : alkaline paper) — ISBN 978-0-253-00987-6 (ebook) 1. Fossils—Hungary. 2. Fossils—Carpathian Mountains Region. 3. Paleontology—Hungary. 4. Paleontology—Carpathian Mountains Region. 5. Geology, Stratigraphic. 6. Geology—Carpathian Mountains Region. 7. Geology—Hungary. I. Szente, István, 1960– II. Dyke, Gareth. III. Title.
QE755.H9F6913 2013
560.9439'9—dc23
2013008540

1 2 3 4 5 19 18 17 16 15 14

TO

Miksa Hantken (1821–1893),
pioneer of micropaleontology

AND

Baron Ferenc Nopcsa (1877–1933),
dinosaur hunter of Transylvania

CONTENTS

DETAILED OUTLINE OF THE TEXT (CHAPTER NUMBERS AS INDICATED)

MAPS

All maps were prepared by László Zentai, cartographer, Eötvös Loránd University, Budapest.

Jul: Kováts Flor. fossil. ad Erdöbénye. Tab. III.

Rohn. lith. Druck. b. Lencsó in Ofen. 1856

1–7 Castanea Kubinyii Kov:

PREFACE

Every single flower awaits a mention
Every handful of dust is remarkable
—Sándor Weöres, "There Is Still So Much to . . ."

Every day we stare in wonder at the gentle vibrations of nature, with perhaps a sentiment similar to the one above in mind. Then we just hurry on. Not so biologists! They seek out, name, and list even the tiniest living creatures and so far have "discovered" about 1.5 million species in the name of science. But this is merely the tip of the iceberg: According to some estimates, the number of still-unknown species may be between 5 and 100 million, and these estimates do not take into account the numbers of individuals within these species. After all, the population of mammals named *Homo sapiens* (one species!) numbers in the billions. Who would now undertake the task of counting "every single flower"?

And so if we know so little about the living world, what can be said about the plants and animals of the past—about the creatures that lived millions, or billions, of years ago? Some researchers think that just 1 percent of the species that have ever lived are alive today, and something on the order of 99 percent are extinct. Nonetheless, the number of extinct (fossil) species so far described can be estimated to be a few hundred thousand at most. We cannot even begin to estimate how many species could have lived in the geological past, or how diverse life once was. Indeed, as the Apostle Paul's simple words teach us, our knowledge is fragmentary. Numberless tiny, microscopic fossils may lie hidden in a single barrowful of sedimentary rock, while in other places spectacular, large fossils wait to be found. Who would now take on the task of mentioning "every remarkable handful of dust"?

This book, therefore, is not a complete inventory of the known fossils from the Carpathian Region. Nor does it provide a complete geological history—although the history of fossils is, of course, tightly interwoven with geological events. The book is not systematic in its treatment of fossils, either, but most of the noteworthy groups are touched upon. Fossils old and young are discussed as part of the long list of ancient plants and animals from the Carpathian Basin: minute species and giants, vertebrates and invertebrates. Some are mentioned for their huge abundance, and others because their discovery was an event. The most famous of all the fossils from this region, the spectacular Mesozoic dinosaurs—*Magyarosarus dacus* and its contemporaries—are treated in detail.

Fossil sites are also included in this book, and although it is not a field guide, we hope that nature lovers and fossil hunters will find concrete and useful information herein. Collecting fossils is a noble passion: let us go out and search for them! As a result of studying fossils we will learn more about nature, the hundreds of millions years that has passed for the living world, and—indeed—about ourselves.

Finally, we give an old greeting common to miners and geologists, and wish every reader good luck!

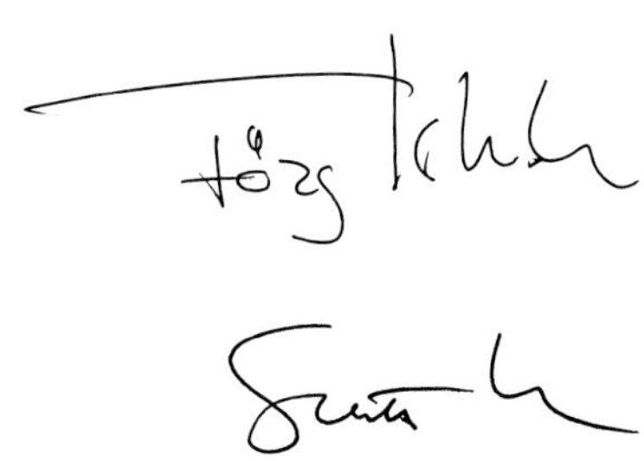

Miocene fossil plants illustrated in an article by Gyula Kováts (1815–1873), "Erdőbényei ásatag virány" (The fossil flora of Erdőbénye), which appeared in 1856 in the first issue of the first Hungarian geological journal. This article, which also included seven lithographic plates of similar beauty, is one of the earliest paleontological publications in Hungarian.

ACKNOWLEDGMENTS

The authors express their sincere thanks to Professors András Galácz and László Kordos, who reviewed the Hungarian version of this manuscript and made useful suggestions that allowed us to improve it. The present English edition has benefited significantly from the comments of academician Barnabás Géczy, who read the first Hungarian version enthusiastically and carefully and provided useful comments. Special thanks are dedicated to Gareth Dyke, the English text editor of this volume, whose work was thorough and who has done much more than could normally be expected. His contribution was essential and is highly appreciated.

The translation of the Hungarian version of this book into English was made possible by financial support generously provided by the following enterprises and organizations: initially Szalai Group SL (Mallorca, Spain), then later TXM, Oil and Gas Exploration Ltd. (Hungary), Geomega Ltd. (Hungary) and the Hantken Miksa Foundation (Hungary).

Most of the fossils figured in this book belong to the collections of the Hungarian Natural History Museum (Budapest) and the Geological Museum of Hungary (Budapest); some are housed at the Eötvös Museum of Natural History (Budapest). We thank the leaders of these institutions for making the collections in their care available to us. Dezső Illés and Bálint Péterdi kindly helped us to explore the rich collections of the Geological Museum of Hungary.

We thank in particular András Szunyoghy for making a number of unique pencil drawings used in this book, and Mariann Bosnakoff and László Zentai for compiling the index and providing locality maps. Eszter Hankó and Tímea Szlepák contributed significantly to the completion of the bibliographic database used in this book.

We thank our colleagues and friends, without whose contributions and support this book would never have been completed. They are, in alphabetical order:

Bada Gábor (Budapest, Hungary); Báldiné Beke Mária (Üröm, Hungary); Barabás András (Pécs, Hungary); Bertalan Tamás (Bonyhád, Hungary); Bókáné Barbacka Mária (Budapest, Hungary); Budai Tamás (Budapest, Hungary); Coralia Maria Jioanu (Deva, Romania); Császár Géza (Budapest, Hungary); Csiki Zoltán (Bucharest, Romania); Czier Zoltán (Oradea, Romania); Czirják Gábor (Budapest, Hungary); Dan Grigorescu (Bucharest, Romania); David B. Weishampel (Baltimore, United States); David Mayhew† (Leiden, Netherlands); Detre Csaba (Budapest, Hungary); Dunai Mihály (Budapest, Hungary); Erdei Boglárka (Budapest, Hungary); Erika Posmoşanu (Oradea, Romania); Fitos Attila (Budapest, Hungary); Fűköh Levente (Gyöngyös, Hungary); Gerhard Mandl (Vienna, Austria); Görög Ágnes (Budapest, Hungary); Hably Lilla (Budapest, Hungary); Hála József, (Budapest, Hungary); Harald Lobitzer (Bad Ischl, Austria); Hír János (Pásztó, Hungary); Jan Sand (Bratislava, Slovakia); Jan Schlögl (Bratislava, Slovakia); John Callomon† (London, United Kingdom); Josef Piller (Stetten, Austria); Kazár Emese (Budapest, Hungary); Kázmér Miklós (Budapest, Hungary); Kocsis László (Lausanne, Switzerland); Kőhler Artúr (Budapest, Hungary); Kovács Sándor† (Budapest, Hungary); Kretzoi Miklós† (Budapest, Hungary); Krolopp Endre† (Budapest, Hungary); Laczkóné Őri Gabriella (Budapest, Hungary); Lantos Zoltán (Budapest, Hungary); Magyar Imre (Budapest, Hungary); Magyari Enikő (Budapest, Hungary); Matskási István (Budapest, Hungary); Mészáros Lukács (Szentendre, Hungary); Mihai Popa (Bucharest, Romania); Miloš Siblík (Prague, Czech Republic); Monostori Miklós (Budapest, Hungary); Ősi Attila (Budapest, Hungary); Ozsvárt Péter (Budapest, Hungary); Pálfy József (Budakeszi, Hungary); Palotás Klára (Budapest, Hungary); Papp Gábor (Budapest, Hungary); Pazonyi Piroska (Budapest, Hungary); Pecsics Tibor (Budapest, Hungary), Pelikán Pál (Budapest, Hungary); Péró Csaba (Budapest, Hungary); Péterdi Bálint (Budapest, Hungary); Piros Olga (Budapest, Hungary); Pongrácz László (Balatonfüred, Hungary); Raucsikné Varga Andrea (Szeged, Hungary); Rodika Ciobanu (Sibiu, Romania); Selmeczi Ildikó (Budapest, Hungary); Stamm Attila (Budapest, Hungary); Szabó János (Budapest, Hungary); Szalai Erika (Mallorca, Spain); T. Bíró Katalin (Budapest, Hungary); Tóth Tamás (Budapest, Hungary); Venczel Márton (Oradea, Romania); Vlad Codrea (Cluj Napoca, Romania); Vörös Attila (Budapest, Hungary); Vörös István (Budapest, Hungary); Vremir Mátyás (Cluj Napoca, Romania); Wanek Ferenc (Cluj Napoca, Romania); Weiszburg Tamás (Budapest, Hungary).

The authors are deeply indebted to Tamás Bertalan, Mihály Dunai, Zoltán Evanics, Zoltán Orbán, and László Sővér—all adept fossil collectors and good friends—for permitting the publication of photos of exceptionally attractive specimens in their collections and the publishers of the *Proceedings of the National Academy of Sciences* (Washington, DC) for permitting reproduction of the figure of *Balaur bondoc*.

Most photos published in this book were taken by the authors, but some were kindly provided by our friends whose help is gratefully acknowledged. People whose photos are used in this book are listed here in alphabetical order, alongside the page number(s) on which the photo(s) in question can be seen (abbreviations: l, left; r, right; a, above; b, below):

Barabás András (8), Bókáné Barbacka Mária (108), John Callomon (96-a), Csiki Zoltán (213, 214-la), Czirják Gábor (196), Donato Di Bari and Roberto Rettori (43-ra), Dulai Alfréd (315-rb, 320-ra), Dunai Mihály (15-a, 18/8, 27, 89, 137-b), Fűköh Levente (424), Gál Lehel (40), Galácz András (95-b), Géczy Barnabás (293-l), Góczán Ferenc and Sieglné Farkas Ágnes (155), Görög Ágnes (99-a,b), Gulyás Péter (141, 148-b), Bernhard Hubmann (110F), Jurcsák Tibor (81, 189, 190), Kákay-Szabó Orsolya (395-r), Kenjiro Sho (385), Kocsis László (256/6, 7), Köhler Artúr (77), Kovács Sándor (49-b), Lantos Zoltán (341-b, 346), Lelkes György (42), Magyari Enikő (376, 379, 383-b), Maria Marino (31, 95-a), David Mayhew (394), Mészáros Lukács (357-b), Monostori Miklós (241-r), Monostori Miklósné (241-l), Müller Pál (317), Nagy Ágoston (138-b), Oraveczné Scheffer Anna (63), Orbán Zoltán (134-b), Mike Orchard (49-a), Ősi Attila (194, 198-a), Ozsvárt Péter (44, 46, 100, 234, 235), Pálfy József (115-a), Pecsics Tibor (53-b), Josef Piller (296), Pintérné Mészáros Ildikó (365), Piros Olga (43-l), PNAS (142, 221), Rákosi László (154), Rédly Szilvia (79, 150-a), Jan Sand (98), Stamm Attila (188-r), Szabó János (87-r), Szinger Balázs (298), Szoláth György (267-a), Szunyoghy András (80, 82, 206-ra, 209-rb, 211-a, 212-a, 359-r, 398-r, 408), Telcs Gábor (412), Uhrin András (20), Venczel Márton (374-b), Vörös Attila (35-a, 37-r), Vremir Mátyás (220-r, l), Kamil Zágoršek (229).

Last but not least we express our sincere thanks to our editors at Indiana University Press, namely Nancy L. Lightfoot, Dawn Ollila, and Robert J. Sloan, for their continuous encouragement, patience, and kindness.

INTRODUCTION: ROCKS, FOSSILS, EVENTS, AGES

The Earth is more than four billion years old. Its history is documented by the rocks that form the Earth's crust, which lies beneath our feet and can be structurally complex in some places. The time that has elapsed since the formation of our planet is infinitely long when compared to the age of the human lineage, and several methods make it possible for us to measure geological time. The study of fossils—the remains of animals and plants preserved in sedimentary rocks—allows us to recognize the order of events that have formed the Earth.

Indeed, for almost a century, the study of fossils was the only way to determine relative geological ages. This method, known as biostratigraphy (stratigraphy is the study of the temporal and spatial relationships of rock bodies), is founded on the irreversible nature of biotic evolution. First, the fossil contents of isolated localities were studied and later, following much debate and many mistakes, it became possible to arrange fossil occurrences according to their geological ages. By about the mid-nineteenth century the relative temporal distribution of most of the important plant and animal groups had been more or less established. This knowledge resulted in a comparative scale that—because it is continuously being developed—has become more and more applicable around the world.

The subdivision of Earth history is also based primarily on the sequence of animal evolution—especially of marine invertebrates, because these are the most often fossilized. Thus, the first (and longest) portion of time in Earth history was named Azoic (i.e., time without animal life) in earlier geological works because the earliest fossil evidence for the several-billion-year history of life had not yet been discovered. The second part of Earth history is called the Phanerozoic Eon (i.e., time of full animal life) and includes rock sequences up until the present day. The dawn of the Phanerozoic was about 542 million years ago, when marine invertebrate animals with hard skeletons first appear in the fossil record. The Phanerozoic is subdivided into shorter eras: the Paleozoic (i.e., time of ancient animal life, from about 542 to 251 million years ago); the Mesozoic (time of middle animal life), which ended about 65 million years ago; and the Cenozoic (time of new animal life). This latest period continues today. Due to developments in dating methods the ages of these time-slices have varied slightly, but their boundaries are most often found at times of crisis or at major mass extinctions characterized by marked changes in fossil assemblages.

The Mihalovits Quarry at Nagyvisnyó (Bükk Mountains, Hungary). These exposed Permian carbonates were deposited in a shallow marine environment about 250 million years ago. The subdivision of the lithologically monotonous succession and the determination of the geological age of the rocks can be done on the basis of its fossil content.

It is also possible to subdivide Earth history using the evolution of plants. The boundaries between eras (Paleophitic, Mesophitic, and Cenophitic) that this evidence leads to, however, do not correspond with the big turnovers in the animal world. For example, the boundary between the Mesophitic and Cenophitic—defined by the replacement of angiosperms with gymnosperms as dominant elements of plant assemblages—is placed in the middle of the Cretaceous period, long before the end of the Mesozoic.

Eras are further subdivided into geological periods, and within periods ages are distinguished. The latter are then further divided into epochs. Periods, ages, and epochs are geochronological categories referring to intervals of time and so correspond in chronostratigraphy to systems, series, and stages. The latter categories form successive parts of rock successions. For example, the Hettangian epoch of the Jurassic (from about 200 to 197 million years ago) is represented in the Bakony Mountains in Hungary by a 150-meter-thick limestone succession of the Hettangian stage. Within geochronological categories early, middle, and late subdivisions can be distinguished, and these correspond to the lower, middle, and upper parts of rocks succession. For example, in the peculiar outcrop at Kálvária Hill in Tata (Hungary) the lowermost limestone beds that are traditionally referred to as the "Gerecse red marble" represent the upper Hettangian, in other words they were deposited during the late Hettangian.

Epochs are subdivided into zones, subzones, and horizons, each defined by characteristic fossil assemblages that represent shorter and shorter spans of time. The area in which they can be recognized, however, becomes increasingly restricted. Each time span, including most recent ones, have characteristic animals and plants whose remains can be fitted into our continuously developing chronostratigraphical frameworks. The precision with which ages can be determined depends on the systematic position of the fossil(s) in question, as each group has its own evolutionary tempo and these can be very different. Another factor that determines precision is the degree of refinement of the biostratigraphic scale being applied. Some groups—famous examples include ammonites and continental mammals—evolved rapidly and therefore their remains make possible detailed biostratigraphic subdivisions of rock successions.

The discovery of radioactivity, as well as large-scale developments in the nuclear industry that took place in the mid-twentieth century, provided the first opportunity to express geological ages using precise dates. Radiometric—often incorrectly called absolute—age determination is based on knowledge of the half-lives of radioactive isotopes and on assumptions about their original ratios. By measuring mass ratios of the end products of decay and the original isotope the date of rock formation can be determined. This method can also be used when fossils are entirely absent—for example, in volcanic rocks—but its applicability is limited to rocks that contain minerals that can be measured (radioactive ones). Another difficulty with this approach is caused by the often very durable nature of the mineral grains that can be dated—they are eroded from coeval original rocks and then redeposited two or more times, resulting in a presumed geological age for the final rock in which they are embedded that can be much older than their actual age. These days, calibrating refined biostratigraphic scales from radiometric age data is a hot research topic.

Events in Earth history can also be revealed by the environments of rock formation. Former environments, and changes within them, are often well reflected by distinctive features of rocks. These features, including color, lithology, mineral composition, and fossil content, are referred to as "facies"—meaning "outer form," from the Latin "visage" or "face." Some rock facies are found throughout Earth history, whereas others are restricted to certain age intervals. Black shale, for example, is known from almost the whole Phanerozoic, whereas nummulite limestone is characteristic of the Eocene.

More than 75 percent of the rocks that are found on the surface of the Earth are of sedimentary origin. The percentage coverage of these rocks in the Carpathian Basin is even larger. The fossils found in almost all sedimentary rocks vary considerably in size, chemical composition, and systematic position. Some fossils can be extracted from their host rocks only by using sophisticated methods and are merely of scientific value; others are attractive and sometimes have a high market value. This book aims to—without any intention of being exhaustive—present a review of all the types of fossils that have so far been described from the Carpathian Region.

The book icon 📖 that appears at the end of each section refers the reader to relevant works that provide source material and further reading; a full list of references would be far beyond the scope of this book. Complete bibliographic information can be found at the end of the book.

Classification of Fossils

Fossil remains vary in several ways. Some have been left behind by "functions" of an animal or plant—such as feeding, moving, resting, and eating—and are preserved as parts, or structures, in sediment that later solidified. These kinds of fossils are called "trace fossils." The other main kind of fossils, the much larger group in terms of numbers of examples, are the remains of parts of former plants and animals referred to as "body fossils." These fossils are distinguished by size—either micro- or macrofossils; study of the former requires a microscope. Fossils are also classified on the basis of other properties including their constituents and relative abundance in the rock record.

FOSSILS OF THE CARPATHIAN REGION

The Paleozoic fossil *Waagenophyllum indicum* (Waagen and Wentzel), a colonial coral from Upper Permian rocks at Nagyvisnyó. The fossil is about 40 cm high. Corals are abundant in the fossil record from the Ordovician onward and mass occurrences of their skeletons frequently make sedimentary formations of considerable economic importance as reservoir rocks. The genus *Waagenophyllum*, characteristic of the Permian, belongs to Rugosa, a large and diverse order of Paleozoic corals that became extinct at the end of the period.

PART ONE

THE PALEOZOIC

Articulated columns of a sea lily (*Poteriocrinus* sp.) from Carboniferous shale exposed in railway cutting no. 2 at Nagyvisnyó in northern Hungary (close to original size). The Paleozoic was the golden age for sea lilies such as these, and they inhabited oceans around the world in huge numbers. Some of the known species had stems reaching 20 meters in length, formed by several thousand disklike or cylindrical columnals. Cups and branches articulated to the end of these stems. These lilies were extremely characteristic elements of the marine biota in the Paleozoic.

1

The Paleozoic

The Precambrian, which makes up about 85 percent of the history of the solid Earth, is represented by very sporadic fossil assemblages in the Carpathian region. A few poorly preserved organic-walled microfossils extracted from "crystalline" metamorphic rocks in a few areas including the Apuseni Mountains of Romania are thought to come from the latest Precambrian. However, due to the scarcity of fossil assemblages of this age, a detailed treatment of the two eons of the Precambrian, the Archaic and Proterozoic, is unjustified given the general scope of this book. Our present knowledge indicates that the living world of the Precambrian was immensely poorer than that of the Paleozoic; there were no organisms possessing hard skeletons at this time, for example, and the most widespread traces of life in the Precambrian are biosedimentary structures called stromatolites. These are mounds of mud and blue-green algae, or cyanobacteria, that have been found on almost all continents, and are particularly characteristic of the Ediacaran (i.e., the period immediately preceding the Phanerozoic).

As noted above, fossils from the Paleozoic are rare in the Carpathian Region. There are, however, a few localities that have yielded attractive fossil assemblages.

Paleozoic sedimentary rocks are much more widespread in the Carpathian region than are those from the Precambrian and, indeed, some of them contain fossil assemblages of scientific value. These fossil floras and faunas, however, are rather isolated in space and time and hardly any can be said to be spectacular or notable. Although many of the known Paleozoic successions were deposited in freshwater environments, sediments yielding exceptionally preserved fossils, such as those from the celebrated Carboniferous Mason Creek biota of North America, are lacking. These fossil assemblages are much less diverse than are those of coeval marine deposits, and many Paleozoic successions in the Carpathian area in general have been metamorphosed by the heat and/or pressure of orogenic processes resulting in the partial or total destruction of fossils. With this in mind, Carpathian Paleozoic assemblages are treated in a single chapter, rather than discussed period by period.

MEMORIALS OF LOST PEOPLES AND FARAWAY COUNTRIES: PALEOZOIC PERIODS

The Paleozoic, also called the "Primary" in older literature, was at least 290 million years long and, as such, was longer than both the Mesozoic and Cenozoic put together. It is subdivided into six periods that can be distinguished in sections all around the world. The earliest of these periods, the Cambrian, was named after the Roman name for North Wales (Cambria). Indeed, the next youngest, the Ordovician and Silurian Periods, are named for tribes that once lived in the area of present-day Wales; the Devonian was named for the county of Devonshire. The name Carboniferous refers to the Latin name for coal (*carbo*) and, as such is, a rare example among geochronological names; finally, the youngest period of the Paleozoic, the Permian, was named after the Perm Province of Russia.

EARTH HISTORY IN A NUTSHELL

Over the course of the almost 300 million years of the Paleozoic, the face of the Earth changed fundamentally. At the beginning of the Cambrian most of the ancient shields forming the so-called core of the present-day continents were concentrated between the 60th latitudes, principally in the Southern Hemisphere. Their arrangement differed markedly from that of today. Some continents (Africa, India, South America, Australia, New Guinea, and Antarctica) formed a huge supercontinent (Gondwanaland) in the Cambrian—the latter three being its northernmost "tongue," lying on the northern hemisphere. North America (Laurentia) and the landmass that would eventually constitute Europe (Baltica) were separated from one another by the Iapetus Ocean. The microcontinents Kazakhstania and Siberia, separated from all other landmasses, were situated near the equator.

Erathem/Era	System/Period	Million Years Ago
Paleozoic	Permian	299
	Carboniferous	359
	Devonian	416
	Silurian	444
	Ordovician	488
	Cambrian	542

Subdivision of the Paleozoic into periods and systems.

The climate of the Cambrian is thought to have been warmer and more balanced than that of the present day and, in contrast to the Precambrian and Ordovician, no traces of glacial sediments have been found. At the very beginning of the Cambrian the biggest event in the history of life is thought to have taken place, the sudden and almost simultaneous appearance of both fossils with hard skeletons and several animal phyla, an event usually described as "the Cambrian explosion." This remains one of the most puzzling enigmas in evolution. In the Cambrian (actually until the end of the Silurian) life was mostly restricted to oceans.

During the Ordovician, the Northern Hemisphere was almost entirely covered with oceans. At the end of this period, considerable areas of the southern continent, including present-day North Africa, became covered with inland ice and glaciers. In the Silurian, huge parts of this region were flooded by ocean; in the tropics evaporite rocks were deposited, while at higher latitudes the ice age persisted. The ice-covered South Pole was situated in present-day Brazil; the Iapetus Ocean became narrower and narrower in the Silurian until, finally, it closed.

Sediments eroding from the folding and uplifting primeval Caledonian chain, whose remains form mountains in the eastern part of North America, on the British Isles, and on the Scandinavian Peninsula, were deposited in the Devonian. The continental succession that resulted from this erosion is widely known as the Old Red Sandstone. A new phase of mountain building began in the Carboniferous and lasted until the end of the Paleozoic and resulted in the formation of the Hercynian or Variscian Chain. The latter has two branches in Europe: the northern one stretches from the southern part of Northern Ireland to the Sudetes Mountains in Poland; the southern branch is traceable across the Iberian Peninsula. Details of the formation of the Hercynian Chain are still unknown, but the approximately 4,000-kilometer-long Ural Mountains, which mark the collision point between the ancient continents of Siberia and Baltica, also belongs to the Variscian Belt. The enormous weight of rock bodies thrust over one another during the Hercynian orogeny even created a flexure of the Earth's crust, which resulted in a series of depressions lying in front this major orogenic belt. These basins provided places for the subsequent deposition of Upper Carboniferous coal measures.

By the mid-Permian, the largest parts of the continental crust assembled into the supercontinent called Pangaea, extending over all climatic belts, and were surrounded by a global ocean called Panthalassa. A huge equatorial ocean divided Pangaea into a northern and a southern part, called Laurasia and Gondwanaland, respectively. The eastward open "embayment" of Panthalassa is named Tethys, after the sister of Oceanus in ancient Greek mythology. This name for the ancient ocean that dominated the surface of our planet for near 200 million years, was coined by influential Viennese geologist and conservative politician Eduard Suess (1831–1914), in his fundamental work *The Face of the Earth* (*Das Anlitz der Erde*). This work, more than 3,000 printed pages, laid long-enduring foundations for thinking about the Earth. In areas in the Permian South Pole glacial deposits were abundant, whereas in Europe red sandstone successions indicate the dominance of a hot and dry climate.

EARLY DEVELOPMENT OF THE LIVING WORLD

In the Paleozoic era there was a tremendous change in the living world. Among the marine invertebrates possessing hard skeletons, whose remains form the majority of the fossil record, three successive assemblages—usually called evolutionary faunas—can be distinguished. The dominant groups within these were especially diverse in an interval of the Paleozoic—if the number of the families is considered—and later on were replaced, gradually or suddenly, by other groups. The Cambrian was, therefore, the time of trilobites. Extinct mollusks known as hyoliths as well as monoplacophorans, inarticulate brachiopods, and primitive echinoderms also comprised the Cambrian evolutionary fauna.

In the Ordovician, articulate brachiopods became dominant in benthic communities and together with massive and lacelike bryozoans, reef-forming stromatoporoideans classified alongside sponges, cephalopods, sea lilies, starfish, and graptolites, they constitute the Paleozoic evolutionary fauna. The leading role of brachiopods persisted until the end of the Permian, when the brachiopods were replaced by different groups of organisms that have remained dominant in modern seas: bivalves, gastropods, vertebrates, arthropods, and bryozoans. The later history of brachiopods, this once abundant group, was quite different—and some managed to survive the decline. The inarticulate brachiopod *Lingula* (small tongue), for example, has persisted for around 500 million years, having changed little, and is today considered a so-called living fossil.

Besides fundamental changes in the composition of marine assemblages, conquest of the continents also occurred in the Paleozoic, at the end of the Silurian and in the Devonian. Spiders and scorpions were the first to inhabit dry lands, and the remains of the first amphibians—the first tetrapod animals—are known from the Upper Devonian. The hot and humid climate of the Carboniferous was especially favorable for the development of life on land, and some plants and insects are known to have reached gigantic sizes. Reptiles became very diverse in the Late Carboniferous, soon after their appearance, and the earliest mammal-like reptiles are known from the Permian.

The marine biota suffered a mass extinction at the end of the Permian and the inhabitants of shallow waters were especially afflicted. About 95 percent of invertebrate families, including some emblematic organisms of the Paleozoic like the fusulinid foraminifers, corals of the orders Rugosa and Tabulata, trilobites, and most of the previously dominant brachiopod groups, disappeared. This extinction was not a rapid event, having lasted several million years; it is interesting that little of the continental biota was affected.

Although all of the groups mentioned above were already in decline in the Permian and their diversity was already strongly reduced, the coincidence of their total disappearance has given scientists plenty to think about and has so far remained an enigma. Today the process of extinction is intensively studied, not just so that we can learn lessons for the future: formation of an anoxic water layer, global cooling, and lethal radiation generated by a supernova explosion that occurred close to Earth have all

Portrait Gallery

Zoltán Schréter (1882–1970)—Eminent Paleozoic fossil researcher who worked in the Hungarian Bükk Mountains

Zoltán Schréter received honors as a student of natural history and geography at the University of Budapest for his hard work and original studies on the former glaciers of the Southern Carpathians. He graduated as a teacher in 1908, received his doctoral degree in geology and paleontology in 1909, and in the same year joined the Hungarian Geological Institute. He worked for this institute until he retired in 1942, after 33 years as vice-director. Schréter was a geologist with a wide range of interests: although his work covered almost the entire area of the former Hungary, most of his studies focused on the geology and paleontology of the mountains of Northeast Hungary. He was elected a corresponding member of the Hungarian Academy of Sciences in 1938, to honor his achievements in the field of raw material exploration. During the period of the academy's management by communist authorities, Schréter, along with many other eminent researchers, was demoted in 1949. Their new status ("consulting member") meant a total loss of their former rights as members, including their pensions (These "consulting members" were reinstated only in 1989, when just 4 of the 122 academicians affected were still alive!).

Schréter was thus forced to return to work at the Geological Institute in 1949, at the age of 67. He carried out geological mapping projects and estimated stocks of Hungarian raw materials until 1958, when he managed to retire again. This marked the beginning of the third period of his productive scientific life, as he was able to entirely devote himself to work with fossils. In addition to a series of papers, he completed a monograph on the Upper Permian brachiopods from the Bükk Mountains (1963); his description of the nautiloids from the same stratigraphic level appeared posthumously (1974).

📖 Schréter 1963, 1974; Balogh 1970

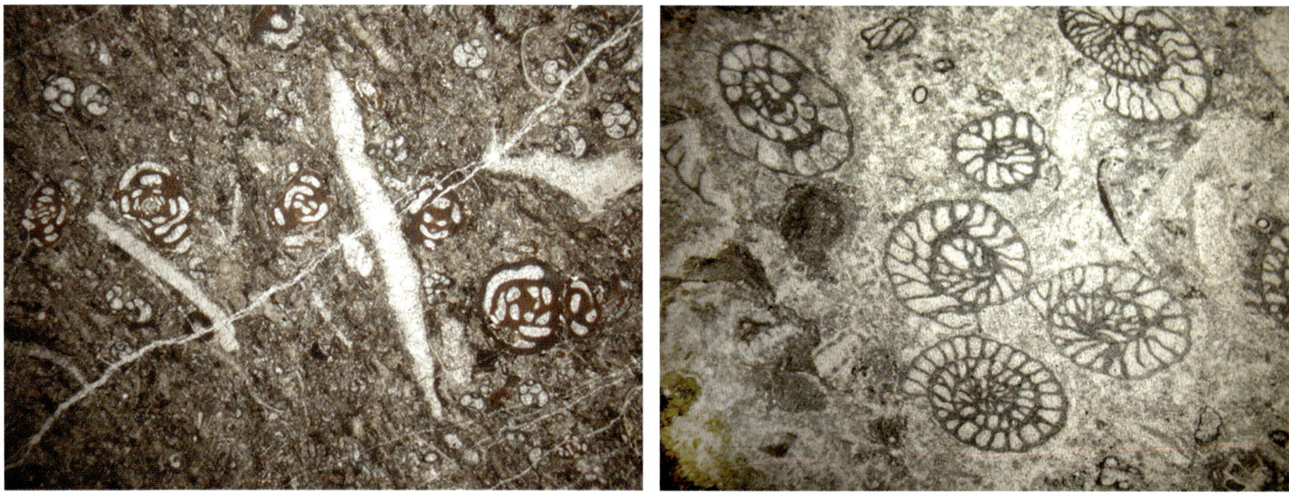

Thin-sections of foraminifera-bearing Paleozoic limestones from the Bükk Mountains. (Left) Specimens of *Glomospirella* in Upper Permian limestones (Szodonka Valley of Lénárddaróc). (Right) An Upper Carboniferous Fusulina limestone (Dédesvár).

been proposed as possible causes for the Permian mass extinction, the most severe catastrophe that has ever befallen life on Earth. Finally, as a consequence of the formation of Pangaea, the area covered with shallow water also decreased significantly at this time and this could have also had a hand in triggering the mass extinction.

CHARACTERISTIC ROCKS

Because the formation of some rock types is confined to certain periods in Earth history, many Paleozoic sediments are characteristic to this era. In general, however, the Carpathian Basin has relatively few rocks of Paleozoic age.

The few successions of this age that are known from this region have restricted areal extent. In Hungary, these rocks are found in the Mecsek Mountains, in the Transdanubian Range and northern Hungary. Successions of considerable extent are also found in Styria in Austria (the Paleozoic of Graz), in the Gemerské (or Slovenské) Rudohorie in Slovakia (the Gemer Paleozoic), and in the Apuseni Mountains and the Banat region of Romania. A review of the Paleozoic geology and paleontology of Hungary, intended to be exhaustive, was published by József Fülöp (1928–1994) the former rector of Eötvös Loránd University in Budapest.

📖 Fülöp 1990, 1994

The Fusulina Limestone

Foraminifera appeared at the beginning of the Cambrian but remained small in size for the next 200 million years. The first larger foraminifera that attained sizes of several centimeters appeared in the Carboniferous and belong to the order Fusulinidae. Fusulinas, like all larger-sized foraminifera, lived in shallow seas in the tropical belt and so their occurrence is indicative of the original peri-equatorial position of their depositional environment. As happened to rocks in the Carpathian Basin, some sedimentary successions have been moved, in some cases several thousands of kilometers, to their present positions by forces of plate tectonics.

The presence of *Fusulina* can even be seen on the weathered surfaces of limestones because these larger forams, similar to other related taxa, usually occur in large, often rock-forming quantities. However, the resistance of different rocks and fossils to weathering can vary; often fusulinids are seen at the surface as rounded or elongate outlines, and they are usually darker than the embedding rock. Since the outer shapes of different species, or even genera, of these forams are often the same, determination requires microscopic study using thin-sections. How these sections are made is explained below.

The order Fusulinidae, and thus Fusulina limestone, is characteristic of the Carboniferous and Permian. The genus *Fusulina* has, however, a much shorter range and is restricted to the Middle and Upper Carboniferous. Fusulina limestones are known to crop out in the Carboniferous of the Bükk Mountains and in Dobšiná, Slovakia, whereas representatives of the genus *Codonofusiella,* also belonging to Fusulinidae, are found in Permian limestones in the Bükk Mountains.

Red Sandstones

Successions predominantly consisting of red, or in some cases gray or green, sandstones, conglomerates, or

A Paleontologist in Action: Let's Make a Thin-Section!

The study of thin-sections is a traditional method in micropaleontology and is based on the translucent nature of very thin (often just tenths of a millimeter in thickness) rock slabs. Rock-forming minerals, as well as minute fossils, can be studied in this way under the microscope. Nowadays machines that make thin-sections are available, but the traditional method for making these, outlined below, is still widely used.

First, a slab—as thin as possible—is cut from the piece of rock to be studied. This is the most dangerous step of the process because, due to the considerable hardness of most rocks, it requires the use of a diamond cutting disk spinning at many revolutions per minute. Usually the rock to be cut is about two to three square centimeters in area and not more than a few millimeters in thickness. One side of the rock slab is smoothed using a thick glass slab covered with wet grinding powder that is harder than the rock, usually limestone. The grinding process itself is simple: the rock slab is gently pressed onto the glass slab and moved in a circle on the mixture of grinding powder and water. Finer and finer powders, usually three or four grits, sprinkled onto different glass slabs are used, resulting in a rock surface of increasing smoothness. Depending on the hardness of the rock, this process may require some minutes and rock slabs must be washed very carefully because contamination of finer powders by larger grains may severely scratch the ground surface. After the surface is completely smooth, the rock slab is then pasted onto a thin glass slide. Traditionally, a strong and translucent natural resin called Canada Balm was used in this stage of the process, but nowadays the use of two-component synthetic resins is expanding. Care is needed to completely remove bubbles from these resins. After solidification of the resin, the rock slab is then thinned again in the same way; frequent examination under the microscope is needed when using the finest powders. Slabs that are too thick are not translucent and are not good under the microscope: micropaleontological thin-sections usually exceed 0.03 mm in thickness which is the exact size for thin-sections of igneous and metamorphic rocks.

Making thin-sections is not difficult work, but it does require experience. Care should be also taken to ensure that the thin-section is of even thickness: If adequate, then fossils can be studied under the microscope. To protect the section, a smaller glass slide is often used as a mount. If properly prepared and well stored, these sections are still usable after decades.

finer-grained rocks that were deposited in continental environments under arid or semiarid climates are widespread throughout the Permian of Europe and are likewise found in the Carpathian Basin. Among them, the most famous is the Rotliegend (the "red underlying succession") that forms the lower unit of the traditionally two-part Permian (equivalent to the "Dyas" of older literature) in Germany. The word "Rotliegend" refers to the stratigraphic position of this succession with respect to the Zechstein succession, which is of economic importance because it contains evaporites and ore deposits such as the Mansfeld Copper Shale. The Val Gardena, or Gröden, Sandstone in the Southern Alps as well as the Verrucano—a rock named for its peculiar weathered surface that resembles a wart (*verruca* in Latin)—also belongs to this group of sedimentary rocks. In Hungary, this particular lithofacies is represented by the Balatonfelvidék Sandstone, uranium-bearing Permian sandstones in the Mecsek Mountains and the Turony Formation in Southern Transdanubia that have been explored using boreholes. A common feature of these successions is that they almost completely lack body fossils, but they are nevertheless worth mentioning because they do contain traces left by amphibians and reptiles. The processes of deposition and diagenesis of these rocks have prevented the preservation of bones and teeth, and so only ichnogenera and ichnospecies have been identified.

In general, most Permian rocks in southern Europe contain ichnofaunas, but in Hungary just a few examples have so far been found. György Majoros, a recognized authority on Permian sedimentology who worked for the former Mecsek Ore Mining Company, was the first to document (in 1964) the occurrence of reptile traces in the Pálköve Quarry at Balatonrendes.

Pentadactyl (five-fingered) traces attributed to the ichnogenus *Korynichnium* were formally described in 1968 by András Kaszap, who was working as an assistant professor at Eötvös Loránd University at the time. He later became the chief geologist on the board of directors for the baths in Budapest. Since this first description, a three-fingered trace has also been found at the same locality and in the 1960s, additional traces were found in cores from borehole Turony-1 drilled during subsurface geological investigations in the Mecsek-Villány region. These violet-brown sandstone beds have yielded two types of traces and some of them were identified by well-known expert on vertebrate traces Hartmut Haubold, now emeritus professor at the University of Halle, Germany. According to Haubold, these traces were produced by amphibians and

Batrachichnus salamandroides (Geinitz 1861) Haubold 1996. Footprint of a small amphibian found 1,220 meters down borehole Turony-1. This species was formerly assigned to the genus *Antichnium* and is considered a good index fossil for the Lower Permian. The specimen was found by Ágnes Barabás-Stuhl, who worked as a palynologist with the Uranium of Pécs (Mecsek Ore Mining Company), and was identified by Hartmut Haubold, now emeritus professor at the University of Halle, Germany, and a recognized expert on fossil footprints.

can be assigned to the ichnospecies *Batrachichnus salamandroides*. From the same borehole, another form was also found and was identified as *Platytherium* by Ágnes Barabásné Stuhl, who was at that time working as a palynologist with the Mecsek Ore Mining Company. Finding tetrapod footprints is not her only talent: she has also resolved fundamental stratigraphic questions relating to the Mecsek Mesozoic.

🕮 Majoros 1964; Kaszap 1968

FOSSILS FROM TRANSDANUBIA

Paleozoic sequences play only a subordinate role in the formation of the Transdanubian Central Range, as their outcrops are usually scattered and are situated far from one another. Indeed, some of these sequences are known only from borehole cores, and the oldest rocks of Lower Paleozoic age in this region were largely metamorphosed during orogenic movements. Some Lower Paleozoic metamorphic rocks are found along the northern shore of Lake Balaton as well as in its northeastern continuation in the Balatonfő area and in the Velence Hills. The western part of the Mecsek Mountains is formed largely from Permian rocks, although various Paleozoic successions have been explored by boreholes drilled in the neighborhood of these mountains. Although nearly all of these successions do contain fossils, most of these organic remains are purely of scientific interest.

THE OLDEST FOSSILS FROM HUNGARY

One of the oldest fossil-bearing successions known in the Carpathian Basin is found at Szár-hegy Hill in Szabadbattyán, near the town of Székesfehérvár. The slate exposed here contains poorly preserved acritarchs. These organic-walled microfossils, reminiscent of armored flagellates of phylum Pyrrhophyta, were first documented in 1985 at this site by Gyöngyi Lelkesné Felvári, a recognized authority on metamorphic rocks who was working jointly with the Italians Roberto Albani and Marco Tongiorgi. Among other forms, they found *Baltisphaeridium* and *Micrhystridium*, which are indicative of Middle Ordovician age, at this site. Thus, this is the oldest confirmed fossil assemblage in Hungary.

Many more fossils are known from the Silurian: the first pioneering paper on fossils of this age was published by János Oravecz (1935–2009), an influential lecturer at Eötvös Loránd University. Black siliceous rocks ("lydite" or Lydian stone) that are embedded in greenish-gray slates found in the vicinities of the villages of Alsóörs and Lovas have yielded microfossils of Silurian age: graptolites dissolved from these rocks with hydrogen fluoride are representatives of the long-ranging and cosmopolitan genus *Monograptus*. Alongside organic-walled fossil sponge spicules, radiolarians and conodonts have also been extracted from these rocks—even though they were previously thought to be insoluble. The later identification of further finds by Ferenc Góczán and Heinz Kozur has contributed to our knowledge of Hungarian Silurian fossils.

🕮 Oravecz 1964; Góczán 1971; Kozur 1984a, 1984b; Albani et al. 1985

DEVONIAN FOSSIL DISCOVERIES FROM RECENT DECADES

Devonian fossiliferous rocks in the Carpathian Basin are almost exclusively found below the surface. For example, a borehole drilled near the village Kékkút revealed a red nodular limestone sequence that was otherwise unknown and invisible from the surface. This type of rock is called *griotte* after its type locality in France (this French word means "cherry colored"). Small remains of mollusks belonging to the extinct group Hyolitha have also been discovered in thin-sections from borehole cores. Hyolithes have conical shells occasionally ornamented with rings on their outer surfaces and are characteristic to pelagic Devonian sediments. Insoluble residues of these rocks also contain conodonts, consistent with their Devonian age. The discovery of, and publications on, this scientifically significant fossil assemblage is due to the work of Gyöngyi Lelkesné Felvári, György Majoros, and Sándor Kovács (1948–2010)—the latter a well-known expert on conodonts.

Several other Hungarian boreholes have also penetrated fossiliferous Devonian rocks, but because they differ from one another they cannot be identified. Fossil assemblages of this age are mostly dominated by conodonts, which themselves are characteristic of different stratigraphic levels in the Devonian.

Supposedly, the Polgárdi Limestone that was for years exploited at a huge quarry in the Szár-hegy Hill is also Devonian in age. This rock was once deposited in a shallow sea and is famous mainly because of the Miocene fossil vertebrates that have been preserved in karstic fissures formed within it. So far the repeated and intensive search for identifiable fossils has remained unsuccessful: only stromatolites have been found thus far.

📖 Lelkesné Felvári et al. 1984

CARBONIFEROUS FOSSILS

In contrast with older microfossil-dominated Paleozoic assemblages, the Carboniferous sequences in the Transdanubian Range contain well-preserved macrofossils. One of these successions is the Lower Carboniferous Szabadbattyán Slate that was once explored in galleries and boreholes under an overthrusted Devonian limestone in the vicinity of Szár-hegy Hill. The subsurface mines that were abandoned several decades ago were left open in order to exploit the lead ore; rich, shallow-water fossil assemblages in dark-gray and black limestones and shales were discovered by Aladár Földvári (1906–1973), professor at the universities in Debrecen and Miskolc. However, it was János Kiss (1921–2005), head of the Department of Mineralogy at Eötvös Loránd University, who was the first to describe the occurrence of these fossiliferous beds (noting the priority of Földvári). Since their discovery, the list of fossils from these rocks has grown longer and longer, thanks to the work of several specialists, and the microfauna and microflora is especially remarkable. Among algae, besides more common Carboniferous forms like *Dvinella* and *Anthracoporella,* well-preserved specimens of the blue-green alga *Girvanella* are of note because in these Szabadbattyán specimens very delicate details that are only rarely visible can also be studied. The rich foraminifera assemblage from these rocks was described by Miklós Monostori, a former head of the Department of Paleontology at Eötvös Loránd University, who noted that species of *Endothyra* frequently found here belong to the group of *Fusulina*-like larger foraminifera. Corals are also known from this locality; the order Tabulata, dominant in the Paleozoic, is represented by *Hexaphyllia* and *Syringopora.* The widely distributed *Dibunophyllum,* and some other forms also recorded here, belong to the order Rugosa, another characteristic group of Paleozoic corals. Occurrence of *Heterophyllum,* a Heterocorallia, is also peculiar to this assemblage; among the brachiopods, a species of *Gigantoproductus,* first described from Szabadbattyán and named *transdanubicum* is the most common.

The other fossil-bearing successions of this age in the Transdanubian Range were deposited in other types of environments. The material eroding from the uplifting Variscian mountain chain filled fluvial basins surrounding these ranges and successions can be studied, although only partially, in small quarries at Kő-hegy Hill near the village of Füle. The successions exposed here range from coarse-grained, even boulder-bearing, conglomerates to finer-grained sandstones; the latter contain a plant assemblage indicative of a Late Carboniferous age. This flora is thus coeval with the large and famous coal measures known across Western Europe. According to the literature on this site the first specimen, the trunk of a horsetail (*Calamites*), was found in 1910 by Ferenc Pávai Vajna (1886–1964), an eminent explorer of the oil and gas fields found in the Zala and Hajdú Counties of Hungary as well as of the thermal-water reservoirs on the Great Hungarian Plain. The Füle flora was then systematically described by Sándor Mihály (1941–1995), one of the very few people to graduate as a paleontologist from Eötvös Loránd University. Until his untimely death, Mihály worked as a researcher at the Hungarian Geological Institute and documented occurrences of—among other forms—the numerous ferns (*Alethopteris, Pecopteris, Neuropteris*), horsetails (*Asterophyllites, Calamites*), and tree sizes of gymnosperms (*Cordaites*) from this site.

📖 Andreánszky 1960; Kiss 1951; Földvári 1952; Detre 1971a; Mihály 1973, 1980; Lelkesné Felvári 1978; Monostori 1978

PERMIAN FOSSIL FINDS

The Permian system in the Carpathian Basin, although represented by peculiar sequences of considerable areal extent, including the Balatonfelvidék Sandstone, is poor in fossils. Besides vertebrate traces only plant remains have so far been found, usually on the bedding planes of fine-grained rocks. Poorly preserved fossils of leaves, trunks, sporangia, and cones can be seen at some sites and were described in 1911 by János Tuzson (1870–1943), a professor at the University of Budapest in one of the famous Balaton Monographs (see chapter 2). According to Tuzson, Permian plants from the Balaton Highlands most closely resemble *Voltzia hungarica,* a fossil described on the

1
2
3
4
5
6

basis of better-preserved specimens from coeval beds in the Mecsek Mountains. Since this early work, only silicified tree trunks from this area have attracted attention: Pál Greguss (1889–1984), professor of botany at Szeged University and a well-known expert on xylotomy, identified *Dadoxylon* and *Arauxylon* from these beds.

More diverse fossil assemblages recently were discovered in borehole cores penetrating subsurface marine Permian beds in the northeastern part of the Transdanubian Range. During the Permian the sea was situated to the northeast of present-day Lake Balaton so the continent and sea were separated by a zone of lagoons in which evaporitic rocks such as dolomite, gypsum, and anhydrite were deposited in a hot and dry climate. The succession that was first drilled near the village of Tabajd near the town of Bicske yielded spores and pollen grains of continental plants (the word "pollen" derives from Latin and means "flour" or "fine powder"). In the direction of the former marine area to the northeast, dolomite beds become more frequent and were first discovered in a borehole near Dinnyés, at Lake Velencei. Cores in this area were found to contain a diverse marine microfossil assemblage in addition to green and red algae and sections of unidentifiable bivalves, *Bellerophon*-like gastropods, ostracods, and a rich and well-preserved fauna of foraminifera. Of the latter, *Collaniella parva* is an important guide fossil for the latest Permian.

📖 Greguss 1961, 1967; Haas et al. 1986

PALEOZOIC FOSSIL ASSEMBLAGES OF THE MECSEK MOUNTAINS AND THE SURROUNDING AREA

The western part of the Mecsek Mountains is largely comprised of Paleozoic sequences that were deposited in continental environments. Many aspects of the geology of these beds, including their fossil contents, have been studied in detail because of intensive searches for Permian uranium ore that was mined here for decades in the second half of the twentieth century. Paleontological research here began well before the discovery of these raw materials, in the years immediately following World War II. While the geology of the town of Pécs (also called Fünfkirchen in older German literature) was being mapped on behalf of the local community to provide a safer water supply, János Böckh (1840–1909), the second director of the Royal Hungarian Geological Institute, wrote a detailed account of these Permian sequences and collected plant fossils. He asked Oswald Heer (1809–1883), a well-known paleobotanist working in Zürich, to study the flora; this resulted in a well-illustrated monograph published in 1877. Unfortunately, however, it is not clear where the fossil remains assigned to the genera *Voltzia, Ulmannia,* and *Carpolithes*—which all belong to pin-leaved plants related to *Araucaria*—were collected within this sequence.

Following the publication of Heer's work, the Permian plants from the Mecsek Mountains received almost no attention for 130 years. The only exceptions to this are mineralized, often silicified, tree trunks frequently encountered in natural exposures as well as in subsurface mines and which were documented by Pál Greguss in a paper published in 1961 in *Palaeontographica*. This German journal, founded in 1846, is the oldest paleontological journal in the world.

Somewhat surprisingly, the Carboniferous floras of Southern Transdanubia have received much more attention; indeed, their discovery is considered to be an important step in the geological recognition of this area and Hungary in general. In 1960, István Soós and Áron Jámbor first documented the occurrence of Carboniferous, plant-bearing shale pebbles in the Miocene conglomerates of the Mecsek Mountains. On the basis of these

Carboniferous plant fossils (all original size). (1) *Asterophyllites* sp.—a representative of the organ genera frequently used in paleobotany. The name refers to a kind of foliage from the treelike ancient horsetail *Calamites*. *Asterophyllites* is characterized by upwardly oriented, pinlike, and dense leaves. Different parts of these plants—including spores and pollen grains, leaves, stems, and seeds—usually fossilize separately, but remains have been traditionally assigned to organ genera and species because it is often unknown whether they belong to the same plant species (from borehole Füle-2, 150.0 m). (2) *Pecopteris* sp.—a very widespread Carboniferous fern characterized by the wide base of its leaflets (from borehole Füle-2, 255.8 m). (3) *Annularia* sp.—another kind of *Calamites* foliage with leaves that originally resembled lancet-like star shapes around the stem. When fossilized, however, only deformed and flattened remains of these once three-dimensional structures can be seen. This genus was distributed worldwide during the Late Carboniferous. The specimen illustrated here is from the coal-bearing sequence at Secul (Southern Carpathians, Romania). (4) *Alethopteris* sp.—a seed fern whose leaves are often abundant in bare rocks in Carboniferous coal-bearing sequences. *Alethopteris* can be distinguished from other similar forms because of its multipinnate leaves in which the bases of the neighboring leaflets are fused (from borehole Füle-2, 257.0 m). (5) *Neuropteris obliqua* Brongniart—a frequently encountered Carboniferous seed fern whose compound leaves are, in contrast to *Alethopteris*, formed from small petiolate leaflets. This species was described by the "father of paleobotany," Adolphe Brongniart (1801–1876), son of the director of the famous porcelain factory at Sèvres, France (from borehole Füle-2, 150.0 m). (6) *Cordaites* sp.—a representative of the tree-like ancient gymnosperms, immediate descendants of psilopsids. The straplike *Cordaites* leaves have characteristic venation patterns that resemble monocotyledons (from borehole Füle-2, 129.0 m).

finds, the authors supposed that some 15 million years ago continental Carboniferous beds, now covered by a thick succession of Neogene sediments, cropped out about 15–35 kilometers to the south of the present-day Mecsek Mountains. Some years later their predictions were borne out as the plant-bearing Carboniferous conglomerate was penetrated by boreholes drilled in this area. In the meantime, new specimens were collected on the surface by Béla Wéber (1932–2003), a geologist at the Uranium Ore Mining Company and author of numerous important papers. This new material was studied by the outstanding paleobotanist Gábor Andreánszky (1895–1967) and, most recently, Csaba Gulyás-Kis summarized the accumulated knowledge on the Carboniferous plants of Southern Transdanubia. According to him, this assemblage is closest in its affinities to coeval assemblages from Upper Silesia of Poland. The Mecsek Flora comprises about 20 species and is dominated by ferns (*Alethopteris, Pecopteris, Neuropteris*), although ancient horsetails (*Calamites, Annularia*) also are present.

As a result of the ongoing search for uranium ores in Permian beds, organic-walled microfossils assemblages have also been discovered. A Triassic age for the spectacular cliffs that make up Jakab Hill, previously thought to be Permian, was determined by Ágnes Barabásné Stuhl using evidence from spores and pollen. Some galleries and surface trenches into these sediments have also yielded fragmentary animal remains, and the fine-grained rocks were also found to contain the remains of freshwater phyllopod clam shrimps.

The discovery and subsequent study of siliceous sandstone and shale sequences penetrated by boreholes north of the Mecsek Mountains has also proved to be important both from geological and paleontological points of view. Core samples from this area were first studied in 1964 by János Oravecz, a specialist in Lower Paleozoic microfossils. As a result of his efforts, these hard, metamorphic rocks believed to be Carboniferous in age yielded the fragmentary remains of marine organisms including graptolites and hystrichosphaerids—evidence for the much older, Silurian age of this succession. Later, Heinz Kozur contributed to the micropaleontological knowledge of these so-called Szalatnak Siliceous Shale beds. In addition to conodonts, he erected a new group he called Muellerisphaeridae to accommodate the numerous species of microfossils encountered in the Szalatnak borehole. Muellerisphaerids are globular forms bearing obtuse thorns somewhat resembling naval mines.

📖 Heer 1878; J. Böckh 1881a; Soós & Jámbor 1960; Greguss 1961; Wéber 1964; Barabásné Stuhl 1981; Kozur 1984a; Gulyás-Kis 2003

FOSSILS FROM NORTHERN HUNGARY

Paleozoic sequences in the northern Hungarian Range are variable and differ considerably from those of Transdanubia. Apart from some isolated outcrops sequences of this age are best exposed in the Bükk Mountains as well as in the Uppony and Szendrő Mountains. Most of rock bodies of this age have been metamorphosed to different extents over the millions of years that have elapsed since their formation.

Bükk Mountains

Kálmán Balogh (1915–1995), an outstanding researcher on the geology of the Carpathians as well as on Triassic fossils, correctly considered the Bükk Mountains to be the most complex geological area in Hungary. The large outcrops of Paleozoic rocks, Carboniferous and Permian, are concentrated at the northwestern margins of these mountains and are built up from folded and thrusted sequences. Lower Carboniferous sandstones and shales in this area represent deposits of a submarine slope covered with deeper water and do not contain fossils. By contrast, Upper Carboniferous shallow-water sequences consisting mainly of shale and limestone beds, the Mályinka Formation, are known to contain rich and diverse fossil assemblages. The Lower Permian, as evidenced by the presence of evaporitic rocks, was deposited in coastal plain and lagoon environments and also contains practically no fossils. Due to the softness of these rocks, this sequence—the Szentlélek Formation—is only rarely exposed, whereas the Upper Permian Nagyvisnyó Formation that is formed predominantly from shallow-water limestones is rich in fossils in a number of places.

The first data on Paleozoic fossils from the Bükk Mountains was provided by Austrian geologists who proved the presence of Carboniferous beds in this area on the basis of relatively frequent brachiopod remains. However, they misidentified these rocks and correlated with the Culm, the lower, marine section of classical Carboniferous in northwest Europe. Construction of the Eger-Putnok Railway in the first decade of the twentieth century gave significant impetus to the recognition of fossiliferous beds, and cuttings made near Nagyvisnyó in Bán Creek valley provided fine exposures of Carboniferous rocks. Cutting no. 2 proved to be the richest in fossils; cutting no. 5, near Nekézseny, revealed richly fossiliferous Upper Permian beds. Although this railway is now abandoned, these cuttings are still the best exposures of fossiliferous Paleozoic rocks in the Bükk Mountains.

The first report on new Carboniferous exposures was published by Elemér Vadász (1885–1970), a highly influential worker and authority on Hungarian geology in the twentieth century. During his long professional career, the

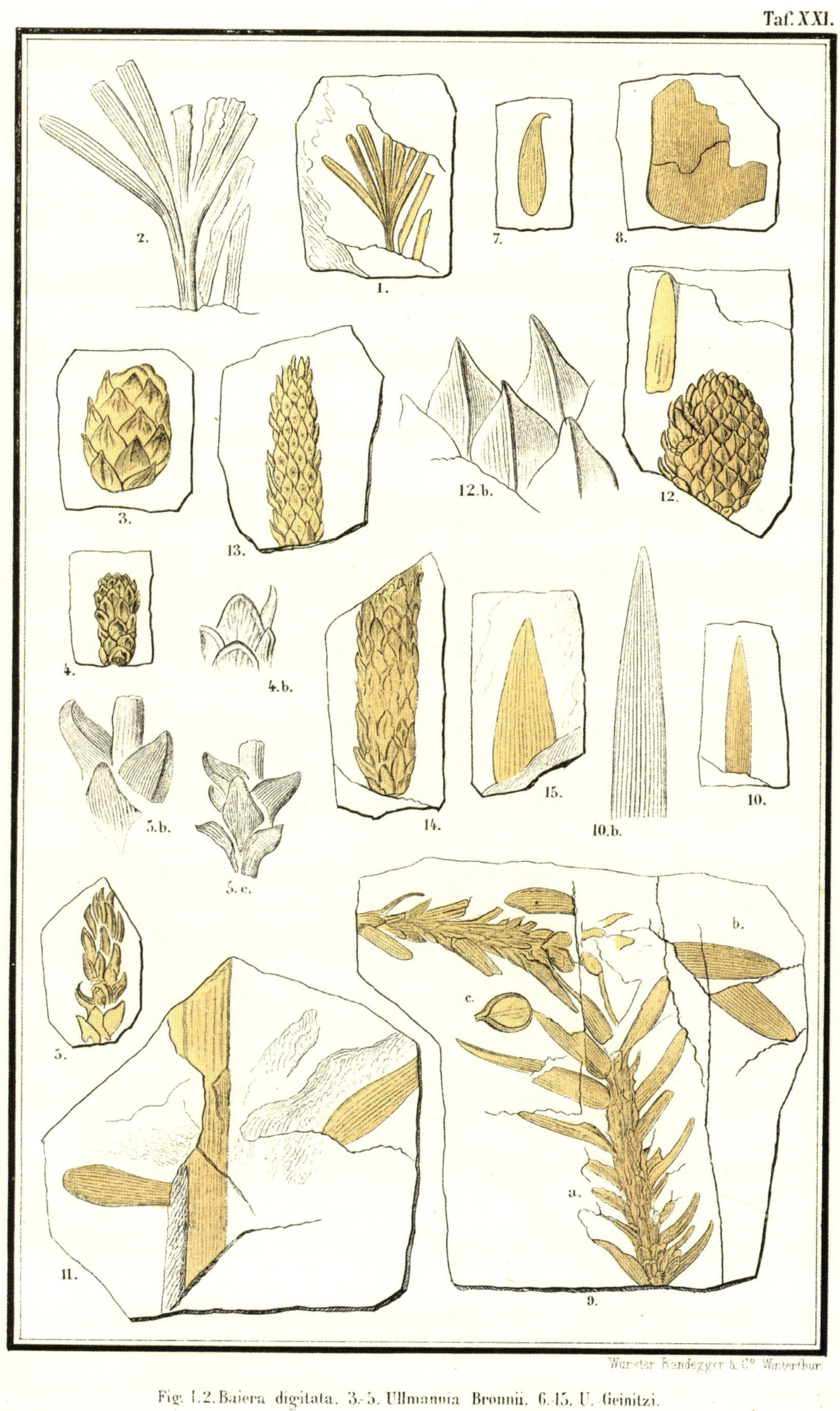

Drawings of Permian plants from the Mecsek Mountains that were published by Oswald Heer.

remarkably versatile Vadász served, among other things, as president of the Society for Hungarian-Soviet Friendship and was also able to recognize that the fossil assemblage at Nagyvisnyó differs considerably from that at Dobšiná, the nearest similar outcrop area. It was left, however, to the geologist Gyula Rakusz (1896–1932)—who, sadly, died young—to document this fauna; his monograph (published posthumously in 1932) dealt with the known fossils known from both these areas. Rakusz's work comprises 220 printed pages and reflects the relative abundances of the fossil groups by then encountered. Brachiopods—associated with corals, gastropods, bivalves, bryozoans, and sea lilies—are by far the most diverse fossils from Nagyvisnyó, whereas the vegetation of the neighboring land mass is evidenced by leaves from the seed fern *Neuropteris*. Since the work of Rakusz, several other Carboniferous exposures have also yielded diverse fossil assemblages.

The next significant event in the history of paleontological research in the Bükk Carboniferous and Permian was the publication of a paper by Zoltán Schréter in 1948. This study first records the presence of trilobites in the area; most specimens were provided by the extravagant private collector Ferenc Legányi (1884–1964). Born into a landowning family, Legányi became a versatile geologist and paleontologist and collected fossils from the Bükk Mountains tirelessly for over 45 years. Verification of a Permian age for some fossiliferous beds in this area on the basis of the aberrant brachiopod *Leptodus* (*Lyttonia* in the older literature) and the trilobite *Pseudophillipsia* is thanks to the efforts of Legányi and Schréter. Permian limestone beds, although exceptionally well exposed in the Mihalovits Quarry at Nagyvisnyó, are usually poor in fossils and the best finds—the only Permian bonanza from this area—are from the Leptodus member exposed in railway cutting no. 5.

After the work of Schréter, geological mapping in the Bükk Mountains was also done by Kálmán Balogh, who then asked well-known experts from Hungary and elsewhere to study the fossils he collected. This cooperation resulted in the 1963 publication of a volume in the Geologica Hungarica Series Palaeontologica devoted entirely to the Bükk Paleozoic. The Carboniferous fusulinid forams collected here were described by the Russian Sofia Yevsseyevna Rosovskaya (1907–1987), the Croatian Milan Herak and Vanda Kochansky (1915–1990) documented Upper Paleozoic calcareous algae, and Schréter described the brachiopods.

The next substantial volume on the Upper Permian fossils from the Bükk Mountains was published in German by Akadémiai Kiadó, at the time the official publisher of the Hungarian Academy of Sciences, in 1974. This publication contains monographic treatments of the forams (by Mária Sidó), ostracods (by Béla Zalányi 1887–1970), and nautiloids (by Schréter).

However, due to the imperfect methods that were used by Zalányi to extract ostracods from these rocks, only a rather poor assemblage was documented. The real diversity of this fauna was discovered later by Heinz Kozur, who used sophisticated chemical extraction techniques. This work resulted in a rich and well-preserved assemblage described in a series of papers and in a 1985 monograph. On the basis of these ostracods, Kozur was able to subdivide this otherwise lithologically monotonous limestone sequence into zones, even though (it must be mentioned) this work was not conducted in accordance with the requirements of the International Code of Zoological Nomenclature—not all the type specimens are housed in public collections. As a result, the validity of the numerous new species described by Kozur is questioned by some experts.

As a consequence of the research that has been done thus far, the Permian fossil assemblages are known to be comparable in their richness to the Carboniferous. Among microfossils, calcareous algae must first be mentioned—as some forms (*Mizzia, Gymnocodium*) identified as early as 1919 remain important index fossils. The Permian macrofauna is dominated by brachiopods, of which some 30 species have been identified. In terms of specimen numbers, *Tschernyschewia* is by far the most abundant—but *Leptodus* and another special cementing form, *Richthofenia,* also occur. Amongst the bivalves, the large-sized *Edmondia* and *Aviculopecten* are characteristic and nautiloids are represented by both orthoconical (*Lopingoceras*) and planispiral (*Tainoceras*) forms. Corals are known on the basis of the widely distributed *Waagenophyllum,* which is especially frequent in a bed of considerable lateral extent around Nagyvisnyó and at a number of other sites. Sections of the gastropod *Bellerophon* are often visible on rock surfaces, although specimens usually cannot be extracted.

In addition to the monographic treatments mentioned above, a series of shorter papers have been published recently that have contributed to our knowledge of the Paleozoic fossils from the Bükk Mountains. A rare Permian fish tooth belonging to an early shark was described in 1983 by Sándor Mihály and Péter Solt and in the same volume of the annual report of the Hungarian Geological Institute another Permian fish paper was also published. Alois Fenninger, a professor at the University of Graz and his coworker J. Nievoll described a find of a leaflike (phyllodont) tooth belonging to a bony fish. New data on Carboniferous bryozoan assemblages were published in 1993 by Kamil Zágoršek, who worked in Bratislava; the ontogeny of a peculiar Carboniferous coral species (*Palaeacis cyclostoma*) was documented in 1999 by Mihály Dunai. A further find of an interesting Upper Carboniferous echinoderm belonging

Colonial corals (*Palaeacis cyclostoma* Phillips) comprising individuals in different numbers from Upper Carboniferous sediment exposed in railway cutting no. 2 at Nagyvisnyó (Bükk Mountains). The hard parts of other organisms, such as the carapaces of trilobites, gastropod shells, or crinoid stems, lying on the muddy and hostile bottom of the Carboniferous Sea provided the only opportunities for coral larvae to settle and develop. This photo shows colonies consisting of individuals in increasing numbers as well as the skeletons of other organisms that make the settlement of corals possible. This interesting form, along with numerous other rare Paleozoic fossils, was documented from Hungary by Mihály Dunai (natural size) 📖 Dunai 1999.

to Ophiocystoidea, also described by Dunai, considerably augmented the known stratigraphic range of this group. Previously, ophiocystoids were thought to have disappeared in the Early Carboniferous.

The most recent results of research on Paleozoic fossils from the Bükk Mountains were published by the young paleontologist Csaba Gulyás-Kis. His MSc thesis was devoted to a revision of the Carboniferous brachiopod fauna and contains descriptions of 36 species belonging to 27 genera. Of these, *Linoproductus, Dielasma, Chonetes, Orthotetes,* and *Choristites* are the most abundant; and he has pointed out that the Bükk assemblage shows close affinities to those known from the Southern Alps, the Karavanke (or Karawanken) Mountains in Slovenia and Austria, and to western Serbia. All these data provide further evidence for the Dinaric relationships of the Bükk Mountains.

📖 Rakusz 1932; Schréter 1936, 1948, 1963, 1974; Herak & Kochansky 1963; Rosovskaya 1963; Balogh 1964; Zalányi 1974; Fenninger & Nievoll 1983; Mihály & Solt 1983; Kozur 1985; Zágoršek 1993; Gulyás-Kis 2001, 2004

Remains (calyx and columnals of different sizes) of a sea lily (*Poteriocrinus* sp.) from Upper Carboniferous sediments exposed in railway cutting no. 2 at Nagyvisnyó (Bükk Mountains). Columnals (parts of stems) are common fossils, whereas calyxes are extremely rare. (Original size.)

Uppony Hills: The Oldest Macrofossils from the Carpathian Basin

Apart from the Upper Cretaceous Nekézseny Conglomerate, which crops out along the southeastern margin of this area, the bulk of the Uppony Hills are composed of Paleozoic rocks. These form, for example, the peculiar Uppony Gorge. All of these rocks are also metamorphosed to some extent and so their geological age, due to the apparent lack of index fossils, was long debated. The first age determinations, based on sound biostratigraphical evidence, were provided by Heinz Kozur and Rudolf Mock (1943–1996). The latter, an expert on conodonts, worked as a professor in the Department of Paleontology at the University of Bratislava until his tragic death, and was also well known as a mountaineer. In a pioneering paper published in 1977, Kozur and Mock described Devonian and Carboniferous conodonts. Their research was carried on by Sándor Kovács, a former member of the Geological Research Group at the Hungarian Academy of Sciences and an outstanding expert on the geology of northern Hungary. Identification of species belonging to *Palmatolepis, Spathognathodus,* and *Idiognathodus* allowed

Kovács to correlate the rock bodies in this area, many of previously unknown or misinterpreted age, with other stratigraphic levels in the Devonian and Carboniferous.

During the detailed geological mapping of this area, numerous research trenches were made in order to collect paleontological samples. In one of the trenches, dug on the top of Strázsa Hill in the village of Nekézseny, blocks of limestone, unknown from other outcrops, were revealed. Two types proved to be rather frequent but Devonian crinoidal limestones are by far the most abundant. The other type, which occasionally forms boulders that exceed one cubic meter, is rarer but much more interesting paleontologically—as its purplish-red, greenish-gray, or greenish-red color contains the relatively frequent shells of orthoconical nautoloideans. These finds were studied and described by Maurizio Gnoli, an Italian expert on Paleozoic nautoliods, and Sándor Kovács in a joint paper published in 1992. Two genera proved to be identifiable: *Michelinoceras*, a nautiloid with one of the longest stratigraphic ranges, and *Kopaninoceras*. Besides other, unidentifiable, cephalopods, brachiopods and bivalves also occur in these limestones and the rich conodont assemblage dissolved from these rocks contains specimens of *Ozarkodina* and *Spathognathodus*, both indicating Silurian age. This means that the Nekézseny nautilodeans are the oldest known macrofossils from Hungary and the whole Carpathian region.

The sedimentary history of the Strázsa Hill section is also worth mentioning. Silurian and lowermost Devonian limestones, already lithified by the Middle Devonian, slipped down as huge blocks into a marine basin where material spilled by neighboring active volcanoes was deposited. The fossil-bearing limestone blocks are thus embedded in volcanoclastic rocks well exposed in an abandoned quarry at the western end of the hill.

📖 Kozur & Mock 1977; Gnoli & Kovács 1992

Szendrő Hills

Paleozoic rocks in the Szendrő Hills are usually strongly folded and shale-like in their appearance as a consequence of the pressure and heat that they have been subjected to since their formation. In addition, the geological age of these rocks—due to a scarcity of fossils—was also unknown for a long time. It was therefore an important step toward the geological characterization of the area when Sándor Mihály began to study the coral remains embedded in the dark gray limestones (known today as the Szendrőlád Limestone) that outcrop in large areas in the southern part of the Szendrő Hills. The results of his work were published in several papers and a 1976 monograph.

By studying orientated thin-sections, Mihály pointed out that these limestone beds also contain a rich colonial coral assemblage. This fauna consists of more than 20 species from the order Tabulata, a diverse group of Paleozoic corals that disappeared at the end of the Permian. About three-quarters of the specimens collected belong to *Favosites*, but *Thamnopora* and *Alveolites* are also worth mentioning. It is surprising that the presumably diverse reef-dwelling communities of Devonian seas have left relatively monotonous fossil assemblages that consist predominantly of tabulates. In addition, only a few sea lilies, sections of rugose corals, and scattered gastropod remains have been found.

Paleozoic fossils (all original size). (1) *Peronidella baloghi* Flügel—a rare calcareous sponge from the Upper Permian of the Bükk Mountains. This species was named in honor of Kálmán Balogh, an eminent Hungarian geologist. (2) *Favosites goldfussi* d'Origny—a morphologically variable (hemispherical, tabular, or tuberous) colonial coral that belongs to the order Tabulata, an important group of reef builders in the Paleozoic. The corallites of this coral are thin, pentagonal in cross section, and closely spaced, forming beehive-like masses. The outer walls of the corallites are perforated and the radial inner walls (septa) are short, thornlike, and often lacking, whereas the horizontal walls (tabulae) characteristic of this order are well developed. The geological age of the metamorphic Devonian limestones in the Szendrő Hills of Hungary (Szendrőlád Limestone) was determined on the basis of remains of Tabulata, in part *Favosites* (from Szendrő, Garadna-puszta). (3) *Thamnopora* sp. With its massive, upward-growing colonies, this genus is an unusual form compared to other Tabulata with colonies formed from short, tightly spaced, and often branching corallites. The septa of *Thamnopora* are short and thornlike with thin tabulae (from Szendrőlád, the valley below Tótharaszt-puszta). (4) *Canina pannonica* (Frech)—a solitary Carboniferous coral. Reduction of the length of septa in this coral has resulted in formation of a wide, inner space late in ontogeny—a characteristic feature of this genus. This space is especially visible on the upper two specimens (from Dobšiná, Biengarten). (5) *Lithostrotion* sp.—a colonial rugose coral from the Carboniferous. Representatives of this genus are frequent worldwide from Carboniferous shallow-water limestones and calcareous shales (Bükk Mountains). (6, 7) *Lopingoceras cyclophorum* (Waagen)—an orthocone-shaped cephalopod from the Upper Permian of the Bükk Mountains. Strongly protrusive rings in the outer shell surface as well as an eccentrically placed septal opening are characteristic features of this species. The Bükk Mountain specimens are usually referred to as *Pseudorthoceras* in older literature (from Nagyvisnyó, Mihalovits Quarry). (8) *Edmondia permiana* Simić—the largest bivalve from the Permian of the Bükk Mountains, specimens occasionally reach 10 centimeters in length. The outline of the valves in *Edmondia* is characteristically rounded and triangular, with the highest region at about mid-length. Co-marginal growth lines form the only ornamentation on the outer surfaces of these bivalves (from Nagyvisnyó, on the hillside above the former railway station).

1
2
3
4
5
6
7
8

1
2
3
4
5
6
7
8
9
10
11
12

Following investigation of these shallow-water limestones, sediments of the coeval but deeper basins in the Szendrő Hills were found to also yield fossils. Conodont assemblages from these basins indicate different stratigraphic levels of the Devonian and Carboniferous, and many finds derive from red limestones infilling fissures of older, shallow-water limestones. These faunas were also studied by Sándor Kovács—who identified, among other forms, the stratigraphically valuable *Palmatolepis* and *Polygnathodus,* both from the Devonian; and *Gnathodus* and *Idiognathoides* from the Lower Carboniferous.

📖 Mihály 1978

FOSSILS FROM SLOVAKIA, ROMANIA, AND CROATIA

Similar to Hungary, the areas immediately adjacent to this country are also poor in attractive and well-known Paleozoic fossils. The nearest, relatively diverse fossil assemblage outside of Hungary is from the Carboniferous of Dobšiná, Slovakia, in the Gemerské Rudohorie—also known as Slovenské Rudohorie (Slovak Ore Mountains).

It is only near the Carpathian Mountains that there is a general paucity of Paleozoic fossils. Some three hundred kilometers to the northwest of the Carpathians, near Prague, is the celebrated region of Barrandium, one of the world's best-known areas for Paleozoic rocks and fossils. Barrandium—named in honor of the eminent French scholar Joachim Barrande (1789–1883), who lived in Prague and published voluminous monographs on the Paleozoic fossils from this country—is a Mecca for paleontologists (incidentally, Barrandov, a southwestern district of Prague famous for its film studios, is also named after Barrande). Rich and diverse Paleozoic fossil assemblages are also known from the Carnic Alps in Austria and Italy, from the Austrian province of Styria (the Graz Paleozoic) and from the Dinaric Alps in Croatia. Of these regions, the Gorski Kotar Range in the Dinaric Alps has yielded Paleodyas (Early Permian) ammonites, which are rare in Europe. This fauna was described by Viktor Vogl (1885–1922), an eminent geologist who worked at the Royal Hungarian Geological Institute.

The first report on fossils found during iron mining at Dobšiná was published by András Kiss, a mining engineer at Roznava, in the mid-nineteenth century. The monographic treatment of this fauna was completed, however, much later by Gyula Rakusz. At the request of Baron Ferenc Nopcsa (1877–1933), director of the Geological Institute in Budapest at the time, this volume contains descriptions of fossils from both Dobšiná and Nagyvisnyó. As outlined in

Paleozoic fossils (all original size except [4]). (1) *Acrodus gaillardoti* Agassiz—a tooth of an ancient shark-like fish from the Permian of the Bükk Mountains. (2) *Stachella* sp.—a *Bellerophon*-like gastropod from the Upper Permian of the Bükk Mountains. This peculiar group of gastropods is characterized by a more or less globular, perfectly symmetrical shell formed by whorls where the last completely envelops all the previous ones (a mode of coiling known as involute). The shell opening also bears a slit situated in the plane of symmetry, and behind this there is a bend—the surface of which is ornamented with creases and rows of pits. Bellerophontoideans are considered by some authors to belong to the class Monoplacophora rather than Gastropoda. Bellerophontoideans appeared in the Silurian and are often frequent in marine fossil assemblages of Paleozoic age. Numerous genera assigned to this peculiar group have been named and Upper Carboniferous and Upper Permian sediments in the Bükk Mountains have yielded numerous, occasionally well preserved *Bellerophon*-like gastropods, most of them still undocumented. Remarkably, the genus *Bellerophon,* the single representative of this previously diverse group, somehow survived the end-Permian mass extinction. Remains of this invertebrate, the so-called Last of the Mohicans, can be found in, among other areas, the Lower Triassic of the Balaton Highlands (Upper Permian, Nagyvisnyó railway cutting no. 5). (3) *Phricodothyris asiatica* Chao—a brachiopod from the Upper Carboniferous of the Bükk Mountains. (4) *Pseudophillipsia hungarica* Schréter—a trilobite from the Upper Permian of the Bükk Mountains. Before the end of the trilobites' long evolutionary history, their diversity was very low compared to the heyday of the group: only five genera are known from the Upper Permian and the number of localities is also strongly reduced. The Bükk Mountains is one of those areas where remains of the last survivors of this group can be encountered, although extremely rarely. The genus *Pseudophillipsia* contains remarkably small—just a few centimeters long—trilobites with semicircular head shields (cephalon), middle sections (thorax) comprising nine segments, and relatively large, rounded tail sections (pygidium). Further characteristics of these trilobites are their big eyes, relatively long and narrow middle segments of the pygidium, and a fine granulation covering the whole shell surface (Nagyvisnyó railway cutting no. 5) (2× magnification). (5–7) *Knightites* sp.—a *Bellerophon*-like gastropod from the Upper Carboniferous of the Bükk Mountains (Nagyvisnyó railway cutting no. 2). (8) *Pseudophillipsia ogivalis* Gauri—a trilobite from the Upper Carboniferous of the Bükk Mountains. These are small forms just a few centimeters in length that have a semicircular cephalon that ends in two long, backward-pointing thorns. The pygidium in these trilobites is strongly sectioned and is surrounded by a wide, smooth margin. The Upper Carboniferous of the Bükk Mountains has yielded many dozens, probably hundreds, of separate pygidia from these trilobites even though the number of articulated, complete specimens (consisting of all three sections) is less than five. The cause of this conspicuous disparity in style of preservation remains an enigma (Nagyvisnyó, railway cutting no. 2). (9) Section of a nautiloid in Upper Permian limestone from the Bükk Mountains (Nagyvisnyó, Mihalovits Quarry). (10, 11) *Bellerophon* sp.—from the Upper Carboniferous of the Bükk Mountains (Dédestapolcsány). (12) *Productus* sp. and *Spirifer sp.*—brachiopods from the Upper Carboniferous of the Bükk Mountains (Nagyvisnyó, Nagyberenás-lápa).

The Strázsa Hill at Nekézseny, locality of the oldest macrofossils known from the Carpathian Basin.

the introduction to Rakusz's work, in addition to the generally poor state of preservation of these fossils, their "unsuitable coloration" also proved to be a major problem. Rusting of iron minerals out of the originally dark gray rocks sometimes replaces calcareous shell materials and leads to yellow, red, or brown patches that makes photography very difficult. At the time, whitening the fossils with ammonium chloride, now a standard method of eliminating such color differences, was unknown. Blackening with graphite powder, another method that was available in Rakusz's time, would have damaged the delicate details of the fossils. In spite of these difficulties, Terézia Dömök (1892–1980), an exceptionally skilled photographer who worked at the Geological Institute, was able to take photos of considerable quality, in which the fossils are highly recognizable.

The Carboniferous fauna from Dobšiná, like that known from Nagyvisnyó, is dominated by brachiopods—but corals, bivalves, and gastropods also occur in significant numbers. A few trilobite specimens have been found in this fauna (the first trilobites from the former territory of Hungary were described from this site in 1902). A few rare ammonites have also been found, and these differ in form from the vast majority of other Paleozoic and Mesozoic taxa because their inner whorls form triangles in lateral view. Later in their ontogeny, shells of these forms become more similar to those of ordinary ammonites (that is, they grow according to a logarithmic spiral). Such ammonites are also known from other Carboniferous faunas, and Rakusz assigned them to the genus *Gastrioceras*. Recently, the name *Dobshinaeceras* has been introduced to accommodate these strange forms. Of the plant remains that occur at Dobšiná, the genera *Calamites* and *Neuropteris* are most frequent.

The continental Carboniferous deposits known from Hungary do not contain coal seams, but in spite of the objections of experts familiar with the paleogeography and paleoenvironment of the country, such as Zoltán Schréter, attempts were made in the past to find coal seams that might be of economic interest in the Bükk Mountains and its surroundings. Not far from the present-day borders of Hungary, in the Banat area of Romania, Carboniferous coals have been exploited since the end of the eighteenth century. The first mines in this region were opened near Reşiţa, and extended mining was carried out in the coal seams of Baia Noua, between Berzasca and Orşova.

Polished section of Silurian limestone from Strázsa Hill displaying several sections of orthoconical nautiloideans (the width of this piece of rock is 34 centimeters).

The fine-grained rocks that form the spoils between these coal seams have yielded well-preserved plant remains. According to the outstanding Hungarian paleontologist Miksa Hantken (1821–1893), these shale-like mudstone beds contain carbonized plant remains in enormous quantities; the first identifications were made by Dionýz Štúr (1827–1893), a well-known Slovakian geologist and paleontologist from the Geologische Reichsanstalt in Vienna. Later, Hantken asked Constantin Ettingshausen (1826–1897), rector and professor of paleobotany of the University of Graz, to study these fossils.

As in other coeval assemblages in Europe, lycopsids (*Lepidodendron*) are frequent in Carboniferous floras from the Southern Carpathians (note that the subsurface part of *Lepidodendron* is called *Stigmaria*). Horsetails are abundant, usually in the form of internal molds of *Calamites* and leaves of *Sphenophyllum;* seed ferns such as *Alethopteris, Neuropteris,* and *Pecopteris* are also common. Specimens of tree-like gymnosperms (i.e., *Cordaites*) also have been found.

As is also the case in the Mecsek Mountains, the Permian rocks of the Southern Carpathians contain uranium ores as well as fossil plants. These plant assemblages, however, are of low diversity compared to their counterparts from the Carboniferous; the pine-like *Walchia* was the most frequent floral element of the semidesert Permian landscape. Nevertheless, these poorly preserved remains have proved to be valuable from a biostratigraphic point of view because they indicate an Early Permian age for the succession.

Carboniferous and Permian fossil plants are also known from the Slovakian sector of the Zemplén Mountains and were first reported by François Sulpice Beudant (1787–1850), a pioneer of geological investigations in the Carpathian Basin. In his 1818 work, Beudant suggested a Carboniferous age for the sandstone beds that outcrop out on the hillside at Luhyňa (Legenye) on the basis of remains of fern-like plants. The plant remains from this succession, exposed all around the village of Trňa (Toronya), were later studied by Austrian, Hungarian, and Slovakian researchers. According to these studies, the Permo-Carboniferous flora of the Zemplén Mountains consists of *Dadoxylon,* represented by silicified stems, as well as horsetails (*Calamites, Spenophyllum*).

In the 1980s, Carboniferous plant remains from the Apuseni Mountains were documented for the first time by eminent Romanian geologist Marcian Bleahu and his colleagues. These fossil-bearing successions comprise conglomerates, sandstones, and shales and are reminiscent of coeval deposits in the Banat.

📖 Vogl 1913; Rakusz 1932; Vaňová 1987; Bitoianu 1988; Popa 2009

1
2
3
4
5
6
7
8
9
10
11

Paleozoic fossils (all original size). (1, 4) *Leptodus nobilis* (Waagen)—a strange, shoehorn-shaped brachiopod from the Permian of the Bükk Mountains. The lower valve of this so-called aberrant form is unusually thick and was cemented tightly to a hard object. The upper valve is thin, with a comblike outline, and the inner structure of the lower valve is also unusual—as it is formed by a central elevation called the median septum and has transverse partitioning walls that can number up to 33. *Leptodus* is often found associated to coral reefs and because it had an upper valve that supposedly did not cover the whole of its soft body, symbiosis with photosynthetic algae was possible. If this was the case, then the conspicuous articulation of this brachiopod would have increased considerably the surface of the mantle in which the algae thrived. Although the *Leptodus* beds from the Permian of the Bükk Mountains were named after this peculiar fossil, well-preserved remains of this fossil are actually quite rare (Nagyvisnyó, railway cutting no. 5). (2) *Avonia echidniformis* (Grabau). A number of the Carboniferous brachiopod species that are known from the Carpathian region were first found and described in China. This is the case for *A. echidniformis,* a taxon introduced by Amadeus William Grabau (1870–1946) a German-American paleontologist also known as the father of Chinese geology (Dobšiná, Mihály Mine). (3) *Spirifer zitteli* Schellwein—one of the most attractive brachiopods from the Carboniferous at Dobšiná. Based partly on this specimen, from micaceous sandstones in the Mihály Mine, Gyula Rakusz erected a new subspecies, *S. zitteli dobsinensis.* (5) *Tschernyschewia typica* Stoyanow. The most frequent brachiopod from the Permian of the Bükk Mountains, this *Productus*-like form has an oval, or rounded, pentagonal outline and can reach three centimeters in width. The distinguishing features of this genus include tiny tubercles on the outer surface, which, depending on mode of preservation, sometimes continue as thorns (Nagyvisnyó, railway cutting no. 5). (6) *Tyloplecta yangtzeensis* (Chao). The largest brachiopod known from the Permian of the Bükk Mountains belongs to the order Productida (Nagyvisnyó, railway cutting no. 5). (7) *Choristites fritschi* (Schellwein)—a relatively large, thick-shelled brachiopod from the Carboniferous. Both the upper and lower valves of this animal are convex and have enrolled beaks and a long and straight hinge margin. Representatives of the genus *Choristites*—frequently encountered in the Bükk Mountains—and some other forms were previously assigned to the genus *Spirifer,* the type genus of the order Spiriferida, a dominant group of brachiopods from the Upper Paleozoic. Recent studies, however, have suggested that although the species found in the Bükk Mountains are really spiriferids, they should correctly be referred to the genera *Choristites* and *Plicatocyrtia,* the latter slightly the smaller of the two (Nagyvisnyó, railway cutting no. 2). (8) *Pseudomonotis* sp.—a Permian representative of this widespread bivalve genus known from the Carboniferous and Permian of the northern hemisphere. The left valve of *Pseudomonotis* is significantly more inflated than the right one and it bears radial ornamentation consisting of ribs of different strengths. The ornamentation of the right valve is usually less developed (Nagyvisnyó, railway cutting no. 5). (9) *Derbya senilis* (Phillips). Discovery of this attractive and characteristic brachiopod at the beginning of the twentieth century provided one of the first pieces of evidence for the presence of Permian rocks in the Bükk Mountains (Nagyvisnyó, railway cutting no. 5). (10) *Taenioceras buekkense* Schréter—a highly incomplete specimen of a nautiloid with whorls of quadrangular cross section. The last whorl of this specimen only slightly covers the previous one—a so-called evolute mode of coiling; the outer, ventral part of the shell bears two rows of blunt tubercles. The hole penetrating the partitioning walls is in a subcentral position. The genus *Taenioceras* was distributed worldwide during the Late Carboniferous and Permian, but these fossils are rare in the Bükk Mountains (Nagyvisnyó, Mihalovits Quarry). (11) *Medlicottia croatica* Vogl—a fragment of a Permian ammonite found by Viktor Vogl near Mrzla Vodica in the Gorski Kotar Range, to the northeast of the town of Rijeka. The species name refers to the locality of the type specimens, Croatia.

Late Mesozoic (Cretaceous) flysch outcrop in the Carpathian Mountains. Following the Paleozoic, the Mesozoic Era comprises the Triassic, the Jurassic and the Cretaceous Periods. The long history of our planet is documented by the rocks and fossils embedded in them. Using this analogy, one can think of successive rock layers as sheets in a textbook on the history of the earth.

PART TWO

THE MESOZOIC

Reitziites reitzi—an important index ammonite from the Middle Triassic of the Balaton Highlands (Felsőörs) (1.3× magnification). Although ammonites had appeared in the fossil record by the Devonian, they became dominant elements of the marine biota only in the Mesozoic. Following repeated bursts of diversification and then decline, they disappeared at the end of the Mesozoic. Triassic, Jurassic, and Cretaceous rocks in the Carpathian region are especially rich in ammonite remains.

2

The Triassic

The approximately 186 million years that constitute the Mesozoic—also known as the secondary, or the middle, age in Earth history—document fundamental changes in the surface of our planet. By the end of the Cretaceous, the supercontinent Pangaea had been largely fragmented and the Tethys Ocean—which once separated the northern and southern continents—closed, resulting in the formation of the Alpine-Himalayan mountain chain that stretched for thousands of kilometers. At the same time, the basins that made up the modern oceans begun to open; the oldest parts of the present-day oceanic crust are Middle Jurassic in age.

The living world also changed considerably. The succession of events aptly named the "Mesozoic Marine Revolution" led to the evolution of the modern marine biota. Scleractinian corals, bivalves, crabs, and bony fishes became dominant; brachiopods, a very diverse group from the Paleozoic, suffered during the extinction at the end of the Permian. In only some periods and environments of the Mesozoic did brachiopods remain frequent. Among ectocochliate cephalopods (that is, those possessing an outer shell), the ammonites attained wonderful diversity in Mesozoic seas but their evolutionary history, lasting from the Devonian, finished at the end of the Cretaceous. The Mesozoic may also be rightly considered the time of reptiles. The celebrated dinosaurs appeared in the Late Triassic and soon became the lords of the continents. Birds, whose oldest representatives lived in the Jurassic, most probably evolved from a group of dinosaurs; the air was also conquered by flying reptiles, pterosaurs, which also went extinct at the end of the Cretaceous—along with the non-avian dinosaurs. Mammals, appearing for the first time in the Early Mesozoic, lived in the shadow of the "frightfully large lizards" for several million years. In the world of plants, the Mesophytic—considered the time of gymnosperms—that began in the Permian lasted until the Mid-Cretaceous, when angiosperms became dominant elements of the continental floras.

In some places Mesozoic rocks form successions several thousands of meters thick; they are widespread in the Carpathian region.

The name Triassic was first introduced in 1834 by Friedrich August von Alberti (1795–1878), who recognized that the peculiar tripartite rock succession cropping out especially in Germany, comprising the lower Buntsandstein or bunter ("variegated sandstone"), the middle Muschelkalk ("shell-bearing limestone"), and the upper Keuper (a succession of mottled beds named after its resemblance to a colorful textile once weaved around the town of Coburg) form a unit—which he named after these three elements (τριάς, the digit 3, or "triad," in Greek).

The Triassic, although one of the shortest periods of Earth history—just 51 million years in duration—is usually subdivided into seven epochs. Of these, the term Scythian, which corresponds to the whole Early Triassic, is sometimes used in place of the terms Induan and Olenekian. As discussed earlier, epochs, or stages, are usually named after the areas in which corresponding successions characteristic of this time were first studied or after the people who once lived or are living in those areas. Thus, Induan refers to the Indus River, near which, in the Salt Range, good marine Permian-Triassic boundary sections were found for the first time. Olenekian bears the name of the Olenëk River in Siberia; the Anisian was named after the Roman name for the Austrian river Enns (Anisus); the Carnian after Carnia, which forms the border region between Austria and Italy; the Norian after former Roman province lying south of the Danube; and the Rhaetian after the Rhaetian Alps that stretch over parts of eastern Switzerland, northern Italy, and western Austria. The Scythian remembers the people who once lived on the East European Platform, and the Ladinian refers to the Ladins, inhabitants of the Italian Dolomite Mountains who adopted the practice of making and consuming sauerkraut from the Germans. These Triassic stages are further subdivided into substages and one of them, the Pelsonian, is named after Lake Balaton (which the Romans called Lacus Pelso) in Hungary.

According to our present-day knowledge, the surface of the Earth in the Triassic was less complex than it is nowadays. The large continents were connected to one another and formed a giant supercontinent named Pangaea,

Erathem/Era	System/ Period	Millions of Years Ago
Mesozoic	Cretaceous	145
	Jurassic	200
	Triassic	251

Subdivision of the Mesozoic into periods.

which was divided by the Tethys Ocean into northern and southern parts, Laurasia and Gondwanaland, respectively. The remaining parts of the globe were covered by a huge ocean called Panthalassa, and this water surface was interrupted by several smaller continents and islands. Some of these islands moved thousands of kilometers on the oceanic plates of the lithosphere and became amalgamated into the orogenic belts of the larger continents. Evidence for their former existence is especially spectacular where the direction of plate movement was more or less longitudinal, since in these cases the remains of fossil organisms can now be found just short distances from one another, even though they lived in different climactic belts in the Triassic. The Rocky Mountains of North America, for example, contains such Triassic pieces of the Earth's crust; these are named "suspect terrains."

Global sea level was also very low at the beginning of the Triassic and a more or less continuous rise in sea level across Europe culminated in the formation of the Muschelkalk Sea. Climate around the globe was balanced and, as evidenced by widespread evaporate formations, large areas lay in arid belts. In the monsoon-influenced eastern areas of Laurasia and Gondwanaland, coal-bearing sequences were deposited and are widespread.

The dawn of the Triassic was also a turning point in the history of the living world. Forests of ancient pines thrived in the northern hemisphere while seed ferns dominated in the south. Among vertebrates, so-called cold-blooded reptiles thought to be better adapted to the warm climate replaced so-called warm-blooded groups as the dominant elements. and near the end of this period dinosaurs and the first true mammals evolved.

The marine biota gradually recovered after the end-Permian catastrophe and the rapid and successful evolution history of modern (Scleractinian) corals began at this time. Bivalves became more and more diverse and replaced the brachiopods as dominant elements of benthic macro-invertebrate assemblages—although in some environments, which acted as border fortresses, brachiopods still played a leading role. Echinoderms evolved into several new groups, as did the ammonites, whose diversity exceeded that attained by Paleozoic forms.

The Triassic sequences of Europe can be classified largely into two types. One, the Triassic system of the areas lying

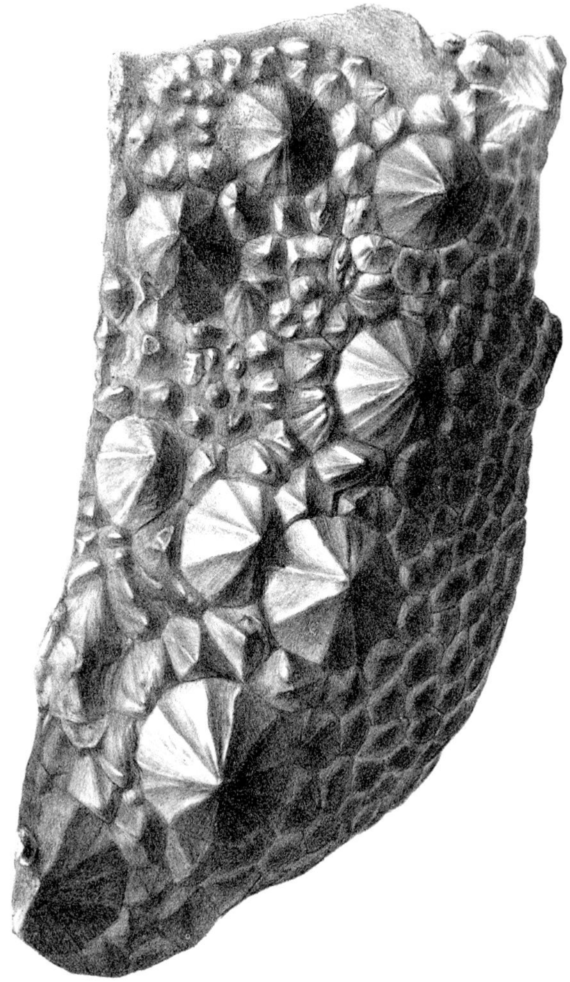

Reptiles thrived in the Mesozoic, and different groups became dominant on land, in the sea, and in the air. In the picture the earliest representation of the Triassic "pebble-toothed pseudo-turtle," one of the most valuable Hungarian fossils, can be seen. Photos of this fossil specimen found about 100 years ago in the town Veszprém and reprepared in the 1930s can be seen at the end of the chapter.

outside the Alpine-Carpathian chain, is more or less similar to the classic sequence in the Germanic Basin. This classic tripartite succession consists of two units of continental origin (the Buntsandstein and the Keuper) and a marine unit (the Muschelkalk) that is sandwiched between them. The other type of succession—the Alpine Triassic—is named for its type areas, the Northern Calcareous Alps and the Southern Alps. The latter is characterized by more or less continuous marine sedimentation that prevails from the earliest Triassic, by the presence of enormously thick platform carbonate successions (see below) and deeper-water deposits, and by fossil assemblages that are much more diverse than those of the Germanic Triassic. Some peculiar deposits from the Alpine Triassic can be traced along the whole of the Tethys Ocean, from the Alps to the island of Timor and

System/Period	Series/Epoch	Stage/Age	Millions of Years Ago
Triassic	Upper/Late	Rhaetian	204
		Norian	216
		Carnian	229
	Middle	Ladinian	237
		Anisian	246
	Lower/Early	Olenekian	249
		Induan	251

Subdivision of the Triassic period into stages.

even onto the Exmouth Plateau off northwestern Australia. Triassic rocks are widespread in the Carpathian region and play an important role in the formation of mountains. For example, Triassic rocks form the main mass of the Transdanubian Central Range, the Bükk and Aggtelek-Rudabánya Mountains and the neighboring Slovak Karst. Thick Triassic sequences are also found in the Mecsek and Apuseni Mountains. The Triassic system in the Hungarian Range represents Alpine development, while that in the Mecsek and Villány Mountains more closely resembles successions in the Germanic Basin.

📖 J. Böckh 1873, 1879; Haas 1993, 2004; A. Vörös 2000

Clever Naming in the Spirit of Respect

According to the rules of stratigraphical nomenclature, geochronological and chronostratigraphical units are not to be named after people. This prohibition, however, did not stop Edward Timothy Tozer (1928–2010), leading expert in Triassic stratigraphy, from creating monuments in the form of substage names to commemorate excellent Triassic workers from around the world. He chose four nameless creeks in Northern Canada, where important Triassic successions crop out, and named them after his predecessors. After doing this, there was no problem in subdividing the Scythian stage into the Griesbachian, Dienerian, Smithian, and Spathian substages because all were named after geographical places and not people.

📖 Tozer 1984

FOSSIL-RICH FORMATIONS

Due to the continuous and irreversible evolution of the living world, fossils provide sedimentary rock deposited in each interval of Earth history with a unique face. Although this applies to each kind of sedimentary rock, carbonates are usually the most characteristic of a period. Although deposition of siliciclastic rocks has largely been controlled by the same physical and chemical rules, changing climatic and oceanographic parameters as well as the interaction of the living world with its environment continuously create new conditions for the formation of carbonate rocks. Thus, the most characteristic and well-known Triassic rocks are limestones and dolomites.

Muschelkalk

The name of the middle unit of this classic Germanic succession is derived from popular misidentification of the brachiopod *Coenothyris vulgaris*. This form, somewhat similar to a bivalve in appearance, is found in large quantities in some beds. Characteristic but varied limestone successions assigned to the Muschelkalk were deposited on an evenly dipping gentle slope (carbonate ramp) covered with shallow water. This relatively shallow depth is indicated by frequent storm deposits (tempestites) full of the remains of bivalves, gastropods, and brachiopods. Fossil assemblages in rocks deposited in lagoons and in near-shore regions tend to be less diversified and almost exclusively consist of bivalves and gastropods. Deeper-water beds, on the other hand, yield abundant articulate brachiopods and rare ammonites. Trace fossils and sea lilies are also characteristic of these sediments. Some Muschelkalk successions in the type area are also justly famous for their reptile fossils. The state of preservation of invertebrates depends on the mineral composition of their original skeleton: calcite shells are often well preserved, whereas organisms with aragonite shells are replaced and often preserved just as internal molds. Overall, the diversity of Muschelkalk faunas is much lower than that seen in the majority of coeval sediments in the Alpine Triassic.

In addition to those in Germany, Muschelkalk successions are known from France, Spain, Switzerland, Poland, and Bulgaria, as well as from the Mecsek and Villány Mountains in Hungary. Such beds form, for example, the main part of the Misina Hill that rises above the town of Pécs and some beds of the Muschelkalk of Southern Transdanubia have long been exploited for their ornamental stone. The attractive olive-green and yellow-patched limestone slabs covering the walls of several public buildings, including the National Széchényi Library in Budapest, come from a quarry in the vicinity of the town of Siklós and are from the deepest marine part of the succession. The cut and polished surfaces of these rocks display abundant brachiopods, usually accumulated in nests, as well as white sections of sea lilies and rare cross sections of ammonites.

The geological relationship between the Middle Triassic successions of Southern Transdanubia and the Germanic Muschelkalk was debated for a long time. Sedimentological

When the "shell-bearing limestone" (Muschelkalk) is really shelly. A piece of limestone, collected on the Misina Hill at Pécs, containing the abundant remains of the burrowing bivalve *Pleuromya* (0.5× magnification).

and paleontological research carried out in the last two decades, however, has provided clear evidence for a Germanic affinity; the main counterargument to this is based on a few ammonite finds, including *Paraceratites,* an Alpine genus. Forms typical of the Germanic Basin are generally absent, but this apparent contradiction is easily explained: the endemic ammonite fauna in the Germanic Basin evolved later in the Ladinian, when freshwater sedimentation prevailed in the Triassic Mecsek Basin. Although it is true that brachiopods and benthic mollusks from the Middle Triassic mountains of Southern Transdanubia have been also documented from the Alpine Triassic, the general paucity of species in these fossil assemblages—albeit represented by huge numbers of specimens—indicates a sedimentary environment that is deviant from a normal marine one. This is another feature that is characteristic of the Germanic Triassic.

Main Dolomite and Dachstein Limestone

In the Late Triassic of the Western Tethys, a vast carbonate platform, most probably the largest in Earth's history, came into existence via deposition of shallow-water carbonates for more than 20 million years. Middle Triassic carbonate platforms of lesser extent had already formed in this area. Products of these older "carbonate factories" can be seen, for example, at Budaörs near Budapest, where a succession traditionally named the *Diplopora* dolomite (nowadays this is called the Budaörs Dolomite) after its characteristic fossil—a green calcareous algae—forms bare hills. A similar sequence seen in the Aggtelek-Rudabánya Mountains is formed largely from well-karstified Wetterstein Limestone. The extent of this Late Triassic platform, however, exceeded that of all earlier ones.

The succession of the Main Dolomite represents sediments formed in tidal flats and in poorly oxygenated lagoons, whereas the Dachstein Limestone was deposited farther from the coast in tidal and slightly deeper environments. The color of these rocks is predominantly light to dark gray because both formed from the remains of calcareous skeletons, which became unrecognizable during diagenesis. Dolomitization occurred early in this process, following the burial of the lime mud, and these Upper Triassic carbonates, several thousand meters thick, were later converted into mountain ranges—including the Dachstein Mountains in the Northern Calcareous Alps. These rocks are also widespread in Hungary: the Papod Hill that rises to the north of the town of Veszprém or the rocky slope of the hill, atop which is the Turul Memorial at Tatabánya, are Dachstein Limestone. The Main Dolomite forms some large masses in the Dolomite Mountains (Italy) and several hills in the Buda Hills where natural dolomite powder has long been mined for use as a scouring powder.

Although different in several respects, these rock types nevertheless share important common features. The most spectacular of these is their cyclic stratification, the regular repetition of characteristic sediment types in the successions. The huge and, at first sight, monotonous Upper Triassic platform was in fact a mosaic of biotic and sedimentary environments that ranged from occasionally dry to lagoons that could be meters deep. Long-lasting subaerial exposure resulted in erosion and in the formation of soils, as evidenced by thin, variegated clay beds known as the "A" members of the cyclothem. Thin beds formed by laminae 1–2 millimeters thick ("B" members) indicate that tidal flats were also frequent and further features of these peritidal beds include abundant pores and shrinkage cracks, interpreted as traces of algal mats. This type of carbonate rocks is known as loferite—named after the village of Lofer, in the Northern Calcareous Alps, where it was first documented. Lagoonal sediments are represented by compact limestone or dolomite beds ("C" members) several meters thick, and the three sediment types together usually form deepening-upward (ABC) sequences, reflecting short-term fluctuations in sea level. Dachstein Limestone sequences of predominantly lagoonal origin are well bedded and occur as thick banks; well-exposed examples can be seen in most rock walls in the Dachstein Mountains of Austria. At the other extreme, fossilized reefs that once separated the platform from the open sea are not stratified, and instead are found as massive rock bodies. Of these, the Gosau Ridge in Austria is probably the most spectacular example.

Identifiable fossils occur almost exclusively in the "C" member, and weathered surfaces of the Dachstein Limestone are often spotted in appearance because of the abundance of gray bodies about 1 mm in diameter. These

Thick beds of the Dachstein Limestone exposed on the Kálvária Hill at Tata, very near the open-air geological museum managed by Eötvös Loránd University.

Cross section of a *Megalodon*-like bivalve in the Upper Triassic Dachstein Limestone at Tata, Kálvária Hill. The minute gray spots on the rock surface are remains of *Triasina hantkeni*, the widely distributed marker foraminifera.

globular, calcite remains are the tests of the peculiar foraminifera *Triasina hantkeni* (see below).

Among the macrofossils found in these rocks, megalodontid bivalves—emblematic fossils of the Upper Triassic platform carbonates—are the most spectacular. Usually occurring as internal molds, these bivalves can reach half a meter in height; specimens often erode from their beds and fall or roll down. People living in the Alps in the past believed that these fossils often found on mountain paths were hearts that became stone. Most megalodontid bivalves lived more or less burrowed into bottom sediments and because of this mode of life are often preserved double-valved. Their sections visible on weathered rock surfaces strongly resemble the traces of cows, leading to the vernacular name "cow track bivalves" (Kuhtrittmuscheln in German) that is used by mountain people.

As noted by the Hungarian geologist Benő Winkler (1835–1915) in his 1883 paper dealing with the geology of the Hungarian Gerecse and Vértes Mountains, these rock surfaces that resemble the trackways of cows were also previously called "stone with puppets" by village people.

Hallstatt Limestone

The huge Late Triassic carbonate platform was separated from the Tethys Ocean by deeper-water areas. According to recent paleogeographical reconstructions, deep "channels" similar to the Tongue of the Ocean in the modern Bahama Platform cut into this platform as well. The celebrated Hallstatt Limestone, a typical Alpine Triassic rock famous for its occasionally well-preserved ammonites, is interpreted as representing one of these environments. According to the most recent, restrictive definition, the Hallstatt Limestone was deposited on top of submarine highs rising in the deep basins. The rock is predominantly red or pink, fine grained, and well bedded, and often occurs as infillings in fissures of older Triassic rocks. Among the macrofossils found in this sediment, ammonites are the most frequent—and specimens, often highly valued objects in private and public collections, are often coated with a thin layer of iron or manganese oxide. As a rule, shells are filled partly with red limestone and partly with white sparry calcite; the latter is especially spectacular because of its complex lobe lines. Specimens preserved in this style are often displayed polished and are quite beautiful. The most famous localities for Hallstatt Limestone are around the town of Hallstatt in the Salzkammergut region of Austria, but this rock is also known to crop out in Tibet and even on the island of Timor! According to museum labels, a number of specimens were collected at a locality called Feuerkogel ("Fire Hill") but this does not mean that there is only one place where spectacular ammonite specimens can be collected. Dozens of such hills where fires were lit on the occasion of feasts are indicated on maps of the Salzkammergut.

At first, the attention of members of the nobility able to buy specimens as curiosities was drawn to these attractive Hallstatt ammonites. For example, Count Klement Lothar von Metternich (1773–1859) had a considerable collection but after his death (he played rather a disagreeable role in the history of Europe, at least from the Hungarian point of view) his collection disappeared. Some years later Edmund von Mojsisovics (1839–1907, see below), a leading expert on Triassic ammonites, attempted to trace Metternich's collection, but without success. Metternich has, however, been commemorated in the species name of

Bivalves, such as *Tutcheria cloacina* (Quenstedt), are frequent elements of the fossil assemblages of the "Kössen Beds." This specimen was found by János Böckh at Akasztó Hill, near the village of Rezi in the Keszthelyi Mountains. Originally, the valves of this bivalve were formed from aragonite, but later recrystallized into calcite. The moisture of the soil as well as the acids produced by the roots of plants could better penetrate into the marly sediment than into the compact valves, so these became naturally prepared. Fossil collectors know that the best way to collect good fossils is to discover and exploit a new locality. In such cases we may rely on finding the products of the natural processes outlined above—at least in the beds closest to the soil—and the hard, time-consuming work of hammering and preparation can be fully, or partly, done by the slow forces of nature. Frequently, all a collector has to do is gather attractive fossils prepared naturally over hundreds of years (0.3× magnification).

a large-sized, disklike ammonite (*Pinacoceras metternichi*) whose specimens are especially attractive if polished.

Early researchers of the Alpine Triassic began to study the rich ammonite fauna of the Hallstatt Limestone in order to establish the chronological order of assemblages of different taxonomic compositions. These attempts were often unsuccessful because of the fissure-infillings, where geologically younger ammonites are frequently preserved below older ones in fissures opened and filled later. Interesting, these successions fly in the face of the law of superposition formulated in Tuscany by the Danish physician Nicolaus Steno (1636–1686), according to which beds formed earlier are always found below those formed later.

In Hungary the Hallstatt Limestone, although very poor in macrofossils, is known to occur in the Aggtelek-Rudabánya Mountains. Some rock types in the Middle Triassic—the "Nemesvámos" or "Tridentinus limestone" (Buchenstein Formation) in the Balaton Highlands—as well as the nature of ammonite preservation are also strongly reminiscent of the typical Hallstatt Limestone.

The "Kössen Facies"

This characteristic type of succession is named for a village in the Northern Calcareous Alps. In the surroundings of Kössen, near the town of Kufstein, a place well known in Hungary for its oubliette, a brachipod- and bivalve-rich dark gray marl and limestone sequence attracted the attention of early researchers in the Northern Calcareous Alps. Since the introduction of the name Kössen into the geological literature, a number of different kinds of rocks have been identified in facies with the type sections, and the term "Kössen Beds" has also been used in a chronostratigraphical sense. The common feature shared by these rocks, however, is the presence of the peculiar bivalve *Rhaetavicula contorta,* an index fossil of the Rhaetian stage of the Upper Triassic. On the basis of lithological characters and fossil content, Swabian, Salzburgian, Carpathian, and other facies have traditionally been distinguished among occurrences of these Kössen Beds and now these successions, as well as the "Kössen" itself, are considered to be different formations. In a modern and restrictive sense, the Kössen Formation was deposited in more or less oxygen-depleted smaller basins formed on the huge carbonate platform of the Dachstein Limestone. However, since older literature usually refers to the "Kössen" in its wider sense, to stop using it completely would be unfeasible.

The water of the Kössen basins was shallow, although deeper than that of surrounding areas. Their faunas and floras were thus a little bit more diverse than was the monotonous fauna of the platform, containing a few more species. Besides the forms mentioned above, colonial corals, especially belonging to the genus *Retiophyllia* (formerly referred to as *Thecosmilia*), and gastropods are common. Indeed, Kössen fossil assemblages can be distinguished easily, on the basis of their composition, from those of the geologically coeval Zlambach Marl that represents deeper-water areas bordering the platform toward the open sea.

Several Upper Triassic successions in the Carpathian region belong to these Kössen Beds in a broad sense. In Hungary, these sediments attain considerable thickness in the Bakony, and especially in the Keszthelyi Mountains and their representative sections and fossil content were documented in a monograph by Sándor Végh (1930–2009), a researcher at the Hungarian Geological Institute.

📖 Végh 1964

REMARKABLE FOSSIL SITES

The most famous Carpathian Basin Triassic localities, the focus of interest since the mid-1850s, are scattered in the hilly areas of the Balaton Highlands. In addition, well-documented faunas are also known from the Buda

Mountains, northern Hungary, the neighboring Slovak Karst, and the Apuseni Mountains of Romania.

Böckh and Mojsisovics: Pioneers of Alpine Triassic Research

When thinking about Triassic localities in Hungary, what first comes to the minds of Hungarian and foreign specialists alike are probably the Balaton Highlands. This gentle landscape is rightly considered a classic and exceptionally well-studied area of the Alpine Triassic, a real El Dorado for well-preserved Triassic fossils. This fame is partly due to fortunate geology: the area lacks the subsequent large-scale tectonic deformation that often makes observation of stratigraphic relationships difficult in the Northern Calcareous Alps. Thrusts and folds disturbing the normal sequence are rare in the Balaton Highlands—and, when present, are easily recognizable—so the chronology of rock bodies and fossil assemblages can be unambiguously established.

Besides these natural conditions, the fame of the Triassic in the Balaton Highlands is also due to the outstanding researchers who have studied and described fossils from this area. The first comprehensive works were published by János Böckh, one of the most eminent Hungarian geologists, in 1872 and 1874.

In these papers, which summarize the results of only two field seasons, several Triassic ammonite and brachiopod species are described. According to the taxonomic practice of that era, however, Böckh assigned the ammonites to only a few genera. Subsequent developments in paleontology, including changes to this taxonomic approach, and more detailed studies of these specimens had meant that new genera have been erected to accommodate the species from this area as their number has continued to increase. One leader in this field for more than 25 years was Edmund von Mojsisovics, an expert on Triassic cephalopods and the stratigraphy of the Alps.

Mojsisovics was a member of a noble family of Slovakian (or Croatian, according to some authors) origin, but spelled his name in Hungarian and spent most of his life in Vienna. He taught law for a while but in 1867 joined the Imperial Geological Institute (Geologische Reichsanstalt) in Vienna. In 1870 he was appointed chief geologist and in 1902 became vice-director. Overall, Mojsisovics's oeuvre is both monumental and fundamental. For example, his monograph that deals with the geology of the country around Hallstatt contains 223 lithographic plates, and his work on the ammonites of the Alpine Triassic contains 94 plates. His 1879 study on the Triassic "reef" in the Dolomites, now considered a classic paper in carbonate sedimentology, was translated into English and republished in 2001.

Although Mojsisovics personally participated only minimally in field research in the Balaton Highlands, he studied many of the Triassic fossils that were collected there. Based on ammonite species first described by Böckh, he erected the genera *Balatonites, Hungarites,* and

Portrait Gallery

János Böckh (1840–1909)—A Pioneering Researcher in the Hungarian Triassic

Böckh was born in Pest, Hungary, and wanted to become an army officer. However, because of an accident he chose geology instead, and in 1858 enrolled at the Academy of Mining at Selmecbánya (present-day Banská Štiavnica, Slovakia). After taking his degree he worked in Vienna for the Imperial Geological Institute (founded in 1849) and then for the Ministry of Agriculture and Industry of Hungary. In 1882 he was elected—"with an imperial education, but with a Hungarian heart"—director of the Royal Hungarian Geological Institute in Budapest. Böckh took part in the geological mapping of several areas in the Carpathian region and discovered, for example, the pelagic, ammonite-rich Triassic sequences in the Apuseni Mountains and in the Banat area. Böckh is rightly considered a pioneer researcher in the Alpine Triassic although his inaugural lecture at the Academy of Sciences in 1876 dealt with the geology and the Middle Jurassic fossils of the Mecsek Mountains. He also played a leading role in the early search for crude oil in Hungary that began in the early 1890s.

In addition to his merits as a scientist, Böckh was an excellent manager. The palace that is the home of the Hungarian Geological Institute, a significant example of Hungarian secessionist architecture designed by Ödön Lechner (1848–1914), was constructed during his directorship.

📖 Schafarzik 1914

Arpadites—the latter named after the Hungarian chieftain Árpád, leader during the Hungarian conquest.

The "Balaton Monograph" Series

Triassic fossils from the Balaton Highlands first came to international interest about one hundred years ago. This was as a result of publication in the early twentieth century of the so-called Balaton monograph series, one of the largest projects in Hungarian science. This series was initiated by Lajos Lóczy, Sr. (1849–1920), a well-known geologist and geographer and the successor to János Böckh as the director of the Hungarian Geological Institute in Budapest. Well-known experts, from Hungary and abroad, accepted Lóczy's invitation to participate in scientific studies of the country around Lake Balaton. These efforts resulted in a series of large volumes that was also reprinted in German, with some papers also appearing in English. The geological monograph written by Lóczy comprises four volumes of paleontological supplements, mostly containing papers about Triassic fossils. The Hungarian edition is over 2,421 pages and includes 128 lithographic and photographic plates.

Authors of these paleontological supplements were not ordinary people. The most productive was Fritz Frech (1861–1917), professor at the University of Breslau (now Wrocław, Poland). Frech, a good friend of Lóczy's, was the tutor to several well-known geologists who later worked in Hungary—including Gyula Prinz (1882–1973), the eminent monographer of Jurassic ammonites from the famous Tűzköves Ravine at Bakonycsernye, and Henrik Taeger (1881–1939), who later worked on Triassic and Eocene fossils from the Vértes Mountains. Frech died in Aleppo at the age of 56, while serving as a war geologist in the German army.

The Conflict between Bittner and Mojsisovics

Another productive author who contributed to these supplements was Alexander Bittner (1850–1902), an expert on the fossils and stratigraphy of the Alpine Triassic. It might be better to now describe Bittner as ill famed because he is also known for initiating a conflict that deeply divided the Alpine geological community in the last years of the nineteenth century. This issue began with a critique written by Bittner about a paper of Mojsisovics dealing with stratigraphic subdivisions of the Alpine Triassic. Mojsisovics was older than Bittner and was also higher up in both the Reichsanstalt and in local society. He had engaged himself for years in finding the right succession for the one and one-half dozen or so ammonite assemblages, or stratigraphic levels, recognized at that time. Owing to the complicated structure of the Alps, as well as the large distances between localities and other causes (as described above), often successions simply cannot be established in the field. Indeed, Mojsisovics, despite modifying his scheme several times, was never able to find the real order in some cases. However, the short, bespectacled, and quick-tempered Bittner accurately judged the sequence of faunas in several cases and did not hesitate to publish his contrary ideas in biting pamphlets.

The debate between the two men soon degenerated into furious attacks, and Bittner did not hesitate to resort to dirty tricks to make his points; he attacked Mojsisovics's manhood. Geologists identified themselves with one or the other and signed open supportive letters as bit by bit the whole geological community became factionalized. Among Bittner's supporters was one troublemaker extraordinaire who printed and distributed letters at his own expense. Although it is not known how the debate ended, it is certain that Mojsisovics retired from the Reichsanstalt. Bittner, by then an old bachelor living with his elder sister, had little time to bask in his glory: On Easter Sunday, 1902, he choked and died of a severe asthma attack. There is another story—amusing, perhaps, to the reader but highly frustrating for one of Bittner's colleagues. This comes, probably, from a story related by Elemér Vadász to András Kaszap, an outstanding historian of Hungarian geology. Gustav Adolf von Arthaber (1864–1943), another author of a Balaton monograph, always took great care of his appearance, a personality trait that tends not to characterize geologists. On one field trip Bittner noticed that Arthaber's rucksack contained shoe-cleaning tools and even a mustache-fixing band. For some reason this made him so angry that he actively hindered Arthaber's application to the Geologische Reichsanstalt. As a result, Arthaber ended up working at the University of Vienna because there were no openings for him in Imperial geology.

A complete bibliography of the Triassic fossils from the Balaton Highlands, including all works available at the beginning of the 1970s, was assembled by Imre Szabó, an expert on Triassic field geology. This bibliography was published as an accompaniment to the Veszprém sheet in the 1:200,000 scale geological map series of Hungary.

📖 I. Szabó 1972

Felsőörs: Forrás Hill (Balaton Highlands, Hungary)

As noted at the beginning of the twenty-first century by Michael Orchard, a Canadian conodont specialist and the president of the Triassic Subcommission of the International Stratigraphic Commission, the slope of Forrás Hill on the northwestern margin of Felsőörs village is

Geological profile of Forrás Hill at Felsőörs, one of the best known Triassic sections in the Balaton Highlands.

a Triassic paleontological Disneyland. This now-classic Middle Triassic sequence in the Balaton Highlands National Park is now protected from the weather by a roof and supplemented with abundant explanatory placards for tourists. The Forrás Hill section is one of the most well known in the Alpine Triassic, and as such is well worth conserving. Its siliceous limestone and marl beds, rich in ammonites, were discovered by János Böckh in 1870. This fauna, including the form nowadays called *Reitziites reitzi,* was described partly by him and partly by Lajos Roth (1841–1928) and József Stürzenbaum (1845–1881). Roth and Stürzenbaum made extensive collections of Felsőörs ammonites at Böckh's request, and Mojsisovics later named this unique assemblage as the type fauna of the "Trachyceras Reitzi Zone."

The Felsőörs sequence reentered the spotlight in the 1980s because a need arose to designate a GSSP (Global Stratotype Section and Point) for the lower boundary of the Ladinian stage. Forrás Hill is the type locality of the Reitzi Zone, which was traditionally regarded the base of the Ladinian, The section became one of the most likely candidates for GSSP status, along with some other sections in the Southern Alps. Although in 2004 the Bagolino section in Italy was chosen as the GSSP for this age, with the base of the Ladinian defined higher in the sequence, the large-scale exploration and fossil collection that has been carried out at Felsőörs makes this Hungarian site exceptionally important among the Triassic sections in the Balaton Highland.

In the lowest fossiliferous beds of this protected section the columnals of sea lilies and brachiopods are dominant. József Pálfy identified 15 species of the latter group, among them the most frequent is *Tetractinella trigonella,* a characteristically triangular form bearing four strong radial ribs. *Caucasorhynchia altaplecta* and *Trigonorhynchella attilina* are also common; stratigraphically higher beds almost exclusively contain ammonites, and *Ptychites*-like forms are dominant.

📖 A. Vörös et al. 1996, 2003

József Pálfy, former head of the Department of Geology and Paleontology at the Natural History Museum of Hungary and professor at Eötvös Loránd University since 2011. Pálfy's scientific career began with a study of Triassic brachiopods from the Balaton Highlands; his later publications have contributed significantly to knowledge of Lower Jurassic stratigraphy as well as the events at and around the Triassic-Jurassic boundary. His interests also include radiometric calibration of the geological timescale.

Aszófő: Farkó-kő Hill (Balaton Highlands, Hungary

As is the case for most good fossil localities, Farkó-kő has been known for a long time. As mentioned by Lajos Lóczy Sr., dark-brown and yellowish-gray rock slabs full of fossils were found first here in 1907, when grapes were planted in the area; the section was studied in detail in the early 1980s, when two trenches about 30 meters in length were dug and numerous fossils were collected. The sequence comprises about 80 beds and has so far yielded nearly 5,000 specimens of ammonites, usually of small size. *Balatonites balatonicus,* as well as species of *Norites, Bulogites,* and *Schreyerites* are the most frequent: most specimens are crushed characteristically and their inner whorls are filled with silica. A few well-preserved specimens make up a diverse nautiolid fauna, of which *Germanonautilus* is especially worthy of mention.

In addition to cephalopods, brachiopods, bivalves, and gastropods are also found at this site; among the bivalves, masses of thin valves of *Daonella* and *Bositra* form whole beds. Specimens of *Solemya* are adapted to oxygen-depleted environments and are considered living fossils; *Pseudocorbula gregaria,* which form monospecific assemblages on bedding planes, are also common.

📖 A. Vörös 1987

Jeruzsálem Hill, Veszprém: The Locality of the "Pebble-Toothed Pseudo-Turtle" (Bakony Mountains, Hungary)

Jeruzsálem Hill is one of the celebrated seven hills of Veszprém. Nowadays, only the names of streets—such as Csákány ("Axe"), Kőbánya ("Quarry"), and Pöröly ("Sledgehammer")—remind us of the times Upper Triassic (Carnian) marls in this area were exploited in pits dug at the top of the hill. This well-bedded marine marl succession has yielded the remains of, among other things, one of the most famous fossils from Hungary, the so-called pebble-toothed pseudo-turtle (*Placochelys placodonta*). The fauna here consists of nearly two hundred species and, thus, is the richest Triassic site in the Balaton Highlands. Indeed, compared to other known Carnian assemblages, only the fauna of the celebrated San Cassiano Formation in the Dolomites is more diverse than these "Veszprém Beds." Fossils from this area are mostly available for study because of the enthusiasm of Dezső Laczkó (1860–1932), a Piarist monk, teacher, and geologist. Laczkó, an outstanding expert on the geology of Veszprém and its surroundings, collected Triassic fossils along with his students for decades and regularly visited the open pits on Jeruzsálem Hill. As he mentioned, however, by the 1890s conditions for collecting there were already unfavorable: "Nobody should believe that this place is an El Dorado of fossils." The fossiliferous beds were accessible in only a very few pits operating at a given time, and were otherwise covered with waste heaps. Due to the exhaustion of the marl reserves and the expansion of the town there were few chances to find new localities even by this time.

Laczkó not only collected there but also made careful notes on the mode of occurrence of the fossils he found. Because of him we know, for example, that huge boulders consisting of sponges and corals derived from the adjacent shallow-water areas were embedded in the marl succession as it was deposited in a relatively deep basin. He also produced a detailed geological description of the surroundings of Veszprém and since 1990 the town's museum, the successor of an organization founded by Laczkó, has borne his name. A sculpture of Laczkó by Ferenc Medgyessy (1881–1958), stands beside the museum building.

📖 Laczkó 1911

Nemesvámos: Katrabóca Hill (Balaton Highlands, Hungary)

According to the enthusiastic, although exaggerated, description given by Dezső Laczkó, this "rocky corner bastion of Katrabócza" is situated south of the village of Nemesvámos. Today this gentle hill is covered with dense forest, but rocks were visible here about a hundred years ago, when a Middle Triassic limestone named the "Vámos Marble" was exploited in several pits. This red, violet, or green flint-bearing limestone is also known as the "Tridentinus limestone," after its characteristic fossil, a globular, ammonite species related to the genus *Arcestes.* These former pits have already collapsed and are covered by vegetation, but they still show well-recognizable patterns along the strike of the rock. Ammonite specimens were mainly collected by Laczkó in working pits and described by well-known experts of that age—such as Carl Diener (1862–1928), rector of the University of Vienna, and Fritz Frech—and are without a doubt the most attractive specimens to come from the Triassic of Hungary. However, the scientific value of these fossils is very limited because of the casual nature in which they were collected (they were not collected bed by bed and little precise positional data is available). Modern reinvestigation of this locality was carried out at the end of the 1990s. After a careful excavation starting off in one of the collapsed pits, two ammonite zones in the Middle Triassic were identified. This more recently collected fauna includes the zonal markers *Protrachyceras gredleri* and *P. archelaus,* as well as the nominal form *Joannites tridentinus.* In addition, several species of the genus *Arpadites* and, in smaller numbers, gastropods and bivalves were also collected.

📖 A. Vörös 1998

Budapest: Fazekas and Remete Hills (Buda Mountains, Hungary)

The Triassic rocks in the Buda Mountains are far less known for their fossils than those of the Balaton Highlands. Indeed, just finding the brachiopods that were described by Károly Hofmann (1839–1891), a pioneer investigator of the geology of the area, was most probably due to the enthusiasm and good luck of this eminent researcher. Although weathered surfaces of the Dachstein Limestone display sections of bivalves, gastropods, and corals in several places, their remains cannot be extracted from the rock. Only two localities have yielded relatively rich fossil assemblages.

Both of these localities are abandoned quarries. One is situated behind the Primary School at Remetekertváros; the other is at Budaliget, at the mouth of the Remete Gorge.

Portrait Gallery

Endre Kutassy (1898–1938)—Researcher of Triassic Fossils

Born in Hajdúböszörmény, Endre Kutassy received his PhD in 1922 from Pázmány Péter (now Eötvös Loránd) University in Budapest. He then spent his professional career in the Department of Geology of that university, beginning in 1924 as an assistant lecturer, becoming a lecturer in 1928, and then Privatdocent in 1929. He was appointed full professor in 1937, shortly before his death. Besides his numerous papers describing Triassic fossils from Hungary and other countries, he wrote a book entitled *Ősmaradványok gyűjtése és konzerválása* (Collecting and conservation of fossils) and completed four volumes dealing with Triassic gastropods, bivalves, and cephalopods, all published in the series Fossilium Catalogus. Although he died prematurely of pulmonary tuberculosis, his life's work is enormous. According to his contemporaries, Kutassy lived very intensively, burning the candle of his life at both ends.

📖 Bogsch 1939

Digging an artificial trench at Farkó-kő Hill near the village of Aszófő in the early 1980s. The Anisian sequence exposed has proven to be especially rich in ammonites, bivalves, and brachiopods. Without artificial exposures such as this one, the Triassic stratigraphy of the Balaton Highlands would be much less well known.

Both sites became known in the 1920s when Mór Pálfy (1871–1930) published an interesting preliminary report on the occurrence of fossils in the Fazekas Hill Quarry. He interpreted the soft, pulverized rock in the fossil-bearing bed as the deposit of a submarine spring. Unfortunately, however, Pálfy did not continue his work. Meanwhile, Elemér Vadász published a paper on the geological age of the Dachstein Limestone in the Buda Mountains, in which he supported his stratigraphic conclusions using the fossils found at the two localities. Finally, a description of a rich fauna comprising nearly one hundred species was published in several papers by Endre Kutassy (1898–1938), also known as Andreas Kutassy, a short-lived but very productive expert on Triassic fossils. The fossil assemblage he described is strongly dominated by gastropods,

Small ammonites, just 1–2 centimeters in diameter, from the Dachstein Limestone at Fazekas Hill.

represented occasionally by relatively large forms, alongside a few bivalves, cephalopods, sea urchins, and corals.

Ammonites from this site were revised in the late 1960s by Anikó Bérczinė Makk, then a student of geology at Eötvös Loránd University and later an acknowledged expert on foraminifera and Mesozoic sediments reached by boreholes drilled in the Great Hungarian Plain. Bérczinė Makk worked until her retirement for the Hungarian National Oil Company. Interestingly, the specimens she studied were all collected by György Buda, who later became the head of the Department of Mineralogy at Eötvös Loránd University. Although later attempts to rediscover the fossiliferous bed were unsuccessful, a multilingual poster was still placed in the abandoned quarry to inform visitors of the importance of the area.

📖 Kutassy 1927, 1933, 1936; Bérczinė Makk 1969; J. Szabó 2011

Csővár, Pokol Valley: A Triassic-Jurassic Boundary Section in the Cserhát Mountains, Hungary

In spite of the ominous name (Pokol means "Hell") of this actually quite friendly valley, the rocks exposed in the stinging bushes on the steep slope of Vár ("Castle") Hill as well as in the quarry in front of it have been of interest to geologists for more than a century. This marly and siliceous limestone sequence differs from all others known from Hungary; recent studies, revealing peculiar fossil assemblages, have confirmed the importance of the Csővár section.

József Szabó (1822–1894), according to some authors the most eminent Hungarian geologist of all time, wrote in 1860 that he believed the whole succession here to be Early Jurassic in age. In the absence of diagnostic fossils, this statement was based on lithological considerations. Nearly half a century later, in 1908, Elemér Vadász proposed an Upper Triassic age for the Csővár Mesozoic, a determination based on the fact that the lower part of this succession is formed by marly limestone beds, superficially similar to the Carnian marls in the Balaton Highlands. Debate as to the age of these rocks continued until the 1973 publication of a paper by Heinz Kozur and Helfried Mostler, the latter a professor at the University of Innsbruck, who used microfossil evidence from conodonts and other invertebrate remains to provide evidence for a Norian, or Latest Triassic, age for the lower part of this sequence.

Later, this age was confirmed by exceptional macrofossil evidence: Viktor Hermann, a talented collector from the Hungarian Geological Institute, found a specimen of *Choristoceras,* a heteromorph ammonite, characteristic of the uppermost Triassic. It is interesting that the lucky find of this form, previously unknown from Hungary, was first published in an amateur periodical of very high quality, the *Ásványgyűjtő Figyelő* (Mineral collectors' observer). *Ásványgyűjtő Figyelő* was edited at Eötvös Loránd University and offered a much shorter time between submission and publication than the official, professional geological periodicals. In 1988, however, the official description of this ammonite by Csaba Detre, Lajos Dosztály (1961–1999), and Viktor Hermann was also published, and then in 1993 the "veteran" Csővár worker Heinz Kozur reported ammonite-bearing Jurassic beds on the slope of the Vár Hill. This was another important discovery because continuous marine ammonite-bearing Triassic-Jurassic boundary sections are rare in Europe. Over the last decade, intensive and detailed paleontological, sedimentological, and geochemical studies have been carried out in this section in order to gain a better understanding of events at the end-Triassic. József Pálfy and co-workers have, for example, been able to locate the Triassic-Jurassic boundary within an interval a few meters thick and to document occurrences of numerous fossils previously not recorded from the locality.

📖 Kozur & Mostler 1973; Detre et al. 1988; Kozur 1993; J. Pálfy, Demény, et al. 2007

Aggtelek-Jósvafő: Baradla Cave (Aggtelek-Rudabánya Mountains, Hungary)

For geologists, caves are not only interesting and sometimes frightening phenomena of nature, but also are places to study rock successions hardly visible on the surface due to the cover of soil and vegetation. Although the Baradla Cave was inhabited by ancient man, geological research in the longest cave in Hungary began only in the 1970s. Systematic sampling resulted in the recognition of several fossils, especially conodonts and calcareous algae, the study of which has contributed significantly to our knowledge of geological structures in the area as well as the depositional history of the Wetterstein Limestone succession into which the cave was carved.

📖 Szentpétery & Less 2006

Hybe (Liptov County, Slovakia)

This small village situated near the town of Liptovský Mikuláš is famous not only because of the church where great Hungarian poet Bálint Balassi (1554–1594) is buried, but also for Upper Triassic outcrops considered to be among the most fossiliferous in the Carpathian Basin. Dark-gray limestone beds here, full of the shells of bivalves and brachiopods, represent the "Kössen Beds" in their broadest sense and are found here to the southeast of the village in the Biely Váh River valley. The fauna from Hybe was described in a 1907 monograph by Walery Goetel (1889–1972), an eminent researcher from Kraków (Poland) who was also a skilled mountaineer active in the High Tatras Mountains. Most Hybe fossils are derived from exposures named Simkovics and Kántor. As revealed in a recent revision of this fossil assemblage by Marián Golej of Bratislava, the bivalve fauna is dominated by the early oyster *Actinostreon* and scallops. Besides bivalves, outcrops at Hybe have yielded a brachiopod assemblage characteristic of the Rhaetian stage. Miloš Siblík, an expert on Triassic and Jurassic brachiopods from the Alps and the Carpathians identified typical forms of the Kössen facies, such as *Fissirhynchia fissicostata, Rhaetina pyriformis,* and *Zugmayerella koessenensis.*

It is interesting to note that gastropods here are rare elements of the fauna. Among them, the genus *Kokenella* is worth mentioning because its disklike shell is strongly reminiscent of ammonites, otherwise lacking in the Hybe fauna. This similarity, associated with poor preservation, even misled Miloš Rakús (1934–2005), a well-known expert on Triassic and Jurassic ammonites.

📖 Goetel 1917; Golej 2005

Drnava: Bleskový Prameň (Slovak Karst, Slovakia)

Leaving the eastern end of Drnava village, one reaches Bleskový Prameň ("Horrible Spring"). This name refers to the large, muddy, and roaring water masses that can suddenly emanate after heavy rains. This water often threatens surrounding buildings with flooding. József Stürzenbaum, from the Geological Institute in Budapest, first discovered fossils in the area, especially abundant in the dark gray crinoidal limestone next to the spring. However, due to his untimely death, his initial 1879 report was never concluded and the fossil assemblage was first studied in detail much later by Austrian experts. A brachiopod fauna consisting of more than 20 species was documented in 1890 by Alexander Bittner; another paper, published in 1896 by Edmund von Mojsisovics, described 12 ammonite species from this locality.

Large-scale, repeated collection work has subsequently proved that the fossil assemblage at this site is more diverse than was previously thought. Results of a restudy of the locality are summarized in a valuable monograph published in 1973 by Vanda Kollárová-Andrusovová (1921–2004), an expert on Triassic ammonites, and Mária Kochanová, a specialist on Triassic and Jurassic bivalves. The presence of 17 ammonites, 16 gastropods, and 56 bivalve species have been documents, making the Norian Bleskový Prameň fauna one of the most diverse Upper Triassic fossil assemblages from the Carpathian Basin.

📖 Siblík 1967; Kollárová-Andrusovová & Kochanová 1973

Braşov: Dealul Melcilor Hill (Southern Carpathians, Romania)

Although this hill, the top of which offers a good view of both the historical center of the town and an area of modern housing developments, is named after modern snails (Dealul Melcilor means "Gastropod Hill"), Triassic beds exposed here have yielded one of the most diverse and gastropod-rich fossil assemblages from the Southern Carpathians. These fossiliferous light-gray limestone beds were first identified as being Štramberk Limestone, an Upper Jurassic shallow-water deposit widespread in the Carpathians, by Erich Jekelius (1889–1970), who was a tireless researcher of the geology of the country around Braşov as well as of the history of the Saxonians in Burzenland (in Romanian, Ţara Bârsei refers to the region around Braşov). Later Jekelius recognized the much older, late-Middle Triassic age of the beds as they cropped out at the foot of the Schneckenberg (the German name for Dealul Melcilor Hill).

This hill, on which traces of an ancient human settlement were discovered, is now a protected area and the Triassic fossils were found near the quarry of the cement factory still in operation at the time. The assemblage consists of small-sized, but well-preserved, remains of abundant gastropods; a few species of bivalves; as well as some ammonites, brachiopods, and sea urchins. An interesting segmented calcareous sponge, *Colospongia,* has also been described from this site.

📖 Jekelius 1935

Triassic Vertebrate Localities from the Bihor Mountains

The study of the Triassic vertebrates known from the Bihor Mountains began in the 1960s under the leadership of Tibor Jurcsák (1926–1992). Exposures scattered around the villages of Alesd, Pestes, and Finiş have yielded the remains of diverse fish as well as aquatic and continental reptiles. Most of these very attractive fossils are now housed in the museum at Oradea.

Portrait Gallery

Tibor Jurcsák (1926–1992)—Expert on the Triassic Reptiles from the Bihor Mountains

Tibor Jurcsák trained first as a biologist, but in 1951 joined the staff of the Museul Ţării Crişurilor (Museum of the Körös River country), where he became a paleontologist. He founded the Department of Natural History at the museum and published several papers on the fossil aquatic and continental reptiles from the Pădurea Craiului ("Kings' Forest") and Bihor Mountains. From 1977 until his death he worked on Cretaceous fossil reptile remains collected from the bauxite mine at Cornet. With his colleague Jenő Kessler, a professor at Babeş-Bolyai University in Cluj-Napoca, he studied supposed bird remains associated with the dinosaurs.

TRIASSIC FOSSILS FROM THE CARPATHIAN REGION

The Carpathian Basin is rightly considered to be one of the richest areas in the world for Triassic fossils, as the successions and fossil assemblages here correspond to both the main facies types of the European Triassic (Germanic and Alpine). Most fossil assemblages are of marine origin; ammonites, bivalves, gastropods, and brachiopods are dominant. However, practically all the known groups of invertebrates can be encountered in this region, along with rare and very valuable fossils of vertebrates. Among the fossil plants, calcareous algae occur in rock-forming quantities in some successions, whereas the remains of vascular plants are much less frequent.

Microfossils

Spores and Pollen Grains

Triassic rocks from the Carpathian region contain a wide spectrum of microfossils; thus, this area is a real treasure trove for such research. In the first place, microfossils of plant origin are worth mentioning because they are known to occur in almost every Triassic sedimentary rock—ranging from shallow lake deposits to deep sea basins.

Being organic-walled fossils, the preservation of spores and pollen requires oxygen-free substrate conditions. Such settings are usually represented by dark-colored rocks, which have long been the focus of palynologists.

Fossil spores and pollen from continental Lower and Upper Triassic successions in Southern Transdanubia were intensively studied because of ongoing searches for uranium ore and black coal. These investigations were done by paleontologists for commercial reasons and not geological ones; the successions in question contain few fossils useful for age determination or stratigraphic correlation. Palynological studies in Middle Triassic marine sequences free of raw materials began only recently.

Spores and pollen from Lower Triassic rocks, as well as older samples, were studied for decades by Ágnes Barabásné Stuhl, who was affiliated to the state uranium ore company. Her efforts resulted in the correct assignment of the Permian-Triassic boundary in the sequence below the Jakabhegy Sandstone previously that was previously believed to be Late Permian in age. Organic microfossil assemblages from these dark-colored continental sediments deposited in reductive conditions consist exclusively of spores and pollen grains. In the lower part of the succession *Densoisporites nejburgi* is the characteristic form, whereas in the middle part *Voltziaesporites heteromorpha* is the dominant element. Index forms extracted from the upper part of the succession (*Triadospora crassa* and *Stellapollenites thiergartii*) indicate that these rocks represent the Anisian stage of the Middle Triassic.

A palynological investigation of the Middle Triassic shell-bearing limestone sequence in the Mecsek Mountains—with participation from Annette Götz, an expert on Triassic palynology from the Technische Universität Darmstadt—is in progress. The initial results are promising: it has been possible, on the basis of organic-walled microfossils, to fit different parts of this thick limestone succession, otherwise poor in index macrofossils, into the framework of Middle Triassic chronostratigraphy as well as to correlate this Hungarian sequence with the Germanic Muschelkalk.

The Upper Triassic organic-walled microfossils from the Mecsek Mountains were first studied by József Bóna, a well-known palynologist as well as the retired head of the now-defunct Geological Laboratory at Komló, an important center for geological investigation in Southern Transdanubia. Numerous grains of *Singulipollenites* indicate a Carnian age for the dark-gray limestone succession known as the Kantavár Limestone; the presence of the marine microplancktonic *Micrhystridium* indicates that the ocean was close at hand. The overlying Karolinavölgy Sandstone has also yielded a rich palynoflora consisting of about 80 species, of which most are spores and pollen, although a few marine planktonic forms are also present. Indeed, the study of these organic microfossils has at long last closed a question—as *Ovalipollis pseudoalatus, Aratrisporites minimus,* and *Triancoraesporites ancorae* equivocally indicate a Late Triassic age for the lower member of this coal-bearing sequence, previously believed to be entirely Early Jurassic in age.

Fossil spores and pollen are also known from nearly all the stratigraphic levels of the Triassic in the Hungarian Central Range, although work on these has focused on Transdanubian occurrences, especially those from the Balaton Highlands. The work of Ferenc Góczán, former head of the also defunct paleontological department of the Hungarian Geological Institute, has resulted in subdivision of the usually fossil-poor Lower Triassic into successive palynozones. Góczán began his work in the early 1960s with a palynological study of the Kössen Beds. Results of these investigations—co-authored with B. S. Venkatachala (1933–2007), director of the Birbal Sahni Institute of Palaeobotany in Lucknow, India—were published in 1964; this paper has become a standard work in Mesozoic palynology.

Góczán was able to subdivide the Triassic successions in the Balaton Highlands into seven zones and acquired resolution comparable to ammonite biochronology, the applicability of which is limited because of the scarcity of index forms in this area. Palynostratigraphy is especially important in the subdivision of the thick marl successions in the Upper Triassic.

Palynological research has also revealed the presence of some interesting and previously unknown forms of organic-walled microfossils from the Upper Triassic at Csővár. Two thick-walled, globular forms of large size (compared to most spores and pollen grains) were assigned from here by Góczán to the new genera *Oraveczia* and *Vadaszia,* named in honor of previous researchers who worked on the Csővár Mesozoic.

📖 Venkatachala & Góczán 1964; Bóna 1995; Góczán 1997; Götz et al. 2003

Calcareous Algae

The Upper Triassic rocks in the Mecsek Mountains have also yielded calcareous plant microfossils. These remains include the calcified reproductive structures of freshwater stoneworts belonging to Charophyta or Chlorophyta (green algae), according to the traditional system.

These tiny globular, or barrel-shaped, bodies ornamented with keels are usually known as "*Chara* oogonia" and even some poorly-preserved specimens of the extinct genus *Atopochara* were recently extracted from the Kantavár Limestone, which has been interpreted as a lacustrine or lagoonal succession.

Astonishing the geologist community of Hungary, charophyte remains were also found by Felicitász Híves Velledits, a hardworking researcher on the Triassic of Northern Hungary, in cores that come from a borehole drilled in the southern part of the Bükk Mountains. The

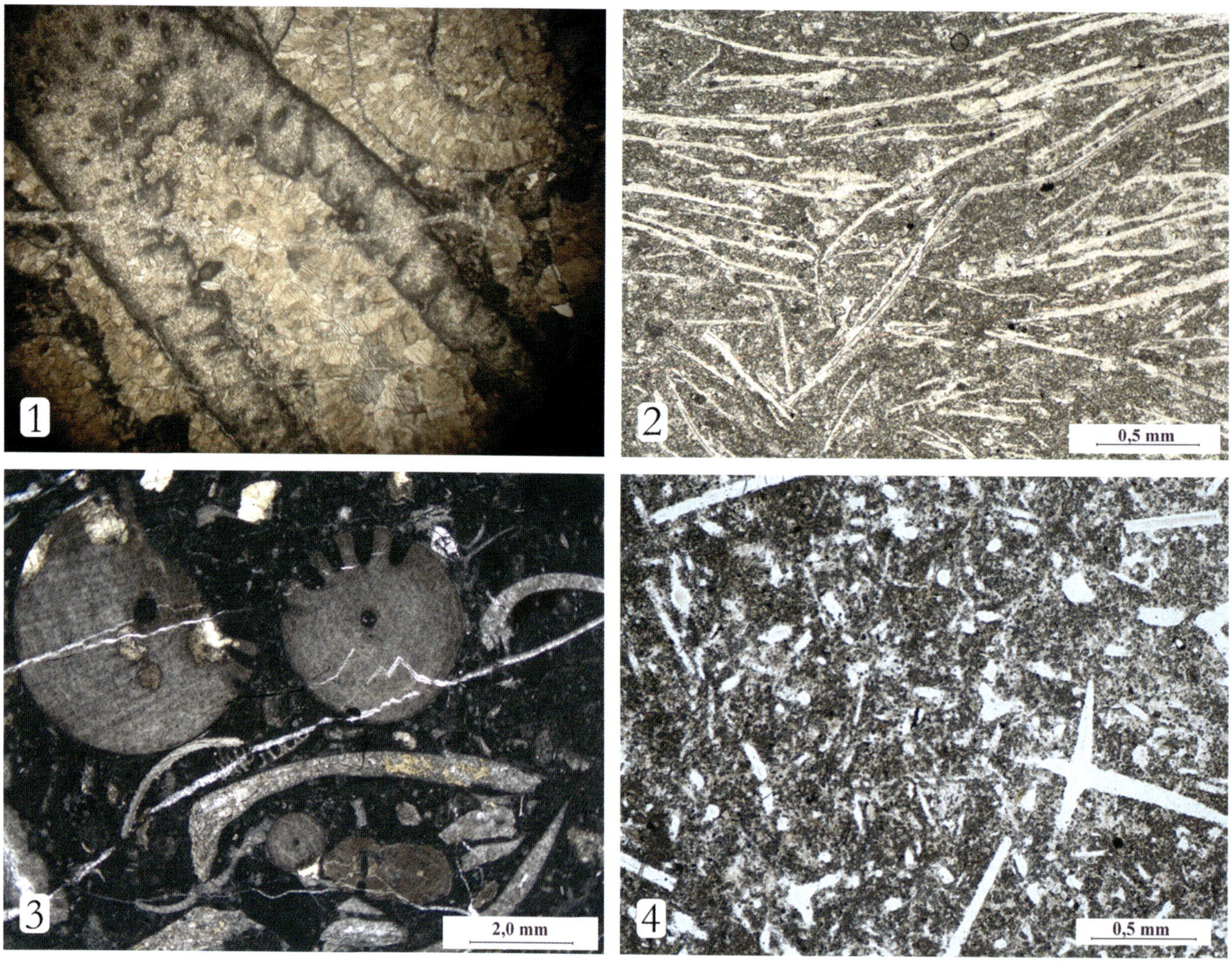

Thin-sections of fossil-rich Middle Triassic rocks. (1) *Diplopora annulata* (Schafhäutl) from Aggtelek. This frequent calcareous green alga appears in thin-sections, usually in the form of calcareous rings or cylinders of some millimeters in diameter; its walls are penetrated by fine radial pores. (2) A deeper-water limestone full of thin bivalve shells ("filaments") from Aszófő. (3) A limestone from Aszófő containing sea lilies, brachiopods, and bivalves. (4) A limestone with abundant sponge spicules (spiculite) from Felsőörs.

occurrence of these fossils proved, for the first time, that these Bükk Triassic deposits were freshwater; they had been thought to be entirely marine in origin. Velledits—along with her colleagues from the Department of Geology at the University of Erlangen, a leading European center of carbonate sedimentology—also published a monograph on the Triassic reef-forming organisms from the Bükk Mountains.

Much more diverse marine calcareous algal floras are also known from platform carbonates in the Hungarian Central Range, the Slovak Karst and in the Apuseni Mountains. All these deposits represent the Alpine Triassic. Algal-bearing sequences, such as the Wetterstein Limestone, are usually several hundred meters thick and are lithologically rather monotonous. Since good index fossils, especially ammonites, are extremely rare in these sequences, calcareous algae are often the only tool for geological age determination and for correlation. The Aggtelek-Rudabánya Mountains and, in particular, the Slovak Karst situated immediately north of them are most probably the richest areas in Triassic calcareous algae in the whole Carpathian region. By the mid-1960s Ján Bystrický (1922–1986) published a thick monograph on the floras of the Slovak Karst containing descriptions of 71 species and subspecies. Modern investigations of these Hungarian occurrences, however, began much later—with the work of Olga Piros, now an internationally acknowledged expert on Triassic algae and, since 2007, director of the Geological Library of Hungary (and last, but not least, efficient and patient editor of Hungarian geological periodicals). Her work has resulted in subdivision of the Anisian, Ladinian, and Carnian Stages, corresponding to a period of about 20 million years, into successive zones defined by fossil algal remains. The most important index fossils are, among others, species of the genera *Physoporella* in the

Remains of calcareous algae on a weathered surface of the Wetterstein Limestone, near Aggtelek (approximately original size).

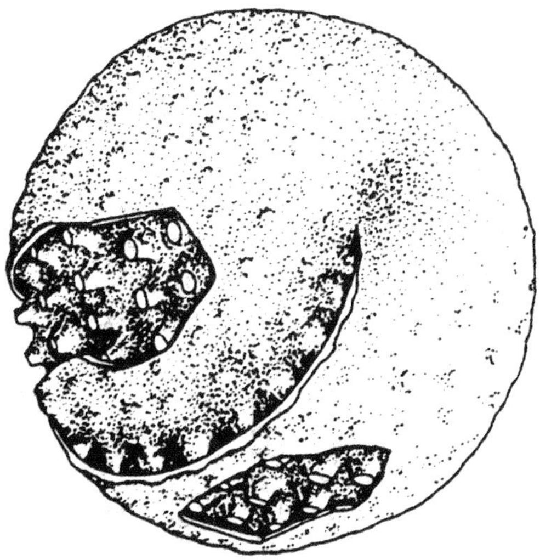

Triasina hantkeni Majzon. This relatively large-sized "small" foraminifera is one of the most widely distributed Triassic fossils. This genus and species were introduced by László Majzon (1904–1973), a well-known expert of foraminifera, on the basis of specimens collected at localities in the Pilis and Gerecse Mountains. *Triasina hantkeni* is a globular—or slightly flattened—form, with younger whorls completely covering older ones. These whorls are connected by radial pillars as shown in the illustration, in which parts of the last whorl have been removed. *Triasina* is an index fossil for the upper Norian and Rhaetian stages. Drawing after Di Bari and Rettori 1996.

Anisian, *Diplopora* in the Ladinian, and *Macroporella* in the Carnian.

A new genus of fossil red algae (*Aggtecella*) was named after the village of Aggtelek, famous for the entrance to the Baradla Cave, by Baba Senowbary-Daryan and Felicitász Híves Velledits. Thanks to the work of Velledits, the oldest Triassic platform margin reef in the Alpine-Carpathian region was also discovered in the Aggtelek-Rudabánya Mountains. Similar, although less diverse, assemblages are also known from the Transdanubian Range, the Bükk Mountains and from the Southern Carpathians.

📖 Bystrický 1964; Flügel et al. 1992; Monostori 1996a; Piros 2002; Senowbari-Daryan & Velledits 2007; Velledits et al. 2011

Foraminifera

Triassic rocks in Southern Transdanubia have yielded just a few animal microfossils. Foraminifera, for example, are known almost exclusively from Middle Triassic limestone successions and have been studied in thin-sections. Most of them are agglutinated forms with tests built up from grains gathered and glued to each other by the animal. The presence of the species *Pilammina* (sometimes also referred to as *Glomospira*) *densa* is worth mentioning because this form resembles a dense skein and is an index fossil for the middle and upper Anisian.

In contrast to the small numbers of foraminifera that are found in the Triassic mountains of Southern Transdanubia, more than one hundred species have been described from the Hungarian Central Range. These species occur in distinct assemblages of more or less the same composition and age-determining significance. The Lower Triassic is characterized by the so-called *Cyclogyra-Rectocornuspira* assemblage, a fauna recorded across almost the whole area of the former Tethys Ocean. For decades, this assemblage was thought to indicate the lower Scythian substage, the Induan stage, but as pointed out in 1996 by Kinga Hips, a carbonate sedimentologist from the Geological, Geophysical and Space Science Research Group at the Hungarian Academy of Sciences and Eötvös Loránd University, this assemblage in fact persists into the upper Scythian (Olenekian). Thus its importance as zonal marker is somewhat decreased. Upward in the sequence, *Meandrospira pusilla*, a species first described from the Sichuan Province of China becomes frequent and indicates a higher level of the Lower Triassic.

Middle and Upper Triassic successions in the Transdanubian and Northern Hungarian Ranges are different. Modern, comprehensive data on these foraminifera are only available for the former area in a monograph published by Anna Oravecszné Scheffer, a retired micropaleontologist who worked at the Hungarian Geological Institute. This study, published in 1987 in *Geologica Hungarica Series Palaeontologica*, summarizes the results of several decades of research. Oravecszné Scheffer showed that deeper basin and shallow-water areas were populated by different foraminifer assemblages; among the deeper-water forms, *Ophtalmidium* was found frequently in the Middle Triassic, whereas *Duostomina* is characteristic to the Upper Triassic. Finally, the widely distributed

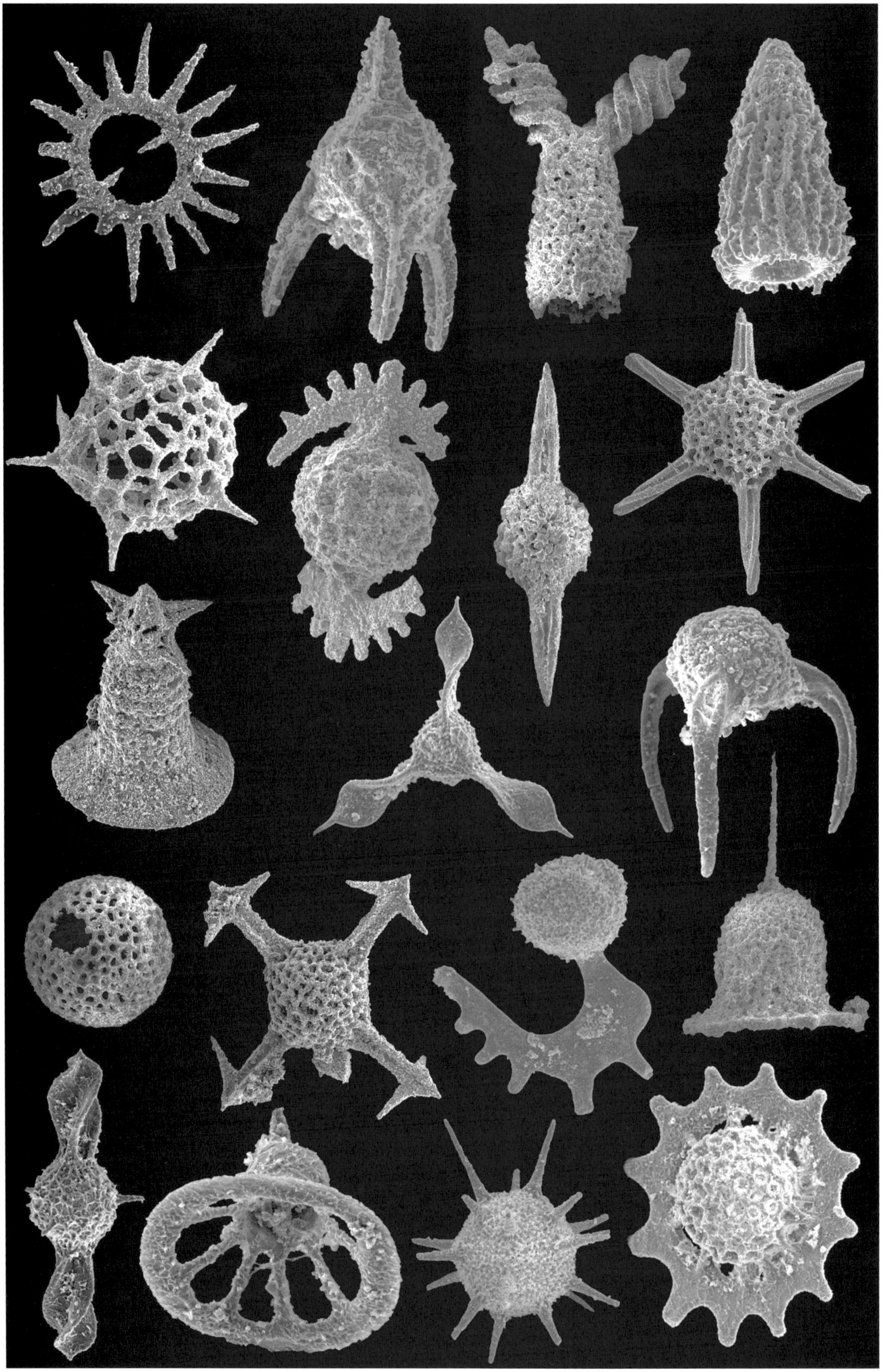

Péter Ozsvárt, a member of the Paleontological Research Group at the Hungarian Academy of Sciences (left) and Heinz Kozur, micropaleontologist, in the library of the Department of Paleontology at the Hungarian Natural History Museum. On the table in front of the two workers are dozens of photographs of radiolarians, the majority of which are new to science. Although radiolarians have long been known, their astonishing diversity was revealed thanks to the use of electron microscopes that allow high-resolution imaging.

foraminifera *Triasina hantkeni* is the marker fossil for the Upper Triassic Dachstein Limestone.

However, Oraveczné Scheffer was not the first worker to study the Triassic foraminifera from the Transdanubian Range. The Balaton monographs include a paper written by M. Elemér Vadász on Triassic foraminifers from the Balaton Highlands. Vadász studied material collected partly by himself and partly by previous researchers, but, unfortunately, he often was not careful enough with other collections and uncritically accepted data written on museum labels and vials concerning localities and geological age of specimens. As a result some of the supposedly Triassic specimens he described turned out to be of much younger—that is, Miocene—age, and did not come from the locality (Felsőörs) indicated on the museum label. Although Vadász did mention that he was not able to find any specimens at Felsőörs and correctly identified this material as belonging to Miocene species—some of them extant today—he nevertheless trusted the collectors and curators and published the fauna as Triassic in age. Later, when persuaded of his mistake, he published a correction.

Electron microscope photos of Middle Triassic (Anisian) radiolarians from the Balaton Highlands (different magnifications).

📖 Oraveczné Scheffer 1987

Mixed-Up Specimen Labels

It is interesting to note that one of the greatest mistakes ever made in micropaleontology is also related to Triassic fossils. German researchers from the mid-twentieth century described a series of conodonts from Cretaceous beds in the heart of Africa, even though most paleontologists at the time were sure than such "cone teeth" had disappeared everywhere at the end of the Triassic. They nevertheless pondered the possible presence of an enigmatic Triassic seaway in which conodonts lived, because it left absolutely no trace

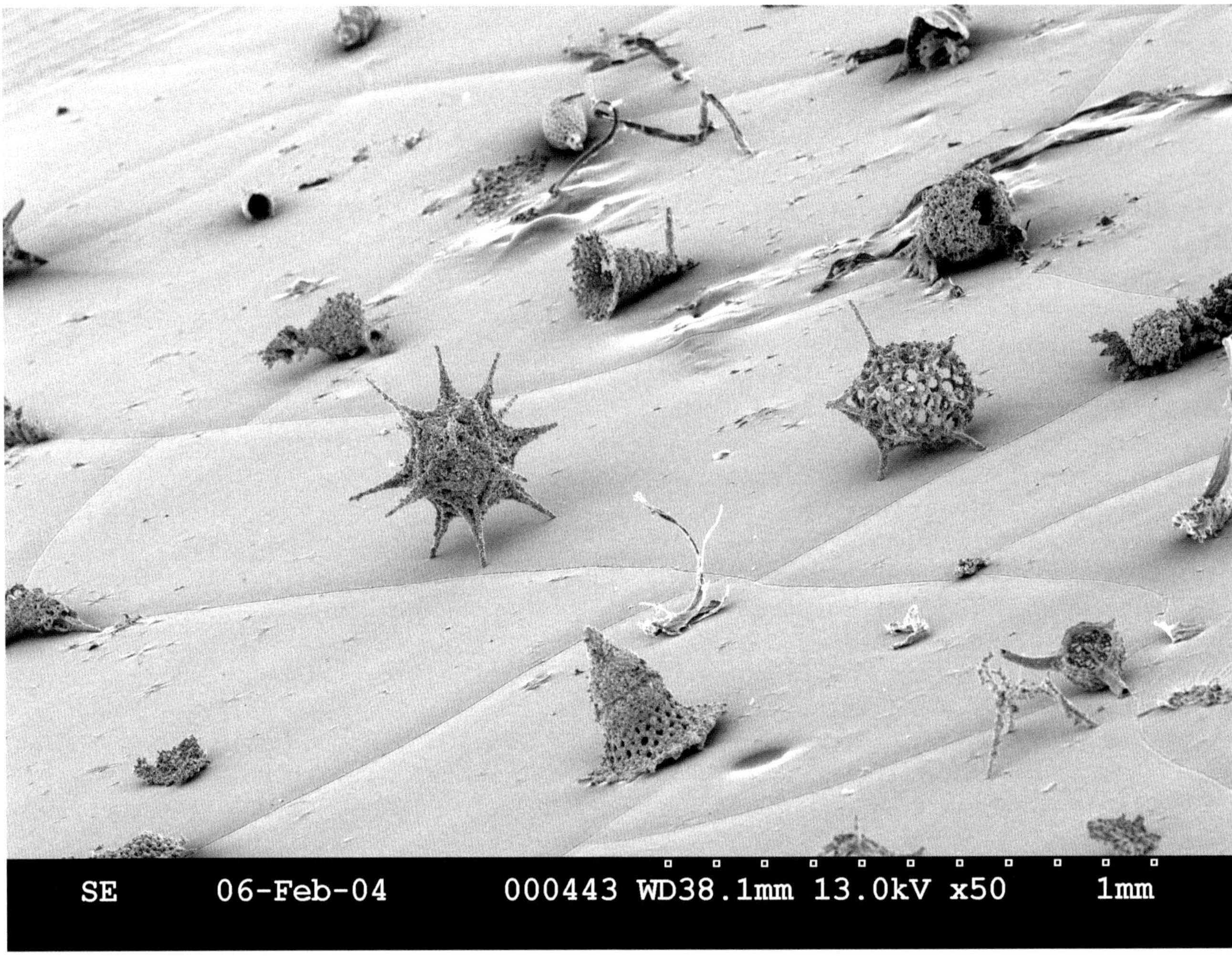

Electron microscope photos of radiolarians stuck onto a supportive sheet (holder).

in the area in question. This problem was resolved only decades later, at the end of the 1970s. As a consequence of a museum disturbance, Middle Triassic limestone samples collected in Spain were put into boxes that were labeled with African localities and a Cretaceous age.

Radiolarians

Radiolarians have only recently enjoyed interest among micropaleontologists, largely thanks to the more general use of modern chemical extraction methods and scanning electron microscopy. Previously these organisms were almost exclusively studied in thin-sections, making the identification of their usually complicated tests possible only in lucky cases in which sections were orientated favorably. The dawn of radiolarian research in Hungary was heralded by Miksa Hantken's seminal 1884 paper, entitled "A magyarországi mész- és szarúkövek górcsövi alkatáról" (On microscopic construction of limestones and cherts from Hungary). The author, considered to be among the pioneers of microscopic study of sedimentary rocks, asked David Rüst (1831–1916), a well-known specialist from Vienna, to study samples he had collected from the Balaton Highlands. The results, including descriptions of several new species, were published in *Palaeontographica* in 1885 and 1891. Rüst's papers are fundamental and still widely used in Mesozoic radiolarian research. Studies of radiolarian-bearing rocks in Hungary then continued with a series of papers published between 1916 and 1937 by Rezső Hojnos (1894–1961), assistant professor at the Paleontological Institute at Pázmány Péter University, later a teacher at Vörösmarty Secondary School in Budapest. The Triassic cherts from the Buda Mountains and the radiolarians they contain were studied in 1936 by Erzsébet Károly.

Because they are linked to deep water, radiolarian remains occur in Hungary only in Middle and Upper Triassic rocks. Modern research on these fossils began with studies of the radiolarites that are widespread in the Bükk Mountains and surroundings, in order to establish their geological age. Later the scope of activities was extended across the whole Hungarian Range; some of the earliest data were produced by Heinz Kozur and Patrick DeWever, the latter a French micropaleontologist, professor at the

Paleontologist in Action: Photographing Microfossils with a Scanning Electron Microscope

The human eye is not able to differentiate two neighboring point-like objects, or details of objects, if their distance from one another is less than 0.1 millimeter. This visual angle can be widened, however, by bringing the objects closer to our eyes. Still, the resulting picture remains fuzzy when viewed with the naked eye if the distance is less than a certain value. Study of tiny objects, or fine details, thus requires a microscope of some kind, and for 400 years the light microscope—consisting of objective, ocular, tube, and lighting equipment—was the only tool available. Resolution that can be achieved with a light microscope is of the magnitude of micrometers and further magnification is limited by the wavelength of visible light (around 500 nanometers).

According to John Callomon (1928–2010), expert on Jurassic ammonites and professor of physical chemistry at University College London, the twentieth century saw two revolutionary innovations in the field of paleontology. One was photocopying, which made paleontological monographs necessary for the identification of fossils widely available, but kept only in relatively few libraries. The other important innovation was the invention and widespread distribution, of the scanning electron microscope, or SEM. In an SEM, an electron beam focused either by a magnetic or electric field scans studied objects that have usually been made conductible with a coating of thin coal or a gold layer. The screen of the microscope then displays the picture formed by the secondary electrons emitted from the object. The back-scattered electrons (BSE), also detectable, reflect any differences in the mineral composition of an object and can also be used in imaging. About 300,000× magnification is achievable with an SEM, but this level is not necessary in paleontology. Lesser magnification is, however, very useful, because this means that most groups of microfossils are now better observable and photographable. Use of the SEM is now routine for paleontologists.

Muséum national d'histoire naturelle de Paris, and former president of the Société géologique de France. In the mid-1980s, Lajos Dosztály joined this research team when he worked at the Geological Institute of Hungary. Dosztály's studies were mainly focused on Middle Triassic radiolarians from the Balaton Highlands. Radiolarians played a particularly important role in research aimed at defining the lower boundary of the Ladinian stage. Because of this, the biostratigraphic value of all fossils occurring alongside ammonites was studied in detail in order to characterize the proposed boundary in as many ways as possible. In terms of radiolarian fauna, the frequent and bizarre *Saturnalis*-like forms that sometimes resemble an eagle are worth mentioning.

Following the premature death of Dosztály, research in this area halted for some years in Hungary. However, Péter Ozsvárt has begun to study Triassic radiolarians, and his promising work has already resulted in additional details on changing compositions of radiolarian faunas at the Triassic-Jurassic boundary.

📖 De Wever 1984; Dosztály 1989, 1993

Ostracods

Because they are widely distributed from freshwater to the deep sea, many Triassic rocks have been found to contain ostracod shells. The dark, marly limestone beds that overlie the marine shell-bearing limestone succession of the Middle Triassic in the Mecsek Mountains, for example, contain valves of the stratigraphically long-ranging freshwater genus *Darwinula* in rock-forming quantities.

Usually less abundant, but much more diverse, ostracod assemblages are known from the Middle and Upper Triassic of the Transdanubian Range. In one Balaton monograph, 44 ostracod species were described by the specialist Gyula Méhes (1881–1959). This fauna derived mostly from Carnian marls exposed when fishponds were dug in the Nosztori Valley situated north of the village of Csopak, as well as from small quarries opened around it. By now, however, all these exposures have completely disappeared and after the 1911 publication of Méhes's work, the Triassic ostracods of the Balaton Highlands were studied only sporadically until the 1970s. At that time Heinz Kozur was able to study and document ostracods from rocks previously not considered promising from a micropaleontological point of view by using chemical extraction methods, including acid dissolution. His pioneering publication on these fossils deals with the fauna of the classical Forrás Hill sequence at Felsőörs, concentrating on the rarer and very interesting deepwater forms, but gives much less attention to the most abundant, common species that form the bulk of the fauna.

With the renewed geological mapping and study of the Balaton Highlands, Miklós Monostori, an acknowledged ostracod expert mainly known for his work on Cenozoic faunas also began to work with Triassic faunas. Ostracods are excellent paleoenvironmental index fossils: each ostracod morphological type has evolved to

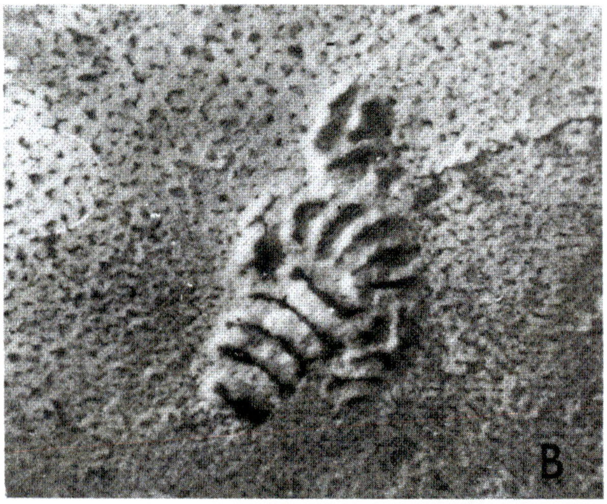

Triadogigantocypris balatonica Monostori—the largest "microfossil" from the Hungarian Triassic is this ostracod, about 13 millimeters in length and resembling a bean in outline (left). Superficially, this specimen might be identified as a bivalve but the characteristic protuberances on it corresponding to sites for the muscle attachments (right) are unambiguous indicators of its systematic position.
📖 Monostori 1991

be characteristic for a specific environment. Thus, the studies carried out by Monostori have contributed in particular to our understanding of the environmental evolution of the Balaton Highlands Triassic Basin. Bed-by-bed sampling and analyses have revealed paleogeographical changes, for example, the uplifting of some segments of the seafloor. The formation of submarine highs, leaving no trace in the lithological characteristics of the rocks deposited in the adjacent deeper basins, is nevertheless reflected in upwardly increasing amounts of characteristic shallow-water ostracods redeposited from the seamounts and mixed with deeper-water forms. Samples collected from the Nosztori Valley section were also found to contain peculiar genera (for example, *Renngartenella* and *Simeonella*) characteristic of hypersaline environments (those consisting of more than 35 grams of salt per liter of water). Their occurrence in the sequence is due to a widespread event called the Carnian salinity crisis, which was presumably triggered by changes in climate. The same hypersaline genera were also found in some clayey beds of the Dachstein Limestone exposed on the Kálvária Hill at Tata.

An artificial research trench dug into Öreg Hill at Vászoly yielded a previously unknown giant ostracod. Relatives of this form, described by Monostori in the respected journal *Neues Jahrbuch für Geologie und Paläontologie* as *Triadogigantocypris balatonica,* a new genus and new species, live floating in modern oceans as well today. Their valves are of chitin and, thus, are not calcified. Large ostracods such as this were common in the Paleozoic, but this Vászoly specimen is by far the largest known from the Mesozoic.

📖 Méhes 1911; Kozur 1970; Monostori 1991, 1994, 1995a

Conodonts

Conodont animals survived the end-Permian mass extinction and persisted up until the end of the Triassic. Their last evolutionary flowering provided a valuable tool, a sort of wonder weapon for Triassic stratigraphy, concerning in particular, geological age determination for rocks that contain few other fossils. Such fossil-poor deepwater sequences are widespread in the Carpathian region, especially in Northern Hungary and in the area of the Slovak Karst.

Although the extraction of calcareous microfossils—for example, ostracods that are often encountered in thin-section—proved to be impossible, the Middle Triassic limestone beds of Southern Transdanubia have often yielded well-preserved conodont faunas. Study of these sequences began in the mid-1970s with the pioneering work of József Bóna on material derived from five outcrops and two boreholes in the Villány Mountains. After samples were dissolved in acetic acid, the residue of these limestones was found to contain, in addition to abundant conodonts, the remains of branches of brittle stars as well as the finely perforated sclerites of sea cucumbers. Coeval shell-bearing limestones from the Mecsek Mountains, studied by Sándor Kovács and the Slovakian Jarmila Papšová, also yielded a similar conodont assemblage that was described in 1986. Both faunas were collected from the Zuhánya Limestone, a succession that represents the deepest marine environment in the sequence and, as evidenced by the presence of the index fossil *Gondolella bulgarica,* is of Pelsonian (middle Anisian) age.

Conodonts were also found to occur in a much wider stratigraphic interval across the Hungarian Range, where several Middle and Upper Triassic rock successions deposited

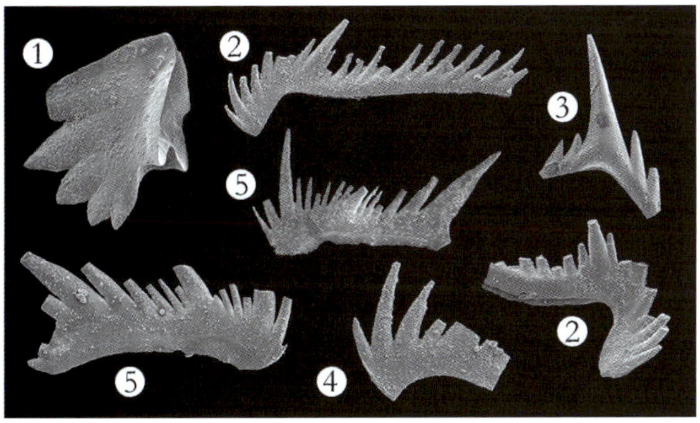

"The Last of the Mohicans" among conodonts—millimeter-sized fossils from Triassic-Jurassic boundary beds at Csővár. This fauna is geologically among the youngest known, because conodonts went extinct at the end of the Triassic. The specimens in this photo were identified by Mike J. Orchard. (1) *Misikella posthernsteini* Kozur et Mock. (2) *"Hindeodella" andrusovi* Kozur et Mostler. (3) *Norigondolella steinbergensis* (Mosher). (4) *"Neohindeodella"* sp. (5) *Neohindeodella rhaetica* (Kozur and Mock).

in deeper marine basins outcrop. Thanks to these discoveries, a Rhaetian age for the dolomite near the village of Rezi in the Keszthelyi Mountains, the folded cherty limestone succession exposed in the corner of the Mátyáshegy Quarry in the Szép Valley of Budapest, and the lower part of the marly limestone sequence at Csővár—all previously thought to be older in the Triassic—was established on the basis of conodont remains. Much of this work was done by Sándor Kovács, but the contributions of Heinz Kozur and Rudolf Mock were also significant. Detailed studies were carried out by Kovács, in the context of the development of a GSSP for the Ladinian stage, on the conodont assemblages of Anisian-Ladinian boundary beds. Kovács's conodont studies from the Triassic of the Aggtelek-Rudabánya Mountains proved to be especially important, because bed-by-bed collecting of basin sediments allowed him to subdivide the Middle Triassic and Carnian into almost 20 successive zones and subzones defined by *Gondolella* species.

📖 Bóna 1976; S. Kovács 1977, 1986; S. Kovács & Papšová 1986; S. Kovács & Rálisch-Felgenhauer 2005; S. Kovács et al. 2006

Fossil Plants

Although the Triassic sequences in the Carpathian Basin are overwhelmingly of marine origin, fossil remains of higher continental plants are very rare. Indeed, even the continental Triassic deposits known from this area are poor in plant remains: there are no plant-rich deposits comparable to the Voltzia Sandstone in the Vosges Mountains of France or to the diverse floras of the Variegated Sandstone (Buntsandstein) and the Keuper in Germany. Most plant remains are of Late Triassic age and come from the mountains of Southern Transdanubia. These floras, however, have received only limited interest and thus far have been documented only as lists. For example, the thick, coal-free sandstone succession in the Mecsek Mountains has yielded poorly preserved plant remains assigned by previous researchers to the genera *Equisetites, Czekanowskia, Podozamites,* and *Clathropteris.* Considering the development of paleobotanical methods and taxonomic philosophies, this flora is in need of revision. The lower part of the coal-bearing sequence in the Mecsek Mountains is also known to contain plant remains, but very little material is available. (Triassic coal seams are thin and separated with thick successions of barren rocks in this area, and so have been exploited only rarely, despite more than 150 years of mining.) Nevertheless, these plant assemblages await study.

Plant remains are not at all uncommon in the variegated clay, sand, and sandstone beds occasionally exposed in the cutting made for the former cable railway at Templom Hill in Villány. One charming anecdote related to the

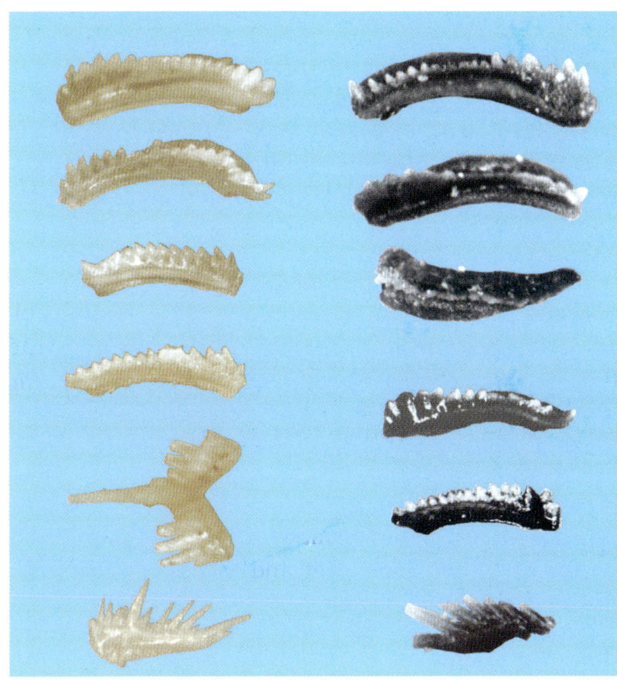

The color of conodonts reflects the measure of pressure and heat experienced by the rock they are embedded in due to subsidence and weight of overlying beds. The greater the depth of the conodont-bearing bed, the darker the color of the specimens. In this picture, specimens from the same geological age are shown from the Villány and Mecsek Mountains (left and right, respectively). The Villány specimens are from rocks that experienced lower pressures and temperatures and have remained light in color, while the Mecsek ones, covered with sediments several thousands of meters thick, are darker. After S. Kovács et al. 2006.

fossils is worth retelling. In 1987, participants in an international geological workshop visited the exposure, which had been cleaned and prepared for the occasion. The Hungarian expert leading the field trip mentioned that a large plant fossil had been found on a similar occasion some years before by a professor from abroad. On hearing this announcement, Krzystof Birkenmajer, a professor from the Jagellonian University in Kraców (Poland) and an expert on the geology of the Pieniny Klippen Belt as well as Spitzbergen and Antarctica spoke up to say that he was that person. Turning back the pages in his field notes, he found the remarks he had made on the previous field trip, then walked up to a bed and hammered out another large plant fossil. Birkenmajer then presented this new fossil to the leader of the field trip, who thanked him and eventually placed it in the same drawer as the first one. However, the fossil plants from this succession have yet to be studied: perhaps a few more visits to the locality by Birkenmajer will be needed to acquire enough specimens to justify a research project.

Alpine Triassic successions in the Carpathian sporadically contain the remains of higher plants. Recently, some marly limestone beds exposed in the Pokol Valley at Csővár have yielded carbonized plant remains including a cone from the early pine-like plant *Voltzia*. Rich and well-studied Triassic plant assemblages are known, however, from the Carnian Lunz Sandstone in the Northern Calcareous Alps, as well as from the coeval Raibl Beds in the Southern Alps.

Invertebrates

Sponges

Sponges are usually regarded as unpleasant fossils by most geologists because their study requires the production of thin-sections or microscope preparations as well as taxonomic expertise. However, Paolo Vinassa de Regny (1871–1957)—an Italian paleontologist similar in character to the all-around personalities of the Renaissance—did not find this group unpleasant, and documented the presence of 27 sponge and hydrozoan species in the Veszprém Beds. As evidenced by letters he sent to Lajos Lóczy Sr., discovered by Péter Vincze in the archive of the Natural History Museum of Hungary, this work was slow because of a lack of comparative material as well the time available. Vinassa de Regny both prepared and drew these specimens himself and so his paper, published in 1911 as a Balaton monograph, took two years to complete. It is also true that he published twenty other papers while he was undertaking this work. Vinassa de Regny named an interesting new species of segmented calcareous sponges in honor of Imre Lőrenthey (1867–1915), the second head of the Department of Paleontology at Pázmány Péter University.

The true home of Hungarian Triassic sponges is, however, the Aggtelek-Rudabánya Mountains, where sections of fossils are often seen on the weathered surface of the Wetterstein Limestone. From the rich fauna identified in thin-sections, the non-segmented, reef-forming *Peronidella* and *Leiospongia*—as well as the segmented *Colospongia,* and especially *Vesicocaulis* and *Stylothalamia*—are worth mentioning. Sponges from Alsó Hill were studied for decades by local expert Sándor Kovács—and recently by Hajnalka Simon as part of her MSc thesis.

📖 Vinassa de Regny 1911; Balogh & Kovács 1976; S. Kovács 1978a, 1978b; Vincze 1987

Cnidarians

Corals are usually uncommon Triassic fossils. Although their recrystallized remains can be found on the weathered rock surfaces of the Dachstein Limestone in places, significant coral assemblages are known only from the Veszprém Marl. The relevant part of the Balaton monograph that contains the descriptions of these 30 species was written by Károly Papp (1873–1963), professor of geology and paleontology at Pázmány Péter University, in the interwar period.

Subordinately, corals can also be reef-forming organisms in the Middle Triassic of the Aggtelek-Rudabánya Mountains. In 1972, Gábor Scholz recorded *Pinacophyllum* and *Retiophyllia* from the fossil reef situated near the Vöröstó entrance of the Baradla Cave.

Reference to Triassic corals can also be found in geological literature on the Mecsek Mountains. In 1958, a paper on Middle Triassic corals from Hungary by Gábor Kolosváry (1901–1968), then the professor of zoology at the University of Szeged, was published in the *Journal of Paleontology.* The paper also treated corals known from the Mecsek Mountains. This report surprised the geologist community of Hungary because no one had found them in the Mecsek shell-bearing limestone before. Interestingly, no corals have been found since—an enigma that was only recently cleared up by Gyula Konrád, head of the Department of Geology of the University of Pécs. He noted that this limestone contains globular structures that consist of radial calcite crystals that formed after the gypsum and that it was these, along with some peculiar stromatolites, that had been incorrectly identified as corals by Kolosváry.

📖 Papp 1911; Kolosváry 1958; Scholz 1972

Gastropods

The Renaissance of Triassic Gastropod Research

Remains of gastropods, bivalves, and cephalopods are common in most Triassic rocks in the Carpathian region and are often the most frequent fossils. As has been seen in other

Marine Upper Triassic beds around the world also occasionally contain globular enigmatic fossils known as *Heterastridium conglobatum* Reuss. The systematic position of this organism is debated and it is usually interpreted as an extinct planktonic hydrozoan. The largest specimens, sometimes reaching the size of a basketball, can be found in the Rhaetian. The specimen shown here comes from Coleşti in the Apuseni Mountains; *H. conglobatum* has also been recorded from the Balaton Highlands (close to original size). 📖 E. Kutassy 1930

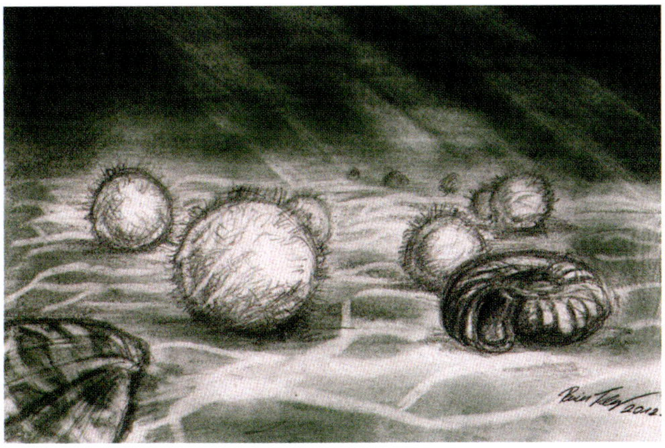

Heterastridium is a mysterious fossil. It is still unclear whether these hydrozoans were gently rolling up and down on the calm seafloor (left) or floating in water (right).

marine invertebrate groups, gastropods were hit hard by the end-Permian mass extinction, although the numbers of taxa that disappeared at this time maybe overestimated. According to Roger L. Batten, expert on fossil gastropods and now emeritus curator at the American Museum of Natural History, some characteristic Permian forms (described by other authors and thus bearing other names) can still be found in the Middle Triassic. It is thus hard to judge whether geologically younger forms really represent Permian taxa or whether superficially similar forms can be encountered in the Triassic. We are getting closer and closer to finding answers, however, as the last two decades have seen a remarkable renewal of interest in the phylogeny and systematics of Triassic gastropods. The revival of this field, somewhat neglected for nearly a century, is partly due to the need to fit the enormous fossil record into the phylogeny of Gastropoda that is otherwise based largely on characters of the soft body, unavailable to paleontologists. The other cause is the recognition that much detailed phylogenetic work can be done based on the protoconch—the shell that forms early in ontogeny—of fossil gastropods. However, the detailed study of protoconchs, which received only limited attention in the past, is now possible thanks to the widespread use of scanning electron microscopes that have made tiny shells just one-tenth of a millimeter in size visible and photographable.

One of the initiators of, and leading personalities in, the study of fossil gastropod protoconchs is Klaus Bandel, professor at the University of Hamburg, Germany. Bandel has also provided a number of modern descriptions

Amblysiphonella loerentheyi Vinassa de Regny. This cylindrical, segmented form is one of the most frequent sponges found in the Veszprém Marl and is also known from the Wetterstein Limestone in the Aggtelek-Rudabánya Mountains. The surface of this sponge is penetrated by fine pores and the wall formed by its simple spicules is a blister-like structure. (Original size.)

of Upper Cretaceous gastropods from Hungary and also rediscovered a Triassic gastropod bonanza, the extremely diverse and well-preserved fauna known from the San Cassiano Formation. Bandel's papers documenting the early ontogenetic shells of numerous Triassic gastropods are of fundamental significance in the field. Studies of this kind, which look at the diverse and well-documented Triassic gastropod faunas from the Carpathian region, have also begun recently.

Low-Diversity Gastropod Assemblages in the Lower Triassic

The Lower Triassic of the Carpathian region—indeed, even the whole Alpine Triassic—contains rather uniform, low diversity fossil assemblages. The base of this succession has yielded bivalve fossils almost exclusively and consists of marl and sandstone beds often named Seisian, after the locality Seis (Siusi in Italian) in the Southern Alps. Besides these fossils, the remains of the gastropod *Bellerophon,* a survivor of the end-Permian catastrophe, can also be found in some places. In the lowest part of these Triassic successions are gastropod-oolite beds, formed from the cemented shells of microscopic gastropods (*Holopella*

Gastropods from the Triassic (1.5× magnification, except [3]). (1) *Werfenella rectecostata* (Hauer)—a guide fossil of the Lower Triassic Campilian beds in the peri-Mediterranean region. The shell of this gastropod is angular in outline and is ornamented with ribs parallel to the axis of the shell. Originally, this taxon was assigned to the genus *Turbo,* which now lives in Southeast Asia, by Franz Ritter von Hauer (1822–1899), a pioneer researcher of the Alpine Triassic. Although this identification was doubted for many years, only in 2004 was the new genus "*Werfenella,*" accommodating the single species "*rectecostata,*" erected by eminent gastropod researcher Alexander Nützel from Germany (Sóly). (2) "*Fusus*" *noricus* Bartkó. Gastropods with apertures ending in a long, tubelike shell part called the sipho occur in great numbers in the Cretaceous and become even more frequent in the Cenozoic. Although very sporadic, forms of modern appearance are also known from the Triassic; one of these is this specimen, collected from the Dachstein Limestone at Remete Hill. It is interesting that this form is represented by a single specimen as it is from material consisting of several hundred specimens. (3) *Worthenia ornata* Kutassy. Representatives of this long-ranging (Carboniferous–Jurassic) genus *Worthenia* are especially frequent in the Triassic. Their shell forms a double cone and consists of four or five whorls ornamented with spiral rows of protuberances. The angular aperture bears a slit, behind which the ornamentation differs from that of the remainder of the surface. This Upper Triassic specimen was found at the rich locality of Coleşti in the Apuseni Mountains (2.5× magnification). (4) *Promathildia winkleri* (Klipstein). The Carnian Veszprém Marl has yielded a diverse gastropod assemblage consisting of nearly 120 species, including numerous examples of the turreted form shown here, first described from the San Cassiano Formation in the Southern Alps. At first glance this taxon is strongly reminiscent of *Turritella,* a genus widespread in the Cretaceous and Cenozoic but, as seen in well-preserved specimens, the axis of the early ontogenetic shell in *P. winkleri* is not coincident with that of the teleoconch, the mature shell. Thus, *Promathildia* has a heterostrophic protoconch, whereas that of *Turritella* is orthostrophic (Veszprém, Jeruzsálem Hill). (5) *Natiria costata* (Münster). This slightly flattened, globular form with an oval outline is a very characteristic gastropod from the Campilian beds. The last whorl of the fast-expanding shell is much larger than earlier ones and the surface of the internal mold is covered with folds parallel to the outer lip. In the Triassic, *Natiria* was already a living fossil because it first appeared in the Carboniferous. This specimen was collected from the Lower Triassic sequence exposed near the railway station at the village of Sóly. (6) *Hungariella kutassyi* Szabó. The genus *Hungariella* was introduced by Kutassy on the basis of *Nerita*-like gastropods that are abundant in the Upper Triassic fauna of the Buda Hills. In 1927, one frequent species from the Fazekas Hill fauna was described by Kutassy as "*Neritopsis spinosa* n. sp.," because of the thornlike protuberances on its shell. However, as János Szabó pointed out in 2007, this name was already preoccupied and so it was replaced by a new one, in this case in honor of Endre Kutassy. (7) *Hungariella stredae* (Kutassy). This species, named in honor of archiepiscopal adviser and zoologist Rezső Streda (?–1960), most of whose collection of several thousand fossils is now housed in the Natural History Museum of Hungary, is the most frequent gastropod in the Remete Hill fauna. (8) *Callotrochus triadicus* (Kutassy). The Upper Triassic gastropod fauna of the Buda Hills consists mostly of ornamented species. This spectacular smooth form with a nicely arched outline was collected from the Dachstein Limestone at Remete Hill.

and *Coelostylina*) and encrusted with a fine-grained lime mud (micrite).

Much more diverse fossil assemblages are known from younger Lower Triassic beds that are traditionally named Campilian, after another locality in the Southern Alps. This Campilian fauna is dominated by bivalves and ammonites, especially *Tirolites* and *Dinarites,* and records a later phase of the recovery of marine life after the Permian extinction. This succession outcrops in several places in the Balaton Highlands, is traditionally named the Tirolites marl (now the Csopak Marl), and typically contains the remains of two gastropod species. Both *Natiria costata,* a relative of the extant nerite snails, and *Werfenella rectecostata* are considered to be index fossils for the Alpine Lower Triassic. Indeed, the first detailed description of these Campilian faunas from the Balaton Highlands was published by Fritz Frech in a Balaton monograph. New, rich localities situated on Iszka Hill near the town of Székesfehérvár were documented in the 1960s by Erzsébet Végh-Neubrandt (1926–2008) and János Oravecz, and from the neighborhood of Balatonfűzfő by Csaba Detre in 1971. A comparative study of Lower Triassic sequences and fossil assemblages in the Dolomites and in the Balaton Highlands was published by Hungarian and Italian authors in 1990 in the *Memorie di Scienze Geologiche,* formerly issued by the University of Padova, Italy. This work of more than 60 pages is accompanied by high-quality photographic plates that illustrate the most frequently occurring fossils.

Because they are less abundant and less attractive, the Lower Triassic fossil assemblages from the Bükk and Aggtelek-Rudabánya Mountains were documented much later. Indeed, the very few localities that are known from

this area have yielded only a few poorly preserved specimens. As documented by Kinga Hips, this characteristic Campilian fauna can be found in a succession called the Szín Marl in the Aggtelek-Rudabánya Mountains.

📖 Detre 1971b; Lenner 1989; Broglio Loriga et al. 1990; Hips 1996; Hips & Pelikán 2002

Sporadic Gastropod Assemblages from the Middle Triassic

Middle Triassic gastropod faunas from the Carpathian region are more diverse than their Lower Triassic counterparts. Poorly preserved internal molds can be collected from the shell-bearing limestone in the Mecsek Mountains at several places, one being on the Lapis Hill at Pécs. Upper Triassic marl and sandstone beds overlying the limestone succession are also known to contain gastropods, occasionally in large numbers. These occurrences have received limited attention over the few last decades—and, according to previous authors, they represent marine genera. Because this sequence has recently been shown to have a freshwater origin, these earlier identifications are obviously in need of revision.

A well-preserved and well-documented gastropod assemblage is known from the Middle Triassic of the Transdanubian Range. The fauna of this red, ammonite-rich limestone from the Ladinian stage is dominated by small forms that were described in one Balaton monograph by Ernst Kittl (1854–1913), a Viennese gastropod specialist. The long-ranging genus *Worthenia* that has a "pagodaform" outline and which is ornamented with spiral keels has proven to be especially common in the assemblage. One new species within this genus was named *Worthenia vamosensis,* after the Nemesvámos locality, by Kittl.

Before the large-scale collection work that has been carried out over the last two decades, only a few poorly preserved gastropod remains were known from the Anisian. Newly explored localities, including Farkó-kő Hill at Aszófő and Cser Hill at Mencshely, have, however, yielded relatively diverse gastropod assemblages that have not yet been studied. An Anisian gastropod faunule consisting of a few species is also known from the Wetterstein reef at Jósvafő.

Diverse Gastropod Faunas of the Upper Triassic

Most of the gastropod species recorded from the Triassic of the Carpathian region are from the Upper Triassic. One assemblage in the Veszprém Marl was studied by Kittl, who documented 117 mostly small species, with the majority of specimens identified to taxa already known from the Cassian fauna. Heterostrophic gastropods, such as representatives of the genus *Promathildia,* proved to be frequent; another gastropod assemblage consisting of small-sized forms is also known from the Rezi Dolomite that is confined to the Keszthely Mountains and the Kössen beds that are also characteristic of this area. Remarkably well preserved specimens were figured from the Keszthely sequence in the Akasztó Hill Quarry at Rezi by Péter Bohn (1937–1998), monographer of the geology of the Keszthely Mountains. In this assemblage, similar to the Kössen beds, smooth forms of the common Triassic genus *Coelostylina* are most frequent, alongside *Promathildia.*

It is worth mentioning that the genus *Coelostylina* is something of a dustbin, accommodating a huge number of typically small-sized, conical, and unornamented Triassic and Jurassic gastropods. The number of species that really belong to this genus is surely smaller. For example, study of the protoconch of a gastropod species that occurs in rock-forming quantities in the Lower Jurassic coal-bearing sequence of the Mecsek Mountains ("*Coelostylina* banks") revealed that these tiny specimens are in fact representatives of *Hydrobia*-like gastropods that lived in fresh and brackish waters. Thus, they are not actually related to *Coelostylina,* whose remains are known only from normal marine sediments.

Formation of the Main Dolomite was not favorable for the preservation of gastropods because their easily dissolvable aragonite shells usually disappear completely during diagenesis leaving an empty space and a corkscrew-like internal mold. Dolomite beds are known to outcrop at several places in the Transdanubian Range, for example at Csákánykő-puszta Quarry in the Vértes Mountains and on the steep slope of Lófingató Hill at Óbarok. Gastropod remains preserved in this way can be identified if the external mold is filled with silicone rubber. This is very rarely done, however, because although these fossils can be attractive, gastropod remains are not considered useful as a rule. This statement applies especially to extinct forms without close living relatives because the mode of life of gastropods can be inferred only in exceptional cases from the shape of the shell.

The Dachstein Limestone has yielded diverse gastropod assemblages in several places, many of which were documented by Endre Kutassy. As a leading Triassic expert on Triassic, Kutassy studied fossils from around the world and published his work in Hungarian, German, Italian, and Dutch. As indicated in papers published in 1927, 1932, and 1936—in which he described the rich faunas of Fazekas and Remete Hills in Budapest—relatively large sized shells of *Purpuroidea* ornamented with striking tubercles are conspicuous. Representatives of *Hungariella,* also relatively large in size and with attractive ornamentation, are the most frequent elements of the Buda Hills faunas.

A single specimen of surprisingly modern appearance, reminiscent of geologically younger gastropods, was also encountered in the Dachstein Limestone at Remete Hill. This peculiar shell ("*Fusus*" *noricus*), with an aperture that ends in a relatively long siphonal canal, was found and described in 1939 by Lajos Bartkó (1911–1988), an eminent geologist who also explored Hungarian medicinal waters—including the Jodaqua site, from which bottled water is now sold—and mineral resources. After retiring, Bartkó completed a geological monograph on Ipolytarnóc, the most famous paleontological site in Hungary. This study was published in the Geologica Hungarica series Geologica.

One of Kutassy's last works was published in 1937, and dealt with the exceptionally rich gastropod faunas of the Codru-Moma Range in the Apuseni Mountains. The specimens he worked on were gathered in the surroundings of the village of Coleşti by mapping geologists—chiefly Pál Rozlozsnik (1880–1940), one of the most versatile Hungarian geologists of the twentieth century. Rozlozsnik is well known, for example, as an expert on Eocene *Nummulites*. The limestone beds exposed near Coleşti represent fossil reefal settings as well as lagoonal settings closeby—an environment that is inhabited today by extremely diverse communities. This was also the case in the Triassic, and the gastropod fauna from this site consists mostly of small-sized forms classified into 70 species in 42 genera. However, for most species described only a single specimen was available to Kutassy for study.

Of the described forms, the genus *Worthenia* proved to be the most frequent, with 10 species attributed by Kutassy. He also had planned to study the bivalve fauna from this site, but was unable to because of his premature death. Bivalves—as well as sponges, corals, and bryozoans—also occur at great diversity in these beds and have largely remained undocumented. Kutassy nevertheless left behind a fascinating body of work: his four volumes for the respected monograph series Fossilium Catalogus, which review the full literature for a given fossil group, are still widely used as important databases. These volumes, however, did not invoke unanimous enthusiasm among contemporaneous researchers living in countries around Hungary. Indeed, the monographs that were written in German and published in the 1930s and 1940s mention only the Hungarian names of localities that—following the Treaty of Trianon—were then considered to belong to countries surrounding Lesser Hungary. One disapproving view of the work of Kutassy can be found at the end of Erich Jekelius's paper dealing with the fauna of the Triassic white limestone from Schneckenberg at Braşov.

Kutassy described a gastropod fauna consisting of relatively large sized forms from the Šentana locality in Slovenia. This paper also provides evidence that he preferred to draw the specimens he worked on instead of photographing them. The artistic work published in this paper was done partly by Terézia Dömök, a talented photographer from the Geological Institute in Budapest, and partly by Sándor Jaskó (1910–1998), another geologist who left a significant body of work.

The years following Kutassy's death of witnessed a gap in the monographic treatment of Triassic fossils, due largely to World War II and the unfavorable conditions that it created for paleontological studies. The first large-scale postwar work on Triassic invertebrates from the Carpathian region, including descriptions of some gastropod species, was published by Vanda Kollárová-Andrusovová and Maria Kochanová in 1973. It dealt with the fauna of the Bleskový Prameň locality on the Slovak Karst. In the meantime, gastropod faunas from the Alpine Triassic were studied intensively by Ferenc Góczán, who delivered a lecture on their stratigraphic use at the 1959 International Mesozoic Conference in Budapest.

As indicated in the footnote to Góczán's voluminous 1961 conference paper, supplemented with a table showing the stratigraphic distributions of several hundred Triassic gastropod species described from dozens of localities, this work was regarded by its author to be an extract from a large-scale study that would contain a revision of the faunas of the Buda Hills as well as discussions and stratigraphic evaluation of all Alpine species. This work was to be published in the Geologica Hungarica series Palaeontologica, but so far has not appeared. Revision of the faunas of the Buda Mountains, now considered classic assemblages of the Alpine Triassic, was begun recently by János Szabó, former head of the Department of Geology and Paleontology at the Natural History Museum of Hungary and an expert on Jurassic gastropods.

📖 Kittl 1912a; A. Kutassy 1934b, 1936, 1937; Góczán 1961; J. Szabó 2011

Bivalves

The Lower Triassic Claraia-*Fauna*

Bivalves are even more common fossils than gastropods in Triassic rocks of the Carpathian region. They occur in enormous numbers in some rocks and give them peculiar appearances. In the stratigraphically oldest beds of the Alpine Lower Triassic, for example, bivalves are by far the largest and most frequent fossils, easily surpassing the rare and usually small gastropods. Indeed, these faunas are of very low diversity: the fossil assemblages of the Seisian beds are dominated almost exclusively by representatives of the genus *Claraia*. The search for the possible reasons for this poverty in older Triassic benthic assemblages is the focus of much geological and paleontological interest. As recent studies carried out along the margins of the former Tethys Ocean—from the Alps to China—have revealed, *Claraia* was among the very few marine invertebrate genera that survived the worldwide formation of an anoxic layer in

Fossils from the graveyard. A slab of friable Lower Triassic sandstone, containing bivalve remains, from the cemetery at Perkupa (0.5× magnification).

the sea. According to several authors, at the beginning of the Triassic the water in oceans around the world became stratified in oxygen content, and an oxygen-depleted water layer containing less than 1 milliliter of dissolved oxygen per liter formed. Formation of this hostile environment is thought to have triggered mass extinction of shallow-marine benthic groups. Indeed, the amount of geological evidence for such an event in terms of basal Triassic sections that contain black shales characteristic of oxygen-depleted bottom conditions is overwhelming.

Although—because of the general lack of ammonites—stratigraphic subdivision and correlation of lowermost Triassic successions is difficult, bivalves have also proved to be useful tools. The wide geographic distribution of some species of *Claraia* over several thousands of kilometers is also associated with a relatively short geochronological range. Because of this, these bivalves are considered useful index fossils.

Three *Claraia* zones, based on successive species, are distinguished in the lowermost Triassic. Of these, the index fossil of the middle zone—based on *Claraia clarae,* the type species of the genus—is the best known. This taxon was described either by Hermann Friedrich Emmrich (1815–1879), a secondary school director in Meiningen, Germany or—according to some—by Franz Ritter von Hauer, one of the leading geologists of the Hapsburg Empire and the Austro-Hungarian monarchy as well as a researcher of the Alpine Mesozoic. The correct spelling of the species name is *clarae,* however, *clarai* can still be seen even in papers and reference books of a high standard. The Arács Marl in the Balaton Highlands, named after the village of Arács (now a district of the town of Balatonfüred), is famous for mass occurrences of *Claraia,* and—although specimens are only preserved as internal molds—sound identification of species is still possible. In addition to *Claraia,* the inbenthic *Unionites,* which has a rather characterless appearance, is also worth mentioning.

In the Aggtelek-Rudabánya Mountains, this stratigraphic level is represented by the reddish-brown, ripple mark–bearing Perkupa Sandstone. In this locality *Claraia* is also associated with *Eumorphotis,* another index fossil. Some of the best-preserved specimens of this bivalve genus were collected—from sediments dug out of graves in the cemetery of the village of Perkupa—by Kálmán Balogh at the beginning of the 1940s.

It is interesting to note that cemeteries overall seem to be good localities for Triassic fossils from Central Europe. Paul Assmann (1881–1967), a German geologist who published two monographs on Triassic fossils from Upper Silesia in the 1930s, also collected well-preserved fossils from the Jewish cemetery in the town of Tarnowskie Góry (Tarnowitz in German), Poland.

📖 Frech 1912; Hips 1996

"Paper Pectens," "Flat Clams," and Co.

The *Claraia* fauna is followed upward in sequence by the more diverse bivalve assemblages of the Campilian beds. Dominant forms in these rocks include *Bakevellia,* a genus of *Gervillia*-like bivalves that are frequent in the Mesozoic; *Eumorphotis;* and species of *"Pseudomonotis"* ("P." *laczkoi* and "P." *loczyi*). The generic attribution of the latter has not yet been established because of their generally poor preservation. Among identifiable forms, representatives of *Costatoria,* ornamented with sharp radial ribs, often occur as pavements on bedding planes. Because the number of ribs increased during phylogeny, *Costatoria* species can also be used as index fossils: *C. subrotundata* occurs in lower beds, whereas *C. costata* characterizes a higher stratigraphic level of the Lower Triassic.

Until recently, only a few bivalves were known from the Anisian of the Balaton Highlands. Indeed, it was often said that these forms remained hidden in the shadow of the abundant brachiopods and ammonites. Besides some forms described by Bittner as "small-sized and characterless," two species were nevertheless found to occur frequently—and both were members of a peculiar group of fossil bivalves known as the "paper pectens" or "flat clams" that once inhabited oxygen-depleted environments. The thin, circular shells of *Bositra* (previously called *"Posidonia"*) and the smooth, round-, and oblong-shaped valves of *Daonella boeckhi,* named to honor János Böckh, were collected in rock-forming quantities from some places. Systematic collection work carried out over the last two decades, however, has changed the picture and has yielded considerable bivalve fauna comprising more than 30 species from the Pelsonian substage. The Farkó-kő Hill section of Aszófő has proven to be particularly rich in bivalves.

The Ladinian comprises mostly deeper-water sedimentary settings and has yielded a low-diversity bivalve assemblage, with *Daonella* as one dominant element. The systematics of the group that contains this genus as well as some other related taxa—including *Halobia* and *Monotis*—are discussed in detail in a voluminous work by Kittl, another publication in the Balaton monographs series. Middle Triassic platform carbonates—especially the white limestone that outcrops in the surroundings of the village of Tagyon—are also known to contain bivalve remains, but it is impossible to extract the fossils from these hard rocks.

The most enigmatic bivalve occurrence in the Balaton Highlands, or even from the whole Carpathian region, is also related to these deeper-water sediments of the Ladinian stage. Slightly north of Main Road no. 8, on a hill near the village of Hajmáskér, a peculiar bed full of large but poorly preserved bivalves crops out. These specimens sometimes reach up to 20 centimeters in length and belong to the extinct genus *Myoconcha*. However, the formation of this *Myoconcha* bed remains an enigma in the Triassic geology of the Balaton Highlands because it is sandwiched between deeper-water, pelagic limestone successions and it shares their characteristic microfacies. Indeed, Triassic and Jurassic deeper-water, basinal limestones and marls are full of embryonic bivalve shells; in thin-section these appear as thin, calcareous threads—hence the name "filaments."

As a result, both the stratigraphic setting and microfacies of this *Myoconcha* bed indicate a deeper-water origin, but such large, thick-shelled bivalves could not have lived in such an environment. It was then suggested that this bed in fact represents the remains of a peculiar benthic community, albeit one fueled with methane dissolved in the water and living at submarine springs, or cold seeps. However, the characteristic values of stable isotopes of oxygen and carbon that would be peculiar to the shells of organisms living in such environments could not be detected in the bed.

This *Myoconcha* bed was treated as a unique feature of the Alpine Triassic. Deciphering the secret of the origin of this bed remains a task for the future.

As is the case for all other benthic groups, the Carnian Veszprém Marl has yielded the most diverse bivalve fauna from the Triassic of the Balaton Highlands, or even the whole Carpathian region. Bittner described more than 90 species from this unit, and based on the characteristic and abundant forms it has been possible to subdivide this thick succession into successive intervals—including the Lima austriaca marl, the Nucula marl, and the "Craspedodon marl." The literature that deals with this marl also refers to the Estheria marl as being part of these Veszprém Beds. *Estheria*—or to give it its correct name, *Cyzicus*—is a genus of phyllopod shrimps that belongs to arthropods; because it must be confined to fresh and slightly brackish waters, mass occurrence of these shrimps in a purely marine succession would be difficult to explain. Further study has, however, revealed that these small-sized and circular "*Estheria*" remains are in fact the larval shells of bivalves.

The most spectacular species in the Veszprém Marl bivalve assemblage is the relatively large sized *Cornucardia hornigii,* which is represented by specimens that reach up to 6–8 centimeters in height and are characterized by strongly twisted umbones. This species, the remains of which are also abundant in the Sándorhegy Limestone overlying the Veszprém Marl, was an early form of the Upper Triassic *Megalodon*-like bivalves that became widespread on the carbonate platforms of the Late Triassic. Besides this form, a diverse assemblage of scallops—including several forms described from this mark for the first time—and frequent occurrences of file clams are also worth mentioning. In addition, an early oyster, *Umbrostrea*? *montiscaprilis* (Klipstein) also occurs. Interestingly, the genus *Cassianella,* so characteristic of the Cassian fauna, is rare in the Veszprém Marl.

In the Aggtelek-Rudabánya Mountains the Middle Triassic comprises in part the Wetterstein Limestone, from which just a very few bivalves are known, and also deeper-water sediments which contain "paper pectens" or "flat clams" as exclusive elements of their bivalve faunas. By the Middle Triassic, in addition to the obligatory filaments (*Bositra*?), *Daonella* is encountered alongside *Halobia* and *Monotis,* which often occur en masse in the Upper Triassic. Representatives of the latter genus include *Monotis salinaria,* an emblematic fossil of the Hallstatt Limestone that was described in 1820 by Ernst Friedrich von Schlotheim (1765–1832), a father of Triassic paleontology. The species name *salinaria* refers to the salt mines (salinas) that have been exploited for thousands of years in the surroundings of Hallstatt.

The flat clams from the northern Hungarian Triassic were documented in a fine 1976 paper by Kálmán Balogh. As the title indicates, the author did intend to publish further parts of a series dealing with bivalves from the South Gemeric Triassic; unfortunately, there was no continuation and all the specimens were left lying in drawers as a task for the future.

Daonella-bearing limestone also occurs in the Bükk Mountains and was first reported in 1989 by Csaba Detre, a philosopher and expert on Triassic fossils. Indeed, the first finds of *Daonella cassiana* proved to be important because the largely metamorphic Triassic in the Bükk Mountains is very poor in fossils, so this occurrence indicates a previously unknown deeper-water interval in the succession.

📖 Kittl 1912b; Bittner 1912a; Balogh 1976; Kochanová 1985, 1987; Detre 1989; Szente & Vörös 2003

Triassic Flat Clams from the Great Hungarian Plain—A Tribute to László Bogsch

In March 1950, László Bogsch (1906–1986), professor in the Department of Paleontology at Eötvös Loránd University, delivered a talk to the Geological Society of Hungary on the presence of *Daonella*-bearing rocks in pre-Cenozoic basement rocks on the Great Hungarian Plain. His announcement, based on fossils found in drill cores, was a sensation and elicited numerous widely differing opinions. According to the paradigm prevailing at the time, the basement of the Carpathian Basin should be composed of a rigid block of crust. This block, often called Tisia, was thought to have consolidated long ago and to be occasionally flooded by the sea. The extents of seaways were thought to correspond to the areas of present-day mountain ranges and their presumed continuation, and the occurrence of *Daonella*-bearing rocks far from all known outcrops could not be fitted into this static picture of the geology of the Carpathian Basin. Because Bogsch did not publish on these conclusive fossils, a significant part of the Hungarian geological community believed that the bivalves in question were in fact Miocene lymnocardiid cockles that had been flattened by pressure from overlying beds.

However, in the course of rearrangement of the collection in the Department of Paleontology at Eötvös Loránd University, specimens from a borehole drilled near the town Tótkomlós again came to light. They acutally pertain to *Daonella;* Bogsch had identified them correctly, and, in light of these finds, this penetrated Middle Triassic sequence must be considered the deeply buried continuation of the belt of *Daonella* shales that crop out in the Apuseni Mountains.

📖 Bogsch 1950

This Middle Triassic flat clam (*Daonella* sp.) comes from a borehole drilled near Tótkomlós on the southeastern flank of the Great Hungarian Plain. In spite of the many doubts, this really is a flat clam and not an artificially flattened one.

Triassic bivalves (all original size). (1) *Hoernesia socialis* (Schlotheim)—one of the most characteristic bivalves of open marine intervals represented by shell-bearing limestones in Southern Transdanubia. The valves of *Hoernesia* are rhomboidal in outline, with the left one (higher on the slab) higher and more inflated than the right (to the right on the slab). In accordance with its name (*socialis,* meaning "social") the valves of this species often occur as pavements on bedding planes (Bükkösd, Megyefa Quarry, Mecsek Mountains). (2) *Umbrostrea cristadifformis* (Schlotheim)—one of the earliest oysters, also characteristic of marine, shell-bearing limestones. On the surface of the left valve in this species a relatively smooth part formed earlier during ontogeny and a ventral part ornamented with radial folds can be seen (Orfű, Mecsek Mountains). (3) *Costatoria goldfussi* (Alberti)—a bivalve that has a rounded pentagonal outline ornamented with sharp radial ribs. The smooth posterior part of the valve, an area characteristic of *Trigonia*-like bivalves, is separated from the remaining parts of the shell by a keel. This species in particular is abundant in shell-bearing limestones of the Mecsek Mountains (Hetvehely, Mecsek Mountains). (4, 5) *Claraia clarae* (Emmrich)—a flat bivalve of circular, oval, or rounded tetragonal outline that is ornamented with folds running parallel to the valve margin. The left valve is more inflated than the right, while the auricles that are situated in front of the umbones usually cannot be studied. The widening area of the posterior margin is only just separated from the remaining part of the valve ([4] Balatonfüred; [5] Hidegkút). (6) *Praechlamys subdivisa* (Bittner). The Veszprém Marl has yielded several species of scallops, including this one, which has valves that bear dense radial ribs (Veszprém, Jeruzsálem Hill). (7) *Granulochlamys margaritifera* (Bittner)—an attractive scallop ornamented with radial ribs that themselves bear small tubercles (*margaritifera* means "pearl bearing" in Latin) (Veszprém, Jeruzsálem Hill). (8) *Antiquilima mytiloides* (Vigh)—a file clam from Upper Triassic dolomite in the Újlaki Hill, Budapest. (9) *Myophoria* sp.—an internal mold of a right valve from Campilian beds at Csopak. (10) *Costatoria chenopus* (Laube)—an attractive bivalve from the Veszprém Marl. (11) *Mysidioptera laczkoi* Bittner—one of the largest bivalves from the Veszprém Marl, it was named after eminent researcher Dezső Laczkó. The outer surface of this shell is ornamented with barely visible radial ribs (Veszprém, Jeruzsálem Hill). (12) *Mysidioptera multicostata* Bittner—another species of the genus *Mysidioptera* that is related to file clams. This one differs from *M. laczkoi* in the presence of striking radial folds (Veszprém, Lánczi Hill). (13) *Rhaetavicula contorta* (Portlock)—a small bivalve with an inflated left valve ornamented by arched ribs and a right valve that is flat and smooth. The largely semicircular outline of this taxon is disrupted only by the acute posterior ear. One curiosity of Triassic paleontology is that this very peculiar and widespread form, unmistakable when compared to any other bivalve, was originally described from Northern Ireland, an area of limited importance in global Mesozoic geology and paleontology. Following first description in 1847, however, *R. contorta* was identified from several other similarly aged localities along the Alpine-Himalayan orogenic belt; its easternmost occurrence is in Myanmar (Szentgál, Bagnyakőpuszta, Bakony Mountains). (14) *Daonella reticulata* Mojsisovics—a flat clam from the Middle Triassic Buchenstein Beds. The oval valves of this taxon are ornamented with spectacular radial ribs (Balatonszőlős, Hegymál).

1
2
3
4
5
6
7
8
9
10
11
12
13
14

The Explosive Success of Collecting

Collecting megalodontid bivalves is often difficult, especially in the case of large specimens that are embedded in rigid rocks. These fossils can be hammered out only after long and painstaking work. Such an obstacle was encountered some decades ago at one locality in the Vértes Mountains by Erzsébet Végh-Neubrandt and Viktor Dank, who was then a student and later an expert on petroleum research and chief geologist at the Hungarian state oil company. Fortunately, Soviet troops were training in the neighborhood and were well supplied with ammunition and explosives. At Végh-Neubrandt's request, large blocks of dolomite were exploded from the rock so that the geologists had only to hammer them to gather the specimens. These, as well as other parts of the valuable "*Megalodon*-collection" made by Végh-Neubrandt over the course of her career, are now housed in the Department of Geology and Paleontology at the Natural History Museum of Hungary.

Megalodontids: Emblematic Bivalves of the Upper Triassic

Upper Triassic bivalve faunas of the Transdanubian Range are more diverse than those of Northern Hungary. The most spectacular forms are megalodontids characteristic of platform carbonates. *Megalodon*-like bivalves appeared about 400 million years ago in the Devonian period. Early representatives of this family lived in the same environments as their descendants and after a long period of stasis they began to evolve rapidly in the Carnian age. From then on, megalodontids, living on the bottom in shallow-water areas in the Tethys Ocean, increased in diversity. Successive species attained larger and larger sizes and as such provide an example of phyletic size increase that is also known as Cope's rule. Some of the largest specimens were up to 42 centimeters high—and specimens of almost that size can be seen, for example, in some road cuttings at Kőris Hill near the village of Bakonybél. Witnesses who visited this locality when the forest road was just being cut into the Dachstein Limestone reported large specimens rolling out from the rocks.

The Transdanubian Range is a real El Dorado for Triassic megalodontids. Mass occurrences of this spectacular *dachsteinbivalv* attracted the attention of early researchers in the area; one of the first reports was published by Antal Koch (1843–1927), professor at the University of Kolozsvár (now Cluj-Napoca) and then-director of the Geo-Paleontological Institute at the University of Budapest. In his pioneering 1875 paper that treats the geology of the northwestern part of the Bakony Mountains, he wrote about these "paleontological inclusions"—fossils from the Dachstein Limestone: "These are almost exclusively confined to internal molds of the characteristic dachstein-bivalve *Megalodus triqueter* Wulf. They are, however, very frequent and specimens of variable sizes can be collected in enormous quantities in some places. Most frequently, one encounters only cross sections. I've found, however, some places where very nice internal molds can be collected in large quantities. I found most specimens on the plain and at the foot of Parrás Hill, where the limestone is full of closely spaced internal molds of megalodons in some places. These can be extracted with a hammer from the rock, but usually they fall to pieces."

Koch also provided a vivid description of the mode of occurrence of megalodontids. In most cases only their sections are seen, but they can be extracted intact from some beds. Specimens that have a peculiar mode of preservation can also be collected, as, for example, from the Aranyosvölgy Quarry in the Táborállás district of Veszprém. This large abandoned quarry from which Carnian dolomite was once exploited is now protected from garbage dumping by a mound formed by rejects from a former prefab construction factory in Veszprém. The specimens that occur in these dolomite beds are internal molds of characteristically small size, but their surfaces are covered with shiny dolomite crystals—similar in appearance to granulated sugar.

These initial observations were then followed by descriptive works, including a 1909 paper by Henrik Taeger on the megalodontids from Vértes, an area largely composed of Upper Triassic dolomite, and Fritz Frech's 1912 chapter on the fauna of the Bakony Mountains. Endre Kutassy also planned a monographic treatment of all species occurring in Hungary, but his premature death prevented him from completing the work. On the basis of his Fossilium Catalogus volume entitled *Pachyodonta mesozoica (Rudistis exclusis)* (Mesozoic pachyodonts [without rudists]) that also deals with megalodontids, Kutassy was prepared to write a comprehensive work.

Some decades later, work on Hungarian megalodontids was completed by Erzsébet Végh-Neubrandt, the former

Triassic megalodontid bivalve (0.5× magnification). (1–3) *Gemmellarodus seccoi* (Parona)—a large megalodontid bivalve that has a left valve that is higher and more inflated than the right one. This genus was named in honor of Gaetano Giorgio Gemmellaro (1832–1904), an outstanding Italian paleontologist. The specimen shown was collected by Lajos Lóczy Sr. near the village of Gyulafirátót and represents the subspecies *G. seccoi seccoi,* which differs from similar forms in its right valve—among other features—which has a markedly rounded outline. This is an index fossil for the upper Norian substage.

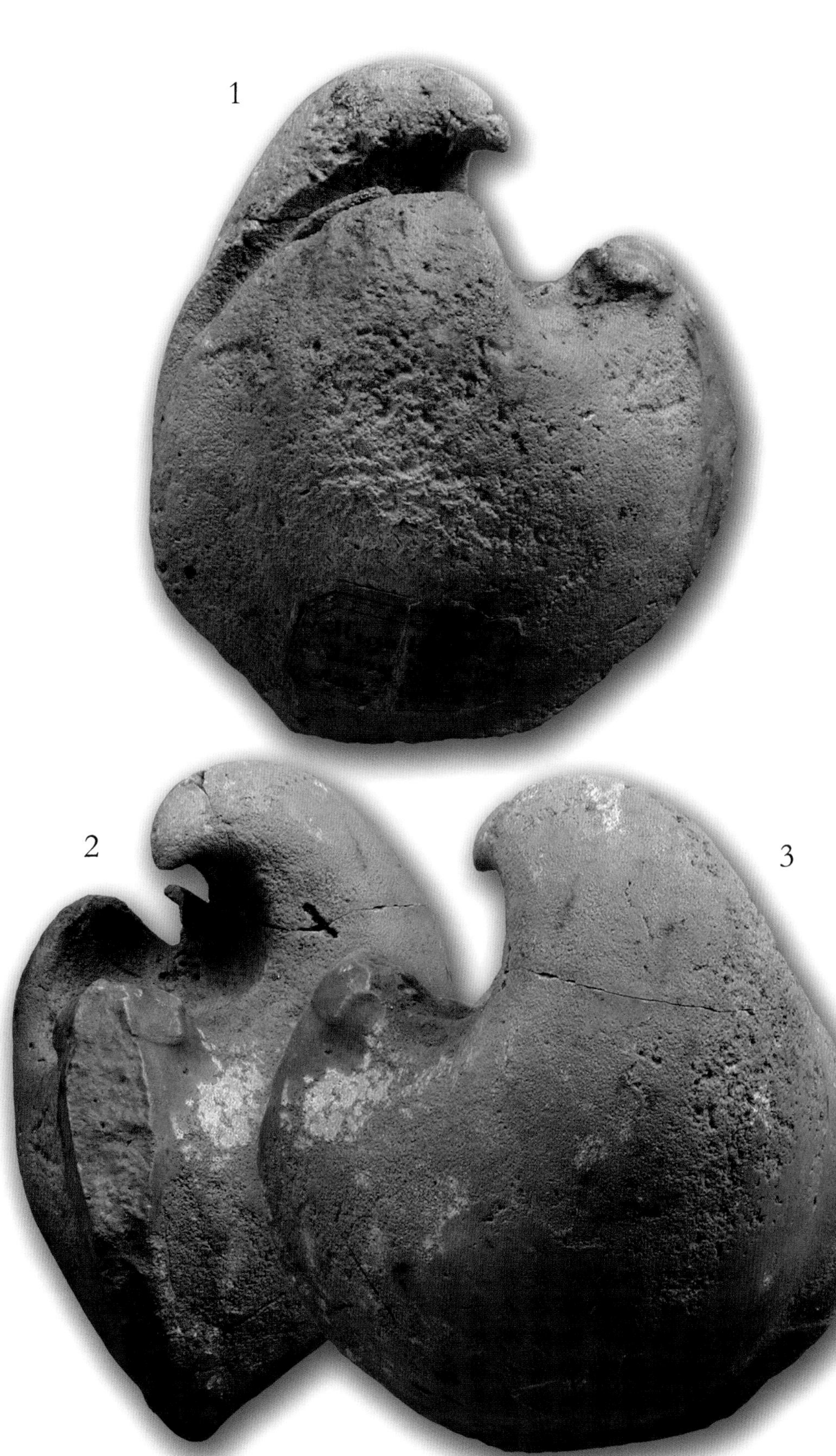
1
2
3

Megalodontid bivalves from the Upper Triassic of the Transdanubian Range (0.6× magnification). (1–3) *Rhaetomegalodon incisus hungaricus* (Végh-Neubrandt). To date this subspecies is known only from its type locality, the huge limestone quarry at Kecskekő Hill to the south of the town of Lábatlan. The largest specimens from this locality are 20 centimeters in height. The genus *Rhaeomegalodon,* as suggested by its name, lived in the Rhaetian epoch. Although Carnian megalodontids did not exceed 10 centimeters in height, those of the Rhaetian could reach up to 40 centimeters. (4, 5) *Neomegalodon arcuatus* Végh-Neubrandt. This figured specimen, collected from the Norian Dachstein Limestone at Várbükk Hill in Csesznek, is the type of this species and has since been found at other localities in the Transdanubian Range as well as in Slovenia and the Dolomites (Italy).

Erzsébet Végh-Neubrandt (1926–2008) and János Oravecz (1935–2009).

Although the professional career of Erzsébet Végh-Neubrandt (called "the Professor" by her students, or "Zizi" or "Zizike" by people close to her) was connected to the Department of Applied Geology at Eötvös Loránd University (which she led for a long time), she also produced an outstanding body of work in paleontology. Her book on the paleobiology of Triassic megalodontid bivalves is widely cited and highly appreciated among Triassic workers of the world.

János Oravecz, affiliated with the Department of Geology at Eötvös Loránd University until his retirement, was remarkable in part for his inquiring focus on the use of chemical extraction methods and their application to rocks otherwise considered hopeless. His activities in this area, for example, resulted in the first verification of the presence of Silurian rocks in Hungary. He also contributed significantly to an understanding of the Triassic geology of the Transdanubian Range with the recognition of the heterochroneity of the Upper Triassic platform carbonates.

head of the Department of Applied Geology at Eötvös Loránd University. Her 1982 volume contains revisions not only of species occurring in Hungary but all species described from the Triassic worldwide. This represented a huge undertaking because megalodontids are known from almost everywhere shallow-water Upper Triassic sediments are found—from Alaska, through the whole Alpine-Himalayan chain, to Japan and New Zealand. It is also interesting that although in the Ladinian and Carnian ages megalodontids lived at relatively high latitudes, in the Norian and Rhaetian they were confined largely to the tropics. Végh-Neubrand's book also offers much more than just systematic descriptions of species and is still in demand—and not only among megalodontid specialists for whom it a useful source of information concerning all aspects of paleobiology, classification, morphology, ontogeny, phylogeny, and biogeography.

Early researchers recognized that megalodontids evolve rapidly, and so their remains may serve as useful index fossils in successions that lack other guide forms such as ammonites. Indeed, the study of megalodontid bivalves made it possible to recognize that the deposition of the Main Dolomite and the Dachstein Limestone began earlier in the northeast of the Transdanubian Range than it did in the southwest. In other words, the base of these formations is heterochronous. In Hungary around 160 localities have yielded megalodontids, and 73 species and subspecies have been distinguished. The Carnian assemblage is characterized, for example, by species of *Neomegalodon* and the genus *Cornucardia,* whereas the genus *Gemmellarodus* and its strange relative *Dicerocardium*—which is reminiscent of a twisted horn—are typical of the Norian, along with species of *Neomegalodon.* In the Rhaetian, the genera *Rhaetomegalodon* and *Conchodon* are index forms.

In addition to their stratigraphic use, megalodontids have provided a good deal of other paleontological interest. For example, when Endre Kutassy was studying large (up to 30 centimeters high) specimens of the species "*Megalodon*" (now called *Gemmellarodus*) *amplus* he had described, he noticed that the surface of one internal mold collected at Felsőgalla (now part of the town of Tatabánya) had regular depressions of some millimeters in diameter. According to Kutassy, these depressions are traces of the oldest known pearls.

With the exception of megalodontids, Carpathian Basin Upper Triassic platform carbonate successions only rarely contain fossil bivalves. In some beds of the Main Dolomite, deposited in shallower water than Dachstein Limestone, barely identifiable internal molds of *Myophoria*-like bivalves do occur in large numbers; this monotonous fauna indicates that some environmental parameters, probably salinity and temperature, deviated considerably

1
2
3
4
5
6
7

from normal values during the deposition of the bed. The Dachstein Limestone, in contrast, is usually richer in non-megalodontid bivalves, with the remains of both burrowing forms (such as *Pleuromya*) and those that lay on the surface of the sediment (file clams and scallops) frequent in some places. The latter groups constitute the bulk of the bivalve fauna found in the Norian gray limestone at the Bleskový Prameň locality.

📖 Kutassy 1934a; Végh-Neubrandt 1982

Bivalves in Deeper Waters

Bivalve assemblages from Upper Triassic basinal deposits differ from those found on the platform. Of the usually small forms that constitute the "Kössen fauna" the peculiar *Rhaetavicula contorta,* an index fossil of the Rhaetian stage, is worth mentioning. The Feketehegy Limestone in the Pilis Mountains, which outcrops over a small area, was in the past grouped with the Kössen Beds because of the misidentification of a bivalve that occurs in large numbers in these rocks. Recent revision of collections from this unit, made chiefly by János Oravecz and Gábor Szilágyi, has revealed an interesting bivalve assemblage consisting of two dominant forms and some accessory faunal elements. *Pteria,* resembling a bird or feather (usually mentioned as *"Avicula"* in the older literature), and *Myoconcha* are the most frequent forms—although some valves of *Praeconia* with preserved traces of their color patterns have also been found.

📖 Haas et al. 2005

Triassic Bivalves from the Mecsek Mountains

Triassic bivalve assemblages from the mountains of Southern Transdanubia differ considerably from those found in the Transdanubian Range. Geologists mapping these areas identified most of the frequent forms, but Elemér Nagy (1934–1998)—an eminent Triassic and Jurassic researcher who worked extensively in the Mecsek Mountains—was the first to illustrate these bivalves in a 1968 monograph.

The lower part of this shell-bearing limestone succession is characterized by mass occurrence of the genus *Costatoria.* This bivalve appears for the first time in the red or green siltstone beds underlying the limestone succession. In the past, these beds were believed to be of Early Triassic age; thus, the bivalve remains abundant on some bedding planes of these rocks—outcropping, for example, in the Sás Valley near the village of Hetvehely—were identified with *Costatoria costata* and thought to be an index fossil of the Lower Triassic. Recently, however, it has been revealed that these fossils represent the species *C. goldfussi* characteristic of the Middle Triassic. *Costatoria*-bearing beds are overlain by a typical Muschelkalk sequence that is constituted by thin limestone beds with wavy bedding planes. This interval has yielded a bivalve fauna of slightly higher diversity, including scallops (*Entolium, Praechlamys*) and abundant burrowing forms (*Pleuromya*).

The appearance of a series of bivalve species in the shell-bearing limestone that are not encountered in older beds marks the time the Mecsek Triassic Sea reached its greatest depth. This interval is known as the Bertalanhegy Limestone and has yielded, in addition to abundant remains of articulate brachiopods (mainly *Coenothyris vulgaris*), the most diverse bivalve assemblage from the Mecsek

Triassic *Megalodon*-like bivalves (all original size). (1) *Rhaetomegalodon incisus* (Frech). This relatively high form is a guide fossil for the Rhaetian Dachstein Limestone. This specimen, displayed in characteristic posterior view, was collected in 1954 by Imre Szabó at the type locality for the species, Kálvária Hill at Tata. (2, 3) *Gemmellarodus paronai praenoricus* Végh-Neubrandt. The dolomite beds in the large abandoned quarry in Aranyos Valley at Veszprém have yielded several specimens of this small megalodontid bivalve. These remains are usually less than 4 centimeters in length and seem to be powdered with shiny dolomite crystals. As seen in anterior view (2), the left valve of this species is characteristically more inflated than the right. The name of this subspecies refers to its Carnian age—*praenoricus* means "prior to the Norian." (4) *Gemmellarodus amplus* (Kutassy). One of several megalodontid species originally described from Hungary, the distinctive features of this taxon include a markedly inflated left valve as well as a posteriorly elongated right one (not visible in the photograph). This specimen was found by Gyula Vigh (1889–1958), who worked extensively on the Mesozoic of the Transdanubian Range, in the quarry at Csákány-puszta near Felsőgalla. (5, 6) *Dicerocardium pannonicum* Oravecz. The genus *Dicerocardium* contains a number of bizarre forms markedly different from other *Megalodon*-like bivalves. The shared features of these forms include right and left valves that are more or less the same size as each other as well as a flat or slightly convex anterior region. Valves of some dicerocardiids are pointed, hornlike, and twisted; those of other species are reminiscent of wings. It is presumed that these bizarre valves served as safe anchors but prevented the animal from sinking into the sediment. The *Dicerocardium* species described from Hungary include *D. pannonicum,* the type specimen of which, taken from Babál Hill near the village of Epöl, is shown here. (7) *Cornucardia hornigii* (Bittner). Representatives of the genus *Cornucardia* are confined to the Carnian stage and have twisted left and right valves that are equal in size. It is assumed that the *Cornucardia* evolutionary lineage deviated from the main stock of megalodontids and gave rise—through *Dicericardium*—to the successful rudists of the Jurassic and Cretaceous. Only a very few species and subspecies are attributed to this genus, and *C. hornigii* is one spectacular and emblematic fossil from the Veszprém Marl. The specimen figured here was found in a quarry at Jeruzsálem Hill in Veszprém.

1
2
3
4

Academician Attila Vörös, former head of the Department of Geology and Paleontology at the Natural History Museum of Hungary and an internationally acknowledged expert working on Triassic ammonites and Jurassic brachiopods.

Triassic. The most spectacular form is *Plagiostoma lineatum,* an emblematic species of the Germanic Muschelkalk that sometimes reaches 10 centimeters in length. In some places, such as the large quarry at Megyefa near the village of Bükkösd, its valves occur as pavements of bedding planes. *Hoernesia socialis,* another bivalve typical to Germanic shell-bearing limestones, is also common, along with representatives of about a dozen species—all, however, are typically subordinate to brachiopods if the number of specimens is considered. At one single locality, the quarry near the village of Váralja in the Northern Mecsek Mountains, a bivalve-dominated fossil assemblage was found. The dominant element of this fauna is *Pseudocorbula gregaria,* a widely distributed and opportunistic species.

This relatively diverse bivalve assemblage, however, disappears suddenly upward in the sequence. The geologically youngest unit of the shell-bearing limestone succession is a peculiar bed, traceable throughout several kilometers, that is full of large, thick-shelled bivalves. Because of their outline, these remains were attributed to the genus *Trigonodus,* characteristic of the Trigonodus Dolomite in the Germanic succession. Unfortunately, no intact specimens with a preserved hinge structure have been collected from this rigid rock, and so this generic assignment still needs to be confirmed. Triassic bivalve assemblages from the Mecsek Mountains show stronger affinities to faunas from the Germanic shell-bearing limestone than to those of the Alpine Triassic.

📖 E. Nagy 1968; Szente 1997

Cephalopods

The First 150 Years of Ammonitology in Hungary

Among the Triassic fossils from the Carpathian region, ammonites have been always a primary focus of scientific interest. This continued attention has largely been because, in addition to their beauty, ammonite shells are important to geologists because they provide a unique tool for the subdivision and correlation of marine successions.

The most diverse ammonite assemblages, the characteristic elements of which were first described in the 1870s by János Böckh, are from the Balaton Highlands. Detailed documentation of these faunas was published by leading ammonite experts including Frech, Diener, and Arthaber, in the paleontological supplements of the Balaton monograph series. Publication of these respected volumes was followed by a half-century hiatus—owing in part to the discovery of a small amount of radioactive minerals in some Middle Triassic beds, which put a stop to Hungarian paleontologists' activities in that stratigraphic interval.

Collecting work, however, was not suspended; indeed, important ammonite collections were made as part of this search for fissile materials. In the 1960s and 1970s these explorations were led by Imre Szabó, a versatile geologist from the Mecsek Ore Company who had worked previously on the classical Jurassic succession at Kálvária Hill in Tata. Several trenches and shafts were dug that resulted in a rich collection of often very well preserved ammonites. These specimens, however, were not made available to the wider geological community for a long time.

In the meantime, interest in Triassic stratigraphy was considerably renewed globally, making it urgently necessary to revise the ammonites and sections known from Balaton Highlands. In the context of geological mapping carried out by the Hungarian Geological Institute, some sections at places often inadequately specified in the older

Triassic ammonites from the Buchenstein Beds in the Balaton Highlands (all original size). (1, 2) *Eoprotrachyceras pseudarchelaus* (Böckh)—large ammonites ornamented with ribs and tubercles. This attractive specimen was found near the village of Szentantalfa by János Böckh, original author of this species that has now been reported to occur at several other localities since its original description. (3, 4) *Proarcestes* sp.—globular form whose last whorls completely cover the previous ones and whose navel is always narrow and sometimes closed. The surface of this ammonite is ornamented with slightly undulating constrictions and occasional radial ribs, while the margin of the aperture is bent gently backward. This ammonite is a rather frequent form in the Ladinian Nemesvámos ("Tridentinus") Limestone in the Balaton Highlands. *Ptychites-* and *Arcestes*-like ammonites are often referred to as "leiostraca" (smooth-shelled) forms in older literature; they inhabited the deeper waters of the Triassic seas (locality unknown).

1
2
3
4
5
6
7
8
9
10
11

geological literature were rediscovered, and several new ones also came to light. The leading player in this revival of the study of Middle Triassic ammonites—a traditional field of paleontology in Hungary—is Attila Vörös, a member of the Hungarian Academy of Sciences and an expert on Jurassic brachiopods. Overall, the Anisian, Ladinian, and Carnian Stages proved to be especially rich in ammonites, and—following the publication of several shorter papers—the first results were summarized in a volume of *Studia Naturalia* issued by the Natural History Museum of Hungary in 1998. This volume documents in 16 photographic plates the most important ammonite species from the aforementioned stratigraphic levels. The most recent monographic treatment of the Triassic ammonites from the Balaton Highlands, the fauna of the Pelsonian substage, was published by Vörös in 2003.

📖 A. Vörös 1998, 2003

Promising Sections and Annoying Difficulties

Although the geology of the Balaton Highlands is very simple when compared, for example, to the Alps, ammonite researchers sometimes encounter difficulties. The first of these can be cover of soil and vegetation that often makes the building of artificial exposures necessary. The

Hungarites, one of the "Hungarian" ammonites from the Balaton Highlands. This attractive specimen has been naturally prepared (0.5× magnification).

considerable thickness of some successions is also a hampering factor; in such cases only one part of them can be explored in a given trench. Ammonite-rich intervals also are usually sandwiched between thick, non-fossiliferous sequences and so studying the fauna of a whole stage at a single exposure is rarely possible. This contrasts with the situation in the Jurassic of the Gerecse Mountains,

Triassic ammonites (original sizes except [7]and [8]). (1) *Tirolites cassianus* (Quenstedt)—a small- to medium-sized, relatively evolute form. The height of the whorls in this species exceeds their width, and the ventral part is either rounded or flattened. Ornamentation consists of tubercles or spines formed at the joining lateral and ventral regions of the shell as well as by wide ribs. Remains of *Tirolites* are characteristic of Lower Triassic rocks across the whole Tethyan region as well as in North America. In Hungary, this ammonite is known to occur all along the Hungarian Range from the Balaton Highlands to the Aggtelek–Rudabánya Mountains (Sóly). (2) *Flexoptychites* sp.—one of the most frequent ammonites from deeper-water Middle Triassic rocks from, for example, the Balaton Highlands. In this species, the last of the rapidly growing whorls covers all the previous ones and ratios of the shell change considerably during ontogeny. Earlier whorls are relatively inflated; those that form later are flattened. The lateral side of the shell bears slightly undulating, rare, and blunt ribs that fade towards the ventral region (Köveskál). (3, 4) *Arpadites* spp.—flattened forms with whorls that only partly overlap previous ones, two keels, and a groove that runs between them that characterizes the ventral region. The lateral regions of this ammonite are ornamented with fine, S-shaped ribs as well as small tubercles situated near the edge of the navel. (3) *A. toldyi* Mojsisovics. (4) *A. szaboi* (Böckh) (Szentantalfa). (5) *Paraceratites trinodosus* (Mojsisovics)—a flattened form that is made up of largely overlapping whorls bearing a crest-like keel on its ventral region. The lateral region in this species is ornamented with bifurcating ribs bearing three tubercles that become flatter during ontogeny. The species name of this stratigraphically important form refers to its ornamentation (*trinodosus* means "bearing three nodes") (Felsőörs, Forrás Hill). (6) *Vandaites stuerzenbaumi* (Mojsisovics). In the Late Triassic the first ammonites that deviate from normal shell morphologies (whorls touching each other, coiled in a plane) are seen. Some of the more unusual forms, known as heteromorphs, even had a snaillike shell; however, most remained coiled in a plane, but with successive whorls that do not touch. Rarely, one part of the shell formed a straight line. Because the onset of this straight shell during ontogeny could be variable, it is often considered a generic character. In *V. stuerzenbaumi*, for example, only the living chamber was straight; the shell had a tetragonal cross section ornamented with strong ribs. The genus *Vandaites* was introduced by E. T. Tozer, in honor of Vanda Kollárová-Andrusovová, an expert on Triassic ammonites. This species was named after József Stürzenbaum the discoverer of the locality (Drnava, Som Hill). (7, 8) *Rhabdoceras suessi* Hauer. In this heteromorph ammonite just the first whorls are coiled; the remainder of the shell, ornamented with ring-like folds or small tubercles, is straight or slightly arched (Pilisszentlélek, Fekete Hill) (slightly enlarged). (9) *Protrachyceras ladinum* (Mojsisovics)—A relative of *Trachyceras* (the type genus of "trachyostraca" ammonites ornamented with numerous tubercles), this species is characterized by whorls that partly overlap previous ones as well as ornamentation that consists of strong, tubercle-bearing ribs (Nemesvámos, Katrabóca). (10, 11) *Norites dieneri* Arthaber. The genus *Norites* comprises ammonites with disklike involute shells. A flat ventral region is another characteristic feature of these forms (Hajmáskér, Berek Hill).

Arpadites sp., a Middle Triassic ammonite from Katrabóca Hill at Nemesvámos. The outline of consecutive whorls forming the ammonite shell is usually a logarithmic (aka equiangular) spiral, in which the distances between the turnings increase in geometric progression. The detail seen in the photo is about 3 cm high.

for example, where the whole system can be studied by just walking a few tens of meters in a red-marble quarry.

Some years ago, researchers hoped that they might be able to study a succession of considerable thickness in a single section connecting two fossil-rich intervals that had been previously exposed. The top of Cser Hill near the village of Mencshely had long been known as a good locality for *"Ceratites" reitzii* and *Hungarites;* on another part of the hill gray limestone beds full of *Balatonites* specimens were also known to outcrop. Research pits and trenches dug across the area, however, provided evidence that the hill is in fact fragmented by geological faults hindering the study of both fossil-rich levels in an undisturbed section.

The Oldest Triassic Ammonites

Ammonites first appear in Triassic sequences in the Carpathian region and form part of the "Campilian" fossil assemblage that corresponds to a higher part of the Lower Triassic, the Olenekian stage. Characteristic elements of this fauna include the thorny form *Tirolites,* the smooth *Dalmatites,* and *Dinarites* that are ornamented with ribs. This assemblage, all of the same composition, can be traced along the Hungarian Range from the Balaton Highlands through the Bükk Mountains to the Aggtelek-Rudabánya Mountains. Preservation of specimens is also uniform, as all are internal molds, and one of the richest localities is at Iszka Hill near the town of Székesfehérvár. The fossil assemblage at this site was first documented by Anikó Bérczinė Makk and the fauna throughout belongs to the *Tirolites cassianus* Zone. None of the seven standard ammonite zones that are geologically older than this have been identified in sections from the Carpathian region.

📖 Bérczinė Makk 1970

Hungarites, Arpadites, Balatonites

In the Balaton Highlands, Middle Triassic basinal sequences have proven to be richest in ammonites. However, because the lower Anisian substage that is represented by shallow-marine sediments (Aszófő Dolomite) has so far yielded no ammonites, study of the sections in this area cannot contribute to knowledge of the faunas in the interval between the *Tirolites* fauna and the *Balatonites balatonicus* Zone, the next ammonite assemblage recorded from the area. Middle Anisian beds, however, are especially well exposed in the Farkó-kő section of Aszófő, and these have proven to be very rich in ammonites. *Balatonites* is the most characteristic element of this fauna, associated with other "trachyostraca" forms such as *Bulogites* and *Acrochordiceras,* and "leiostraca" forms such as *Norites* and *Discoptychites.* This assemblage indicates the Pelsonian substage.

Higher beds in the Anisian have yielded an assemblage characteristic of the Illyrian substage. Among the genera encountered in several sections, *Paraceratites* is considered to be an index fossil and occurs along with *Lardaroceras,* a peculiar ammonite ornamented with tubercles and fine-arched ribs. The diversity and number of specimens of *Ptychites*-like forms, supposedly indicating depositional environments deeper than those in older beds, is higher in the Illyrian beds. Overall, the upper Anisian has yielded ammonite faunas at several places in the Carpathian Basin; this stratigraphic level is represented, for example, by the assemblage from on the top of Dunna Hill in the Aggtelek-Rudabánya Mountains. Here the lilac or red nodular embedding rock is strongly reminiscent of the Han Bulog Limestone described near Sarajevo in Bosnia and Herzegovina. A slightly older geological age is indicated by the faunule from the shell-bearing limestone in the Mecsek Mountains. The first identifiable find of the zonal marker *Paraceratites binodosus* was documented in 1973 by Csaba Detre; interestingly, ammonites, usually very rare in this thick succession, can usually be found at certain localities. These sites tend to be concentrated in the western part of the mountains—most probably indicating the position of the Middle Triassic Sea.

The Fassanian substage, for decades unofficially considered the base of the Ladinian stage, has yielded the most diverse ammonite assemblage from the Triassic of the Balaton Highlands. This richness was the basis on which a proposal was submitted to the International Commission on Stratigraphy to designate the Forrás Hill section as the GSSP for the Ladinian stage defined at the base of the Reitzi Zone, which corresponds to the base of the Fassaian. The zonal marker for the Reitzi Zone is *"Ceratites reitzii"* (long known as *Xenoprotrachyceras reitzi* and now called *Reitziites reitzi*), described by János Böckh and spectacular because of two rows of spines on its ventral surface. According to the GSSP proposal the first appearance of the species *Kellnerites felsoeoersensis,* originally described by József Stürzenbaum, should be used to mark the base of the Ladinian.

A typical Reitzi assemblage contains, besides the name-giving form, species of *Hyparpadites* and *Hungarites* as well as several other genera. If the number of specimens alone is considered, then *Flexoptychites* is the most frequent; the ammonites, preserved in rocks containing grains of volcanic origin, are often variegated and different shades of red, lilac, brown, or green. The Forrás Hill section has yielded most of the available material, but several well-preserved specimens have also been collected from the section at Mencshely and the trench dug on Öreg Hill at Vászoly.

Higher beds of the Ladinian stage (in the traditional sense) have also yielded diverse and well-preserved ammonite assemblages, including specimens considered to be among the most attractive from the whole of the

1
2
3
4
5
6
7
8
9
10
11
12

Balaton Highlands. Most of these were collected from the so-called Nemesvámos marble, a peculiar red limestone reminiscent of Hallstatt Limestone at Katrabóca Hill. From this locality, the genus *Protrachyceras* is represented by several specimens in the fauna; among "leiostraca" ammonites, *Ptychites*-like forms are replaced by globular arcestids (*Joannites, Proarcestes*), which also have weakly ornamented shells. Unfortunately, all ten upper Ladinian localities are from the same stratigraphic level, and so their scientific importance is subordinate to the aforementioned sections.

📖 Arthaber 1911; Frech 1911a, 1911b; Detre 1973; A. Vörös 1998, 2010

Upper Triassic Ammonites

As in the upper Ladinian, only sporadic ammonite finds come from the Carnian stage. In the Balaton Highlands, the pelagic, cherty Füred Limestone overlies volcano-sediment successions of the Buchenstein Beds. Although well exposed, occurring as steeply dipping beds in a spectacular abandoned quarry situated at the southern end of the Nosztori Valley near the village Csopak, Füred Limestone is rather poor in fossils. The specimens known from this rock, which is widely used as a building stone, are small in size but usually well preserved. Of the diverse assemblage, representatives of *Trachyceras* are encountered at several exposures—as well as those of *Dittmarites,* which is ornamented with fine, arched ribs. A member of the overlying Veszprém Marl is referred to using the traditional name "Austriacum limestone" because of the presence of densely ornamented *Trachyceras*-like ammonites. This succession, however, is not too fossiliferous and scarcely yields a single specimen in a century. The upper beds of this thick marl succession are also poor in ammonites: In his "Balaton monograph," Lajos Lóczy, Sr., mentioned that the genera *Carnites, Lobites, Lecanites,* and *Trachyceras* are known from this sequence but have been sampled and studied only a few times over the last century.

Carnian ammonites are also known from Budapest. An assemblage of dwarfed specimens, including the important index fossil *Anolcites,* was discovered by Károly Hofmann (1839–1891) in dolomite rocks outcropping near Apáthy Cliff in the Hűvös Valley. Hofmann was a pioneering researcher on the geology of the Buda Mountains. Ammonite assemblages from the brachiopod-rich, crinoidal "Szádvárborsa Limestone" (officially called the Silická Brezová Limestone) collected at localities in present-day Slovakia also testify to the presence of the Carnian stage. Descriptions of the small forms were published in 1940 by Kálmán Balogh, a lifelong researcher of the geology of Northeastern Hungary and adjacent areas. Carnian ammonites were also found in nearby parts of present-day Hungary—with *Sirenites* and *Austrotrachyceras,* for example, identified from a faunule found near the village of Szőlősardó by Leopold Krystin, a Viennese worker who specialized in the conodonts and ammonites of the Tethyan Triassic and who contributed significantly to our knowledge of these groups from across the Northern Calcareous Alps to the Himalayas.

Triassic ammonites from the Balaton Highlands (all original size). (1–3) *Neoprotrachyceras* spp.—Carnian ammonites from the Veszprém Marl at Veszprém ([1] *N. attila* [Mojsisovics] [Jeruzsálem Hill];, [2, 3] *N. baconicum* [Mojsisovics]. [Sintér Hill]). (4) *Paratrachyceras hofmanni* (Böckh). This finely ornamented specimen was collected from the fossil-poor Füred Limestone that represents the lower Carnian substage. This species was named by János Böckh for his brother-in-law Károly Hofmann (Sóly, Fenyves Forest). (5) *Lecanites loczyi* Frech. Several specimens of this small smooth form were found in the Veszprém Marl and based on them, Frech named a new species in honor of Lajos Lóczy Sr. The type specimen, also shown here, was refigured—probably because of its accurately painted simple lobes that are reminiscent of those of ancient ammonites in several books, including those by Elemér Vadász and András Tasnádi-Kubacska (1902–1977) (Veszprém, Szalay Hill). (6, 7) *Arpadites telleri* Mojsisovics—an ammonite from the Buchenstein Beds. (Szentkirályszabadja). (8, 10:) *Protrachyceras* spp.—stratigraphically important ammonites from the Ladinian ([8] *P. archelaus* [Laube] [Felsőörs, Forrás Hill]; [10] *P. longobardicum* [Mojsisovics] [Balatoncsicsó]). (9) *Hyparpadites liepoldti* (Mojsisovics). This attractive ammonite, ornamented with ribs and nodes, is a subzonal index fossil for the Reitzi Zone that is now assigned to the Anisian (Felsőörs, Forrás Hill). (11) *Balatonites balatonicus* Mojsisovics. This evolute and flat form is the index fossil for the Pelsonian substage and is named after Lake Balaton. The ventral surface of this fossil looks like the top of a house in cross section, while the lateral region of the shell is ornamented with radial and occasionally bifurcating ribs that bear three rows of tubercles. *Balatonites* was distributed worldwide in the Anisian age of the Middle Triassic and the specimen shown here is the type of the species (Mencshely). (12) *Austrotrachyceras austriacum* (Mojsisovics). Some fossil species become famous for their numbers of specimens, and others because of their rarity. This species, "*austriacum,*" possesses a moderately involute shell comprising whorls—the height of which well exceeds their width—and a rounded ventral region, and belongs to the latter group. The specimen figured here was found by Dezső Laczkó, in a small quarry in the Nosztori Valley at Csopak at the beginning of the 1900s. After his lucky find he named the rock in the quarry the "Austriacum limestone," an alluring name that was subsequently widely distributed in the geological literature dealing with the Balaton Highlands that led to large numbers of people visiting the quarry—hidden in the bushes some meters from busy Road no. 73—in hopes of finding some attractive specimens. Visitors, however, almost always leave disappointed because this succession, now known as the Nosztor Limestone Member, is extremely poor in fossils.

The Norian stage has also yielded relatively few ammonites, although the Late Triassic marks the first appearance of heteromorph ammonites. Besides common forms, such as tiny specimens of *Paraplacites* and *Megaphyllites,* straight-shelled fragments of *Rhabdoceras* (considered an index fossil) were found by János Oravecz in the bivalve-dominated assemblage of the Feketehegy Limestone. The same so-called normal ammonites were also identified from the unattractive—although stratigraphically important—Remetehegy faunal assemblage, which is known from the now-abandoned quarry at the mouth of the Remete Gorge at Budaliget. A dwarfed fauna known from another Buda Mountains locality, Fazekas Hill, is also partly Norian in age. In Northern Hungary, presumably Norian ammonites have also been found on Haragistya Hill near the village of Jósvafő, in beds thought to represent the Zlambach Marl, a classical formation described from the Northern Calcareous Alps.

The uppermost stage of the Triassic, the Rhaetian, has proven to be even poorer in ammonites than earlier stages. These occurrences, due in part to their rarity, have become famous. In particular, the fauna from the Bleskový Prameň locality near Drnava has been known since the end of the nineteenth century; its most recent revision, published in 1973, described the presence of 12 species. The favorable state of preservation of fossils from this site has made possible the study of hitherto unknown early ontogenetic stages of shells, although the majority of recorded species belong to normal ammonites. The most interesting of the forms collected are heteromorphs, including *Vandaites stuerzenbaumi,* which is a zonal marker for the Rhaetian.

The Rhaetian ammonite fauna of Csővár was discovered much later than the aforementioned assemblage. Since the late 1980s, when a flattened specimen of *Choristoceras marshi* built up from loosely coiled whorls was found, this section has yielded a good deal of Upper Triassic and lowermost Jurassic ammonites—including the abovementioned peculiar species from Drnava. Two heteromorphs from this locality are indicators of the two uppermost standard ammonite subzones (Stuerzenbaumi and Marshi) of the Rhaetian.

📖 Balogh 1940; Bérczíné Makk 1969; Detre et al. 1988; J. Pálfy, Demény, et al. 2007

Triassic Nautiloids and Coleoids

In addition to ammonites, the remains of other cephalopods can also be found in the Triassic rocks of the Carpathian Basin. Although never among the most frequent fossils, nautiloids have been collected from several localities, especially in the Middle Triassic, and some species were described by Fritz Frech in his Balaton monograph. However, because previous finds were so sporadic, the discovery of a diverse and well-preserved assemblage in Pelsonian beds at Aszófő was unexpected. This fauna consists of about a dozen species and was documented by Attila Vörös in a paper published in the *Fragmenta Palaeontologica Hungarica,* issued by the Natural History Museum of Hungary. The most common genera in this assemblage—including *Pleuronautilus,* which is ornamented with blunt ribs, and *Germanonautilus,* which is characterized by a smooth shell and a tetragonal whorl cross section—have also been found in several localities in the Balaton Highlands. Indeed, an attractive specimen of *Germanonautilus* was also found in the Mecsek Mountains. Ortothoconical cephalopods, including both ectocochliate forms such as the *Orthoceras*-like *Michelinoceras* and endocochliate coleoideans (*Breviconoteuthis* and *Mojsisovicsteuthis*), also occur rarely.

📖 A. Vörös 2001

Bryozoans

During the Paleozoic, bryozoans were important elements of shallow-water benthic communities. However, most genera went extinct at the end of the Permian and only a very few forms are known from the Lower Triassic. In general, bryozoan diversity gradually increased until the Carnian age and was then followed by a decline toward the end of the Triassic, when bryozoan faunas are known from only a few localities around the world. Likewise, in the Carpathian region, one Carnian succession in particular, the Veszprém Marl, has yielded a considerable bryozoan fauna that was described by Vinassa de Regny.

Brachiopods

"Lamp Shells" from Southern Transdanubia

Brachiopods and bryozoans are related to each other but have not shared the same fate. Although bryozoans flourish today in some environments, brachiopods play a very subordinate role in extant marine communities.

The dominant Paleozoic brachiopod groups, including *Productus*-like forms, went extinct at the end of the Permian. From the Lower Triassic onward remains of the environmentally very tolerant inarticulate brachiopods, including forms related to *Lingula,* are known—as these were able to survive the crisis that generally afflicted organisms living in shallow seas. By the Middle Triassic, as abundant remains of shells found in some places testify, brachiopods reattained considerable diversity; the Middle and Upper Triassic sequences in the Carpathian region are known to contain well-preserved assemblages that have been the focus of much paleontological research over the years. In Southern Transdanubia, however, brachiopods are confined to the middle marine interval of the Triassic system: limestones, dolomites, and sometimes siltstones contain *Lingula*-like inarticulate brachiopods throughout

Triassic brachiopods (all original size). (1) *Lingula* sp. The genus *Lingula* ("little tongue" in Latin) is one of the best known "living fossils." Recently, however, important differences in morphology between the known extinct and living forms have been emphasized by Christian C. Emig, an expert on lingulid brachiopods. Phosphatic shells of *Lingula*-like brachiopods, also known as "tongue shells," are not uncommon in the Triassic limestones and dolomites of the Mecsek and Villány Mountains. One dolomite succession that is well exposed in the "Winery Quarry" on the northern slope of Templom Hill at Villány is also traditionally called the "*Lingula* Dolomite." Because the valves of *Lingula* are not articulated to one another they are often found separately (Bükkösd, Megyefa Quarry, Mecsek Mountains). (2) *Punctospirella fragilis* (Schlotheim). Besides *Coenothyris vulgaris* the shell-bearing limestone also contains other brachiopods, including this small spiriferid. *P. fragilis* is subordinate to the aforementioned species in benthic assemblages of Hungary. It also occurs in the upper shell-bearing limestone in the Germanic Basin forming monospecific shell beds traceable across tens of kilometers. It is thus also used for the correlation of sections (locality as above). (3) *Adygella julica* (Bittner). This smooth-shelled form is the most common component in the brachiopod fauna of the Veszprém Marl. Valves that are less than 10 millimeters in length are joined along a commissure line that is straight in its anterior part; the lines of larger forms, as shown here, bear two dorsally curved folds (Veszprém, Jeruzsálem Hill). (4, 5) *Spiriferina fortis* (Bittner)—another common brachiopod from the Veszprém Marl that can be distinguished from *P. fragilis* because of the different length of its inner shell parts as well as its outline and the number of radial folds in the valves (locality as above). (6–8) *Fissirhynchia fissicostata* (Suess)—a characteristic brachiopod from the Kössen Beds. The genus and species of this taxon are named for the mode of occurrence of the ribs ornamenting the valves (Hybe). (9) *Coenothyris vulgaris* (Schlotheim). The most abundant brachiopod in the shell-bearing limestone occurs in large quantities in some Middle Triassic beds in Southern Transdanubia. This taxon is, however, restricted to a relatively thin interval, which is interpreted to represent the deepest marine settings of the sequence. The species name refers to its mode of occurrence (*vulgaris* means "common" in Latin), and as seen in the specimen shown here, the shape of *C. vulgaris* is rather insignificant—as is common among the *Terebratula*-like brachiopods. One of the distinguishing outer characters of this taxon is the low fold of the commissure line on the anterior part of the shell (Bükkösd, Megyefa Quarry, Mecsek Mountains). (10) *Rhaetina pyriformis* (Suess)—another characteristic brachiopod from the Kössen Beds that is often found associated with colonial corals. The genus *Rhaetina,* as the name suggests, is a characteristic brachiopod from the Rhaetian stage of the Upper Triassic. The species name refers to the resemblance between these fossils and pears (*pyrus* means "pear" in Latin) (Cresuia, Apuseni Mountains). (11) *Tetractinella trigonella* (Schlotheim). One of the most recognizable of the Triassic brachiopods, the genus and species name of this form refer to its four spectacular ribs and the triangular shape of the valves. This genus is among the most famous of all brachiopod genera cited in paleontology textbooks because, together with the Upper Jurassic *Cheirothyris,* it provides a clear example of the phenomenon of homeomorphy. In this case, the outer shape of the two genera is identical but their inner features as well as their shell structures are quite different. In the past *T. trigonella* specimens could be collected in large numbers weathered out of the Recoaro beds in the Forrás Hill section at Felsőörs. Since this locality is now protected, further collecting is prohibited (Felsőörs, Forrás Hill).

the sequence. Articulate forms, considered the only decent brachiopods by some experts, are only known from the Zuhánya Limestone, which was deposited in a relatively open sea environment. As documented in several papers and a voluminous doctoral thesis by Csaba Detre and, more recently, in works by Ákos Török, the valves of *Coenothyris vulgaris* as expected because of its name occur in rock-forming quantities in some limestone beds.

One interesting problem in this area arose in the mid 1980s, when József Pálfy and Ákos Török—both students working on their MSc theses in the Department of Paleontology at Eötvös Loránd University—were researching the inner features of Anisian brachiopods from the Balaton Highlands and the Mecsek Mountains. This sort of work had become standard in order to recognize cases of homeomorphy, a common phenomenon in brachiopods that occurs when shells with identical outer shapes are seen in unrelated species. The internal morphology of *Coenothyris vulgaris* had not yet been studied, and comparisons of drawings made from serial sections of specimens derived from the two areas yielded a surprising result: the internal features of *C. vulgaris* from the Balaton Highlands were different from those from the Mecsek Mountains. Thus, the species *C. vulgaris,* identified on the basis of external features, may be polyspecific.

Other brachiopods, including the attractive *Punctospirella fragilis* and the triangle-shaped *Tetractinella trigonella,* were also recorded from the Middle Triassic of the Mecsek and Villány Mountains. The species, however, number not more than five.

Brachiopods from the Hungarian Range

Much more diverse brachiopod assemblages are known from the Anisian in the Balaton Highlands, and species from the Alpine shell-bearing limestone called "Recoaro" were the first to attract the attention of researchers in the area. Some of the most frequent forms were documented by János Böckh in 1872, and the first comprehensive study was published as a Balaton monograph. Alexander Bittner and Fritz Frech, both Triassic invertebrate specialists, described about 50 brachiopod species from the Middle and Upper Triassic—especially from Anisian and Carnian beds. Following a gap in study, their work was continued by József Pálfy; the results of his activity in this field so far have shed light on paleobiogeographic affinities as well as the environmental and evolutionary significance of the fauna. Systematic descriptions of 34 species from the Pelsonian substage were published in 2003.

In contrast to the Balaton Highlands, studies on Triassic brachiopods continued uninterrupted in other parts of Europe. One leading personality in this field was the Lithuanian Algirdas Dagys (1933–2000), who worked on Triassic assemblages from the Caucasus and introduced several new genera based on species previously assigned to other classical genera. One of them, *Koeveskalina,* is based on the taxon "*Spiriferina Köveskalyensis*"—originally described by Dionýz Štúr—from the Balaton Highlands. The genus and species names of this organism, although in Latinized form, refer to the small village of Köveskál that is situated in the pretty Káli Basin not far from Lake Balaton.

As overviewed by Csaba Detre on the occasion of the 1992 Regional Field Symposium on Mesozoic Brachiopods held near Lake Balaton, the Carnian of Hungary is also rich in brachiopods. However, because of the lack of accessible localities, the fauna of the Veszprém Marl first described by Bittner in 1900 has not been studied since. A new occurrence was discovered in the early 1990s by Gábor Csillag and László Gyalog, both geologists mapping in the Vértes Mountains. Dolomite beds outcropping over a relatively large area near the village of Gánt contain an assemblage that is dominated by the genus *Cruratula.* Another similar, peculiar assemblage is also known from the Maly Karpaty Mountains in Slovakia; Carnian brachiopod assemblages are known—although they are documented mainly in faunal lists—from the Buda Mountains and from the Triassic inlier at Nézsa that is situated on the east side of the Danube. A coeval fauna known from the Bükk Mountains is also little studied; limestone and dolomite successions from Norian and Rhaetian stages are also known to contain brachiopods occasionally but, with the exception of a few localities, play only a subordinate role in ongoing studies.

An Anisian assemblage similar in composition to those seen in the Balaton Highlands was described in 1972 by Gábor Scholz from a reef succession exposed near the village of Jósvafő. As seen also at the Forrás Hill locality in Felsőörs, *Tetractinella trigonella* is the most spectacular element of this fauna, and assemblages of more or less the same composition have been documented by Miloš Siblík from localities on the Slovak Karst. Carnian brachiopods were described in 1940 from the same area for the first time by Kálmán Balogh; these derive from the pink, crinoidal Silická Brezová Limestone.

Although Rhaetian Kössen Beds also make up parts of the geology of the Transdanubian Range, and in some places (such as near Szent Miklós Spring in the Keszthely Mountains) they also contain brachiopods, the richest occurrences are from areas outside Hungary. The locality at Hybe has proved to be especially noteworthy from this point of view and other, typical Kössen brachiopods are also known from some localities in the Apuseni Mountains of Romania.

📖 Bittner 1912b; E. Nagy 1968; Siblík 1986; Detre 1993, Detre, Szentes, & Szente 1986; Detre, Lantos, & Ó-Kovács 1992; J. Pálfy 1988, 2003; J. Pálfy & Török 1992

Ophiolepis balatonica Detre and Mihály, a brittle star from the Lower Triassic at Sóly in the Balaton Highlands. (Approximately original size.)

Echinoderms

Sea Lilies

Nearly all of the marine Triassic successions in the Carpathian region contain echinoderm remains, of which fossil sea lilies are the most common. Their calcite skeletal parts are easily recognizable on freshly broken rock surfaces and are characteristically shiny. Often remains embedded in rocks—with the rare exception of highly characteristic sections—cannot be identified precisely; there are, nevertheless, rock types from which parts of the stem (columnals) and arms (brachials)—as well as much less frequently the calyxes—can be extracted using either chemical or physical methods.

In the Transdanubian Range some Triassic successions in the Balaton Highlands have long been known to contain the identifiable remains of sea lilies—frequently in large quantities. Often seen in thin-sections, the skeletal parts of these fossils are also present in Lower Triassic limestones, but these occurrences have not yet been studied.

The much better studied and documented sea lilies are known from Middle Triassic successions often referred to as "Alpine shell-bearing limestones" in older literature. The fossil assemblages of some beds are dominated by brachiopods and sea lilies—the latter represented by just a few species, but usually a huge number of specimens. Taxa such as *Dadocrinus gracilis* and *Encrinus liliiformis*—both described at the beginning of the nineteenth century, when work was just starting in the Triassic—are encountered frequently in the geological literature and in the area. Indeed, hundreds of columnals are literally falling out of the rock in some places, including at the Forrás Hill section in Felsőörs. The aforementioned two species, so characteristic of their particular stratigraphic interval—along with the brachiopod "*Terebratula*" *decurtata* that is now attributed to the genus *Decurtella*—were in the past considered index fossils of the so-called Decurtata level.

The crinoid assemblage of the Veszprém Marl is much more diverse and is found in soft and friable sediments. The sea lily assemblage, which alone comprises 23 species, was recorded in a 1911 Balaton monograph by Francis Arthur Bather (1863–1934), a British expert on fossil echinoderms who was then president of the Geological Society of London. The abundance of crinoid remains—as well as the amount of the work that went into this monograph—can be gauged from the fact that Bather studied 11,000 specimens in total, of which the genus *Isocrinus* proved to be the most common.

Although found relatively late, this succession has also yielded a good number of interesting crinoid remains that had not previously been recorded. In the course of her work on foraminifera, Anna Oraveczné Scheffer discovered some skeletal parts of the planktonic crinoid *Roveacrinus* when studying washing residue from the Veszprém Marl. As she discussed in a 1979 paper, these pelagic forms had a wide geographic distribution, but they are also of some biostratigraphic value and indicate a lower Carnian age.

Recently the Middle Triassic sea lilies from the Mecsek Mountains were studied in detail. Although recorded by earlier researchers who worked on the geology of this area, the crinoid assemblage remained undocumented in detail for many years before the task was taken on by an expert from abroad. Hans Hagdorn—director of a secondary school for commerce as well as the owner of the Muschelkalmuseum Hagdorn in Ingelfingen, Germany—is an acknowledged expert on Triassic crinoids; his studies resulted in a 1997 paper written jointly with Gyula Konrád and Ákos Török, a renowned carbonate sedimentologist and professor at the Budapest University of Technology and Economics. They state that the Middle

3
1
2
4

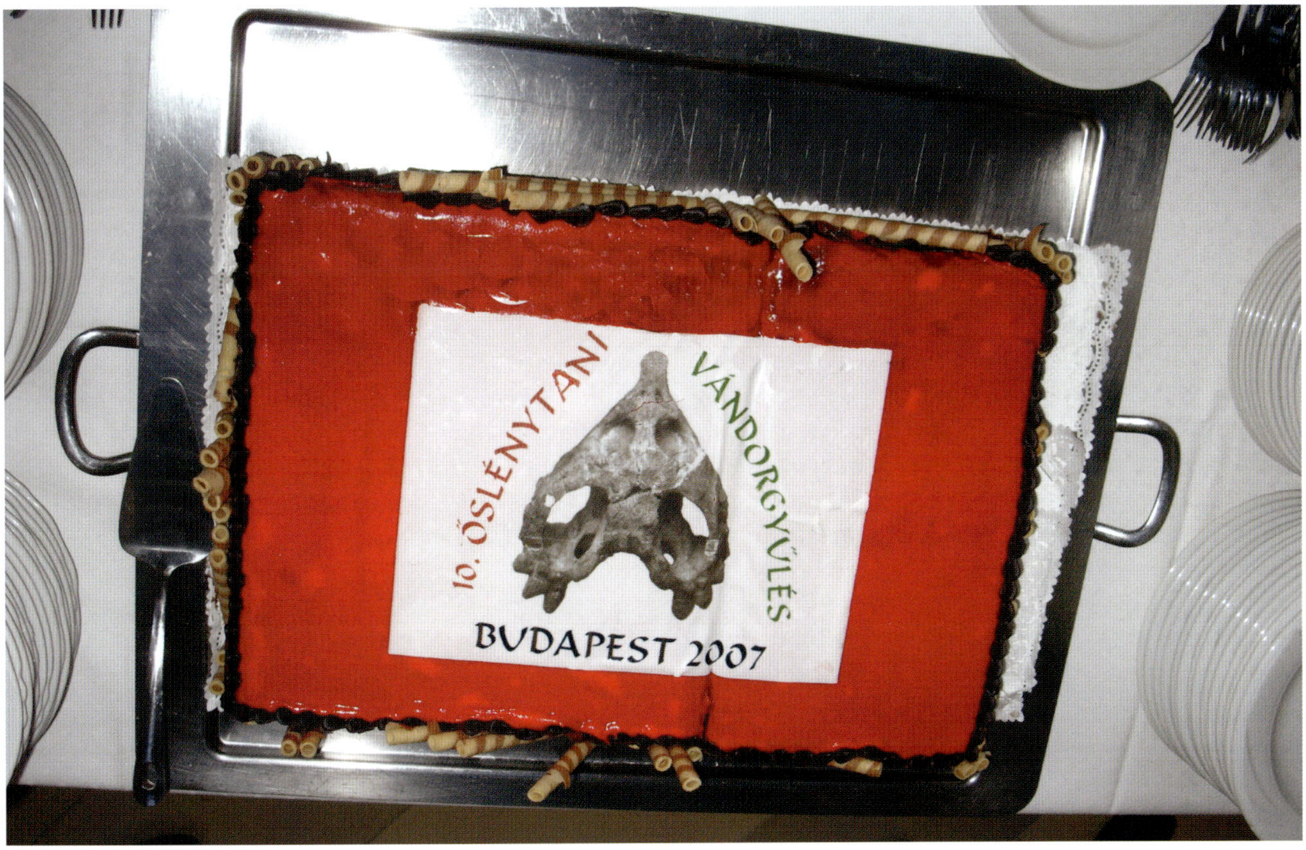

Ancient reptile from Marzipan. The figure of this pebble-toothed pseudo-turtle has appeared in innumerable different renderings since its remains were discovered more than 100 years ago by Dezső Laczkó. These include images on the pages of popular scientific books and even on postage stamps. Here, the characteristic skull of *Placochelys* is seen on the top of a cake prepared for the ceremonial dinner of the 10th Jubilee Hungarian Regional Paleontological Meeting in spring 2007.

Triassic crinoid assemblage from the Mecsek Mountains comprises nine species, and as such is more diverse than previously supposed. It also proved possible to subdivide this monotonous limestone succession, which contains age-determining fossils in just a single level, into successive zones based on the sea lilies. A final point of interest is that the Mecsek fauna shows affinities with coeval assemblages in Upper Silesia as well as in the eastern part of the Germanic Basin.

Besides numerous stem elements, a single calyx of *Dadocrinus* with articulated arms was collected recently from the shell-bearing limestone in the Mecsek Mountains. This very exceptional find was exhibited at the University of Pécs by Kriszta Sebe, who works on the geology of the Mecsek Mountains, and her supervisor Gyula Konrád. Sadly, this fossil was so appealing that someone stole it.

Middle Triassic limestones of the Aggtelek-Rudabánya Mountains are also known to contain crinoid remains.

One of the most valuable fossils known from Hungary, the "pebble-toothed pseudo-turtle," found at Jeruzsálem Hill in Veszprém (0.5× magnification). The skull is the most important part of the skeleton of this animal that had "button-like" teeth, which it is thought to have used to crush the hard shells of bivalves and brachiopods that lived on the bottom of the sea. The known fossil remains of *P. placodonta* are housed in the Geological Museum of Hungary in the Hungarian Geological Institute, Budapest. (1) Upper view of the skull. The small outer nasal openings as well as the large orbits and wide upper temporal fenestrae of this reptile are visible behind the elongated tip of the nose. (2) Lower view of the skull showing the flat grinding teeth that are attached to the roof of the mouth. (3) Upper view of the lower jaw bearing four teeth. (4) A fragment of the attractive carapace of *Placochelys*, with more or less regularly arranged larger cones surrounded by smaller ones. Extraction of this large piece of the armor of *Placochelys* from the surrounding rock ended up revealing the remainder of the almost undamaged fossil. Like true turtles, the armor of these pebble-toothed pseudo-turtles was formed by an inner bone layer derived from the skin during ontogeny and then covered by a horny layer from the outside. The armor of these Placodont-like reptiles is composed of bony plates, which never formed belts joined along regular sutures as they do in true turtles. The most important difference of note is that, in contrast with true turtles, the armor of these Triassic animals was not fused with the vertebrae and ribs.

The faunule recently described by Hans Hagdorn and Felicitász Velledits consists of six taxa, most of them identifiable only at the family level.

📖 Bather 1911; Hagdorn et al. 1997; Hagdorn & Velledits 2006

Sea Urchins, Starfish, and Sea Cucumbers

The remains of sea urchins, so abundant and spectacular in some rocks of the Cretaceous and so-called Tertiary, are uncommon in the Triassic of the Carpathian Basin. The only considerable known fauna, published by Bather in his Balaton monograph, comes from the Veszprém Marl. This assemblage comprises 20 species of regular *Cidaris*-like forms; although the species is usually represented by isolated spines and fragments of corona, the assemblage is nonetheless one of the most diverse known from the Triassic.

Starfishes also are rare fossils in the Triassic of the Carpathian Basin. One single stratigraphic level, the highest part of the Lower Triassic, is known to contain relatively frequent and articulated remains. The Sóly section in the Balaton Highlands has yielded numerous specimens that were assigned in 1983 and 1987 to the extant brittle star genus *Ophiolepis* by Csaba Detre and Sándor Mihály. These finds are also interesting because there are only three other areas—China, Pakistan, and Northern Italy—in which Lower Triassic brittle stars are known to occur. Indeed, species found in the Dolomites are apparently indistinguishable from those found in the Balaton Highlands. Carmela Broglio Loriga, an expert of the geology and paleontology of Northern Italy, and her colleague Anna Berti Cavicchi interpreted the Dolomites specimens as representatives of a new genus called *Praeaplocoma,* which is confined to the Triassic.

Isolated, tiny skeletal elements (sclerites) of sea cucumbers are often found in dissolution residues of Triassic marine sediments, and some of them have considerable stratigraphical value. For example, the Late Triassic age of the limestone succession of the Csővár Quarry was evidenced by Kozur and Mostler on the basis of microfossils including peculiar, wheel-like remains of the sea cucumber genus *Theelia.*

📖 Bather 1911; Broglio Loriga & Berti Cavicchi 1972; Kozur & Mostler 1973; Detre & Mihály 1987; Broglio Loriga et al. 1990

Vertebrates

Vertebrate localities are exceptionally rare in the Triassic of the Carpathian Basin. One well-known and unique fossil is the skeleton of *Placochelys placodonta,* widely known as the "pebble-toothed pseudo-turtle," found by Dezső

Reconstruction of *Placochelys placodonta.* This animal might have moved in a similar way to recent sea turtles, swimming and maneuvering with its shovel-like limbs. It was approximately 90 centimeters long. Pencil drawing by András Szunyoghy.

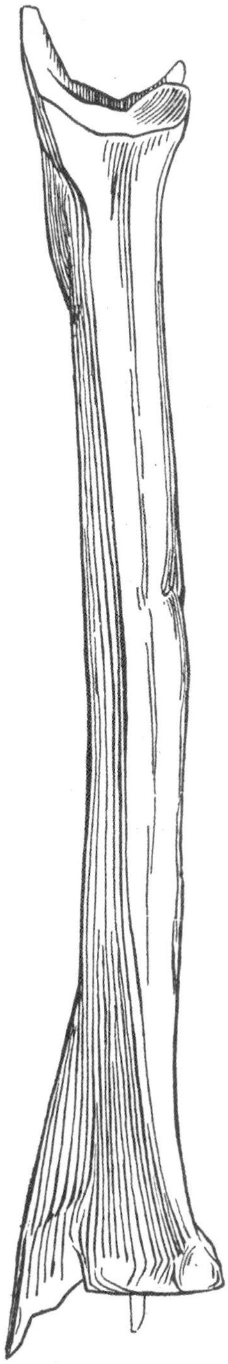

The massively elongate cervical vertebra of the "giraffe-necked" fossil reptile *Tanystropheus biharicus*. This special vertebra, 116 millimeters in length—and collected in the surroundings of Lugaşu de Sus, near Aleşd—was one of the first finds of the Triassic reptiles of the Bihor Mountains. It proved to belong to a giraffe-necked aquatic ancient reptile—a *Tanystropheus*, described as a new species (*T. biharicus*) by Tibor Jurcsák.
📖 Jurcsák 1975

Laczkó at Jeruzsálem Hill in Veszprém. This is very likely one of the most important and valuable fossils ever found in present-day Hungary.

Across the Hungarian border, fossil vertebrates that have been found around the margins of the Apuseni Mountains are also worth mentioning.

The "Pebble-Toothed Pseudo-Turtle"

Lessons Learned from a Double-Prepared Skull

In 1899, Dezső Laczkó, a teacher from Veszprém and an enthusiastic fossil collector and geologist found a number of remarkably well preserved bones in Upper Triassic marl beds at Jeruzsálem Hill in Veszprém. He carefully prepared these bones from the tough Veszprém Marl, and they proved to belong to a "pebble-toothed pseudo-turtle," a placodontomorph. This group of extinct reptiles includes marine turtle-like animals that are not, however, directly ancestral to present-day chelonians. The skeleton of this Veszprém specimen consists of a complete skull as well as fragments of armor, ribs, and limbs and was described in 1907 by Otto Jaekel (1863–1929), professor at the University of Greifswald in Germany. To enable further study, the valuable remains were transported to Frankfurt where they were reprepared "very carefully and cleverly" by Strunz, one of the curators, under the supervision of Fritz Drevermann (1875–1932), another fossil reptile expert. The rock infilling the skull was also removed, and this careful work meant that several smaller bones were revealed inside the skull and many previously hidden details became visible. Results of this repreparation of the skull were published by Baron Friedrich von Huene (1875–1969), a close friend of Baron Franz Nopcsa's.

Study of the skull of the Veszprém specimen has contributed significantly to knowledge of the phylogeny of placodontomorphs, and has meant that this group is now considered more closely related to therapsids than to other groups of reptiles. Recently, the skull was morphologically reevaluated by Olivier Rieppel of the Field Museum in Chicago; the paleoecology of placodontomorphs has been discussed by Jean-Michael Mazin of the University Claude Bernard in Lyon, and Giovanni Pinna, director of the Natural History Museum in Milan.

Placodontomorphs were a group that comprised a few genera and species and were not widely distributed either geochronologically or biogeographically. Remains of these reptiles are only known from Europe, especially from the Middle and Upper Triassic of the Alpine-Carpathian region. *Placodus,* a well-known genus form Germany, is thought to be a close relative of the Hungarian *Placochelys.*

📖 Jaekel 1911; Huene 1931; Mazin & Pinna 1993; Rieppel 2001

Vertebrates from the Bihor Mountains

Bihor Localities and Their Diverse Fauna

Remains of marine and coastal reptiles have been collected from a series of Middle Triassic limestones that outcrop in patches around the villages of Aleşd, Peştiş, Finiş, and Lugaşu de Sus because of the painstaking work of Tibor Jurcsák, a paleontologist from Oradea. Systematic collecting that he has carried out for more than a decade has resulted in a fauna that includes the remains of sharks (*Hybodus, Acrodus*), "semi-bony fish" (*Colobodus, Birgeria*),

ancient turtles with wide carapaces (*Proganochelys* and *Triassochelys*), "pseudo-turtles" (Placodonts [*Placochelys, Placodus*]) and a small ichthyosaur called *Mixosaurus*. The most interesting finds from this area, however, include the "giraffe-necked" reptile *Tanystropheus biharicus,* which has a debated systematic position, and the "marvelous" ancient reptile *Nothosaurus mirabilis,* which belongs to nothosaurs.

Tanystropheus

This "giraffe-necked" ancient reptile is most likely interpreted as a highly specialized predator that was adapted to a coastal and marine way of life. It had 12 cervical vertebrae that made up a long, whip-like neck more than half the length of the animal. The fossil remains of representatives of this genus of animals 3–6 meters long are known mainly from Europe.

Jurcsák 1975, 1976

The Marvelous Reptile

Because of the steady collecting work that has been carried out in the Bihor Mountains, the remains of Triassic fossil reptiles that have been found include vertebrae, ribs, teeth, and limbs; they fill an entire cabinet. The most attractive are on display in the Museum of Crisul Rivers Country in Oradea and include numerous marine, turtle-like placodonts that are likely the same, or similar to, the "pebble-toothed pseudo-turtle" *Placochelys placodonta* from Veszprém. Besides these pseudo-turtles and the aforementioned *Tanystropheus,* the Bihor finds of *Nothosaurus* are also most interesting. This genus contains medium-sized shallow-water predators that do not exceed 4 meters in length and that are characterized by flat and elongate skulls. The Bihor finds have been classified partly into the previously described species *N. mirabilis* and partly into the newly erected species *N. transsylvanicus*. The species name of the latter refers to its locality in Transylvania; that of the former means "marvelous."

Determining the geological age of these rare reptile bones is a difficult task. Associated with *Nothosaurus* and *Tanystropheus,* remains of *Ceratites*-like ammonites have also been found in the Bihor Mountains and indicate a Middle Triassic age for the succession. Nothosaurians are otherwise known from several Middle Triassic localities across Europe and North America; recently some bones that pertain to this were also found in the Balaton Highlands, along with shells of the index fossil ammonite *Balatonites balatonicus*.

Jurcsák 1978

Tanystropheus very probably lived in shallow, near-shore waters. However, this image shows it as an ambush hunter. Pencil drawing by András Szunyoghy.

A Middle Jurassic ammonite (Choffatia sp.) about 30 centimeters in diameter from the Rózsa Quarry at Siklós (Villány Mountains, Hungary). Ammonites nearly went extinct around the Triassic–Jurassic boundary, but the group managed to survive and radiate again in the Jurassic, at which point they became very diverse.

3

The Jurassic

The middle part of the Mesozoic era, which lasted about 55 million years, is called the Jurassic (denoted with a J in geological abbreviations) after the Jura Mountains, which lie on the border between Switzerland and France and where surface rocks of this age are widespread. The Lower, Middle, and Upper Jurassic are also called Liassic, Dogger, and Malm, respectively. These series, as in the Triassic, are also subdivided into stages, again named after French, German, and English towns and villages where rocks of these types are common. There is one exception: the Tithonian stage, whose name has a mythological origin. Tithonos was the son of the Trojan king Laomedon and lover of Eos, the goddess of the dawn. According to the Greek myth, when Eos asked Zeus to make Tithonus immortal, she forgot to ask for eternal youth. Tithonus, indeed, lived forever—but, as he got older and older, he shrank and eventually turned into a cicada, or cricket; in this form, he was able to stay with the goddess of the dawn forever. Indeed, the events in the last stage of the Jurassic (Tithonian) are inseparable from the dawn of the subsequent (Cretaceous) period—this is how Tithonos has gained immortality in the earth sciences. Jurassic rocks—full of fossils in most cases—are widespread throughout the Carpathian Basin.

If we had been able to look down on the Earth from space about 150–200 million years ago, we would have seen a single, huge supercontinent, called Pangaea by geologists, surrounded by water, the Panthalassa Ocean. The giant continent of Pangaea had a bay close to the equator that was a few thousand kilometers long, stretching in an east–west direction. The ocean within this huge bay (already in existence in the Triassic) is called Tethys; it was bordered by the Eurasian landmasses and tectonic plates in the north and by Gondwanaland (principally comprising the African Plate) to the south.

Strictly speaking, the Carpathian Basin did not yet exist in the Jurassic because the Alp-Carpathian mountain chain was formed only later, during collision of the European and African plates, which also resulted in the slow closure of the Tethys Ocean. However, taking a closer look at European paleogeographic and plate tectonic reconstructions, it is clear that the local situation was rather more complex: in the western corner of Tethys, where the ocean opened like a pair of scissors, a series of microplates can be seen submerged, or elevated, with respect to sea level. In this diverse environment, separated by basins and elevated highs on the seafloor, life was abundant and fossils that testify to this can be found in many different kinds of rocks. Of course, in the shallow epicontinental sea above ancient Europe and in the deep oceanic environment far from the shore, different types of rocks that preserve different fossils also formed. The first of these rock types is generally referred to as the "northwest European facies"; a later type is called "Tethyan" or "Mediterranean."

We also know that the climate in the Jurassic was warmer and more humid, with fewer fluctuations in temperature than we see today. The polar regions were cooler, of course, like those of lower latitudes, but very probably there were no ice caps on the poles. Global sea levels remained permanently high and marine depositional environments were widespread throughout the whole Carpathian Region and across the larger territory of present-day Europe.

FOSSIL-RICH FORMATIONS

The Jurassic formations of the Carpathian Region were deposited in a wide range of different environments, from freshwater lakes to the deep sea. Around the seashores, coal-bearing successions were formed which preserve the remains of the plants and land animals living in marshes and swamps. Far from the shores, in the open marine basins, ammonite-rich calcareous deposits or radiolarian ooze accumulated. Sedimentation in these areas was slow, but more or less continuous; however, on the elevated topographic highs of the seafloor, as on seamounts, for example, sediment formation was more ephemeral and episodic. Here strong bottom currents swept away the mud, and much of the sediment was dissolved—leading to the formation of successions with repeated periods of

System/ Period	Series/Epoch	Stage/Age	Millions of Years Ago
Jurassic	Upper/Late	Tithonian	151
		Kimmeridgian	156
		Oxfordian	161
	Middle	Callovian	165
		Bathonian	168
		Bajocian	172
		Aalenian	176
	Lower/Early	Toarcian	183
		Pliensbachian	190
		Sinemurian	196
		Hettangian	200

Subdivision of the Jurassic period into stages.

non-deposition. At the foot and in fissures of some of these seamounts, the peculiar Hierlatz Limestone—especially rich in brachiopods, ammonites, and gastropods—was formed. During the Late Jurassic, on the shallow, continuously subsiding seafloor, thick, reefal Štramberk-type limestone was deposited.

Rosso Ammonitico

One of the most characteristic Jurassic rocks seen in the Carpathian region was formed far from the shore and is referred to by its Italian name: *Rosso ammonitico*. This name clearly refers to its appearance: *rosso* means "red," and the most common fossils found in it are ammonites. In Hungary, this rock is also often called the Gerecse Marble, since its compact, massive variety was intensively quarried in the Gerecse Mountains and has been used in the building industry since Roman times. The Gerecse Marble is, however, not a marble but just a red, ammonite-rich limestone, generally laid down to a thickness of a few, or a few dozen, meters. The main part of the rock is largely composed of very fine-grained $CaCO_3$ precipitated from seawater and by tiny skeletal fragments of different organisms. Since the shores were far away, this limestone is very poor in terrigenous elements—those derived from the erosion of rocks on land—and its reddish color is due to the presence in the limestone of small amounts of iron oxide.

Of course, the most characteristic megafossils in this kind of rock are ammonites. Because sedimentation was very slow, only a thin rock layer was formed after a very long period, and empty ammonite shells accumulated on the seafloor. Other cephalopods, such as nautiloids and belemnites, also appear frequently; other benthic elements—such as brachiopods, gastropods, bivalves, and sea urchins—are generally rare. Vertebrates are extremely rare or practically absent, probably for two reasons: first, marine vertebrates (including ichthyosaurs and crocodiles) were always the top predators in ecosystems and as such were a priori rare; and secondly, the Rosso ammonitico facies does not favor preservation of fossil bones. During lithification, bones were easily destroyed; as a consequence, the few that have been found, such as the Gerecse crocodile (see below), are particularly unique and sensational.

These red, ammonite-rich rocks have another feature that has yet to be explained: their nodularity. This could be a result of lithification, but in most cases it is more likely to do with the fact that benthic animals mixed and reworked the once-soft sediment—a process called bioturbation. In most cases, the mediators of bioturbation had no solid skeletal elements and so the only evidence of their presence is the result of their activity: burrows and nodules. In the Rosso ammonitico limestone, crustaceans are generally suspected to be the cause of most bioturbation and nodularity, but it is difficult to exclude annelid worms or any of the other infaunal or epifaunal organisms. These taxa usually do not fossilize, and leave behind only traces generally parallel to bedding planes, reticulated and filled with clayey (softer) sediment within the harder limestone.

Rosso ammonitico–type rocks can be found in many parts of the Transdanubian Range. In the Bakony Mountains, they are found outcropping at the bottom of shallow valleys and forming cliffs a few meters high in some places. For building stone, the largest active quarries are found in the Gerecse Mountains, in the vicinity of Tardos.

This formation is also known in the Middle and Late Jurassic of the Mecsek Mountains, as well as from many other Carpathian region outcrops. Indeed, in many places it is common for the Rosso ammonitico to appear twice in Jurassic successions; the older unit represents the lower and middle part of the period, whereas the younger one is Late Jurassic in age. The two rock bodies may have a similar appearance, but their faunal content is quite different; between them there is often a siliceous formation (radiolarite) that corresponds to the deepest part of the Jurassic Sea.

Silica Rocks

In the deepest basins of the Jurassic ocean, the seafloor was far below carbonate compensation depth—meaning that the calcareous plankton and sediment particles that sank into deeper waters were dissolved and that no carbonate was preserved on the seafloor. The fine mud accumulated on the ocean bottom mainly contains small particles of siliceous skeletal elements that are thought to come from radiolarians that settled on the ocean bottom. Following

The name "Rosso ammonitico" is used for several different rock types. This image shows minute fossil shells on the surface of a Jurassic limestone block, the surface of which was eaten away by dilute hydrochloric acid. The calcite shells of mollusks were substituted by silica in this case and remained intact after the acidic treatment (slightly enlarged).

lithification, this ooze became a sedimentary rock called radiolarite, a variety of chert, but its fossils cannot be studied in most cases. The strange-shaped, complex, and fragile skeletons of the radiolarian were dissolved and precipitated as silica gel. For paleontological studies on radiolarians, the most suitable sediments are those that formed close to the carbonate compensation depth: in these rocks, radiolarians are very abundant (because they are the main rock builders) and their primary skeletal structures are well preserved within the still calcareous sediment.

Similar in appearance to radiolarite, another type of chert found in the Carpathian region is called spiculite, because it is formed from the dissolution and subsequent precipitation of sponge siliceous spicules. (These spicules are the structural elements that form the skeleton of the sponge.) Although radiolarite cherts are found in units up to a few meters thick in the Carpathian region, in most cases spiculite cherts are known only to form lenses, or thin cherty layers.

We might expect cherty formations in areas where Jurassic rocks are found, but these sediments almost never form spectacular cliffs like the Rosso ammonitico. Cherts—although very hard—are not very resistant to erosion: this brittle rock is very rigid and crumbles easily, so it is often found as debris within soils.

The Hierlatz Limestone: A Peculiar Austro-Hungarian Facies

The so-called Hierlatz Limestone—named after its type locality, the Austrian Hierlatz Berg (Hirlatz Hill in modern maps) in the Dachstein Group—is one of the most characteristic Jurassic rocks in the Carpathian Basin and is generally very rich in fossils. Indeed, the famous Austrian alpinist and geographer Friedrich Simony (1813–1896), born in Bohemia but presumed to be of Hungarian origin, first found the rich fossil locality at Hierlatz Hill. He noted that "from the brachiopod limestone . . . at least 50 different species of brachiopods, gastropods, cephalopods, and crinoids have been found, of which the first group are by far the most frequent." This rock has a conspicuous appearance: the intermediate cavities among the well-preserved brachiopod and ammonite skeletons are partly filled with pink or yellow micrite and partly by snow-white, fibrous, sparry calcite. Fossils can be very frequent in this light-colored rock and its most typical feature is the brachiopod-lumachelle character.

For years, different opinions have abounded regarding the formation of the Hierlatz Limestone. Because of its lumachelle appearance, some workers have considered it to be a coastal deposit of shallow water origin, while others have favored a deeper-water environment of formation. A detailed restudy of the formation has revealed that the Hierlatz Limestone is attached to Early Jurassic submarine fault zones and so was deposited in submarine fissure fills (also called Neptunian dykes) and usually is associated with breccias around submarine escarpments. Organisms, mostly brachiopods, lived on the rocky walls of these fissures and their shells accumulated and were preserved in the Hierlatz Limestone at the foot of these submarine hills.

Typical Hierlatz Limestone occurs only in the Northern Calcareous Alps and in the Transdanubian Range (the Bakony and Gerecse Mountains), which are disjunct areas of the former Austro-Hungarian Empire. Furthermore, although the classical Hierlatz Limestone is a Lower Jurassic formation, similar rocks are also known from the

A mass occurrence of small ammonites, gastropods, and brachiopods in the Hierlatz Limestone at Lókút (close to original size).

Prepared brachiopods sorted by species from the Hierlatz Limestone (0.7× magnification). (Above) *Linguithyris aspasia, Prionorhychia? forticostata.* (Below) *Bakonyithyris pedemontana, Securina partschi.*

Middle and Late Jurassic. These sediments are thought to have been deposited in similar paleonvironments and paleotectonic settings (fault scarps).

📖 Simony 1850; A. Vörös 1991

The Gresten Facies

During the Early and Middle Jurassic, a thick succession of paralic deposits (formed on seashores), including coal seams and fossil plants, was deposited close to the European continent on the shallow, subsiding northern shelf of the Tethys Ocean. This facies is named after its type locality, the Austrian village of Gresten. Gresten-type rocks are found in many places across the Carpathian Basin, including the Mecsek Mountains in southern Hungary.

At certain levels within this facies fossils are frequent, especially the remains of ancient vegetation and the invertebrates (mainly gastropods and bivalves) that lived along the seashore. In the Mecsek Mountains, coal-bearing deposits are covered with a coarse-grained sandstone of shallow marine origin, followed by a thick, spotty marl (also known as the Fleckenmergel, its German name) that was deposited in a deeper environment. This marl is characterized by fossils of both planktonic and nektonic creatures, including ammonites and belemnites, as well as abundant large-sized, three-dimensional trace fossils. Of these traces, *Zoophycos* is thought to have been produced by feeding worms.

Semi-Reef Deposits

Štramberk-type limestones are also characteristic and frequent in the Jurassic of the Carpathian Basin. These rocks are light colored, more or less bedded, and often massive and thick, deposited in the vicinity of reefs or lagoons. The type locality for this limestone facies is near Štramberk—in the Moravian-Silesian Region of the Czech Republic, where it has been quarried for many years. The real Štramberk Limestone yielded some very fine fossils, especially bivalves, gastropods, and ammonites, which were worked on by early paleontologists. In contrast, similar large rock bodies in the Eastern and Southern Carpathians, and in the Apuseni Mountains of Romania, are also referred to as Štramberk limestones—but these are usually very poor in macrofossils.

REMARKABLE FOSSIL SITES

Within the Carpathian Basin, fossil-rich Jurassic rocks are found at the surface in the Transdanubian Range and in the Mecsek and Villány Mountains. A large part of the Bükk Mountains in northern Hungary is also built from Jurassic rocks, but these monotonous shales do not contain megafossils, only rare microscopic remains. Very different Jurassic rocks can be found all along the Carpathian Arc, and their fossil assemblages have been discussed in detail in different papers and monographs: this is one reason many Carpathian Basin localities are familiar to Jurassic workers worldwide. Among these sites are abandoned, or still active, quarries as well as unremarkable ravines and cliffs that are often hard to identify on maps. In Hungary, both the discovery and early research on most Jurassic outcrops is a result of the work of Jenő Noszky Jr. (1909–1970), and large-scale collecting at many sites was supervised by his academic protégé József Konda (1929–1995). These Noszky-Konda collecting campaigns resulted in tens of thousands of fossils, often sampled bed by bed, which have given unique insights into the Jurassic stratigraphy of the Carpathian Basin.

Mihály Dunai—geologist, engineer, and fossil collector who discovered *Temnodontosaurus* remains in the Gerecse Mountains. This photo was taken at the famous ammonite locality in the Tűzköves Ravine (Bakony Mountains), where thousands of cephalopods have been collected.

Bakonycsernye: The "Production Site" of Cartwheel-Sized Ammonites

Officially affiliated to Isztimér and otherwise known in the geological literature as the Tűzköves Ravine of Bakonycsernye (Tűzkő means "chert"), this locality is one of the best known and most intensively studied from the Hungarian Jurassic. It was first found in 1859, when Flóris Rómer (1815–1889)—Benedictine monk, archaeologist, art historian, and later professor and member of the Hungarian Academy of Sciences—missed his road in the vicinity of Csernye. After a short walk he noticed a column of smoke billowing from a lime kiln that "led the way to a marble quarry." He talked with a young Slovak man who lay alone close to the kiln and then, with great composure, he walked away—hoping that he would "again enjoy the pleasure of the miraculously large ammonites and orthoceratids [an old term for *Atractites* and related forms] that he had seen." In his 1860 monograph *A Bakony* (The Bakony), Rómer recalls that he "fantasized about the fossils" that he saw "for more than a year and dreamt about them every night for a week." In spite of the hints provided at the lime kiln—or perhaps because of them—Rómer completely lost his way in the forest and found the quarry again only the following day, when he had company. He was totally overjoyed and immediately collected two large cases of fossils!

Following in the footsteps of Rómer, Miksa Hantken, one of the greatest Hungarian paleontologists, also visited the Bakonycsernye locality and collected fossils, including 44 ammonite specimens that were presented at the World Expo in 1873. Sadly, Hantken was still working on a monograph about these fossils when he died, and so his work was never completed.

The first comprehensive evaluation of this fossil fauna appeared much later, in 1904. Gyula Prinz reported 111 cephalopod species and "mutations"—as he called them. However, the Bakonycsernye ammonites finally became well known not because of Prinz's work, but because of two volumes published in 1966 and 1967. These thick volumes appeared in the series Geologica Hungarica and were written by Barnabás Géczy—later head of the Department of Paleontology at Eötvös Loránd University—who had collected and cleaned the fossils for years. Géczy carefully collected bed by bed at Bakonycsernye, using explosives where necessary to remove fossils from the harder layers. His hard work led to spectacular results: all in all, some 8,000 ammonite specimens were collected from the Tűzköves Ravine; these were classified by Géczy into 345 species and subspecies, 92 of which were new to science. All of the Jurassic stages and most of the substages are documented at this site; the Middle Jurassic Aalenian stage is represented by an especially rich fauna. Up the succession, the Bajocian stage is represented still by ammonites, but even higher, because siliceous deposits start to dominate and preservation state becomes poorer, no ammonites have been collected. Nevertheless, from a paleontological point of view, Bakonycsernye is one of a kind: The name of this locality has even been immortalized in the name of one ammonite genus, *Csernyeiceras*.

📖 Rómer 1860; Gy. Prinz 1904; Géczy 1961a, 1966, 1967; Géczy & Meister 1998; Galácz et al. 2008

The Upper Jurassic cliff at Prédikálószék ("The Pulpit"), west of Hárskút, in the Bakony Mountains. In the red, Rosso ammonitico–type limestone ammonites are generally abundant.

In the Vicinity of Lókút and Hárskút

These two little villages are found in the heart of the Bakony Mountains, close to the city of Zirc. Localities in their vicinities have yielded rich Jurassic assemblages composed mainly of brachiopods and ammonites.

Especially well known among geologists is the Jurassic succession at Lókút Hill, east of the village that bears the same name. Indeed, this place is a Mecca for Hungarian Jurassic workers: within a more or less continuous succession, 7 of the 11 Jurassic stages can be documented on the basis of ammonites, and the presence of the remaining 4 can be inferred on the basis of lithological changes. Nearby Kávás Hill—situated to the east of the village—is also a classical locality for Early Jurassic ammonites. The fauna from this site was first described by Lajos Kovács (1908–1978) in the *Bulletin of the Hungarian Geological Society* in 1934.

To the southwest of Lókút, the foot of Papod Hill is also a well-known fossil locality. The hillside here is rather steep and was probably similarly steep during the Jurassic, when it was an escarpment of a submarine high. On this ragged slope, among boulders of different ages falling down the escarpment, brachiopods lived in great numbers. Generic names, including *Papodina* and *Lokutella,* introduced by brachiopod researcher Attila Vörös, refer to these remarkable localities.

The vicinity of Hárskút is also rich in ammonites and the natural outcrops and artificial exposures have yielded a rich and well-preserved fauna. Interestingly, however, Lower and Middle Jurassic beds are rather condensed here, and only certain short periods of time are represented. These "missing" beds either were never deposited or were swept away by bottom currents; as a result, the Upper Jurassic succession is thicker and more complete. Late Jurassic ammonites can be collected further away as well, close to a series of low cliffs called Rendkő, and the nearby Gyenespuszta site is famous for its unique and well-preserved Bathonian (Middle Jurassic) ammonite fauna. This fauna was described by András Galácz, the former head of the Department of Paleontology at Eötvös Loránd University.

📖 Galácz 1980; A. Vörös 2009; Főzy et al. 2011

Csárda Hill at Úrkút

Csárda Hill in the vicinity of Úrkút is a nature conservation area that is protected because of the importance and natural beauty of the Lower Jurassic rocky formations that outcrop there. This limestone surface, often referred to as a "paleokarst," first became visible between 1926 and 1933, when quarry workers using simple hand tools scraped out oxidized manganese ore (also Jurassic in age) from the karstic fissures and holes. This spectacular karst landscape was formed much later, when the limestone body emerged and became dry land, following the Jurassic period. Subsequently, manganese ores, which have a submarine origin, were redeposited and infilled the karst.

Although the geology and mineralogy of this site is of interest, fossils were also found here in large numbers. Most of the rocks at Csárda Hill are the Lower Jurassic crinoid and brachiopod-rich Hierlatz Limestone: whereas crinoids can often be seen only as glistening calcite on fresh, broken rock surfaces, brachiopods often form "nests" in great abundance.

The precise age of the Hierlatz Limestone was constrained by determination of the brachiopod fauna, which is late Sinemurian. These invertebrates lived in great numbers along the submarine escarpment, around seamounts, and accumulated at the foot of the hills, where the topographic high meets the real basin. Outside the nature conservation area, covering Eocene limestone beds can be seen above the manganese ore. These rocks are also rich in fossils, especially *Nummulites* and other larger foraminifera.

📖 Pocsai & Sasvári 2005

Csárda Hill near Úrkút, one of the finest localities for Sinemurian (Lower Jurassic) brachiopods.

Gerecse Localities

The main part of the Gerecse Mountains, between Lábatlan and Tatabánya, is composed of Triassic carbonates but Jurassic rocks also occur—especially on the hilltops. The so-called Gerecse Marble—a Rosso ammonitico–type limestone—has been quarried here since Roman times, and the oldest quarries can be found on Nagypisznice Hill. (Famous crocodile fossil remains, discussed in a later section, were found in one of these quarries.) In the middle of the old quarry yards trees now grow; these abandoned places have a unique atmosphere.

Quarries that are still active can be found above the village of Tardos. Using explosives, huge limestone blocks are removed and large slabs are sawed on the spot. Outside the sawmill on the scrap heap, searching among the fallen pieces, cut and polished *Atracatites* and ammonite fossil remains can often be found.

The building industry mainly uses this thick-bedded, massive Lower Jurassic limestone: it can be seen, for example, in Budapest quay and in many public buildings and footways. It was the raw material for the famous Lion Well at the palace of King Matthias at Visegrád and was also used to pave many of the halls in the Hungarian Royal Palace at Buda. The red limestone fades fast in open air, but fossil cephalopods are easily recognized—even on weathered surfaces.

The Tölgyhát Quarry can be found in the northern part of the Gerecse Mountains. Standing on top of the Jurassic formations in this area it is possible to enjoy the wonderful scene toward the north, facing the Danube and the plain that fades into the distance. The relatively thin (just tens of meters) Jurassic succession below is the most complete series of this age in the Gerecse Mountains. The rock sequence comprises a thickly bedded, massive Lower Jurassic limestone, poor in ammonites and covered by a Toarcian Rosso ammonitico–type marl with more frequent fossil cephalopods. Above these units sits a red, nodular Rosso ammonitico–type limestone of Middle Jurassic age, followed by a radiolarite and then another example of a nodular Rosso ammontico–type limestone. These sediments complete the Jurassic sequence.

Beside these localities, specialists on the rocks and fossils of this age from this area will know a series of further sites, also in the Gerecse Mountains. For example, at Teke Hill many beautiful fossil gastropods, otherwise rare in the Hungarian Jurassic, have been collected from a Lower Jurassic fissure. A limestone with an unique white appearance exposed in an artificial shaft on the northwestern edge of Szél Hill has yielded small and rare Tithonian ammonites (such as *Cyrtosiceras*); another diverse ammonite fauna, with some superbly preserved shell material, is known from an extremely hard Upper Jurassic purple limestone in the Paprét Ravine.

A canyon-like abandoned limestone quarry in the Gerecse Mountains that exposes the Lower Jurassic succession capped with a Toarcian marl.

📖 Főzy 1993a; Főzy et al. 1994; Géczy & Szente 2006

Tata and Kálvária Hill

One particularly notable locality is situated inside the city of Tata and can be found within an old, abandoned limestone quarry complex that is now an open-air geological museum. Since this site is protected by law, all collecting—and even hammering on the rock—is forbidden. Nevertheless, its Jurassic beds are rich in fossils, especially cephalopods, and one can still study both the underlying Triassic rocks and the Cretaceous sediments above. Work at this site goes back almost a century; Nándor Koch (1885–1961) first published a three-page Jurassic faunal list for this locality. The Late Jurassic is especially well-represented by ammonites of the Oxfordian, Kimmeridgian, and Tithonian. Several short papers deal with

"Lying Stones": Fake Fossils

It seems in science, as in all things controlled by people, there are frauds from time to time. There will always be people who will alter, falsify, and deliberately misinterpret fossils. Of all fake fossils, however, some of the most famous are the pseudo-fossils made by the students and co-workers of Würzburg University professor Johann Bartholomeus Adam Beringer (ca. 1667–1738).

Beringer's colleagues carved limestone into the shapes of animals such as lizards, frogs, spiders on their webs, and bees above a flower. This process is thought to have started as a joke, but as the hoaxers got better and better, more and more outrageous fakes were made and then it all got out of hand. Finally, Latin, Arabic, and Hebrew characters were carved in the same manner and planted at a place where Beringer frequently went to search for fossils. Because at this time the real nature of fossils and how they formed in rocks was not known to scientists, people, including Beringer, accepted the fakes. He truly believed that most of the strange things he was finding were the "capricious fabrications of God" and decided to publish his "scientific" results. When his book appeared, a scandal broke and both Beringer and the hoaxers were discredited. The professor tried to buy back all the books that had been published but this project was never completed: the copies that remain are precious artifacts in a number of big libraries, and the stones became known as Lügensteine, or the "lying stones."

The stories of Aradi and Beringer are, of course, different: the latter was a victim of a joke, whereas Aradi actually tried to mislead his contemporaries. There is no known reason for his actions, but perhaps he wanted to achieve instant fame without doing any hard work.

It is quite common, unfortunately, for real fossils to be manipulated—carved or falsely reassembled. For example, the Lower Jurassic ammonites *Dactylioceras*—also known as snakestones—often were augmented with the curved head of a snake. In this case, the sculptors did not mean to deceive but to make something that fit the legend suggested by the coiled shape of the shell.

In contrast, the infamous Piltdown Man was a deliberate forgery that was discovered only decades after the sensational find of a supposed early human. Thought for many years to be a missing link between apes and humans, it turned out that Piltdown Man was in fact a chimera made of a combination of an orangutan jaw and the skull of a modern human. This fake is often used by creationists to discredit paleontologists.

More recently, a similar forgery involved some Chinese feathered dinosaurs. A unique find called *"Archaeoraptor"* was reported to be a linking fossil between birds and dinosaurs, but turned out to be fabricated from two specimens. The case was particularly embarrassing for the National Geographic Society, which publicized the false fossil in its magazine.

the geology and the Mesozoic fossils of the site; the most comprehensive publication on Tata is a monograph by József Fülöp.

📖 Fülöp 1976; Haas 2007

A Jurassic Fossil Site in Budapest That Never Existed

On November 2, 1904, a young man called Viktor Aradi Jr. (1883–1937?), gave a remarkable talk in a session of the Hungarian Geological Society. He reported that he had found Jurassic ammonites, brachiopods, belemnites, and echinoid spines at sites on Buda Hill, within the cherty limestones at Szépvölgy and Farkasvölgy. This was of interest to the assembled geologists because no one previously had been able to document any Jurassic deposits within the Buda Hills, the gentle mountains around the Hungarian capital. Later, in 1905, Aradi published his results in the *Bulletin of the Hungarian Geological Society* as a paper entitled "Liász és dogger a budai hegységben" (Liassic and Dogger in the Buda Hills). According to the widely accepted rules of the era, Aradi finished his article with a polite sentence: "Finishing this preliminary note, I express my sincere thanks to János Böckh, min[isterial] coun[selor]; Dr. Ferencz Schafarzik, prof. [of the University of Engineering]; Dr. Sándor J. Krenner, a court counselor; and Drs. Antal Koch and Imre Lőrenthey, university professors, for their kind support."

Although he had previously supported the young colleague, Lőrenthey became suspicious and he visited the sites that Aradi mentioned to make some personal field observations. He also suggested a follow-up collection campaign using explosives, if needed, but when the Hungarian Geological Society organized a field trip to the site, the numerous fossils that were collected belonged mostly to the brachiopod genus *Lingula*. Not a single Jurassic fossil was found, only Triassic ones, and the rock matrix surrounding many of Aradi's specimens was observably quite different from the local geology.

A fake fossil that resembles an ammonite—one of the famous "lying stones" of Beringer, from the collection of the University of Karlsruhe (Germany).

Of these dubious specimens, the ammonite, which would have been of special importance for dating the sequence, was miraculously lost. It has subsequently come to light that faded registration numbers could be seen even on some of the fossils shown by Aradi, suggesting that he took fossils from a museum to bolster his work, but this would be hard to prove now. When the full nature of Aradi's hoax came to light, in 1907 Lőrenthey published a sharp-toned paper entitled "Vannak-e juraidőszaki rétegek Budapesten?" (Are there any Jurassic rocks in Budapest?) in order to establish the age of the beds in question. The respected professor was then more polite and generous when he later simply regarded Aradi's report as "a mistake": as he pointed out, "from the literature the whole fauna of Aradi, as well as the Jurassic of Budapest, must be deleted."

There is a subscript to this strange historical episode as much later it turned out that some slightly deformed rocks in the western quarry of Mátyás Hill, close to Aradi's sites, are much younger than thought before: Heinz Kozur and Rudolf Mock established a latest Triassic, Rhaetian, age for these sediments using micropaleontology. Finally, today it is accepted that a lowermost Jurassic age for some of the beds in this area that remain covered cannot be excluded.

To finish the story of Aradi, it is interesting to note that following the discovery of his fraud he changed both his home country and profession. After World War I, he acted as chief censor of the occupying Romanian troops in Hungary, and later worked as a sociologist in Romania. It seems that he was unable to stay in one place, however, since in 1933 he went to the Soviet Union—a decision that probably had something to do with the fact that twice he was arrested as a spy while living in Romania. Aradi, the habitual troublemaker, ended up in the paradise of the wage-earning class: in 1937 his comrades arrested him—supposedly not because of paleontological tricks—and finally he disappeared in the Soviet Union.

Fossil Localities in the Mecsek Mountains

There are quite a number of well-studied, classic Jurassic outcrops in the eastern parts of the Mecsek Mountains. Some, like the spectacular large opencast mines at Pécs-Vasas or in the Karolina Valley, expose only the lowermost parts of the Jurassic succession, in which paralic black coal seams occur. At these sites, mostly pyritized remains of mollusks that lived in these former saltwater swamps, carbonized plant remains of the former vegetation, and the footprints of dinosaurs that once walked around on the marshes can be collected side by side. Indeed, the so-called Basa Pit in the vicinity of Hosszúhetény is a classical locality from which Lower Jurassic bivalves have been collected for many years.

With a little luck it is possible to collect ammonites and belemnites from the thick marl successions (also Lower Jurassic in age) that overlie these coal deposits. *Zoophycos*-type trace fossils are also frequent in these marls while upsequence, into the Liassic, the Toarcian is in many places represented by a black shale with high organic content. The only known surface exposure of this facies is in the remote Réka Valley, where dark shale fish fossils (including teeth, isolated scales, and sometimes whole skeletons) can be found.

In the Middle Jurassic, nodular marl and limestone ammonites are the most frequently encountered fossils—but occasionally echinoids, brachiopods, and (from the Bathonian) sponge remains can also be collected. The latter, in particular, are rare fossils in the Hungarian Jurassic. The best Middle Jurassic outcrops are found naturally exposed in hillsides, in ravines (such as the Hidasi Valley, Somosicsörge, and the Takanyó Valley), or occasionally in road cuttings (such as that along the road from Zobákpuszta to Kisújbánya). Older localities can be found also at Ófalu (Kohltal, Kalktal).

The Upper Jurassic of the Mecsek Mountains is generally poor in megafossils: the most complete succession known to yield Oxfordian, Kimmeridgian, and Tithonian ammonites is found close to Zengővárkony, not far from a long-abandoned lime kiln.

📖 Peters 1862; Főzy 1993b; Galácz 1995

Localities in the Villány Mountains

The Villány Mountains are situated in southern Hungary, close to the Croatian border. The most famous Jurassic locality in this region is Templom ("Church") Hill near the

Templom Hill at Villány, a classic locality for Callovian ammonites and Pleistocene small mammals.

village of Villány, where a half-meter-thick bed contains a very rich fauna, including abundant cephalopods—particularly ammonites—in an abandoned limestone quarry. The fauna of this stromatolitic bed, which also contain bivalves, is Callovian in age, but older fossils from the same locality are also known. Paleontology professor Barnabás Géczy, who published a paper on the site in 1984, describes its discovery thus: "The Pliensbachian beds at this locality were recognized by British researchers, professors D. Ager, J. Callomon, and D. Donovan. Villány was, and will always remain, a good example of international geological cooperation for it was Lenz, who worked in Prague, who first noticed the site at all and it was Till who completed a paleontological monograph on the locality while based in Vienna. Lóczy then studied this area, along with Rollier who was based in Zurich, while the stromatolitic character of the ammonite bed was first noted by two Polish researchers, Radwansky and Szulczewski."

The stromatolites at this site, which are laminated structures of mostly biological origin, form a series of mattress-like bedding structures of fine laminae that encase the

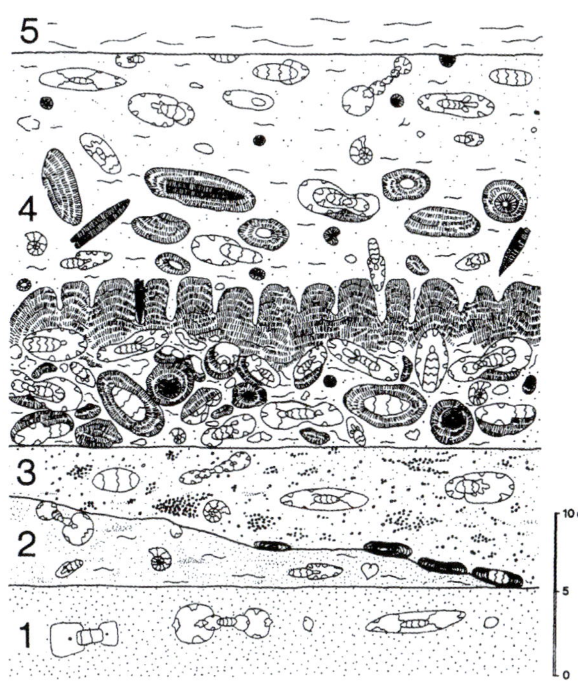

A drawing of the stromatolite bed at Villány. Crusted ammonite and belemnite fossils can be seen as cross sections; in the middle of this image, domal structures form a continuous mattress. The overwhelming component of the rich ammonite fauna in this bed (containing about 180 species) is middle Callovian in age, but some fossils are also from the late Callovian. (1) Pliensbachian limestone; (2) Bathonian sandy limestone; (3) Lower Callovian limestone with iron oolite; (4) Callovian stromatolite bed; (5) Oxfordian limestone. After Géczy & Galácz 1999.

Lower Jurassic beds exposed in a trench on the Somssich Hill at Villány. This fauna mostly comprises large specimens; this photo was taken by the late John Callomon, who participated in the Mediterranean Jurassic Colloquium of 1969.

fossils in this limestone. Indeed, the stromatolite beds cover ammonites and belemnites like the crispy coating on fried chicken: a characteristic limy covering that is thought to be due to the activities of cyanobacteria (commonly known as blue-green bacteria). Empty cephalopod shells, gently rolling around on the seafloor, were eventually completely overgrown with stromatolitic laminae. The fossil content of the ammonite bed indicates a deeper-water origin of these stromatolites.

Extraordinarily well-preserved Middle Jurassic ammonites, coated with a shiny ferro-manganese oxide crust, are known from the Rózsabánya Quarry at Siklós. The first fossils from this locality were found by the multifaceted geologist András Kaszap.

📖 Till 1910–1911; Lóczy 1915; Kaszap 1961; Ager & Callomon 1971; Géczy 1984, 1998; Géczy & Galácz 1998; A. Vörös 2011

Jurassic Outcrops in the Carpathian Mountains and in the Apuseni Mountains

The Carpathian Mountain chain borders the Carpathian Basin for about 1,000 kilometers, and Jurassic outcrops are found in many places. A list of some of the classic and important localities is provided here, but a complete list is far beyond the scope of this volume. In the Western Carpathians, the Lower Jurassic (Toarcian) outcrop of Marianka and the Middle Jurassic localities of Oravský Podzámok (Slovakia) and Wielka Świstówska (Poland) are worthy of mention because of their abundant ammonite faunas. Of these sites, the latter is especially interesting because Bathonian cephalopods—otherwise rare in the Carpathian Basin—are known in great numbers and have great species diversity. Most of these specimens were collected by Edward Passendorfer (1894–1984), the tireless and long-lived High Tatras researcher. Another nice Bathonian ammonite collection is known from Vadu Crişului (formerly Vad, Pădurea Craiului Mountains).

From the Eastern Carpathians, localities around Lacul Roşu and Hăşmaşul Mare ("Nagyhagymás" in the old geological literature) must also be discussed. The most detailed descriptions of these sites date to the time of the Austro-Hungarian monarchy and are known by their Hungarian names. Ferenc Herbich (1821–1887) and Melchior Neumayr (1845–1890)—Hungarian and German geologists, respectively—are to be credited with describing beautiful and abundant Lower, Middle, and Upper Jurassic ammonites from these sites, some of which have turned out to be new to science.

It was also Herbich who published the first detailed description of the fauna from Cetea in 1886. The limestone cliffs at this site contain a very diverse Late Jurassic (Tithonian) coral and gastropod assemblage with the

View of the Rózsabánya Quarry at Siklós. Although this thick-bedded limestone contains quite a lot of fossils, they are impossible to extract from the compact and very hard rock. Polished slabs of this limestone are common in buildings. They were used to cover the walls of the Deák Square Station in Budapest, for example. The vivid red iron oolite, which yields some of the most spectacular Jurassic ammonites known from Hungary, can be seen cropping out in the left corner.

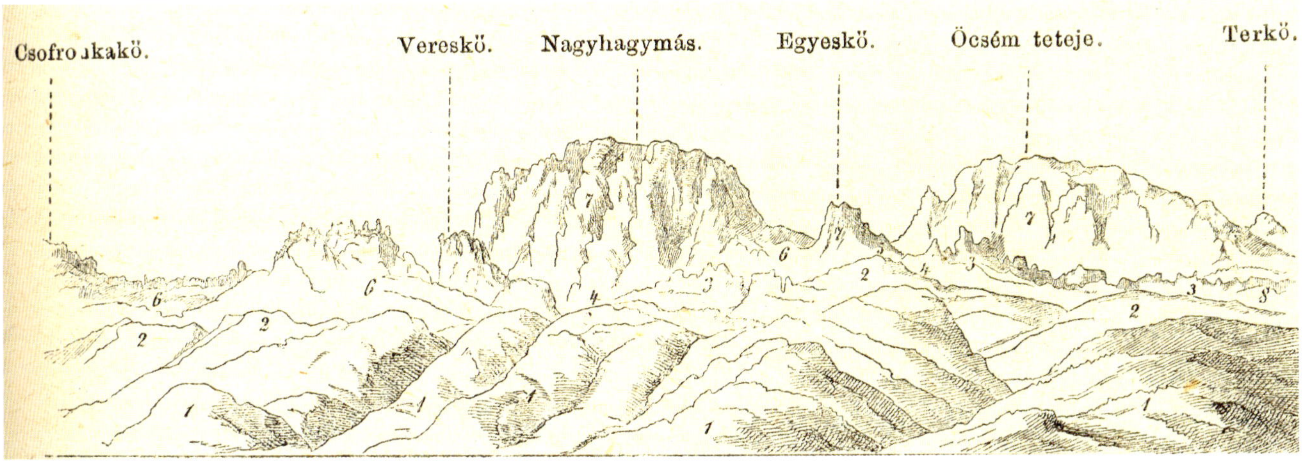

Jurassic rocks in the Hăşmaş Mountains of Romania as viewed from the Georgheni Basin—a wispy drawing from Herbich's 1878 monograph accompanied by its original explanation: (1) mica-schist, (2) gneiss, (3) Triassic lime, (4) Hallstatt Limestone, (5) Adnethian Beds, (6) Acanthicum Beds, (7) Štramberk Limestone. The fossiliferous Jurassic strata (5, 6, and 7), which belong to the Transylvanian Nappe were emplaced by gravity and today rest on the crystalline basement (1, 2) above the Triassic (3). The spectacular cliff also named in this sketch was built up from uppermost Jurassic (Tithonian) reefal Štramberk limestones. Ammonite specimens collected by Neumayr and Herbich come from deeper levels in the saddles between the cliffs.

The Late Jurassic ammonite *"Oppelia compsa"*—known today as *Taramelliceras compsum*—from Herbich's 1877 monograph. The original figure caption informed the reader that "after nature it was carved into rock by J. Wittinger." Before photographs were in general use, paleontological papers were illustrated with carefully executed lithographs. These were sometimes of very good quality, and actually much better than a poor photographic image. However, it sometimes was the case that fossils in these hand-carved images were misdrawn, or sometimes "completed," so that the image ended up being a compilation of two or even more specimens. As a consequence, some illustrations are of limited scientific value. This specimen is from the base of the cliff called Gyilkoskő, as noted by Herbich: "Its original is deposited in Kolozsvár (Cluj)"—where it is still kept today.

Flattened Toarcian ammonites from the Marianka black shale.

latter represented by no fewer than 72 species, of which 64 belong to *Nerinea* and related forms. From the hills around Braşov, the Cristian locality is also justly famous for its pretty Lower Jurassic fauna. The well-preserved ammonites from Racoş can be found in one of the geologically oldest occurrence of the Rosso ammonitico facies. This fossil site is located in the Perşani Mountains in the Tepő Creek valley (also known as Töpe), a tributary of the River Olt. Ammonites similar to those found at Racoş are also known from the vicinity of the Orbán Balázs Cave at Mereşti; Doman, Steierdorf-Anina, and Holbav in the Southern Carpathians are famous for Lower Jurassic plant remains.

Farther to the south, on the riverbank at Duna Sviniţa (a site that is better known by its German name, Swinitza), is one of the most important of the Carpathian Jurassic fossil localities. Here the best-preserved ammonites come from a single 30-centimeter-thick iron-oolite bed; some of the best-preserved internal molds may reach, or exceed, the size of a soup plate. Some also have fine suture lines preserved and are precious constituents of historical collections.

📖 Neumayr 1873; Herbich 1878; Jekelius 1915, 1925; Rakús 1994; Tomas & Pálfy 2007

JURASSIC FOSSILS FROM THE CARPATHIAN REGION

Of the Jurassic fossils known from the Carpathian Basin, the most important are marine invertebrates—the remains of gastropods, bivalves, and especially cephalopods (ammonites, nautiloids). In some rocks brachiopods are also common, and at certain specific sites echinoderms (mostly echinoids and sea lilies) can also be found. Thanks to mass occurrences of these fossil invertebrates, it is often possible to reconstruct the ancient environments in which these animals lived and died. On the other hand, the ages of many Jurassic marine rocks can also be precisely constrained by their constituent ammonites, which had a very rapid evolutionary rate.

Microfossils

Jurassic formations in general are represented by hard limestone across the Carpathian Basin; their microfossil content was traditionally studied using thin-sections (in two dimensions only). In contrast, newly developed techniques (such as maceration with undiluted acid) mean that microfossils extracted from rocks can be studied in three dimensions; this has breathed new life into micropaleontological research.

Foraminifera

Jurassic rocks across the Carpathian Basin contain relatively abundant foraminifera. Indeed, László Majzon, in his voluminous handbook *Foraminifera Studies,* published in Hungarian in 1966, noted 44 forms from the lower part of the Liassic and 41 forms from the upper part, which altogether number 87 known from the Lower Jurassic. Majzon's book accurately reviewed all available data from the Paleozoic to the Quaternary, but listed few publications dealing with Jurassic-aged rocks. In the 1960s, little work had been done on this subject.

Among the early papers in this field, only the publications of Elemér Vadász and Mária Sidó are worth mentioning. Vadász gave faunal lists from Lower, Middle, and Upper Jurassic samples collected in the Gerecse Mountains, and determined microfossils from thin-sections and, occasionally, washed residues. In these latter cases, samples were taken from the upper, weathered rock surfaces. Sidó should also be acknowledged for her pioneering work on the Liassic successions that overlie the manganese ore at Úrkút in the Bakony Mountains. She recognized many species that belong to different families—including Lagenidae, Polymorphynidae, and Reophacidae—in radiolarian-rich marl formations.

Majzon was able to foresee that many things were left to do in the Jurassic foraminifera research arena. In accordance with this, new data were later published by Anikó Bérczіné Makk, who studied thin-sections from the Bükk Mountains and also borehole samples drilled deep into the Great Hungarian Plain. As a result of her work, many formations of uncertain or unknown age were added to the Hungarian Jurassic stratigraphy.

Compared to the method of making thin-sections, widespread application of a concentrated acetic acid solution led to a leap forward in research quality. It has also meant that many forms, long undetermined, can now be precisely observed and identified.

📖 Majzon 1966; Bérczіné Makk 1999

More and More Data

Due in large part to intensive research in recent decades and successes using an acetic acid solution, many new Jurassic foram faunas from Transdanubia are now known. One of the earliest researchers to attain spectacular results was Ágnes Görög: when one compares her data to those summarized by Majzon in 1966, one sees the minimum number of known forms doubled. From the Mecsek Mountains alone, for example, 59 Sinemurian, 130 Pliensbachian, and 49 Toarcian taxa are now distinguished. However, because the evolutionary rate of Jurassic forams is rather slow, certain species have a long time range—so the group as a whole is rather limited in biostratigraphic value.

This characteristic of the fossil record also means that the faunas known from certain stages overlap to a considerable extent. As far as we know, only a single species is restricted to the Sinemurian, another to the Pliensbachian, and four to the Toarcian. In contrast, 25 taxa are characteristic of the Sinemurian–Plienbachian, 28 are known in the Pliensbachian–Toarcian, and 53 occur from the Sinemurian into the Toarcian. The most diverse known forms are members of the genera *Dentalina, Nodosaria,* and *Astacolus*—which are represented by more than a dozen species—but the genera *Planularia, Vaginulina,* and *Lenticulina* are also abundant. A separate monograph dedicated to the rich benthic assemblage of the Mecsek Mountains Bathonian lists 82 species.

Ágnes Görög, professor of micropaleontology at Eötvös Loránd University. Görög described the earliest known planktonic foraminifera (*Globuligerina geczyi*) from Lower Jurassic beds in the Gerecse Mountains.

The feeding habits of forams (such as suspension filtering and detritus eating), their living positions (on the surface of sediments or within them), and environmental requirements (such as water depth and organic matter supply) are reflected in the diversity of their test shapes and by the materials that comprise these external shells (agglutinated, porcelain, or hyaline). This range of diversity also raises the possibility that different morphogroups can be distinguished on the basis of their appearances; thus, their degrees of dominance in ancient environments can be inferred. In certain specific situations, forams can provide an effective tool for environmental reconstruction.

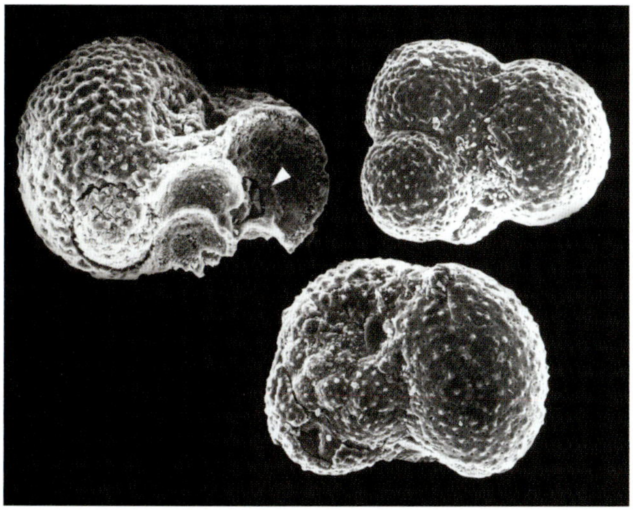

Globuligerina geczyi Görög, an early planktonic foraminifera from the Early Jurassic of the Gerecse Mountains (approximately 150× magnification).

Importantly, in addition to benthic foraminiferal assemblages, more and more data have now been gathered about the early planktonic forms as well. One group in particular, the so-called protoglobigerinids, were almost completely unknown.

📖 Görög 1995

Beginnings of a Success Story:
The Appearance of the First Planktonic Forams

One of the greatest foram "inventions" and one of their most important evolutionary milestones occurred when one or more members of this group first became planktonic. The Cretaceous and Cenozoic seas contained diverse and rich planktonic foram assemblages, so researchers

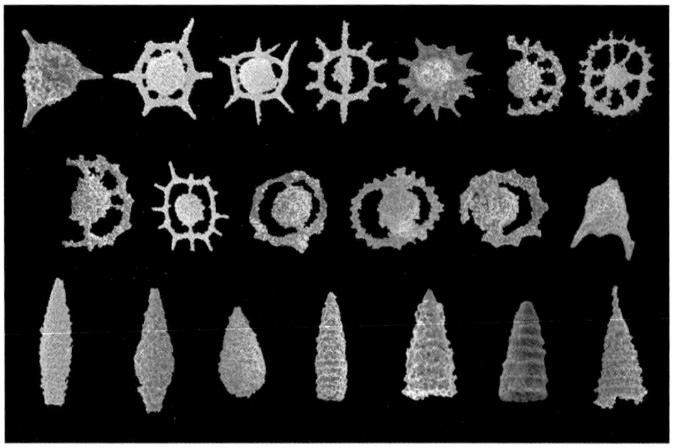

Lower Jurassic (Hettangian) radiolarians are known from the Csővár section, not far from Vác. This locality is famous because here it is possible to study successions from both the latest Triassic and the earliest Jurassic, encompassing the Triassic-Jurassic boundary, continuously. Of particular importance here, both macrofossils (mostly ammonites) and microfossils (radiolarians and spores) can be collected and studied in parallel. Similar fossiliferous radiolarian localities of this age are rare globally. Comparable boundary beds have so far been found only in Japan and Canada (50–100× magnification).

always suspected that their earliest forerunners might be found in the earliest stages of the Jurassic. Ágnes Görög also published important data on this subject when she discovered a new species in samples from the huge quarry at Nagypisznice Hill (Gerecse Mountains). She named this new form in honor of Barnabás Géczy, and the holotype of the species (on which the description was based) is from the late Hettangian, or early Sinemurian, part of the Rosso ammonitico–type Pisznice Limestone Formation. This is one of the earliest planktonic forams ever recorded. It is worth remembering that foraminifera that have a floating mode of life differ from their benthic counterparts in shape and size, and also in the structure of their tests. Planktonic forams are usually small, their chambers overlap considerably, and their outer walls are thin and have a peculiar structure. In subsequent studies, Ágnes Görög also revealed that these early planktonic forms (often referred to as protoglobigerinids), although extremely rare in the lowermost Jurassic, are much more common in the Middle Jurassic. After thorough analysis of their characteristic features (including the aperture, umbilical swelling, structure of the wall and coiling), Görög was able to distinguish seven different taxa in total (at the species and subspecies levels). She also applied an interesting approach to identifying forms that were hitherto known only from thin-sections: she put the separated, already determined, specimens she had obtained using acetic acid solution into resin to make orientated thin-sections. This enabled her to see the wall structure and the inner part of the test and allowed the production of a determination key to identify protoglobigerinid foraminifera from thin-sections.

📖 Görög 1994, 1995

Radiolarians

The minute, microscopic remains of radiolarians can be very abundant in some Jurassic rocks—especially those that were deposited in deeper-water environments. The tiny tests of radiolarians accumulate in the deep ocean, and if the material of this radiolaritic ooze is resolved and silica later precipitates, then a so-called radiolarite is formed. Some silica formations in the Transdanubian Range and in the Mecsek Mountains were known long ago: archaeological sites in these areas often yield ancient artifacts made of radiolarite, or other silica-rich rocks.

The first results to be published on Jurassic radiolarians—like those on Triassic forms—were completed by pioneer Miksa Hantken in 1884. Hantken had collected samples from the Bakony, Gerecse, and Mecsek Mountains; the radiolarians he extracted were later studied and described by David Rüst.

After a long hiatus, this research was continued by András Barabás—under the supervision of Heinz Kozur—who started to work on Late Jurassic radiolarians from the Mecsek Mountains. His results were summarized in his excellent thesis and also in a review paper. At the same time, Lajos Dosztály, a young geologist from the Hungarian Geological Institute, also started research on the radiaolarians of the Bakony and Gerecse Mountains. His first, very promising results were published in 1988, but soon after—because of his untimely death—Jurassic radiolarian studies ground to a halt in Hungary. Recently, following in the footsteps of Dosztály, Péter Ozsvárt is preparing to publish new results on the Triassic and lowermost Jurassic radiolarians from the Csővár section.

📖 Barabás 1995; Dosztály 1998; J. Pálfy, Demény, et al. 2007; Götz et al. 2009

Calpionellids: Tiny Parachutes from the Jurassic-Cretaceous Boundary

One group of ciliated protists—known as tintinnids—has been around since the Ordovician, but became abundant in the fossil record only during the Late Jurassic and Early Cretaceous. Mesozoic forms of these protists, also known as calpionellids, have a characteristic amphora or vase-shaped shells called loricae, which are mostly organic (proteinaceous) but may also incorporate minute pieces of minerals, particles of coccoliths (individual plates of calcium carbonate formed by certain single-celled algae), or diatoms. Loricae are very small, generally less than 0.2 millimeters in length, and ancient forms of these were likely calcified and so are well represented in the fossil record.

Tintinnids are aquatic (mainly marine) organisms and, orientated tip down, they float in the water column like tiny parachutes, feeding mostly on photosynthetic algae and bacteria. They used hairlike projections (cilia) that point out from the top of their loricae to filter their planktonic food.

The most important of the calpionellid genera are *Calpionella, Calpionellopsis,* and *Tintinopsella.* These all have narrow stratigraphic ranges around the Jurassic-Cretaceous boundary—especially within the Early Cretaceous (Berriasian and Valanginian Stages)—and play a prominent biostratigraphic role.

Even today, identification of calpionellids is based on studies of thin-sections because the successful extraction of their tiny shells from rock is nearly impossible. Indeed, because the plane of a thin-section may cut the loricae at different angles, the same shape may end up having a very different appearance when viewed in cross section, whereas different shapes may be thought to have similar appearances. Although this can make determination difficult, these shells tend to appear in mass accumulations, so there is a high probability that one thin-section will contain a nearly longitudinal cut that allows precise determination. The size of the shell, its thickness, and the existence or absence of a collar are all characteristic features. One example, the descriptions of dozens of new forms by István Nagy (1936–2003) from the Mecsek Mountains, is thought to be invalid. This seems to have been one case in which differently orientated sections of formerly known species were misinterpreted and mistakenly identified as new ones.

A recently developed technique based on the differential solubility of test minerals allows extraction of entire calpionellid specimens from solid rock. The initial results are very promising; however, using thin-sections to determine these fossils will remain a quick, easy solution for the near future.

In Hungary, calpionellid limestones occur in the Transdanubian Range and in the Mecsek Mountains. When a Romanian micropaleontologist described *Tintinopsella carpathica* from Lower Cretaceous beds in the Carpathian Mountains, the same species was subsequently recognized in Hungary all around the Carpathian Basin. Indeed, calpionellids are common in calcareous rocks deposited in pelagic environments—and, because of their wide geographical range, they are also an important tool for correlating rocks between widely spaced regions.

📖 I. Nagy 1986

Ostracods

Ostracods have long been known to micropaleontologists from the Jurassic formations of the Carpathian Basin. Their presence can be proven by a characteristic oval, or bean-shaped, cross section that can often be seen in thin-sections taken from limestones and marls; but, again, it was the application of acetic acid solutions to rock samples that led to a breakthrough in this field. Early results using this approach were published by Miklós Monostori, then head of the Department of Paleontology of Eötvös Loránd University in Budapest, who worked on samples from the Pliensbachian of the Bakony Mountains and from the Bathonian of the Mecsek Mountains.

This Bathonian nodular marl is a condensed formation, which means that just a small amount of sediment was deposited over a long period of time. One might expect ostracod shells to be preserved in large amounts in such a rock, but, surprisingly, the Mecsek Bathonian marl is relatively poor in ostracods: one kilogram of this marl contains just 28 specimens on average. It is nevertheless quite diverse: 13 species in six genera have been described; the genus *Cytherella,* represented by four species, is most frequent. The genera *Bairdia, Anisobairdia, Cardobairdia, Pontocyprella,* and *Paracypris* are also relatively common, but some of Bathonian ostracods described from the Mecsek Mountains have turned out to be new to science. According to recent work, although Jurassic ostracods were not very good age markers, they are accurate indicators of ancient environments. For example, comparing the Bathonian ostracod fauna known from the Mecsek Mountains with other European faunas of similar age, it can be concluded that this marl was deposited in a pelagic ocean environment either in the deepest part of the sublitoral zone or in a shallow bathyal environment.

📖 Monostori 1995b

Rare Finds: Moss Animals (Bryozoans) in the Hungarian Jurassic

One interesting faunal element in the Bathonian of the Mecsek Mountains is a small bryozoan that was found as an encrusting colony on an *Eopecten* bivalve. The mollusk specimen was found by fossil collector Zoltán Orbán (Bonyhád), who kindly donated it to István Szente. Later, Szente noticed that the bivalve was encrusted with a bryozoan. This was described by a specialist from Bratislava, Kamil Zagoršek. Although this particular fossil bryozoan is neither an age or paleoenvironmental marker, it is the first known record of this group from the Hungarian Jurassic. This discovery is also a fine example of cooperation between fossil collectors, professional paleontologists, and taxonomic specialists. It also shows that there is always something new under the sun.

📖 Zágoršek 1995

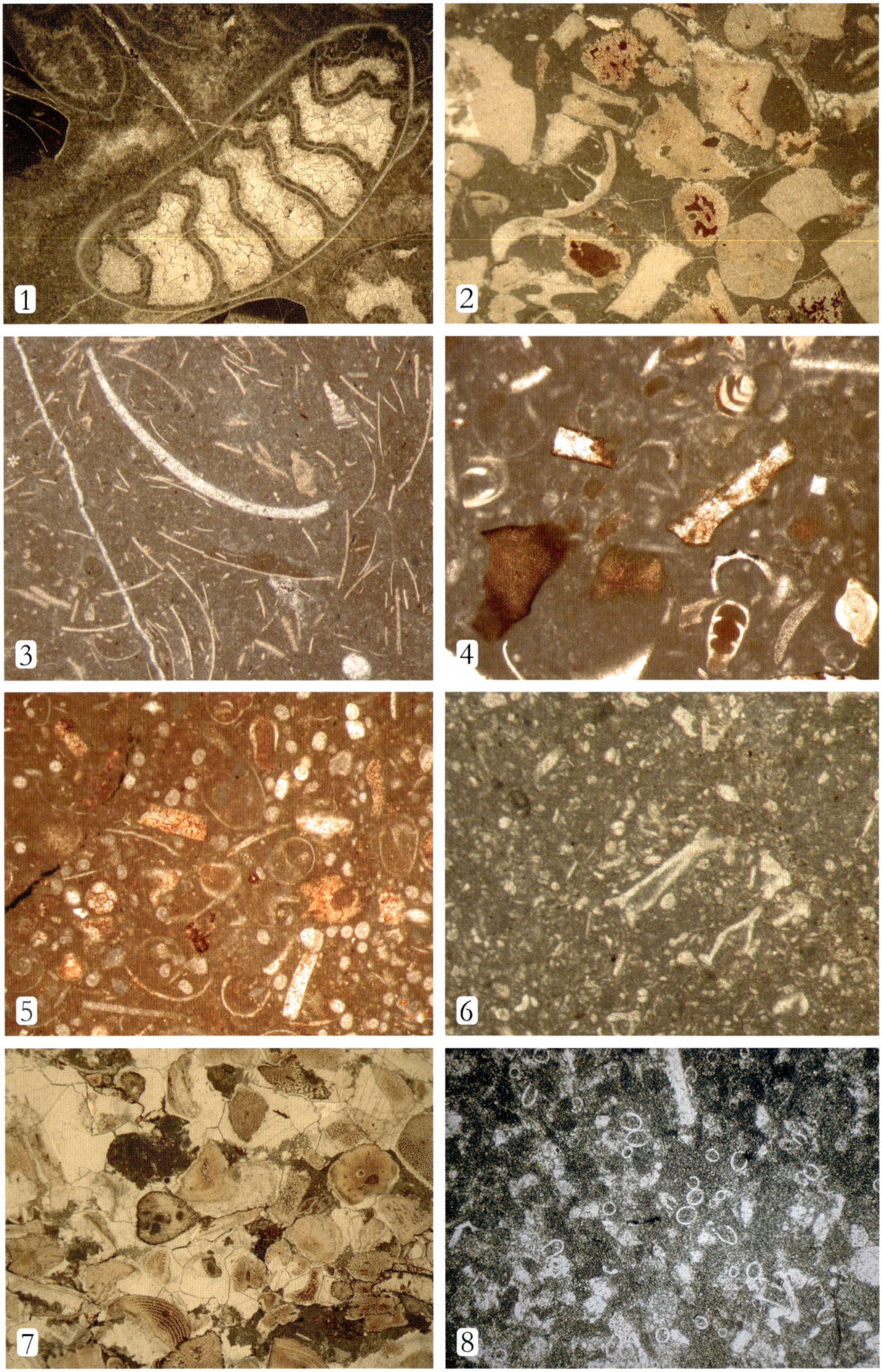
1
2
3
4
5
6
7
8

Fossil Plants

An excellent image of the Early Jurassic flora can be based on the diverse fossil records that have come to light from coal mines in southern Hungary and Romania. In contrast, the Middle and Late Jurassic floras of the Carpathian Basin are lesser known because most of these sediments are marine.

Localities and Paleontologists

Fossil plants of Jurassic age are found in three areas of the Carpathian-Pannonian region: in the Mecsek Mountains in southern Hungary, and in the Partium and Banat (Bánság) regions of western Romania. At all three locations, Lower Jurassic coal is mined and an abundance of fossil plants are also found in layers overlying the seams. Indeed, the fossiliferous sequence can be seen in the large opencast mines near Pécs (Pécs-Vasas, Pécsbánya, and Rücker). Many spectacular plant fossils have also come to light when sifting through spoil banks adjacent to the subsurface mines (Zobák, Béta, and Kossuth shafts). Four particularly notable fossil sites in the Crişul region of Partium are known: Dumbrava, Şuncuiuş-Recea, Bălnaca, and Recea. These sites lie within a circle that only slightly exceeds 10 kilometers in diameter. South of this area, in the Banat region—from the village of Doman down along the lower course of the Danube to the village of Sviniţa—nearly 20 fossil plant localities are mentioned in the geological literature. The most famous of these is probably Steierdorf-Anina, where coal deposits were discovered along the András streamlet by woodsman Miklós Hammer in 1790.

Coal has been exploited for nearly 200 years from the mines in the Banat and Mecsek, and over this long period countless beautiful fossil plant specimens have been collected by miners, geologists, and paleobotanists. As the mines have been closed one after the other, this continuous supply of fossils has gradually ceased because in the spoil heaps leaf and twig remains are degraded by weathering of the thin, layered shale. This coal, and thus the plant fossils, is of lower Liassic age—Hettangian and lower Sinemurian. The fossils that have persisted intact in these deeply buried sediments for 190 million years are strongly weathered after just a couple of years of surface exposure; once the last mine is closed, the only sources remaining to document this ancient vegetation will be the remains stored in museums and other collections. Apart from the significant fossil plant localities mentioned above, few fossil plant remains of Jurassic age are known from other parts of this region.

One rare exception to this is the manganese ore mine at Úrkút: Toarcian layers in this mine include huge silicified tree trunks that are thought to belong to relatives of *Araucaria*. Over the last 150 years numerous researchers have discussed the Jurassic plants from the Carpathian Basin. In early work, the fossil floras of the Mecsek Mountains were reviewed by Miksa Hantken, Móricz Staub (1842–1904), and István Zoltán Nagy. The most recent scientific results have been published by Mária Barbacka; at the same time, the fossil plants of the Körös and Banat regions have been studied by nearly a dozen geologists and paleobotanists. A detailed and thorough survey of the history of research in this area was provided in 1998 by Zoltán Czier, a worker at the Museum of Nagyvárad (Oradea).

Paleobotanical research on Romanian localities is currently being done by Zoltán Czier and Mihai Popa, who works in Bucharest. Both have already published a number of valuable papers on this topic.

📖 Popa 2009

Where Dinosaurs Were Grazing: Plants of the Swamps in the Mecsek Mountains

Along the narrow shoreline, salt-tolerant plants including ferns (probably tree ferns such as *Clathropteris* and

Fossils—skeletal remains—in thin-section from Jurassic rocks in the Bakony Mountains (under different magnifications). (1) A Lower Jurassic limestone with minute, differently oriented ammonite shells. These empty tests, accumulated on the seafloor, were buried rapidly and so fine mud was unable to enter the shells. The white calcite (sparite) between undulating suture lines was formed during diagenesis and infill spaces, including the inner parts of these tiny ammonites (from Kericser in the vicinity of Lókút). (2) Mass occurrence of sea lily fragments in a Lower Jurassic (Pliensbachian) limestone. In the middle of some of these crinoid pieces, small channels—where the soft body of the echinoderm was placed—can be seen (from Fenyveskút in the vicinity of Lókút). (3) Mass occurrence of thin-shelled bivalves (*Bositra*) in a Middle Jurassic (Bajocian) limestone (from Kericser in the vicinity of Lókút). (4) Benthic foraminifera in a Middle Jurassic limestone (from Közöskút Ravine in the vicinity of Hárskút). (5) A Middle Jurassic limestone with small sections of different mollusks and recrystallized test of radiolarians appearing as white spots. The fragile scales and spines of the test—important taxonomic features—cannot be studied in thin-section (from Közöskút Ravine in the vicinity of Hárskút). (6) An Upper Jurassic (Kimmeridgian) limestone with small particles of miniature sea lilies (*Saccocoma*) which—according to widely accepted opinion—were buoyant in life. These fossil crinoids, which often appear as a split bone in thin-sections, were referred to as Lombardia in older paleontological literature. (7) A Lower Jurassic limestone rich in skeletal debris. The relatively big, occasionally perforated, fragments are sea lily particles that were corroded by moving water or partly eaten by other organisms (from Mohoskő, in the vicinity of Lókút). (8) An Upper Jurassic (Tithonian) limestone containing the variously orientated cup-like tests of calcareous protists (calpionellids) (from Rendkő in the vicinity of Hárskút).

A Paleontologist in Action: Extracting Microscopic Fossils from Hard Rocks

Although microfossils can be seen in thin-section using light microscopes, researchers are keen to extract them and study them in three dimensions. For some groups—such as the calpionellids that are especially abundant in marine Upper Jurassic–Lower Cretaceous sediments, or certain planktonic foraminifera—a lucky cross section can provide enough information for a precise determination. Other fossils, like the larger forams (see chapters 1 and 5), can be determined only in orientated thin-section because their appearance is uninformative, and only their inner structure is characteristic.

Most microfossils are difficult—if not impossible—to determine in thin-section and are also often hard to extract from surrounding rocks using traditional preparation methods. These fossils are far too small for the use of chisels, hammers, or needles. In order to extract them, completely new methods that work at the level of atoms and molecules are needed: chemicals! Which kind of chemical can be used depends on the mother rock and on the material of the fossils themselves.

A Paleontologist in Action: Maceration with Acetic Acid

The most frequent microfossils, forams and ostracods, have calcium carbonate shells that are often preserved in calcareous rocks. These rocks are mainly limestones that have almost the same chemical composition as the fossils themselves. Microfossils were first extracted from these sediments in the 1980s, when French ostracod researchers developed a method applicable in most cases. This process was subsequently also applied successfully in the laboratory of the Department of Paleontology of Eötvös Loránd University, where many kinds of Jurassic rocks and microfossils—hitherto impossible to study—were investigated.

This so-called maceration procedure uses both physical and chemical processes at the same time and takes advantage of differences in the microstructure of the fossils and their surrounding matrix. First, the rock is broken into pieces about one cubic centimeter in size, which must be kept completely dry. Next their container is filled with undiluted (minimum 98 percent) acetic acid (CH_3COOH) that enters only into the rock and slowly dissolves it. (The use of dilute acid leads to a rapid reaction and dissolution of the microfossils.) As a consequence of repeated heating and cooling of samples, the reaction products form into crystals and the fine laminae of the rock foliate and weather away. However, because the skeletal remains of the fossils generally have a more solid inner structure then the surrounding rock, the acid has less effect on them. Sometimes the fossils are coated with a fine metal oxide, which also protects them. Unfortunately, fossils that have inner cavities filled with sediments will be lost to this process, as the crystals destroy them from inside as they grow.

In general, the time needed for acetic acid maceration depends on the purity of the limestone. A high clay content will result in a more rapid reaction, but in the case of generally lighter colored, very pure limestones, this process may take weeks—and even then results can be dubious.

As crystals form, they cement the sample into a single slab that must be taken out of its plastic container and again broken into pieces. The acid must be washed away quickly; by adding water it becomes dilute and the reaction is speeded up. The procedure can be repeated with any non-disintegrated rock pieces, and all acid and calcium acetate must be washed off the residue—otherwise the microfossils will be damaged. Usually, a base of some kind (such as ammonium hydroxide [NH_4OH]) is used to neutralize the acid.

There are many additional ways to extract other kinds of microfossils. Coccoliths (a type of single-celled alga with a calcium carbonate test), for example, are collected by drying the fine suspension which contains their separate algal plates, no more than 10 micrometers in size, and making smear slides. Other non-calcareous microfossils (like pollen and spores, for example) are extracted from hard rock by using other acids of different kinds (including hydrochloric acid [HCl], formic acid [HCOOH], or hydrofluoric acid [HF]). These methods are treated in more detail in chapter 4.

Dictyophyllum) that live in swamps may have flourished. Indeed, the enormous palmate dissected leaves of one of these fossil ferns exceeded 1.5 meters in length. Although the whole of this giant leaf was not preserved, its great many fragments serve as a jigsaw puzzle so that its original form—including its non-dissected basal parts and the fingerlike, serrated long leaf sections that branched from it—can be pieced back together.

Along with giant forms, there were pteridophytes of all sizes—including representatives of the genus *Todites*—that

Paleobotanist Mihai Popa of the University of Bucharest. His interests are wide ranging, but he focuses in particular on Lower Jurassic plant remains from the Banat region.

had pinnately compound leaves and resembled some modern ferns. Swampy areas adjacent to forests were dominated by horsetails (*Equisetites, Equisetum, Neocalamites*) having stems the thickness of an arm and primitive leaf segments arranged in whorls on their straight stems. Jurassic medium-sized horsetails were not considerably different from their giant Carboniferous ancestors, and are also quite similar to the small-sized genus *Equisetum* that is alive today.

Conifers and pteridosperms were also important elements of Jurassic forests, members of the genus *Komlopteris* being the most frequently encountered of the latter group. Indeed, this name pays tribute to one former mining town, Komló (as, incidentally, does *Komlosaurus,* a dinosaur known from the Mecsek Mountains). *Komlopteris* had large, pinnately compound leaves along with a special structure, a scale-covered cupule that surrounds the seeds of pteridosperms, which has also been reported. Additional pteridosperms from the Mecsek Mountains include *Sagenopteris, Ctenozamites,* and *Pachypteris,* alongside lianas and conifers (such as *Elatocladus, Elatides,* and *Palissya*) known from both fossil leaves and well-preserved cones.

In addition to all these arborescent forms, shrubs also occurred in Jurassic forests and include ginkgos (relatives of one living fossil, *Ginkgo biloba*), members of the genera *Ginkgoites,* as well as *Sphenobaiera* and *Baiera* that grew in drier soil habitats. Both the leaves and the seeds of these plants are well preserved in the fine-grained mud. Occasionally a mass occurrence of fossil remains has been recorded: "Compressed several layers of leaf-mass often form deposits of considerable thickness among the thin shale layers" (Barbacka 2000). Cycads (*Nilssonia* and *Bjuvia,* or earlier *Macrotaeniopteris*) and Bennettitales (*Pterophyllum, Anomozamites*), the latter similar to the former, also occur rarely. In addition to thick, coriaceous leaves, the seeds and sporophylls of these plants have been found.

📖 I. Z. Nagy 1961; Barbacka 1994a, 1994b, 2000; Barbacka & Konijnenburg-van Cittert 1998

Invertebrates

Sponges

Across the northwest European shallow-water facies zone, beautiful fossils of Jurassic sponges are common. For example, in late Oxfordian and Kimmeridgian times a huge sponge reef existed on the northern margin of the Tethys Ocean. Remains of this reef have been found on the Iberian Peninsula, through Poland and Dobrogea or Dobruja (southeastern Romania), and into the Doneck Basin (eastern Ukraine). Jurassic sponge remains from the Carpathian Basin are rather rare.

The Sponges of Philipp Počta

Although rare, the Middle Jurassic (Bathonian) sponges from the Mecsek Mountains merit mention. These fossils were first collected by pioneer geologists Károly Hofmann and János Böckh. They are often just a few centimeters long and are cylindrical or cone-shaped, resembling calcareous nodules; they were first noted by the early Bohemian paleontologist Philipp Počta (1859–1924) in an 1866 paper in the *Annales of the Hungarian Royal Geological Institute.* These fossils have since been identified as representatives of hexactinellid sponges, a group with skeletons made up of four or six-pointed siliceous spicules (the structural elements found in most sponges). Hexactinellids are also often referred to as glass sponges.

In these sponges, siliceous spicules ("needles") of different sizes are either fused to one another or occur individually and are embedded into the body of the sponge—which itself consist of collagen, a protein. These spicules are arranged more or less systematically—and in certain places they perforate the body of the sponge, which most often has a shape like a calyx. Following the death of these sponges, the organic body decays and only the remains of the microscopic spicules are left on the seafloor. In these cases the original shape of the sponge animal cannot be reconstructed, but in the Mecsek fossils the entire skeleton is preserved, although silica has been replaced with calcium carbonate. On the surface of these fossils a characteristic reticular structure can be seen.

The old Mecsek fauna, collected about 120 years ago, was revisited recently by Polish sponge expert Andrzej

1
2
3
4
5

A lithograph of a pteridosperm from the Mecsek Mountains, as published by Móricz Staub in the 1882 volume of *Földtani Közlöny* (Bulletin of the Hungarian Geological Society). This fossil, originally described as *Ctenopterys cycadea*, is currently assigned to the genus *Ctenozamites*. At the time of Staub's work, investigations on fossil plants were limited to macroscopic observation and descriptions were typically documented by lithograph illustrations. Nowadays modern microscopic techniques provide high resolution photographs.

Pisera. According to his preliminary unpublished results, new generic names are needed for some sponges in this fauna to replace the old names used by Počta. These mainly Bathonian and partly lower Callovian sponges comprise a dozen species—most of them probably new to science. Besides the most common hexactinellids, representatives of the Lychniscosida and Lithistida ("stone sponges") groups are also present in the Mecsek fauna.

📖 Počta 1886

From Sponge Spicules into Flint

Living sponges consist of organic material and calcareous, or siliceous, spicules. The size of these spicules can vary from just a few to a few hundred micrometers—indeed, most of them are very small and cannot be seen with the naked eye. The materials that make up these spicules as well as their type of symmetry are important taxonomic features.

Spicules of inorganic material are surrounded by the organic body of the sponge; the meshing of many spicules together creates a skeleton. When the sponge dies and all the organic material has decayed, only these very tiny, resistant spicules remain. Indeed, it is very common for these siliceous spicules (themselves composed nearly exclusively of silicon dioxide) to dissolve as well, and the gelatinous silica that remains in the porous sediment then precipitates to fill cavities. As a result of diagenesis, the soft sediment turns into limestone and this gelatinous silica becomes flint—a kind of chert—forming nodules, lenses, or even whole beds. The original shape of the spicules cannot be seen in the flint; only the origin of this siliceous material can be traced back to the sponge. Spiculite (a type of rock that originates from siliceous sponges) can be found, for example, in the Lower Jurassic Isztimér Limestone Formation in the Transdanubian Range.

Gastropods and Bivalves

Jurassic Bivalves from the Villány Mountains

In Jurassic formations in the Villány Mountains, bivalves are the most common fossils after ammonites and brachiopods. Bivalves are abundant in Pliensbachian, Bathonian, and Callovian beds, whereas gastropods are represented by just a very few specimens. The history of research on these fossils, like the history of work on all Jurassic successions in the Carpathians, was characterized by a series of lucky and not-so-lucky accidents.

The first report on the bivalves from this area was provided by Lajos Lóczy Jr. (1891–1980), who listed 11 forms. Following this, research ground to a halt for more than half a century and was renewed only when Attila Vörös redescribed the fauna in his PhD thesis. Vörös studied more than 200 specimens, most from old museum collections, and was able to identify 12 taxa (5 to the species level) from Pliensbachian beds alone (the precise age of these beds had just been determined a few years before). Within this fauna, the most common elements are free-living scallops, some of which are smooth (*Entolium*) and some of which are ornamented with radial ribs (*Pseudopecten*). At Villány, just below the ammonite-rich limestone bed, there is thin (8 centimeters thick at the maximum) sandy limestone that also contains about a dozen different bivalve species. Some of them lived on the surface of the sediment, while others belonged to the infauna—living

Early Jurassic plant remains from a strip mine spoil bank in the Mecsek Mountains. Fossil plants suitable for scientific investigation are always preserved in the otherwise barren beds that intercalate with the coal seams (magnification is 0.5×, except [5]). (1) *Sagenopteris* sp.—Leaf of a seed fern similar to the four-leaf clover. This genus is thought to include arboreal forms as well as lianas. (2) *Phlebopteris* sp.—leaf of an extinct fern. (3) *Komlopteris nordenskioeldii* (Nathorst) Barbacka—slender and small-sized sunleaf of a seed fern. This type of leaf was developed from twigs exposed to direct sunlight and this species is one dominant element of the Early Jurassic flora of the Mecsek Mountains. A coalified and compressed cupule of this species (the outer cover of the seeds in seed ferns), has also been found. (4) *Komlopteris nordenskioeldii* (Nathorst) Barbacka—broad and larger leaves of a seed fern developed by twigs exposed to indirect sunlight. (5) *Todites* sp.—fragment of an ancient fern leaf (original size).

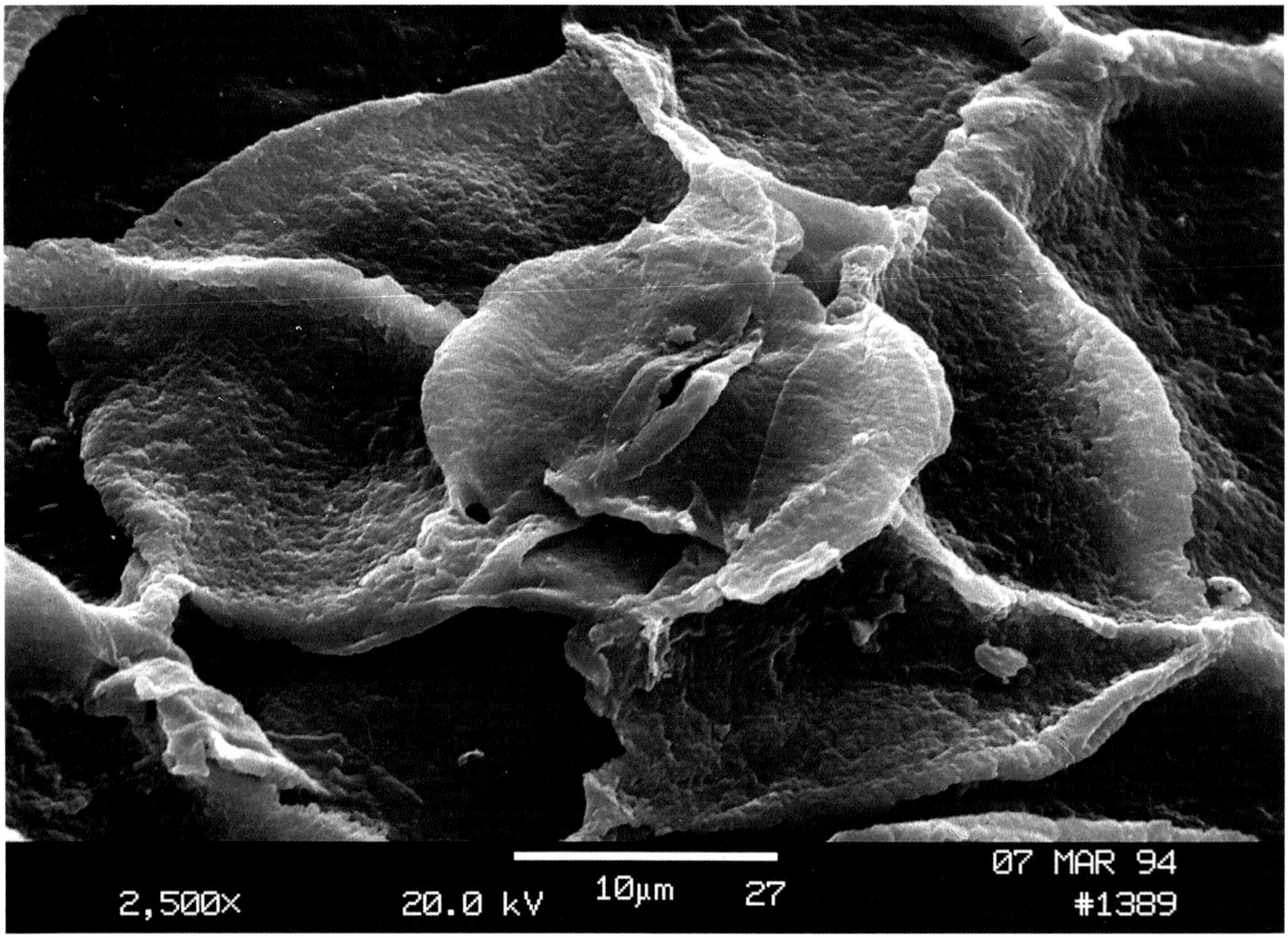

This photograph obtained by electronmicroscopy shows the stomata of *Ctenozamites cycadea* (Berger) Schenk, a fossil pteridosperm from the Mecsek Mountains. Plant remains are usually flattened, highly coalified, and preserved as a very thin filmlike envelope. However, they are sometimes so excellently preserved that it is possible to make preparations of the outermost protective layer of the leaves, the cuticle that covers the epidermis. Using this microscopic method, 196-million-year-old cells, stomata, and other details are recognizable. In this case, the structure of the stomata is indicative of a plant presumably adapted to humid conditions.

just in the sediment. All in all, these ammonite-rich beds have yielded about 70 specimens—again representing 11 species, but which, in contrast to the aforementioned older bivalves, lived in a deeper environment. Because most of these forms are completely extinct, we do not have a clear understanding of their lifestyles even today. This is especially true of *Inoceramus*-related forms, even though they are the most common elements of this bivalve fauna.

The best-preserved and most diverse bivalve assemblage—containing 26 species—was collected from a brown-colored Bathonian limestone that has a characteristic chocolate-like appearance. Indeed, this rock type is no longer found in the Villány Mountains, either in situ or as debris. This rock is known just from museum specimens, although their labels clearly indicate that it used to be exposed in the Villány Mountains. We know that some of the Jurassic formations in this region had limited lateral extension, so it is possible to imagine that the fossils from this "chocolate rock" were collected from a facies that no longer exists. Another quite likely alternative, however, is that (according to Attila Vörös) these specimens were collected from the Middle Jurassic in the Pădurea Craiului Mountains of Romania and were mislabeled.

More information about the Jurassic bivalve faunas of the Villány Mountains was obtained when a temporary site was found by geology students in 1990. This group was led by university lecturer Csaba Péró, who is deeply knowledgeable about the geology of the Carpathians, and visited Szobros ("Statue") Quarry near Villány—where Upper Jurassic limestone was once mined, but today it is an open-air sculpture park. In the vicinity of this quarry the group found a recently made pit that was likely intended to be a wine cellar. In this pit, below several meters of loess, a previously unknown Jurassic sequence

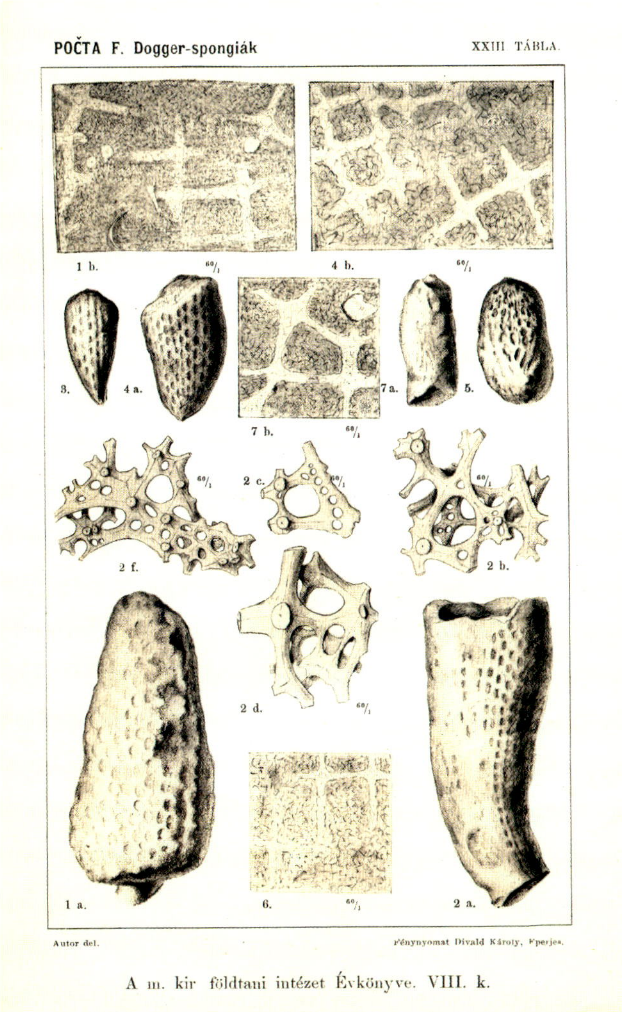

Middle Jurassic sponge remains from the Mecsek Mountains. A lithographic print from an 1886 paper by Philipp Počta. Another interesting feature of the Mecsek sponge fauna is that it comes from deep water Rosso ammonitico type sediments, while the faunas known from northwestern Europe are from shallower marine facies.

composed of cherty, siliceous limestone beds was found. Here the rock contained many fossils, including bivalves, belemnites, and brachiopods; the group collected about 100 bivalves alone—all of which suggested an Early Jurassic age. More precisely, these fossils are from the upper part of the Pliensbachian stage, which is also known informally as the Domerian, after Monte Domaro in the Southern Alps where these beds were first recognized. Until this discovery, this substage had not been documented in the Villány Mountains.

The bivalves that were collected from this site comprise seven species, with most belonging to a large scallop called *Pseudopecten aequivalvis* and to *Gryphaea gigantea*. This latter species is the last known member of the Early Jurassic *Gryphaea* evolutionary line; the genus had a worldwide distribution at that time. All the fossils collected were also remarkably silicified—likely because of dissolution from dissolved and precipitated sponge spicules. Concentric rings on the surfaces of the thick-shelled bivalves, called beekite, are the results of repeated dissolution and precipitation of silica.

A similar bivalve assemblage can be collected near Şuncuiuş, in the Apuseni Mountains of Romania, and it is no surprise because this locality is on the other side of the Villány Zone, a tectonic belt that can be traced below the surface across the basement of the Great Hungarian Plain. All in all, the discovery of these Domerian fossils at Villány is entirely thanks to the interest and experienced eyes of the geology students and their professor. Their eagerness was justifiable, since temporary exposures (like the pit they found) can disappear within a few days—but this turned out not to be the case. Perhaps because of the hard rock beneath the soft loess, the builder gave up on this cellar. Twelve years later, the pit can still be found; it was visited recently by the authors.

The subsequent Jurassic bivalve assemblages from the Villány Mountains are of different paleobiogeographic aspects. Understanding this change has provided deep insight into the movement of the Tisia Terrane, a small microcontinent, a part of which is now represented by the Villány Zone. As the microplate moved, Tisia was slowly distanced from the rest of Europe; as a result, Pliensbachian and Bathonian bivalve assemblages from this area have strong affinities to those known from time-equivalent beds in northwest Europe. However, this similarity is less expressed in Callovian fauna.

📖 A. Vörös 1971; Szente & Vörös 1992

Jurassic Bivalves and Gastropods from the Mecsek Mountains

The oldest member of the Jurassic succession in the Mecsek Mountains, the so-called Mecsek Coal Formation, began to form in the Triassic in a system of lakes and rivers. The oldest mollusk remains are found only higher up in the formation, in a marine delta facies that was formed as a result of global transgression during the Early Jurassic. Otherwise barren beds among the coal seams have yielded about 20 gastropods, and about the same number of species of bivalves. These mollusks were living on the delta plain, which was not covered with water all the time. Fossils other than mollusks are extremely rare in this fauna. With the exception of one frequent bivalve (*Unio vizeri*) and a very rare, small-sized gastropod (*Viviparus verinorum*)—both restricted to freshwater environments—all other taxa are marine and must have been able to tolerate excursions from normal salinities and form seven distinct communities, six of which are dominated

Portrait Gallery

Karl Ferdinand Peters (1825–1881)—Physician, Teacher, and Geologist

The work of Karl Peters is a good example of just how multifaceted the best researchers of the second half of the nineteenth century really were. Peters first completed his studies as a doctor, then as a teacher, and then finally he became a geologist focusing mostly on mineralogy. Between 1855 and 1861, when he was appointed professor of mineralogy and geology at the University of Pest by the Austrian government, he made a tremendous effort to save and organize the mineralogical collection that was at the time on the verge of destruction. His work was highly appreciated by his predecessor (and his successor), József Szabó, the father of Hungarian geology. Peters's fundamental paper on the geology of the Mecsek Mountains was published toward the end of his stay in Hungary. He also completed a paleontological monograph and was one of the first people to study fossils and carbonate rocks in thin-section—a method we now consider routine.

by bivalves. Differences between these communities can be understood as a consequence of the environmental diversity of the delta plain; fossils are amply accumulated in shell beds, often more than 10 centimeters thick. They are even of occasional use in helping correlate the folded and buckled coal seams in deep mines and boreholes.

This coal formation has abundant organic material and pyrite; therefore, it disintegrates and weathers rapidly at the surface. Because of this it has almost no natural outcrop and can be seen only in mines and in waste dumps and piles. Similar to this coal succession, the overlying marls, also known as the "Gryphaea marl," contain many bivalves. *Gryphaea* itself is believed to have been an unattached recumbent recliner on the seafloor, whereas other bivalves (such as *Palaeoneilo elliptica, Cardinia listeri,* and *Hippopodium ponderosum*) lived within the sediment on the seafloor. Still others (including *Pseudopecten aequivalvis* and *Antiquilima succincta*) were even able to swim short distances by quickly clapping their valves together, whereas some (*Modiolus ventricosus, Cuneigervillia hagenowi,* and *Myoconcha psilonoti*) were specialized to live partly embedded in sediment. In contrast, the easily weathered—and, therefore, poorly exposed—marls contain very few gastropods.

The best-known bivalve locality of this age is Basa Pit, beneath Köves Hill, a place already known to Peters. Fossils from this locality occasionally appear as rare, scattered, but whole specimens, or in shell beds that are packed mainly with broken specimens, which accumulated during heavy storms. Besides *Gryphaea,* pectinids are the most frequent bivalves at these sites and are also found in the sandstone beds that are intercalated into the lower part of this marly succession. These beds are best known from Ófalu.

By the end of the Sinemurian, because of the gradual deepening of the sea, once-flourishing benthic life became less abundant. In deeper-water sediments *Inoceramus*-like bivalves (a group that had worldwide distribution in the Jurassic and Cretaceous) became dominant, whereas in calcareous marls remains of *Oxytoma inequivalve,* probably the most often mentioned Jurassic bivalve, are also frequent.

Deposition of the Fleckenmergel Formation started in the Pliensbachian. Gastropods are not common in this unit, and bivalves are even rarer—with one exception: the fauna of the black shale in the lower part of the Toarcian, which records a global anoxic event and is characterized by abundance of the bivalve *Pseudomytiloides dubius.* This fossil mollusk had the ability to secrete sticky threads, known as byssal threads, much as the living zebra mussel (*Dreissena*) does in order to stick itself to substrate. Indeed, *Dreissena* is able to stick to any hard surface, although it adheres mostly to rocks. The fossil bivalve *Pseudomytiloides* is often found attached to ammonite shells, which raises the question of whether the latter received their "fellow traveler" while still alive, floating and swimming in the water, or the bivalve larvae settled on the empty shells that form stable islands on the seafloor. The soft deepwater sediment would otherwise have been lethal to the larvae.

Strangely enough, there are even deep infaunal bivalves (*Solemya, Goniomya*) preserved as fossils in this black shale, proving that for short periods of time, the dissolved oxygen content of the bottom waters exceeded 0.1 milligram per liter, the minimum value necessary for multicellular animals to survive. Fossils provide the only evidence for these changes in oxygen content, as there are no geochemical or lithological signals in these basically anoxic black shales.

Peters, the first researcher to work in the Jurassic of the Mecsek Mountains, noticed that this black foliating shale, found almost exclusively in the Réka Valley (spelled "Recca" in the old geological literature), is very similar to a formation of the same age in southern Germany—the

Holzmaden Shale. The most characteristic fossil in the German sediment is a small bivalve, called *Posidonia* in older literature and known today as *Bositra*. This shale, otherwise famous for its vertebrate fossil remains, is one of the richest fossil localities in Europe and is regarded internationally as one of the Fossil Lagerstätten—the Posidonia Shale (Posidonienschiefer). In the Mecsek Mountains, however, *Posidonia* is missing from the black shale; it was reported only once—by Jenő Noszky Jr., who reported a mass occurrence of these fossil bivalves during a talk on the 1959 Hungarian Jurassic in the Mesozoic Conference, held in Budapest. Why he did this remains a mystery, because *Posidonia* is not present.

From the Aalenian, the bivalve fauna of the typical Fleckenmergel is dominated by *Bositra buchii*. Other faunal elements are forms that can attach to stable substrates and include certain types of scallops (*Entolium, Eopecten,* and *Camptonectes*) and *Plagiostoma,* which is closely related to the genus *Lima*. Up the sequence, in the Bathonian, the bivalve assemblage becomes more diverse: in addition to earlier forms such as *Inoceramus*-related species, other members of the infauna—including taxa close to *Malletia* and *Anisocardia*?—can also be found. However, because these bivalves are preserved only as internal molds, they cannot be determined with certainty. From Upper Jurassic formations only a few bivalves related to *Inoceramus* and representatives of the genus *Cuspidaria* are found.

📖 Peters 1862; Szente 1995

The First Rudist

Rudist bivalves reached their evolutionary zenith during the Cretaceous (see later sections) but the earliest representatives of this group appear in the Late Jurassic. Although there are no traces of rudists in Hungarian deeper-water Jurassic deposits, there are other places around the Carpathian Arc they have been collected. These fossil sites tend to occur in allochtonous tectonic position and often are isolated cliffs (Klippen, in German) or blocks surrounded by or embedded in younger deposits. Softer younger deposits usually have been eroded around masses of hard shallow-water Upper Jurassic limestones, resulting in a picturesque landscape. Indeed, one of the most famous Jurassic rudist assemblages is found in the light-colored Ernstbrunn Limestone, which forms klippen in the geologically complicated Waschberg Zone north of Vienna. In the Carpathians, another white limestone—the Štramberk Limestone, named after the northern Moravian city of Štramberk, which contains a huge quarry opened into a large klippe—has yielded some Jurassic rudists. These rocks are rich in fossils, especially bivalves, gastropods, and corals that were deposited in a reefal environment; in the Northern Carpathians, the most famous rudist localities are Polish klippen near the cities of Wadowice and Przemyśl (Woźniki, Inwałd, and Kruhel Wielki).

In the Apuseni Mountains, as part of the rich fauna figured in the monograph by Ferenc Herbich, the rudist *Diceras* was found in the Cetea Klippe. This site—called Csáklya-szirt in older geological literature—is rather picturesque, and is the subject of a poem by Hungarian poet Lajos Áprily (1887–1967).

📖 Herbich 1886

Jurassic Gastropods from the Transdanubian Range

Gastropods are generally rare in the Mediterranean Jurassic and their remains are mostly found in sediments that formed on topographic highs on the seafloor. One such deposit is the Hierlatz Limestone, formed almost entirely from fossils. In the Bakony Mountains, for example, the mainly Pliensbachian occurrence of this rock has yielded hundreds of specimens representing around 40 species of gastropods; from another Sinemurian site on the Nagy-Teke Hill in the Gerecse Mountains, a similarly large and even more diverse fauna has also been collected. This fauna is of particular interest because it is close in composition to another locality—the type locality of the Hierlatz Limestone—in the Northern Calcareous Alps. Gastropods with an apertural slit (forms related to *Pleurotomaria* in a broad sense), which is reminiscent of Paleozoic faunas, are quite common here—alongside high, conical gastropods related to the living genus *Trochus*. Overall, the most conspicuous faunal element is *Discohelix,* a typical genus found in the Mediterranean Jurassic and represented by numerous species. Among forms with mainly flat and disk-shaped tests, some are also pentagonal in shape, and are separated into the distinct genus *Pentagonodiscus*.

Another gastropod-rich rock type of this age formed in fissures on seamounts (topographic highs on the seafloor). From these more or less horizontal fissures very well-preserved gastropods have been collected. A fissure at a Bajocian site discovered in the 1960s in the vicinity of Bakonybél has yielded more than 1,000 specimens, classified into more than 50 species. Other localities of similar age are known from around the Mediterranean Sea (Southern Alps, Central Apennines, and Sicily) and have yielded similar gastropod faunas. Indeed, some of the faunal elements in the Bakony sites were unknown before they were collected here, and so their generic or species names refer to the site or to the region in which they were found (*Trochotomaria somhegyensis, Zircia zircensis,* and *Bakonyia planapex*). Many of these fossils were later recorded outside Hungary as well.

The younger Jurassic of the Transdanubian Range is very poor in gastropods as compared to the Sinemurian

1
2
3
4
5
6
7
8
9
10
11
12
13
14
15
16

Jurassic bivalves and gastropods (original size). (1, 2) *Pleurotomaria* sp. Representatives of this common Jurassic gastropod genus are often collected from the Transdanubian Central Range and from the Lower Jurassic of the Mecsek Mountains. *Pleurotomaria* belongs to the group of slit-bearing gastropods, which are so named because of a longitudinal cut on the last whorl. Today these gastropods are restricted to calm environments in the deep sea and possess common ornamental features, such as longitudinal ribs and fine nodules (Kárász). (3) *Ptychomphalus rotellaeformis* (Dunker). This relatively rare gastropod occasionally occurs in barren beds in the Mecsek Coal Formation. It is also a member of the gastropod lineage that possesses an apertural slit (Pécsbánya). (4) *Discohelix* sp. Hundreds of specimens belonging to this genus, an index fossil in the Mediterranean region, have been collected from Jurassic rocks in the Transdanubian Central Range (Lókút). (5) *Tretospira angulata* (Dunker). The most conspicuous gastropod from the Mecsek Coal Formation. The outline of the shell has a break: above it, the relief is plain and leaning outwards, and below it, it is bulging. The break forms an edge. Outside the Mecsek Mountains the species is known only from the Hettangian deposits of northeastern France, Luxembourg, and northwestern Germany (Pécsbánya). (6, 10) *Cardinia* sp. The genus *Cardinia* appeared in the Late Triassic, and by the Early Jurassic had become a very common member of the bivalve fauna. Remains of these taxa can be found worldwide. The outline of the shell varies considerably, from triangular to rectangular to oval. Because of this high variability, in the past 200 years it was placed into dozens of different species; however, the number of denominated species is much higher than necessary. (6) *C. quadrata* Agassiz. This species is known from those layers of the Mecsek Coal Formation that were deposited under normal salinity or close to it. It differs from *C. listeri* because it has a more rectangular outline (Pécsbánya). (10) *C. listeri* (Sowerby). The species *listeri,* abundant in the "Gryphaea marl" in the Mecsek, is one of the earliest described species and also one of the few that is distinct and distinguishable from others (Hosszúhetény). (7) *Bakevellia waltoni* (Lycett). This bivalve has a very unequal, triangular shape, with the left valve slightly more convex then the right. This is the most common fossil from the Gervillea Beds, often explored during coal mining (Pécsbánya). (8) *Eomiodon menkei* (Dunker). This is the most common bivalve in the Mecsek Coal Formation and has a medium-sized shell—differently elongated, rounded, or trapezoidal in shape. The species belongs to Neomiodontidae, a family that had worldwide distribution during the Jurassic and Cretaceous and was typical of brackish environments (Pécsbánya). (9) *Weyla hungarica* (Jekelius). *Weyla* is one of the most often cited Jurassic bivalve genera. The reason for its popularity is its peculiar paleobiogeographical distribution, first analyzed by Argentine researchers Susana Damborena and Miguel Manceñido in a paper now considered a milestone in Jurassic bivalve paleobiogeography. During the Hettangian and Sinemurian, *Weyla* was restricted to the eastern part of the ancient Pacific Ocean; therefore, remains of these bivalves are emblematic and well-known fossils from the Rocky Mountains and the Andes. Subsequently, during the Pliensbachian, the area occupied by this bivalve increased and it reached Europe, an invasion made possible by the Hispanic Corridor, an ancient seaway that connected the former Pacific Ocean and the western edge of the Tethys. However, this bivalve never become abundant in Europe and the number of known occurrences is low. One that is known is in the Southern Carpathians, where Erich Jekelius originally described *W. hungarica* using the generic name *Janira.* The single figured specimen remains the only one known, and is the type specimen of this species (Cristian). (11) *Modiolus hillanus* Sowerby. *Modiolus* first appeared in the Devonian and by the Jurassic was already a living fossil. It still exists as a peculiar faunal element of shallow marine bivalve assemblages (Pécsbánya). (12) *Bositra buchii* (Roemer). Remains of *Bositra* can be found almost everywhere Middle Jurassic sediments are represented by deeper-water deposits, and sometimes in rock-forming quantities. In spite of this, the interpretation of its mode of life—as is true of all the other "paper pectens," or "flat scallops"—is one of the most difficult issues in Paleo-Mesozoic facies analyses. Once it was supposed that they belonged to the pseudoplankton, or they spent all of their life floating and swimming in the pelagic water masses. Today, it is widely accepted that they lived on the seafloor. The specimen was collected by János Böckh in the vicinity of Szentgál. (13) *Unio vizeri* (Nagy)—an early representative of the genus *Unio,* which still exists. On the basis of the distribution of the recent *Unio* species, it is very likely that the relatively "thick" *U. vizeri* inhabited slow rivers or lakes. The appearance of the genus is unique in the European Early Jurassic bivalve assemblages, which are nearly exclusively composed of marine elements (Pécsbánya). (14, 15) *Gryphaea mucculochi* Sowerby. *Gryphaea* is aptly called "the *Drosophila* of paleontology" because there is hardly any other extinct form to which so much attention has been paid. (*Drosophila* is the fruit fly widely used in experiments in order to demonstrate genetic processes.) *G. mucculochi* represents the second member of the classical evolutionary line, and occurs in rock-forming quantity in the Lower Jurassic Vasas ("Gryphaea") Marl Formation in the Mecsek Mountains (Nagymányok). (16) *Parainoceramus fuscus* Quenstedt. Not counting the representatives of the genus *Bositra,* this is nearly the only bivalve in the Middle Jurassic of the Transdanubian Range that occurs almost everywhere; although it is never abundant, it is not rare. This beautiful specimen was collected from the Bakonycsernye Tűzköves Ravine, and it was Miksa Hantken who intended to figure it in a lithograph. Finally, the figure—as the only non-cephalopod macrofosssil from the Aalenian beds—appeared on the first plate of the monograph by Gyula Prinz.

János Szabó, a specialist on Jurassic gastropods and the former head of the Department of Paleontology at the Hungarian Natural History Museum, saying goodbye to colleagues and the former site of the department. For almost 200 years the Hungarian Natural History Museum was located within the landmark building of the Hungarian National Museum. It moved to a new site in 1996.

and Pliensbachian. The only exception to this is the Bajocian stage.

📖 J. Szabó 1979, 1980, 1981, 1982, 1983

Jurassic Bivalves in the Transdanubian Range

Although fossil bivalves are found in nearly every stage, they are typically not in abundance. Exceptions to this include some Toarcian and Middle Jurassic formations (the "filament-bearing" rocks) that contain numerous fragments of thin-shelled bivalves in rock-forming quantities, generally thought to belong to the genus *Bositra*. This has been confirmed using thin-sections, and this rock type tends to give off a clanking sound when hammered and splits into thin sheets.

One of the oldest (perhaps the oldest) and most mysterious of the bivalve faunas from the Transdanubian Range was reported by János Böckh from a series of boulders cropping out in the middle of a forest in the vicinity of Szentgál. In these boulders, he recognized taxa such as *Gryphaea* and *Cardinia*, which are otherwise characteristic of the Northwestern European Lower Jurassic. Oddly, these fossils have never been found again, but the samples that Böckh collected can still be examined; they contain poorly preserved specimens of some long-ranged taxa. The surrounding rock, however, differs from the typical development of the so-called Dachstein-type Liassic—also known as the Kardosrét Limestone Formation, the oldest member of the Jurassic in this region. Geological mapping of this area also shows that Triassic, Jurassic, and also Cretaceous formations—although sometimes difficult to distinguish—can be found close together. Rediscovery of Böckh's site, which yielded a fauna of special paleobiogeographic importance, is an exciting task for the future.

Overall, however, the most diverse bivalve assemblage from the Transdanubian Range is found in the Hierlatz Limestone. Among more than 30 species, those that lived on the surface of the sediment—either as free swimmers or attached to the substrate by byssal threads—are dominant. Most common of all are pectinids, mainly the delicately ornamented genus *Praechlamys* and also taxa related to *Lima*. Of particular interest is the observation that infaunal elements, quite common in the Alpine type locality, are much rarer in Hungarian collections. Finally, it is worth mentioning the mass occurrences of the bivalve *Caenodiotis janus* at some Pliensbachian localities. This peculiar form was previously documented only from eastern side of the Southern Alps and from the Central Apennines outside the Transdanubian Range. Fossil occurrences show that these regions were once geographically close to one another.

Bositra beds (called the *Posidonia* Beds in earlier literature) were first reported from the Southern Bakony Mountains by János Böckh, who figured a well-preserved specimen of "*P. alpina*" (*Bositra buchii*). Because of this occurrence Böckh correctly concluded that the age of these fossil-bearing rocks is Middle Jurassic. Subsequently, Elemér Vadász came to a different conclusion, assuming a stratigraphic gap between the Lower and Upper Jurassic formations. Vadász published a paper in the prestigious Balaton monograph series and dedicated a long discussion to *Posidonia*-related species and to a rebuttal of Böckh's ideas about the age of these rocks. His position later turned out to be incorrect.

Besides *Bositra, Inoceramus*-like bivalves are characteristic of the Middle Jurassic, often appearing in monospecific occurrences. This group of bivalves includes generally small and smooth taxa that are very different from their younger Cretaceous counterparts. Often, because these bivalves are preserved as internal molds in nodular facies, the inner features of their shells cannot be studied and thus in many cases their generic identification remains uncertain. Bivalve faunas of this age are more diverse in formations deposited within fissures in seamounts, as is also the case for gastropods. From the Bakonybél locality, for example, 80 specimens representing 15 species have been collected—more than a third of them attributable to the genus *Isoarca,* which was a widespread taxon in the Middle and Late Jurassic but which has an uncertain systematic position and an unknown mode of life.

In the Upper Jurassic bivalves became more varied again but their fossils are nowhere common. Over the last several decades only about 200 specimens have been

A happy moment. The 80-year-old Barnabás Géczy receives a present of an Aalenian ammonite from his students. This celebration was part of a scientific festival organized at Eötvös Loránd University. Professor Géczy led the Department of Paleontology for a long time and wrote several textbooks and handbooks on paleozoology and paleobotany. His main research was based around the study of Jurassic ammonites, and his 1966–1967 monograph on the Lower Jurassic cephalopods of Bakonycsernye comprises two thick volumes and 637 pages.

gathered from a dozen fossil sites in the Transdanubian Range, a very small number compared to the thousands of ammonites which are known from these formations. All in all, this fauna comprises about 20 species, the most marked of which belong to the extinct genus *Rhynchomytilus*. Species of *Cuspidaria* (called *Neaera* in older literature) are also frequent, as are their extant relatives in modern deep-sea bivalve faunas. *Placunopsis? tatrica*, which is often reported from the Mediterranean realm, is another common Upper Jurassic bivalve, but it is badly in need of confirmation.

📖 Szente 1996, 2003

Cephalopods

Cephalopods are the most characteristic Jurassic fossils from the Carpathian region. Within this large group, of special interest are the extinct ammonites. In certain beds, extinct belemnites are common, as are fossils of other kinds of cephalopods with internal skeletons. Jurassic nautiloids, forerunners of the extant *Nautilus*, are nowhere abundant but are found in all kinds of pelagic sediments.

If Jurassic, Then Ammonite

In Jurassic seas, ammonites, which are cephalopods with outer shells, were abundant all over the world. This group nearly went extinct around the Triassic-Jurassic boundary when an important mass extinction occurred, but it managed to survive and radiate again in the Jurassic when the group was most diverse. Hundreds of genera represented by thousands of species have been distinguished from the Jurassic, although it is hard to give an exact number, as this depends on the taxonomic philosophy of the researchers. Although fossils are tangible and concrete, paleontology is always subjective to some degree.

Jurassic Ammonite Workers

The Jurassic of the Carpathian Basin is relatively rich in ammonites and so the earliest studies on these fossils were done by pioneer paleontologists. One of the first was the German worker Melchior Neumayr, who focused mainly on Upper Jurassic ammonites collected from the Eastern Carpathians. Neumayr wrote one of the first paleobiogeographical papers, today regarded as a classic, recognizing that the European ammonite faunas represent two different types. One of these faunal types can be found around the Mediterranean Sea, representing the former Tethyan realm, and is rich in phylloceratids and lytoceratids. The

András Galácz, professor in the Department of Paleontology at Eötvös Loránd University and specialist on Middle Jurassic ammonites. His monograph on the Bathonian ammonites from Gyenespuszta (Bakony Mountains) was published as part of the Geologica Hungarica series, and his popular pocket book on living fossils is a must read for anyone interested in paleontology. Galácz also wrote a handbook on different fossil groups, co-authored with Miklós Monostori, which is widely used in university settings.

1
2
3
4
5
6
7
8

other faunal type, Northern European, is poor in *Phylloceras* and *Lytoceras*-related ammonites and represents the Boreal realm. Another German-speaking paleontologist, the Austrian Alfred Till, was the first to describe the Callovian ammonites from Villány Hill.

Over the years, many of the directors of the Hungarian Geological Institute continued with work on Jurassic ammonites—a nice tradition. Miksa Hantken, János Böckh, Lajos Lóczy Jr., and Gyula Vígh (1889–1958) all studied ammonites from Bakonycsernye, the Mecsek Mountains, the Villány Hills, and the Gerecse Mountains. Jenő Noszky Jr., carried out large collection campaigns in the Bakony Mountains but published very little, whereas Gyula Prinz, the author of the first monograph on the Bakonycsernye fossils, was not even active as an ammonite specialist for much of his career. Later Prinz focused on geography and paleogeography, developing his Tisia theory and ideas about the formation of the Carpathian Arc, which determined the way people thought about the formation of this basin for many years. Lajos Kovács's paper on Lower Jurassic ammonites from the Bakony Mountains received much international recognition; Gusztáv Vígh (1920–1984), following in the footsteps of his father, published a series of papers on Upper Jurassic ammonites from the Bakony and Gerecse Mountains as well as from Tata. In the last two decades Christian Meister, a curator at the Museum d'histoire naturelle in Geneva, and Zoltán Kovács, a lecturer at the Liszt Academy of Music, contributed significantly to the knowledge of Lower Jurassic ammonites of the Transdanubian Range. Lower and Middle Jurassic ammonites of the Northern Carpathians have been recently intensively studied by Jan Schlögl of the Comenius University in Bratislava.

The Ammonite Clock: Better than Isotope Dating

From a stratigraphic point of view, ammonites are the most important fossils from marine deposits. Because of the fast evolutionary rate seen in this group, subsequent taxa (genera and species) follow each other rapidly and so it is possible to use these fossils to subdivide the Jurassic into formations. For example, in the Transdanubian Range 8 of the 11 Jurassic stages can be documented using ammonites; within these stages, 42 ammonite zones have been distinguished. If we divide the duration of the whole Jurassic with the number of ammonite zones, then each lasts for a little more than 1,000,000 years, and in certain cases even shorter subzones can be distinguished down to around 100,000 years. This means that Jurassic ammonites can be used with precision to date rocks to the region of 150,000–200,000 years—very accurate for this age of sediments. For comparison, radiometric dating, the technique normally used to date rocks based on comparisons of observed abundance of a naturally occurring radioactive isotope and its decay products, has a margin of error of a few percent—a few million years in Jurassic-aged rocks.

Big and Small Forms: Males and Females?

Often in the same bed, or beds close to each other, similar ammonites are found together—one larger, slightly more involute, with well-developed ornamentation and a regular adult body chamber. The other is often smaller, more evolute and with a similar, but often less complex, ornamentation and an irregular adult body chamber and characteristic aperture. In this sense, irregular means eccentric in coiling or contraction on the final part of the body chamber, while the aperture may have lateral lappets (like *Otoites*) or may have a cap-like prolongation (like *Sphaeroceras*). However, the inner whorls of the two ammonite types are nearly identical.

In these cases, larger ammonite forms are called "macroconchs" (M) and the smaller forms are called "microconchs" (m). It is thought that the macroconch forms were the females and the microconchs were the males, because the females would have needed to be larger for egg

Lower and Middle Jurassic ammonites from the Villány Mountains (all original size except for [1] and [8]). (1) *Villania galaczi* Géczy. *Villania* is a fossil genus that was documented for the first time from Hungary, and later became much more widely known. Alfred Till recognized that this ammonite was new to science in 1911; for decades it was thought to be endemic to the Pliensbachian of the Carpathians. Recently, however, it was also found in Turkey and the United Kingdom. *Villania galaczi* was described by Professor Géczy in honor of his colleague and student András Galácz (0.5× magnification). (2, 3) *Oecotraustes (Thraxites) thrax* Stephanov. This strangely shaped eccentric ammonite was collected from Bathonian beds below the famous Villány ammonite bed. This taxon has a test built up from smooth, regularly coiled inner whorls and a body chamber that is consistently thick and bears ventro-lateral nodules on a short section. (4, 5) *Prohecticoceras* spp. These ammonites are common elements in the Bathonian beds at Villány, and their characteristic shape and lateral ornamentation, with open V-shaped ribs, resemble "pilot biscuits," popular cookies in Hungary. In (4) *P. retrocostatum* (De Grossouvre), an important zonal index fossil, is figured; (5) features *P. angulicostatum,* a form originally described from Villány by Lajos Lóczy Jr. (6, 7) *Bomburites globuliforme* (Gemmellaro). This ammonite is a rare constituent in the Callovian at Villány. In the early phases of its ontogenetic development, the whorl height and width grow evenly and cover the inner whorls to create an inflated appearance; the final stage of the adult body chamber is particularly narrow. (8) *Oxycerites tilli* (Lóczy). This ammonite is one of the most spectacular elements in the Callovian fauna. The external part of its high whorls is ornamented by shallow, very feebly curved ribs that can be seen only if illuminated with low-angle light (0.5× magnification).

1
2
3
4
5
6
7
8
9
10
11

production. However, the story is not so simple, because in some cases the stratigraphic ranges of the two forms are not quite the same. This could either mean that the paleontological record is incomplete or that two supposed pairs of ammonites are in fact different species.

Phylloceras, Lytoceras, and Related Ammonites

Jurassic ammonites are classified into three orders: Phylloceratoida, Lytoceratoida, and Ammonitoidea. The name of the first order incorporates the Greek word *phyllo,* which means "leaf," in reference to the shapes of the complex suture lines in these ammonites. The whole of this group are also characterized by their involute shells, in which the last whorls cover earlier ones nearly completely. This type of coiling gives a rather special shape to these ammonites, which can also be flat or inflated. The genus *Phylloceras* can be smooth or ornamented with very fine growth lines, whereas *Hantkeniceras* is a phylloceratid with a slightly rectangular cross section and strong connection to Hungary; it was named by Lajos Kovács, who worked on Lower Jurassic ammonites from the Bakony Mountains, in honor of the great geology professor Miksa Hantken.

Lytoceratids are usually moderately big, and sometimes large, ammonites although the Middle Jurassic *Nannolytoceras* is rather small. All *Lytoceras*-related ammonites are also referred to as "loosely coiled" as their subsequent whorls do not cover one another, just touch each other. In general, phylloceratids and lytocertids had a very slow rate of evolution and as a result are not all that useful for biostratigraphy. They are nevertheless important because they comprise the bulk (up to two-thirds or more) of Jurassic ammonite faunas from the Mediterranean (Tethyan) realm.

Order Ammonitoidea

This order comprises ammonites of very different shapes, sizes, and ornamentation. Some of them are no larger than a coin in their adult stages, whereas others can reach the size of a truck tire and subsequent whorls can cover earlier ones at markedly different rates. Because of these differences in growth, some ammonitoids are evolute and some are involute, and their varied ornamentation can comprise ribs or tubercles. In these ammonites, ribs can project forward or backward, and can be simple or branching. It is also common for ornamentation to change during ontogeny, with earlier whorls having simple ornamentation and later ones becoming more complex. The adult body chamber can also often be differently ornamented, as can the phragmocone, and many adult microconchs can have a special aperture with lappets. At the largest sizes, the big macroconchs have a simple aperture without lappets.

Jurassic ammonites were numerous and diverse: hundreds and thousands of species and subspecies have been

Lower Jurassic ammonites from the Transdanubian Range and the Mecsek Mountains (all original size). (1) *Hildoceras* sp.—midsized ammonites with a wide umbilicus and square cross sections. These forms also have two grooves and three keels on the ventral side and ornamentation that consists of crescent-shaped ribs above a lateral groove. The inner part of their lateral sides is smooth. *Hildoceras* is a frequent and very characteristic Toarcian ammonite (Gerecse Mountains). (2, 3) *Frechiella* sp.—small-sized ammonites with a narrow umbilicus and inflated whorls. The venter bears two, often weak, grooves and has three keels; the latter side is smooth or weakly ornamented by simple ribs. The suture line of this ammonite is simple and resembles more closely the simple lines seen in taxa like the Triassic *Ceratites* than it does the sometimes extremely complicated sutures of many Jurassic ammonites. *Frechiella* was named for the geologist F. Frech by his student Gyula Prinz, and is an uncommon but often typical element in the Toarcian beds of the Transdanubian Range (Gerecse Mountains). (4) *Juraphyllites* sp.—midsized, moderately evolute ammonites with rectangular cross sections, flat sides, and rounded ventral regions. Characteristic features include prominent prorsiradiate ribbing on the final and ventral parts of the adult body chamber; younger specimens are smooth. *Juraphyllites* is a common faunal element in Sinemurian and Pliensbachian assemblages (Bakony Mountains). (5, 6) *Mercaticeras* sp. This genus is related to *Hildoceras,* but has grooves and keels on the ventral part of the inner whorls that become smoother during later ontogenetic stages. The ribs of *Mercaticeras* are also strong and flexed; this ammonite is a typical Toarcian genus (Gerecse Mountains). (7) *Echioceras* sp.—a very evolute ammonite with a rounded quadrangular cross section and a single ventral keel. The ribs in this taxon are strong, straight, and simple. The figured specimen has been pyritized so it is metallic in appearance. (8) *Pleuroceras* sp.—midsized evolute forms with rectangular cross sections and a serrated keel on the venter. Ribbings in this genus consist of very strong radial ribs that bear forward-facing ventrolateral nodules. *Pleuroceras* is characteristic to the Pliensbachian and, in certain levels, it is very frequent in Liassic formations in the Mecsek Mountains. (9) *Grammoceras* sp.—related to *Hildoceras.* The umbilical edge of this ammonite is flat and the falcoid ribs are dense. *Grammoceras* is an important element of late Toarcian assemblages (Mecsek Mountains). (10) ?*Microderoceras* sp.—evolute ammonites with rectangular cross sections and strong, simple ribs with two rows of tubercles. This genus occurs in Sinemurian and Pliensbachian beds. (11) *Amaltheus* sp.—small- and medium-sized ammonites with a shape and cross section that resembles *Oxynocticeras.* Very characteristic features include a prominent serrated keel and slightly sigmoid (S-shaped) ribs that have strong lateral nodules on inner whorls. This especially attractive ammonite was named after the goat in Greek mythology that fed the young Zeus in a cave on Mount Ida, Crete. The genus *Amaltheus* is a rare element in the Pliensbachian in the Transdanubian Range, but can be relatively common in some levels at the same age in the Mecsek Mountains (Mecsek Mountains).

1
2
3
M 66 664
4
5
6
7
8

described. These forms also evolved very rapidly, as evidenced by the changing ornamentation of their shells. Ammonite biostratigraphy is based on the subsequent appearance of different forms, genera, and species—most of which can be determined at least to genus and are characteristic of the age of the beds from which they were collected.

Operculum or Jaw Apparatus?

The paired aptychi and the unpaired (much rarer) anaptychus, which resembles a bivalve's valve, are the hard plates of ammonites. These plates are made of calcite and some organic material and often fall out of the shell after death and embed into the sediment separately. They are rarely preserved in place and as a result aptychi and ammonites are often found apart and are classified into distinct groups. Aptychi, for example, are distinguished based on their constituent material and surface ornamentation—those with granular, lamellar, or punctuated surfaces are called *Granulaptychus, Lamellaptychus,* and *Laevaptychus,* respectively.

Because of their shape, aptychi were long thought to be operculi, doors that hinge on the aperture of the shell, an idea supported by the fact that these paired organs cover more or less the entire aperture of the ammonite. Indeed, the extant *Nautilus* possesses such a caplike cover on its shell, similar to an operculum. However, the preservation of some exceptional fossils has shown that at least some aptychi functioned instead as part of the complex jaw apparatus of ammonites. Occasionally, as is true of some German Upper Jurassic ammonites, aptychi have been preserved in life position inside the shell and besides their basic calcitic structure a counterpart of mainly organic material can be found. Thus, these elements are thought to have functioned as both upper and lower jaws—or even as operculum and jaw apparatus at the same time. Calcite is harder to dissolve and is more resistant in seawater than is an aragonitic shell, and so we might expect to find aptychi preserved in deeper water sediments when ammonite shells have dissolved. It also seems to have been the case that some closely related ammonites had extremely similar, practically indistinguishable, aptychi. Despite this they do have some limited value in biostratigraphy, although to date no detailed research has been done on the Jurassic aptychi of the Carpathian Basin. The known Cretaceous record of these structures has been studied in more detail.

The Young Researcher and the Senile Ammonite

In 1908 the young Elemér Vadász, who was later to become a highly regarded professor, found and described a supposedly interesting ammonite under the name *Lytoceras evolutum.* The specimen was rather large—41 centimeters in diameter—and its last whorls were detached, so Vadász regarded it as an uncoiled ammonite with a free body chamber. Uncoiling of the last whorl(s) is typical for some Cretaceous ammonites, but never occurs in lytoceratids. It seems that Vadász had simply overlooked the fact that his specimen was rather dissolved and this had led to an uncoiled appearance. Although he mentions that the ammonite was "corroded" over its plane of symmetry, he explained the "free" body chamber as a result of aging in this "individual" ammonite, regarding the last uncoiled whorl as indicative of a senile growth stage. Although at the time the author was rather young, an informed reader

Middle Jurassic ammonites from the Bakony Mountains (0.8× magnification). (1) *Spiroceras orbigny* (Baugier and Sauzé)—This is a midsized, initially loosely coiled but subsequently nearly straight, heteromorph ammonite with a cross section that is close to circular and ornamentation that consists of simple ribs and two rows of tubercles. *Spiroceras* is a typical Bajocian species; in Hungary some nice specimens have been collected from the Transdanubian Range and from the Mecsek Mountains. (2) *Tmetoceras scissum* (Benecke)—a midsized evolute ammonite with an oval cross section and a deep ventral furrow. Ribbing in this species consists of simple ribs ending in tubercle-like swellings on the ventrolateral edge. *Tmetoceras* is an important Aalenian zonal index fossil. (3) *Ludwigia* sp.—a midsized ammonite with a rooflike ventral region. Its sides are flat and its bending ribs are strong, although some of them are intercalated or branching. The genus is characteristic for the Aalenian, and many specimens have been found—especially at Bakonycsernye. (4) *Holcophylloceras* sp.—a medium- to large-sized form with a high oval cross section and a narrow, point-like umbilicus. Another characteristic of this genus is strong, broken constrictions that are visible on internal molds. *Holcophylloceras* is one of the most common ammonites in the Hungarian Jurassic. (5) *Erycites* sp. Similar to the frequent *Hammatoceras,* this midsized ammonite has no tubercles, flat sides, and a blunt, low venter. The end of the adult body chamber is also slightly constricted and becomes smooth. *Erycites* was named after the Sicilian Monte Erice, and in Hungary it is a frequent element in Aalenian faunas. (6) *Parkinsonia parkinsoni* (Sowerby). This genus unites midsized ammonites with relatively high cross sections and those with biplicate ribs that are projected forward on the upper third of the flank. Ribs are interrupted on the middle of the rounded venter and different species occur in Bajocian and Bathonian beds. The species figured here is a zonal index fossil for the uppermost Bajocian. (7, 8) *Labyrinthoceras* sp. This is a characteristically eccentric ammonite because it is not spirally coiled like most of its counterparts and it has a constricted adult body chamber. The umbilicus is narrow and the whorls are ribbed and inflated. The name *Labyrinthoceras* refers to the extremely complicated suture line in this genus. A microconch specimen is figured here with characteristic lateral lappets. These macroconchs can sometimes be larger than a fist.

1

2

3

4

5

6

7

may take the opinion that in this case it was not the ammonite that was senile!

📖 M. E. Vadász 1908

Nautiloids

In addition to ammonites, fossil nautiloids, distant allies of the extant *Nautilus* and *Allonautilus* can also be collected. Their remains are never frequent, but can be found in all kinds of pelagic formations (Rosso ammonitico). It is quite common that the phragmocone, collapsed into pieces along its simple, arched chambered walls, and the body chamber are found separately. However, there are few published articles about Jurassic nautiloids. In 1906, Gyula Prinz published on some specimens, including a few that were new to science, that were collected from Bakonycsernye and from the Gerecse Mountains. One of these new forms (*Nautilus kochi*) was named after Antal Koch, a professor in Kolozsvár; another (*Nautilus semsey*) was named after the great benefactor of Hungarian science Andor Semsey (1833–1923).

Nautiloids had a low evolutionary rate. Indeed, the extant *Nautilus* is considered a living fossil since it basically has not changed since the Oligocene. Jurassic nautiloids, although also very similar in appearance to recent ones, are generally classified into the genus *Cenoceras*. One new subspecies of *Cenoceras* was described by B. Géczy from Tűzköves Hill (Szentgál) and given the name *C. truncatus vadaszi*.

The calcified jaw apparatus of nautiloids, the so-called rhyncholit, is a more or less arrowlike structure situated within the beak of the animal. Since it is found very rarely in situ in fossils (that is, inside the shell) these structures are most often collected and treated separately, like the aptychi of ammonites. Rhyncholits tend to be about one centimeter in length, are slightly curved, and possess a sharp ridge along their midline. They are found only occasionally, but are relatively common, for example, in the Bathonian marl in the Mecsek Mountains.

📖 J. Prinz 1906; Géczy 1961b

Cephalopods with Internal Skeleton

An Attractive Fossil: Atractites

In the Lower Jurassic formations of the Transdanubian Range, the cylindrical remains of a cephalopod with an internal skeleton called *Atractites* are relatively common. In most cases, only the phragmocone of these animals is found and these slightly conical fossils, sometimes as thick as an arm, resemble a drill core. Their septa are rare and concave, and in this genus prolongation of the phragmocone, the telum, reached about one meter (although it is rarely preserved). The entire, full-grown animal could reach 4–6 meters and, as its name suggests, *Atractites* is a pleasing looking fossil. In earlier paleontological literature, this genus was often named *Ausseites*. Because the Lower Jurassic Rosso ammonitico–type limestone was and is widely used as building stone, remains of these fossils are common in floors, stairs, and footings of Hungarian public buildings.

Belemnites

Belemnites, belonging to the order Belemnitoidea, are geologically the most important group of cephalopods that possess an inner shell, although normally only the bullet-like back part of their shells (the guard, or rostrum) is commonly found as fossils. Belemnite systematics is based on morphological features, including symmetry,

Lower and Middle Jurassic Ammonites from the Mecsek and Villány Mountains (original size). (1) *Leioceras* sp.—a midsized moderately involute ammonite with a strong ventral keel. The inner part of the largely flat and smooth flanks is ornamented by bending, folding ribs. *Leioceras* is common in Aalenian faunas (Mecsek Mountains). (2) *Hildaites* sp.—a midsized evolute ammonite with flat flanks and a high, rectangular cross section. On its ventral region there are two ridges and three furrows; ribbing consists of strong, simple, and outermost falcoid ribs. This genus was named after Saint Hilda (c. 614–680), who lived in the abbey at Whitby—which was so heavily infested with snakes that the nuns were afraid to leave their cells even to urinate. According to the legend, Saint Hilda turned these snakes into stone—specimens of the serpenticone ammonite *Dactylioceras ammonites*. *Hildaites,* together with its close ally *Hildoceras,* are characteristic elements of Toarcian ammonite assemblages. The fossil figured here is a completely flattened specimen (Mecsek Mountains). (3) *Strenoceras niortense* (d'Orbigny)—evolute ammonites with very clear, sharp, straight, and mainly simple ribs and two rows of tubercles. *Strenoceras* is a characteristic element in Bajocian faunas and is of value for zonal indices. (4, 5) *Oxycerites* sp.—a medium- to large-sized ammonite with a very high cross section. The umbilical wall of this genus bends gently and is unexpressed, and the venter is very sharp. *Oxycerites* is characteristic of the Callovian in the Villány Mountains and has very weak and blunt ribs that smooth the adult body chamber. (Villány Mountains) (6) *Stephanoceras* sp.—a midsized ammonite with a wide umbilicus and a cross section that is close to circular. The ribbing in this genus is comprised of short primaries that end in small nodules and which give rise to two, or occasionally three, secondary ribs which cross the rounded venter without interruption. The microconchs of these ammonites have lateral lappets. *Stephanoceras* is typical and common in Bajocian ammonite faunas—from Hungary, beautiful specimens have been collected from the Bakony, Gerecse, and Mecsek Mountains (Mecsek Mountains). (7) *Reineckeia* sp.—a midsized ammonite with a rounded cross section and very strong, distant and mainly biplicate ribs. Branching positions are marked with nodules in this important Callovian genus (Villány Mountains).

1
2
3
4
5
6
7

Calliphylloceras and *Lytoceras* (0.5× magnification). These ammonites are among the most common genera from the Jurassic of the Transdanubian Range and eventually from the peri-Mediterranean region. Phylloceratids are very involute (the later whorls cover the inner ones nearly entirely), and lytoceratids are very evolute (the subsequent whorls just touch each other). Representatives of the order Ammonitoidea can be characterized by an intermediate type of coiling (the later whorls cover the earlier to a different extent).

size, and proportions as well as the shape of the conical inner hole, the alveolus. Belemnites are common fossils in rocks worldwide; sometimes their guards are washed together and so these cephalopods can be found en masse on the surface of layers. These are the so-called belemnite battlefields, where the fossils may indicate the direction of paleocurrents above the seafloor. In the Jurassic formations of the Carpathian Basin belemnite remains are generally not common, but they may appear in a number of rock types. The guard in belemnites is mostly composed of calcite and thus tends to preserve well: this explains why, in some deepwater sediments, basically aragonitic ammonite shells are dissolved but belemites often remain. For this reason, belemnites are often index fossils in rocks that lack ammonites.

The rostra of belemnites may accumulate when sedimentation rates were very slow, so only a thin succession of sediments was deposited over a long period of time. One example of such a condensed bed is the "Callovian ammonite bed" that was described by Villány: the belemnites from this unit were studied and published by A. Galácz and A. Vörös, who were able to distinguish 10 species within 4 genera (*Hastites, Belemnopsis, Hibolites,* and *Rhopaloteuthis*).

📖 Galácz & Vörös 1969

Brachiopods

In the words of Attila Vörös, "Brachiopods are the most common Jurassic fossils all over the world. In Hungary, however, in terms of their frequency and scientific importance, brachiopods are secondary to ammonites. Except in siliceous rocks, deposited in the deepest environments, these fossils are known from practically all kinds of marine deposits, and are therefore important in biostratigraphy as well as for environmental studies."

Early Studies

The French traveler and naturalist François Sulpice Beudant (1787–1850) first mentioned Carpathian Basin Jurassic brachiopods in his 1822 book. Flóris Rómer also wrote with great enthusiasm on the fun associated with collecting brachiopods in his 1860 book on the natural history

Late Jurassic ammonites from the Transdanubian Range (original size). (1) *Lemencia* sp. This is a perisphinctid genus and includes some midsized, moderately evolute ammonites with more or less oval cross sections and strongly biplicate ribs. *Lemencia* is a characteristic element of early Tithonian faunas (Mogyorós-domb, Sümeg). (2) *Gregoryceras* aff. *fouquei* (Kilian). *Gregoryceras* are midsized, moderately evolute ammonites with rectangular or trapezoidal cross sections and strong blunt ribs that bend backward on the adult stages of shells. Recognition of the genus is easy, but because of the numerous transitional forms that have been named, determination of different species is difficult. *Gregoryceras* is a very typical Oxfordian ammonite (Domoszló, Gerecse Mountains). (3) *Semiformiceras semiforme* (Oppel)—a small ammonite genus with a very narrow umbilicus and high, oval cross section. The venter of the phragmocone in this form has a serrated keel, replaced by an expressed furrow on the venter of the adult body chamber. Inner whorls are ornamented by weak periumbilical ribs and small nodules; the ornamentation fades away on the final part of the shell. The eccentric adult body chamber gives a peculiar appearance to this typical early Tithonian genus, which includes three successive zonal index fossils (Hárskút, Bakony). (4, 5) *Ptychophylloceras ptychoicum* (Quenstedt)—small to midsized ammonites that are closely related to *Phylloceras.* Very involute, the outer whorls in this genus cover the inner whorls entirely and the cross section is rectangular with rounded edges. Flanks are unornamented and the wide venter in this genus has expressed, distant wrinkles (*ptychos* means "wrinkle" in Greek). *Ptychophylloceras* is especially common in the Tithonian (Hárskút, Bakony). (6) *Volanoceras aesinense* (Meneghini)—midsized, very evolute ammonites with rectangular cross sections and with concave ventral regions. The strong umbilical tubercles and the elongate (clavate) outer tubercles are connected by simple, broad ribs. *Volanoceras* and related genera (including, for example, *Simoceras*) are important ammonites in Tithonian (mainly early Tithonian) fossils assemblages (Lókút Hill, Bakony Mountains). (7) *Haploceras verruciferum* (Zittel)—a small, smooth ammonite with a high oval cross section. Another characteristic feature is the ventral flare (*verruca* means "wart" in Latin) on the end of the adult body chamber. Possibly a microconch pair of a similar, but bigger (macroconch) form, *H. verruciferum* is an important zonal index fossil in the early Tithonian (Lókút Hill, Bakony Mountains).

Complicated suture lines (intersection between the septum and the outer shell) in a Jurassic ammonite (*Labyrinthoceras*). The septa, which were intensively folded on their periphery, strengthened the otherwise thin and relatively fragile shell from inside, making it resistant to pressure. The precise function of these often extremely folded suture lines remains unclear. Indeed, ammonites developed more and more complex suture patterns from the Paleozoic up to the end of the Jurassic. They could be a sign of more and more refined adaptations to marine life and buoyancy in the deep sea; following the Jurassic, during the Cretaceous, some ammonites developed less complex and reduced suture lines. Elements of the suture that point backwards are called lobes; those pointing towards the aperture are called saddles.

of the Bakony Mountains: "Often I took home boulders and rocks by carriage in order to chop them into pieces of hazel size in the shadow of my garden and to extract the prettiest terebratulids of different types." These words reflect not only his deep love for these fossils but also explain how to collect brachiopods: patient work to break hard rock into hazelnut-sized pieces is still the only way to get a decent brachiopod collection.

The subsequent, and brightest, period in Hungarian Jurassic brachiopod research is characterized by the work of János Böckh, who published on 34 species—12 of which were new to science—from the Liassic (Lower Jurassic) in one important 1874 paper on the geology of the Southern Bakony Mountains. Böckh also handled brachiopods from the Mecsek Mountains. These he described in an 1881 monograph that was mainly focused on ammonites. This period, however, ended with a paper by Károly Hofmann who published on the first Lower Jurassic brachiopods from the Gerecse Mountains in 1884.

Over the next 25 years, no new data were published on Hungarian Jurassic brachiopods, and much later Vadász, Koch, and Lóczy contributed to knowledge with their work on taxa of this age from the Southern Bakony Mountains, Kálvária Hill (Tata), and Villány Mountains. Later still, Gyula Vígh and his son, Gusztáv, published on a series of new brachiopod faunas from the Gerecse and Bakony Mountains.

The quest for Hungarian Jurassic brachiopods was picked up again in the early 1970s by Attila Vörös, who has published numerous papers on the systematics, biostratigraphy, and paleobiogeography of these fossils over the last 40 years. Indeed, much of the Jurassic brachiopod section in this book is based on his results. Over his long career Vörös has handled almost the entire Jurassic brachiopod fauna but has placed special emphasis on the diverse faunas of the Lower Jurassic. Brachiopods of this age (Hettangian–early Sinemurian) are now studied and have been described in detail by one of his students—later a co-worker—Alfréd Dulai.

📖 Beudant 1822; Rómer 1822; J. Böckh 1874, 1881c; Hofmann 1884; A. Vörös 1997

Best-Known Localities

Jurassic brachiopods can be found at a number of different fossil sites but there are some places, and especially some rock types, that are particularly rich in their remains. This is because brachiopods are benthic animals but they inhabit the seafloor unevenly. They prefer certain places and so their remains have accumulated in certain rock types rather than others.

In Hungary, we know about important brachiopod assemblages from a series of different localities in the Transdanubian Range. In the Bakony Mountains the best-known brachiopod fossil sites are Mogyorósdomb (Sümeg); Csárda Hill (Úrkút); Tűzköves Hill (Szentgál); Középső-Hajag, Fenyveskút, Lókút Hill, and Kericser (all Lókút); Kőris Hill and Som Hill (Bakonybél); Borzavár; Hamuháza near Tés. Rich faunas are also known from the

Internal mold of a nautiloid (*Cenoceras*) from the upper Liassic in the Bakony Mountains. The slightly curved suture lines in the chambered section (phragmocone) can be seen in lateral view, the ventral view clearly shows the position of the siphuncle, and the concave septum is pierced by a duct in the middle (0.5× magnification).

Orientated *Atractites* remains can be seen in the basal blocks used to build the Museum of Applied Arts in Budapest. These rocks are from the Gerecse Mountains, possibly from the huge quarries near Tardos, and contain specimens up to 40 centimeters in length.

western flank of the Vértes Mountains (Csóka Hill, Mór), from Tata, from many Gerecse Mountain localities (including Asszony Hill, Nagysomlyó, Hosszúvontató, Teke Hill, and Szél Hill), as well as from the northwestern edge of the Pilis Mountains (Velka Skala).

Some of these important Transdanubian brachiopod sites were immortalized by Attila Vörös, who introduced the following generic names: *Bakonyithyris, Papodina, Kericserella,* and *Lokutella.* In Southern Hungary, Jurassic brachiopod sites are known from the Eastern Mecsek Mountains (sites include Szászvár, Váralja, Márévár Valley, Hidasi Valley, Vasas, Pécsvárad, Zengővárkony, Pusztakisfalu, and Ófalu) as well as from the Villány Mountains (Templom Hill, Szársomlyó).

📖 A. Vörös 1997, 2009

Changing Diversity through the Jurassic

Although brachiopods suffered greatly as a result of the crisis at the Triassic-Jurassic boundary, faunas recovered surprisingly rapidly: Hungarian brachiopod faunal diversity gradually increased in diversity from the Hettangian up until the Pliensbachian as the increasingly fragmented Jurassic seafloor provided more and more favorable habitats for benthic communities. The Toarcian anoxic event then led to a drastic decrease in the numbers of benthic faunal elements, including brachiopods. Because these invertebrates filter organic particles from seawater they were unable to tolerate bottom waters with low oxygen content, as was the case during the Toarcian. The Aalenian that followed was, thus, practically free of brachiopods; later, during the Bajocian, tectonic movements resumed and resulted in submarine slopes with scarp breccias that

A guard of a Lower Jurassic belemnite from the Mecsek Mountains (Ófalu) (0.5× magnification).

A rare fossil from the Mecsek Mountains. Apart from belemnites, fossils of cephalopods with internal skeletons from the whole of the Mesozoic are rare. For some reason, remains of these animals are more frequent in the Tertiary and so a small fossil, found in 1874 by János Böckh in Toacian shales in the Réka Valley north of Pusztakisfalu, is of special interest. This find is a 35-millimeter, shiny, and slightly pyritized piece of shell that was named by István Zoltán Nagy as a new species, *Teudopsis subacuta.* This species consists of just the rachis, the arrow-like end of the proostracum (the elongated bladelike structure that is attached to the dorsal part of the septated phragmocone). 🕮 I. Z. Nagy 1958

again furnished living space for brachiopods. This resulted in a second acme for the group, although from the Bathonian onward brachiopods are again absent from most Hungarian sections. This gap in the fossil record can easily be explained by a dramatic deepening of the Jurassic ocean. Because of the considerable water depth that was achieved during this time, calcareous mud and shells (including brachiopod valves) could not accumulate and only radiolarian ooze remained on the deep seafloor. In other deep basins that lacked a hard, rocky basement brachiopods were unable to settle and so the third (Late Jurassic) peak in their diversity is likely due to resumed tectonic movements in the Tithonian. During this time, on steep edged submarine slopes and among huge fallen rocks, brachiopods flourished again.

🕮 A. Vörös & Dulai 2007

The Most Important Genera

In recent decades a large amount of brachiopods have been collected and found their way into Hungarian public collections. The published works of Attila Vörös, and most recently of Alfréd Dulai, are based on studies of 20,000 specimens. Most of these fossils were collected alongside ammonites, and so are extremely age constrained. Here, the most important brachiopod genera from the Transdanubian Range are listed stage by stage.

In the Bakony Mountains, the Hettangian is comprised of monotonous, thickly bedded, oolitic limestones of shallow marine origin. Thanks to the work of Alfréd Dulai, from these rocks we know about 15 species that make up 5 genera. Of this age, the most numerous are the terebratulids *Lobothyris* and *Zeilleria.*

The Sinemurian is comprised of deeper water sediments, including cherty and crinoideal limestones and Hierlatz-type limestones. The brachiopod fauna from this age, especially in the late Sinemurian, is extremely rich and about 80 species in nearly two dozen genera have been distinguished. The characteristic genera at this age are *Prionorhynchia, Cuneirhynchia, Salgirella, Bakonyithyris, Zeilleria,* and *Securina.*

Jurassic brachiopod workers. In the middle is Miloš Siblík, a retired member of the Geological Institute of the Czech Academy of Sciences. On the left is Alfréd Dulai, head of the Department of Paleontology of the Hungarian Natural History Museum, Budapest. On the right is Harald Lobitzer, a former member of the Geological Institute in Vienna and an honorary member of the Hungarian Geological Society. Lobitzer always initiated and supported cooperation among the geoscientists of the former Austro-Hungarian Empire.

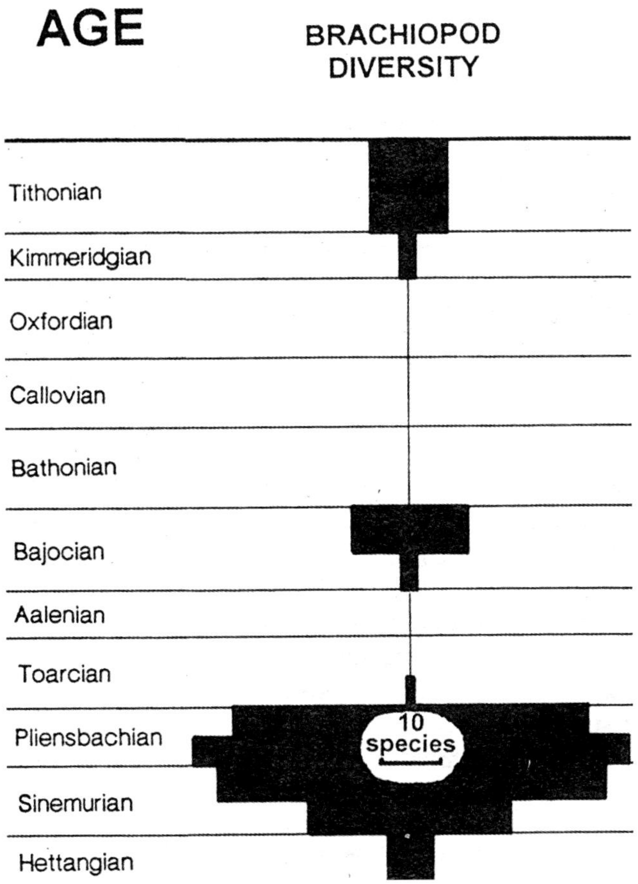

Changes in the diversity of the brachiopod fauna throughout the Jurassic in the Transdanubian Range. The subsequent rich faunas seen in this area are connected strongly to extensive tectonic phases—when tension resulted in fragmentation of the seafloor, where brachiopods were flourishing. In the muddy anoxic basins of the Toarcian and Aalenian, and in the very deep basins across the Bathonian–Oxfordian interval, brachiopod larvae were simply unable to settle and colonize. After A. Vörös 1997.

The red limestones and Hierlatz Limestones of Pliensbachian age in the Bakony Mountains have yielded the richest Jurassic brachiopod fauna known from the whole Carpathian Basin—the entire fauna consists of nearly 100 species. Among the most important genera are *Pisirhynchia, Lokutella, Apringia, Liospiriferina, Securithyris, Linguithyris,* and *Bakonyithyris. Lokutella kondai* is named after József Konda (1929–1995), a noted researcher on the Jurassic of the Transdanubian Range and once the head of the Hungarian Geological Institute.

There are practically no brachiopods known from Toarcian and Aalenian rocks and the Middle Jurassic fauna is also less diverse than its Lower Jurassic counterpart. Only about 20 species in a handful of genera have been documented from rocks of this age.

In the red nodular Tithonian limestones, as well as in the late Jurassic Hierlatz-type rocks, brachiopods are relatively common again, but the fauna still is not very diverse. Characteristic taxa include *Pygope,* which is shaped like buttocks, and the triangular *Triangope.*

📖 A. Vörös 1983, 2009; Kázmér 1998; Dulai 2002, 2003

What Is Hidden within Brachiopod Valves?

The valves of brachiopods, because of their massive hinge, do not fall apart easily after death like the valves of bivalves do. If separate valves are found as fossils then this means that strong currents, wave action, and redeposition have had a part to play. The unopened shells of dead brachiopods can also often be infilled with sediments—through the little hole where the cylindrical pedicle (a kind of stalk, an extension of the body wall that attaches the animal to hard substrates) used to be. If sedimentation is slow, then fine calcareous mud can fill the shell completely, but if it is quick then spaces are left behind. Sometimes these spaces can be filled with sparry calcite, either completely or just partially. The boundary between the lime mud and the sparry calcite was originally more or less horizontal. If these features, called paleolibella or geopetal structures, are parallel to one another and the calcite fills are in an upper position, the remains have not been reworked and the bed containing the remains is not overturned. Different orientations of paleolibella structures indicate reworking. If the mud fills are consequently in an upper position, the bed in question is overturned and most probably forms part of a recumbent fold.

The very fine mud that can enter into the closed valves of brachiopods often also preserves the lophophore, a fragile inner organ that filters nutrients from surrounding seawater. Lophophores are diagnostic features, key elements for brachiopod classification, but can never be seen from the exterior of the animal. Indeed, it is common for two brachipods to be very similar on the outside but to have extremely different lophophores. In fossils, of course, these fragile features are almost impossible to study: How can we see them if they are embedded inside the hard shell? This problem can be solved using skill and experience: a serial section of the fossil brachiopod is needed.

Echinoderms

The nice looking and well-preserved remains of Jurassic echinoderms are rather rare throughout the Carpathian region, because this geological period is represented mainly by pelagic and other rather deep marine sediments. This is true in spite of the fact that crinoidal limestones, made up from ground and sand grain–sized particles of crinoids (which are, of course, also echinoderms) are characteristic and common Jurassic rocks.

Characteristic Sinemurian brachiopods from the Bakony Mountains. (1) *Cuneirhynchia.* (2) *Zeilleria.* (3) *Linguithyris.* (4) *Prionorhynchia.* (5) *Liospiriferina.* (6) *Cirpa.* (7) *Securina.* (Close to original size.)

What Value a Sea Urchin without a Shell?

Eight of the fourteen most important extant echinoid groups appeared in the Jurassic and so this period is often called the "explosion of sea urchins."

Most often, echinoderms are preserved without their test—as internal molds or the remaining part of their shell is recrystallized, or even permineralized in most cases. Since most of the important systematic features (such as the form and configuration of the plates of the test) cannot be studied on internal molds, these fossils are often very difficult to determine. As a result, it is no surprise that the number of articles that have been published on these fossils from the Carpathian Basin is rather limited. The only relevant work is by echinoid specialist Erzsébet Szörényi (1907–1987) who described a large sea urchin (*Laticlypus giganteus,* a new species in a new genus) from the late Middle–Upper Jurassic of Pálihálás in the Bakony Mountains.

Explanatory notes to geological maps and occasionally other papers dealing with Jurassic fossils and stratigraphy may contain data on sea urchins. The Middle and Late Jurassic echinoids are usually mentioned under the names *Collyrites* and *Metaporinus.* It is likely that many of the Upper Jurassic echinoids actually belong to the genera *Tithonia, Cardiolampas,* and *Cyclolampas. Tithonia* can be small, cylindrical, or cordiform (heart-shaped); *Cardiolampas* is moderately large and cordiform; and *Cyclolampas* has a test ovate in outline.

All abovementioned echinoids belong to the large group of irregular sea urchins. The rare regular echinoids of the Upper Jurassic formations belong to the family Cidaridae. Commonly the roll-shaped test is not collected, but only the spines. These spines—although they can be thin or stout, and ornamented with serrated or thorned ribs—are very characteristic. Many of the cidarids of the Bakony Mountains may belong to genera *Rhabdocidaris* and *Cyathocidaris.*

Unlike the Upper Jurassic beds, the Lower and Middle Jurassic contain very few echinoids.

📖 Szörényi 1966

Submarine Forests

Sea lilies—their name and shape quote the flower—were common echinoderms of the Jurassic marine ecosystem. After the first Paleozoic acme, this archaic group reached its second peak in diversity in the Mesozoic. Most of them lived attached to the seafloor in some way. Their stalk was slender and long—sometimes very long—built up by tiny cylindrical or polygonal elements. On the top of the stem there was the calyx, in which most of the organs were placed. The fragile arms were attached to the calyx and were equipped with cirri, which facilitate feeding by moving the organic media down the arm and into the mouth.

Sea lilies lived en masse—the slowly feathery crinoid forest above the seafloor must have been a spectacular sight. Where current or wave action disturbed the water

A Paleontologist in Action: Making Serial Sections from Fossil Brachiopods

The aim of making serial sections is to study the inner skeletal elements of the brachiopod that are enclosed within the rock. This procedure is destructive, so it is advisable to make a copy of the whole fossil in plaster of Paris before beginning the sectioning. The properly fastened (i.e., fixed in a gypsum cylinder) fossil is first oriented tip down, perpendicular to its plane of symmetry, held firmly (preferably with the help of equipment), and then carefully ground. The smooth surface that results should be checked regularly with a microscope as the fine elements of the lophophore will gradually appear in cross section. After several tenths of a millimeter, the surface has to be checked repeatedly, as subsequent drawings of these cross sections will reveal the entire three-dimensional nature of the lophophore (this method resembles CT scanning, often used to visualize human embryos, as we can recognize only a section of the body onscreen). Often it is the case that once the fossil is sectioned, or even in the middle of the work, it becomes clear than the vulnerable lophophore is not preserved: it was either replaced, broken, or completely destroyed because of sediment infilling the valves.

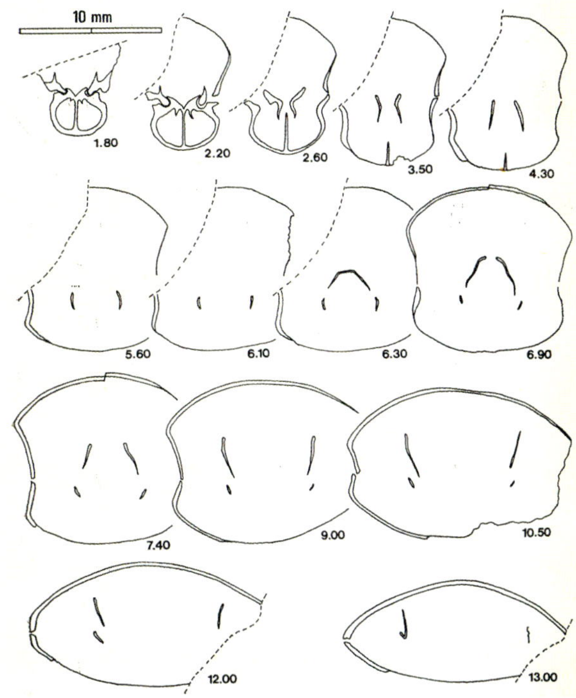

Fourteen successive serial sections of the Sinemurian brachiopod *Securina securiformis* (Gemmellaro). The length of this specimen was 15.7 millimeters; its lophophore was 13.8 millimeters and nearly reached the opposite wall of the shell. In many cases, making serial sections such as this is the only way to accurately classify brachiopods. After A. Vörös 1983.

the larvae were unable to settle. There were also forms belonging to the plankton as well.

Sea lilies are the only extant Crinozoa subphylum. The diversity of today's crinoids is far less than the Mesozoic (and Paleozoic) diversity of the group. Studying only the recent crinoids does not facilitate an understanding of the mode of life of the extinct forms.

When Crinoids Die

When a sea lily died, the soft ligamentary tissue inside the cylindrical stem, the arms, and also the cirri decayed, and the highly porous columnals and other types of ossicles fell apart, like a pearl necklace when the string breaks. The disintegrated elements were easily transported away by the weakest currents. The rootlike holdfasts—if there were any—remained on the spot. But the rolling columnals were redeposited and accumulated far from where the sea lily lived, at the foot of the elevated highs or trapped inside the submarine fissures. The already unrecognizable crinoid elements turned into a calcareous sand, and finally, after the diagenesis, into crinoidal limestone, also known as encrinite. In the Hungarian Jurassic this type of rock is typical. If we hammer a crinoid limestone on the freshly broken surface, we will see shining spots—each of them is a piece of broken crinoid element. But the parts of the sea lily cannot be determined from these ground particles.

Crinoids from the Bakony and Mecsek Mountains

In the Hungarian Jurassic well-preserved crinoids—that is, suitable for paleontological study—are rather rare, and therefore the number of the relevant publications is also low. The best-known fauna is from the Eperkés Hill (Bakony Mountains). The well-preserved fossils, mainly columnals and calyxes, were found here in red, nodular, calcareous marl. The remains can be easily recognized and picked up from the loose rock. For determining species calyxes, which have a rather complex structure composed of many elements, are especially suitable. Today nine species are recognized—one of them was new to science—belonging to the genera *Proholopus, Gammarocrinites, Psalidocrinus, Phyllocrinus, Lonchocrinus, Hemicrinus,* and *Saccocoma.* The first five genera unite regular sea lilies and display calyxes with pentaradial symmetry and with shorter or longer stems. *Hemicrinus* is a special spoon-shaped form with a modified extremely short stalk; *Saccocoma* is a stemless crinoidea, possibly belonging to the plankton.

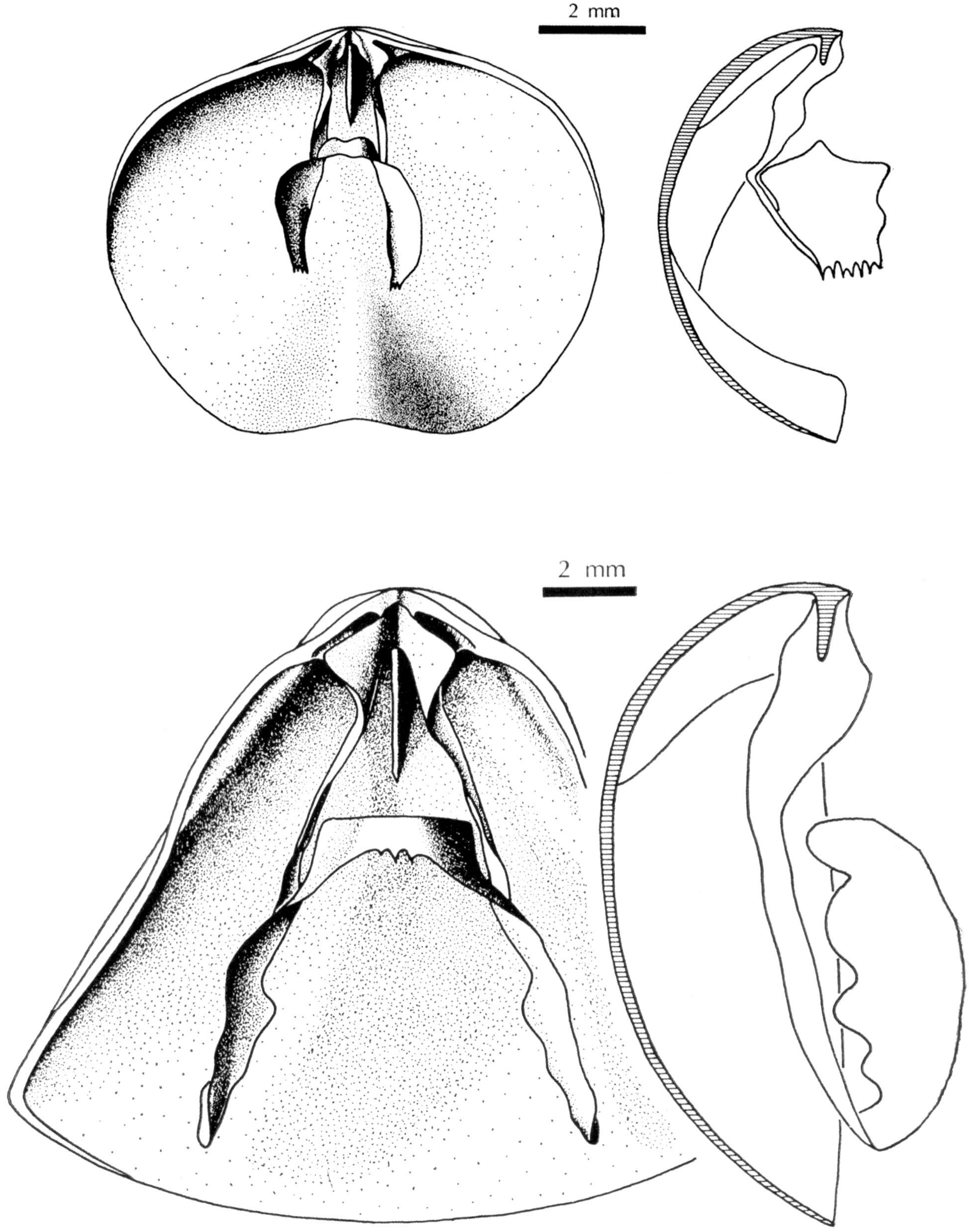

Ventral and lateral views of the brachial valves of the brachiopods *Bakonyithyris pedemontana* (Parona) (above) and *Securina securiformis* (Gemmellaro) (below). The fragile lophophore in these specimens was reconstructed by making serial sections. After A. Vörös 1983.

Pygope diphya (Buch)—a characteristic Late Jurassic brachiopod (original size).

It is strange that the Eperkés Hill crinoidea fauna, which was collected alongside some superb Kimmeridgian ammonites, was described as Tithonian.

Jurassic crinoids are also known from the Bajocian and Bathonian marls of the Mecsek Mountains. On the basis of different stem columnals, eight species were distinguished. The most common genus was *Balanocrinus,* which was represented by different species. Close allies, like *Margiocrinus* and *Chariocrinus,* were also recognized. The common cylindrical stem elements possibly represent the genus *Psalidocrinus* or *Apsidocrinus.* On the basis of the calyxes another eight species were described; most of these represent *Phyllocrinus.* In determining crinoids, the main problem is that the different elements of the complex test are often fossilized separately and rarely preserved together. Therefore it may happen that different names are applied to the calyx and the columnals of the same species.

📖 Manni et al. 1992; Bódy 2000

Vertebrates

In the Jurassic of the Carpathian Basin, remains of vertebrates are only rarely preserved. In the marine sediments, occasionally teeth of fishes or bones of reptiles can be found. Of special interest are the dinosaur footprints of the Mecsek Mountains. Of course, these were preserved not in a marine facies, but in the muddy floodplain of the former marches.

Fish

Although many types of cartilaginous and bony fishes might have been present in the Jurassic marine environments, their fossils are rather rare—each shark tooth is a lucky find. Only the Toarcian yielded more body fossils.

When the Oxygen Is Gone

During the Toarcian anoxic water masses—water bodies with low or very low oxygen content—appeared in the Jurassic ocean. This phenomenon is known by geologists as the "Toarcian anoxic event." The word "event" refers to the fact that it happened within a short period of time, and it had a large geographical distribution. The anoxia is reflected in the special geochemistry and petrology of the Toarcian formations. During the event, dark shale rich in organic matter and manganese-rich carbonate of particular stable isotope composition were formed. If the oxygen-depleted water layer impinged the bottom, life had a very limited chance of survival in the sediment on the seafloor, so even the mud-dwelling creatures and scavengers were unable to survive. Carcasses of animals that occasionally drifted down from the higher water layers—which had a healthy enough oxygen content to maintain life—remained intact on the seafloor and were buried in one piece. Especially famous are the localities around Holzmaden (Germany) where hundreds of entire, extremely well-preserved fish, crocodiles, and Ichthyosaurs have been found in the Toarcian black shale.

Securithyris adnethensis (Suess)—the largest brachiopod from the Lower Jurassic in the Bakony Mountains (original size).

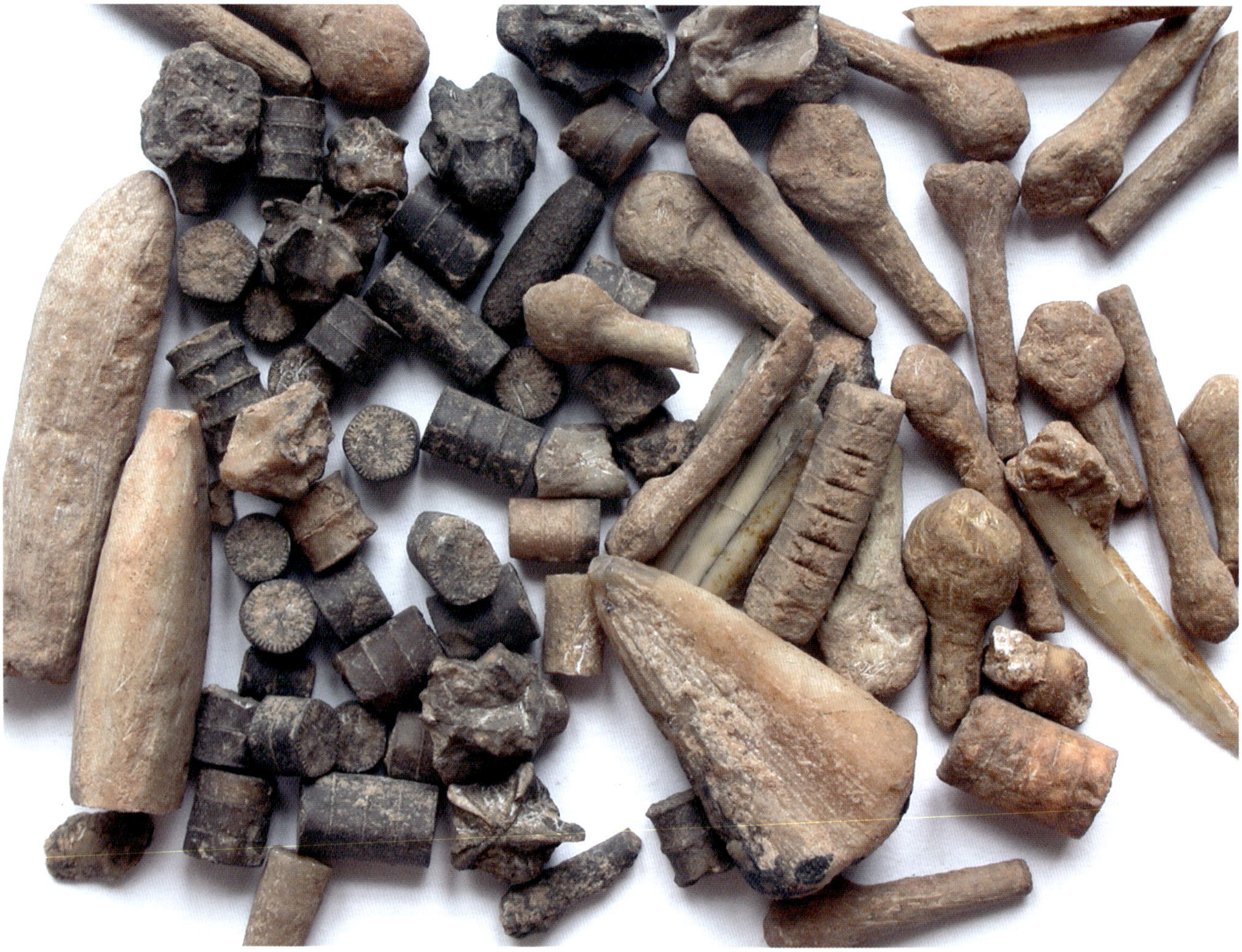

Small fossils from the Kimmeridgian calcareous marl of the Eperkés Hill (Bakony Mountains): remains of sea lilies coated by black manganese oxide (columnals and calyxes), short stalked crinoids resembling a kitchen spoon (*Torynocrinus*), fragmentary belemnites (left), echinoid spines (lower right) and shark teeth (2× magnification).

Toarcian black shale cropping out in the Réka Valley (Mecsek Mountains).

The Hungarian Toarcian is not as famous as the German one: only a single crocodile and Ichthyosaur were found. Some smaller finds—namely pycnodontid fishes—were collected from manganese nodules from Úrkút (Bakony Mountains). Some fine specimens are on exhibit in the Hungarian Geological Institute in Budapest.

In southern Hungary, in the Mecsek Mountains, no manganese ore was left behind after the Toarcian event, but due to the anoxic environment a dark, foliating shale rich in organic matter was deposited. The best—and practically the only—surface exposure of this formation can be found in the Réka Valley. Turning over the "pages" of this shale we can find fishes—rarely exceeding 10 centimeters in length—preserved from the tip of the head to the end of the caudal fin. Their closer study is badly needed, until now only the species *Leptolepis normandica* was determined.

Marine Reptiles

The Crocodile of the Gerecse Mountains

On the right bank of the River Danube, south of Lábatlan, the "Gerecse red marble" has been quarried for centuries. This is a Rosso ammonitico type of limestone deposited

As yet undetermined bony fishes from the Toarcian black shale of the Réka Valley (original size).

The Gerecse crocodile was found a few meters away from this artificial trench cut into the steep wall of the abandoned quarry at the Pisznice Hill. The site is a nature conservation area and therefore strictly protected—any collection activity requires a special permit. However, there are many other quarries in the Gerecse Mountains where anybody may collect fossils.

in a pelagic environment; fossils of cephalopods can be common, but vertebrate remains are very rare. That is why it was a sensation, when a young student, Attila Fitos, discovered a large crocodile in one of the abandoned quarries

The roughly cleaned bones of the crocodile reassembled, more or less, as it was in the rock. The animal was about 4–5 meters long. After preliminary study of the skeleton, the whole find was given to the Hungarian Natural History Museum.

on Pisznice Hill in August 1996. Unfortunately the skull was only partially preserved, but many vertebrae, ribs, limb bones, the pelvis, and some dermal plates were found in a relatively good state of preservation. Altogether about 200 pieces were found. The Gerecse crocodile belongs to the suborder Mesosuchia. Beautiful fossil crocodiles of the same age have been found in Germany and in the United Kingdom, but nothing like them was previously documented from Eastern Europe.

The Ichthyosaur of the Gerecse Mountains

In 2010, after the unique find of the Gerecse crocodile, another great fossil—a Jurassic ichthyosaur—was discovered in the red nodular Toarcian marl of the Gerecse Mountains. The bones are still waiting for thorough cleaning, but it has been established that they represent a *Temnodontosaurus*. These torpedo-shaped predators of the Jurassic exceeded 10–12 meters; the Hungarian specimen was about 8 meters long. A detailed study and description of the find has yet to be completed.

Fused, carefully cleaned dermal plates of the Gerecse crocodile. The reptile had tough armor, as many extant crocodiles have (0.7× magnification).

Jurassic Dinosaurs

Since the Jurassic period was dominated by marine sedimentation within the Carpathian Basin, remains of dinosaurs are rare and can be found only in places where deposits of ancient terrestrial environments are preserved. Hundreds of mainly three-toed (tridactyl) footprints, arranged in long tracks, have been found near Pécs in southern Hungary. More recently some large, poorly preserved sauropod tracks were recorded in the lowermost Jurassic of the Southern Carpathian. Since large dinosaurs were landlocked, the appearance of their footprints has paleogeographic implications.

The Dinosaurs from Pécs

Although the Jurassic of the Transdanubian Range is fully marine and pelagic, a very thick Jurassic succession of the Mecsek Mountains was deposited in a "half graben" that describes an extended deltaic floodplain where coaly deposits several hundred meters thick formed in the Early Jurassic (see also the fossil plants section of this chapter). Outcrops of coal form a zone shaped like a reverse S between Pécs and Komló. These were intensively mined using vertical shafts and intricate networks

Dogger Quarry in the Gerecse Mountains, where a *Temnodontosaurus* skeleton was found. The location of the discovery is marked with a figure in red.

Jurassic vertebrate fossils (original size). (1) Bony fish in nodule, found in the Toarcian manganese ore complex at Úrkút (Bakony Mountains). (2) Teeth of a pycnodontid fish from the Tithonian limestone at Steierdorf-Anina (Southern Carpathians). (3, 4) Hettangian reptile vertebra from coal seam no. 11 at Pécsbánya (Mecsek Mountains). The poorly preserved fossils were collected by pioneer János Böckh and determined as *Ichthyosaurus* by Friedrich von Huene.

Some of the Gerecse *Temnodontosaurus* vertebrae are larger than a cup.

of underground tunnels as well as large open-cast mines. In 1966, György Wein (1912–1976), a notable geologist from the Hungarian Geological Institute, discovered the first fossil reptile footprints in Pécs-Vasas II, an open-cast mine. The hand-sized prints were found in the barren beds between coal seams. András Tasnádi-Kubacska supervised the making of plaster copies and the original block was taken to the Geological Institute. Tasnádi-Kubacska reported how the footprints were collected and transported to the Hungarian Geological Institute. Three footprints could be seen on a bedding surface, two in front of each other, and a third going across. Local geologist Zoltán Csörnyei found a fourth when he happened to split a loose rock. The main find was a splendid example, perfect for a museum, but it lay in bedrock. Fortunately the mine's staff helped to solve the problem: they laid a kilometer of air tube from a pneumatic jackhammer and the manager of the mine, the local Communist Party secretary, visiting geologists, and miners all took turns. Eventually the block bearing the prints was ready to be pried from the ground.

By then, retrieving the block had become a major project. A large scraper was called in to make a road down to the find and then a house-sized excavator crept down its

Dinosaur tracks in the Karolina Valley open-cast pit. Local miners call them "petrified chicken tracks" and, that is not so very far from the truth, since dinosaurs are closely related to birds.

1

2

3

4

Footprints in the museum. Reassembled slabs bearing *Komlosaurus* footprints waiting for final exhibition in the old ceremonial hall of the Hungarian Natural History Museum, before its restoration.

steep slope. It squared up to the block, and seized with the jaws of its bucket. In spite of the massive size of the hydraulic jaws, three attempts were needed to lift the block. By the time the excavator carried the block to a waiting lorry, it was late afternoon. The trip back to the Geological Institute was extremely slow and long.

🕮 Tasnádi-Kubacska 1967

More Footprints

In 1980 more "dragon tracks" were discovered, this time at Komló. László Kordos, paleontologist at the Geological Institute, studied these footprints, but when he came to compare them to the specimens found in 1966, he found that the originals had been damaged. Apparently vibration during the transport had caused the seemingly solid block to crack along the bedding planes and literally fall to pieces. Fortunately, some of the pieces revealed new footprints! Kordos described the dinosaur of the Mecsek Mountains as *Komlosaurus carbonis,* a new species and genus. The name refers to the mining city of Komló, and to the Latin name for coal (*carbo*).

More footprints were found in summer 1988, when geology students Kinga Hips, Ágoston Nagy, and Zsolt Pataki had a university practicum around Pécs. They discovered hundreds of *Komlosaurus* footprint arranged into pathways in the large strip mine of the Karolina Valley. At the same time new footprints were found in Pécs-Vasas, in the upper part of the middle coal seam set. The footprints, which formed tracks, occurred in several levels on bedding planes of fine-grained sediment. Hundreds of these footprints were transported to the Hungarian Natural History Museum and others are stored in the local museum at Komló. A series of tracks was assembled and mounted in the corridor of Eötvös Loránd University.

🕮 Kordos 1983; Ősi, Barbacka, & Szente 2005; Ősi et al. 2011

The Age of the Footprints

Initially the whole coal seam series was regarded as Lower Jurassic, but palynological studies have revealed that the deepest beds represent the upper reaches of the Triassic (Rhaetian). Higher sequences include some marine layers

It is usually impossible to match fossil prints with the dinosaur species that left them. However, we are sure that the track makers were bipedal. *Komlosaurus* was midsized (2–3 meters long), and very likely a predatory dinosaur that left these tracks when walking around on the seashore. Its body was balanced by its muscular tail. Reconstruction by András Szunyoghy.

Komlosaurus footprints are characteristically three-toed (tridactyl) and range from 12 to 20 centimeters long. Its step was about 60–70 centimeters. Although most of the tracks show only three toes, the track makers likely had four or five toes on their feet. When walking or running only the second, third, and fourth toes touched the ground. If the animal was moving very slowly, or "elbowed," then it left three- or four-toed prints.

that have yielded ammonites (*Arietites, Coroniceras*) that suggest early Sinemurian age. The majority of the coal formation, which contains the *Komlosaurus* tracks, is Hettangian in age, but footprints also occur in the lower Sinemurian.

Giant Footprints of the Southern Carpathians

More than 40 years after of the discovery of the first dinosaur footprint near Pécs, during a field excursion conducted as part of a Jurassic meeting held in Romania, sauropod footprints were found in Hettangian (earliest Jurassic) alluvial deposits in Colonia Cehă Quarry, Anina (Reşita Basin). The track-bearing sandstone surface was densely trampled by sauropods, which makes identification of individual tracks and trackways difficult. However, of special interest is a two-step trackway, showing pes and manus footprints of relatively small size (the pes is 40 centimeters long and 20 centimeters wide; the manus 15 centimeters long and 10 centimeters wide). The material is consistent in shape and size with known Hettangian *Parabontopodus* ichnospecies (a species defined by footprints only) trackways reported from Northern Italy and Central Poland. The Romanian footprints are deeply impressed; the first pes is surrounded by a high displacement rim, and the second was filled by upwelling, saturated sediment. The presence of the footprint indicates that during the earliest Jurassic, the "Moesian Island" (a territory that comprises the present-day Anina region) was at least temporarily attached to the Pangaean (Eurasian) mainland; thus, sauropods arrived. Moreover, the normal size (for Hettangian times) of the Romanian footprints speaks against insular dwarfism of this animal, unlike the Late Cretaceous dinosaurs in Romania and Hungary. Based on the Anina footprints and those of other sauropod tracks in the region, these huge animals roamed both the northern and southern rim of the Western Tethys.

📖 Pienkowski et al. 2009

How Can a Footprint Be Preserved for a Million Years?

Most footprints that animals leave behind are eaily ruined or lost. However, prints made in soft mud become dry and harden in sunlight, and so subsequent marine or fluvial flooding will not necessarily destroy them. As millions of years pass, the mud gradually turns into hard rock and preserves the prints, which become fossils.

The leg pushed into the sediment makes a real print—a negative. The sediment-filled print is a positive track—a natural cast, a mold. Heavy animals may deform the soft sediment below the actual surface; this is how "ghost prints" are formed in different levels.

This is how the *Komlosaurus* footprints were formed. Among the coal layers, their ghost prints also are frequent.

The Mesozoic—and especially the Cretaceous—may rightly be considered the time of reptiles, with dinosaurs preeminent. This image shows the left hind limb of *Balaur bondoc,* a recently discovered Transylvanian carnivorous dinosaur, a close ally of *Velociraptor* (close to original size). *Balaur* had a big toe with a large claw that could be hyperextended and was presumably used to slash prey, as well as a large claw on the second toe. It had short and stocky feet and legs with fused bones, indicating that it was adapted for strength rather than speed. Compared to *Velociraptor, Balaur* was probably more of a kickboxer than a sprinter.

4

The Cretaceous

The last period of the Mesozoic era (denoted with a K in geological abbreviations), lasted for about 80 million years and was named after a special rock, "chalk" (from the German word *Kreide*), that formed in large amounts at this time. Although the specific Cretaceous "writing chalk" formed from the remains of masses of chrysophyte algae familiar across northwestern Europe is not found in the Carpathian Basin, formations of other Cretaceous-aged deposits are nevertheless common.

This originally bi-, or tripartite (recently declared officially bipartite) period is usually divided into 12 stages by geologists. Most stages—just like those of the Jurassic—were named after (usually French) villages. The Cretaceous, following the relatively calm Jurassic, was one of Earth history's most eventful periods from a tectonic point of view. Because of this activity, the Cretaceous layers in the Carpathian Basin are very diverse, and include rocks of continental and marine (both nearshore and offshore) origin.

The biotas of the Cretaceous, thanks to this environmental variation, were also very rich and are evidenced in the fossil assemblages known from this period. The massive extinction that marks the end of this period was a milestone in the evolution of life on Earth. It marks not only the end of the Cretaceous and the end of the Mesozoic, but the beginning of the Cenozoic.

During the Cretaceous, the drift and breakup of continents increased in speed. Parallel with the opening of a southern basin in the Atlantic Ocean, Africa began moving and rotating in a counterclockwise direction. Because of this movement, the Tethys seaway gradually became narrower and the formation of mountains in the Alp-Carpathian region also began: thick layers of rock laid down at earlier times were shortened, and thrusting (formation of geological structures called nappes) began in the Carpathian Basin. Also as a result of these movements, oceanic sea levels changed several times during the Cretaceous. With such diversity in environmental conditions, many kinds of sediment were deposited encompassing continental, shallow marine and deep marine (pelagic). The climate of the Cretaceous, although similar to that of the Jurassic, was warmer and experienced less fluctuation in temperature than we see today.

THE MAJOR EVENTS OF AN EVENTFUL PERIOD

The events described above resulted in spectacular changes to the western basin of the Tethys Ocean, which practically closed during the Cretaceous. In its place—the Western and Central European landmasses of today—a shallow ocean formed with an archipelago. As a result of global lowering of the sea level in the early Cretaceous, shallow marine platform sediments resurfaced in several places leading to the formation of bauxite deposits in the Villány Mountains of Hungary and in the Pădurea Craiului region of present-day Romania. This period is also characterized by a series of terrestrial and shallow marine environments, followed by rising sea levels and deposition of pelagic sediments. The remnants of the latter are found expressed, for example, in the thick, ammonite-bearing sedimentary sequences of the Gerecse Mountains of the Transdanubian Range. The so-called Urgon-type limestone, characterized by thick-shelled bivalves and gastropods, was also deposited during such a transgression in the Bakony Mountains, Transylvania, and the Carpathian region, for example.

In some areas the Upper Cretaceous is represented by a thick succession of layers that were formed in a third cycle, which includes rocks (for example, bauxite) deposited on land, in coal beds formed in swamps and marshes, shallow marine reefal limestones, or pelagic marls and limestones. The most spectacular development of this third cycle—the Upper Cretaceous—found in Hungary is in the area of the Southern Bakony, between Sümeg and Magyarpolány.

The relatively fast tectonic events of the Cretaceous that altered the Earth's surface, drifting of continents and the formation of mountains, were also often accompanied by intensive volcanic activity. The most marked examples of this activity within the Carpathian Basin can be found in the Metaliferi Mountains and the Lower Cretaceous of the

System/ Period	Series/Epoch	Stage/Age	Millions of Years Ago
Cretaceous	Upper/Late	Maastrichtian	71
		Campanian	83
		Santonian	86
		Coniacian	89
		Turonian	94
		Cenomanian	100
	Lower/Early	Albian	112
		Aptian	125
		Barremian	130
		Hauterivian	134
		Valanginian	140
		Berriasian	145

Subdivision of the Cretaceous period into stages.

Mecsek Mountains. In the Mecsek, basaltic lava poured to the surface underwater leading to the formation of well-preserved pillow structures.

The circular reefs that evolved around the volcanic island—or islands—of the Mecsek were similar to modern atolls and hosted a rich fossil fauna. Parallel with thrusting and the formation of mountains, the erosion of the uplifting Carpathian Basin also began at this time; as a result, the Carpathian sandstone (also known as flysch) began to form, mostly in the Upper Cretaceous.

FOSSIL-RICH FORMATIONS

The different kinds of sedimentary rocks of the Cretaceous were differentiated, and named, by geologists in the past mostly for the characteristic fossils they contained. Terms such as "Turrilites marl," "Hippurit limestone," "Gryphaea marl," and "Inoceramus marl" are found in geological literature that dates back almost a hundred years. In the next section we will discuss the most important fossiliferous rocks and facies of the Carpathian Cretaeous, approximately in accordance with the age of their formation.

Aptychus Marl

The aptychi of ammonites usually fossilize separately from the rest of the animals' bodies and are often found in Lower Cretaceous marls. There are numerous examples of typical localities, including the Gerecse Mountains and different regions of the Carpathians. These aptychi, similar to the valves of living clams, are thought to have functioned either as operculum or as jaw apparatus. In

Aptychus from the Early Cretaceous Bersek Marl (original size).

addition, because they were composed of calcite, they often fossilize in layers, whereas the aragonitic shell of the rest of the ammonite would dissolve. The term "Aptychus marl" was first introduced to Hungary by Miksa Hantken, the first professor of the Department of Paleontology at the University of Budapest, in reference to the occurrence of rocks containing aptychi in the Gerecse Mountains. The same formation is also often called the Bersek Marl, after the most important outcrop at Bersek Hill. We know that the Bersek Marl was deposited in an offshore environment, and apart from its aptychi, it also contains the abundant fossils of ammonites and belemnites as well as the less common remains of echinoids and corals. This rock is a gray and monotonous marl succession that represents the Berriasian, Valanginian, and Hauterivian stages of the Cretaceous almost completely, as evidenced by its rich fossil content. This part of the Lower Cretaceous—using an older terminology—is also often called the Neocomian, thus the Aptychus marl is also sometimes referred to as "Neocomian marl" in the Carpathian Basin.

Crinoidal Limestone

In the different periods of the Mesozoic, rocks that are basically built from, or at least contain masses of, the remains of crinoids (sea lilies) were formed and are commonly encountered. The term "encrinites" is often used for these kinds of rocks. In Hungary, however, the name "crinoidal limestone" as an independent lithostratigraphic term is generally applied to a rock body known from the Middle Cretaceous of the Transdanubian Range. The official name of this unit, the "Tata Limestone," refers to the town of Tata, where one of the best-known and most heavily studied exposures of this rock type occurs. Tata Limestone can be found on the surface all the way from Tata to Sümeg; indeed, the castle hill of Sümeg is composed of this rock (note that it is not a volcanic cone, as it is described in several tourist brochures). This light, pure, and often coarse-grained limestone was deposited in a mid-depth offshore environment and consists of crinoid fragments ground to fine calcareous sand. Independent crinoidal elements, including the calyx, columnals, stem, or arm ossicles, are rarely seen. In a fresh break in the rock, sparkling particles are seen, each of which is a distinct crinoid remnant. It is this sporadic sparkling quality that tells us that the crinoids are present in a rock-forming quantity.

In case this is not proof enough, just make a thin-section! When a thin-section of Tata Limestone is placed under a light microscope large numbers of eroded crinoid fragments can easily be seen sitting closely together. However, despite this abundance of crinoid fragments, the Tata Limestone is relatively poor in macrofossils. Even if something is found it is usually a brachiopod, although a significant and strongly condensed ammonite fauna has been found in the bottom layer, at the base of this unit, at certain outcrops, including Tata. The specimens that were collected from the base of the Tata Limestone were first thought to be Aptian in age, but recent investigations have shown that some are younger. As a result, the actual age of the Tata Limestone, previously referred to as the "Aptian crinoidal limestone" is most likely somewhere between the Aptian and Albian ages.

Sea lilies exist today, of course, but their role in nature compared to that of their Paleozoic and Mesozoic counterparts is extremely minor. It is interesting that limestones formed exclusively from the mass accumulation of sea lily remains are unknown in periods that follow the Cretaceous. In English, we use the term "vanished" or "dead facies" when describing rocks that were once quite common but are no longer formed due to present environmental conditions. Crinoidal limestone is one good example of such a facies; there are no known occurrences younger than the Tata Limestone anywhere in the world.

Munieria Marl

Munieria-bearing beds can be traced into the axis of the Transdanubian Range to within a relatively small area that encompasses Úrkút, Zirc, Tés, and Környe. Because it was formed in a range of rather different depositional environments—fresh water (rivers and marshes), brackish, coastal, and shallow sea conditions—the lithology of this marl is also very diverse. One formation, also known as the Tés Marl, comprises clays, marls, and occasionally intercalated limestone beds. There are no spectacular surface outcrops of this lithology in the Carpathian Basin and the best localities are found in the Zirc Basin and in the vicinity of the village of Tés. In the Pig Pasture (Disznólegelő) meadow at Zirc, a waterlogged muddy area indicates the presence of the impermeable clay horizon of the formation under the surface. Due to the range of ways in which this rock can form, it may contain many different kinds of fossils. For example, in his 1934 publication, Jenő Noszky refers to it as the "ostrea-brachiopoda-ostracoda-calcareous algae clay and marl group." The most commonly found calcareous alga in this lithology is *Munieria baconica,* sometimes present in rock-forming quantities. *Munieria* belongs to the calcareous green algae; its generic name refers to the French paleontologist Ernest Munier-Chlamas (1843–1903), whereas its species name was given in reference to the Hungarian Bakony Mountains. When the rich and diverse ostracod assemblage found in the Munieria marl is examined, most workers have concluded that it is most likely middle Albian in age.

📖 Noszky 1934

Orbitolina Limestone

This rock was named after one characteristic: larger foraminifera from the Cretaceous, *Orbitolina,* which occurs in large amounts in some marine rocks. The geologists who mapped the Bakony region usually referred to one part of the Albian sequence (today known to be part of the Zirc Limestone) using the term "Orbitolina limestone." The Bakony region is, however, not the only place where rocks abundant in *Orbitolina* can be found: there are Orbitolina limestones in the Villány Mountains (the "Nagyharsányi Limestone") and, as evidenced by borings from the foothills of the Vértes Mountains, some rocks that make up this range are built solely from this type of foraminifera.

Turrilites Marl

Turrilites is a heteromorph tower-like Albian-Cenomanian ammonite. The Turrilites marl, rich in *Turrilites* and lithostratigraphically often referred to as the Pénzeskút Marl, is a hemipelagic, mostly gray, more or less silty, sometimes sandy, dolomitic marl that is rich in diverse fossils. This characteristic lithology in the Transdanubian Range can be seen between Csehbánya and Oroszlány. The most spectacular surface exposures can be found around the villages of Jásd, Szápár, and Pénzesgyőr in the Bakony Mountains and the base of this formation is especially rich in fossils; it is usually packed with ammonites, gastropods, bivalves, and the remains of echinoids. The microfauna of the Turrilites marl is also rich and diverse, and the age of its base—derived from examination of stratigraphically important ammonites—is late Albian; the upper part of the succession is most likely Cenomanian.

Gryphaea Marl

Gryphaea is an interestingly shaped marine bivalve, closely related to *Ostrea* and known mostly from the Jurassic. In older Hungarian literature, clams that are now called *Pycnodonte* and that are found at certain stratigraphic levels in the Bakony Mountains were also considered to be

Requienia limestone cliffs around the manganese mud waste dump deposit in the vicinity of Úrkút.

Gryphaea. The most spectacular occurrences of this rock are known only from drilling and can be found in the area surrounding Sümeg. Apart from *Pycnodonte,* other mollusks, such as *Ceratostreon* (once called *Exogyra*), are also found in another Upper Cretaceous (Santonian) Gryphaea marl—the Jákó Marl, which was named after the village of Bakonyjákó. The term "Exogyra marl" is also encountered in older paleontological literature. This rock is usually gray, and is considered to be a marl (specifically, a calcareous marl or a silty marl) from a lithological perspective. The lower part of this unit was deposited in a shallow marine environment, whereas its upper layers were formed in deeper sea. The fossils that this rock type contains (bivalves, gastropods, and solitary corals—most commonly *Cunnolites*) were generally deposited in specific layers that contain fossil debris nearly exclusively. Geologists consider these massive accumulations of shells to represent storm deposits.

Rudist Limestone

Rudists (*Hippurites* and related genera) are large bivalves with thick, strange looking horn-shaped shells that can be abundant in Cretaceous reefal limestone. Sometimes these rocks are also referred to as "Hippurit limestones" and can be found commonly in the Transdanubian Range as well as at Pădurea Craiului and Fruška Gora Mountains. The occurrence of this lithology in the Bakony Mountains has long been known; Antal Koch referred to this rock type in 1875 as "hippuritmeszek" ("Hippuritkalk").

The rudist limestone of the Bakony is a light grayish, thickly bedded, and well-stratified rock. When it is examined in thin-section, rudists and the skeletal debris of other mollusks are present in rock-forming amounts. The most spetacular surface outcrops of this lithology are found near Sümeg and Ugod. Named after the latter village, it is also sometimes called the Ugod Limestone. The rudist limestones of the Bakony are considered Upper Cretaceous, more specifically Campanian in age, based on their rich mollusk and foraminiferal fauna.

Inoceramus Marl

Inoceramus are marine bivalves known from Jurassic to Late Cretaceous marine deposits that are characterized by their mostly large, flat valves and concentric ribs around their umbos. In the Carpathian region and in the Transdanubian Range, these bivalves are most commonly known

Rudist bivalves embedded and preserved in close to life position in the vicinity of Sümeg, Sintérlap Quarry (close to original size).

Vertical flysch layers in the Southern Carpathians.

from Upper Cretaceous sediments. Hungarian geologists mostly know this gray, clayey limestone as a well- or not-so-well-bedded calcareous marl that crops out in the southwestern part of the Bakony, in the vicinity of Sümeg and Magyarpolány. For the latter town, this rock is also sometimes called the Polány Marl. This rock was deposited on a reef slope and in the neighboring offshore basin. It reaches its greatest thickness in the foreground of the Bakony; in the vicinity of Devecser, for example, about 800 meters of this lithology was drilled. However, the name of this unit—the "Inoceramus marl"—is misleading: the fossils of this bivalve are never abundant in the formation; they are, in fact, only relatively common in the upper part of the succession. Because this marl is relatively poor in its preserved fossil macrofauna it was named after a quite rare, but characteristic and easily recognizable fossil.

Flysch with Abundant Trace Fossils

The very specific rock type called flysch is widespread in the Carpathian region. It consists of alternating sandy and calcareous layers formed into often very thick successions, sometimes reaching several thousand meters, that were deposited on Cretaceous (or younger) continental marine slopes. Flysch is characterized by specific sedimentological characteristics such as flute casts, current ripples, and ichnofossils. However, a surprisingly small number of actual body fossils are found in this rock type: it is a rare and happy moment for the geologist when an ammonite or a bivalve—often flat or deformed—appears in the rock. Even so, the common occurrence of trace fossils in flysch is compensation for the lack of actual fossils, even though in most cases it is not known what kind of animal, or even what type of activity, left the traces behind. Surfacing layers of flysch can be found in tectonic zones that follow the bend of the Carpathians. These zones can be differentiated according to rock type and age and have outcrop along long and narrow nappes. There are more than a dozen different such nappes along the arc of the Carpathians and all have been separately named by local geologists.

REMARKABLE FOSSIL SITES

In the Mesozoic, Cretaceous rocks in the Carpathian Basin were deposited under diverse environmental conditions. As a consequence, different raw materials, including coal, bauxite, iron ore, marl, and limestone were formed, and all have been mined and quarried over the years. The most intensive exploitation took place in the early part of the twentieth century. Deep mines and huge quarries produced a large number of fossils (mainly invertebrates) which were unknown before, although many smaller, fossiliferous outcrops of the different Cretaceous formations were already known to pioneer geologists.

Paleodictyon, a characteristic trace fossil from Carpathian flysch (0.3× magnification).

Collecting cephalopods from the ammonite-rich layers of the Tilos Forest, near Pénzesgyőr, Bakony Mountains. In this region, the Lower Cretaceous is represented by relatively thin successions, which is why most of the fossil sites, like this one, are small natural cliffs or minor quarries serving local purposes only. On the left is Ottilia Szives, who contributed to our knowledge on the Cretaceous ammonites of Hungary.

Marine Invertebrate Localities in the Bakony

Cretaceous deposits of the Bakony can be found in the axes of the mountain range, in a relatively narrow zone that runs in a north-northeast–southwestern direction between Tés and Sümeg. Several places, such as Zirc (Márvány Quarry), Hárskút (Közöskút Ravine), Bakonynána (Zsidó Hill) and Pénzesgyőr (Tilos Forest), are significant localities for Lower Cretaceous invertebrates—especially ammonites. Gastropods and rudists, all well preserved, can be collected from the cliffs protruding from the side of the manganese waste dump deposit in the vicinity of Úrkút. Near the small town of Ajka, the spoil bank of the coal mines that were in use for over a century has become internationally known as a site yielding Upper Cretaceous, freshwater gastropods—mainly the genus *Pyrgulifera*. With a little luck, Upper Cretaceous clams, even large ammonites (*Pachydiscus*), can be found in the huge quarries in the area surrounding Sümeg. It was in a small, wet field, in the vicinity of the same town, where outcrops of the Jákó Marl, abundant in *Cunnolites* corals, were first found. More detailed descriptions of the location and geological conditions of these outcrops and sites are provided in the different monographs (as seen in the box), all devoted to the Cretaceous of the Transdanubian Range.

Iharkut: The Site for Fossil Reptiles

In 2000, Attila Ősi, then a third-year student in geology, along with his friend András Torma, went to the Bakony Mountains to search of dinosaur remains. What they found was beyond their wildest dreams: not only bones, but an entire fossil assemblage. The students had discovered the first Cretaceous dinosaur remains known from modern-day Hungary. The success of their endeavor had more to do with hard work and research than dumb luck, as they had been looking for places where terrestrial sediments had been deposited in Hungary during the Cretaceous. Logically, they knew that marine deposits would not contain the remains of dinosaurs that lived on dry land. That is how they happened upon the Iharkút site and the area surrounding it, where there was a large open air bauxite pit. Here, the bauxite had been buried under a Late Cretaceous terrestrial deposit, protecting it from later erosion. Luckily, the clayey, aleurolitic layers belonging to the Csehbánya Formation not only concealed the ore of alumina below, but also preserved the bones of Late Cretaceous fossil reptiles. As a result of several field campaigns over the years, thousands of fossils have been found by Ősi and Torma; we now know that the vertebrate assemblage of Iharkút comprises numerous fossil remains from fish, amphibians,

The Iharkút bauxite mine in the early years (2000) after discovery of the first bones. The ore was accumulated in the deep karstic hole—once a tropical tower karst pit. The site and the collected fossils are good evidence that as a result of strenuous research we can make remarkable discoveries—even in a small country like Hungary, which has been intensively mapped and studied by geologists for the past 150 years.

The Iharkút pit 10 years after the discoveries of the first fossils. The bauxite mining is over and the pit is barely refilled. Where the exploitation took place, there is now a green lake. But exploitation of the bones is not yet finished and the regularly organized field campaigns each year have yielded more and more new fossils—some of them representing completely unknown species.

turtles, crocodiles, serpents, dinosaurs, flying pterosaurs, and birds. The study and description of this rich fossil assemblage is still in progress, but the results so far clearly indicate that Iharkút is one of the most important fossil sites of its kind—not only in Hungary, but also worldwide. Both the age and geographic position of the Iharkút are important to our understanding of vertebrate evolution in the early Late Cretaceous.

The Bersek Hill of the Gerecse Mountains

Cretaceous rocks in the Gerecse Mountains can be found in the northern part of this region, on the side of the Danube. The biggest and most spectacular outcrop is the huge quarry at Bersek Hill, where the lower Necomian marl (Bersek Marl Formation) and the upper so-called Lábatlan Sandstone are both quarried—the marl is transported to the nearby cement factory. The lower, Valanginian part of the marl is thick, gray, and quite poor in megafossils; the upper, reddish Hauterivian part and also the higher Barremian layers are rich in mollusks, especially cephalopods. Thanks to long years of persistent and systematic collecting and research in this area, a very rich ammonite and belemnite fauna had been found, but the presence of bivalves and echinoids is worth mentioning as well.

Rocks similar to those found at Bersek Hill also crop out nearby, in the small quarry at Póckő, and a little further away in the Paprét Ravine. Beautiful Valanginian ammonites are found in the lush creek of the Nyagda Valley that runs along the edge of Lábatlan. Rocks in the Köszörűkő Quarry, near the Danube, are younger than the youngest part of the Bersek Hill sequence: they represent the Albian stage. Mostly colonial animals—in particular, corals—were found here when the quarry was still open.

📖 Főzy & Janssen 2009

Mecsek and Villány

A few Cretaceous layers are also found in southern Hungary, but for the most part they lack any spectacular macrofossils.

The western yard of the Bersek Quarry in the Gerecse Mountains, a well-known site for Early Cretaceous fossil cephalopods.

From Mecsek, mainly ammonites found in outcrops of the Valanginian marl near Kisújbánya are worth mentioning, along with corals and mollusks from another Lower Cretaceous formation known as the Magyaregregy Conglomerate. The Cretaceous in Villány is identified first and foremost on the basis of its microfossils; the few elements of macrofauna are known only from drillings.

The Fruška Gora

The most important feature of a notable fossil site is not the fact that there is something that can be found only in this place; it is much more a question of the care with which the fossils were first collected, the precision of their preparation, and the level of professionalism in their processing. This is the case when considering the Cretaceous of the Fruška Gora—a mountainous range between the Danube and the Sava in present-day Serbia—often called the Pétervárad Mountains in older Hungarian literature. The rich invertebrate fauna of this area—comprising gastropods, bivalves, ammonites, and corals—was studied in detail by Gyula Pethő (1848–1902). He inherited the collection of Antal Koch, who had discovered the site by the Čerević stream, and augmented it with his own collection—which also comprised a good deal of well-preserved and nice-looking fossil material. Later Pethő took the

Trace fossil (*Zoophycos*) from Lower Cretaceous marl in the Bersek Quarry. These kinds of trace fossils are known from the Cambrian and are still found today—for a long time, their characteristic form was thought to represent the remains of a plant but, in fact, *Zoophycos* turned out to be the lobed feeding traces of animals, most likely worms.

Cretaceous Monographs

Cretaceous research in Hungary has a long tradition and has yielded impressive results. Early Cretaceous localities, often rich in fossils, have been intensively collected and studied over many years. In particular, the research of a team of specialists was supervised by József Fülöp, once the head of the Hungarian Geological Institute, who produced separate monographs on the Lower Cretaceous of the Bakony, Gerecse, and Villány localities. The volume on the Tata horst is not exclusively on the Cretaceous, but deals with the entire Mesozoic at Kálvária Hill—a site that is protected because of its geological importance and natural beauty. Detailed descriptions of the mainly Upper Cretaceous formations and fossils near Sümeg were provided by János Haas, a close co-worker of Fülöp's, in a separate monograph. The so-called Urgon formations and the Middle Cretaceous sediments of the Transdanubian Range were thoroughly described by Géza Császár, another co-worker of Fülöp's, in two separate volumes.

Besides these monographs, focused mainly on general geology and stratigraphy, special papers dedicated to Cretaceous paleontology have also been published. Monographic descriptions of Aptian spores and pollen grains were given by Margit Deák; Béla Zalányi described the Aptian ostracods; Lenke Czabalay, the gastropod fauna of the Aptian–Cenomanian of the Bakony and the rudists from Sümeg; and Erzsébet Szörényi, the Cretaceous echinoids from Hungary (the "Aptian" beds studied by the abovementioned authors are now considered mostly to be Albian in age). Mária Beke published a pioneering paper on Cretaceous nannoplankton remains, and Ilona Bodrogi reviewed the planktonic forams.

Most of these volumes appeared in the Geologica Hungarica series produced by the Hungarian Geological Institute. An often-cited 1979 monograph by Gábor Scholz deals with the Albian ammonites of the Transdanubian Range. This work was published—for personal reasons—not in Hungary, but in the German journal *Palaeontographica*. For a long time, it was the only Cretaceous ammonite monograph from Hungary, but this situation changed when Ottilia Szives published her volume *Aptian-Campanian Ammonites of Hungary* in 2007.

📖 Fülöp 1964, 1966; Benkőné Czabalay 1965; Deák 1965; Haas et al. 1985; Császár 2002

whole collection to Munich, where he began to study all the material alongside Karl Alfred Zittel (1839–1904), one of the most prominent paleontologists of the time. After years of painstaking work, Pethő described the mollusks and the brachiopods from Fruška Gora; Zittel gave the coral fauna to another co-worker, Eduard Pratz, for study.

As the years passed, the number of specimens known from from Fruška Gora increased, and more and more problems appeared in the evaluation of the fauna. Pethő suffered from serious setbacks and was hampered by an extended illness in completing his monographic work on the subject he had first taken up as a young man. His publication on the Cretaceous invertebrates of the Pétervárad Mountains finally appeared five years after his death, in German, in the well-known journal *Palaeontographica* (then edited by Zittel). This monograph—even though it is incomplete—contributed enormously to our knowledge of the Cretaceous fossils from the region. Indeed, Pethő's former colleagues, working back in Budapest, considered it important to publish his paper in Hungarian: the Royal Hungarian Natural History Society decided to fund and print the paper—"the writing of the late chief geologist, our most illustrious one." Imre Lőrenthey was entrusted with the editorial work; the Hungarian version of Pethő's monograph, complete with 24 beautifully illustrated plates, appeared four years after the German edition in 1910.

📖 Pethő 1906

Protected Outcrops at Perşani Mountains

The Perşani Mountains are situated in the elbow of the Carpathian arc in Romania. There are two noteworthy Cretaceous sites in this region that are protected because of their geological value. One is the Upper Jurassic–Lower Cretaceous ammonite-rich succession found along the Carhago Creek, running in a northerly direction into the main gorge of the Olt River.

The other protected site of this age is found along the Falu Creek—a small tributary of the Olt—which flows through the village of Ormeniş. The outcrop is in the vicinity of the last houses of the village, and here Upper Cretaceous (Turonian–lower Coniacian) fossil-rich strata can be found cropping out in a 3-meter-high and 10-meter-long section. The site was discovered by Herbich in the second part of the nineteenth century and detailed descriptions of the fauna were given first by Ion Simioniescu (1873–1944). The Carhago Creek and Ormeniş localities were first protected by law in 1954 because of their scientific importance. Their area covers 1.6 and 4 hectares, respectively.

📖 Herbich 1878; Dénes et al. 2005

Additional Transylvanian Fossil Sites

In addition to the localities listed above, there are of course numerous other outcrops and localities in Transylvania that

A rain-washed, weathered surface of a slab packed with *Acteonella*-like gastropods in the vicinity of Ohaba Ponor.

are rich in the fossil remains of invertebrates but are not protected. From the Southern Carpathians, a nice-looking collection of Cenomanian ammonites (mainly *Calycoceras* and related forms), gastropods, and bivalves (*Acteonella, Transilvanella, Hippurites*) was obtained from Ohaba Ponor. The Upper Cretaceous mollusk fauna from the vicinity of Vinţu de Jos also became well known when Móric Pálfy (1871–1930) published his first results on the locality.

📖 M. Pálfy 1902

Vertebrate Fossils Sites of the Bihor Mountains

Early and Late Cretaceous beds occur in equal numbers in the Bihor Mountains. Some of these localities also yield—or yielded—the remains of fossil vertebrates. Although Baron Ferenc Nopcsa mentioned a *Megalosaurus* tooth, found in Borod around the end of the nineteenth century, the sites that have provided the finest and richest vertebrate remains

The broad "Valley of Dinosaurs" and the bordering low hills above the village of Sînpetru. The landscape in this area has not changed much since the first dinosaur bones were found here more than 100 years ago. Amid the vegetation, clambering on the hillside, light-colored patches can be seen caused by the intensive trampling of herds of sheep. In these places, where Cretaceous sandstone crops out, new finds of fossil reptiles can be expected even today.

This spectacular locality is called Râpa Roşie because of the reddish color of the rocks and is located 2 kilometers north of the city of Sebeş. The thick, pebbly, and sandy series seen here was deposited in a continental environment and was long regarded as Oligocene, or even Miocene, in age. Indeed, the fossil reptile bones that occur in this succession were thought to have been reworked—that is, later transported into the sediment as pebbles. Most recent finds, including dinosaur footprints, show that the majority of this succession was in fact laid down during the Maastrichtian (latest Cretaceous) and that the numerous dinosaur, crocodile, and pterosaur remains found here are not reworked but preserved in situ.

were not discovered until 1978. In the region of Pădurea Craiului, not far from the small village of Cornet, dinosaur bones were found deep in a bauxite mine, 40 meters beneath the surface. According to stories told about this discovery, when a miner urinated on the dark rock the fossil bones—also gray in color, but lighter—became visible. Among the thousands of fossil bones collected were dinosaur remains, pterosaurs, and bird bones. The Cornet bauxite mine was later closed, and today the fossil-rich layers are below the water level inside the flooded mine, which makes further collecting nearly impossible. The mine entrance has already collapsed, so the site may be closed forever.

🕮 Benton et al. 1997, Posmoşanu & Cook 2000

Dinosaur Localities in the Haţeg Basin

Some of the most famous dinosaur localities in Europe are those situated north of the Southern Carpathians, on the border of the Haţeg Basin, in the foothills of the Retezat Mountains. Many of these sites were discovered and monographed by the legendary scientist, adventurer, and spy Baron Ferenc Nopcsa. Most of the known fossil reptile remains were found in the vicinity of Sînpetru, but as Nopcsa himself wrote in 1899, "besides Sînpetru, fossils bones were also collected from Demsus, above the church, from Szacsál, along the forest road which goes to Plostina, and also at Boldogfalva, at the end of the garden of the Kendeffy family, additionally on the right riverbank of the Sibişel, NE from the village that bears the same name as the river." Further precious pieces of bone were also collected from Valioara: the bones at this locality are found in gullies and are dark, nearly black, and shiny. Nopcsa's dinosaur localities and the fossil vertebrates (including dinosaurs, turtles, crocodiles, pterosaurs, and ancient mammals) that have been collected from these sites will be discussed in more detail in later sections.

🕮 Weishampel et al. 1991; Nopcsa 1899

Cretaceous Fossils from the Carpathian Region

The Cretaceous deposits of the Alp-Carpathian region were formed under a variety of paleogeographical circumstances. Creatures of this time inhibited deep and shallow seas as well as near and offshore waters, islands, and the mainland. The richness and diversity of life in this period is reflected in the variety of Cretaceous fossils that have been found. Marine invertebrates (mainly gastropods, bivalves, cephalopods, and corals) are most abundant in the Cretaceous rocks of the Carpathian Basin, but there are sites from where rich vertebrate faunas have also been collected.

Microfossils

Cretaceous micropaleontology covers a wide range of different subjects, including the study of *Nannoconus,* a plankton organism that occurs in rock-forming quantities in certain places in the Carpathians. When it comes to certain continental and near-shore successions, knowledge of spores and pollen is highly necessary, whereas when working with

A Paleontologist in Action: How to Prepare a Nannoconus Smear

The "recipe" below is taken from the early monograph on Hungarian nannoplankton that was published by the pioneer Báldiné Beke in 1965:

> In order to prepare a sample for study, the larger portion of the rocks—which cannot be washed through a strainer—was abraded by mixing into coarse-grained sand. This fraction, along with the finer grained, washable portion, was covered with tap water and the resulting suspension boiled for a few minutes (boiling helps to avoid unwanted adherence between particles). Next, the suspension was blended with a glass stick and, after a few seconds, a few drops were taken and put onto a slide. By this time, the larger components of the suspension were already falling out, but not yet the smaller nannofossils. Drops were then evaporated by slowly warming the slide over a flame and the whole sample was covered with a coverslip affixed with Canada balsam. This study was conducted using just a standard biological microscope with magnification from 600 to 3000 times, and, of course, with oil immersion if necessary.

This is the traditional way to make a nannoplankton smear. Additionally, it is worth mentioning that more recently the widespread use of scanning electron microscope techniques (which achieve much higher resolutions than light microscopes) have resulted in new insights into nannoplankton research.

shallow marine deposits, knowledge of specific calcareous algae and ostracods is recommended. The benthic foraminifera play an important role in the Lower Cretaceous, and the planktonic forms are more important in Upper Cretaceous rocks. There is, for example, extensive literature on the Cretaceous larger foraminifera, called *Orbitolina*.

The Tiny Nannoconus and Related Forms: The Nannoplankton

Suborder Chrysophyta, within the diverse planktonic algae, are forms in which the outside slimy skin produces

Oogonium (egglike female reproductive organs, often referred to as "fruit") from the alga *Microchara* found in Late Cretaceous deposits in the Bakony Mountains (SEM images). After Rákosi 1989.

calcareous plates. These flat plates, called coccoliths, wrap around the spheroid algae and as a result, coccolithophores are a major contributor to carbonate production and rock formation in certain deep-sea sediments.

This group is of special stratigraphic importance in the Tertiary, but the chalk from the Upper Cretaceous of Northwestern Europe is also built mainly—nearly exclusively—from coccoliths. After the death of the algae, the spheres that comprise the calcareous plates fall to pieces. These tiny creatures and their remains are also classified as "nanno" (meaning tiny) planktons. In the Cretaceous layers of the Carpathians, representatives of *Nannoconus*—a genus belonging to the coccolithophores—are frequent and may even be present in rock-forming amounts in the Lower Cretaceous, contributing to the "biancone" type—white, compact limestones seen at this time. *Nannoconus* is only 5–30 microns long, formed from peg-like calcite crystals arranged into a cone-like spiral. Within this structure are tiny tubes, opened on both sides. The shapes of these cones and tubes represent characteristic systematic features.

Mária Báldiné Beke was the first to examine the Hungarian Cretaceous nannoconids. Her monograph about *Nannoconus*-bearing rocks and their facies-defining remains was considered a milestone in her field when it was published. Later, András Nagymarosy, László Félegyházy, and Attila Fogarasi also completed considerable research on Hungarian Cretaceous nannoplankton.

Based on the nannoplankton research of these workers and others a precise biostratigraphic subdivision for

Pollen grains (*Hungaropollis* spp.) from the Upper Cretaceous of the Bakony Mountains, under a light microscope (approximately 300× magnification). After Góczán & Siegl-Farkas 1989.

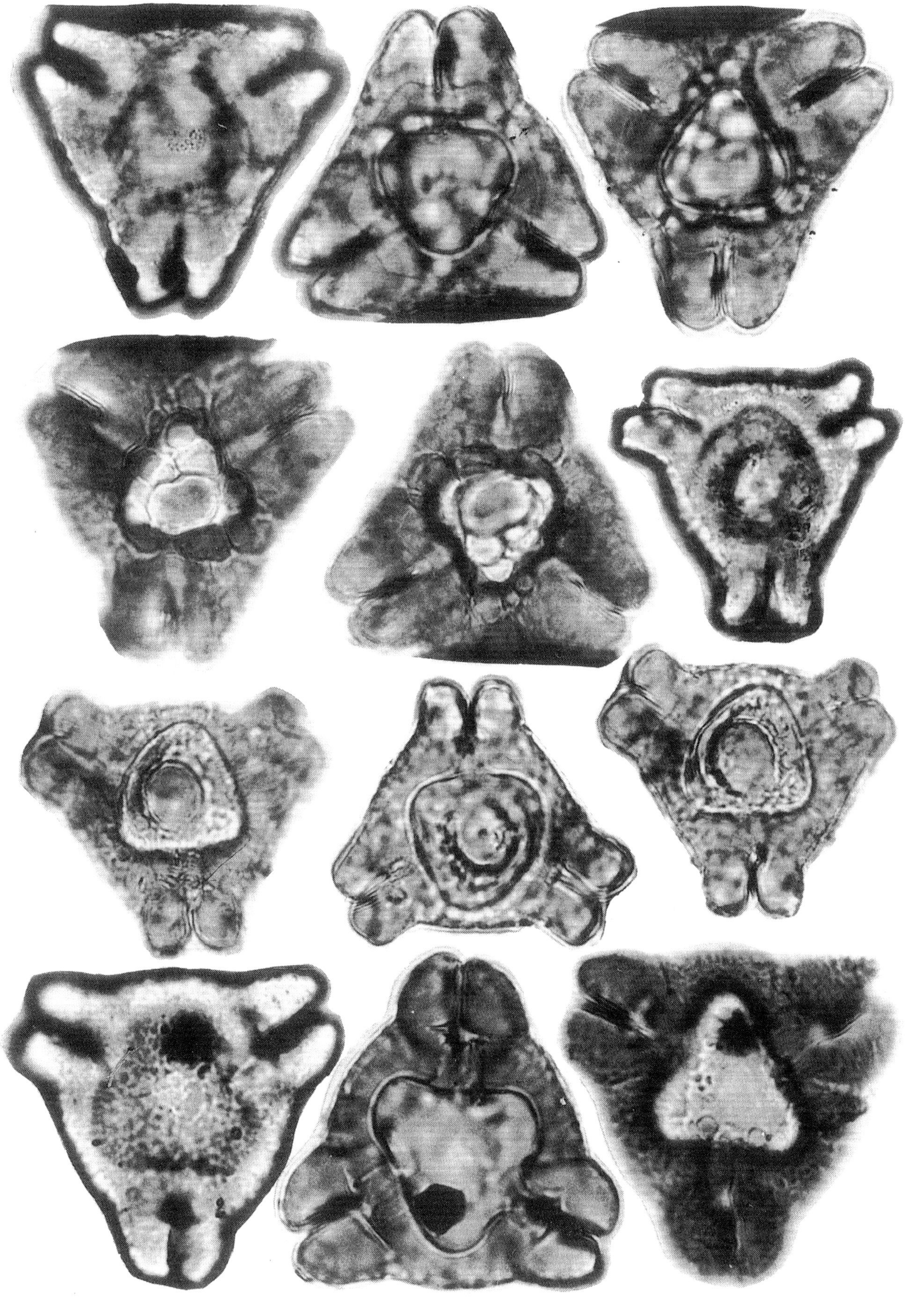

A Paleontologist in Action: How to Study Fossil Spores and Pollen

To study organic skeletal remains—such as fossil pollen grains—in hard rocks, we must take advantage of the different resistances that different substances have against acids. For example, if we dissolve the carbonate fraction of a rock with hydrochloric acid (HCl), the organic components, including any skeletal remains, will accumulate together with the other, insoluble, mineral grains in the residue. The amount of hydrochloric acid–resistant components can be large, or even overwhelming, if we prepare a sample from a marl or a sandstone. To dissolve out quartz grains and silicate minerals, hydrofluoric acid (HF) can also be used. This highly corrosive acid can dissolve even the glass containers! It is also important that neither hydrogen peroxide (H_2O_2)—an efficient chemical often used to screen rocks—nor another oxidative material is used to try to break down organic skeletal remains from rocks, since both these agents will dissolve everything that has an organic component (see also discussion of screening and washing fossils in chapter 6). After acidic treatment of a rock, resistant, heavy minerals can easily be separated from the lighter skeletal debris of biological origin by using differences in their specific gravities. According to the most recent and widely accepted terminology, "palynology" covers the study of all this skeletal debris of organic origin. Therefore, irrespective of the taxonomy of the skeletal remains, what is important is their substance and the selected preparation method.

In palynological samples there are no inorganic skeletal remains—if there were any, they would have been dissolved by the acid during treatment. The study of these remains, the inorganic component, is called micropaleontology.

certain rocks at the substage level, or even to higher resolution, can be achieved.

📖 Báldiné Beke 1965

Green Algae

The best-known algal group from a paleobotanical perspective belongs to the green algae (chlorophyta). The genus *Munieria,* a representative of the order Dasycladacea, is one of the most characteristic forms and is extremely frequent in certain layers of the Lower Cretaceous Tés Formation in the Transdanubian Range. The species *Munieria baconica* (named after the Bakony Mountains) is found in so-called Middle Cretaceous strata from Spain to the Middle East.

The genus *Chara* (commonly stonewort), belonging to Charophyceae, are also representatives of green algae. These forms resemble land plants because of their stemlike and leaflike structures, although their branching system is more complex. The axes of these algae, reminiscent of vascular plants in many ways, may be calcified and preserved; but are the fossilized spherical, or egg-shaped, remains of the female reproductive organs are found most frequently. The tiny *Chara* "fruits," 5 millimeters in length at most and usually with a twisted surface, are frequently found in the Munieria marls and the Tés Formation. Klára Rásky (1908–1971), for example, studied this group in detail and, in a 1945 article, published descriptions of seven species new to science from the Bakony Mountain areas of Kisgyón, Bakonynána, Herend, Zirc, Tés, and Hárságypuszta.

📖 Rásky 1945; Rákosi 1989

Spores and Pollen

In order to examine the reproductive cells of certain fossil plants, including the spores and pollen, rocks must be prepared in a special manner (see text box). Spores and pollen (considered collectively as microspores) can be seen using either a light microscope or an electron microscope. Specialists have developed a thorough nomenclature to describe the diverse morphologies of these reproductive structures, found as fossils most often in continental and shallow marine deposits. Spores and pollen can vary in shape and size, in their symmetry, and in the decorations on their surface. Surface decoration can comprise ridges, grooves, protruding hooks, spines, and depressions. The number and positioning of airbags is also an important character.

Despite their huge diversity in morphological structure, the possibility of classifying spores and pollen in a natural system, based on their evolution, is limited. Thus, the systematics of Mesozoic spores and pollen is essentially artificial.

Several researchers have studied Cretaceous spores and pollen from Hungary. The work of Margit Deák was dedicated to the flora of the Lower Cretaceous Muneria marl. Ferenc Góczán and Ágnes Sieglné Farkas published on palynomorphs from the Upper Cretaceous of the Bakony Mountains, adding valuable data on the microflora of Senonian layers from outcrops in northern Hungary and from boreholes into the basement of the Great Hungarian Plain.

The Importance of Calpionellids

Calpionellids are calcareous microplanktonic protists that characterise the Late Jurassic–Early Cretaceous interval in

A Paleontologist in Action: Preparing Orientated Sections from Orbitolina

To understand the taxonomy and evolution of the larger Cretaceous foraminifera, it is essential to study their so-called embryonic apparatus, the first part of their tiny tests to evolve. The size and shape of the first chamber (proloculus) and the number and arrangement of the surrounding subsequent chambers are also important taxonomic features that can be studied in an orientated section of *Orbitolina*.

First, melted resin, or Canada balsam, is dropped onto a slide. Then, an *Orbitolina* specimen is pushed into the congealing material. The more or less disk-shaped fossil must be set on its edge, perpendicular to the slide; then, using abrasive powder (emery), about half of the specimen is cut away. Approaching its plane of symmetry, work must continue with great care, regularly controlling the fossil under the light microscope. With luck, the embryonic apparatus can be seen—the first chambers of the foram in the middle of the apical part of the test. Next, the sample is heated gently to remove the fossil, so that it can be glued into a reverse position, attached to the glass by its already smooth side. In this way a standard thin-section can now be made, using progressively finer abrasive grit until the sample is only 30 microns in thickness. If accurate enough, the orientated cut will be in the middle of the proloculus—and the embryonic apparatus of the foram can be studied, drawn, or photographed.

the Tethyan realm. They are especially useful from a biostratigraphic point of view but their study is basically only possible using a thin-section method. A brief description of the structure of these tiny, calcareous creatures—including the most important genera—is provided in chapter 3.

Cretaceous Foraminifera

Lászlo Majzon, the most prominent scholar of Hungarian Cretaceous foraminifera, wrote the following about Cretaceous forams:

> Foraminifera had an important role during the Cretaceous period. Some genera appear for the first time in Cretaceous deposits and this makes a difference with respect to the previous—Jurassic—sediments. In general terms, we can say that the Cretaceous calcareous forms were bigger than their Jurassic forerunners. But among the agglutinated types, there are specific well-developed forms too. In certain Lower Cretaceous sediments, the species of *Orbitolina* occur in rock-forming quantities while in the Upper Cretaceous *Gryphea* Marl, *Quinqueloculina* is especially common. Similar observations can be made based on certain Hungarian *Globotruncana* species in the Senonian.

In his voluminous 1966 handbook on foraminiferal research, Majzon listed nearly 200 different Cretaceous forms.

Following his pioneering work, Hungarian micropalaeontologists contributed much to our knowledge of the smaller Cretaceous foraminifers. Anikó Bércziné Makk studied mainly Cretaceous forams from boreholes in the Great Hungarian Plain—an important topic related to hydrocarbon exploration; and Ilona Bodrogi—among others—produced detailed descriptions of the foraminiferal fauna of the Albian–Cenomanian Pénzeskút Marl Formation in the Bakony Mountains.

📖 Majzon 1966; Bodrogi 1989

Larger Foraminifera of the Cretaceous: Genus Orbitolina and Allied Forms

Orbitolina is a Cretaceous genus of foraminifera that is readily visible to the naked eye. This agglutinated form

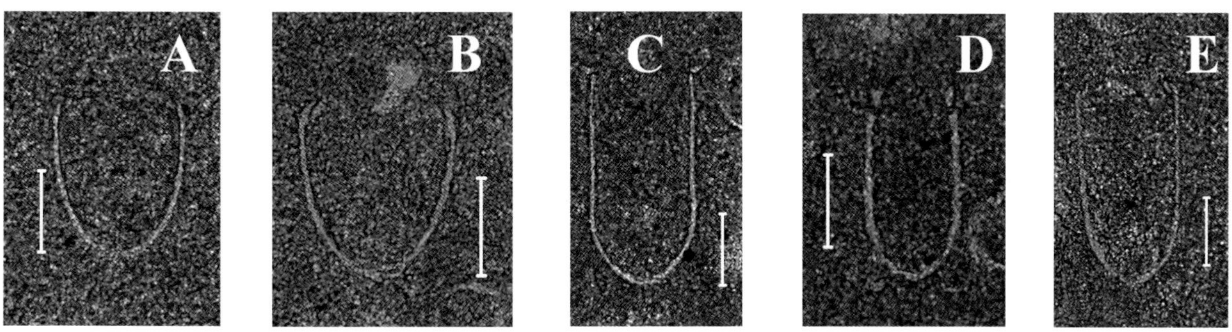

A series of different calpionellid cross sections seen in thin-section. The tiny remains of these planktonic creatures are abundant in sediments from across the former Tethys Ocean. The specimens illustrated here are from an Early Cretaceous section near Hárskút (Bakony Mountains). (A, B) *Calpionellites darderi* (Colom). (C) *Praecalpionellites dadayi* (Knauer). (D) *Calpionellopsis simplex* (Colom). (E). *Praecalpionellites siriniaensis* Pop (scale = 0.05 mm).

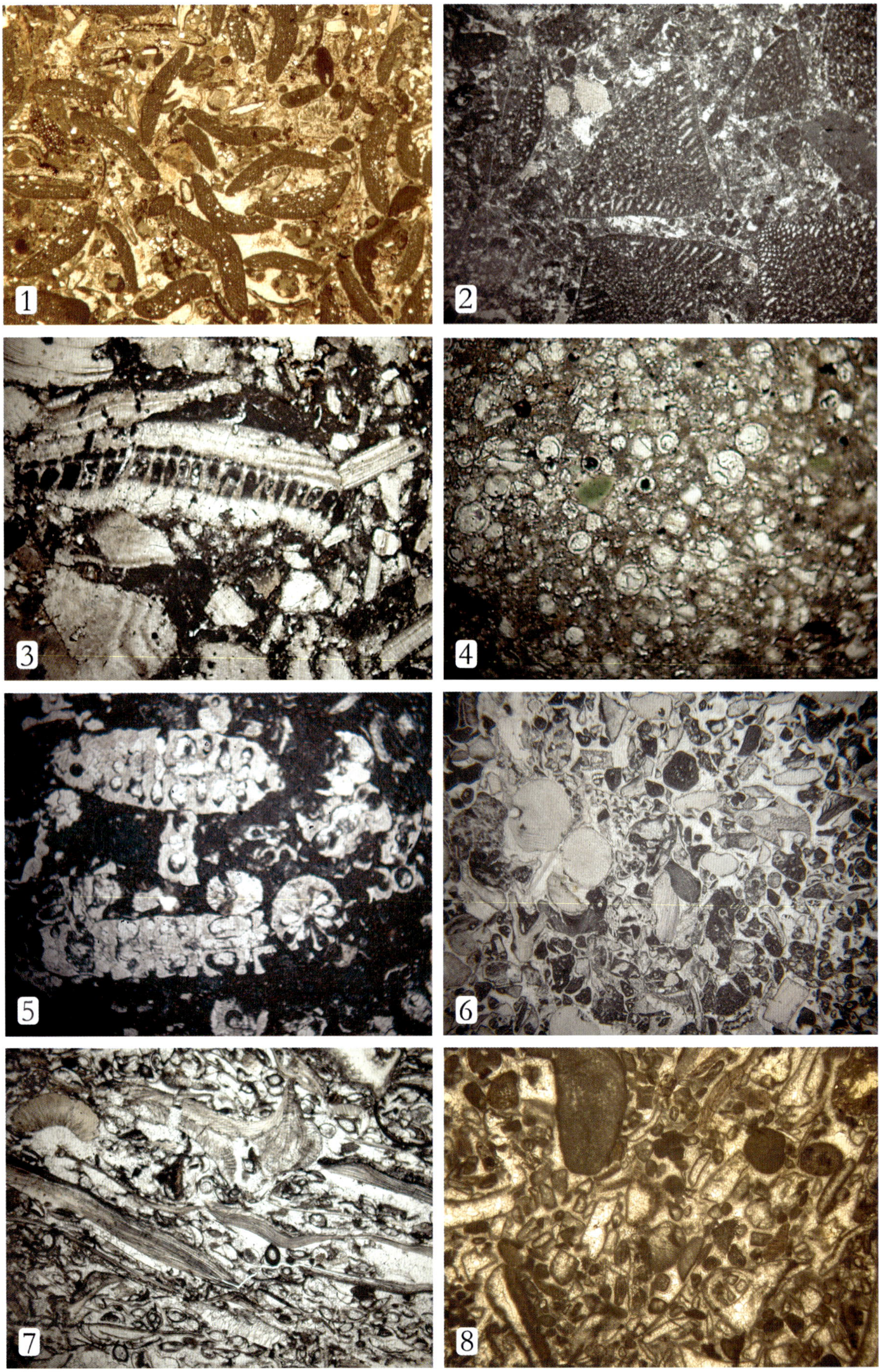
1
2
3
4
5
6
7
8

generally reaches a few millimeters, and occasionally a few centimeters, in diameter. Specimens of *Orbitolina* have a characteristic flat or conical shape, and locally may appear in masses, such as in the Aptian–Cenomanian deposits of the Transdanubian Range and in the Villány Mountains. Orbitolina limestone is also known from boreholes from the foothills of the Vértes Mountains. *Dicyclina, Cuneolina,* and some other genera are closely related: some species, in particular, have very short ranges and therefore are of important biostratigraphic value.

In Hungary, Kálmán Méhes (1911–1988) dealt with this genus for the first time. He was one of the first workers to emphasize the importance of the study of the embryonic apparatus during taxonomic work. Later, Ágnes Görög and Felix Schlagintweit published work on Hungarian *Orbitolina*-rich deposits.

📖 K. Méhes 1964; Görög 1993

Ostracods

In certain Cretaceous deposits, tiny—mostly disintegrated—shells of ostracods occur quite often. Béla Zalányi, for example, described 51 species, including 39 new to science, from the Munieria marl in the Bakony Mountains in his 1959 monograph. The most common faunal element in this marl is *Gomphocythere baconica* Zalányi, which appears in rock-forming quantities only occasionally. On the basis of the ostracod fauna, Zalányi distinguished three successive intervals within this Aptian marl: a lower, freshwater part, a middle brackish portion, and an upper marine section. Generally speaking, ostracod faunas, similar to benthic foraminifera, provide good information about paleoenvironment. Ostracods, in particular, reveal information about water depth, distance from shore, changes in salinity, and oxygen content through a succession.

The ostracod fauna of the Cenomanian Pénzeskút Marl (also known as the Turrilites marl in old Hungarian literature) is also rich and well documented. Based on samples from boreholes and surface outcrops, Miklós Monostori determined 15 different species present in this rock. The fauna is characterized by high abundance of the genus *Cytherella*. Many Cretaceous ostracod faunas are waiting for detailed study.

📖 Zalányi 1959; Monostori 2000a

Fossil Plants

The last important change in the evolution of the Earth's flora took place in the Cretaceous: during this time, flowering plants (angiosperms) first appear, and have ever since played a key role in modern ecosystems. In the Early Cretaceous, gymnosperms (conifers and cycads) were dominant in global floras, but this changed as angiosperms took over by the second half of the period, about 100 million years ago. From Hungary, and also the Carpathian Basin, mostly primitive plants (algae) are known from this time. Beautiful and spectacular plant macrofossils are considered rare.

A Few Finds from Hungary

Among the varied Cretaceous rocks that are found in the Carpathians, some were deposited in shallow marine environments as well as near shore and also on dry land: it would be logical to think that the fossil remains of plants would be frequently present. Surprisingly, this is not the case. If we do not take into consideration the sporadic presence of masses of primitive plants (algae), and the fairly common occurrence of spores and pollen in continental deposits, we can say that the remains of plants in the Cretaceous of Hungary and across the Carpathian Basin are very uncommon.

It is hard to say for sure why Cretaceous fossil plants are so rare from the Carpathians. Luck is often an important factor in the fossilization process, and this manifests itself via quick embedding. Developed terrestrial plants usually

Fossil-rich Cretaceous rocks in thin-section (at different magnifications). (1) The mass occurrence of flattened, croissant-like *Orbitolina* in an Albian calcareous marl (Tés Marl Formation, Bakony Mountans) at Zirc. (2) A limestone densely packed with conical *Orbitolina* remains. The minute, elongated chambers, larger on the edge of the test and smaller inside the *Orbitolina*, can clearly be seen. The embryonic apparatus—an important feature for taxonomy—is in the plane of symmetry and cannot be seen in these specimens because they were not cut exactly in the middle. (3) Mass occurrence of rudist bivalve fragments from the Zirc Limestone Formation (from 4.5 meters down the borehole Tés-27). A characteristic feature is the inner gridlike structure of the thick bivalve valves. (4) An Albian marl, rich in Calcispherulidae remains, from 444 meters down the borehole Jásd-42. Calcispherulidae are small-sized, spherical remains of uncertain taxonomic affiliation—in this case probably the reproductive organs of sea algae. In thin-section their relatively thick shells can usually be seen; this is a distinctive feature because similar-looking (also spherical, and hollow) radiolarian remains are most often filled with compact, recrystallized calcite. The green spots in this thin-section are fillings of the mineral glauconite. (5) Longitudinal and transverse cross sections of calcareous green algae (*Munieria*) remains from the borehole Zt-61 in the Albian Tés Marl Formation. (6) Echinoderms in rock-forming quantity: sea lily fragments closely packed together in the Aptian–Albian Tata Limestone. (7) Thick-shelled *Ostrea*-like bivalves surrounded by tiny ostracods (with both valves preserved in most cases) from 40.8 meters down the borehole Tés-27 in the Tés Marl Formation. (8) Eroded fragments of sea lilies and mollusks in the Tata Limestone.

A carbonized tree trunk from the forest of dinosaurs. In the Late Cretaceous Iharkút deposits, now widely known because of their fossil reptile remains, fossil plants are also fairly common. Most of them are indeterminate tiny pieces of branches, leaves, or fruits, but this fossil is a giant! The visible part of the slightly curved, tapering trunk is about 8 meters in length. Showing no branching, and highly flattened because of the pressure, this trunk is from an ancient gymnosperm. Photographed from above, the lower part of the trunk continues into the wall of the pit, therefore its complete size remains unknown. The two parallel dark strips are not branches, but the track of a huge lorry that drove over the tree trunk.

lack a hard skeleton (in contrast to calcareous algae) and, thus, are difficult to bury and preserve as fossils.

Among the few remains of fossil plants that are known, those from the Lower Cretaceous marl of the Bersek Hill at Gerecse should be mentioned, as well as those preserved in bauxite at Iharkút. The finds from Bersek Hill have long been known; they are strongly flattened and suffer from carbonization. In contrast, the flora from Iharkút consists of beautiful, large leaves that were found only recently, following the discovery of dinosaur remains and as a result of mapping work at the fossil site. Processing the Iharkút fossil plants has not yet been completed.

📖 Andreánszky 1948; Rásky 1954

A Palm Leaf from the Valley of the River Jiu

A beautiful fossil palm leaf (*Juranyia hemiflabellata*), from the Jiu valley in Translyvania, was described by János Tuzson. This gigantic fossil leaf, collected from the Upper Cretaceous, was named after professor of botany Lajos Jurányi (1837–1897). Palm trees appear in the Upper Cretaceous, so *Juranyia* is an early representative of this group.

Invertebrates

Cnidarians

Research on Cretaceous Corals

The Hungarian Cretaceous is not very rich in coral remains, but a nice-looking and relatively diverse fauna has been found in some localities. Only a few researchers, however, have studied this topic in any detail: Gábor Kolosváry has produced the most publications on the Hungarian fossil coral fauna, including that known from the Cretaceous. Kolosváry collected fossils himself and

Late Cretaceous landscape. In the foreground is a reconstruction of *Juranyia*. *Juranyia hemiflabellata* is a relatively complete fossil palm: its leaves may have reached 2 meters in length and had a 4–6-centimeter-thick, 60-centimeter-long stem. This fossil is an intermediate between *Pandanus*-like and *Palmetto*-like forms.

Silicified wood trunks near Lancrăm. In spite of its name—Sebeş means "swift" in Hungarian—this tributary of the Mureş meanders leisurely in the vicinity the village of Lancrăm. The pebbly, clayey riverbank deposits were considered to be Miocene until October 2000, when Mátyás Vremir and his colleagues from the University of Cluj Napoca (Romania) discovered the first Transylvanian dinosaur tracks on a bedding surface. This proved a Cretaceous age for these sediments. The huge crag-like blocks on the right and left bank of the river are also fossils: these are pieces of petrified—more precisely, silicified—tree trunks. Some of them reach more than 10–20 meters in size. While the smaller fragments found along the shore are rolled away by the river, the bigger specimens are passed round by the running water and are slowly abraded. During the Late Cretaceous it is very likely a similarly sized river accumulated these logs, which subsequently became silicified.

also studied the collections of former geologists. His most important contribution to the subject is a paper on the Cretaceous corals of Hungary, published as part of the Yearbook of the Hungarian Geological Institute in 1954. Interestingly, in the same year another monograph on Cretaceous coral faunas was published by a very young scholar, Barnabás Géczy. He was later to contribute a great deal toward a better understanding of Jurassic ammonites, but his outstanding paper on corals dealt exclusively with the genus *Cunnolites* (earlier called *Cyclolites*)—of special importance amongst Cretaceous solitary forms. A description of the coral fauna from the Fruška Gora Mountain was published by Eduard Pratz in 1910, and Slovenian specialist Dragica Turnšek authored a modern description of Early Cretaceous corals from the Mecsek Mountains.

📖 Géczy 1954; Kolosváry 1954; Császár and Turnšek 1996

Some Important Coral Localities

Most of the Cretaceous coral known from the Bakony Mountains (mainly *Cunnolites*) comes from strata referred to the Csingervölgy Member of the Jákó Marl Formation, in the vicinity of Sümeg and Ajka. Additional specimens are known from the older (Albian–Cenomanian) Pénzeskút Marl, which outcrops near Zirc, Pénzesgyőr, and Olaszfalu. A fairly diverse coral assemblage, including solitary and colonial forms, is known from the Lábatlan Köszörűkő Quarry, which is in the northern foothills of the Gerecse Mountains. Kolosváry also gave a detailed account of the Upper Cretaceous coral faunas from different localities in the Uppony Mountains (Northern Hungary) and listed corals from northwest of Bántapolcsány and also from the vicinity of Nekézseny (Jőcsös Valley and railway cutting). From the Mecsek Mountains we know about Early Cretaceous corals from localities at Mecsekjánosi, Gadány Hill, Korhadtfás Ravine, Márévár, Magyaregregy (Hodács Valley), and Zengővárkony.

Instead of providing a long faunal list of the coral species that have been determined, here is a summary of the Hungarian fauna as given by Kolosváry in his 1954 paper: "We know 35 species form the Senonian of Nekézseny and Bántapolcsány, 19 from the Senonian of Ajka, 14 from the Senonian of Sümeg, 1 or 2 species from the Cenomanian of Pénzeskút and Olaszfalu, 5 Aptian forms from Lókút and Pénzeskút, 15 from the Neocomian of Lábatlan, and 12 from the Early Cretaceous of the Mecsekjánosi." Indeed, only reevaluation of the Mecsek fauna has contributed new results to this list. Cretaceous corals—mainly *Cunnolites* and related forms—are known from many other non-Hungarian localities as well: including Borodul Mare, Vinţu de Jos, Alba Iulia, Inuri, and Čerević.

Results from the Study of Cunnolites from Sümeg

Barnabás Géczy studied more than 4,500 *Cunnolites* specimens—most of them collected from Sümeg, and nearby localities—and distinguished 17 taxa (species and subspecies) within the fauna. As part of his work, he also developed a key for species determination in which the convexity of the calyx, the length of the depression in the middle of the calyx, the shape of the skeleton, and the base and the presence or lack of the epitheca were considered the most distinctive features.

The range of *Cunnolites,* even within the Late Cretaceous, is relatively narrow: this group, so diverse toward the end of the Mesozoic, became extinct by the Eocene. The possible closest ally of *Cunnolites* is the genus *Fungia* (and related forms), which appears in the Tertiary. These mushroom corals (Fungiidae) are generally solitary marine animals, but occasional colonial forms also exist. In favorable conditions these corals cover the seafloor en masse, at a maximum depth of 80 meters. Most species remain fully detached from the substrate in adulthood and all can reproduce both asexually—either via gemmation (budding) or division—and sexually.

Cunnolites specimens from Sümeg. On specimens photographed from above, the septa and the elongated depression (calyx) can be seen. Concentric growth lines and radial ribs appear on the base (1.5× magnification).

The *Cunnolites* specimens from Sümeg can be found in the marl in a rather chaotic way, suggesting that they were living free on the surface of the muddy sediment and were washed together by occasionally stronger currents. The wide base of these corals prevents them from sinking into the soft sediment. This kind of adaptation to soft-bottom conditions is known as a "snowshoe strategy." Surprisingly, in spite of the large numbers of the studies specimens, no hint of asexual reproduction has been proven. The different irregularities on the skeletons could be explained as evolutionary anomalies.

Géczy paid special attention to the overgrowing organism as well, an interesting subject from a paleobiological point of view. In many cases, worm tubes were found attached to these corals, even during the life of the cnidarian. The calcareous worm tube was growing around the coral, whose skeleton became distorted (deformed). Two new parasitic species of worm (*Serpula cyclolitophyla, Serpula hemisipunclida*) were described from Sümeg.

📖 Géczy 1954; Kolosváry 1957

Gastropods and Bivalves

Gastropods and bivalves are conspicuous fossils in many Cretaceous deposits throughout the Carpathian region. Gastropods are frequently encountered in freshwater and shallow marine sediments, whereas bivalves occur—often in rock-forming quantities—in nearly every aquatic formation. Many of the traditional formation names in the Carpathian region (for example, the Gryphaea marl and the Inoceramus marl) refer to the characteristic bivalve found in the sediment.

Gastropods

Remains of gastropods are found in nearly every sedimentary Cretaceous formation—from terrestrial sediments to deep marine deposits. However, because it is hard to collect these fossils from limestones, well-preserved, spectacular gastropod faunas are rare and mostly come from marls and clayey sediments. On the surfaces of Cretaceous limestones, the cross sections of gastropods are often recognizable, but are very hard to hammer out. However, this is often not a problem because they can be identified in cross section. Nerinellidae (*Nerinea* and related forms) is one peculiar group that flourished during the Cretaceous: these

A spectacular-looking, fan-shaped colonial coral (*Diploctenium*) from the Upper Cretaceous of Sümeg (2× magnification).

gastropods differ from others because they have longitudinal inner wrinkles, which narrow the inner space of the shell. Nerinellidae first appeared in the Early Jurassic but by the Cretaceous they inhabited the same habitats as rudists and went extinct alongside them. Most were elongate, tall, and slender—but some were stockier, almost egg shaped. Unlike most other gastropods, which can be identified on the basis of the outer ornamentation of their shells, these forms can only be sorted from their internal anatomy. It is thus necessary to make a cut across, or close to, the axis of the spiral, because the important systematic features are often the inner wrinkles. Representatives of different species, or even genera, may have very similar, or even identical, outer appearances.

In Hungary, almost all shallow marine Cretaceous deposits contain fossil gastropods. *Nerinea*-related forms can be recognized on the surfaces of Early Cretaceous limestone cliffs in the Villány Mountains; the poorly known Early Cretaceous gastropod fauna of the Mecsek Mountains is also composed mainly of these fossils. In contrast, the gastropod fauna of the Bakony Mountains is better known and has been well studied; the oldest mass occurrence of these gastropods is from the basal layers of the Aptian Tata Limestone. Unfortunately, this rich assemblage is so far known only from faunal lists. Albian gastropods are much better documented: although the gastropod fauna of the "Munieria marl" (Tés Marl Formation) was studied by Elemér Vadász, this rich assemblage, containing about 70 species, is still not fully described. Lenke Czabalay published on 7 *Nerinea* and 4 *Nerinella* species from the rudist-bearing Zirc Limestone Formation, and suggested the presence of nearly 20 other rarer taxa. She also published a monograph on the relatively poorly preserved—and consequently difficult to determine—gastropod fauna of the Pénzeskút Marl Formation.

The Late Cretaceous gastropod fauna of the Bakony Mountains is richer and better known than those of the Lower Cretaceous. The base of this succession is formed

A Paleontologist in Action: Making Serial Sections from Fossil Corals

Making serial cuts is a traditional way to study the ontogenetic evolution of fossil corals. This method—introduced about 100 years ago—led to interesting results, especially in Paleozoic corals. In this way, the evolutionary relationships among different ancient groups were revealed and the mode of life of some already extinct groups was suggested. Unfortunately, in the case of the highly recrystallized corals from Sümeg, this method is not applicable; but, quite surprisingly, making cuts from the embryonic portions of these specimens worked well. In the words of Géczy:

> Before polishing a specimen, I described the fossil and made a sketch. Then the upper surface was attached to a slide with melted Canadian balsam. The horizontal position of the fossil was manipulated under a light microscope and abraded with a sheet of glass. No abrasive powder was used. After polishing the coral on one side, the balsam was heated; the specimen was removed and re-attached by its basal surface. In some cases both transverse and oblique sections were prepared. The inner structure of the coral was studied during the process and both sides of the thin-section were observed under transmitted light so that each form could be drawn.
>
> One ideal requirement—a serial photograph of the same specimen—was not achieved because of technical reasons, although the high number of studied specimens (109) allowed us to make a detailed reconstruction of its ontogeny.

Finally, young *Cunnolites* specimens from Sümeg were arranged into four groups, revealing the relationship between different species groups.

📖 Géczy 1954

1
2
3
4
5
6
7
8
9

from freshwater deposits, which also contain a coal seam in the vicinity of Ajka. These barren beds, and also the coal itself, contain a rich mollusk fauna overwhelmingly dominated by gastropods. Because of their compaction, these coaly layers have yielded some heavily flattened fossils; only the interbedded gray marls contain well-preserved specimens. By far the most common of these are the markedly beautiful shells of the genus *Pyrgulifera,* the focus of paleontological interest following the start of mining activity in this area in the nineteenth century. The most common forms were illustrated by Miksa Hantken in a comprehensive 1878 book that reviewed the coal seams and coal mining in the countries of the Hungarian crown. Hantken noticed that among the gastropods found here not only were forms that fitted with the name of the genus (*pyrgulifera* means "little tower bearing"), but also some relatively smooth forms that have only a minute spiral shell.

Later Hantken was invited to head the paleontological department at the University of Pest, founded in 1882, and so the first monograph on the mollusk fauna of this coal-bearing succession was published by the Austrian geologist Leopold Tausch (1858–1899). Tausch found that the 6 bivalves and nearly 80 gastropod species and subspecies recorded from this site were living in a non-marine environment. It is also clear from his publication that he was able to compare very few species of the Ajka fauna with known forms from other freshwater localities of more or less the same age, such as the Gosau Beds of the northern Alps or some successions of South Eastern France. Tausch supposed instead that the Cretaceous gastropods of Ajka were the fossil relatives of faunal elements from the present-day Lake Tanganyika.

The bewildering diversity of the Ajka *Pyrgulifera* has not been lost on later researchers. In 1892, Paul Oppenheim (1863–1934), a specialist on Cretaceous and Eocene mollusks, described a new taxon and reviewed the fauna. Erzsébet Neubrandt, later a professor at Eötvös Loránd University and an expert on Triassic megalodontid bivalves, reported in 1949 on some extremely large specimens that probably differ from other known *Pyrgulifera.*

The next publication on this topic is from 1962 and was written by Ferenc Bartha, best known as a specialist on Pannonian mollusks. Bartha tried to explain why in the lower part of the coal-bearing succession relatively smooth, thin-shelled phenotypes occur, and thick-shelled taxa that are ornamented by ribs and spines appear only later in the sequence. Bartha suggested that this must be a result of increasing salinity, due to a marine transgression.

The most recent description of this gastropod assemblage was provided by German scientists Klaus Bandel and Frank Riedel, who published their lavishly illustrated paper in the annual report of the Natural History Museum of Vienna. Because of the exceptional preservation of the fauna, they were able to study the embryonic stage of gastropod shells, which is often the only reliable taxonomic feature for fitting an extinct genus into a higher taxonomic rank. In the assemblage, formerly well known on the basis of adult shells (the so-called teleoconch), they distinguished six new genera, of which two (*Czabalaya* and *Szaboella*) were named for Hungarian gastropod researchers. Bandel and Riedel also came to an interesting conclusion regarding the most common taxon in the assemblage, *Pyrgulifera* itself: although the fauna is very diverse, it is impossible to distinguish different species. They also pointed out that living genera in Lake Tanganyika (*Lavigeria* and *Paramelania*), albeit similar in shape, have markedly different juvenile stages. These are not related to the *Pyrgulifera* fauna of Ajka and can be clearly distinguished.

Above this marine deposit, a coaly succession also contains gastropods. Most characteristic of these is *Cassiope,*

Cretaceous gastropods (original size). (1) *Cimolithium inauguratum* (Stoliczka)—The Upper Cretaceous "Inoceramus marl" in the Bakony Mountains, deposited in a deeper-marine environment, is poor in gastropod remains. This image shows one exceptional fossil find. This large-sized form is known not only from the Bakony Mountains and from the Alps, but also from India, Madagascar, Iran, and Libya. (Sümeg, Haraszt Quarry) (2) *Cassiope* sp. Late Cretaceous lagoons were inhabited by a range of gastropods, among them *Cassiope,* which is known as *Glauconia* in older paleontological literature. *Cassiope* is one of the most common and abundant forms (Sümeg, Hárs Well). (3, 4) *Nerinella* sp. *Nerinella* and related gastropods were typical faunal elements in tropical, shallow marine environments in Jurassic and Cretaceous times. One feature common to these gastropods is a longitudinal wrinkle that makes their inner space tighter and may have served as protection against predators. In addition, the number and shape of these wrinkles are characteristic systematic features: as a result, these gastropods can be identified only in cross section: (3) shows the rather smooth outer surface of the shell, and (4) shows a longitudinal cut, close to the axis of coiling of the gastropod. The inner space is painted in this image (Grădistea de Munte, Hunedeoara County, Cenomanian stage). (5, 6) *Pyrgulifera* sp. Among the diverse representatives of this genus, there are a number of both rather smooth and heavily ornamented types. Extremities are connected by a series of intermediate forms (Ajka). (7–9) *Acteonella* sp.—*Acteonella*-type gastropods, like *Nerinella,* were quite common in warm, tropical Late Cretaceous seas. Occasionally beds are full of these often large-sized shells, and it has been supposed that some forms may have had photosymbiontic algae living within their mantles. (7) *Acteonella* sp. (Hunedeoara County). (8) *Acteonella goldfussi* (Cherges̗, Hunedeoara County). (9) *Acteonella* (*Trochactaeon*) *gigantea* d'Orbigny (Ohaba-Ponor, Hunedeoara County).

formerly referred to as *Glauconia*—which inhabited Cretaceous lagoons all over the world, in large numbers and with great diversity in shape. From the "Mollusk marl" at the Hárskút ("Hárs Well") locality, near Sümeg, 5 species and subspecies of nearly 40 different phenotypes were distinguished by Lenke Czabalay in 1964. In contrast to the richness of the shallow marine deposits, this deepwater "Inoceramus marl" seldom contains gastropods.

📖 F. Bartha 1962; Benkőné Czabalay 1964b; Bandel & Riedel 1994; Tausch 1886

Rudists: The Most Peculiar Bivalves of All Time

Although generally ordinary looking fossils, bivalves have experimented with a number of bizarrely shaped valves over their nearly 600-million-year history. Of these experiments, even among the most unusual, the rudists were the most prominent. Because of their often very large size and bizarre shape, they are occasionally referred to as the "dinosaurs of the bivalves."

Named for their often huge and heavy shells, rudists originated from within the Late Triassic megalodontid bivalves. Their other traditional name, "pachyodonta," refers to the few, but large, differentiated hinge teeth that hold together the valves of their shells. Rudists began their evolution during the Late Jurassic; it is thought that their direct forerunner was a *Diceras*-like bivalve. The Jurassic forms were elongate, with both valves similarly shaped and markedly twisted. Until they mysteriously went extinct at the end of the Cretaceous period, rudist bivalves inhabited warm and shallow marine environments all over the world.

Some rudist taxa were free living on the seafloor, and had a curved shape, but the classic morphology of these bivalves consisted of a lower, more or less conical valve that was attached to the seafloor, and a smaller, upper valve that served as a kind of lid for the organism. The small upper valve took a variety of interesting forms, including a simple flat lid, a low cone, a spiral, and even a star shape. The curved rudists stood close together on the seafloor like "carrots in the earth," as noted by the German geologist Johannes Walther (1860–1937). His valuable *A Föld és az élet története* (The history of Earth and life), was published by the Hungarian Royal Society of Natural History in 1911.

One reason for the success of the rudists is thought to be that during their evolution the lower valve of a typical bivalve twisted and formed a cone-like shape, and as a result the animal was able to avoid burial by rapid sedimentation and was able to grow upward. Further evolution resulted in the replacement of the ligament (hinge)—a fibrous, flexible organic material, generally on the outside of the clam—which opened and held together the bivalve shell. In the case of rudists, the ligament developed instead within the shell, and was replaced by an inner channel. As a consequence, torsion of the valve became unnecessary, and at the same time, secretion of the valve all along the mantle became possible. This resulted in a regularly growing cone-like funnel shape, or horn-like lower valve. These latter shapes were dominant and are exemplified by the Late Cretaceous genus *Hippurites* and allied forms. These animals supported each other on the seafloor: by doing so they were able to maximally exploit the possibilities offered by a nutrient-rich environment.

In addition to the shape of the valve, rudists were also characterized by another innovation. Their thick shell evolved into a hollow, cellular structure, meaning that a bigger, higher, thicker, and more efficient and more resistant shell could be produced from the same amount of calcium carbonate. *Radiolites* and related genera had a very thin upper valve, which is generally accepted to have been translucent during the life of the animal. This is thought to have allowed symbiotic algae to flourish inside the bivalve, similar to those seen in the giant clam alive today (*Tridacna gigas*).

It has often been written that rudist bivalves built reef structures; they did, but only occasionally. These bivalves developed structures mainly parallel to bedding surfaces, whereas real reefs are characterized by growth perpendicular to bedding.

Since rudists mainly inhabited very shallow marine environments, and because ammonites are mostly missing from this type of facies, age estimates for the formations that contain these bivalves are generally based on microfossils, or even on the rudists themselves. However, age determination has proven difficult in many cases, especially for forms described from the Alps and from the Carpathian region, and has resulted in a number of incorrect interpretations. These days, thanks to a sophisticated geochemical process—measuring the ratios of different stable strontium isotopes—it is possible to accurately determine the geological ages of the different Alpine rudist-bearing successions.

Rudists in, and around, the Carpathian Basin

In the Jurassic formations of the Carpathian Region rudists are quite rare, but in the Cretaceous they are more abundant. Forms found in Cretaceous-aged rocks can be subdivided into two types, which differ in their age and type localities. The first of these types is Urgonian—Early and Middle Cretaceous rudist facies that are named after a rock that forms a spectacular cliff in southeastern France, in the vicinity of Orgon. The second type is the Late Cretaceous rudist-bearing facies, called the Gosau type after a locality in the Northern Calcareous Alps where it was first studied in detail.

Urgon-type formations, although discontinuous, can be found at the surface all along the Carpathian arc. They also crop out in the Mecsek and Villány Mountains and along the Transdanubian Range. In this rock type gastropods and bivalves are very frequent, although the precise fossil content changes from bed to bed. This variable fossil content reflects continuous changes in the environment at the time the rock was forming, and indicates changes in the salinity, oxygen content, and temperature of these Cretaceous lagoons.

On weathered surfaces of Lower Cretaceous limestone in the Villány Mountain, rudists are conspicuous fossils. In the older literature this formation was described as "Valletia limestone"; however, the genus *Valletia* does not occur here. Representatives of the genera *Agriopleura, Toucasia, Caprina, Requienia, Eoradiolites, Monopleura,* and *Praecaprina* are found instead. Although it is impossible to extract the fossils from the hard rock, it is nevertheless possible to determine them, since rudist taxa can be determined from cross-sections. Indeed, in most cases, making sections perpendicular to the growing direction of the rudists is not only helpful but necessary in order to determine the different genera and species.

The Lower Cretaceous rudist faunas of the Mecsek Mountains have yielded only a few species; in spite of this, they are very interesting. The most striking feature of these faunas is their strongly endemic character: four of the seven species known from these mountains were new to science when described. Indeed, in one case (*Bicornucopina*), it was necessary to introduce a new genus name. The Mecsek taxa have not been recorded from younger rocks at other localities: it seems that their fragile valves, which slipped down the slope of this ancient atoll and were subsequently buried, provide a unique snapshot of rudist evolution not preserved in other places.

The Urgonian facies of the Transdanubian Range is younger than those found in the Mecsek Mountains, and also has a very different faunal content. In Transdanubia, the richest assemblages come from the rock known today as the Zirc Limestone Formation. This rock was known in the past as either the "Caprotina limestone," or the "Requenia or Agria limestone" after its typical faunal elements. The fossil shell found in this rock is sienna in color, which contrasts with the yellow-gray limestone. The most frequently encountered taxa are the strongly twisted rudists *Toucasia* and *Pseudotoucasia,* although the funnel-shaped *Agriopleura* is also relatively common. Besides the dominant *Requenia*-related forms, certain species of *Radiolites* (*Eoradiolites*) also occur; but because of the hard surrounding rock and the complex shape of the bivalve, complete specimens are only occasionally collected. Similar taxa were also documented from limestone blocks in the Köszörűkő Quarry near Lábatlan, and there are numerous Urgonian occurrences of similar age all over the Carpathian region. Of these, and of special interest, is one that contains dozens of species that was recently described from Ohaba, near Haţeg.

Rudists flourished toward the end of the Cretaceous period, and rocks of this age in the Carpathian region are generally rich in these fossils, which have been described in voluminous monographs. One such work was completed by Gyula Pethő, who described the diverse fauna of Fruška Gora. At this locality, the rudists are from the latest Maastrichtian and therefore represent one of the very last moments—in geological terms, of course—before the whole group went extinct. It is also worth mentioning that three of the species described for the first time from this site are representatives of the genus *Pironaea,* characterized by numerous inner pillars within the shell.

Rudists were also recorded long ago in the Late Cretaceous of the Bakony Mountains. It is interesting to note that for a long time these bivalves were known only from faunal lists and from short papers, containing figures, produced by Sándor Jaskó and Kálmán Barnabás (1910–1980)—the latter became a successful bauxite researcher—as their monographic description was completed only much later in 1982. It was Lenke Czabalay (1927–2010), a noted specialist working on Cretaceous gastropods and bivalves, who determined more than 50 species from the "Hippurit limestone" (today known as the Ugod Limestone Formation) in the vicinity of Sümeg. Nearly half of these taxa belongs the genus *Radiolites,* while almost all of the other half represent the genus *Hippurites.* Strange *Plagyoptychus*-like forms, which have peculiar channels within the shell, are represented in this fossil fauna by only two specimens. The most conspicuous faunal elements are *Vaccinites*—which may reach the size of a human thigh—and *Lapeirouseia,* which grows rapidly in diameter and has a thick cellular wall structure. On the basis of this rudist fauna, the Hippurit limestone at Sümeg is now thought to be from the Campanian stage.

A rudist fauna more or less the same age, but less diverse, was documented from limestone blocks deriving from an Upper Cretaceous conglomerate in the southeastern part of the Uppony Mountains (northeastern Hungary). Geology student Gabriella Szlatki identified seven taxa in her 2007 thesis; it is extraordinary that although hundreds of specimens were collected, very few belonged to *Radiolites.* This unexplained scarcity, combined with the relatively low diversity of the fauna, is reminiscent of the rudist reefs in the Northern Calcareous Alps.

Rudist occurrences in Upper Cretaceous rocks in the Transylvanian Mid-Mountains (Bihor Mountains) are even more diverse then those of the Transdanubian Range. Their stratigraphic range is also relatively long, spanning from the Santonian to the Maastrchtian. More then 50 localities

Tightly standing *Radiolites*-like rudists from the Upper Cretaceous of Lapusnik (Caraş-Severin County). This specimen was collected by János Böckh (0.75× magnification).

have been documented, most of them are around Oradea, with the best-known outcrops in the vicinity of Borod and Roşia. The rudist faunas of these sites were examined by a series of scientists, among them Dan Patrulius (1921–1982), who was originally an oncologist and later a renowned geologist, and Denisa Lupu, who published a monograph on the rich fauna.

📖 Benkőné Czabalay 1964a; Czabalay 1971, 1982, 1994; Patrulius 1974; Lupu 1976, 1992

Rudist bivalves (0.66× magnification). (1) *Pironaea polystyla* (Pirona). Representatives of this peculiar genus lived toward the end of the Cretaceous, during the Campanian and Maastrichtian stages. *Pironaea* has a horn-shaped lower (right) valve, which has numerous pillars (they can be clearly seen in the photograph) in contrast to other *Hippurites*-like bivalves, which have a maximum of three of these pillars. The upper ends of the pillars were firmly fitted into the upper (left) valve when the clam was closed. Possibly this system allowed the bivalve to circulate more water and therefore to filter their food more efficiently. The specimen figured here was collected by Gyula Pethő from Čerević Creek in the Pétervárad Mountains (now Fruška Gora Mountains, Serbia). (2) *Bicornucopina petersi* Hofmann and Vadász. This rudist is endemic to the Mecsek Mountains; it has not been found anywhere else. This is the type species of the new genus dedicated by Károly Hofmann to pioneer Austrian geologist Karl Ferdinand Peters, who contributed greatly to our knowledge of the Jurassic in the Mecsek Mountains (Magyaregregy). (3–5) *Radiolites* sp. In the Upper Cretaceous "rudist limestone" of the Bakony Mountains, both *Hippurites* and *Radiolarites*-like forms are common. Of these, the latter is frequently represented by small-sized, unique specimens, similar to those figured here. Large, smooth forms, up to the size of a palm, also occur (Sümeg). (6) *Toucasia carinata* (Matheron). The "Requienia limestone" of the Bakony Mountain was named after this fossil. Of the two twisted valves, the left is bigger. Also characteristic is the sienna color of the shell, which helps to identify even small fragments. This rudist is a close ally of *Requenia,* which was attached to the ground—in contrast with other rudists—by the left valve, which was the lower one in these forms (Bakonybél). (7–10) *Vaccinites* spp.—the most frequently encountered *Hippurites*-like rudist from the Upper Cretaceous of the Bakony Mountains to show the right valve in lateral view (7) and a cross section of the same valve (10). The three inner pillars are characteristic of this genus: the different species have these pillars at distinct angles. (Sümeg, Kecskevár Quarry). (8, 9) *Valletia germani* (Pictet and Capiche). This small sized rudist is so far known only from the south of France, from the Jura Mountains, and from the Mecsek Mountains (Magyaregregy).

1
2
6
3
4
5
8
9
7
10

Bivalves in the Shadow of Rudists

Fossil bivalves are also frequent in most Cretaceous formations. On the weathered, rain-washed Early Cretaceous cliffs at the Szársomlyó and Tenkes Hills, cross sections of a large, flat bivalve called *Chondrodonta* can be observed. In the Mecsek Mountains, in conglomerate layers that evidence the former richness of this Early Cretaceous atoll, bivalves are by far the most common fossils. Besides the numerous and radially ribbed *Ctenostreon,* thick-shelled oysters (*Aetostreon*), the beautiful *Neithea,* and the "truss"-ornamented *Fimbria* are characteristic faunal elements.

In the deeper marine Early Cretaceous formations of the Transdanubian Range, bivalves are relatively rare; however, more than 300 specimens have been found over the last 100 years. This fauna comprises about 30 species, among them, and most common, is the extinct *Rhynchomytilus,* documented at nearly all localities. In the marl and sandstone successions of the Gerecse Mountains, *Nuculana, Acesta,* and *Cuspidaria* are common forms that document deeper-water environments. The basal beds of the Aptian–Albian encrinite at Tata have also yielded a low-diversity bivalve fauna, including some genera formerly believed to be extinct (*Praechlamys, Spondylopecten*).

Albian formations generally contain diverse bivalve assemblages. In the early Albian Vértessomló Aleurite Formation, known nearly exclusively from boreholes, *Protocardia,* which indicates the low oxygen content of seawater, is the most common. Besides rudists, non-rudist bivalves are fairly common in the so-called lower rudist limestone (Zirc Limestone Formation) near Úrkút. The most conspicuous faunal elements are the fan-shaped bivalves *Chondrodonta.* Their shells, often 30 centimeters in size, stand close to each other, and can be seen in section in most cases; they are nearly impossible to extract from the hard rock. Different

Cretaceous bivalves (original size). (1) *Endocostea decipiens* (Zittel). The Inoceramus marl (Polány Marl) in the Bakony Mountains contains *Inoceramus* remains nearly everywhere. Among these remains this small-sized form is common and was originally described from the Gosau Beds in the Alps (Magyarpolány). (2) *Pycnodonte vesicularis* (Lamarck). Oysters—and related forms—have a more then 200-million-year evolutionary history. Most of them have an irregular shell as a result of attachment to a substrate, but during their long evolution species often developed that lived free on the seafloor during some, or all, of their lifetime. Balance was maintained by the thick, heavy left shell, often partly subsided into the muddy substrate. The most widely known genus of this oyster is the Jurassic *Gryphaea,* which was cosmopolitan at this time; the specimen figured here is from the Late Cretaceous of the Bakony Mountains. Indeed, the so-called Gryphaea marl was named after this fossil, which is similar in shape to its Jurassic counterpart but differs in inner features and shell structure. The generic name *Pycnodonte* means "strong (or thick) tooth," which characterizes the appearance of this fossil; the species name refers to the cellular structure visible to the naked eye on the shell of worn specimens. (Bakonyjákó) (3) *Ostrea* sp. In the Munieria marl (Tés Formation) of the Bakony Mountains oyster valves are common. This specimen was collected by Archduke József (1833–1905) at Tés, and was donated to the Hungarian National Museum. (4) *Neithea stefanoi* (Choffat). The genus *Neithea* is represented by several species that occur in the Hungarian fauna. They are characterized by a convex right and a flat, or concave, left valve (the latter is figured here). Specimens belonging to this relatively large species are fairly common in Lower Cretaceous limestones between Ajka and Úrkút (Bocskor Ravine, Ajka). (5) *Nuculana scapha* (d'Orbigny). In the Lower Cretaceous formations of the Gerecse Mountains, which were deposited in a deep sea, there are also forms, besides extinct bivalves, the relatives of which can still be found in deeper marine environments today. One such form is the genus *Nuculana* (Lábatlan, Bersek Quarry). (6) *Cucullaea glabra* (Parkinson). Generally, *Arca*-type bivalves are common in the Cretaceous of the Bakony Mountains. The specimens shown on this photo make up one frequent faunal element in the Pénzeskút Marl, but like other bivalves in this formation, their shells are rarely extracted from the rock and only their brown, glossy, phosphate internal molds are collected (Pénzeskút). (7) *Rhynchostreon suborbiculatum* (Lamarck). This oyster, similar in form to *Gryphaea,* is locally abundant in the Albian-aged Pénzsekút Marl of the Bakony Mountains, and was referred to as *Exogyra columba* in older paleontological literature (Jásd). (8) *Buchia keyserlingi* (Trautschold). Representatives of the genus *Buchia* inhabited cold, boreal seas close to the North Pole during Late Jurassic–Early Cretaceous times. Because there are numerous species belonging to this genus, *Buchia* is important to biostratigraphy in this region. In certain short periods, when the Boreal Sea and the Tethys Ocean were connected, the habitat available to this bivalve increased considerably toward the south. This unique specimen was collected from the Valanginian marl in the Gerecse Mountains, likely witness to such an incursion (Lábatlan, Bersek Quarry). (9) *Ceratostreon pliciferum* (Dujardin). This Late Cretaceous oyster from the Bakony Mountains is generally represented by specimens 2–3 centimeters long, but occasionally—especially in limestone beds—individuals may reach three or four times this size. In older literature this bivalve was referred to as *Exogyra matheroni* (Sümeg). (10) *Inoceramus cuvieri* Sowerby. This *Inoceramus,* with strong concentric ribs, lived during the Turonian and Conacian stages of the Cretaceous. This specimen came from Ormeniş, one famous locality in the Perşani Mountains. (11) *Limaria marticensis* (Matheron). The Late Cretaceous Jákó Marl in the Bakony Mountains was formerly called the Lima marl because of the frequent occurrence of this fossil (Sümeg). (12) *Ctenostreon pseudoproboscideum* (Loriol). This is the most common bivalve in the Early Cretaceous of the Mecsek Mountains. Strongly ribbed specimens, like this one, are found mainly as internal molds (Kisújbánya).

Portrait Gallery

Ion Simionescu (1873–1944)—Romanian Geologist and Paleontologist

Ion Simionescu received his doctorate from Vienna University, but his career was mainly connected with the University of Iaşi in Romania, where he was first a professor and then (in 1922–1923) the rector. Simionescu published around 100 paleontological and stratigraphical papers, most (but not all) related to the territories of Moldova and Dobrogea. He described, for example, Cretaceous *Inoceramus* bivalves from the Ormeniş locality in the Southern Carpathians, and also worked with Middle Jurassic ammonites from Monte Strunga in the Bucegi Mountians. Thanks to his work, these sites are now regarded as classics. In addition to his scientific activities, Simionescu was also an active teacher who published several textbooks and popular papers: in recognition of his scientific work, he was elected a member of the Romanian Academy of Sciences and became its president in 1941.

bivalves can also be found in certain parts of the Albian Pénzeskút Marl Formation, where alongside *Rhynchostreon* (an oyster, again resembling the Jurassic *Gryphaea*), flat and dark-brown shells of *Plicatula* are the most frequent.

Cephalopods

The geological and paleontological importance of Cretaceous cephalopods can be compared with that of the Jurassic forms. Indeed, for the subdivision of marine Cretaceous successions, ammonites are of particular importance. Belemnites—characterized by a generally lower rate of evolution—are also frequent in certain rocks, whereas nautilids, which have only restricted biostratigraphic usefulness, are interesting but not very common components of Cretaceous faunas.

Ammonites

One important characteristic of Cretaceous ammonite faunas, besides the suborders Phylloceratina, Lytoceratina, and Ammonitina that were already present in the Jurassic, is the radiation of Ancyloceratina, a group of strange-shaped forms. These taxa were also present during the Late Jurassic, but in the Cretaceous they rapidly diversified to become one of the most distinctive components of ammonite faunas. The most typical feature of the majority of the Ancyloceratina is a tendency to have shells that are not regular spirals like those of most other ammonites. These irregularly coiled ammonites are called heteromorphs.

In the Carpathian region, the Cretaceous ammonite faunas are especially rich, diverse, and occasionally well preserved. This important group of cephalopods flourished

Early Cretaceous Ammonites from the Bersek Quarry in the Gerecse Mountains (0.5× magnification). (1, 2) *Moutoniceras moutonianum* (d'Orbigny). The often large shell of *Moutoniceras* is built up from very loosely coiled inner and middle whorls, followed by a body chamber that has a straight shaft and a final, back-turned hook. Only a few species within this genus have been described on the basis of ornamentation (ribs and weak tubercles), yet this genus is characteristic to the Barremian. The species shown here, represented by a fragment of a middle whorl, is an index form for the early Barremian. (3) *Subpulchellia changarnieri* (Sayn). This ammonite within the family Pulchellidae is relatively large in size. Its umbilicus is very tight and it has flat walls. Broad, flat ribs are seen only on the outer part of the shell while the venter is flat and rounded on the adult body chamber. This species is often common in lower Barremian beds. (4, 5) *Subpulchellia didayana* (d'Orbigny). This species belong to the *Pulchellia* group in its broadest sense, as this comprises small to medium-sized, variously ornamented ammonites that have a variety of cross sections. Recently, numerous new taxa (including genera and subgenera) have been described and in terms of determination, characters from the ventral region are of importance. Of these, the position of the ribs, rows of tubercles, as well as their shape are often used. Related forms first appear in the Hauterivian, but real pulchellids occur only in the Barremian, and many have a value as zonal index fossils. The species shown here is rare, but characteristic for the early Barremian. (6) *Toxancyloceras vandenheckii* (Astier)—a medium-sized typical ancyloceratid ammonite that has heteromorph coiling. A young animal has crioceratid coiling—a very loose spiral that continues in a straight tube and ends in a hook-like bend. Ornamentation includes the ribs and rows of tubercles, and shown here are the positions of former spines. This species is also an early Barremian zonal index fossil. (7) *Jeanthieuloyites* sp.—Related to the better-known genus *Olcostephanus*, these ammonites have conspicuous constrictions and a characteristic ornamentation that comprises straight, branching ribs. The rounded ventral region has a weak, smooth bend, as its ribs are interrupted. This form is quite common in the Valanginian and Hauterivian strata of the Carpathian region, but is otherwise not very well known from other localities.

1
2
3
4
5
6
7

for hundreds of millions of years, finally going extinct at the end of the Cretaceous.

Cretaceous Ammonite Workers

Like the study of their Triassic and Jurassic counterparts, research on Cretaceous ammonites has a long tradition in the Carpathian region. The first mention and illustration of Cretaceous ammonites in Hungary was provided by the Viennese geologist Franz von Hauer, who reported Lower Cretaceous (Albian) specimens from the Bakony Mountains. Miksa Hantken first handled Early Cretaceous ammonites from the Gerecse Mountains. After this early work, however, interest in Cretaceous ammonites in Hungary was renewed only in the mid-twentieth century, when Jenő Noszky Jr., István Zoltán Nagy, Anna Horváth (1915–2000), and Gábor Scholz started their research. Subsequently, Emőke Miszlivecz, István Főzy, László Bujtor, and Ottilia Szives have contributed to this topic to different degrees. This long list of names is, however, somewhat misleading because some of these workers left geology soon after they started. The list is also at least partly related to the controversial activities of József Fülöp, once considered the authority on Hungarian geology. As head of the Hungarian Geological Institute and later director of the Hungarian Office for Mining and Geology, Fülöp oversaw the collection of thousands of Cretaceous ammonites and so was able to exert great influence over who got to work on them. Fülöp also left behind cupboards of unstudied Cretaceous ammonites, which have provided work for paleontologist for decades.

In Romania, Early Cretaceous ammonites from the Carpathians were studied by Ion Simionescu, distinguished professor at the University of Iaşi, and Emil Avram from the Geological Institute of Bucharest—among others. Research on Late Cretaceous ammonites (and *Inoceramus* bivalves) from this region was also published by László Szász (1939–2008), who spent his active years in the Bucharest Institute before settling down in Budapest when he retired.

A Short History of Work on the Bersek Hill Ammonites

The Bersek Hill above Lábatlan and its vicinity, including the nearby Nyagda Valley, is the classical site for Hungarian Cretaceous ammonite research. Miksa Hantken first pointed out the Cretaceous age of these rocks, before then erroneously regarded as Eocene, on the basis of their fossil content. Most of Hantken's ammonite collection was identified by Urban Schlönbach in Vienna.

After the pioneering work of Hantken, many geologists and paleontologists have visited these Cretaceous outcrops in the Gerecse Mountains, but a major leap forward was to take nearly a century, when Cretaceous research in Hungary, initiated by József Fülöp, found a new focus. Fülöp, the general director of the Hungarian Geological Institute since 1958, had significant financial and human resources under his control.

Fülöp also coordinated the work of specialists across a number of different fields to edit and publish a series of Cretaceous monographs, all under his name. For the monograph on the Cretaceous of the Gerecse Mountains, his university colleague and former roommate István Zoltán Nagy provided ammonite data, including faunal lists and biostratigraphic data. Nagy had wanted to publish his results on the Bersek Hill ammonites in a separate volume under his own name, but Fülöp would not allow it. Indeed, as a result of this pressure, Nagy left the Institute and found a position in the quieter Hungarian Natural History Museum in Budapest. In the paleontology department, far away from Fülöp, Nagy was able to publish some short descriptive papers on the Early Cretaceous Gerecse ammonites.

At the same time, also under the supervision of Fülöp, a new large-scale collecting effort was undertaken at the Bersek Hill, and thousands of ammonites were carefully collected bed by bed. It is very probable that Nagy would not have wanted to work with these specimens in any case, but it was Fülöp who would describe and publish these ammonites himself. But as director of the Institute, later head of the Office for Mining and Geology, and then rector of Eötvös Loránd University of Budapest, Fülöp was always too busy to dedicate sufficient time and energy to this scientific work. After his death in 1994, some 11,000 ammonites kept in 100 drawers in 4 large oak cupboards remained undescribed; they were later donated to the Hungarian Natural History Museum. However, the scientific value of the entire collection was called into question because the original field notes were missing. The ammonites had been collected from five different overlapping sections in the Bersek Quarry, and the only reliable information on sampling was the bed numbers written on the labels of each fossil. After a year of work in the collection and in the field, using the indicated bed numbers as a starting point, it proved possible to reconstruct the location, sequence, and bed numbers used by the original collectors. Tibor Steiner, the former leader of the original collecting group and later the caretaker of the Tata open-air geological museum, provided valuable help with this research.

After overcoming these initial difficulties, it eventually became possible to begin a new scientific work on the Bersek ammonites. The first results based on Fülöp's 11,000 specimens were published more than 40 years after the fossils were first collected. The Bersek Quarry ammonites are from the Early Cretaceous Valanginian, Hauterivian, and Barremian Stages, and are more than 120 million years old. What is 40 years of waiting compared to 120 million?

📖 Fülöp 1958; I. Z. Nagy 1967, 1968; Főzy & Janssen 2009

What Happened to Ammonites at the Jurassic-Cretaceous Boundary?

It is well known that geological stages are recognized and differentiated based on the different faunal associations seen in subsequent rock bodies, and that the end of some geological periods are marked by mass extinctions. In this context, it is worth asking this question: What happened to the geologically important ammonites at the Jurassic-Cretaceous boundary? Taking a large-scale view, we can say that the first appearances of important ammonite suborders subdivide the Mesozoic era: Phylloceratina appeared at the Permian-Triassic boundary; Lytoceratina and Ammonitina appeared at the Triassic-Jurassic boundary; and finally, Ancyloceratina, so characteristic to the Cretaceous, appeared close to the Jurassic-Cretaceous boundary. In this context, "close" means around the boundary rather than exactly at the boundary: indeed, the early representatives of Ancyloceratina had already in the fossil record by the last stage of the Jurassic, the Tithonian.

A similar conclusion is also reached based on a survey of lower taxonomic ranks (in this case, families). Not a single new family appeared after the Tithonian-Berriasian boundary, which also marks the boundary between the Jurassic and Cretaceous periods. Of course, an important faunal change took place, but it was rather slow. Toward the end of the Tithonian, Simoceratidae (typical to the Tethyan realm) became less and less important and finally went extinct; the same happened to a Boreal ammonite family, Virgatitidae. Aspidoceratids, ataxiocedratids, and himalayitids also became rare and disappeared after the boundary. In general, latest Jurassic (Tithonian and Volgian) ammonite faunas can be characterized by a great degree of variety, which somewhat diminished during the earliest Cretaceous (Berriasian). These changes are likely related to the continuous fall in global sea levels, a trend that started from its highest level ever in the Tithonian.

From the Berriasian onward, the first representatives of the family Neocomitidae play an important role. More and more families of Ancyloceratina appeared from the Early Cretaceous onward, with the most important faunal change taking place around the time of the Barremian-Aptian boundary—the traditional, but no longer officially used, boundary between the Lower and Middle Cretaceous.

In other words, although a faunal change did take place at the Jurassic-Cretaceous boundary, this change was not sudden or dramatic. The change was, thus, less important than those that took place at the beginning of the Jurassic, and it cannot be compared to the mass extinction that marks the end of the Cretaceous, when the whole global ammonite fauna went extinct. The question remains this: If ammonites are so essential to Mesozoic stratigraphy, why was the Jurassic-Cretaceous boundary first drawn the way it was? The answer lies in the history of earth sciences.

In the Northwestern European countries (mainly England), where the first biostratigraphical observations were made almost 200 years ago, there is a very important environmental change toward the end of the Jurassic. Above the marine Late Jurassic Portlandian facies, earliest Cretaceous rocks vary, but basically consist of terrestrial Purbeckien layers. This change is abrupt, easy to recognize on the basis of rocks as well as on the basis of their faunal content, and so this change was chosen as the marker for the Jurassic-Cretaceous boundary. Unfortunately, defining the boundary in this way does not coincide with any of the evolutionary events that are seen in ammonites.

📖 Főzy 2003

Tata: Pockets in Time

The "Aptian crinoidal limestone," long mentioned in the Hungarian geological literature, is very poor in fossils. However, there are some encrinite beds toward the base of the formation that have locally yielded cephalopods, brachiopods, echinoids, and gastropods—among other fossils. The best examples of this rare fauna were gathered at Tata, where fossils were collected from pockets within the uneven surface of this latest Jurassic–earliest Cretaceous (Tithonian–Berriasian) limestone. Here, ammonites are sometimes preserved without shells as phosphatic, or rarely as glauconitic, internal molds sorted by size—although calcitic shells are also found. Because microfossils are also practically absent, these taphonomic features suggest that fossils accumulated as a result of multiple reworking. The empty shells of dead animals lay on the sea bottom for a long time, as evidenced by warm tube encrustations on their insides, and were finally buried. The more or less filled shells were then excavated, rolled away, sorted by size, and then finally deposited and reburied by the bottom currents. This process may have been repeated several times; many of the remaining reworked fossils were trapped by the uneven surface of the seafloor, whose pockets were packed with skeletal remains and internal molds. Formation of these pockets was very slow—therefore, their fossil contents represent long periods of time; the rocks are condensed, in other words.

Ammonites from these basal encrinite beds were listed by Fülöp in his monograph on The Mesozoic basement horst blocs of Tata. He regarded the fauna to be late Aptian in

1
2
3
4
5
6
7
8
9
10
11
12
13
14
15
16

age. Twenty years later Ottilia Szives revised this ammonite collection and showed that the fauna also contains some Albian elements. According to Szives, these pockets have condensed some 11 million years in the basal layers of the Tata Limestone, and preserved ammonites out of time. It is also interesting to note that these basal layers, only a few centimeters in thick, represent more time than the rest of the encrinite beds, which are often tens of meters thick.

📖 Fülöp 1976; Szives 1999

The Stolen Ammonites of Tilos-erdő ("the Forbidden Forest")

The fossils of the Turrilites marl are some of the earliest recorded from the Bakony Mountains. A knight, Franz von Hauer, first mentioned these remains in 1862. The basal bed of this marl is extremely rich in well-preserved fossils, especially ammonites. Some of the specimens are particularly beautiful, as their iridescent shells are—at least partly—preserved. However, the first detailed evaluation of this fauna, written by young geologist Gábor Scholz, did not appear until 1979. This name will be familiar to fossil collectors, as Scholz is the author of the popular small-format book *Kövületek* (Fossils), part of the Búvár Zsebkönyvek series.

This ambitious and talented geologist was then working in the Hungarian Geological Institute, where his main subject of interest was the fauna of the Albian–Cenomanian Turrilites marl. Scholz reviewed the collection of the Institute and collected new material from different localities. As part of this work, he made a long trip to France to study collections of similar age from foreign outcrops and completed his manuscript on the rich ammonite fauna of the Turrilites marl when he returned home. One of his most important localities was Tilos-erdő, near Pénzesgyőr, where he found a particularly well-preserved fossil assemblage. According to some former colleagues, on finishing his work Scholz was called in by Fülöp, all-powerful leader of Hungarian geology. Fülöp asked for a copy of his manuscript as well as for the specimens, but Scholz would not give them to him, as he was afraid that his boss would publish his results without him. As he did not want to wait in line for new research, he faced the music and it was made clear to him that he could

Early Cretaceous (Aptian) ammonites from Kálvária Hill, Tata (original size). (1) *Tetragonites heterosulcatum* (Anthula). This genus, a close ally of the Jurassic *Lytoceras,* unites some midsized Early Cretaceous forms with a moderately wide umbilicus, and a square or rounded cross section. The strong, straight, prorsiradiate constrictions on the internal mold appear as collar-like swellings on the shell. (2, 3) *Jauberticeras jauberti* (d'Orbigny). Also belonging to lytoceratids, and a close ally of *Tetragonites,* this small ammonite has a unique appearance. The whorl section is much wider than it is high, and its slightly arched ventral region forms a distinctive keel when it meets the lateral wall. This genus is a typical late Aptian ammonite. (4) *Holcophylloceras guettardi* (Raspail). This species is one Cretaceous representative of a genus already well known from the Jurassic. *Holcophylloceras* is a phylloceratid with a minute umbilicus and with strong, slightly broken constrictions on a high, flat wall. (5, 6) *Diadochoceras nodosocostatum* (d'Orbigny). This genus includes some relatively evolute ammonites with rounded to polygonal cross sections. Their ornamentation is robust: on the main ribs there are three strong rows of tubercles, while on the intercalatories the tubercles are weak or missing. The known Hungarian material is extraordinary rich and diverse and these ammonites are typical faunal elements in Aptian strata. (7) *Tetragonites duvalianus* (d'Orbigny) differs from the ammonite in (1) at the species level only; the cross section of its whorls increase at a lower rate, while its constrictions are less steep and are higher in number. The fact that they are also less prominent, however, is related to preservation. (8) *Parahoplites melchioris* (Anthula). *Parahoplites* are midsized ammonites with high oval cross sections. The ventral region is rounded; ribs are strong, slightly bent, simple or bifurcating. Ribbing may be smoothed on the middle whorls, but is well expressed on the adult last whorl. (9, 10) *Valdedorsella getulina* (Coquand). Members of the genus *Valdedorsella* are midsized, moderately inflated forms with rounded or slightly rectangular cross sections. Most of the species have slightly curved constrictions, which appear like swollen ribs on the shell. The numerous related forms (genera and subfamilies) are united within the family Desmoceratidae. *Desmoceras* itself appears in the late Aptian, and remains frequent into the Late Cretaceous. Representatives of *Valdedorsella* are common in the late Hauterivian–Aptian interval worldwide. (11) *Cheloniceras cornuelianum* (d'Orbigny). *Cheloniceras* are moderately evolute, heavily ornamented late Aptian ammonites. Their sculpture is composed of ribs and tubercles and in this way many subgenera are distinguished. (12) *Diadochoceras nodosocostatum* (d'Orbigny)—the same species shown in (5) and (6). It differs only by having a slightly weaker ornamentation. (13) *Hypacanthoplites elegans* Fritel. The form belongs to a group, which includes a large number of Aptian–Albian ammonites, similar in shape and ribbing, which were called "hoplitids" in the old paleontological literature. (14) *Procheloniceras albrechtiaustriae* (Uhlig). This genus amalgamates some moderately evolute ammonites with rounded cross section and strong ornamentation. Ribbing is slightly retroverse, composed of thick, simples and intercalatories, having swollen tubercles. (15) *Cheloniceras* (*Paracheloniceras*) *rerati* Collignon. This subgenus is so far known only from Madagascar and Hungary. Although ornamentation is strong and characteristic, the position of ribs and tubercles is difficult to observe because these specimens were flattened during fossilization. (16) *Silesitoides tatricus* Passendorfer—a little known ammonite with characteristic prorsiradiate ribs. Formerly this species was known only from Poland and North Africa, but the known Hungarian material is extremely rich.

1
2
3
4
5
6
7
8
9
10
11
12
13
14
15
16

not continue with his work in an effective way in Hungary. So he posted the ammonites from the Forbidden Forest to Germany—at the expense of the Institute—and followed the boxes himself. He soon found a position in Tübingen with the support and help of another ammonite worker, Jost Wiedmann (1931–1993). Wiedmann was also a refugee from the German Democratic Republic. As a result of this, the Bakony ammonites were finally published in two huge tomes in the classic German series Palaeontographica. Scholz's career ended suddenly, after so many earlier difficulties. He left Tübingen and never again contributed to paleontology.

📖 Scholz 1979, Szives 2007

Ammonites of the Márvány Quarry: Remembering a Vanished Outcrop

In the heart of the Bakony Mountains, in the vicinity of Zirc and on the right side of the main road to Borzavár, there is a little abandoned quarry called Márvány ("Marble") Quarry. Its name refers to the reddish, Late Jurassic (Tithonian) ammonitico rosso-type limestone that was quarried a long time ago. Indeed, this place is well known among Hungarian geologists not just because of these Jurassic strata, but because of the overlying Cretaceous deposits that are sometimes found. This Cretaceous formation, often referred as "bank" in the Hungarian literature, had a maximum thickness of 50 centimeters and a rather limited extension. It was no more then 2 or 3 square meters when it was discovered during a university excursion led by the young György Wein in 1934. (Wein went on to become an excellent geologist, and 32 years later he discovered the first dinosaur footprints from the Mecsek Mountains.) Cretaceous pelagic limestone beds above the Jurassic at this site once had a vast extension when they formed, but only a small portion remained in the Marvány Quarry. These beds yielded a very rich and well preserved Early Cretaceous cephalopod fauna. The first collecting was done by Wein and his colleague Jenő Noszky Jr., who soon afterward published a long list of some 30 cephalopod species and argued that the limestone was late Hauterivian in age. Because the Early Cretaceous was thought to be a period of terrestrial bauxite deposition throughout the Bakony Mountains, this marine limestone bank attracted much attention from early geologists.

A few years later, Fülöp published his monograph on the Cretaceous of the Bakony Mountains, and suggested

Early Cretaceous (late Albian) ammonites from the vicinity of Pénzesgyőr in the Bakony Mountains (original size).
(1, 2) *Mortoniceras* sp. These are middle sized, moderately evolute ammonites with a cross section that is close to square or trapezoid in shape and with a low or high, but always well expressed, lateral keel. Ornamentation in these taxa is made up of very strong ribs that bear numerous rows of tubercles. A number of subgenera have been described, and these ammonites are good markers as they occur only in middle and late Albian assemblages. (3, 4) *Beudanticeras beudanti* (Brongniart). This genus comprises ammonites with moderately or very involute coiling. Their walls are nearly flat and the ventral region is highly arched. Certain species may have shallow, prorsiradiate constrictions and there are no ribs or tubercles. The genus is known from Albian ammonite faunas worldwide. (5) *Anisoceras armatum* (Sowerby). These are midsized ammonites that have a helicoid (tower-like) iniical spiral that continues into a flush straight part and ends in a hook. These ammonites make up the family Anisoceratidae and have ornamentation that typically contains ribs and tubercles. It is also quite common that the "flawed" double ribs in these taxa run from tubercle to tubercle. (6) *Mariella bergeri* (Brongniart)—heteromorph ammonites with high, helical shells and ribs that bear rows of tubercles. These taxa are representatives of the family Turrilitidae (which includes about a dozen genera, such as *Turrilites, Mariella* and *Ostlingoceras*) and are frequent in Albian and Cenomanian strata. *Mariella* is common to the fauna of the Bakony Mountains and is also known from outcrops across the Carpathian Basin. (7) *Ostlingoceras puzosianum* (d'Orbigny)—a close ally of *Turrilites* with a very high, tower-like shell. Ornamentation in this taxon consists of regular, simple ribs that bear two, or sometimes three rows of tubercles.
(8, 9) *Scaphites hugardianus* (d'Orbigny). These ammonites have minute or midsized inner whorls with a very narrow umbilicus and a short, uncoiled, hook-like outer whorl. Inner whorls are ornamented with dense ribbing and small tubercles; on the uncoiled, final part of the shell, the ribs are either more widely spaced or the whorl itself is smooth. (10, 11) *Salaziceras salazacense* (Hébert and Munier-Chalmas)—*Salaziceras* is a small ammonite with a narrow and deep umbilicus and thick, "inflated" whorls. Ornamentation includes thick ribs that originate from strong periumbilical swellings. The known Hungarian material of these ammonites is unique both in species and number. *Salaziceras* is one characteristic faunal element in late Albian assemblages. (12, 13) *Stoliczkaia notha* (Seeley). The genus *Stoliczkaia* unites midsized ammonites with a rather narrow umbilicus and high oval whorl sections. Their strong ornamentation consists of ribs that are especially thick above the venter as well as secondary or intercalatory ribs. These ammonites are quite common in the late Albian of the Bakony Mountains. (14, 15) *Stoliczkaia dispar* (d'Orbigny)—a characteristic element of late Albian ammonite faunas that also has a value to zonal indices. (16) *Puzosia mayoriana* (d'Orbigny). The genus *Puzoisa* comprises midsized or large, moderately evolute ammonites with rounded or oval cross sections. Ribbing in these forms is generally weak and fades out in later ontogenetic stages. These ammonites typically have strong and deep constrictions that are parallel to their ribbing. Very large macroconchs are occasionally found as fossils, but generally the smaller inner whorls or young specimens occur.

that the limestone bank was in fact Barremian in age. Subsequently, the locality was visited by many geologists and this now-famous cephalopod limestone was always included in student excursions and field trips. Indeed, because of its unique features and special importance, this Early Cretaceous bank was also intensively collected—and, finally, was almost completely destroyed.

What remains of the Márvány Quarry was revisited recently and most of the ammonites collected over the last 70 years from these beds were restudied. This work confirms the age of the ammonite fauna as late Hauterivian; a number of specimens erroneously determined to be Barremian were found. Nevertheless, there is still a slight discrepancy between these ammonite data and more recently obtained belemnite results—determined by a specialist from the Netherlands, N. M. M. Janssen. Although most of the belemnites from this quarry are indeed Hauterivian, some of them are also characteristic to the early Barremian. Perhaps these problematic specimens were actually collected from the uppermost (youngest) beds of the Cretaceous limestone.

All that remains is for us to ask this question: Why spend so much time and energy to determine the age of four almost completely vanished beds? Instead of dwelling on the scientific merit of this issue, let us paraphrase words attributed to Agatha Christie: No reason. That is why it is so important.

📖 Főzy & Janssen 2006

A New Group Appears on Life's Stage: The Ancyloceratid Ammonites

Ancyloceras and allied heteromorph ammonites are very typical elements in Cretaceous faunas. The first representatives of this group, which belong to the family Bochianitidae, had straight (*Bochianites*) or slightly curved (*Protancyloceras*) shells, and appear in the fossil record in the Late Jurassic, during the Tithonian. Ancyloceratids are of special significance from the Valanginian and younger, when the very diverse group comprising *Crioceratites* and its close allies appeared. These ammonites were coiled very loosely, so that their whorls never touched each other. Although paleontologists have distinguished dozens of subgenera and hundreds of species within this group, the evolution of the *Crioceratites* lineage remains complex and is still not clearly understood. Indeed, the taxonomy of these ammonites is a headache, given that *Ancyloceras* and its closest relatives are restricted to the Barremian and Aptian. The shells of these taxa have a loosely coiled phase that ends in a kind of hook. They are generally heavily ornamented with both ribs and tubercles, which were originally spines in most cases.

The genus *Ptychoceras* and related forms have straight shells made up of three or four shafts squashed together and connected with tight, 180-degree bends. *Hamulina* and *Anahamulina* make up a small but unique family that has U-shaped shells. From the Aptian and Albian many regular-shaped ancyloceratids are also known, including the genera *Deshayesites, Mathoceras, Parahoplites, Hypacanthoplites,* and *Cheloniceras.* All of these ammonite genera have been found in Hungary from the Albian onward—and include *Turrulites* and its close allies that have high, helicoid spirals reminiscent of gastropods.

In the Late Cretaceous ancyloceratid ammonites were also very diverse, and numerous morphotypes are represented by many genera. Among these forms, the straight-shelled *Baculites* and the loosely coiled *Glyptoxoceras* are worthy of mention. Both are rare but characteristic elements in the Late Cretaceous faunas of the Bakony Mountains. The most extreme form of all is *Nipponites.* This ammonite is known only outside Europe and has a shell that resembles the Gordian knot.

Uncoiled heteromorph ammonites were long regarded as degenerate forms, taxa that had started down the path towards extinction. However, we now know that it is much more likely that their strange shapes are the result of unique specializations. These ammonites were probably poor swimmers but masters of buoyancy; many of them are thought to have either drifted in the plankton, collecting small animals on their long tentacles like modern jellyfish, or crawled along the seafloor feeding on sessile or slow-moving animals such as clams. There is also good reason to believe that the mode of life of these creatures also changed as their extreme shells grew, and that different morphologies resulted in different centers of gravity and balance distributions. As a result, the position of the ammonite within the water body may have changed. In any case, the long duration in the fossil record of Cretaceous heteromorph ammonites appears to support the hypothesis that

Late Cretaceous ammonites (0.75× magnification). Specimens (1), (2), and (4) were collected from the Campanian Polány Marl Formation in the vicinity of Sümeg, a sequence that is otherwise poor in ammonites. These old quarries to the north of the city do not yield fossils anymore; they were recently turned into waste dumps. Specimen (3) is from near Ohaba Ponor in Hunyad County, Romania. (1) *Pachydiscus* sp. Generally big, moderately involute ammonites with high and oval cross sections. Ribbing in this genus comprises mainly simple ribs that are generally more marked around the umbilical shoulder and above the venter, but often fade out on the flat walls. (2) *Menabites suemegensis* Főzy—a moderately evolute ammonite with whorls that are much higher than they are wide. Ribs are strong, widely spaced, and bear rows of tubercles. The number of these rows, however, as is the case in all allied forms, changes through ontogeny. Some forms related to this genus reach the size of tires. (3) *Calycoceras* sp.—medium or large-sized ammonites that have a relatively narrow umbilicus. The cross section in these forms is thick and high, rounded or rectangular. Inner whorls bear rows of tubercles that vary in number and position from periumbilical through lateral to ventral. Many subgenera within this genus have been distinguished, but all are characteristic to the Cenomanian. (4) *Glyptoxoceras* sp. The initial part of the shell in this taxon forms a helicoid spiral, followed by a loose, elliptical portion and a straight region towards the very end that may or may not bend backward.

László Szász (1939–2008), geologist at the Geological Institute of Romania, did mapping in the Carpathians and also published numerous papers on the Late Cretaceous ammonite and bivalve faunas of Romania. The photograph was taken in his home in Budapest, where he settled after retirement.

they were a successful group. Their extinction at the end of the Mesozoic cannot be associated with the appearance of so-called strange heteromorph shapes, as similarly shaped forms are also known from the Triassic and Jurassic.

The Last Aptychi

Aptychi were bivalve-like fossils, but actually belong to ammonites. These are considered to be either a closing hatch on ammonite shells or a double-plated jaw piece (mandible) similar to that seen in some modern cephalopods. Since these remains are composed mainly of calcite, they accumulate mostly in sediments where the aragonitic shells of ammonites have dissolved and are therefore not preserved

Zdeněk Vašíček, a Czech engineer and paleontologist who teaches and works at the Technical University of Ostrava. Vašíček has published several papers on Early Cretaceous cephalopod (ammonite and belemnite) faunas from the Western Carpathians.

An internal mold of a Cretaceous nautiloid (*Heminautilus* sp.) from Barremian deposits in the Bersek Quarry, Gerecse Mountains. Lateral and ventral views (original size).

(see also discussion of Aptychus marl, above). Because different ammonites had different aptychi, these fossils are consequently useful for biostratigraphy. Indeed, in some Lower Cretaceous formations stratigraphy based on these remains may lead to resolution close to that achieved with ammonite zonation. Unfortunately, however, aptychi are almost unknown from deposits that are younger than late Hauterivian, so their use at these ages is impossible.

The Last Ammonites from the Carpathian Region

In Hungary, Cretaceous formations younger than Cenomanian are poor in ammonite remains. From the Santonian, for example, just a single specimen is known from a borehole. The Campanian is better represented, although all the known specimens were collected from the same formation (Inoceramus marl) in the vicinity of Sümeg. Most of them were collected in 1944 by Jenő Noszky Jr. from the now abandoned Haraszt quarries north of the city. Among these Sümeg ammonites, representatives of the genus *Pachydiscus* are most common. This genus is closely related to the largest ammonites that ever evolved, some reaching 2 meters in diameter. The known Hungarian specimens are smaller, but some of them reach 40 centimeters in diameter. Texanitid ammonites (forms closely related to the genus *Texanites*) are also significant elements of the fauna around Sümeg, especially from the point of view of biostratigraphy.

From the Carpathian Mountains, ammonite faunas younger than the Campanian are also known and were mostly described by László Szász. These are from the Maastrichtian, the last stage of the Cretaceous, and when they went extinct they marked the end of ammonites as a group. Only traces of their former dominance remain—as fossils, of course.

📖 Szász 1981, Főzy 2001

Cretaceous Nautiloids from Hungary

Where ammonites are found nautiloids may also occur. Interestingly, however, they were never as abundant as ammonites and are known from far fewer genera and species. The number of fossil specimens is also much smaller: in most cases thousands of ammonites can be collected, while just one or two nautiloids are found.

In Hungary, these rare cephalopods are most common in the Bersek Marl in the Gerecse Mountains and in the Turrilites marl in the Bakony Mountains. Specimens are also found here and there in different collections, but published data is scare. Indeed, the only article dedicated entirely to this subject was written by Nagy, who reported *Eutrephoceras, Cymatoceras, Eucymatoceras,* and *Angulithes* from the

1
2
3
4
5
6
7
8
9
10
11
12
13
14

Early Cretaceous in the Gerecse Mountains. From the same place, *Heminautilus* has also been found, and *Angulites* is now known from the Late Cretaceous of Sümeg.

📖 I. Z. Nagy 1963; Főzy 2001

Belemnites

Belemnites are not easy to find, identify, or collect. These fragile, elongated, often bullet-shaped fossils—the "rostrum" or "guard" of the animal—are difficult to extract from rocks. In addition, different belemnites from different geological stages may also have very similar appearances, and the ontogeny of the different species is little understood. Consequently, there is chaos within the already published paleontological literature that deals with these cephalopods: the number of available taxonomic names is far too high and the real meaning of many of them remains uncertain (although this is certainly true of many fossil groups!).

Probably for these reasons, little attention was paid in Hungary to Cretaceous belemnites until recently. In most cases researchers simply listed them as present in faunas or sometimes figured the best specimens. However, this situation changed recently when N. M. M. Janssen, a specialist from the Netherlands, made a thorough review of the known Early Cretaceous assemblages. He obtained spectacular results from a study of the more than 200 Gerecse Mountains belemnites that were collected bed by bed from the Bersek Quarry. Janssen was able to determine 36 species and correlate the successive belemnite associations he observed with the available ammonite zones. Because it is thought that belemnites had a lower rate of evolution than ammonites, there are approximately two times more belemnite associations than ammonite zones within a given Early Cretaceous stage.

The Bersek Quarry belemnite fauna is late Valanginian to Barremian in age and contains both flattened representatives of the family Duvalidae and elongated, circular Mesohibolitidae. Within the first group the genera *Duvalia, Pseudobelus,* and *Pseudoduvalia* have been determined; Mesohibolitidae is represented by *Belemnites, "Combemorelites," Curtohibolites, Hibolithes,* and *"Mesohibolites."*

📖 Janssen & Főzy 2003, 2004, 2005; Vašiček et al. 1994

Early Cretaceous (Valanginian–Barremian) belemnites from the Transdanubian Range (1.1× magnification). It is normally the case with fossil belemnites that only the back part of the skeleton—the guard—is preserved, therefore their taxonomy is more often than not based on this part of the fossil. These calcite remains may occur in rocks in which the aragonitic shells of ammonites have dissolved. Specimens (4) and (5) in this figure were collected in the vicinity of Hárskút (Bakony Mountains); the remainder are from the Bersek Quarry (Gerecse Mountains). (1) *"Belemnites" pistilliformis* Raspail—ventral view. (2) *Duvalia silesica* Uhlig—lateral view. (3) *"Mesohibolites"* aff. *elegans* (Shvetsov)—ventral view. (4) *Duvalia* sp.—lateral view. (5) *Pseudobelus brevis* Paquier—ventral view. (6) *Hibolithes jaculiformis* Shvetsov—lateral view. (7) *Hibolithes* gr. *subfusiformis* Raspail—lateral view. (8) *Conohibolites escragnollensis* (Delattre)—ventral view. (9) *Duvalia dilatata* (de Blainville)—lateral view. (10) *Adiakritobelus* sp.—ventral view. (11–13) *Duvalia grasiana* (Duval-Jouve)—lateral, ventral, and dorsal views. (14) *Conohibolites* gr. *gladiiformis* (Uhlig)—ventral view.

Brachiopods

Bottoms from the Jurassic

Cretaceous brachiopods were less diverse than their counterparts in the Jurassic. Indeed, in the Cretaceous representatives of the group Terebratulida, which have mostly smooth shells with few characteristic visible taxonomic features, are most common. Most of these forms can be distinguished and determined only by making serial sections so that the lophophore can be studied. In some cases, when the valves have fallen apart or the inner space of the lamp shell has been filled with coarse-grained sediments that destroyed the lophophore, these fossils cannot be determined precisely. Among the Early Cretaceous forms, *Pygope* and related genera (*Pygites* and *Triangope*) are the most common. These brachiopods are easy to recognize because their appearance is very unusual (their name comes from the Greek *pygos,* which means "buttocks"). One look at their shell and the reason becomes clear. These midsized (at most 8 centimeters), rounded, and triangular shells have an elongate hole in the middle, best expressed in young specimens.

Pygopid brachiopods are known from the latest Jurassic–earliest Cretaceous and are easy to recognize and determine because there are just a few related species. The most common of these is *Pygope diphya,* which bears a species name that refers to "Diphyakalk," a term common in older paleontological literature. Nowadays, this is the Tithonian–Early Cretaceous *diphya*-bearing limestone, also known as the Szentivánhegy Formation in the Transdanubian Range. *Pygope* and related forms have also been reported from the Mecsek Mountains as well as from many other Carpathian localities.

📖 Kázmér 1998

Echinoderms

Sea Urchins (Echinoids)

In many of the Cretaceous formations across the Carpathian region, echinoids are quite common. In general, it is true that in deposits younger than Cenomanian these remains become less frequent. Because these fossils are often preserved as internal molds only, with little or no

1
2
3
4
5
6
7
8
9
10

Club-like spines of regular echinoids from one of the quarries betwen Zirc and Borzavár. These egg-shaped, sometimes cylindrical and flattened echinoid spines are composed from a single calcite crystal and were attached by a short stem to the test, itself composed of many distinct plates. After the death of the animals, the spines detached and were transported by the currents above the seafloor. As a consequence, they were embedded and fossilized in a different place from where they had lived (0.6× magnification).

Reconstruction of a regular echinoid with club-like spines.

shell remains, and because sea urchin taxonomy is based mainly on the arrangement and characteristics of individual skeletal plates, these fossils are often very difficult, even impossible, to determine precisely. Spines detach after the animal dies and are buried separately, and the fine, often hairy spines of irregular echinoids are not generally not preserved at all. The bigger barrel- or club-shaped spines of regular sea urchins are, however, easily found at some sites.

There are just a few Hungarian publications that deal specifically with Cretaceous echinoids: these invertebrates are generally included only at the ends of faunal lists. All of the relevant papers were written by Erzsébet Szörényi, a widely recognized echinoid specialist, and her 1965 *Échinides du Crétacé inférieur de la Hongrie* (Early Cretaceous echinoids of Hungary) is based on the detailed study and description of 685 specimens gathered by geologists mapping the Transdanubian Range. Szörényi described many new species during the course of her work, including *Corthya plana* from the Bersek Quarry in the Gerecse Mountains. This species is the only representative of *Corthya* known from Hungary.

In the Bersek Quarry, even though the echinoids are flattened, they can be determined precisely because their tests

Cretaceous echinoids from the Albian glauconitic marl in the Bakony Mountains (original size). (1, 2, 8) *Discoides* spp.—regular sea urchins with high, domed skeletons that were often referred to as *Discoidea* in older literature. Representatives of this genus are common throughout in sandy facies of mainly Cenomanian age all across Northwestern Europe, and species are distinguished based on details of their apical plates. The specimen shown in (8) was found by M. Hantken at Akli farm, near Zirc, on one of his trips into the Bakony Mountains. (3) *Allomma kalon* Szörényi—test flattened below, weakly domed above. A rare but easy to recognize fossil common in the glauconitic marl. This genus is little known and has an uncertain taxonomic position, possibly related to the more frequent *Hemicidaris.* Originally described from France, this Bakony occurrence represents the only further locality it has been found. This illustrated specimen was collected at Bakonynána. (4) *Holaster hungaricus* Szörényi—The heart-shaped (cordiform) genus *Holaster* is one of the most diverse and one of the longest-lived genera of Cretaceous echinoids. It ranges from the Early Cretaceous to Late Cretaceous (Cenomanian). The lower side of the test (base) is flat, the anterior part is sulcate, and the posterior is truncate. *H. hungaricus* is a rare taxon. Szörényi studied some 5,000 echinoids from the Bakony Mountains, and only one of them was ranged into this species. It was found in Olaszfalu. (5) *Pseudoholaster baconicus* Szörényi. The genus comprises forms with cordate test with deep anterior sulcus with sharp groove. Lower surface flat but with keeled plastron. The species is very common in the glauconitic marl of the Bakony. It is especially abundant in the minute quarries in the vicinity of Olaszfalu. The figured specimens also were collected there. (6, 7, 9, 10) *Epiaster* sp.—cordiform, rounded or sometimes elongated test with shallow anterior sulcus. Posterior face obliquely truncate, lower surface flat. Representatives of the genus are frequent elements of the glauconitic marl. About a dozen new species were described from different Bakony localities: (6) is *E. pseudodistinctus* Szörényi from Jásd; (7) is *E. angulosus* Szörényi from Olaszfalu; (9) and (10) are *E. hemiastriformis* Szörényi from Bakonynána.

A crinoid calyx (*Psalidocrinus*) from a quarry near Borzavár. Overhead and lateral views (4× magnification).

are also preserved. The "Aptian crinoidal limestone" at Tata has also yielded a very diverse, but poorly preserved, echinoid fauna; even more varied is the fauna known from the Bakony Mountains. However, some of the best localities to collect these invertebrates are small, abandoned quarries along the road between Zirc and Borzavár, where not only echinoids but also an abundant crinoid fauna are known. Szörényi listed *Cidaris, Balanocidaris, Rhabdocidaris, Acrocidaris, Magnosia,* and *Pyrina* present at these outcrops. The small (maximum 3 centimeters) irregular echinoids that are relatively common in the lowermost Cretaceous limestone facies of the Bakony Mountains are called *Collyropsis* in Hungarian paleontological literature; the smaller, cylindrical forms are referred to as *Metaporinus*.

📖 Szörényi 1955, 1965

Sea Lilies: Lots of Debris but Few Fossils

We are not very familiar with crinoids because they are not abundant in modern oceans. In contrast, in certain environments in Mesozoic seas crinoids were very abundant: their disarticulated, ground and accumulated skeletal remains make up the so-called crinoidal limestone (encrinite). However, their intact—articulated—remains are very rare.

Nevertheless, some of the best localities for collecting crinoids are the small quarries along the road between Zirc and Borzavár, where echinoids are also found in great abundance (see above). German researcher Herta Sieverts-Doreck (1899–1991) listed 28 species of crinoids in 19 genera at these sites, including 16 that were new to science. Of these, at least *Pyramidocrinus* and *Apsidocrinus* are known only from here and the site at Štramberk.

The fauna includes some minute, strange-looking, and very specialized forms, called *Torynocrinus*. These crinoids were about 2 centimeters long and had a short, thick stem attached to the ground and an oblique, spherical calyx with short arms that look like clamped fingers. Like other crinoids, *Torynocrinus* would have filtered small particles of food from seawater using these arms, which were equipped with feather-like organs.

📖 Sieverts-Doreck 1961

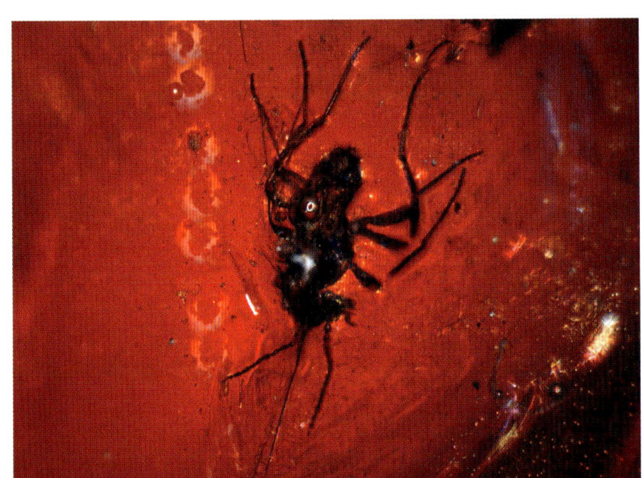

A 1-millimeter-long insect preserved in fossil amber from the Ajka Coal Formation.

Insects

Cretaceous Insects Preserved in Amber

In the Late Cretaceous Ajka Coal Formation pieces of fossil amber of all sizes were relatively common. Although all the coal mines in this area are now abandoned, many fossil amber specimens survive in collections, and recently fossil insects were discovered in some amber pieces, which is also informally known as "Ajkait." Fossil insects have also been found in the dinosaur bone–bearing sedimentary succession at Iharkút.

Vertebrates

Reptiles and Birds from the Bihor Mountains

In 1978, in the region of Pădurea Craiului (Romania), dinosaur bones were found deep inside the Cornet bauxite mine. Although most of these fossil bones were dark-colored fragments only, and proved very difficult to identify, some vertebrae and other identifiable pieces of limb bones were also found (but no skull remains so far). In total, after intensive collection efforts over a number of years, more than 10,000 fossils were recovered and are deposited in the museum at Oradea (Muzeul Ţării Crişurilor).

Dinosaurs in the Bauxite

Bauxite is one ore of aluminum and it is often found above carbonate rocks (such as limestones and dolomites), where it forms by lateritic weathering, residual accumulation of intercalated clays, or by clay dissolution as a residue of the limestone. Since plant and animal remains decay completely during the formation of the ore, fossils are found in bauxite only very rarely. This is why the site at Cornet that preserves fossil bones is so interesting: no similar environment anywhere in the world is known to preserve dinosaur remains.

The Cornet fossil reptile bones were transported and accumulated together into clayey deposits rich in aluminum minerals. Bones were washed together and then redeposited by heavy showers into karstic holes and as a consequence all the skeletal remains are mixed up. Typically, the bones that are found side by side are never articulated and represent different specimens and often species. When the first findings arrived at the Museum of Oradea, biologist and paleontologist Tibor Jurcsák started to work with them. His study was supported by technician Elisabeta Popa, who hand-prepared all of the dark bones form the dark and hard bauxite. A large proportion of the fossils still wait to be cleaned. Although Jurcsák focused on the dinosaurs known from the site, his colleague Eugen Kessler made a study of the birds and pterosaurs. After the death of Jurcsák, Erika Posmoşanu continued research on the fossil reptiles of the Bihor Mountains; subsequently British paleontologists joined in with the revision of the fossil assemblage.

Bone fragments in the dark rock. About 40% of the Bihor specimens are more or less smooth vertebrae although elongate, fragmentary bones (like metatarsal bones and phalanges) are also frequent. Flat or strangely shaped bones (such as hipbones) and skull remains are very scarce, or are damaged so heavily that they are very difficult to determine (0.25× magnification).

📖 Jurcsák & Popa 1978, 1979; Jurcsák 1982; Jurcsák & Popa 1984; Jurcsák & Kessler 1991; Benton et al. 1997; Posmoşanu 2003; Galton 2009; Dyke et al. 2011

Dinosaurs of the Bihor Mountains

According to Tibor Jurcsák saurischian dinosaurs are represented in the Cornet fauna by the small theropod *Aristosuchus,* whereas ornithischians are more diverse and include the following genera: *Dryosaurus, Valdosaurus, Hypsilophodon, Iguanodon, Vectisaurus,* and *Hylaeosaurus.* Of these, the last is an ankylosaur (an armored dinosaur that walked on four legs), and the others are small—or medium-sized, bipedal herbivores. Although similar fossil reptiles are also known from England (from Sussex and the Isle of Wight, for example) and from sites in Spain and Belgium, the dinosaurs from the Bihor Mountains are very difficult—indeed, sometimes impossible—to refer to already known species. Perhaps there are some small species within this fauna that are only known from this site.

The chaotic coexistence of young animals and adult, male and female, specimens in the fauna makes these eroded, fragmentary bones even more difficult to determine. Revision of the fossil reptile fauna from the Cornet site suggests that these dinosaurs represent fewer species than was thought before. It seems that the most common element of the fauna was a small ornithopod, called *Camptosaurus;* remains of *Valdasaurus*—a close ally of *Dryosaurus*—are also frequent. This latter species, however, is poorly known, therefore the badly preserved Cornet specimens are very difficult to interpret. Besides these animals, armored and carnivorous theropod dinosaurs were present, although remains of the latter are extremely rare. In any case, the existence of carnivorous dinosaurs can also be proved by the presence of puncture marks on some bones—these are the bite marks of ancient predators.

📖 Jurcsák & Kessler 1991; Benton et al. 1997; Posmoşanu 2003

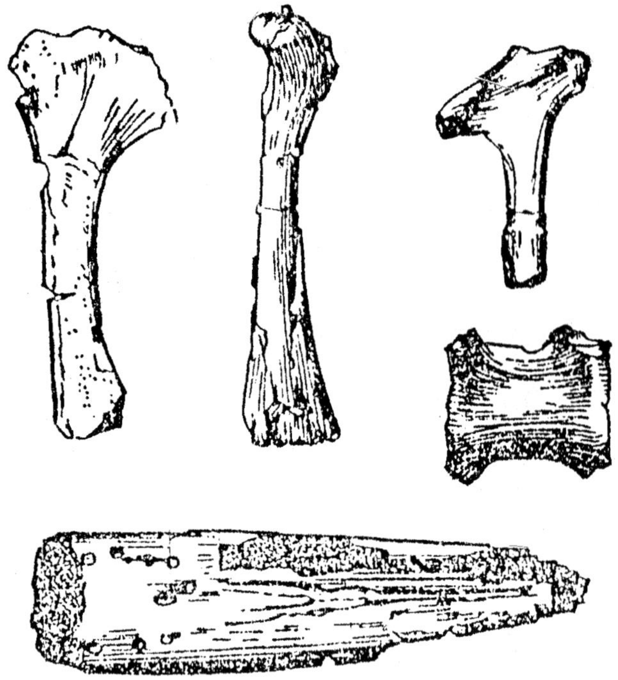

Different pterosaur remains (limb bones, vertebrae, and skull fragments) from the figures in the work of Jurcsák & Kessler, in different magnifications. None of these bones are more than a few centimeters long.

Unidentified Flying Creatures: Pterosaurs from the Cornet Mine

Alongside robust and stout dinosaur bones, tiny and fragile elements have also been found at Cornet. These fossils were first thought to be the remains of flying reptiles, pterosaurs, but are very fragmentary and poorly preserved. As a result, exactly what kinds of pterosaurs were present at Cornet is somewhat uncertain: the Late Jurassic *Cycnorhamphus* and the Early Cretaceous, otherwise Asian form *Dsungaripterus* were reported from the assemblage by Tibor Jurcsák, a curator at the Ţării Crişurilor Museum, and his co-workers. Recent revision of these fossils has shown, however, that most of the pterosaur bones from Cornet cannot be conclusively identified beyond the level of Pterosauria or Pterodactyloidea, and claims that specific taxa are present, such as *Cycnorhamphus,* cannot be supported because the fossils are too poor.

We do know that all of the Cornet pterosaurs were small, as their bones are all only a few centimeters long. These small reptiles hovering over the present-day Pădurea Craiului Mountains would have looked like toys compared to many Cretaceous pterosaurs, which had wingspans of several meters.

📖 Jurcsák & Popa 1982; Jurcsák & Kessler 1986; Dyke et al. 2011

Evolution or Revolution? The Ancient Birds of Cornet

The fossils originally described as ancient birds are the most intriguing remains from the Early Cretaceous Bihor Mountains vertebrate fauna. Originally a few distinct forms were recognized, but, as is true of the pterosaurs, the fossil remains on which identifications were based are not great.

A fragmentary femur that was given the name *Palaeocursornis biharicus* was first identified as a Cretaceous paleognathous (ratite) bird, and a partial humerus referred to as *Eurolimnornis corneti* was initially supposed to be an ancient, grebe-like neognathous bird. (Both of these names were originally attached to a set of remains first thought to belong to a bird known as *Palaeolimnornis,* and this name was itself used a replacement for an earlier one, *Limnornis.* Paleornithology is not an easy job.)

But this is not just a question of taxonomy and nomenclature; the described birds from Cornet were supposed to belong to Neornithes, a large group that includes all living avians. This is significant because, if correct, neornithines were present in the earliest Cretaceous at the Berriasian Cornet locality. Modern birds evolved, diversified, and specialized rapidly soon after the first representatives of this group appeared in the Late Jurassic. Some specialists have suggested that at least some of the so-called birds described from Cornet may really represent small carnivorous dinosaurs with fragile skeletons, or may belong instead to pterosaurs. It has been known to paleontologists for some time that theropods and birds are closely related, so it is not a surprise that their remains are hard to distinguish in many cases. From an evolutionary point of view, although pterosaurs and birds are not closely related to one another, they were adapted to the same mode of life and so have similar aspects to their skeletons. Birds, non-avian theropods, and pterosaurs all have hollow bones, for example.

The most recent revision of the Cornet fossils has concluded that although there is some evidence for birds in the assemblage, none of the fossils can be referred to Neornithes. In other words, there was no revolution by the Berriasian, just slow evolution. At least one of the Cornet fossil birds is thought to belong to the genus *Enaliornis,* a close ally of the better-known and often-cited *Hesperornis.* This latter species was flightless and was a foot-propelled water bird, while *Enaliornis* was more primitive and may have retained large wings.

📖 Kessler & Jurcsák 1984a, 1984b, 1984c; Dyke et al. 2011

Archaeopteryx *at Cornet?*

Among the Lower Cretaceous fossil reptile bones found at Cornet, some remains were originally identified as belonging to the earliest fossil bird, *Archaeopteryx.* This bird is otherwise known only from Upper Jurassic lithographic limestones near Solnhofen, in southern Germany,

Teeth of pycnodontid fish from the Upper Cretaceous near Sümeg (close to original size).

and these fossils are among the most intriguing, best preserved, and perhaps the most valuable fossils from anywhere in the world. Even the fine details of their feathers remain, in contrast to the claimed Romanian records that consist of two or three fragmentary bones. Restudy of these tiny bones has revealed that they certainly do not belong to *Archaeopteryx;* at best they can be considered to be from an indeterminable archosaurian—they might be placed anywhere within the large group that comprises all birds, dinosaurs, pterosaurs, crocodiles, and a range of other taxa—both living and extinct.

📖 Kessler & Jurcsák 1984a, 1984b, 1984c; Dyke et al. 2011

Terrestrial Fauna of the Bakony Mountains

For a very long time, a single fossil turtle from Sümeg remained the only Cretaceous reptile to be found in the Bakony Mountains. Finally, however, in 2000 the ice broke and huge numbers of reptile fossils began to be discovered in the vicinity of Iharkút, including many species of dinosaurs, pterosaurs, lizards, crocodiles, turtles, and birds new to science. Today the Iharkút fossil fauna is of international interest and has contributed a great deal to our understanding of the world of Cretaceous reptiles.

The Boszorkány-hegy ("Witch Hill") Crocodile Tooth

In 1951, a half-page paper was published in the *Bulletin of the Hungarian Geological Society.* This short and curt article reported that during bauxite explorations, bone fragments and a "better preserved tooth-like fossil" had been found in the vicinity of Olaszfalu, embedded in the fractured pisolite bauxite. This tooth later turned out to belong to a fossil crocodile.

The original author of the 1951 paper emphasized that this find alone was worthy of mention because bauxites and other similar deposits are always so poor in megafossils. The fragments, then, had special importance and might also suggest that more detailed examination of such deposits was worthwhile in the future. Maybe the lack of such "devoted attention" is the reason we had to wait 50 years after the discovery of the Boszorkány-hegy crocodile tooth for the sensational discoveries at Iharkút.

📖 Kretzoi & Noszky 1951

The Sümeg Turtle

Although fragmentary, another pretty fossil of a marine turtle was found in 1963 in a reef facies of Upper Cretaceous limestone (that also contains colonial corals, calcareous algae, echinoids, and rudist bivalves) in the Kecskevár Quarry near Sümeg. The skull, tail, and limbs of this specimen were missing, and the fossil was sent to the Hungarian Geological Museum by Lajos Kocsis, who was an enthusiastic collector of the Sümeg fossils. This 34-centimeter fossil was described as *Senonemys suemegensis,* a new species within a new genus, by Péter Bohn, a geologist who also classified it into the family Emyidae, within the suborder Cryptodira. At the time of its discovery this fossil was the oldest certain turtle fossil known from Hungary. Nobody expected that 36 years later, thousands of Cretaceous fossil reptile bones would be found in the Bakony Mountains.

📖 Bohn 1966

The History of the Discovery of the Iharkút Fossil Site

As mentioned above, the now widely known fossil vertebrate site of Iharkút was discovered by Attila Ősi in 2000. In 2003 the enthusiastic young researcher later recounted the moment of the discovery: "The remains were embedded on the top of a sandstone layer and were covered by 1–2 cm thick clay. We removed a large piece of the covering sandstone and revealed the clayey layer that we lifted up like a carpet and had a look beneath. The bones were taking a rest there, if they were waiting for us for 85 million years."

Following their first successful field campaign, the fossil collectors were eager to begin a second one in summer 2001. It was during this two-week exploration period that they first discovered bone-bearing rock material on the spoil of the bauxite pit. Huge amounts of rock yielded hundreds of bones and teeth, and later the fossil-rich clay

Digging for dinosaurs in the vicinity of Iharkút. As a result of successive field collecting campaigns organized yearly since 2000, thousands of Cretaceous terrestrial vertebrate fossils have been found. The fossil material collected from this site has attracted international interest.

layer was also located in situ, 25 meters below the surface, as it was exposed on the eastern side of the pit. Fossils were found in the finer grained sandy part of this layer, yet in order to continue to hunt for fossils, the researchers needed the help of the mine, since it seemed impossible to remove the thick covering sandstones using just bare hands and simple tools. Fortunately, the management of the bauxite mine supported the the first Hungarian dinosaur expedition—and by the beginning of the next field season, digging machines supplied by the mining company had done most of the excavation work. As the bedding surface of the fossil-rich layer was cleaned the quest for bones started afresh, but immediately new difficulties were encountered. Fieldwork at Iharkút was hampered by the extreme heat and occasional heavy summer showers. On very hot days the young paleontologists working in the open-air bauxite pit felt like they were in the Gobi Desert or the American Badlands. On rainy days the gooey mud tacked to their tools made excavation work impossible.

From the very beginning, collecting was supervised and managed by the discoverer of the site, Attila Ősi. Year by year volunteers, mainly biology and geology students, were involved in the excavation work, which was occasionally supported by the National Geographic Society, the Hungarian Scientific Research Fund, and the Hungarian Natural History Museum.

📖 Makádi et al. 2006; Ősi 2003; Ősi & Rabi 2006

The Amphibians from Iharkút

From time to time, during excavations at Iharkút, as a result of regular and careful sediment screening, the tiny bones of fossil amphibians have been collected. Among these bones, some of special interest were described as *Hungarobatrachus szukacsi;* if classified correctly, this fossil is the earliest representative of the order Neobatrachia, a group that contains most extant anurans—more than 500 species. Previously only the much more ancient Archeobatrichia and Mesobatrichia lineages had been described from the European Mesozoic, and so the discovery of *Hungarobatrachus* suggests that neobatrachids—and also the lineage leading to true frogs—appeared earlier in Europe than was previously thought. It is possible that these

An afternoon siesta at Iharkút during the excavation campaign in 2009. Attila Ősi examining an extremely well preserved bird bone found and cleaned that morning.

ancient frogs arrived by rafting on plant debris from Africa or—less probable, but difficult to rule out—were attached as eggs to the feet of birds. During the Late Cretaceous Iharkút was located on an archipelago between these two continents, and so frogs may have eventually conquered Europe by island hopping.

🕮 Szentesi & Venczel 2010

Hungarian Dinosaurs

Previously only Jurassic footprints from the Mecsek Mountains and some poorly preserved, never identified or described fossil bones from the vicinity of Ajka showed that dinosaurs once existed in present-day Hungary. This situation changed only when the excavations at Iharkút yielded thousands of new finds. Most of the fossils that have been found at this site (including bones, teeth, and armor elements) belong to dinosaurs—animals that are always the focus of public interest—both herbivorous and carnivorous. Of these finds, the most complete fossil material belongs to a lightly armored species within the family Nodosauridae (see the bone map below). This dinosaur is very close to *Struthiosaurus,* which was described from Transylvania and was later also found in Austria, but also differed considerably and needed a new name. This animal was described by Attila Ősi as *Hungarosaurus tormai,* a new species in a new genus after András Torma, a friend of Attila's and an ardent member of the first Hungarian dinosaur expedition.

It was also clear from the earliest stages of the Iharkút excavations that, besides the armored *Hungarosaurus,* another herbivorous dinosaur was also present in this Cretaceous menagerie. This second herbivore is a close ally of *Rhabdodon*—another Transylvanian dinosaur—but very probably represents a new, as yet undescribed, species.

🕮 Ősi 2005b; Ősi et al. 2003; Ősi & Rabi 2006

The quest for minute fossil remains. Zoltán Szentesi, a member of the Iharkút team, and his co-worker are screen washing the bone bed on the site. In the basins is a bubbling mixture of clayey rock containing fossils and hydrogen peroxide. This strong oxidant disaggregates the pieces of rock and makes small bones and teeth easier to wash through strainers. The sticky, clayey deposit is washed and dried. In what remains, small organic particles—including bone fragments, minute fossils fruits, and amber drops—can be found. Although screen washing usually destroys the fragile bones held together by the matrix, occasionally determinable fossils—including the remains of little amphibians, and small, but resistant, dinosaur teeth—can also be found.

Carnivorous Dinosaurs from Iharkút

Carnivorous dinosaurs are represented in the Iharkút fossil fauna—but mainly on the basis of minute, often serrated teeth. These remains are thought to belong to Dromeosauridae, one lineage of maniraptoran theropods that are among the lesser-known members of the Iharkút menagerie. So far, all the fossils known from these small and medium-sized animals can be placed in a single museum drawer.

Most numerous of these specimens are about 60 slightly flattened, curved, and finely serrated teeth that are just a few centimeters long. Some are comparable with others described from Muthmannsdorf in Austria as *Megalosaurus pannoniensis* Seeley. In addition to these teeth, a few

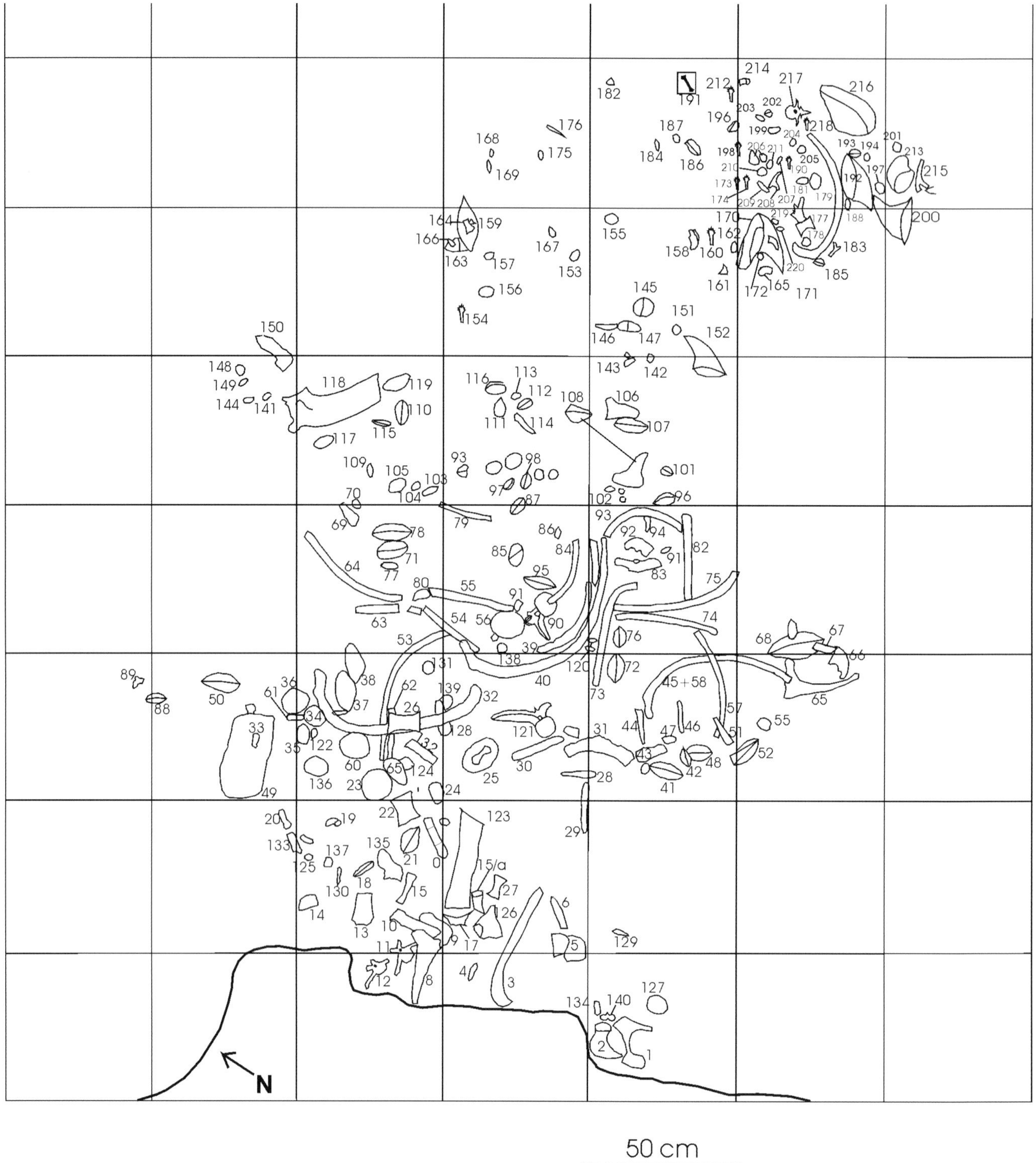

A detailed "bone map" of the Iharkút fossil site. This area of just a few square meters has yielded hundreds of separate bones or fragments of bones, 98% of which belonged to an armored dinosaur. The remaining 2% comprise turtle remains and crocodile teeth. Since all the dinosaur bones are different, it seems likely that they represent the remains of a single specimen whose skeleton was scattered over a relatively small area. There were fragments of ribs, vertebrae, parts of forelimbs and hind limbs, varied armor plates, and even skull fragments within the bone material; taken together they represent the type material of *Hungarosaurus*. This is not only the best-preserved specimen of the species, but it is also the most beautiful and one of the most complete Late Cretaceous armored dinosaur known from all Europe. Besides the type, five additional *Hungarosaurus* specimens were found in Iharkút; unfortunately, they are far less complete.

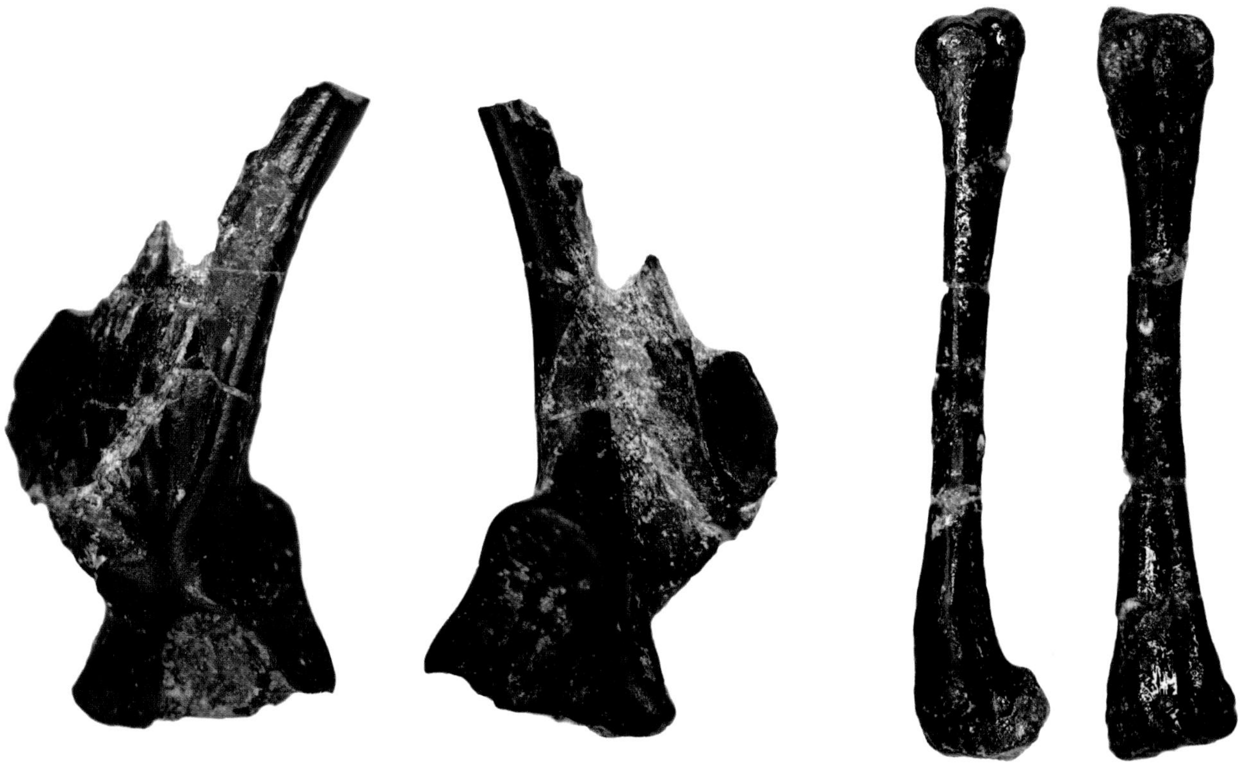

Hungarobatrachus (the "Hungarian frog") was described as an extinct genus of advanced frog on the basis of isolated ilia (left) and tibiofibulae (right) from the Iharkút locality. This frog had an extremely high iliac crest, which would have been the point of attachment for the jumping muscles (9× magnification).

disarticulated bones have also been found that cannot so far be precisely determined, and to date just a single new carnivorous species (*Pneumatoraptor fodori*) has been described. As the name suggests, this animal had pneumatic bones, meaning that airsacs penetrated some of its bones—as is common in birds and certain types of dinosaurs.

📖 Ősi, Apesteguía, & Kowalewski 2010

Ajkaceratops kozmai: The First European Horned Dinosaur

From the beginning of the Iharkút excavation, some strange-looking small bones were found from time to time; their surfaces resembled broken seeds. Because these bones were interpreted tip down and were thought to be dinosaur claws, not much attention was paid to them. However, there was a breakthrough in summer 2009, when a more complete example that looked like a beak was found. Some ornithopods, such as the European *Rhabdodon,* have similarly shaped premaxillae at the front of their skulls, and so suspicion fell on this dinosaur.

However, a thorough study of these bones proved that they represent a completely different group—the horned dinosaurs, also known as ceratopsians. The mystery of the "peach pits" was also solved: these bones should be orientated tip up, as they represent predentary fragments (the bone of the lower beak).

Ceratopsians were a highly diverse and abundant radiation known primarily from the Cretaceous of Asia and western North America, with only controversial remains reported from other continents. The new material from Iharkút unambiguously demonstrates that ceratopsians were present in the Late Cretaceous of Europe, and these exceptional fossils were published in *Nature.* The new genus was named after the city of Ajka, and the species name refers to Károly Kozma, geologist of the bauxite mine.

📖 Ősi, Butler, & Weishampel 2010

Turtles in the Mine

Turtle remains are the most abundant fossils at Iharkút. Since 2000, three partial skulls, two isolated lower jaws, several hind limbs, two partial plastrons, and hundreds of shell fragments and other postcranial bones have been unearthed. The skulls and lower jaws belong to a currently unnamed new bothremydid taxon, as the massively built turtle skull generally shows a great deal of diagnostic features and thus can be identified. In contrast, the postcranial remains—including the plastron and shell—are less diagnostic. Specialists regard these isolated postcranial

Skeleton of *Hungarosaurus* in the exhibition hall of the Hungarian Natural History Museum.

bones (hind limbs, pectoral and pelvic elements, and vertebrae) to be only indirectly referrable to bothremydids simply because no other groups of turtles have so far been identified in the Iharkút fossil collection.

The family Bothremydidae is a group of well-known freshwater pleurodiran turtles in Late Cretaceous herpetofaunas. Because of their unspecialized ecological function the group was widespread in Late Cretaceous ecosystems. The Iharkút bothremydid is a new taxon and a member of the so-called *Bothremys* group. According to a preliminary cladistic analysis, this animal is a close relative of *Foxemys mechinorum,* known from Late Cretaceous deposits in southern France. The newly discovered material suggests that members of the *Bothremys* group first appeared in the Mediterranean and that the appearance of bothremydids in North America at the same period can be explained by a sudden migrational event from Europe.

📖 Botfalvai 2004; Botfalvai & Rabi 2006

Teamwork. Members of the Iharkút excavation try to remove a large block of the bone bed in one piece so that the larger bones are less damaged and smaller ones can also be collected. Even the tiniest fossils can be extracted by screening the sediment.

The First Cretaceous Crocodile Remains from Hungary

Fossil crocodile remains are very scarce from Hungary; this is another reason Iharkút material has special importance. In addition to the unique Jurassic crocodile remains found in the Gerecse Mountains, the Iharkút crocodiles represent the second Mesozoic appearance of this group in Hungary. In addition to fragmentary bones, numerous well-preserved, almost complete skulls, mandibles, teeth, whole limb bones, and vertebrae have been collected; so far three different kinds of crocodile have been identified. One belongs to the suborder Euscuhia, a group that includes all living crocodiles as well as many extinct forms. The fossil is a species of Alligatoroidea, most likely new to science. Because it is the earliest member of the group it cannot be allied with already-described primitive species. In addition, because the new Iharkút species is older than those already known from North America, it may suggest that the whole group originated in Europe.

The second crocodile known from Iharkút is a relatively little known form, evolutionarily far from its abovementioned counterpart, belonging to the extinct suborder

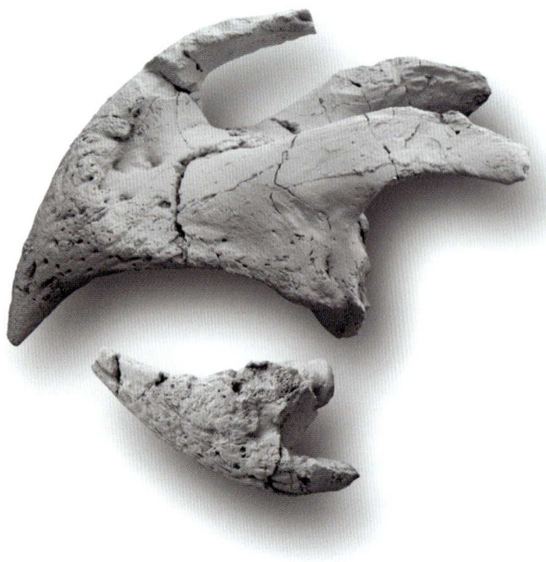

The beak of a dinosaur (close to original size). Type material of the sensational new *Ajkaceratops kozmai*. Although the known bone material is very scarce, these characteristic skull elements provide a solid case for the presence of ceratopsian dinosaurs in the Late Cretaceous of Europe. The figured bones held a real horny beak, which would have given the dinosaur a characteristic appearance. The closest allies of *Ajkaceratops* are the Asian *Bagaceratops* and related forms.

Ziphosuchia. Based mainly on dental fragments and teeth that were found separately, this animal was determined as *Doratodon carcharides*.

The teeth of the Iharkút *Doratodon* are flattened and wrinkled, resembling the teeth of some theropod dinosaurs. They were more or less triangular in shape and had a serrated cutting edge. In contrast, teeth of the Alligatoroidea are "globidont"—not sharp, but globular; this suggests that *Doratodon* had a different feeding habit and mode of life. The final and most distinctive member of the Iharkút crocodile club is another Eusuchia that had very special teeth, like those of a seal. When the first teeth of this animal were found researchers ruled out every other possibility before turning to crocodiles.

📖 Rabi 2004

Yet Another Crocodile?

In the very early stages of the Iharkút excavations, minute toothlike fossils just a few millimeters in length were found. These fossils had flattened crowns and a chewing surface that was covered with more or less eroded cones. To begin with, the researchers were unable to identify these teeth, and later multituberculata—an ancient mammalian group—was suspected. Indeed, these early mammals had even smaller teeth, and something was wrong with their structure: mammal teeth have roots, whereas the Iharkút specimens do not. Finally, the sensible conclusion seemed to be that these strange teeth were closest to rootless, acrodont fish teeth, although after a year it became obvious that this idea too was wrong. Attila Ősi (2005a) was cleaning a crocodile skull when he finally discovered the answer:

> The solution was waiting for us in the autumn of 2003, when during preparation work I noticed place on the inner side of the skull, at the end of the row of teeth where no more were expected, I found two pitch-black teeth. They had cones and grooves just like the teeth we had already found but could not identify and I couldn't believe my eyes. The last tooth was nearly 1.5 cm in size, twice the size of the last one and, as proved by the empty place for another tooth, four times larger than those in the front. What kind of crocodile was this? What were these strange shaped teeth used for? We now started to search for this new creature, previously unknown to everyone.

Later investigations revealed that the peculiar animal was really new to science and it was named *Iharkutosuchus makadii*, after the locality and after László Makádi, a key person in the excavations. This unique crocodile very probably lived in freshwater streams and rivers and searched for food along the muddy bottom—its mode of life somewhat similar to the duck-billed platypus, that strange Australian semiaquatic mammal that lays eggs. It is even the case that the adult platypus loses its teeth, and the remaining hard and flat grinding surfaces that are left behind resemble the teeth of *Iharkutosuchus*. However, the platypus is carnivorous, feeding mainly on annelid worms, larvae and freshwater crabs, and *Iharkutosuchus* is thought likely to have been a vegetarian, searching for fruits and seeds.

📖 Ősi 2005a, 2008a; Ősi et al. 2007; Ősi & Makádi 2009; Ősi & Weishampel 2009

A hind tooth of the bizarre Cretaceous crocodile from the Bakony Mountains. Cones and ditches aligned into rows—rare features for a crocodile—are visible.

Giant Freshwater Mosasaurs

Of special interest within the Iharkút fauna are the remains of lizard-like creatures. So far mainly vertebrae (more than 80 specimens have been found), fragments of the skull (including postorbitals, postfrontals, and squamosals) and teeth have been collected. These lizard-like fossils represent a minimum of 4 different animals, with about 90 percent of them belonging to 1 large species, thought at first to be *Varanus* like. This explains why László Makádi wrote the following about the formidable dimensions of this animal: "Comparing the size of the vertebrae with those of a living species, the water monitor (*Varanus salvator*), we can conclude, that the Iharkút species may have reached five to six meters in length. In this way it greatly exceeds the Komodo dragon (*Varanus komodoensis*) the largest living lizard which reaches about three meters, and is close to the seven meters long *Megalania prisca,* known from the Pleistocene of Australia."

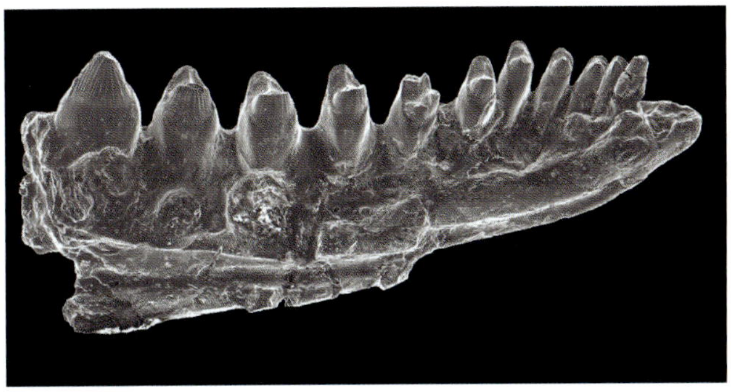

Left dentary of *Bicuspidon* aff. *hatzegiensis* Folie and Codrea, from Iharkút. Besides the large fossil reptile remains collected from this locality, some small lizards have also been found. This tiny jaw belongs to a small lizard closely related to extant Teiids, and has a simple (monocuspid) hind tooth and a single-peaked front tooth. In the middle of the jaw, the flawed (bicuspid) teeth of this lizard can be seen. SEM image from Makádi 2006.

The making of the beast. László Makádi, a member of the Iharkút team, prepares a scientifically accurate reconstruction of the mosasaurid that was discovered at the site. One by one, each bone of this animal was copied, molded, and cast and, finally, the whole skeleton was assembled (although reconstructing the skull, known from very incomplete fossil remains, was a real challenge). This reconstruction, along with other fossils from Iharkút, is exhibited at the Hungarian Natural History Museum in Budapest.

However, this view of a huge terrestrial lizard wriggling across the fluvial plain—the reconstructed environment of Iharkút—was soon revised by Makádi when further skull remains showed that these reptiles were very well adapted to living in water! It seems that they were in fact closer to the aquatic mosasaurs than to the mainly terrestrial varanid lizards. (However, this was not an enormous mistake; mosasaurs and varanids are close allies.) Of special interest is the fact that mosasaurs were powerful swimmers and inhabited warm, shallow epicontinental seas. The Iharkút species probably lived in a freshwater environment, which is a unique feature for the whole group.

📖 Makádi 2005

Bakonydraco galaczi, *the Fishing Dragon of the Bakony Mountains*

Discovery of the first Hungarian pterosaur remains in 2001 was outstanding, even given the series of discoveries from Iharkút. One of the first finds was made by Ferenc Szőke, a member of the first Hungarian dinosaur expedition, during a summer field collecting campaign. Thanks to painstakingly accurate preparation, the hollow and fragmentary pterosaur limb bone was entirely removed from the matrix. (Cleaning and extracting these fossils is not easy, since the thickness of the hollow bones can be less than 1 millimeter.) Luckily, one end of the bone was preserved nearly entirely, so the original size of the piece could be estimated. After a comparison with other similar fossils, found in other places, it became clear that this 16.5-centimeter bone was a proximal part (which is closer to the body of the pterosaurs) of a digit of a fourth finger. This finger—which stretched the flying membrane of the reptile—was about 30 centimeters long.

Dozens of lower jaw fragments of *Bakonydraco*, each representing a separate specimen. Flying reptiles were very common in this Late Cretaceous ecosystem, and in fact this locality has yielded some of the most abundant pterosaur remains from anywhere in Europe.

In the next field season, some peg-shaped pieces of bone a few centimeters in length were found with little holes in their sides; it was soon discovered that these were jaw fragments of pterodactyloid pterosaurs. Subsequently dozens of bones of this kind have been found, among them a beautiful lower jaw. Each belongs to a separate individual, which means that pterosaurs were very common in the Iharkút ecosystem.

To begin with, the flying reptiles from the Bakony Mountains were identified as representatives of the genus Azhdarcho, also known from Spain and Uzbekistan, but are important because they show that members of this group were sometimes toothless. Azhdarchid pterosaurs also include some of the largest flying reptiles of all time. The huge *Quetzalcoatlus*, named after the feathered Mexican snake god, had a wingspan of about 10 meters and belonged to this group. Comparing the corresponding bones of different animals, described from different localities, we can estimate how big an animal was. Applying this reasoning, the jaw of of the Iharkút pterosaur is about one-third the size of the corresponding bone in *Quetzalcoatlus*. Thus, we can estimate that the Ihrakút species was proportionally smaller and may have had a wingspan of around 3.5 meters.

The final conclusion that was reached based on the Iharkút pterosaurs is that these fossils represent a new genus and species within azhdarchids named *Bakonydraco galaczi*. The genus name refers to the dragon (*draco*) of the Bakony Mountains, and the species name refers to paleontologist András Galácz, who strongly supported Iharkút research from the beginning.

Researchers suppose that *Bakonydraco* used its strange, elongate beak as do graceful skimmer birds. These birds fly with their knife-edged lower mandibles submerged in the water, and when they find something they snap their

Late Cretaceous reptiles from Iharkút. (1, 2) *Iharkutosuchus makadii* Ősi, Clark, and Weishampel—skull of a eusuchia crocodile from above and below. The flat teeth, covered by wrinkles and bumps, suggest that this animal was probably feeding on plants as well as animals (0.75× magnification). (3, 4) *Bakonydraco galaczi* Ősi—an excellently preserved lower jaw of the flying reptile (pterosaur) from above and from the side. In the front of this bone two rows of little holes can be seen: these are the places where nerves and blood vessels left the bone. This unique specimen was found by László Makádi, a key member of the Iharkút team (0.75× magnification). (5) *Hungarosaurus tormai* Ősi—an entirely preserved small tooth of the armored dinosaur of Iharkút (original size). (6) Tooth of an herbivorous dinosaur related to the well-known Late Cretaceous European *Rhabdodon*. The root of this tooth is partly broken (original size). (7) Finely serrated tooth of a midsized carnivorous dinosaur. The root is not preserved (original size).

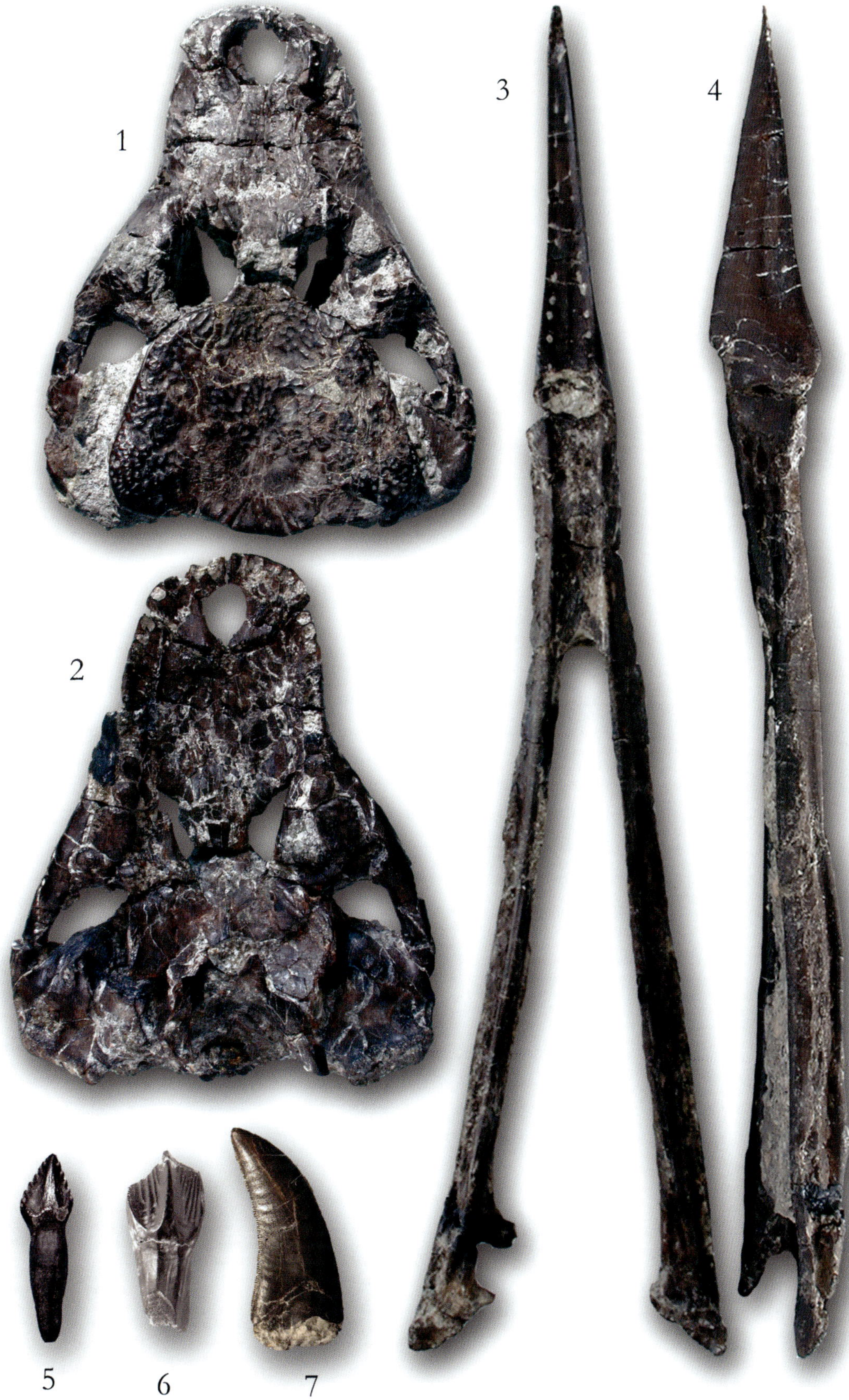
1
2
3
4
5
6
7

Portrait Gallery

Baron Ferenc Nopcsa (1877–1933)—The Dinosaur Baron

Baron Nopcsa was a geographer and geologist, ethnographer, soldier, and spy. A many-faceted scientist and a researcher who paid close attention to even the smallest details, yet a man of broad vision, Nopcsa was capable of great scientific insight and synthesis. He was a Hungarian patriot who feared for his country and took many political gambles as well as a respected member of academic bodies and aristocratic circles. Most unusually, the baron was not only a great traveler but was equally at home among warring Albanian tribes, Csobans of Transylvania, or at the emperor's court in Vienna. In addition, he was almost elected king of Albania, and once hijacked an airplane.

From time to time the baron would grow his hair long and vanish from his friends to travel in mountain regions under the pseudonym Peter Gorlopán. He sent back reports directly to the chief of the monarchy's ruling elite, Franz Conrad von Hötzendorf. His character was a unique combination of a Renaissance-like openness and the romantic idealism of a Jókai novel. This adventurous existence came to a shocking end—in a murder-suicide.

Nopcsa's main interest was the remains of dinosaurs, including those found on his family's own estates. Among these was a duck-billed dinosaur, an armored reptile, a sauropod with a long neck and tail, and several other exciting findings, including a predatory theropod and a flying reptile. These are the dinosaurs of Nopcsa: the *Magyarosaurus dacus* and others.

From a scientific point of view, the significance of these priceless relics lies in their link to other well-known dinosaurs of similar age from North America and Asia. The Transylvanian dinosaurs appear to have lived on an island—which may explain their dwarfed dimensions—and they were present some 67 million years ago, just prior to the extinction at the end of the Cretaceous. They may even have witnessed the fatal meteorite.

In addition to dinosaur research, Nopcsa made important contributions to reptile systematics and to the question of the origin of flight. He always looked to biology when attempting questions in paleontology and in this respect, alongside his friend the Viennese scientist Othenio Abel, he is considered one of the founders of paleobiology.

🕮 Főzy 2004

heads down and clamp onto a fish. *Bakonydraco,* however, could not have done this because its lower jaw was too fragile. Perhaps instead it had a throat pouch, like that of a modern pelican, and fished in shallow seas. An image of a floating and snorkeling pterosaur is not favored by scientists; indeed, the locomotion and feeding habits of flying reptiles are still incompletely understood.

🕮 Ősi, Weishampel, & Jianu 2005

Little Birds

Some tiny, extremely fragile bones from Iharkút have been identified as bird remains and represent not only the first Mesozoic bird bones from the Bakony but also from Hungary. Limb bones, including femora and a tarsometatarsus of a robin-sized enantiornithine bird as well as a larger species, have been discovered. They have been compared with similar European remains found in southern France and in Spain.

Birds appeared during the Late Jurassic and by the Cretaceous the lineage was already characterized by a diverse fauna. Fossil bird bones, however, are rare everywhere because they are so fragile and can fossilize only under certain conditions. For this reason even small amounts of fossil data, like that from Iharkút, take on special importance. We also know that during the Late Cretaceous birds and pterosaurs coexisted: the Iharkút findings might be used to support the hypotheses that ecological niches, once occupied by small pterosaurs, were inhabited by birds, while the larger-sized azdarchid pterosaurs were adapted to life over open land, flying over marshes and the seashore.

🕮 Ősi 2004, 2008b

Where, and When, Did the Iharkút Menagerie Live?

The dinosaurs, lizards, mosasaurs, turtles, crocodiles, pterosaurs, and birds of the Bakony Mountains all lived during the Santonian stage of the Late Cretaceous, about 85 million years ago. Dating these bone beds is done using palynology and suggests that the Iharkút reptiles are slightly older than those known from the Haţeg Basin (see below) and much younger than those known from

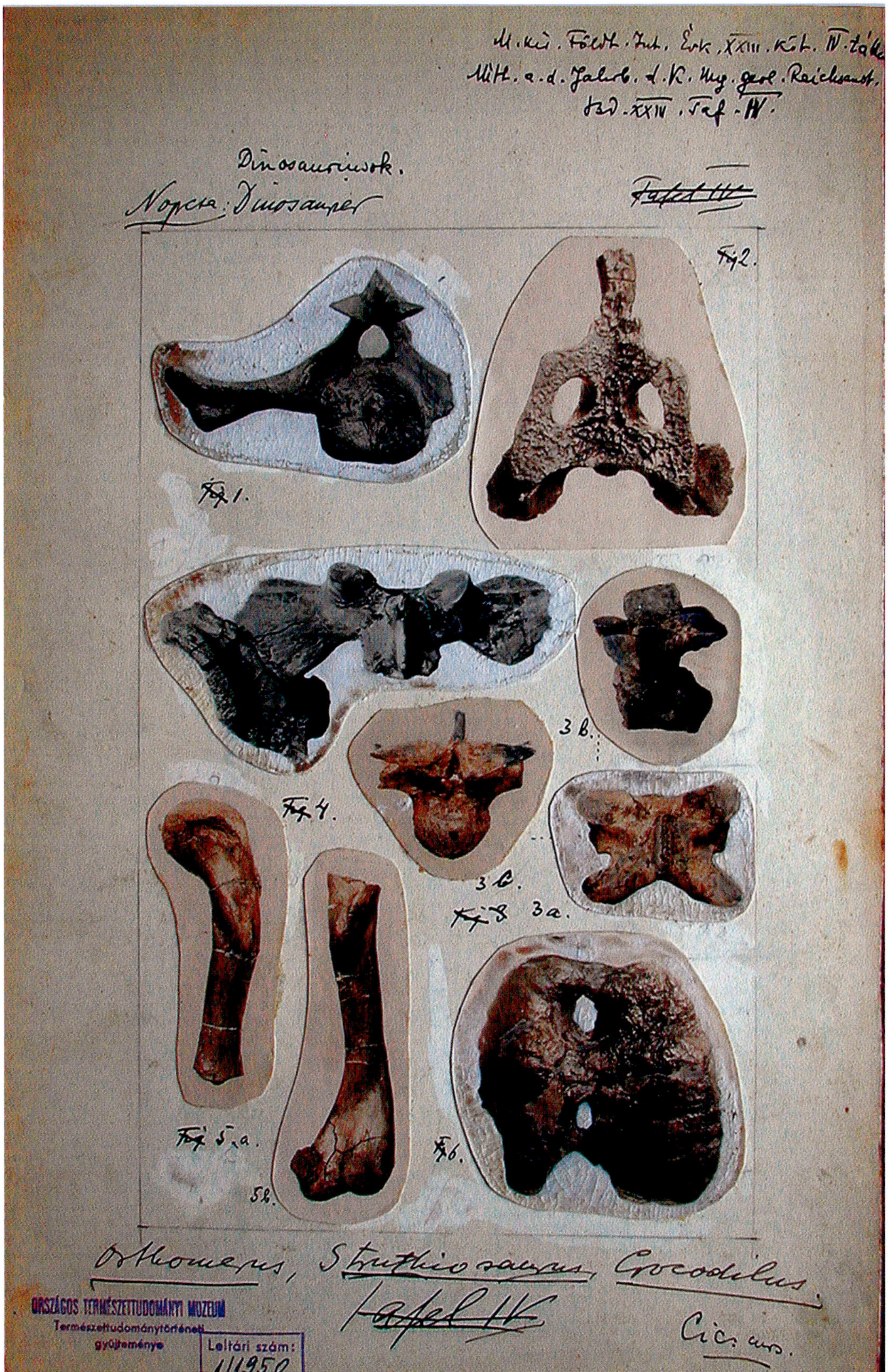

One of the original photo plates from an article on the dinosaurs of Transylvania marked with the energetic handwriting of Baron Nopcsa. The article was published in the *Annals of the Hungarian Geological Institute* in 1915.

In the footsteps of Baron Nopcsa. In 1997 colleagues from the Deva Museum organized a vertebrate symposium at which the latest results on Haţeg reptiles were presented. This shot was taken on the meeting's field trip, in the vicinity of Sînpetru—where dinosaur bones can still be found today. From left to right: Fabio dalla Vecchia, Zoltán Csiki, László Kordos, Coralia-Maria Jianu, Natalie Bardet, Xavier Pereda-Superbiola, Vlad Codrea, Costin Rădulescu, Mátyás Vremir, Dezső Illés, and István Főzy.

the Bihor Mountains. On the basis of the currently accepted paleobiogeographical framework, we can assume similarly between the Haţeg and Cornet sites and conclude that Iharkút was also part of a chain of islands that stretched from around present-day Spain into the Caucasus. Parts of this island arc were temporarily connected by land bridges to the mainland, allowing animals to migrate and mix. Long periods of isolation of islands, however, explain the appearance of many unique forms.

The ancient environments on these islands can be understood by examining the sediment in the bone beds and by analyzing the composition of the flora and fauna. It is generally supposed that the Iharkút fauna lived in and around a vast floodplain, with meandering rivers, shallow lakes, and swampy areas all around. Sand, silt, clay, and dolomite pebbles, more or less cemented, are evidence for a high-energy depositional system; the occasional remains of turtles found in siltstones are representative of lower-energy systems, such as swamps or lake deposits. In a wider context, the whole area was likely surrounded by a hilly highland of a few hundred meters, from which sediment was derived.

Vertebrates from the Haţeg Basin

Discovery of dinosaur sites in the Haţeg Basin, in the foreground of the Southern Carpathians, is thanks to the early work of the "dinosaur baron," Ferenc Nopcsa. This man—who otherwise had an adventurous and eccentric life—described five dinosaur species, one turtle, and one crocodile—and discussed the presence of pterosaurs from Haţeg after many years of work. Among his dinosaurs, there was a sauropod (*Titanosaurus dacus*), one carnivore (*Megalosaurus hungaricus*), one ally of *Iguanodon* (*Rhabdodon priscus*), a duck-billed (hadrosaurid) dinosaur (*Telmatosaurus transsylvanicus*), and an armored form (*Struthiosaurus transsylvanicus*). The turtle he named after his Albanian secretary, *Kallokibotium bajazidi*, and the crocodile was given the name *Allodaposuchus precedens*. The

A lucky find: a nearly complete skull of *Telmatosaurus.* Although the bones were heavily flattened under enormous pressure, only a small part of this skull is missing and most of the bones are present. This fossil is the so-called holotype, on which the description of the animal is based; the fossil is stored in the Natural History Museum of London under the inventory number R 3386. Because Baron Nopcsa collected his dinosaurs at the turn of the twentieth century, he was very lucky that they did not need any preparation or consolidation: he found bones in the places the animals had died, and had only to pick them up from hillsides. Today researchers generally have to be prepared for hard digging and less immediately promising results.

baron's pterosaur remains, however, were neither named nor studied in detail, and so the list above represents the work of later researchers. In addition to fossil reptiles, the remains of mammals are now known from Haţeg.

🕮 Grigorescu 2010

The First Transylvanian "Dragon Bones"

The famous Transylvanian dinosaur localities can be found in the foothills of the picturesque Retezat Mountains in the Southern Carpathians. Most of the known fossils were found in the vicinity of Sînpetru, Săcel, and Sântămăria-Orlea—at one point almost all on land owned by the Nopcsa family. At that time, it was customary for the children of wealthy families to be sent to study to Vienna, the capital of the Austro-Hungarian Empire, and to spend summers back on their estates. During one such holiday, Nopcsa's 12-year-old sister Ilona found the first of the fossil reptile bones. When he was back in Vienna in the autumn, the baron took these bones and showed them to Eduard Suess, professor of geology at the Vienna Academy of Sciences. It was already obvious to the young man, however, that these bones were from "dragons"—dinosaur remains—and Suess was able to confirm this. The fossils, indeed, were the first to come from Transylvania and so were a scientific sensation. The Academy soon made the decision to send Gustav Arthaber, Suess's assistant, back to the site on the Nopcsa family estate to investigate, collect more fossils, and return them to Vienna. The story, however, does not end well: Arthaber was not able to find any more bones at Sînpetru, in part perhaps because the Academy in Vienna kept refusing to fund his expedition to Transylvania.

The young baron again visited Suess, who encouraged him to study the fossil bones. Nopcsa is supposed to have claimed that he could not because he had no appropriate knowledge in paleontology. "Then get some," the professor is reported to have said. So Nopcsa did exactly that!

This experience was the reason Nopcsa decided to study paleontology after his years in the Viennese school for nobles, the Theresianum. Instead of choosing military service or land management on the family estate, he entered the university and decided to be a geoscientist. He was a tough, devoted researcher who took on an enormous workload, and very soon had collected the necessary geological and anatomical information about his bones and their context. He was little more than 22 years old when he gave his first talk to the Viennese Academy of Sciences; his lecture was published and this made the young man quite well known among colleagues. Suess, however, remained his tutor and mentor as his interest focused on the fossil reptile bones from Transylvania. Nopcsa traveled widely, learning and collecting fossils, and completed his thesis back on his family's estate in time to celebrate his graduation on July 21, 1903. By this time, of course, he was already highly regarded as a paleontologist.

Old Localities and New Discoveries

Following the death of Nopcsa, the quest for dinosaurs stopped in Transylvania until it was resumed in the middle of the 1970s under the supervision of Dan Grigorescu from Bucharest. As this new project developed, American, English, French, and Belgian colleagues joined the team; many new discoveries were made at classic localities, and of course, new fossil sites were found.

Newly collected bone material was sent to the paleontological departments at Bucharest and Cluj Napoca Universities and to the Museum in Deva (Muzeul Civilizatiei Dacice şi Romane). In Bucharest, the name of Grigurescu and his younger colleague Zoltán Csiki are associated with this research, alongside their counterparts Vlad Codrea and, later, Mátyás Vremir from Cluj Napoca and Coralia-Maria Jianu from Deva. David Weishampel, a specialist on Eastern European dinosaurs at Johns Hopkins University, should also be mentioned. Weishampel participated in many excavations in Transylvania, and is author or coauthor on many papers in this area.

🕮 Grigorescu 1981; Smith et al. 2002; Grigorescu et al. 1990, 2010

Teeth cluttered up. The crowded teeth of *Telmatosaurus* form three rows behind each other and were replaced continuously by new ones during the life of the animal. Unlike sharks, whose teeth roll forward to replace broken or missing ones, in *Telamatosaurus* subsequent members grew from small holes (alveolus) and the new teeth gradually replaced the old, worn ones. At the same time, different members of the tooth family were in use. As Nopcsa wrote in his 1915 article, the dentition of this dinosaur clearly relates to the feeding habits of the animal: "Taking into consideration how weakly fixed the teeth are that are exposed to abrasion, we can immediately admit that this animal was chewing mainly soft aquatic plants." Indeed, all hadrosaurid teeth are somewhat similar: their narrow, leaf-shaped teeth are gently serrated toward the tips, and have longitudinal ridges in the middle. Very likely, as suggested by the baron, this dinosaur lived in marshy flat areas and fed on soft vegetation. Its teeth were weakly fastened in the jaw and were unable to chop tough plants or hard branches, so the animal used them for grinding rather than for cutting. However, the exact menu for *Telmatosaurus* remains unknown.

Transylvanian duckbilled dinosaur in a pencil drawing by András Szunyoghy. *Telmatosaurus* is one of the smallest known hadrosaurs: it was barely 5 meters long and 500 kilograms. The specimens from Haţeg clearly show fusion of certain bones in the skull and also some vertebrae, indicating clearly that they were fully grown adults and not young animals when they died. Very probably, *Telmatosaurus,* like many other hadrosaurids, was equally good at moving on four legs as it was on two, but we will never know exactly what this animal looked like. Living reptiles are very colorful and patterned: perhaps *Telmatosaurus* was like this as well.

The Duckbilled Telmatosaurus transsylvanicus
What Became of the Transylvanian Hadrosaurid?

The first dinosaur to be found and described from Transylvania was a hadrosaurid, which Nopcsa called *Telmatosaurus transsylvanicus*. This animal has gone on to become the best-known Romanian dinosaur; nearly its entire skull is known, as well as many vertebrae and larger limb bones. The fossils belong to several different specimens of different sizes, and *Telmatosaurus* is probably the smallest of all the known duckbilled dinosaurs. Interestingly, although this dinosaur lived just before the mass extinction at the end of the Cretaceous, it shows some ancient, primitive skeletal features.

📖 Weishampel et al. 1993

The Eggs of the Duckbilled Dinosaur

Almost 100 years after the first dinosaur remains were found in Transylvania, a few kilometers to the south of well-known localities, magnificent finds—including the first dinosaur eggs from this region—were discovered. To begin with, these spherical remains—similar to cannonballs—were thought to be from the Transylvanian sauropod dinosaur *Magyarosaurus dacus*. However, some tiny, toothpick-like embryonic bones were found, and these proved that the fossil nests had belonged to the duckbilled dinosaur *Telmatosaurus*. Dinosaur eggs have been found in several places around the world, but embryonic bones are

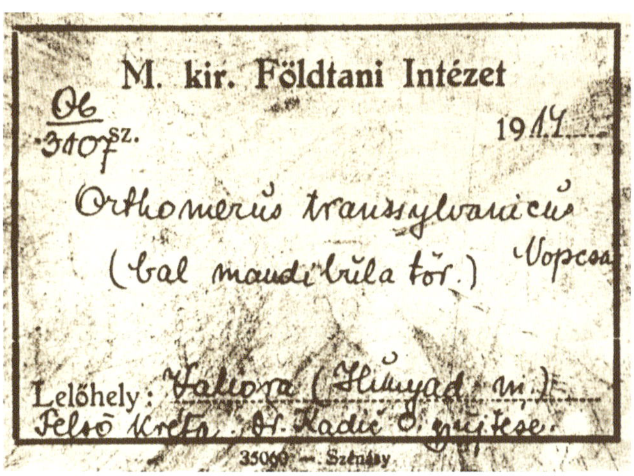

M. kir. Földtani Intézet
Ob
3107 sz. 1911
Orthomerus transsylvanicus
(bal mandibula tör.) Nopcsa
Lelőhely: Valiora (Hunyad m.)
Felső Kréta. Ot. Kadić O. gyűjtése.
35000 — Szénásy

A yellow, worn-out index card inscribed with the name *Orthomerus:* a duckbilled dinosaur from Transylvania. Nopcsa originally described this Transylvanian hadrosaur as *Limnosaurus*. However, it later turned out that this name had already been used for an American crocodile, so it was occupied. As a result, Nopcsa renamed the dinosaur *Telmatosaurus.* Later the baron thought that the duckbilled remains he had found were the same as those of another fossil reptile already named *Orthomerus,* and so he again renamed this dinosaur. According to the latest work, however, Nopcsa's fossil reptile differs from all other forms previously described, and so the name *Telmatosaurus* should once again be used. According to the bottom line of this index card, the fossil was collected by Ottokár Kadić (1876–1957) at Valiora.

The photo shows the natural collapse of the hillside at Tuştea, where the first Transylvanian dinosaur eggs were found. Subsequent to the discovery of the Tuştea fossil egg locality, remains of eggs were found at several other places (such as Pui, Nălaţ-Vad) in the vicinity. In spring 2001 a team made up of Belgian and Romanian experts excavated more than 40 eggs from 11 nests in the central part of the Haţeg Basin, at the Toteşti-baraj locality. At the same time, beautiful remains of fossil mammals were also discovered.

considered extremely rare. Altogether just a few localities are known where the remains of tiny baby dinosaurs have been found.

The eggs of birds are made up almost exclusively of calcium carbonate ($CaCO_3$), whereas eggs of modern reptiles are usually soft and leathery. We do not know exactly what the eggs of fossil reptiles looked like; however, some groups at least probably were calcareous and had thick shells. One thing is certain: their composition and structure may have been changed as a result of subsequent geological processes (dissolution and recrystallization).

Eggs at the Bottom of the Wall

Tuştea is located in the northwestern corner of the Haţeg Basin, 4.5 kilometers east of the village of Valiora. Dinosaur eggs were found in one of the hills at the border of the village. These hills in the vicinity were saturated by water because of unusually heavy rains, and the loose Late Cretaceous pebbly load collapsed under its own weight to create a vertical wall. Dinosaur eggs were found in a dark-red bed made up of finer grains (sand, silt, and clay). During the first field investigations, 14 roundish, cannonball-like eggs were found in the upper, 50 centimeter-thick, section of this red layer. The eggs were approximately 14 centimeters in diameter and would have had a volume of 0.8–1.1 liters. They were found in groups of 3 or 4, not far from each other, and the 8 most complete eggs had their bases intact, although their tops were cracked and shell fragments were also found in the neighboring sediments. The beds containing these exceptional remains are referred to the Densuş-Ciula Formation by geologists. The rocks were deposited in a fluvial, limnic-paludal environment; thus, it is likely that the nests of *Telmatosaurus* were buried by an unexpected flood.

A few years after the discovery of these first eggs, the upper, coarse-pebbled beds were removed with a bulldozer; many fragmentary, and some well-preserved, eggs were found. Based on the position of these eggs in the nest and the presence of dinosaurs of different growth stages, it was assumed that these Transylvanian duckbilled dinosaurs protected their young and took care of them in much the same way as their North American relatives are thought to have done.

📖 Grigorescu et al. 1990; Weishampel et al. 1991; Grigorescu et al. 2010

Zalmoxes robustus: *The Transylvanian Ornithopod Dinosaur Mochlodon, Rhabdodon,* or *Zalmoxes?*

The second dinosaur from Haţeg (Hátszeg) described by Nopcsa was originally referred to as *Mochlodon.* Since its description, however, there has been much uncertainty regarding its precise classification within ornithischian dinosaurs and related groups. Because of several ancestral skeletal characteristics, Nopcsa first thought that *Mochlodon* was a close relative of Hypsilophodontidae, but later this animal was considered close to *Camptosaurus,* a relative of *Iguanodon.* Ultimately, because of additional fossil discoveries, Nopcsa came to the conclusion that

An eggshell under magnification. The thickness of these shells is not more than 2.3–2.4 millimeters. Their outer surface has an irregularly bumpy appearance, and so it cannot be confused with anything else. This photo shows the surface of a fragment magnified a few times: the protuberances on the eggshell's surface are the outer terminations of crystal prisms; the smaller pores among the bumps—through which the growing embryo was able to get air to breathe—were later filled with crystalline calcite.

Cross section of a dinosaur egg, clearly showing prismatic calcite crystal bundles packed side by side.

Mochlodon should be classified with the genus *Rhabdodon,* which had been described earlier from southern France. Subsequently, the name "*Rhabdodon*" was used for this dinosaur, but according to one view this Transylvanian dinosaur—initially described as two species (*Mochlodon,* and later *Rhabdodon suessi* and *robustus*)—is in fact distinct and differs from the French *Rhabdodon priscus.*

Finally, a paper published in 2003 appears to have laid to rest the issue of the taxonomic status of this enigmatic Transylvanian dinosaur. Considering all the available data, David Weishampel and his co-authors proposed a new genus name (*Zalmoxes*) for the Haţeg dinosaur and distinguished two species within it, *robustus* (already known from the work of Nopcsa) and a new species, which they called "*shqiperorum.*" Based on cladisic phylogenetic analysis, these two species were referred to Euornithopoda within Rhabdodontidae (which does belong in the Iguanodontia).

📖 Nopcsa 1902, 1904

Dan Grigorescu from the University of Bucharest handles a flattened dinosaur egg collected from Tuştea.

The Strange History of Zalmoxes

Zalmoxes was supposed to have been the slave of Pythagoras's. After he was set free he went to Dacia, where he was active as a teacher, a doctor, a vegetarian, and prelate. Subsequently, *Zalmoxes* was venerated as a deity of eternity, mystery, and ecstasy by the local population, and he was thought to have power over life in the next world. This Transylvanian dinosaur—similar to the emancipated slave—lived in the land at one time called Dacia and was immortalized thanks to Nopcsa and the paleontologists who followed him. The name *Zalmoxes* will forever remain in the paleontological literature.

As mentioned above, the species *Zalmoxes robustus*—which is accepted these days—was named by Baron Nopcsa. The species name "*Zalmoxes shqiperorum,*" described simultaneously with the establishment of the new genus name, refers to the land of Albania, which is called Shqiperia (the land of the eagles) by the locals, who are called Skipetars. Nopcsa traveled a great deal in Albania, knew the country well, and loved its simple folk. Thus, this new name is closely connected with the baron, whose paleontological activity was so important to our understanding of European fossil reptiles.

📖 Weishampel et al. 2003

Zalmoxes Remains

Zalmoxes remains were found at 24 sites across the Haţeg Basin, as well as some other Late Cretaceous Romanian localities. The total number of bones that belong to this animal is estimated to be of the order of a few hundred pieces. Indeed, about 80 percent of the skull is known, and the rest of the body—the postcrania—is fairly complete. Although vertebrae, ribs, and bones from the shoulder, pelvic girdles, and limbs are not completely known, they have been reconstructed.

The Lamb-Headed Dinosaur
Struthiosaurus transsylvanicus

In 1912, Baron Nopcsa's private secretary, the Albanian Elmas Doda Bajazid found some fossil bones close to Sînpetru (Szentpéterfalva). Among his finds—which may belong to the same animal—skull fragments, ribs, and vertebrae were recovered as well as a shoulder blade, a fragment of a coracoid, and some elements of dermal armor. These bones were collected together and named the holotype of an armored dinosaur called *Struthiosaurus transsylvanicus* by the baron in 1915. Nopcsa followed up his first, rather concise paper on *Struthiosaurus* with a detailed, monographic description that dealt with the anatomy, geographical distribution, biology, and evolutionary relationships of the dinosaur. *Struthiosaurus* had ornamented spines along its back and is certainly the most bizarre animal known from the Haţeg fauna. Its other remarkable characteristic is its dwarfed body size: the whole animal did not exceed 2 or 3 meters in length, making it one of the smallest dinosaurs in its group. Indeed, based

A *Zalmoxes* herd looking for food. These animals—which could walk easily on either two or four legs—were about 3–4 m long. They balanced their bodies by their long, strongly muscled, horizontal tails which would have allowed them to lean forward. DRAWING BY ANDRÁS SZUNYOGHY.

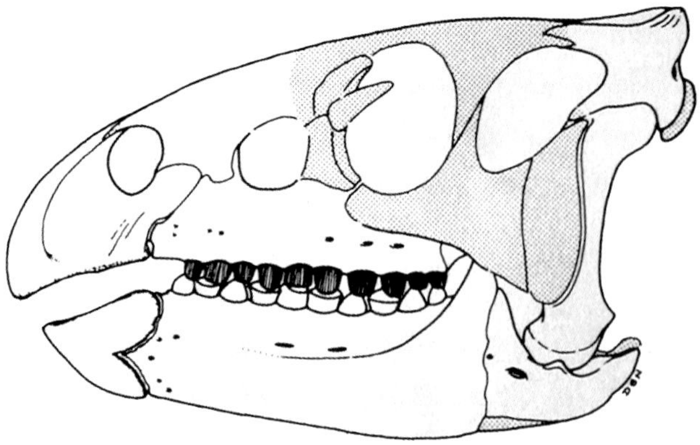

Lateral view of the skull of *Zalmoxes* (formerly *Rhabdodon*) from Haţeg. The hatched regions, next to the eyesocket, are less well known, but based on several separate bones, the skull of this animal has been reconstructed. The bones all correspond with one another in size and their preservation is so similar that researchers have assumed that they belong to one animal. Overall, the skull is stubby, and at its front it terminates in a remarkable pointed and toothless beak. According to Nopcsa's 1915 article, "We may presume that *Rhabdodon* . . . could use this beak to crop hard vegetation which it then chopped up with its teeth . . . instead of chewing and grinding."

on its small size and other osteological characteristics, the Transylvanian *Struthiosaurus,* although similar to the duckbilled *Telmatosaurus,* is considered to be a representative of a basal lineage. This is especially unusual because this dinosaur was one of the last of its kind, living just before the extinction at the end of the Cretaceous.

📖 Nopcsa 1929

To Which Group Does Struthiosaurus Belong?

Nopcsa thought that all dinosaurs characterized by armor or spines had originated from a common ancestor, and so he classified them all in a taxonomic group of his own, which he called Thyreophoroidea (armored). This name (which can hardly be pronounced) was not used by other paleontologists for a long time; it lay buried under increasing piles of literature. However, it later (in the mid-1970s) turned out that every armored (Ankylosauria), plated (Stegosauria) and hardheaded (Pachycephalosauria) dinosaur also shares a special feature: eyelids covered by a movable, bony plate. This common character is now thought to have come from a common ancestor, and so the Thyreophoroidea was resurrected to be used for the group. Originally *Struthiosaurus,* along with *Stegosaurus* (which has plates on its back), was classified by Nopcsa in another family, which he called Acanthopholididae; subsequently the status of this animal was determined to be within the armored dinosaurs (Ankylosauria). According to current paleontological opinions, the Transylvanian

The first reconstruction of the Transylvanian armored *Struthiosaurus*—by Géza Toborffy, following Nopcsa's instructions. As the baron wrote in 1915, "This figure does not lay claim to being exact; it rather serves for a general orientation."

armored *Struthiosaurus* belongs in Nodosauridae alongside ankylosaurs—also small dinosaurs, but with lighter and less-articulated armor.

Teeth and Diet

The teeth of *Struthiosaurus transsylvanicus* have not yet been found, unless we count one small (not more than 7 millimeters in size) find often regarded as the premaxillary tooth of an armored dinosaur and first considered to be from *Rhabdodon* by Nopcsa. If this tooth is really from *Struthiosaurus,* then the presence of premaxillary teeth in Nodosauridae must be a primitive, ancestral character. Although the teeth and the structure of the jaw usually reflect an animal's diet, this cannot be taken for granted. Armored dinosaurs such as *Struthiosaurus* are considered herbivores by paleontologists, but it is nevertheless possible that the vegetarian diet of some species was complemented, or even replaced, with meat. Certainly Baron

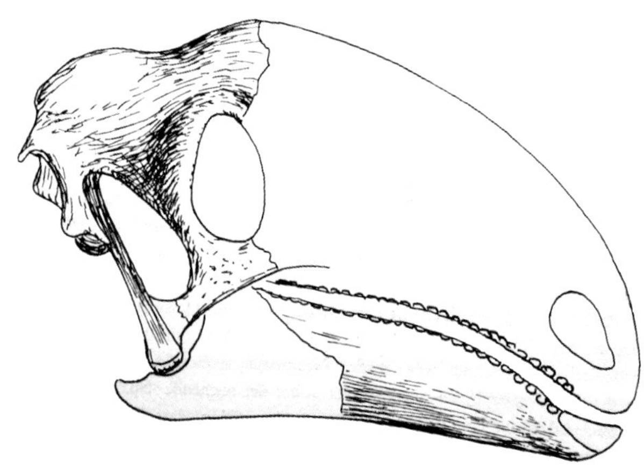

Reconstructed skull of *Struthiosaurus transsylvanicus.* The supratemporal fenestrae in this dinosaur were fully closed, whereas the lower temporal ones were slit-like and cranial orbital cavities were large. Ossification of skull sutures shows that this specimen is probably from an adult. The fragmentary small bones have been carefully glued together; the missing parts have been replaced with plaster.

A recent reconstruction of the Transylvanian armored dinosaur by András Szunyoghy.

Nopcsa had a very clear idea about the dentition and diet of this animal when he wrote, "The small, uniform teeth in the lower jaw of *Struthiosaurus* are found in separate dental sockets, and considering that the animal is a dinosaur, they are present in relatively small numbers." He also thought that tooth replacement must have been a slow process, since visible tooth buds are rare; the teeth—which were found—lack definite occlusal surfaces and were more suitable for shearing rather of grinding:

> Thus, the structure of the jaws indicates that this animal ate foods that were small, soft, and required little chewing. Due to its heavy armour, we also know it could move slowly, therefore we can determine the quality of food in more detail. . . . Considering the nutritive value of soft food and that it could not run away from the *Struthiosaurus,* food such as larger insects and their larvae, caterpillars, slugs, or soft fruits seem likely. . . . Since at that time plants had no fruits, it can be precluded that soft fruits would have served as food and thus only soft and short land animals can come into consideration. . . . Taking all this into account, *Struthiosaurus* must be regarded as a slug and insect eater, although this remains a hypothesis until we get acquainted with the missing parts of the body. . . . However, all this seems to be the most proper.

📖 Nopcsa 1915

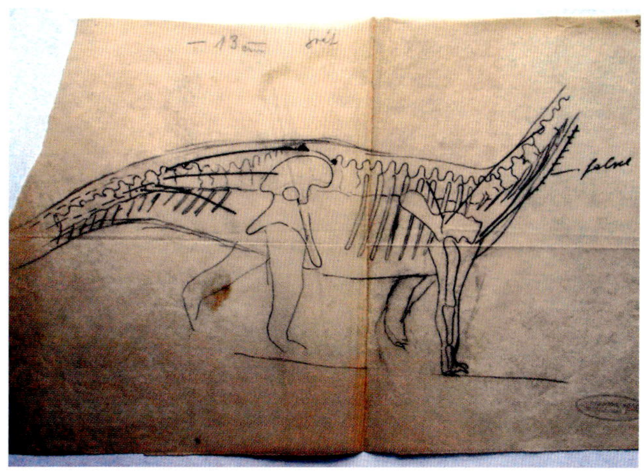

Baron Nopcsa's drawing of the Transylvanian sauropod, made after detailed evaluation of the finds. This original illustration, which can be seen in the Historical Collection of the Hungarian Natural History Museum, was made on thin tracing paper and attached to a letter to Huene. The baron's idea was to explain some features of anatomy to his friend, including the forward position of the shoulder blade and the strong cervical and rearward muscle bundles, which he marked with lines. Nopcsa emphasized that to the rear of the sixth to eighth caudal vertebrae, the tail of this dinosaur was able to make a whiplike motion. It is interesting to note that a more detailed reconstruction has never been made of *Magyarosaurus dacus,* and that the length of this animal probably did not exceed 6 to 8 meters (including its long neck and tail).

The Dermal Scales of Struthiosaurus

The back, neck, and tail of *Struthiosaurus* were covered with dermal scales. Although the fusion of these scales with the bones themselves cannot be discerned (as, for

A reconstruction of a sauropod dinosaur from Transylvania. Pencil drawing by András Szunyoghy.

instance, a turtle's shell is fused to its ribs), they were either found separately, or were fused together, within the skin. After the death of the animal the armor probably fell apart, and so it is hard to reconstruct it: Only one tiny (3.6 centimeters) armor element is present among all the Transylvanian finds described by Nopcsa; more recently found dermal pieces have yet to be described in detail.

📖 Nopcsa 1929

Limb bones and vertebrae of Transylvanian dinosaurs from the collections of the Geological Institute of Hungary.

A Dwarf Sauropod Dinosaur: The **Magyarosaurus dacus**

The largest animal amongst the fossil reptiles of Haţeg is a dinosaur originally described as *Titanosaurus*. This taxon belongs to the sauropod dinosaurs, a group of animals well known from cartoons and storybooks, notable for their long necks, long tails, large size and robust legs. This giant herbivorous dinosaur can be reconstructed thanks to dozens of bones from several specimens, including separate vertebrae and limbs. Baron Nopcsa always intended to write a monograph on the Transylvanian sauropods, but kept postponing the project; it was never completed. In the end he ceded the examination of these remains to co-worker and friend Friedrich von Huene, who eventually named *Magyarosaurus*. Remains of this animal are known from several sites, some located far from each other, but are all from large-sized, herbivorous animals. *Magyarosaurus,* itself a small representative of a sauropod dinosaur, would have seemed like a juvenile when compared to some of the known forms.

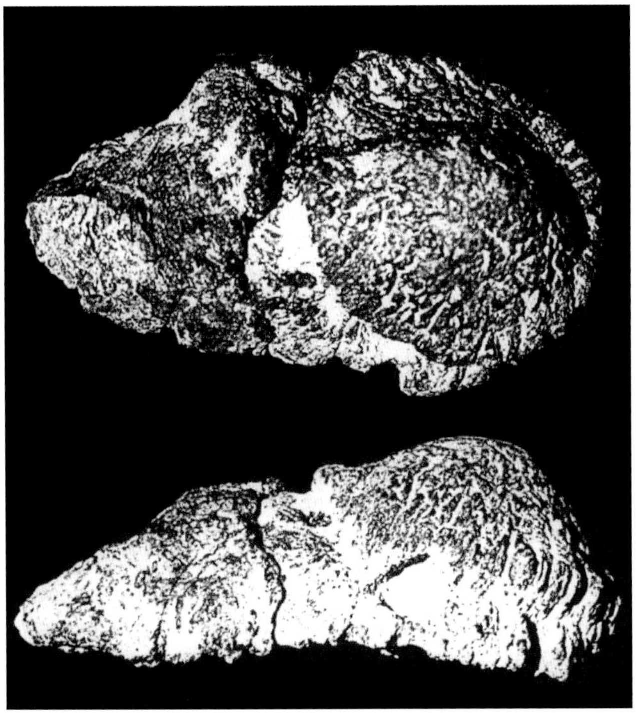

The backs of some titanosaurid dinosaurs were covered by scale-like osteoderms. This 13-centimeter osteoderm that was recently found presumably belonged to *Magyarosaurus*. After Csiki 1999.

The History of a Strange Name

One portion of the bones that come from the beds at Sînpetru (Szentpéterfalva) do belong to a sauropod dinosaur. In fact, in 1915 Nopcsa considered that these remains should be classified within the genus *Titanosaurus,* already known at that time, but as a new species: "I suggest using the name *T. dacus* for the designation of the Transylvanian species." The animal was to be named after the ancient Dacian people, who at one time lived in this part of Romania. However, Huene thought that these finds were so different from any of the known remains of *Titanosaurus* that they should be classified as a new, distinct genus, which he called *Magyarosaurus*. This explains how the name *Magyarosaurus dacus* came into being in 1932. To a Hungarian, this name sounds weird, because it combines the word for their nationality (Magyar) with the Roman name for a tribe that once lived in present-day Romania. Conflict between Hungary and Romania over the region of Transylvania (where a huge number of ethnic Hungarians still live) has been ongoing for centuries. Based on the remains from Sînpetru, Huene distinguished two further species and called them *Magyarosaurus transsylvanicus* and *Magyarosaurus hungaricus.*

📖 Nopcsa 1915

The "Island of Dinosaurs"

It turns out that each dinosaur collected from near the village of Haţeg is characterized by very basal osteological features and is considered to be a small form within its group. Nopcsa noticed this first and came up with one explanation for the strange phenomenon: the dinosaurs were living on an island. We know from living faunas that evolution on an island is often characterized by the survival of ancestral characters accompanied by dwarfed body sizes. Indeed, the latter feature could be caused by unfavorable conditions, lack of food, and inbreeding over millions of years. Miocene elephants that lived on islands in the Mediterranean Sea are considered to be another example of this phenomenon; their skeletons, often less than 1 meter in length, have been found in Crete, Malta, and Sicily. Indeed, Nopcsa was far ahead of his time: recent results have confirmed the existence of an oceanic island arc that stretched from Southern Europe into the

What Happened to Nopcsa's Dinosaurs?

The baron collected a considerable amount of fossil reptile bones in his time, most of which are now housed in the Natural History Museum, London. His precious collection was purchased by the English piece by piece between 1906 and 1924; Nopcsa himself spent a lot of time in this institution, which housed the largest fossil reptile collection of the time. He probably thought that it was important to keep his published specimens near relevant comparative material. However, the purchase price he received was not unimportant to him, as this is how he covered his regular travel expenses. The majority of the specimens in the possession of the Geological Institute of Hungary were in 1915 collected by Ottokár Kadić, a famous paleontologist and cave researcher, although some may have been excavated by Nopcsa. Only two, not very impressive, bones from the Haţeg Basin (Romania) are held today in the Department of Geology and Paleontology of the Hungarian Natural History Museum. We know from correspondence between Nopcsa and Huene that some of the Transylvanian sauropod fossils were given to the museum in Munich, they were probably destroyed during World War II. Several bones also remain in the Natural History Museum in Paris—but no one knows how they got there, long after the death of the baron, in 1944. New work at Nopcsa's old sites was resumed in the 1970s and newly collected fossils were transported mainly to the museum in Deva (Romania) and to the paleontological laboratory of the University of Bucharest.

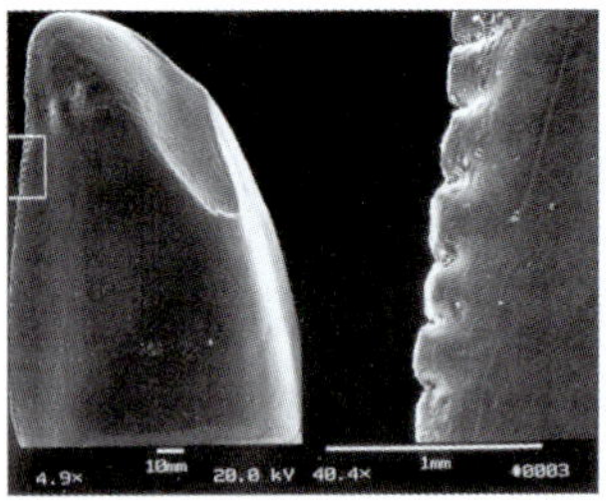

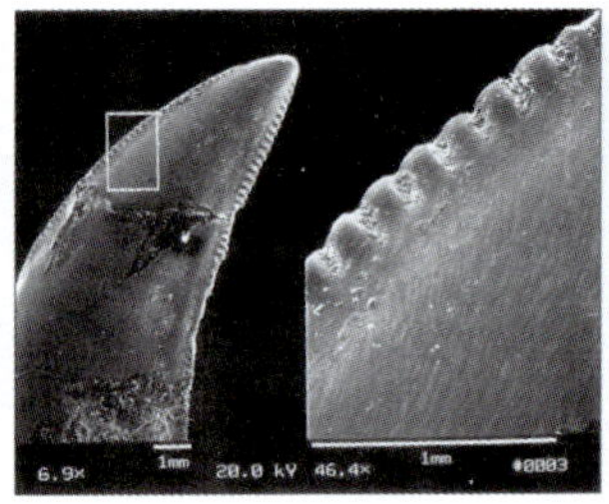

Electron microscope images of tiny predatory dinosaur teeth from the Haţeg Basin: a tooth of a *Troodon*-like theropod (left) and a tooth from a *Velociraptor*-like dinosaur (right) (high magnification). AFTER CSIKI & GRIGORESCU 1998.

Caucasus region. Nopcsa's small dinosaurs, dwarfs compared to related forms, likely lived on one island within this arc that is now often referred to as Haţeg Island.

Predators from Haţeg

A Few Remains from a Small Number of Dinosaurs

The duckbilled dinosaurs *Telmatosaurus, Zalmoxes,* and *Magyarosaurus* were all herbivores from Haţeg. They fed on the fibrous leaves of palms and ferns, coastal and land plants. *Struthiosaurus,* which was protected by body armor, may have also eaten protein-rich animals when it had the chance, but it cannot be definitively called a predator. It is certain that predatory dinosaurs must have existed in the Haţeg area, but in low numbers, as herds of herbivores can feed only a few carnivores. As a result, few fossils are likely to have remained from carnivores, and so little is known about the predatory reptiles from Transylvania.

🕮 Csiki & Grigorescu 1998

The Alleged Megalosaurus Vertebrae

Nopcsa published descriptions of two tail vertebrae that he thought belonged to an ancient predator. At this time, however, all fragmentary fossils of large-sized Jurassic and Cretaceous carnivorous dinosaurs were supposed to belong to the genus *Megalosaurus;* this is what the baron thought, too, although he could not identify the fossils with any formerly described species: "It has not been decided yet, whether the animal is a *M. pannoniensis* Seeley or a *M. hungaricus* Nopcsa, of which only teeth are known, or perhaps a *Megalosaurus? breday* Seeley." Because the two vertebrae have been lost, further examinations are impossible.

🕮 Nopcsa 1915

Pelican, Owl, or Small Dinosaur?

The remains of fragile, thin-boned, birdlike dinosaurs are very similar to the fossils of real birds. (This is, of course, not incidental because the two groups are closely related.) Paleontologists are in real trouble if the fossils of these groups are incomplete; this is often the case, as usually only separate bones are found and different opinions are reached by

The rediscovered forehead bone. Lizard-hipped predatory dinosaurs are generally called theropods. Giant theropods also occurred among the fossil reptiles at Haţeg, but their remains are little known. This fossil, the forehead bone from a large theropod, was found in the collection of the Geological Institute of Hungary. Nopcsa originally thought that this bone fragment belonged to an herbivorous dinosaur, but more recent work has shown that he was mistaken; the bone belongs instead to a predator. The baron thought that the crest on the top part of the fragment was a secondary sexual character, like the antler of a stag, and so he supposed that the specimen was male. This might still be true, even if the bone is from a theropod. 🕮 Weishampel & Jianu 1996

different workers. Some of the Transylvanian bones that Nopcsa had given to the Natural History Museum in London were described and named first as *Elopteryx nopcsai* by Lambrecht and then later as *Bradycneme draculae* and *Heptasteornis andrewi* by two English scientists, Colin Harrison and Cyril Walker. All three of these animals were originally imagined to have been giant birds: *Elopteryx* was thought to be a cormorant-like bird or a relative of the pelicans; the latter two genera were thought to represent the oldest owls. Most subsequent researchers, however, were dead against these ideas and committed themselves to the notion that

Late Cretaceous fossil reptiles from the Haţeg Basin. (1) *Magyarosaurus dacus* (Nopcsa)—shin bone of a sauropod dinosaur from Transylvania (0.5× magnification). (2) *Allodaposuchus precedens* Nopcsa—a well-preserved but fragmentary skull of a crocodile from Valiora, nicely complete on its right side (0.6× magnification). (3) *Kallokibotium bajazidi* Nopcsa—squat thighbone of a tortoise from Haţeg (0.8× magnification). (4) *Zalmoxes* sp.—vertebra of an ornithopod dinosaur from Haţeg (0.4× magnification). (5) *Allodaposuchus precedens* Nopcsa—thighbone of a crocodile from Valiora (0.5× magnification). (6) *Struthiosaurus transsylvanicus* Nopcsa—a fragment of a characteristically T-shaped cross section from a rib of an armored dinosaur (Nodosauridae) discovered at Sînpetru (0.5× magnification). (7, 8) *Zalmoxes* sp.—Variously worn leaf-shaped teeth from an ornithopod from Haţeg (original size). (9, 10) *Allodaposuchus precedens* Nopcsa—vertebra of a crocodile from Valiora (0.7× magnification).

1
2
3
4
5
6
7
8
9
10

Cretaceous flying reptiles. Pencil drawing by András Szunyoghy.

these species were actually small theropods. According to some opinions, the fossil material referred to *Heptasteornis* is an alvarezsaurid, a small dinosaur closely related to birds but which had extremely short forelimbs. Some experts think the three species represent only two or even only one species. This debate can be closed only if much more complete fossils are found that can be compared to these earlier ones.

📖 Grigorescu 1984; Naish & Dyke 2004

Flying Reptiles from Haţeg

The Lost Reptile

The reptile fossils from Haţeg that were studied by Nopcsa also include some pterosaur bones that subsequently have been mentioned a few times in the paleontological literature. However, these specimens have never been described in detail and were long thought lost. Although some of this material has been found, it is still not known which kinds of flying reptiles existed at Haţeg in the Late Cretaceous.

The "Monster of Valiora"

For a long time, Nopcsa's rather unclear data presented the only indication that the Cretaceous strata of the Haţeg Basin contained fossil flying reptiles. But some new bones were found at Valiora relatively recently—including a frontal, the fragmentary end of an upper arm bone, and a few connected but incomplete skull bones that are nevertheless well preserved. These fossils were first thought to be from large theropods, but it has later turned out that they are from flying reptiles.

These intriguing bones were described as *Hatzegopteryx thambema,* a new species within a new genus, placed within the family Azhdarchidae, within pterosaurs. The name of this species is of Greek origin and means "monster." The explanation for this is the giant size of the ancient animal. According to the work that was done on this fossil, the wing tips of the "Monster of Valiora" were at least 12 meters apart and the skull of the animal was about 3 meters long. If all this is correct, then *Hatzegopteryx,* together with an American pterosaur called *Quetzalcoatlus,* was one of the largest flying animals of all time.

A huge azdarchid pterosaur cervical vertebra from Râpa Roşie. It is hard to say how big this animal was, but on the basis of the size of this vertebra it is probable that it was comparable with the largest known flying reptiles, taxa like *Quetzalquatlus* and *Hatzegopteryx,* that attained wingspans of 10–11 meters.

The inner structure of the skull bones from Valiora are characterized by elongate and very thin walled cells. This made it possible for the bones of this pterosaur to be strong, unbending, and powerful but still light.

📖 Buffetaut et al. 2002

Kallokibotium bajazidi:

The Tortoise Named after a Secretary

The Beautiful Behind of Bajazid

One of the most interesting specimens among the fossil reptiles from Haţeg is a land tortoise. Its existence was first mentioned by the young baron in his first scientific study in 1897, yet he finished the description of this fossil only in 1923. He named it after his secretary, Bajazid—which might seem strange, but in fact the baron used the fossil to make a delicate reference to the nicely rounded shell of the tortoise. One might have thought that Nopcsa would name a tortoise after Bajazid because of his working speed, but no: this name is a compound of Greek words meaning the "beautiful behind of Bajazid."

Tortoise remains are fairly common in the terrestrial "dinosaur layers" at Haţeg, and so now the skull, dorsal

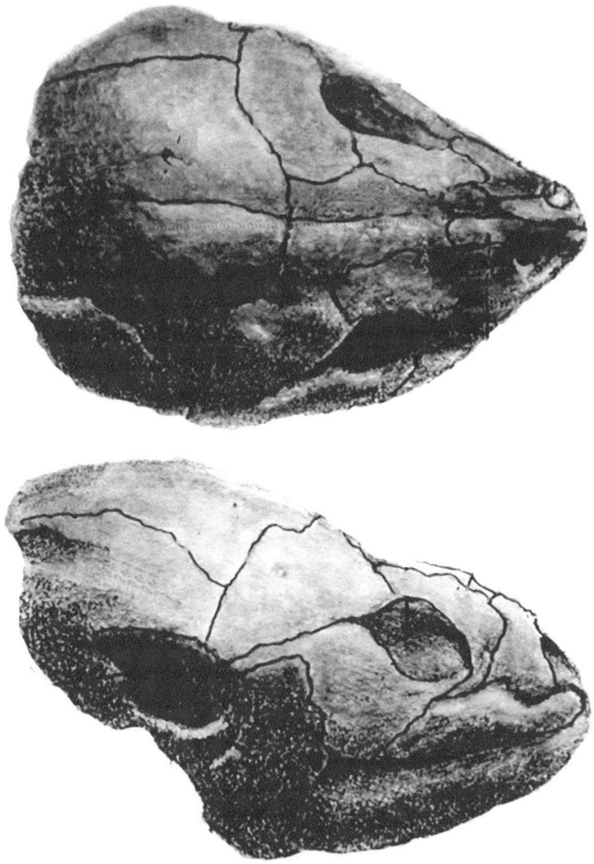

Skull of *Kallokibotium bajazidi* (0.6× magnification). "[I]t is quite rounded, its back is circular, its front is sharp. Its nose holes look ahead, the eye-sockets at about the middle of the skull look obliquely upwards. The septums of the incisive bone as a division of the nasal cavities are important; they distinguish *Kallokibotium* from any living or extinct tortoises, so they are very significant taxonomically" (Nopcsa 1923).

and ventral shells, vertebrae, thighbone, and certain parts of the pelvis of *Kallokibotium* are well known. The remains of many smaller and larger, older and younger specimens have also been found at Sînpetru and in its wider neighborhood. However, *Kallokibotium,* similar to the dinosaurs from Haţeg, is a fossil reptile of debated taxonomic position. All the fossils of this animal that were collected by Nopcsa are housed today in the Natural History Museum in London.

📖 Nopcsa 1923

Where Does It Belong?

The bones of *Kallokibotium* are so unique within its group that it has been necessary for paleontologists to classify this extinct animal in an independent family (Kallokibotidae), as was first done by Nopcsa. The big remaining question is this: To which order does this family belong within tortoises? Nopcsa originally thought *Kallokibotium* was a member of the suborder Amphychelya, although he emphasized that certain of its bones (for example, its hips) suggest rather a relationship with the other important group of tortoises (suborder Cryptodira). Present-day scientists are split about this question as well, and generally the taxonomy of tortoises remains very unsettled.

The Shell

The most characteristic feature of tortoises is their shell, which consists of two parts: the top, usually rounded, is called the carapace, and the bottom, generally flat, is called the plastron. Both parts are made up of dermal bones and covered with horny plates. The horn is shed over time, so that only the bony parts remain, although the outlines of the horny plates remain. They are not identical, however, to the sutures of the dermal bone plates, an anatomical feature that serves to increase shell strength in these animals. Unique within vertebrates, the elements of the bony shell fuse with flattened ribs, the spinous processes of the vertebrae, and bones from the shoulder and the pelvis.

The first real tortoises are known from the end of the Triassic period in the Mesozoic era and this group flourished up until the end of the Cretaceous, the time of

The upper shell (carapace) of *Kallokibotium bajazidi* (0.4× magnification). "The vertebral nerve laminae of the dorsal shell are very wide, which is to be considered a primitive feature. This is also supported by the fact that the nerve laminae of many living tortoises are wider in youngsters than in adults" (Nopcsa 1923).

A cheerful company. Paleontologists rest in a cool brook near Sînpetru in the summer heat. On the right is David B. Weishampel of Johns Hopkins University, and on the left is Dan Grigorescu. The researchers soak the crumbling, clayey sediment that contains dinosaur bones in water and wash it through a fine metal sieve. The washed residual matter retained in the sieve is separated with the help of a magnifier or a microscope so that new plant and animal fossils around Haţeg—including fossilized seeds, fish scales and teeth, fragile bones of small frogs, tiny reptile remains, eggshell fragments, and mammal teeth—can be found. In Haţeg, scientists were very lucky: they discovered remarkable fossils after washing just a few hundred kilograms of sediment. Tons of sediments often have to be sieved at known North American and Asian fossil sites to find worthwhile specimens.

Kallokibotium. Although these reptiles survived the extinction 65 million years ago, their numbers diminished and the group did not begin to diversify again until the Eocene. More than half of the around 30 described families of these reptiles are known only from fossils alongside the remains of the "tablet toothed" reptiles (Placodonts)—such as *Placochelys.* These animals are thought to be the relatives of tortoises and have been found in Triassic rocks.

📖 Nopcsa 1923

The Crocodile of Haţeg

A Crocodile Near a Village

Nopcsa reported on finding a fossil crocodile near Valiora in his 1915 study on Transylvanian dinosaurs. He also published pictures of a fragmentary, but well-preserved, skull and a few other bones of this animal. According to the baron's preliminarily identification, the specimen from Valiora very much resembled a reptile called *Crocodilus affulevensis,* which had been described from the southeastern part of France at the end of the nineteenth century. In 1928 Nopcsa revised his earlier views and pointed out differences between the Transylvanian and French crocodiles. He also proposed that the fossil from Haţeg Basin be named *Allodaposuchus precedens.* This taxon—although it lived at the end of the Cretaceous—is much more similar to fossil crocodiles known from the beginning of this period. Indeed, it belongs, together with several other fossil reptiles, to the suborder Eusuchia, as do all living species of crocodilians (crocodiles, alligators, and gavials).

📖 Nopcsa 1915, 1928

Recent Discoveries from Transylvania

"Microvertebrates" from Haţeg: Fish, Amphibians, and Mammals

Fossil reptiles, dinosaurs, tortoises and crocodiles are beyond doubt the most spectacular elements of the Cretaceous fauna from Haţeg. Indeed, as a result of systematic collections in recent years, new vertebrates, including fish, amphibian, and mammal fossils, have been found at many

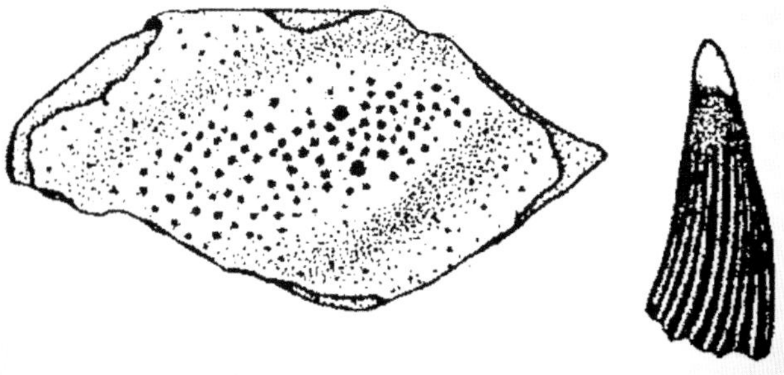

A scale and a tooth of a Transylvanian alligator gar. This small rhombus-shaped scale and the tiny tooth are probably from an alligator gar—which belongs in the genus *Lepisosteus,* according to paleontologists. Alligator gars are very ancient, and have looked pretty much the same since the time of the dinosaurs. Late Cretaceous *Lepisosteus* remains are known from France and Portugal as well, but they are much larger than these Transylvanian fossils (although it is true that only small pieces are likely to be retained in the sieve). Alligator gars today live in freshwater in Asia and Central America, but because they can also tolerate changes in salinity, they can swim into the ocean. Some living specimens can reach sizes of several meters (4× magnification).

of Nopcsa's old sites. The most significant of these finds are mammal remains, since hardly any similar fossils of this age have been found in Europe. However, the fossil mammals from Haţeg are not the immediate ancestors of developed living mammals, from which *Homo sapiens* has risen; rather, they are representatives of a long-extinct group. Nevertheless, these important European fossils shed light upon the very early period in the extremely complicated history of mammals.

📖 Grigorescu et al. 1999

Fragile Bones

Tailless amphibians, including frogs, have thin, toothpick-like bones. Their fossils are rare and little known yet quite a number, including new frog remains (skull fragments, jaws, pelvis, and limb bones) had been found by washing Transylvanian Cretaceous sediments. Experts have classified these fossils into the genus *Eodiscoglossus,* within frogs. Fossils of tailed amphibians have also been found in the recent past in addition to tailless ones.

Tiny Jaws

The few millimeter-sized teeth that were discovered in Transylvania years ago are thought to belong to lizards, which are extremely rare in the Cretaceous fossil record. However, these incomplete remains have yet to be identified. Strictly speaking, although lizards are known from the end of the Jurassic, by the Late Cretaceous a variety of forms were running about under the feet of the Transylvanian dinosaurs. Interestingly, however, no snake fossils have yet been found in the Transylvanian washed sediment, although they are thought to be common in other European fossil sites of the same age. Another single tooth of a small crocodile has also been found, but it cannot be identified with the remains of Nopcsa's *Allodaposuchus.*

Ancient Mammals

The most interesting microfossils that have appeared in the washed dinosaur-bearing sediments are small teeth and bones of mammals, coeval with ancient reptiles. Paleontologists well know that the history of mammals was about as long as that of dinosaurs in the Mesozoic, and the first remains of these animals are known from the Triassic. The Transylvanian finds are of particular importance scientifically because they belong to the order Multituberculata. This ancient, extinct mammal group is well known from North America and Asia, but its remains were not known from Europe until these discoveries. Multituberculatas were small herbivores with special teeth that bear multiple tiny cones—hence the name of the group. Other records from European mammal faunas of similar ages all belong to the much better known placental mammals.

📖 Csiki & Grigorescu 2001

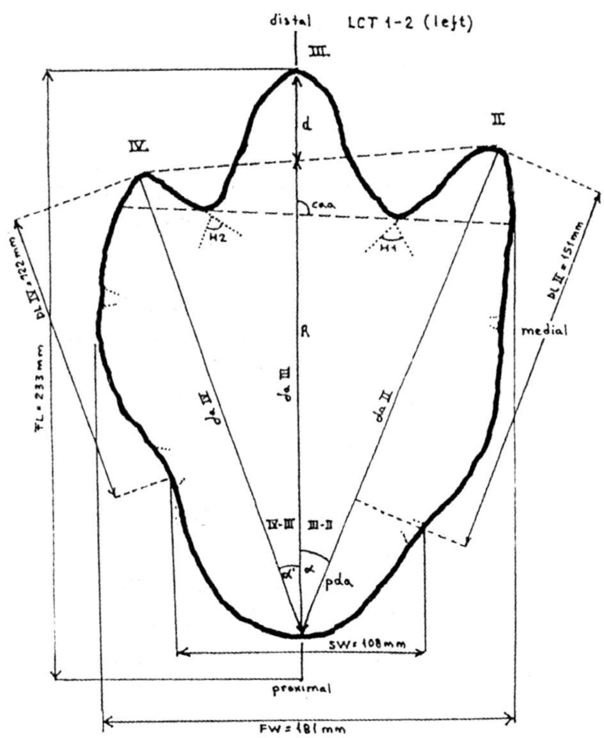

The footprints found near Lancrăm are 23 centimeters long and 18 centimeters wide, and their stride length is 103 centimeters. It is not known which kind of dinosaur made these footprints, although the most likely answer is that they were made by an *Iguanodon*-like animal such as *Zalmoxes.*

Mátyás Vremír—discoverer of many important fossil reptiles, including the theropod *Balaur bondoc*—collecting bones.

Type locality of *Balaur bondoc.* The continental red beds of the Sebeş Formation are dominated by conglomerates—pebbly sandstones with sandstone and mudstone interbeds that are occasionally flooded by the river. During high water levels there is no access to the outcrops. Late Cretaceous red beds that are found in the vicinity of Sebeş (Sebeş and Sárd Formations) can be correlated with the dinosaur-bearing beds of the Haţeg Basin (Sînpetru and Densuş-Ciula Formation).

More Than Just Nopcsa's Dinosaur: A Look Outside the Haţeg Basin

As already discussed, the first Transylvanian dinosaurs were reported from Late Cretaceous rocks in the Haţeg Basin by the notorious, pioneer paleontologist Francis Nopcsa more then 100 years ago. Soon after he reported on these first discoveries in his PhD thesis, Nopcsa extended his research to all dinosaur-yielding formations in Transylvania and predicted that they should continue a northerly direction. Following this prediction, recent surveys have indeed resulted in the discovery of many new fossil localities in Alba, Cluj, and Salaj Counties. The dinosaurs found in these areas show the same peculiar small size, as emphasized by Nopcsa when he defined the so-called Haţeg Island. This island stretched far beyond the present limits of the Haţeg Basin, but even after a century of research, it still cannot be precisely defined: at least part of the continental deposits documenting it has been removed by subsequent erosion.

Upper Cretaceous vertebrate remains outside the Haţeg Basin are uncommon, but their presence in often rather thick and complex red beds is obvious. Fossils that have been found include fish, amphibians, lacertilians, turtles, crocodiles, dinosaurs, and footprints. Although most of the taxa found in these areas were already known from Haţeg, some new and exciting discoveries have still been made. Among them is a new dinosaur, a "stocky dragon" from Romania.

📖 Codrea et al. 2010; Vremir 2010

Lancrăm Footprints

The coarse-pebbled beds—which crop out on both sides of the Sebeş River, next to Lancrăm (located north of the Haţeg Basin)—were long considered Oligocene or Miocene in age. A few fragmentary and worn-down dinosaur bones that were found in this area were thought to have been redeposited into the younger succession from Cretaceous rocks below. However, in October 2000 geologists found a three-toed dinosaur footprint in one sandstone bedding plane exposed on the river bank. This find unambiguously confirmed the age of the succession as Cretaceous, rather than Tertiary, during the time of dinosaurs. This was the first dinosaur footprint found in Transylvania.

📖 Vremir & Codrea 2002

A 2009 Discovery in Romania: Balaur bondoc—*The Stocky Dragon*

A wonderfully preserved new type of dinosaur—similar to the frightening *Velociraptor,* but with two sharp claws on each foot instead of just one—was discovered by geologist and paleontologist Mátyás Vremir of the Transylvanian Museum Society of Cluj Napoca. The new specimen was collected from red floodplain mudstones of the lower-middle part of the Late Cretaceous (Maastrichtian) Sebeş Formation, exposed at the Sebeş Glod locality near Sebeş in Alba County, Romania. The type specimen of this new dinosaur is an articulated skeleton containing many vertebrae, a scapulocoracoid, part of the pelvis, and numerous limb bones. The most spectacular part, however, is an exceptionally well-preserved left hindlimb with superb claws. This predator was named *Balaur bondoc,* which means "stocky dragon" in the Romanian folk language. The animal was a maximum of 2 meters in length and is thought to represent a young specimen that died

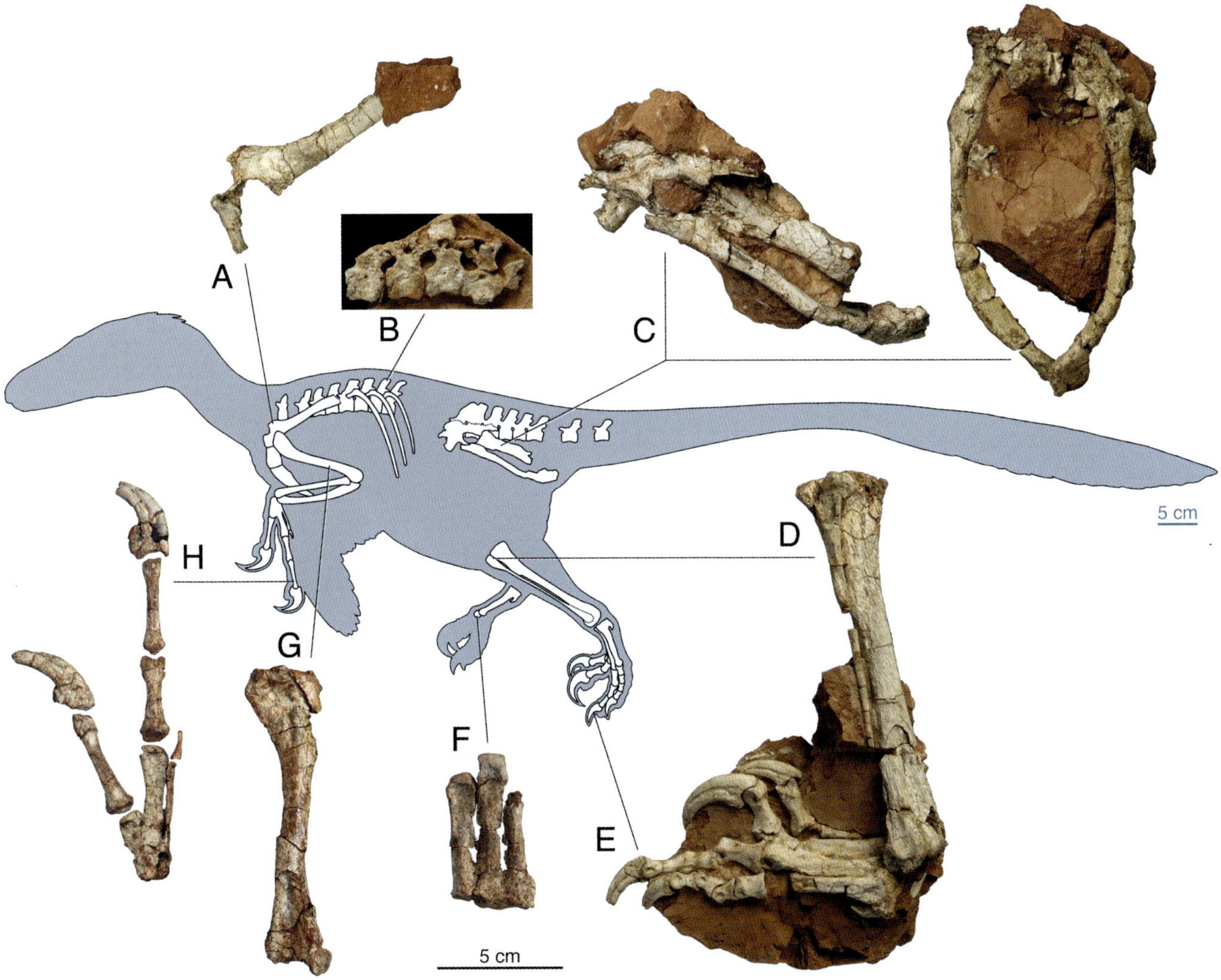

Skeletal anatomy of *Balaur bondoc* with photos of individual bones. (A) left scapulocoracoid. (B) Middorsal vertebrae. (C) Pelvis. (D, E) Left hindlimb and pes. (F, G, H) right limb bones.

accidently during a heavy monsoonal shower. Its carcass was transported and buried relatively quickly in the red floodplain mud, explaining why it remained articulated and is relatively complete. Besides the holotype, another referred specimen had been found in 1997 about 100 kilometers from the type locality, at the Tuştea dinosaur nesting site, by a collecting party led by Grigorescu.

Balaur might turn out to be one of the largest predator in its ecosystem, and possibly preyed on (at least partly) animals such as *Zalmoxes, Struthiosaurus, Telmatosaurus,* as well as on other dinosaurs. The precious fossils were described jointly by Romanian and American paleontologists and were donated by their discoverer to the Transylvanian Museum Society in Cluj Napoca (EME) in order to help start to rebuild the society's once-famous paleontological collection.

🕮 Csiki et al. 2010

During the Cenozoic the biology of the planet became more and more similar to that of the present day. Angiosperms, although appearing earlier, became more dominant in the flora. Miocene floras comprise representatives of several genera that can be found in present-day floras. The photo above shows a leaf of a peculiar shape (*Pungiphyllum cruciatum* [Al. Braun] Frankenhäuser and Wilde), which belonged to an extinct plant of unclear taxonomic status. This fossil leaf—from the Miocene at Ipolytarnóc—was regarded as "*Quercus*" (oak) in older literature (close to original size).

PART THREE

THE CENOZOIC

Shells of the gastropod *Tympanotonos* from the Eocene beds exposed at Csákvár. Eocene and Oligocene successions in the Carpathian region frequently contain limestones and marls deposited in marine lagoons. Species of the extinct genus *Tympanotonos* are thought to have been well adapted to the changing conditions of these special environments, which were hostile for most marine animals. The shells average 5–6 cm in height.

5

The Paleocene and Eocene

The Cenozoic, meaning "recent life," is the next major era in the history of the Earth, and covers the period from 65 million years ago until the present. The name comes from two Greek words, which can be translated as "new animal life." This is intended to indicate that the animal life of this era was similar to that seen today (in contrast to the Mesozoic and the much older Paleozoic). During the Cenozoic the continents drifted into their present—and, for us, seemingly final—positions.

Remarkable changes also occurred in the climatic conditions of the Earth: the balanced climate of the Mesozoic was followed by a more extreme and colder one. This climactic change was gradual, and probably started in the Late Eocene. In the polar regions—at first presumably in the Antarctic—a continental ice sheet slowly developed. Indeed, the last 2 million years of the Cenozoic—the epoch called the Pleistocene—are referred to as an Ice Age in the strictest sense.

Parallel to this change in climate and paleogeography, a change in the living world on Earth also occurred. After the extinction event at the end of the Cretaceous, vacant niches were filled by recovering lineages of animals and plants. The evolution of flora is principally characterized by the explosive expansion of angiosperms. At the end of the Cretaceous empty marine niches were also filled by representatives of different groups. Ammonites and belemnites—together with several other creatures that are characteristic of the Mesozoic oceans—became extinct, and the diversity of the formerly significant brachiopods declined. Their position in the living world has been assumed by bivalves. This group has existed for a long time, and thrived in the Tertiary. Moreover, new groups of planktonic foraminifera appeared that now play an important role in defining Tertiary stratigraphic subdivisions.

The evolution of vertebrates at this time was characterized by the expansion of mammalian lineages. If the Mesozoic can be called the age of reptiles, then the Tertiary should be known as the age of mammals. Due to one small, but significant, moment in the evolution of this group, one branch of the primate tree was pushed in a specific direction, and for the first time in Earth's 4-billion-year history a creature evolved that was able to acquaint itself with that history: Man.

The long unit of geological time called the Cenozoic is divided into the Tertiary and the Quaternary. The Tertiary—a no longer official, but still widely used, subdivision that encompasses more than 62 million years—is represented by two periods: the Paleogene and the Neogene. The Quaternary spans the period from 2.6 million years ago to the present. From this last period we know about many rocks and sediments that were left behind across the Carpathian Basin. Further subdivision of the Tertiary is based on correspondence between past and recent life, based in large part upon the similarities between fossil and recent bivalve and gastropod faunas. The names of the epochs of the Tertiary—the Paleocene, Eocene, Oligocene, Miocene, and Pliocene—emphasize ever-increasing similarity with recent faunas.

From the Oligocene onward, the Carpathian Basin was covered by the Paratethys Sea, which spread over the depression north of the Alp-Carpathian range and extended from present-day Switzerland as far east as the Aral Sea. Until the end of the Middle Miocene, completely marine conditions prevailed. The Late Miocene is the age of the brackish-water Lake Pannon. During the Pliocene, the entire area of the Carpathian Basin became a continental terrain. The varied Tertiary sediments and rocks of the Carpathian region and their fossil contents reflect these paleoenvironmental changes.

Paleocene

The end of the Cretaceous period was characterized by strong orogenic movements, including thrusting (nappe formation) and uplift in the Carpathian-Pannonian region. This area became a continental terrain which was to last for a long period, and erosion was predominant at the beginning of the Tertiary. Sedimentation took place sporadically, and only a few of the rock formations from

Erathem/Era		System/Period	Millions of Years Ago
Cenozoic	Quaternary	Quaternary	2.6
	Tertiary	Neogene	23
		Paleogene	65

Subdivision of the Cenozoic into periods and epochs, with approximate ages.

	Series/Epoch	Millions of Years Ago
Tertiary	Pliocene	5
	Miocene	23
	Oligocene	34
	Eocene	56
	Paleocene	65

Subdivision of the Tertiary into epochs, with approximate ages.

this time period have remained undisturbed: rocks in the Carpathian region that date to the Paleocene (the first 10 million years of the Tertiary) are found only rarely. Layers of probable Paleocene age are known only from deep boreholes in Hungary, and the only fossil locality of importance in the Carpathian region is at Jibău, Transylvania. A rich association of fossil vertebrates has recently been found here.

🕮 Gheerbrant et al. 1999

Remains of Freshwater Turtles

The first remains of freshwater turtles—found in the Paleocene beds of Jibău, more recently known as the Rona Limestone Formation—were collected by Antal Koch, a famous professor who worked at the University of Cluj. Koch classified the bones that he found at this site into the genus *Emys* in 1900. More than 100 years since the discovery of these first remains, new finds from this site were described by colleagues from Cluj. These fossils are now considered to belong to the genus *Palaeochelys,* which is in the family comprising freshwater turtles (Emydidae). The genus *Palaeochelys* is endemic to Europe; their fossils so far have been found only on this continent. However, the other known forms that have been classified in this group are all younger, of Oligo-Miocene age, and were all found in other areas (England, France, Spain, and Georgia). This means that the freshwater turtle remains from Jibău are the oldest fossils of this group, and thus suggest a link between European fossil sites located far from each other.

🕮 Koch 1900; Vremir & Codrea 1996; Vremir 2004

Eocene

The Paleocene was followed by the Eocene epoch, which spanned about 22 million years. In the Carpathian Basin rocks dating to this period are found mostly in the areas of the Hungarian Range and in the Transylvanian Basin. During the Early Eocene, the legacy of the Paleocene, particularly extensive continental erosion, remained predominant. As a result only a few deposits of this age are known. In contrast, the Middle and Late Eocene were characterized by gradual and repeated marine transgressions, which resulted in increasingly thicker successions of shallow-marine sediments overlain by deeper basinal ones; however, the total thickness of these successions do not exceed a few hundred meters. Of these deposits, the shallow-marine ones are especially rich in fossils, particularly the remains of invertebrates. The most characteristic fossils from the Eocene are *Nummulites,* which are larger foraminifera. These fossils frequently occur in rock-forming quantities.

Due to the collision of Africa and Europe, the closure of ocean basins—which occupied the area of the present-day Inner Carpathians and the Transylvanian Basin—took place during the Cretaceous. Therefore, only remnants of the once enormous Tethys seaway can be seen at this time, including the Magura Ocean (which existed in the area of the Outer Carpathians) and the deep basin of the Dinarids. A piece of crust—forming the present-day Carpathian Basin—was located between these two oceans and was divided by the northeast-to-southwest-trending Mid-Hungarian Lineament. Sedimentation took place in four distinct marine basins, whereas in the area of present-day Mecsek a continental succession was deposited in another small, separate basin. The evidence for this latter continental sequence was revealed by deep boreholes. Based on these sediments, and their fossil contents, we know that the climate during the Eocene was warm: from tropical to subtropical.

The most relevant Eocene successions were deposited in the area called Hungarian Paleogene Buda Basin, where sedimentation commenced after a hiatus of several million years. Connection via the Tethys to the World

Series /Epoch	Stage/Age	Millions of Years Ago
Eocene	Priabonian	37
	Bartonian	40
	Lutetian	49
	Ypresian	56

Subdivision of the Eocene into stages, with their approximate ages.

Loose limestone made up almost exclusively of *Nummulites* on the outskirts of Copşa Mică.

Ocean may have been provided by the Slovenian Corridor. In many places successions start with terrigeneous sediments, bauxite and coal, which overlie older rocks. Transgression reached the area from the west. The development of the basins was characterized by a continuous shift from the southwest towards the northeast; thus, in this direction sedimentation started later and later. At first fossil-rich shallow-marine sediments deposited, and these are overlain by deepwater successions.

FOSSIL-RICH FORMATIONS

The successions in Hungary that contain the richest fossil faunas (especially mollusks) are Eocene marine sediments. The Nummulites limestone is the most significant of these rocks that contains fossils in rock-forming quantities. There are also marls from the Upper Eocene in which bryozoans—occurring in large quantities—can be determined. However, in terrigenous beds of Early Eocene age vertebrate fossils are very rarely found.

Nummulites Limestone

The most famous rock facies of Eocene age is limestone containing the protozoan *Nummulites* in rock-forming quantities. Indeed, it is common for this fossil, and only this fossil, to comprise the Nummulites limestone, which is strange because it means that in this tropical sea there were hardly any other benthic organisms. It is also astonishing just how high the carbonate productivity of this sea must have been: its calcium carbonate budget was almost entirely due to production by protozoans.

Occurrences of this rock type are well known all over the Tethyan realm. For example, the pyramids of Giza are covered with blocks of this stone, which can be easily carved. Nummulites limestone can be found at several places in the Carpathian Basin, where it is referred to as either the Szőc Limestone (formerly the "Main Nummulites limestone") in Hungary (which is also rich in other fossils) or the Szépvölgy Limestone (which contains other larger foraminifers in large quantities). This latter limestone can be observed, for example, in the quarry at Mátyás Hill in Budapest: several public facilities and fenceposts in Budapest are built from the easily workable stone from this quarry.

Besides these limestones there are plenty of other argillaceous or marly sediments across the Carpathian Basin, which also contain *Nummulites*. These rocks have been well studied since they were penetrated by several boreholes during the course of exploration of coal seams at Tokod and Dorog.

The upper horizon of the 10–30-meter-thick Kapus Formation, which crops out at several places in the Transylvanian Basin, also contains *Nummulites*. in rock-forming quantities in the vicinity of Cluj. However, this rock cannot be regarded as a real limestone, because *N. perforatus* specimens, even though they occur in large quantities, are not cemented together locally. Some kinds of limestone are made up purely of fossils.

The upper horizon of the so-called Cluj Limestone, which also crops out in the surroundings of the town (in the Plecska Valley), can also be regarded as a Nummulites limestone. This rock—from a higher stratigraphic horizon than the Kapus Formation—is rich in the species *N. fabianii*.

Shallow-Marine Limestones and Marls Rich in Mollusk Fossils

Limestones containing the skeletons of mollusks, predominantly gastropods and bivalves, in large quantities are also characteristic of Eocene-aged deposits in the Carpathians. These rocks often comprise corals, crustaceans, and echinoderms—especially echinoids (sea urchins) and mollusks. Fossil-rich limestones of shallow-marine facies in Hungary include the Szőc Limestone in the Southern Bakony Mountains and the Szépvölgy Limestone, which crops out in the surroundings of Budapest. Since these facies also contain larger foraminifera in considerable quantities, they have already been discussed in connection with the Nummulites limestone, above.

It is common that Middle Eocene sediments in Hungary are not pure limestones; often these rocks are clayey and marly. Nevertheless, in many cases, the fossils of mollusks and invertebrates are common in these rocks. The most famous of these are the Forna beds near Gánt (presently called the Forna Formation), which will be treated in more detail in our discussion of Eocene mollusks. Similar marly

Marl containing the internal mold of mollusks in large quantities in the abandoned coal mine near Dudar.

beds can also be found in the Tatabánya Basin, in the Vértes and Gerecse Hills and in the Dorog Basin, although they have been classified as different lithostratigraphic units.

In Transylvania, the upper, massive section of the Cluj Limestone—the *Nummulites fabianii* horizon of Priabonian age—is the richest in fossils. From the vicinity of the city of Cluj alone almost 40 foraminifera, 8 corals, 2 polychaete annelids, 4 cephalopods, nearly 70 gastropods, more than 80 bivalves and 150 ostracods, 2 decapod crustaceous, 7 brachiopods, 22 echinoderms, and 28 species of vertebrates have been described. The first paleontological data that relate to this rock were published by Antal Koch in 1894; during the second part of the twentieth century Hungarian and Romanian experts working at this site (including Miklós Mészáros, Niţa Vlaicu-Tătărîm and George Bombiţa) thoroughly investigated the fauna.

The Bryozoan Marl

The Upper Eocene Buda Marl was deposited in a deeper environment than earlier formations. One would search for thick-shelled bivalves and gastropods in this rock in vain; instead of the species which occur in shallow-marine limestones in large quantities, only a few remains of the quite thin shelled bivalve (*Propeamussium*) are found. However, the Buda Marl has one specific facies, the so-called bryozoan marl, which—as its name suggests—contains the remains of these one-time moss animals in large quantities. A well-known occurrence of this rock, comprising the fragments of bryozoan colonies in rock-forming quantities, is the upper horizon in the eastern quarry at Mátyás Hill. The rich fauna of this marl will be treated in later discussions of fossils.

The Brebi Marl occurring in the Transylvanian Basin is similar to the Buda Marl both in facies and age.

🕮 Zágoršek & Kázmér 2001

REMARKABLE FOSSIL SITES

Eocene rocks contain beautiful fossils in several areas. However, among the localities known to be rich in fossils there are several that are considered to be highly remarkable fossil sites even in professional circles. These include bauxite and coal mine dumps in the Transdanubian Range (Szőc, Dudar, Iszkaszentgyörgy, and Gánt) and quarries where Eocene marine limestones had been excavated. Good examples of the latter include the Kis-Sváb Hill in Budapest and the large number of quarries in and around Cluj. Sadly, many of the remaining mine dumps are overgrown with weeds, and as quarries get older waste is—legally or illegally—dumped into them. Many older localities, well known from literature, are currently not accessible.

Due to mining of raw materials, Eocene-aged fossils could be found at several places in the Carpathian Region until recently. Such sites include the dumps of coal mines—which exploited this treasure as it was discovered—stretching from Kisgyón in the Bakony Mountains to Kosd, at the foot of the Naszály. However, most of these mines are now closed—either due to depleted reserves or because they were operating at a loss. In some areas (regarded as classical sites from a paleontological point of view), such as in the Tatabánya Basin, in the Pilis Mountains, or at Kosd (see above), mining operations stopped decades ago. The Kosd mine recently suffered from karst water inrush, but in 1909 Elemér Vadász described a rich fossil assemblage, that included several new taxa, from its barren layers.

Unfortunately, rocks containing really well-preserved fossils are usually soft, not very resistant to weather conditions, and have few natural exposures. Since mining operations often started on outcrops of coal seams, remaining natural exposures have usually been buried by mine dumps. These can be recognized by their shapes—very different from the surrounding landscape, even though they are often overgrown with plants. Humidity triggers the decay of pyrite in the rocks, and may induce an exothermic process that leads rocks rich in organic material to start to glow. Even the deeper zones of the dumps, made up of blocks of all sizes, are affected by oxygen, of which there is often enough for combustion. The fossil hunter on a dump might feel as if he were trekking on the hillside of a smoking volcano! Indeed, such similarities are even increased by the smell of sulfur, and the vivid yellow sulfur encrusting on the rock surfaces is similar to that of the fumarola fields. In contrast to such old dump sites, abandoned pits of open-cast bauxite mines—which still exist in large numbers all over the Transdanubian

Some millimeter-sized fragments of bryozoan colonies from the Eocene at Mátyás Hill. After Zágoršek & Kázmér 2001.

1
2
3
4
5
6
7
8
9

Beds of the Szőc Limestone, yielding marine invertebrate fossils in large quantities in the abandoned open-pit coal mine close to Dudar. The classic locality, the mine dump of the underground mine, which had been expanding for decades, has been overgrown with grass.

Range—provide another excellent opportunity to collect fossils.

🕮 Vadász 1909

Szőc, Malom Valley

There are huge, crater-like, abandoned open-pit bauxite mines at several places in the Bakony Mountains that are considered excellent Eocene fossil localities. Remains be collected not from the terrigenous bauxite, but from the overlying—mostly marine—beds.

Such fossil-rich beds are found, for example, next to the road that leads from Padragkút to Öcs, east of the village of Szőc. The overlying succession of the almost totally exploited bauxite lens no. 11 comprises meter-thick coaly clay overlain by a thick sequence of Szőc Limestone. Microfacies of this limestone were examined by Gábor Vakarcs (later a hydrocarbon expert in the United States) and Péter Várnai in 1984, during their university studies. This pit is probably the best accessible locality for echinoids (sea urchins). Information on the fauna was provided in the dissertation of Levente Szabó in 2005, in which 18 species are described. The most prominent elements of this fauna are huge, occasionally 10-centimeter-high (albeit always slightly deformed) *Conoclypus* specimens. In terms of numbers of specimens, *Echinolampas* is the most frequently encountered taxon, but in addition to echinoids, calcareous algae, corals, gastropods, bivalves, and, rarely, the casts of nautiloids can be found in this rock, even though it is predominantly made up of *Nummulites* skeletons.

Ajka and Köleskepe Ditch

Next to the road, a few hundred meters from the less and less discernible remains of the completely demolished buildings of Jókai Mine—which was closed at the beginning of 2000—grayish-yellow marl beds form a sort of a bluff. This rock contains specimens of the mostly large-sized *Nummulites* species, *"millecaput"* ("thousand headed") with a diameter reaching 10 centimeters. The sedimentary environment of this succession might have been a sheer submarine desert where sand grains were made from Saint Ladislaus's coins (fossils of the flat *N. millecaput*). Other faunal elements, such as bivalves and mud-dwelling echinoids, occur only occasionally in the fine-grained layers. This plentiful locality is a regular stop on geological nature trails.

The Pénz-lyuk ("Money Pit") at Pénzesgyőr

On the way from Pénzesgyőr to Bakonybél, where the road bends sharply to the right, a small Eocene limestone pit, the Pénz-lyuk, can be found right next to the road. This site preserves an extremely rich in fossil fauna: the name of this village, Pénzesgyőr ("coin stones"), comes from the fossils that can be found in the surroundings. The limestone in the pit consists almost exclusively of *Nummulites* (*N. perforatus*): indeed, this locality may even have been known to Hantken. Both small and large (macrospheric and microspheric) forms can be observed together, as they are in many other places. In the vicinity of Pénzesgyőr there are several other localities where one can also find *Nummulites*. For example, along the road from the center of the village toward Hárskút, on the outskirts, *N. perforatus* fossils can be picked up in large quantities, as these coin stones have been washed together by rainwater. Specimens are usually slightly bloated, rounded forms; the "real" Saint Ladislaus's coins (*N. millecaput*) cannot be found here, as this species occurs in another stratigraphic horizon.

Dudar

The locality at the abandoned coal mine at Dudar existed as a fossil collecting site for decades and provided fossils of the Csernye Formation, which is characterized by a succession made up of clay, marl, and sandstone beds. Fossils accumulated in a one-time inlet, where deposits became disturbed during storms. Several indications point to a rapid burial by catastrophic deposition. The sandy marl contains both fragmentary and complete fossils.

Gastropods are the most frequent in the fossil assemblage. The 1966 monograph by László Strausz (1901–1988) contains the description of 173 species and subspecies. The large-sized *Velates,* having a cap-like shell and frequently retaining its original color pattern, can be regarded as the heraldic animal of the locality. Specimens of the *Campanile* genus—more than half a meter in size—derived from this locality are also noteworthy. Unfortunately, the description of the greatly diverse assemblage of bivalves has not yet been written.

📖 Strausz 1966a

Gánt, Bagoly Hill

The abandoned open-pit bauxite mine—at present a geological educational site—provides the best exposure of the Forna Beds, which contain a rich and very well-preserved fossil fauna. This profile is one of the most studied successions of the Eocene in the Bakony Mountains. According to Flóris Rómer, Móric Májer—a teacher from Pécs—was the first to find fossils in the area. Karl Alfred Zittel from Munich was also among the first to do so. He was one of the most prominent persons in the history of paleontology, and described 30 mollusk species from this site—11 of these species were new. The bauxite was discovered in the 1920s; investigation of the overlying Eocene beds started simultaneously.

Description of the fossil mollusks—made by Endre Szőcs—was published in 1953. The monograph comprises descriptions of 163 gastropods, 37 bivalve species, and 1 cephalopod species. Remnants of gastropods belonging to the *Cerithium, Tympanotonos,* and *Ampullina* genera are the most frequently occurring fossils.

📖 Szőts 1953

Tokod, the Turnoff to Sárisáp

While waiting for the bus, we may collect corals to make the best use of our time. At the foot of the Római-szőlőhegy ("Roman Vineyard Hill"), in the steep wall behind the bus stop, yellowish-gray marl is exposed. At first sight it seems an outcrop of no importance, however, it is one of the richest, or even the richest Eocene coral collecting sites in the Carpathian Basin. Most of the specimens belong to the solitary *Circophyllia* genus, which sometimes may reach 10 centimeters. The other frequent form is the colonial coral *Euphyllia,* which can be found only in fragmentary preservation. A solitary coral, the hornlike *Ceratotrochus,* is also frequent. Besides corals, gastropods and bivalves can be observed; among the latter, the large-sized *Crassatella* is noteworthy.

Solymár, Várerdő Hill

The hill rising above the Jegenye Creek valley was considered one of the best Eocene fossil collecting sites in the Buda Hills. The Eocene succession is exposed by narrow gullies on the southern side of the hill. The odd thing about the outcrop is that it comprises rock types that have no other exposures. Some of them contain echinoids in large numbers. These outcrops are things of the past; the gullies have been partly buried or hidden by thorny, impervious bushes. Thus, the 1966 dissertation by Miklós Monostori and publications made on the basis of his work represent the last description of this locality.

Subsequently, the echinoid fauna of the Várerdő Hill was studied by Attila Bartha, who later became an oil exploration geologist. His 1988 dissertation deals with the 1,500 specimens collected mostly in the nineteenth or at the beginning of the twentieth century. The most frequent genera are represented by *Echinolampas* and *Echinanthus.*

Kis-Sváb Hill: The "Lost Fossil Paradise" in the Heart of Budapest

Kis-Sváb Hill (at one time Martinovics Hill) was a noted fossil site in the old geological literature. Quarries in this area became well known predominantly of their fossil plants and decapod remains. Of the fossil plants, the nutlike fruits of *Nipa* and *Juglans* should be mentioned; they were found in large numbers in those days.

The collecting work of fossil crustaceans was started by Miksa Hantken, who decided to elaborate them. However, both preparation process and scientific investigation were ceded to Imre Lőrenthey. The latter became enthusiastic about the beauty and diversity of the Kis-Sváb Hill crustaceans, delved into the theme, and became an expert on decapods. His multivolume work *(Paleontológiai tanulmányok a harmadkori rákok köréből* [Paleontological studies of Tertiary crustaceans]) deals not only with the fossil crustaceans of the surroundings of Budapest, but remains derived from the countries of the Hungarian Kingdom and from several foreign countries.

Let us return to the Kis-Sváb Hill, where Hantken showed a preference for collecting crustaceans, because their remains were extremely well preserved there. Lőrenthey described 39 species from this locality; there were many new ones among them. However, the Kis-Sváb Hill is remarkable not only from a paleontological point of view, but also from a mineralogical point of view. Exceptionally beautiful honey-colored crystals of baryte and amazing calcite crystals were found in the fissures of the quarry. After the quarry was closed, the area was consumed by the growing city. By the time the site received well-deserved official protection, the lower quarry—which was famous of its fossils—had been filled with building rubble. The upper horizons of the quarry can still be seen.

📖 Schafarzik & Vendl 1929

Eocene beds cropping out in the Plecska Valley near Cluj. Rocks exposed here are relatively poor in fossils; in the forest nearby, however, some hundreds of meters from here, horizons characterized by the richest macro- and microfauna of the Cluj Limestone can be observed.

Classic Localities in the Vicinity of Cluj

The Eocene beds of the Transylvanian Basin have been known since the time of Antal Koch. An especially beautiful and rich fossil assemblage was found in Cluj and its vicinity, in several horizons. Many of the specimen index cards contain the names the localities Cluj-Monastir, Plecska Valley, and Bácsi-torok. Middle and Upper Eocene limestones of the quarries—which had not been worked regularly, and can be found even today—yielded a lot of beautiful fossils (remains of mollusks, echinoids, and turtles, and bones of sirenians).

Căpuşu Mic, which can be regarded as the type locality of *Nummulites perforatus,* is located not far from Cluj. In its vicinity *Nummulites* can be found in extremely large numbers. Sometimes the material, quarried from the hillside for road reconstruction and building material, consists purely of *Nummulites*. Perforatus beds have several outcrops here. The other, easily accessible exposure along the

Portrait Gallery

Antal Koch (1843–1927)—Professor and Paleontologist

Antal Koch, an epoch-making person and master of Hungarian geology and geological education, was born in Sombor (Zombor), Serbia—formerly the southern part of Hungary. At first he was educated to be a teacher at Pécs, and from spring 1862 he attended lectures on mathematics and science at the University of Budapest. In compliance with his original qualification, he taught at grammar schools; however, his greatness in teaching and research was proven later, during his years at the university.

In 1872 he was appointed to the Department of Mineralogy and Geology of Cluj University. "He was young, nevertheless he took up his post at the department and he was well prepared for it in every respect, and he obtained it due to his work and diligence," remembered Elemér Vadász in 1943. In the person of Koch, an open-minded, versatile expert capable of hard work occupied the chair; he was committed both to education and research. He spent more than two decades in Cluj, where he reorganized the education and the rich collection of the Transylvanian Museum Society, which served the aims of the university, and which he enriched with several valuable fossils.

In 1893 Miksa Hantken, professor of the Department of Paleontology of the University of Budapest, died. In 1894 József Szabó, professor of mineralogy and geology, also died. Thus, Antal Koch was invited to Budapest, where he started to reorganize the department and the collection. Most of his publications in paleontology and geology (works based on his numerous field observations) were written in the course his activity in Budapest. "His work dealing with the Tertiary formations of the Transylvanian Basin based on his own investigations compared with those of others provided the first description of the development and paleogeographic evolution of the Carpathian Basin"—noted one of his colleagues, Elemér Vadász.

Among his paleontological works, those dealing with vertebrate fossils should be particularly emphasized. His affection for nature and his profession is indicated by his coat of arms: in the talons of the rearing gryphon above the helmet is the long-handled geology hammer of old-time geologists, and in the middle of the shield a fossil Tertiary echinoid, illustrated with scientific accuracy from the dorsal and ventral views.

📖 E. Vadász 1943

Coat of arms of Antal Koch. At the bottom of the shield there is Aranyi Hill; on the upper part are two Eocene echinoids with a geology hammer between them.

road is located along the way to Monastireni, 1 kilometer from the village, in the road cutting.

Rădaia (Andrásháza): An Exceptional Locality of Eocene Fossil Vertebrates

Remains of Eocene land vertebrates are very scarce in the Carpathian Basin. For that very reason, the locality at Rădaia is particularly important: two especially important fossil land vertebrates were found here. One is *Brachydiastematherium transilvanicum,* which belongs to the Titanotheriids—that is, to an ancient family of odd-toed ungulates, already extinct in the Oligocene. The other is *Prohyracodon orientale,* which represents one of the most ancient Rhinocerotids ever found.

📖 Pávay 1873; J. Böckh 1881b

EOCENE FOSSILS FROM THE CARPATHIAN REGION

Instead of "Eocene," the term "nummuliticum" can be found in earlier literature. The term refers to the most frequently found fossils of this period, the *Nummulites.* Besides the larger foraminifers occurring in rock-forming quantities at several places mollusks are also frequent; in certain localities mollusks and the remains of other marine invertebrates, such as echinoids, can often be found. Vertebrate fossils, especially the remains of land animals, are unique.

Microfossils

Because of exploitation of raw materials (such as coal and bauxite), Eocene successions (more than any others) have been transected by numerous boreholes. The samples derived from these boreholes, and from surface exposures, have made complex micropaleontological investigations possible. In particular, detailed investigation and documentation of organic-walled (spores and pollen) and inorganic-walled Eocene microfossils (nannoplankton, foraminifers, and ostracods) have been completed.

Spores and Pollen

Eocene successions in the Carpathian Basin include plenty of dark-colored sediments that suggest the remains of well-preserved spores and pollen. Because of this, successions in the Transdanubian Range have been studied by László Rákosi for several decades. According to one of his research papers, completed in 1991, in the Bakony Mountains and in the northeastern part of the Transdanubian Range are four successive spore-pollen assemblages. Based on his work and evidence that comes from other fossils these assemblages partly overlap one another, but each of them is present only in some places.

The Late Eocene age of one thick continental succession, the Szentlőrinc Formation—unique to the Carpathian Basin—revealed spores and pollen from boreholes south of the Mecsek Hills. According to work done by József

The remarkable locality at Rădaia, in an old photo.

The reddish, russet-colored terrigenous beds at Rădaia can be studied at the present time as it was in the past, in the days of Koch and Böckh, when the unique fossil vertebrates were found. The terrigenous beds of probable Eocene age (formerly considered Oligocene) has often been visited by paleontologists looking for further remains—with no success in finding fossils so far.

Bóna, this assemblage resembles the Borken pollen association found in Germany.

📖 Rákosi 1991

Nannoplankton

By the end of the Cretaceous, planktonic coccolithophorids (belonging to Chrysophyta) underwent an explosion in their evolution. Several genera belonging to this group are known from the Eocene—and on the basis of these successions can be subdivided and correlated. The development of nannoplankton zones, based on these tiny calcareous remains, has been accelerated considerably by deep-sea drilling. Thus, Eocene beds that overlie basement rocks and that are rich in nannoplankton have been transected over considerable areas of the ocean. One of the most productive and respected experts to work on Eocene nannoplankton from the Transdanubian Range was Mária Báldiné Beke. She summarized her results in a 1985 monograph that showed that forms important from a stratigraphic point of view belong mainly to the genera *Discoaster, Reticulofenestra,* and *Chiasmolithus*. Báldiné Beke showed that five of the seven global nannoplankton zones can be distinguished in the Carpathian Basin, with two impossible to identify.

📖 Báldiné Beke 1972, 1984; Báldiné Beke & Nagymarosy 1986

Small Foraminifera

In addition to larger foraminifera, on the other end of the spectrum, small-sized planktonic foraminifera are also found in great abundance in Eocene beds. For periods of varying duration, these foraminifera have been studied by several researchers. One special genus, characterized by chambers that end in single spines, is called *Hantkenina*.

Examination of these smaller foraminifera was first initiated by the demand for more precise subdivision and dating of deeper marine successions. Katalin Horváthné Kollányi summarized examinations carried out by former workers and the results of her own studies in 1983. Based on the genera of globigerinids (in a broad sense) she was able

Thin-section preparation of a foraminifera and two sectioned foraminifera. Making orientated thin-sections of smaller foraminifera is not usually done; sometimes, however, this is necessary for studying of the internal parts of the tests. The Micropaleontological Collection of the Hungarian Natural History Museum includes several preparations of this kind that date back to the nineteenth century. This valuable collection was rediscovered and has been worked on and popularized by Péter Ozsvárt.
📖 Ozsvárt 2005

to distinguish seven zones in the Transdanubian Eocene. From these, the lowermost is characterized by *Hantkenina,* while the uppermost features *Turborotalia*. Taxa within the *Globigerinatheka* genus are also significant.

In the first place, small foraminifera are significant because of their paleoecological importance. However, in some places they also determine the features of certain rocks. In the lower part of the Transdanubian Eocene succession, for example, the Miliolina limestone is found in several places. This rock contains an assemblage comprising few species but huge numbers of fossils, including large numbers of the porcelaineous Miliolinas, which indicates hypersaline waters. Other characteristic forms include *Pyrgo, Triloculina,* and *Quinqueloculina:* the latter two differ from each other in the numbers of chambers found in each coil. Small foraminifera from the classic (unfortunately now lost) locality at Dudar were described in a small monograph published in 1988. Knowledge of this fauna was augmented with new data by Gábor Halupka in his 1996 dissertation. Halupka studied miliolinid foraminifera extracted from the sediment infilling gastropod shells; Péter Ozsvárt evaluated the foraminifera fauna of the Eocene Csordakút open-pit mine in his doctoral dissertation.

📖 Sztrákos 1979, 1982; Horváthné Kollányi 1988; Halupka 1996; Ozsvárt 2007

Larger Foraminifera

Eocene is characterized by the frequent occurrence of extraordinarily large microfossils. The so-called larger foraminifera belong to the Rotaliina order (comprising forms characterized by hyaline, perforated tests [external shells]) and the porcelaineous Miliolina order (named after the glossy white, microgranular-walled tests). Locally, their remains occur in rock-forming quantities.

Nummulites***: The Best-Known Fossils of the Eocene***
That the stone objects now known as *Nummulites* were not coins but animals was first realized by Franz Ernst Brückmann (1697–1753), a doctor from Wolfenbüttel who devoted a monograph to this group. At present they are classified into the Rotaliina order. The abundance of the so-called coin stones, overshadowing all other fossils, is so characteristic of the Paleocene, Eocene, and Early Oligocene sediments of the tropical Tethys Ocean that Eugène

Saint Ladislaus's Gold Coins

The origins of the legend related to *Nummulites,* a widely known fossil of the Carpathian Basin are unknown. It is first mentioned in a 1499 work written in Latin by Franciscan monk Pelbárt Temesvári (ca. 1435–1504). However, the story of the coin stones can be read in Hungarian in the 1527 Érdy codex. One thing is certain—namely, the legend dates to the Middle Ages. The legend has appeared in several versions. These are about Saint Ladislaus (1040–1095) or Saint Stephen (975–1038) who scattered money before the pagan enemy, and the gold turned into stone. According to another version—which gives a less favorable account of the Hungarians—the pagans being pursued scattered the money, and the king wanted his soldiers to keep fighting rather than stop to pick up it. Therefore, he changed the gold coins into stone. The latter story is indicative of perfect knowledge of mankind, as well. It is most likely that the legend is of German origin, and reached Hungary by the followers of Princess Gisela of Bavaria (ca. 980–ca. 1059). However, it cannot be ruled out that the story was invented in the Carpathian Basin. The names of the neighboring villages Pénzesgyőr (*Pénz* literally means "money") and Bakonyszentlászló ("Bakony Saint Ladislaus") refer to the fossils that can be found in large quantities there.

The strange shape of *Nummulites* made people think, and not only those in the Carpathian Basin. The origin of the remains is treated in legends and tales everywhere: in Spain they were considered the money of the Saracens on the run; in France they are called Saint Peter's, or Saint Boniface's, money. Legends referring to money turned into stone and to other strange things derived from the rocks were collected compiled by József Hála and his companions.

📖Géczy 1994, Hála et al. 2004

Fossil of legends—*Nummulites millecaput*, the largest among the larger foraminifera in the Carpathian Basin and all over the world.

Renevier (1831–1906), the great Swiss geologist, proposed the name "Nummuliticum" for this period.

Wherever Eocene marine beds occur in the Carpathian Basin, *Nummulites* can be found. The most famous areas are the Transdanubian Range and the Transylvanian Basin, sites of the first findings. It is worth mentioning that Miksa Hantken (1821–1893)—who was one of the epoch-making scientists dealing with *Nummulites* and Cenozoic foraminifers in general—worked in the area of present-day Hungary.

Types of *Nummulites*

The appearance of *Nummulites* tests, although similar to those of other larger foraminifera, does not add up to much: some are striated and some are perforated. The most important characteristics, such as the size of the initial chamber and the pattern of septa, can be seen only in sections that show the largest test-diameter (equatorial section), perpendicular to the axis of coiling.

Portrait Gallery

Miksa Hantken (1821–1893)—The Pioneer of Micropaleontology

Miksa Hantken was born in Jablunka, Austrian Silesia. From 1840 to 1842 he conducted his university studies in Vienna, and graduated from the Academy of Mining at Banská Štiavnica (Selmecbánya) in 1846 as a mining engineer. Subsequently, he studied for another two years in Vienna, where he attended lectures on physical geology. In 1852 he was employed as a mining engineer at the Dorog coal mine. One of his main professional goals was to establish scientific research on coal. He started to examine foraminifers, the remains of which occur in large quantities and in high abundance in the Eocene barren rocks.

Based on knowledge obtained in the course of his industrial research, he became a prominent person in micropaleontology. He perceived the dimorphism between the two generations of *Nummulites*. His work was honored with his admission to the Hungarian Academy of Sciences. After spending some years as a mining engineer, from 1861 Hantken taught natural history at the Commercial Academy of Budapest, and from 1866 he worked as a researcher at the Botanical and Mineralogical Collection of the Hungarian Natural History Museum. In 1869 he was appointed director of the Royal Geological Institute of Hungary. Besides fulfilling this function, he wrote his monumental book on Eocene and Oligocene foraminifers. His two other important works, a, 1871 monograph describing brown coal occurrences in Dorog and an 1878 book providing a review of coal seams and coal mining Hungary, were published during this period.

In 1882 Hantken was appointed the first professor of the Department of Paleontology of the University of Sciences at Budapest. He established the paleontological collection and library. He carried out complex research as well; description of American *Nummulites* found in Florida (in 1886) and microscopic analysis of limestones (which was a pioneering initiative) are also to Hantken's credit. For his services, in 1888 the University of Bologna granted him the honorary degree of Doctor Honoris Causa. He died in Budapest on June 26, 1893, due to illness he contracted in the course of his research in connection with the earthquake at Zagreb. In addition to his precious works, his memory is preserved in numerous species and generic names of fossil animals.

📖 Tasnádi-Kubacska 1969

Preparation of *Nummulites* from Miksa Hantken's famous collection. This beautiful collection of green specimen boxes contains samples both from Hungary and—with a view to making comparisons—from foreign countries. Hantken's collection won a gold medal at the International Exhibition held in Vienna in 1873.

A Paleontologist in Action: Examination of Nummulites by Splitting the Specimen

To achieve a more precise determination for *Nummulites*—as we would for other larger foraminifera—we must become acquainted with the structure of its shell, especially the form and size of the initial chambers and their arrangement. Although all these features can be studied using thin-sections, these are not easy to make because of the thin and coin-like shape of *Nummulites*. Instead of wasting time trying to make thin-sections, a splitting method is often used instead.

In order to split it, the fossil should be gripped by pincers or laboratory tweezers and placed over a gas flame. Once warmed for a few minutes, the *Nummulites* should be thrown into a glass of cold water. It should then split in two—but if it does not do so immediately, then it can be tapped on its edge with a small hammer. If we are lucky, the fossil will split along the plane of symmetry of its disk-shaped shell, revealing the inner initial chamber. If we are unlucky, then we might need to try again with another specimen!

Chambers can be seen even with the naked eye on the fracture plane; they can be studied under a magnifying glass or with a low-power microscope. We can sketch the chambers, which can then be compared with illustrations already present in the literature.

Assilina, Operculina, *and* Discocyclina

Assilina, another larger foraminifer and a relative of *Nummulites,* is also frequently encountered in some Eocene beds in the Bakony Mountains. One difference between these two genera is that throughout the ontogeny of *Assilina* the diameter of its tests (which are not larger than 5 centimeters) increases, whereas their thickness only slightly increases. The other key difference is in the pattern of coils and septal pillars that form ridges on the surface of the shell. From this genus, the species *Assilina spira* is frequent in the Middle Eocene beds of the Bakony Mountains: the species name indicates that its spiral can be seen clearly from the outside.

Operculina is also related to the two genera mentioned above, but it differs in that its later whorls do not usually overlap earlier ones. Although *Operculina* can be found in limestones, this foraminfer is most frequent in successions of the so-called Operculina marl, which can be found in several places in the Transdanubian Range.

Yet another, larger foraminifer is found in rock-forming quantities—mainly in sediments of the slightly deeper, but still shallow-marine, facies of the Upper Eocene. This form appears under the name *Discocyclina* in the literature; however, because this name has been used to refer to several similar genera, some experts have argued that it is more reasonable to call this foraminifer by the alternative names *Orthophragmina* or *Orbitoides.*

Whatever its correct name, *Discocyclina* is also a flattened form, but with a selliform shell. On rock surfaces, specimens can appear as nail-like bodies, although its internal structure—which can be observed only in sections under the microscope—considerably differs from *Nummulites.* In this foraminifera there are two types of chambers: one is made up of concentric rings (equatorial chambers), and the other joins these laterally.

The microscopic images of this genus show its characteristic cone-like pillars (crystalline cones) intruding into the shell, which is made up of calcite. The functionality of these morphological elements has long remained an enigma; however, it has recently been realized that they could have acted as lenses to concentrate light in the interior of the test and keep symbiont algae alive. As is seen in extant larger foraminifera, algae may have lived inside the test and could have enabled *Discocyclina* to live in deeper areas of the photic zone.

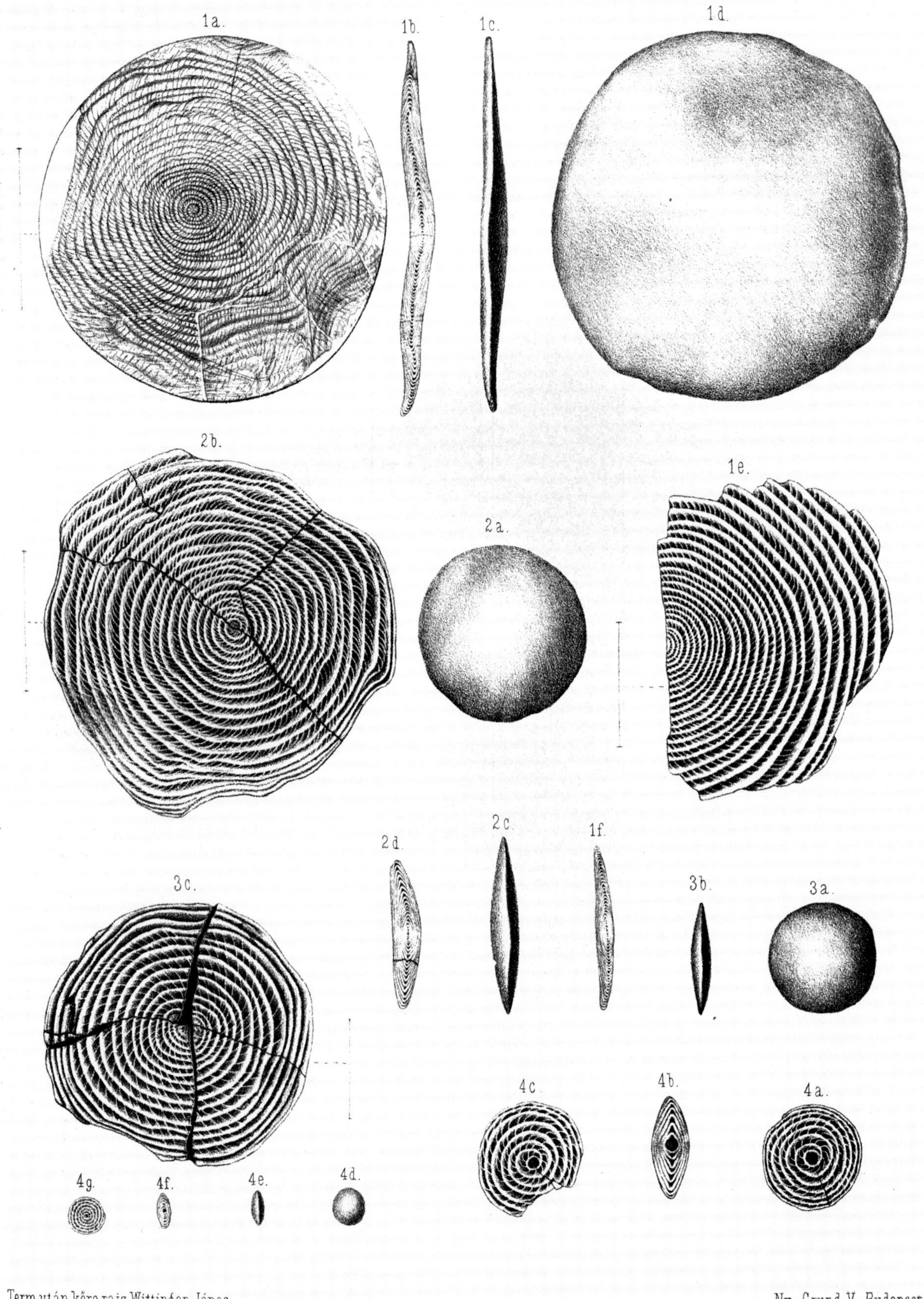
1a.
1b.
1c.
1d.
2b.
1e.
2a.
2c.
1f.
2d.
3c.
3b.
3a.
4c.
4b.
4a.
4g.
4f.
4e.
4d.
Term. után kőre rajz. Wittinger János.
Ny. Grund V. Budapest.

For a long time, *Discocyclina* in Hungary received less attention than *Nummulites,* but in the 1980s György Less published a substantial monograph that deals with these taxa.

📖 Less 1987

Alveolina

Eocene beds in the Transdanubian Range overlie much older formations, especially Triassic rocks that are extremely thick and cover large areas. Accumulation of these sediments mainly commenced in continental fluvial and lacustrine environments. At first, transgression resulted in the development of lagoons and inlets, in which miliolinids proliferated in large numbers (just as they do today); this group is able to tolerate salinity conditions different from those found in seawater and comprises the spindle- or cigar-like elongated alveolinids, which are another group of larger foraminifera.

Alveolina is found frequently and in large quantities in the lower part of Eocene successions. The internal structure of the tests of these foraminifera—which occasionally reach a few centimeters in size—is also characteristic: chambers that form coils and are arranged parallel to the axis are subdivided by secondary septa. As well as observing microscopic thin-section images from the often-mentioned Alveolina limestone under transmitted light, it is also possible to recognize that, due to their microcrystalline structure, the dark material of the shell predominates—contrary to the inner space of the chambers, which can be seen as rows of small light spots. At one time—due to nomenclatural considerations—forms now known as *Alveolina* were mentioned under the name *Fasciolites;* however, the former name is still considered valid. *Alveolina* remains from Hungary were described in a monograph by Mária Jámbor-Kness, who worked on the Eocene larger foraminifera for decades.

Alveolina belongs to the group of so-called porcelaineous foraminifera. Even in these successions they are found alongside hyaline foraminifera, which later became

A lithographic plate from the work of Miksa Hantken and Zsigmond Ede Madarász entitled "Nummulinák Magyarország óharmadkori rétegeiből" (Nummulites from the older Tertiary beds of Hungary), which was posthumously published in 1924 by Pál Rozlozsnik, Royal Hungarian chief geologist, who was also a successful expert on foraminifera. This work is extremely important because it introduced the subdivision of Eocene successions based on *Nummulites.* Besides their stratigraphic importance, Hantken recognized dimorphism in *Nummulites,* in particular that the tests of the sexual and asexual generations of a species are different. The tests of the sexual generation are smaller and their initial chamber is larger, thus, they are called macrospheric forms; however—considering the size of embryonic chambers—the larger-sized asexual generation comprises microspheric forms.

Tibor Kecskeméti, retired deputy director general of the Hungarian Natural History Museum, is a well-known expert on *Nummulites.* Over the course of his long research career Kecskeméti determined the presence of six distinct *Nummulites* assemblages in the Transdanubian Range, which have since served as a definitive basis for stratigraphic correlation.

the dominant forms—with the most common being *Nummulites.*

📖 Jámborné Kness 1981, 1988

Ostracods

The tiny skeletons of ostracods can be separated from most of Eocene sediments by washing. The valves of ostracods are excellent indicators of environment because their individual morphotypes—represented by several heterochronous genera—have not changed in habitat throughout Earth's history.

In the Carpathian Basin, research on ostracods was begun by Gyula Méhes (1881–1961), who published a monograph dealing with the forms found in the vicinity of Budapest. After a break of several decades, Miklós Monostori continued this work. His choice of topic (which he has never once regretted), however, was partly forced. Originally, in the 1960s, he wished to work on the mollusks, which can also be found in large numbers in Eocene beds, but he was strongly advised not to pursue this work. The all-powerful professor of geology of that time wanted someone else to work on mollusks. This person has since passed away, leaving much malacological work still to be done.

At the time, coal exploration in the Dorog Basin provided the basis for all this work, and as a result several boreholes were dug in the area. Several thousand specimens were collected from these boreholes and from open-pit mining operations: when examined, they reveal a rich fauna comprising almost 50 species. Description of this association was published in a single volume in English

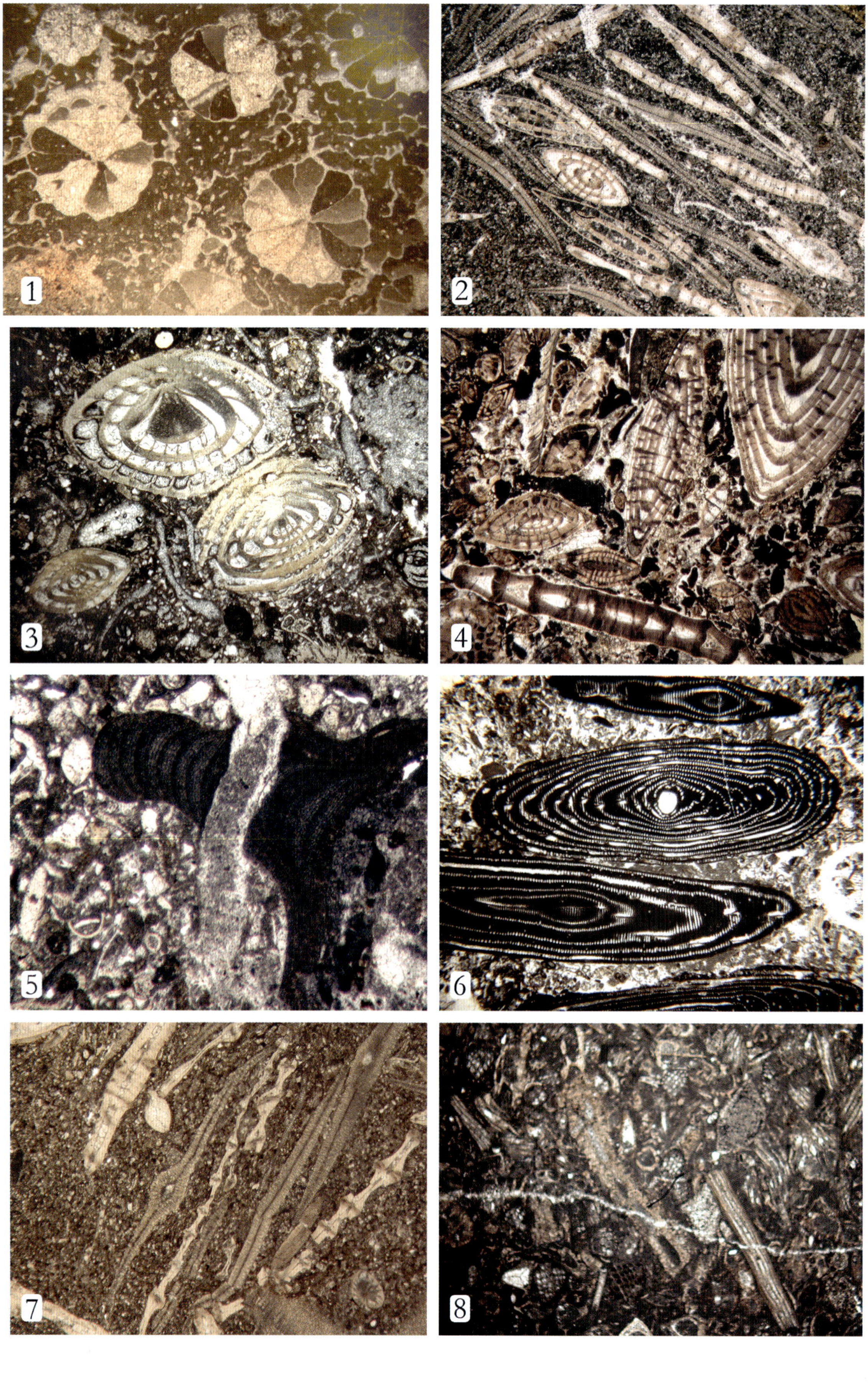
1
2
3
4
5
6
7
8

Miklós Monostori—professor in the Department of Paleontology of Eötvös Loránd University and a well-known expert on fossil ostracods—starting for work. As an ostracod expert he formerly dealt mainly with the fauna of the Tertiary; nowadays he mostly works on Mesozoic assemblages.

by the Hungarian Academy of Sciences in 1985. The book, comprising (among other topics) a detailed paleoecological evaluation of the Dorog area, is an important work for research on Eocene ostracods.

Unfortunately, the scanning electron microscope (SEM), which makes the high-quality representation of small specimens possible, only became widely used just after putting the book went to press. In light of this, the photos in the volume, made using lower-magnification light microscopes, seem pretty simple. Monostori has recently published a series of articles featuring photos taken using an SEM. Besides the ostracods from the Dorog Basin, he has also worked on specimens from other areas; one study of samples taken from material infilling some large gastropod shells led to surprising results. In such protected microenvironments, ostracod assemblages that differ from those of the surrounding host sediments have been preserved. Moreover, different species of gastropods contained different microfaunas, indicating that the deposition of the bed containing gastropods took place over a longer period, whereas the bed containing fossils reflects an environment of different salinity, temperature, or bottom water conditions.

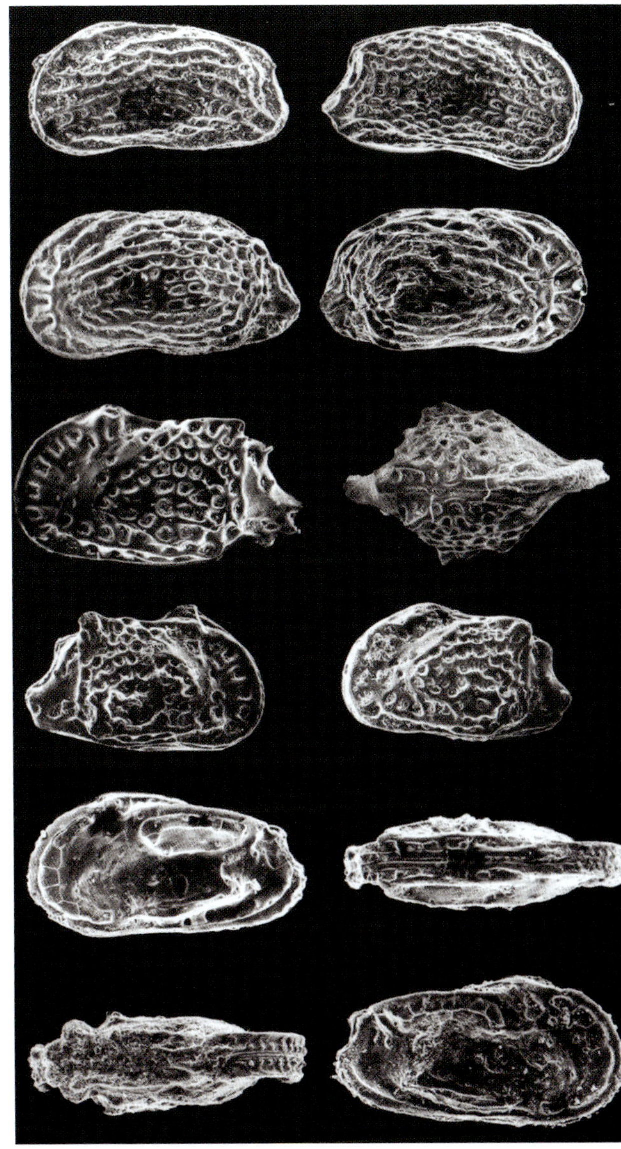

Eocene ostracods from the Transdanubian Range (approximately 50× magnification). Specimens shown in the upper eight photos are *Pokornyella* sp.; in the lower four photos are *Hornibrookella* sp.

Strangely, although numerous freshwater ostracod associations have not yet been found, certain freshwater forms occur among marine associations. The association of the deepwater Buda Marl is characterized by the genera *Bairdia, Krithe,* and *Cytherella.* These faunas show similarities with

Fossils in thin-section from Eocene rocks from the Transdanubian Range (2–5× magnification). (1) A colonial coral—spaces between radial septa are filled with spathic calcite and lithified calcareous mud and therefore have different colors. (2) Large foraminifers appear slightly left of center in this photo; near the bottom right, a *Nummulite* can be seen. The narrow, elongated bodies made up of two parts, subdivided by a darker stripe, are discocyclinids; the elongated sections of different thicknesses belong to specimens of *Operculina.* (3) *Nummulites.* (4) *Nummulites* and *Assilina* (the latter is at the bottom of the photo, ascending obliquely from the right toward the left). (5) Red algae (the dark body stretching upward from the right to the left of the photo, with curved internal surfaces and honeycomb-like structures) penetrated by a crack filled with calcite. This crack formed after lithification. (6) *Alveolina*—the white spot in the upper, large specimen is the initial chamber. (7) *Discocyclina* and nummulitids. (8) Bryozoans (moss animals).

A *Nipa* fruit from Eocene beds at the Kis-Sváb Hill (0.6× magnification).

assemblages described from the United Kingdom, France, and Italy.

📖 Gy. Méhes 1936; Monostori 1985, 1996b, 1998, 2000b

Fossil Plants

Although the Eocene formations of the Carpathian Basin comprise brown coal seams suitable for industrial exploitation, relatively few fossil plants have been described from these deposits. The most diverse assemblage from the Carpathian Basin, however, was discovered in the neighborhood of Lábatlan. One of the most characteristic plant remains of these Eocene fossil assemblages is the fruit of the palm *Nypa*. Another well-known locality for fossil fruits and woody nuts (sometimes arranged in ball-like clusters) was found in the main quarry of the Kis-Sváb Hill in Budapest; later ribbed fruits were also found near Dudar, in the Bakony Mountains. *Nypa* was one constituent of the mangrove vegetation that fringed the Eocene seashore and the fossil fruits of another palm, *Juglans eocenica,* described originally as a walnut, have also been recorded from Kis-Sváb Hill.

The fairly rich fossil flora from the fine-grained sediments of the Upper Eocene sequence in Budapest known as the Buda Marl was published by Klára Rásky. This sequence is formed from basinal sediments which contain relatively well-preserved leaves, fruits, and (rarely) flowers that drifted in from surrounding terrestrial areas. However, the most diverse Eocene flora from the Transylvanian Basin is known from the vicinity of Girbou. The fossil fruits (*Cedrelospermum,* described originally as

Remains of calcareous tube-worms (*Rotularia*) from Kálvária Hill at Bakonybél.

The Eocene beds of the Carpathian Basin rarely yield complete sponge remains, but traces of the activity of boring sponges (*Cliona*) can often be observed on mollusk shells. *Cliona*, and other organisms that live in such a way, play an important role in maintaining the marine calcium carbonate cycle by the destruction of shells. This photo shows a semi-destructed, *Cliona*-bored shell of a gastropod (*Ampullina*) from Gánt (close to original size). Besides sponges, several other organisms play a part in the disintegration of calcium carbonate shells. The first traces of the shell-destroying activity of algae and fungi, however, can only be observed using a microscope. As the process goes on, the originally crystalline structure of the shell becomes unidentifiable as it undergoes micritization. Traces of boring sponges and boring clams as well as bryozoans that live in calcareous shells can be seen even with the naked eye. As the shells are weakened by borings they become fragmentary, and due to wave action these fragments then disintegrate.

Embothrites) from the overlying layers of the gypsum mine near Leghia are also noteworthy.

Invertebrates

Worms

Although worms that secrete calcareous tubes are not frequent in large, rock-forming quantities (such as serpulite), in certain places in the Carpathian Basin, Eocene beds frequently contain the tubes of the worm *Rotularia*. These worms have spirally coiled tubes in the early phases of the ontogeny; later in their development their tubes are straight. This characteristic fossil is referred to as *Tubulostium* in older literature. The dwelling tubes of these worms—which are often identified as gastropod shells by amateurs—are especially frequent in the Bakony Mountains, at the boundary of the nummulitic limestone and the overlying marly succession.

Corals

Fossil corals are not uncommon in certain Eocene successions; however, detailed, up-to-date descriptions are not available. The latest overview was written by Gábor Kolosváry in 1949; he mentions several dozen species, including numerous new ones from the Eocene beds of the Transdanubian area. This improbably large number of species (in light of other works by the author) suggest that far-reaching conclusions should not be drawn from this work.

Corals are mentioned as coming from several places along the Transdanubian Range in the literature. The richest locality can (or could) be found at the foot of the Római-szőlőhegy, near the village of Tokod. Both solitary (*Circophyllia, Ceratotrochus*) and colonial forms (*Euphyllia, Stylocoenia*) can be found here. Coral remains were also frequent in the mine dump at the former coal mine at Dudar, which can be also regarded as a one-off locality. Unfortunately, however, this fauna has not been systematically collected and studied.

A unique occurrence of corals was documented by Sándor Mihály, who presented descriptions of specimens attached to shells of *Cerithium subcorvinum*, a large gastropod. However, general conditions in the lagoon at this time were likely largely unfavorable to corals, which can tolerate only normal marine salinity conditions. Specimens that have been found are scarce and smaller than usual. Suitable substrate was also lacking, and corals could settle only on the shells of some large mollusks.

📖 Hantken 1871; Kolosváry 1949a; Mihály 1975; Mihály & Vincze 1984

Small gastropod shells (*Melania distincta* Zittel) from a freshwater calcareous marl at the base of the Eocene succession at Gánt. Lake habitats like this one came into being because of the transgression of the sea in the Eocene and an increase in groundwater levels (1.2× magnification).

Gastropods and Bivalves

Pioneering Studies

According to the 1987 study by András Galácz, Eocene fossil assemblages are the most abundant and most significant in Hungary. This is especially true of mollusks, and their abundance in Hungary is comparable only to classical localities in the Paris Basin and in northern Italy. This huge variety is emphasized when one reads about the 1,300 species and subspecies mentioned in the voluminous general work by Endre Szőts (1914–1984), which deals with the Hungarian Eocene. This number of species might be an exaggeration, but there is no doubt that Eocene beds comprise the most abundant Cenozoic assemblages across the Carpathian Basin.

Apart from larger foraminifera, mollusks comprise the largest numbers of specimens. Indeed, their generally well-preserved remains have been a focus of interest among Hungarian and other researchers for more than 150 years. In 1853 Miksa Hantken was the first to publish an article on Carpathian fossil mollusks, using the determinations of Moritz Hörnes (1852–1917), who was then the director of the museum in Vienna. Subsequently, discovery of the Forna beds gave a further boost to these studies. Forna-puszta is located close to Gánt, in the southern foothills of the Vértes. In 1876, Edmond Hébert (1812–1890)—geology professor at the Sorbonne and one of the biggest names in European paleontological research—visited Hungary as the guest of Miksa Hantken to study the fossils from this area. He was accompanied by Ernest Munier-Chalmas (1843–1903), a self-taught paleontologist who also left behind a considerable body of work. The stratigraphic conclusions that Hébert and Munier-Chalmas reached on the basis of these fossils have influenced Hungarian researchers dealing with the Eocene ever since. Indeed, the relationship between the two French scientists and their Hungarian colleagues, and their collective role in the study of Eocene mollusks, was appreciated when Michel Durand-Delga (a recognized authority on the tectonics of the Circum-Mediterranean region) gave a talk on the occasion of receiving an honorary membership in the Hungarian Academy of Sciences in 1998.

Seraphs sopitum (Solander) a characteristic fossil gastropod from the Eocene of the Bakony Mountains. In older literature this taxon is often referred to as *Terebellum* (original size).

📖 Szőts 1956; Galácz 1987; Bárdossy et al. 2000

Eocene gastropods from the Transdanubian Range (original size). (1) *Cerithium subcorvinum* Oppenheim. This gastropod, with a long, narrow shell reaching a length of 10 centimeters, is the most dominant form in the Forna fauna. For the most part, its shell lacks ornamentation and only the early whorls show dense spiral and axial lines (Gánt). (2) *Strombus tournoueri* Bayan—a large gastropod with a dorsally concave body whorl that dominates the shell. In the past, exceptionally beautiful specimens could be found near Bajót (Nyergesújfalu). (3) *Clavilithes noae* (Chemnitz). This nice-looking gastropod is characterized by an aperture terminating in a prolongation (siphonal canal). Its shell is ornamented with axial wrinkles, which are absent on the last whorl of mature specimens. (Zámoly) (4) *Faunus fornensis* (Zittel)—a long, narrow shell. Except for the last whorl of adult specimens, this species has a spiral ridge in the upper part of the body whorl. Its name comes from Forna-puszta, the famous locality for Eocene fossils near Csákvár (Csákvár). (5) *Voluta subspinosa baconica* Strausz—a large-sized gastropod with a narrow aperture. The upper part of the whorls in this species is ornamented with blunt ribs ending in spine-like projections (Dudar). (6) *Ampullina perusta* (Defrance)—a gastropod characterized by a roundish, thick-walled shell. The surface of the shell is sculptured with fine spiral and axial grooves, and this species is usually referred to as *Natica vulcani* in older literature (Gánt). (7) *Campanile parisiense urkutense* (Munier-Chalmas). The largest gastropod known from Eocene beds in the Bakony Mountains, these species can reach lengths of 50 centimeters (the length of coeval Caribbean specimens of this genus may have exceeded 1 meter). In marls and sandstones well-preserved shells are rare, but almost complete specimens are often collected from limestone beds, although preserved as internal casts (Dudar).

1
2
3
4
5
6
7

The "Eocene Program"

Exploration investments and the opening of mines—as part of the so-called Eocene Program in the second half of the 1970s—were the result of political decisions and led to considerable economic losses for Hungary. However, these explorations were a good thing for paleontology, and especially for the study of Eocene fossils.

As a result of some of these political decisions, it was deemed necessary to provide the miners in the already exhausted Tatabánya coal basin with work for the future, and moreover, work close to where they lived. To do this, geologists were asked to find more coal seams than were already known, even though this was impossible without detailed drilling exploration. This led to the exploitation of seams that lay below the karst water level: in several cases, a single stroke of a pickaxe triggered karst water inrush. When inrush flow happened, political leaders shifted the responsibility onto the geologists.

Recent Work on Mollusks

Several experts have studied fossil mollusks recently. One is László Strausz, who prepared a monograph based on his examinations of an assemblage near Neszmély comprising very well-preserved, small-sized taxa. Anna Kecskeméti-Körmendy has also gained recognition as a result of her descriptions of Eocene fossils; her monograph dealing with the mollusks from the Dorog Basin was published in 1972, her work dealing with fossil mollusks from fine-grained successions in the Northeastern Bakony was published in 1980. She also wrote a paper with Miklós Mészáros (1927–2000) that discusses fossils from limestone successions (reef-archipelago facies) that was published at the same time, and in 1990 she published on additional well-preserved remains from the vicinity of Nagyegyháza-Mány-Csordakút. These fossils were collected as a result of an intensive drilling project, the so-called Eocene Program.

🕮 Szőts 1953; Strausz 1966a, 1974; Kecskemétiné Körmendy 1972, 1980, 1990; Kecskemétiné Körmendy & Mészáros 1980

Gastropods

At almost all localities, the remains of gastropods are the fossils most frequently found. Limestones deposited in freshwater are characterized by *Pyrgulifera* (which is known from the Cretaceous as well), whereas marls are characterized by accumulations of specimens in large quantities, most often one species of the *Melania* genus. An abundant gastropod assemblage can be found in these shallow-marine marly-sandy beds; the genus *Tympanotonos* is represented by the highest number of species. The ornamentation of this gastropod includes spiral ridges and spines that were thought to have developed as a response to the salinity of its environment. Although this explanation is attractive, a comprehensive analysis of these complex assemblages has yet to be carried out.

Among the known characteristic forms, *Cerithium* (which often exceeds a size of 10 centimeters, and is the emblematic animal of the Eocene at Gánt) and *Turritella* ("having a turreted shell") must be mentioned. The *tokodensis* species of the latter may reach the same size as *Cerithium; Velates* (which often is preserved with its original color patterns) and *Editharus* (formerly *Cantharus*), as well as *Ampullina perusta* (called *Natica vulcani* by earlier authors) are also frequent. At the rich fossil locality of Ivókút, near the village of Bajót, well-preserved, cone-shaped specimens of these gastropods can be collected in large numbers. *Clavilithes*—characterized by a long siphonal canal—and *Faunus* are also frequently occurring forms. Indeed, the species *"fornensis"* of the latter was named after the famous locality of Forna-puszta by Zittel.

In limestone sequences mollusks are generally preserved as internal casts. Specimens—sometimes up to 50 centimeters and belonging to the *Campanile* genus—can be collected from overlying successions in certain bauxite openpit mines in the Bakony Mountains or in Upper Eocene exposures in the vicinity of Cluj, alongside other large gastropods (*Hippochrenes* and *Thericium*). The spindle-shaped *Seraphs* (formerly *Terebellum*) and the tall-shelled *Tibia* (formerly *Rostellaria*) are also characteristic of the nummulic limestone and can be found as internal casts.

Bivalves

Remains of bivalves are among the most common Eocene fossils, and their appearance also depends on salinity and bottom conditions, similar to gastropods. Assemblages that consist of just a few species occurring in large numbers indicate considerably decreased salinity or salinity fluctuations. These layers are often light-colored marl beds, containing almost exclusively *Modiolus,* or dark, coaly clay beds containing the shells of *Anomia gregaria* and *Brachidontes corrugatus.* These latter two taxa almost

Velates schmidelianus (Chemnitz) (original size). A gastropod with a unique shell form from the Eocene at Dudar in lateral view (1), apical view (2), and apertural view (3). This species is characterized by a large body whorl that completely wraps over previous whorls and determines the cap-like form of the shell. Another striking characteristic of Dudar specimens is their preserved zigzag color pattern.

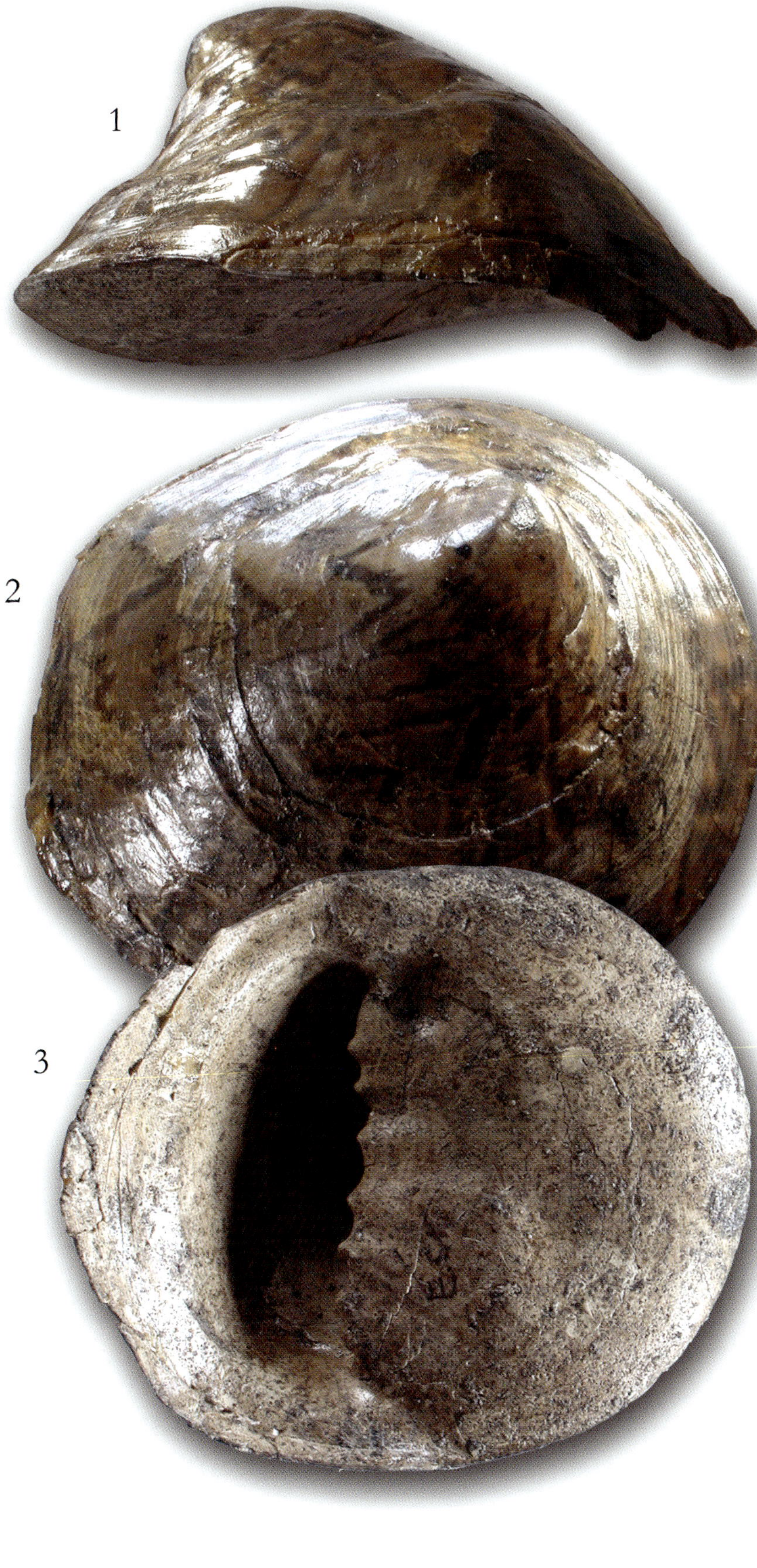
1
2
3

Propeamussium sp. The Buda Marl is usually poor in macrofossils. Among the rare fossils that are sometimes found, however, this small-sized, thin-shelled bivalve is most frequent. Its present-day relatives, similar to these ancestral forms, live in deeper-marine environments (original size).

always occur together and were considered to be indicative of coal because they characterize the layer often found immediately above coal seams.

Shallow-marine marls and sandstones contain the greatest numbers of the best-preserved and largest bivalves. Among these, *Fimbria* (formerly *Corbis*) is the most prominent and has a palm-sized thick shell that is ornamented with ridges parallel to the edge as well as radial ridges that are most pronounced at either end. *Crassatella* is also a large form with a straight posterior margin; this bivalve is usually fossilized with closed, double valves, whereas the shells of *Corbula* (which is one of the most tolerant bivalves known from an ecological point of view) are usually exfoliated. The large *Striostrea* (which belongs to the group that includes oysters) is also worth mentioning: shell beds made up from this taxon may be as thick as 10 centimeters and can be found in the vicinity of Kincsesbánya in some abandoned bauxite pits.

In limestones, the shells of bivalves composed of calcite are often preserved. Of the known taxa, fossil scallops are frequent: the shells of these bivalves can be smooth (*Lentipecten, Propeamussium*) or can be ornamented with radial ribs (*Chlamys*). According to the recent and comprehensive work of Katalin Bodó dealing with Upper Eocene bivalves from the Buda Hills, besides the taxa mentioned above, most specimens belong to either the spiny *Spondylus* or to *Plicatula,* which is ornamented with wavy, radial folds.

📖 Bodó 1992

Transylvania's Famous Bivalve

Besides the Transdanubian Range, the Transylvanian Basin is the other area in the Carpathian Basin that is rich in Eocene fossil mollusks. This territory was invaded twice by the sea, and in both periods fossil-rich successions were deposited.

Fossil remains from this region aroused the curiosity of scientists long ago. Johann Ehrenreich Fichtel (1732–1795), known as the father of geology and paleontology in Transylvania, described and illustrated the most remarkable of these Eocene fossils in 1780, under the name "Gryphit." In his book, he used the vernacular name "foal-hooves" to denote these thick bivalve shells. However, their scientific description was published a hundred years later in 1871, when Elek Pávay (1821–1864) introduced this taxon (which belongs to the Ostreidae) into the literature and gave it the name *Gryphaea esterhazyi.*

Sokolowia buchsii (widely known as *Gryphaea esterhazyi*) is a characteristic Eocene bivalve from Transylvania (0.5× magnification).

This unique bivalve has a height of 14 centimeters and a width of 18 centimeters, and at first sight resembles the more famous Jurassic *Gryphaea* genus. However, significant differences between the shells of these two bivalves came to light in the first decades of the twentieth century. Therefore, the first genus name for the Eocene form, *Gryphaea,* had to be replaced—first by *Fatina* and later by *Sokolowia.* Subsequently, the species name *"esterhazyi"* had to be added to the list of invalid species names, as it turned out that almost twenty years before Pávay's description, this very characteristic bivalve had been described by Constantin Caspar Grewingk (1819–1887), professor at Tartu (then Dorpat) University, using the name *Gryphaea buchsii.* Grewingk had been working on the Eocene fauna in the Fergana Basin.

This remarkable bivalve, correctly known as *Sokolowia buchsii,* can be found in Middle Eocene beds across Transylvania that are thin but can be traced over large areas. At the same time, Transylvania is also the westernmost occurrence of this species. It is surprising—and indicative of the considerable differences between Eocene and present-day paleogeographic positions—that no specimens of this bivalve (which has a large geographic extent) have so far been found in the Transdanubian Range, although these rocks are quite close to the Transylvanian localities.

Miklós Mészáros, professor at Cluj University and who played an important role in the description of the Eocene fossil mollusks of the Transdanubian Range, summarized his studies on Eocene mollusks of the Transylvanian Basin in a 1957 monograph. This work contains the descriptions of 21 gastropod species, 31 bivalves, and a cephalopod—all from the Middle Eocene; from the much richer fauna of the Upper Eocene, 41 gastropod species, 50 bivalve species, and a cephalopod species were recorded.

📖 Koch 1896; N. Mészáros 1957

Cephalopods

Cephalopods are far from being as frequent in the Eocene as they were in the Mesozoic era; nevertheless, they do also occur in Eocene beds and nautiloids are the most common group, known from several localities. Nautiloids in fact were found in especially large numbers at Piszke, in the vicinity of Dunaalmás, during the digging of a railway cutting. According to Viktor Vogl's studies, Piszke was one of the few places in the world where Eocene nautiloids could be collected. Scarce fossils of Coleoids can also be found in some Eocene beds and include the strange, curved *Belosepia* and *Spirulorostra* (the latter resembles the Mesozoic belemnites) as well as *Archaeosepia,* which resembles the recent cuttlefish bone and was described by Erzsébet Szörényi as a new genus from the Buda Hills.

📖 Vogl 1908; Szörényi 1933; Galácz 2004

Bryozoans

The Upper Eocene bryozoan marl in the Buda Hills, found between the Discocyclina (Szépvölgy) limestone and the deepwater Buda Marl, is very rich in fossil bryozoans, containing them in rock-forming quantities. Although the Belgian Eduard Pergens (1862–1917) described 51 species from this formation in 1887, little work has been done over the last 100 years. Just a few short works have dealt with these remains, despite the fact that they are available in large quantities. At the end of the 1990s, Miklós Kázmér, associate professor in the Department of Paleontology at the Eötvös Loránd University and a well-known expert on the Eocene limestones of the Buda Hills, and Kamil Zágoršek, a bryozoan expert from Bratislava (Pozsony), carried out a reexamination of this fauna.

However, since these workers did not know the fate of the specimens studied by Pergens, they carried out new collecting work out at Mátyás Hill in Buda, in the Lapos Quarry at Üröm, and at Úrhida, close to Lake Balaton. As a result of this work, Kázmér and Zágoršek described an exceptionally variable fauna containing 131 species that may have lived at a depth of some hundreds of meters in a publication of the Senckenberg Research Institute. A similarly abundant bryozoan assemblage can also be found in the Upper Eocene of the Transylvanian Basin as well as in the vicinity of Liptov (Slovakia).

📖 Zágoršek & Kázmér 2001

Brachiopods

The nummulites limestone in the Bakony Mountains has yielded brachiopod remains as well. The abandoned limestone quarry in the village Csabrendek north of the town Sümeg and the bauxite pit at Nyíres-puszta near Nagytárkány have proven to be an especially rich brachiopod localities. The assemblage is of markedly low diversity, especially if compared to geologically older occurrences. Only three smooth-shelled forms, all representing the order Terebratulida, have been hitherto recorded. A new species assigned to the genus *Magellania* (*M. hantkeni*) was described in 1944 by Ilona Meznerics (later Ilona Csepreghyné Meznerics [1906–1977]), an acknowledged specialist on Miocene invertebrate fossils. Recent studies proved, however, the need of erecting a new genus (*Mezneric sia*) to accommodate this small-sized, globular form. Stratigraphically slightly younger deeper-water marly deposits were found contain a more diverse assemblage dominated by micromorphic elements, among which representatives of the genera *Terebratulina* and *Argyrotheca* make up more than 90 percent of the specimens.

📖 Meznerics 1944; Bitner & Dulai 2008; Bitner et al. 2010

1
2
3
4
5
6
7
8
9
10
11

Coleoid (a kind of cuttlefish) remains from the Transdanubian Range (original size).

Crustaceans

The oldest sediments that contain decapods in large quantities are Eocene in age. Most of the known remains come from the lower (already buried) layers of the quarry on the northern side of the Kis-Sváb Hill (which was at one time called the Martinovics Hill) as it rises above the Városmajor district. According to Imre Lőrenthey's (1867–1917) posthumously published monograph that evaluated the fossil decapods known from the countries that made up the Hungarian Monarchy, this locality yielded 35 species in 20 genera. Lőrenthey's book was copyedited by the German Karl Beurlen (1901–1985) in 1929—and as far as can be estimated from the number of specimens in collections, the most frequent remains are carapaces of *Lophoranina* (which are ornamented with transverse rows of prominences lined with tubercles) as well as almost complete specimens of the large-sized *Palaeocarpilius.*

Pál Mihály Müller became famous for his work on Badenian decapods and has published valuable data on Eocene forms, including peculiar ones that occur together with corals. In the course of a reevaluation of the decapods of the Late Eocene Szépvölgy Limestone (which outcrops mainly in the area of Budapest), it turned out that the fauna is much more varied than was previously thought. Müller and his co-author carefully made an inventory of all the tiny forms that had escaped the attention of former collectors and showed that 58 decapod species are derived from the coral-bearing facies of this limestone. Among the rich fauna, 35 species from 18 genera were recorded as being new to science.

Fossil crustaceans are also mentioned in the literature as being present at several other sites, but based on descriptions there is no known locality that compares with the Kis-Sváb Hill. Worthy of mention, however, is a hidden fossil site in the Sűrű Hills near Dudar, in the Bakony Mountains, which also seems to be rich. Eocene fossils were studied by János Tomor-Thirring (1910–1979) who mentioned four species belonging to the genus *Harpactocarcinus,* characterized by a carapace with a jagged edge. Two of these species were new.

📖Lőrenthey & Beurlen 1929; Tomor-Thirring 1934; Müller & Collins 1991

Echinoids

Spiny-Skinned Creatures

Fossil echinoids are often found in Eocene beds as both nummulitic limestones and marly layers may contain their remains. Indeed, fossils of these invertebrates are often well preserved and beautiful, so people get much pleasure from collecting them. Indeed, in his 1860 volume

Eocene gastropods and bivalves from the Transdanubian Range (original size). (1) *Strombus* sp.—A gastropod from Dudar that has not been determined in detail. (2) *Brachidontes corrugatus* (Brongniart)—a beautiful bivalve ornamented with folds that usually occurs together with specimens of *Anomia gregaria* (Bayan) (Tatabánya). (3) *Meretrix vertesensis* (Taeger). This rounded, triangular-shelled bivalve lived in near-shore habitats during the Eocene. Occasionally, its shells are found in large numbers in sandstone or marl beds (Mór). (4) *Chlamys biarritzensis* (D'Archiac). This bivalve is sculptured with radial folds and is one of the characteristic fossils from the Upper Eocene limestones of the Buda Hills. In the "bryozoan" marl, the carbonate that makes up the fossil shells is often replaced by silica, as in the case of the figured specimen (Budapest, Szép Valley). (5–7) *Tympanotonos* spp.—Turreted shells of this taxon with different sculptures are frequent in the Hungarian Eocene. According to the traditional interpretation, the sculpture on these shells varies depending on salinity conditions. (5) *T. calcaratus* (Brongniart). (6) *T. hungaricus* (Zittel). (7) *T. rozlozsniki* Szőts (Gánt). (8) *Editharus brongniarti* (d'Orbigny). The most beautiful and most common gastropod from the Forna beds, this taxon can easily be distinguished from the somewhat similar *Clavilithes noae* because of its shorter siphonal canal (Gánt). (9) *Corbula exarata* Deshayes. These convex shells—ornamented with strong ribs parallel to the edge—are frequent in Eocene shallow-marine marls in the Bakony Mountains. The right valve of *Corbula* is considerably larger than the left one, and because these calcareous shells also commonly contain layers of organic matter that is not preserved, they usually fall apart along these layers (Tatabánya). (10) *Lentipecten corneus* (Sowerby)—This smooth-shelled bivalve with symmetrical valves can be found in several Eocene sediments deposited in environments of normal salinity (Budapest, Mátyás Hill). (11) *Fimbria maior* (Bayan). The largest bivalve from the Transdanubian Range Eocene, *Fimbria* is ornamented with ribs parallel to its edge that are interrupted by radial grooves found mainly on the posterior part of its thick shell. In older literature this bivalve is often referred to as *Corbis* (invalid name) instead of the correct *Fimbria* (Dudar).

Portrait Gallery

László Strausz (1901–1988)—Paleontologist and Hydrocarbon Geologist

László Strausz, one of the most creative Hungarian paleontologists, was born in Budapest in 1901. From 1920 to 1924 he was a student of natural sciences and geography at the University of Budapest, as a member of the Eötvös College. He took his doctorate in geology in 1924 and then studied for two years as a scholarship student at the Collegium Hungaricum in Berlin. In 1928 he got his teaching diploma, and for a year he taught geography at the teacher-training grammar school in Budapest. Between 1931 and 1933 he worked twice as a research fellow at the Geological Institute in Budapest.

Strausz became associated with the hydrocarbon industry in 1933, when he was employed as a hydrocarbon geologist at the British-American EUROGASCO Company. Subsequently, he worked for its legal successors (MAORT, MASZOLAJ) until 1962, when he retired from the Hungarian National Oil and Gas Trust, where he was a chief geologist and department head. However, he continued research at his last place of employment, SzKFI (the Research Institute of Hydrocarbon Exploration and Development), from 1980 until his death. His studies focused on the Tertiary (especially Miocene) sediments of the Transdanubian area and partly on the Cserhát Hills. His biostratigraphical and paleoecological interpretations of successions were based predominantly on fossil mollusks and in the 1940s, his interest was aroused by the stratigraphy of the Pannonian (then called Pliocene) Stage, the tectonism of the recent past in geological history, and the use of facies analyses in oil exploration.

Strausz published almost 70 scientific publications, including 9 monographs, in which he described mainly Eocene and Miocene mollusk faunas. The latter were published mostly in the Geologica Hungarica series Palaeontologica; his work dealing with Badenian gastropods in Hungary was published by the Hungarian Academy of Sciences in the form of a handbook and a monograph in German. One still much used work is the 1928 handbook *Geologische Fazieskunde,* which was published by the Geological Institute of Hungary in 1928. In 1953 his life work was honored with the Kossuth Prize and in the same year he received the title Doctor of Geology; in 1986 he received an honorary membership to the Hungarian Geological Society.

📖Németh 1989

Flóris Rómer, lover of the Bakony Mountains, wrote enthusiastically about his experience doing just this: "En route to Halimba I was surprised at the innumerable, very well-preserved, and excellently-sized remains of larger oysters and echinoids which can be collected in some minutes in the hillside gullies."

The first information about and illustrations of these fossils from the Carpathians was given by Elek Pávay (1820–1874) in an 1874 paper ("A budai márga ásatag tüskönczei" [The fossil echinoids of the Buda Marl]). For the most part, these faunas are well examined across the basin in large part thanks to the work of Erzsébet Szörényi, a prominent echinoid expert. Her 1929 paper contains an evaluation of the assemblage of the Buda Marl and her 1973 monograph deals with the regular echinoids of the Bakony Mountains and the faunas from former localities at Iszkaszentgyörgy, which were created as a result of bauxite mining. An additional description of the assemblage of the Buda Hills discusses 42 species and was written by Attila Bartha, both in a 1988 dissertation and a 1992 overview paper. So far a monographic evaluation of the irregular echinoids from the Bakony Mountains has yet to be written. In the Buda Hills, representatives of the genera *Echinolampas* and *Echinanthus* are the most frequent fossils; in the slightly older faunas of the Bakony, *Schizaster* and *Conoclypus* are characteristic.

Eocene beds in the Transylvanian Basin are also rich in fossil echinoids. The first (and also the last) monographic evaluation of them was the 1884 work of Antal Koch—who was director of the Geo-Paleontological Institute of the University of Budapest—*Die alttertiären Echiniden Siebenbürgens* (The Early Tertiary echinoids of Transylvania).

📖 Pávay 1874, 1879; Koch 1884; Szörényi 1973; A. Bartha 1992

Vertebrates

The overwhelming majority of fossil vertebrates found in Hungary and across the Carpathian area are of marine origin. Of these, the shiny teeth of chondrichthyans (sharks

Remains of Eocene decapod crustaceans from the Sűrű Hills at Dudar (1–3, 5) and from Kis-Sváb Hill (4). (1) A well-preserved decapod that has not been determined in detail. (2) *Lophoranina reussi* (Woodward). (3, 5) *Harpactocarcinus hungaricus* Tomor-Thirring. (4) *Micromaia* sp. (original size).

and rays) are the most frequently encountered—but if we are lucky, besides fish remains, fossil marine mammals such as sirenians (dugongs and sea cows) can also be found in the variable, shallow-marine beds. However, fossil vertebrates are really rare!

Marine Vertebrates

Sharks

Shark Teeth in Transylvania

Sharks belong to the group of cartilaginous fish (chondrichthyans) that do not have typical bony skeletons. Since cartilage does not fossilize well, or requires special conditions, teeth are the only parts of sharks that are commonly preserved as fossils. Although shark teeth have no true roots (they are just modified placoid scales) and occur in many different marine sediments, these fossils are not common anywhere across the Carpathians. Collecting these remains is a troublesome business, even if they are known to occur at several places.

Only one locality in the Carpathian region is known to yield Eocene shark and ray teeth in exceptionally large numbers. This site is in the village of Porceşti (Porcsesd in Hungarian, and today Turnu Roşu) in the northern foothills of the Fagaras Mountains. Fish remains have been collected here for 150 years and most of them have been found in the calcareous layers (of Eocene age) which cover Upper Eocene (Priabonian) conglomerate beds. The exceptionally rich and well-preserved fish tooth assemblage from this area was first described by Ludwig Johann Neugeboren (1806–1887)—and his collection, comprising several thousand specimens, is the valuable possession of the Sibiu Museum. Rodica Ciobanu, the current museum attendant, is also an enthusiastic researcher of fossil sharks, and has described remains belonging to 28 shark and ray genera in a monograph that deals with the Paleogene cartilaginous fish fauna of Romania. Twenty-two of these species also occur at Turnu Roşu, and so this site is considered to be the richest locality. In addition, a number of older localities that had yielded teeth in large numbers no longer exist; they

1
2
3
4
5
6
7
8
9
10
11

Remains of an Eocene echinoid in the riverbed at Someşul Mic, below the dam at Cluj.

Shark teeth from the Paleontology Collection of the Sibiu Museum.

have been destroyed, buried, or built on. It is also possible that the original pockets that contained fossil teeth simply became exhausted. Although at the border of Porceşti fossil shark teeth are much less frequent than they were formerly, the exposures of the fish tooth–bearing beds have been declared protected, and fossil collecting requires permission.

The Someşul Mic riverbed at Cluj was also noted as a remarkable fossil shark tooth locality because beds here, exposed in the riverbed, yielded teeth. There was one horizon in the succession (which also contained bivalve and echinoid remains in large numbers) from which a rich assemblage was obtained, but after the reconstruction of the dam located within the borders of the town, the "best" shark tooth–bearing beds were flooded. This site is not accessible at the moment.

Beautiful shark teeth are known from sites at the border of Albeşti Muscel (7 kilometers to the Northeast of Cîmpulung in County Argeş), from Eocene beds (the so-called Rákóczi Sandstone) at Jibău and from Middle Eocene (Bartonian) layers at Leghia.

Finally, chondrichthyan remains can be found outside Transylvania in other marine Eocene beds across Hungary. Exceptionally beautiful shark (and ray) teeth have, for example, been found in the vicinity of Szépvizér, Csordakút, and Tata.

📖 Neugeboren 1851; Ciobanu 2002

The Shark's Jaws

A shark may have several hundred teeth during its lifetime; these are arranged in rows, one behind another, like the grandstand seats in a stadium. However, only the front set of teeth does the work: As a front tooth is worn down (for example, as it breaks while feeding) it is replaced by another tooth from the next row that shifts forward into position—like a bullet in a revolver. Thus, the teeth of sharks, often referred to as revolver dentition, consists of teeth of different sizes and types, with marginal teeth generally smaller than those in the center. The size and (more or less) shape of teeth depend on the age of the shark—but they are all usually sharp and pointed, and have one or more tips (referred to as the main cusp and cusplets) and a slightly serrated cutting edge.

The jaws of chondrichthyans rapidly fall apart after death so most often their teeth are buried and fossilized separately. Historically, paleontologists have given lots of

Eocene echinoids (original size). (1, 2) *Atelospatangus gardinalei* (Oppenheim). One characteristic element of the Upper Eocene echinoid assemblage in the Buda Hills, this genus is ornamented with remarkable warts (Solymár). (3) *Echinanthus scutella* (Lamarck)—a frequent species in Middle and Upper Eocene faunas. Attila Bartha proved that there was a relationship between the height of these semi-infaunal echinoids and the grain size of seabed sediments; in coarser sediments, the tests of these echinoids are higher while in fine-grained sediments specimens tend to have lower tests. (4) *Porocidaris schmiedeli* (Münster)—an ornamented spine of an echinoid (Csákberény). (5) *Stereocidaris nummulitica* (Sismonda)—a rare representative of a regular (round) echinoid; just a few specimens of these animals have been found so far in the Bakony Mountains (Csehbánya). (6, 7) *Coelopleurus nopcsai* Vogl. This beautiful, regular form is one of the most prominent taxa comprising the rich echinoid fauna of the Transylvanian Basin. The name of this species pays tribute to Baron Ferenc Nopcsa (Cluj, Suceag Quarry). (8, 9) *Macropneustes brissoides* (Leske). This species, found in large numbers in the Bakony Mountains, is characterized by an almond-shaped periproct (Úrkút). (10) *Schizaster lorioli* Pávay. This echinoid is characteristic of the "bryozoan" marl. Silicified, cracked fossil specimens are often found (Budapest, Szép Valley). (11) *Echinolampas giganteus* Pávay. This spectacular form—described first from the Transylvanian Basin—is also derived from the Buda Hills (Cluj-Mănăştir).

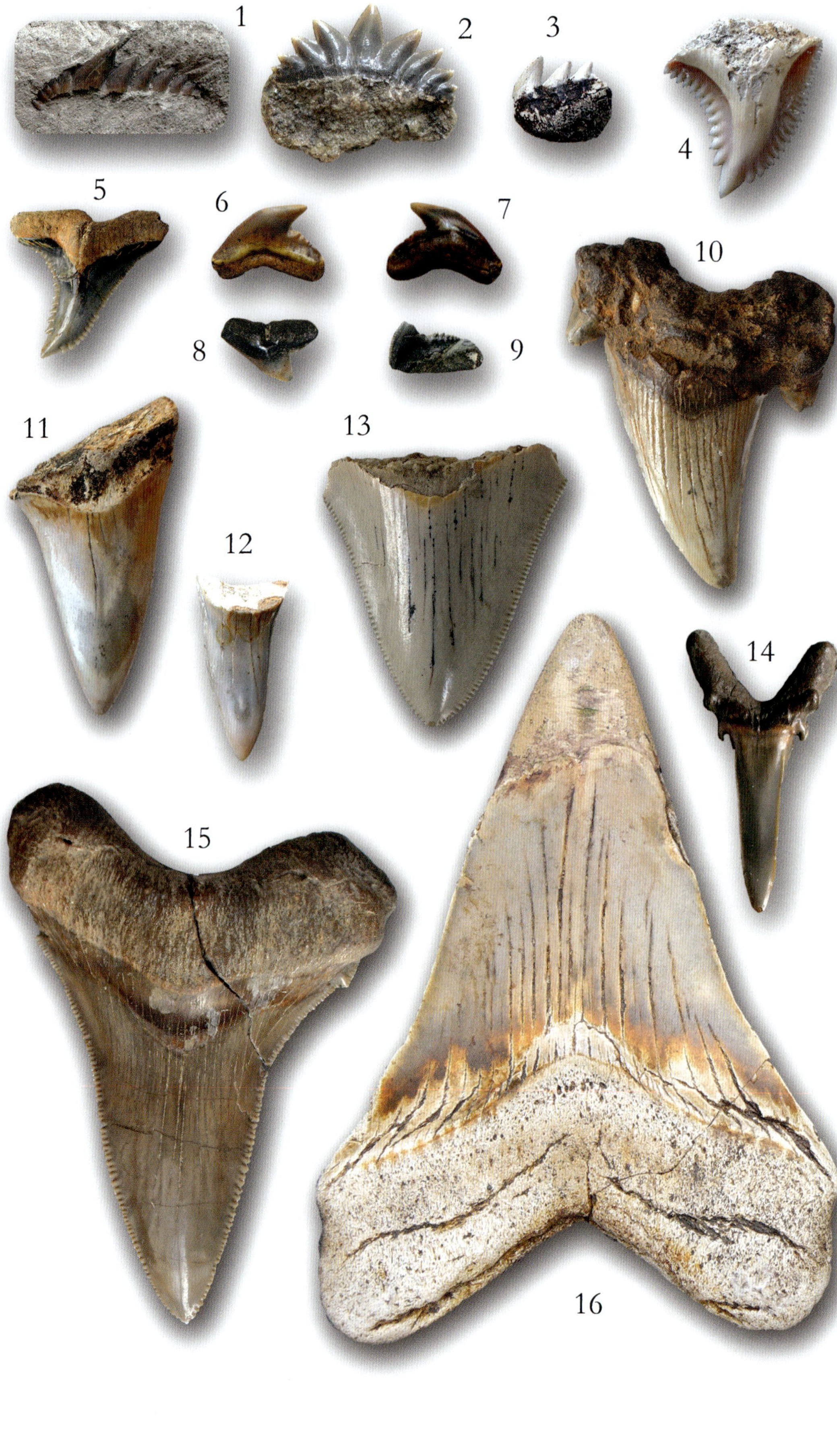
1
2
3
4
5
6
7
8
9
10
11
12
13
14
15
16

different names over the years to teeth of different shapes and sizes, so often it is the case that teeth belonging to the jaw of the same species are listed under a number of species names. One goal of research on sharks is the reconstruction of the complete set of teeth and thus take superfluous names out of use.

Eocene Sharks from Porceşti

Sharks are an ancient type of fish and they have changed little over millions of years. This explains why several genera—known from the Eocene—are alive today in the world's oceans. Most sharks are open-water predators; however, their diet and eating habits are variable and different species have different kinds of teeth. Here, mainly based on the work Rodica Ciobanu, we review some of the most important remains of shark teeth that are known from the Carpathian Region. All the taxa discussed come from Porceşti (Turnu Roşu).

Notorhynchus is a representative of the extant Hexanchidae family (cow sharks) that includes the six- (hexa-) and seven- (hepta-) gilled sharks (hence their family name). "Hexanchid" teeth—especially those that are attached to the lower jaw—are composed of oblique, comb-like denticles. The size of the teeth progressively decreases in the direction of the inclination of denticles and forms belonging to this family are often referred to under the genus name "*Notidanus*" in older literature. The recent bluntnose sixgill shark, or cow shark (*Hexanchus griseus*), occurs in the Mediterranean Sea; although it may reach a few meters in size, it is not dangerous to humans.

The genera *Ginglymostoma* and *Nebrius* are relatives of modern nurse sharks. The more or less symmetrical, curved crown of their teeth is made up of tiny denticles; those in the center larger are than those at the shoulders of the blade.

The genera *Striatolamia* and *Carcharias* belong to the Odontaspididae family. Their teeth are long, pointed, and awl-like, with secondary cusplets that may be present or absent. The modern grey nurse shark (*Carcharias taurus*), often called a maneater, is a relative of these Eocene species and occurs today in warm seas.

The genera *Isurus, Xiphodolamia, Lamna,* and *Isurolamna* are all relatives of mackerel sharks (Lamnidae family). The genera *Otodus* and *Carcharocles* are considered representatives of the fossil Otodontidae family, which also includes forms close to mackerel sharks but of debated taxonomic status. Broadly, because there is no accepted opinion about what kinds of genetic information are carried by certain morphological characteristics, taxonomic classifications used by experts who deal with recent or fossil remains may be different. Because of this, experts often refer to the same kind of shark but use different generic names.

Teeth classified in the Otodontidae family are rather different: they can be straight or curved, symmetrical or asymmetrical. Of these, *Carcharocles* (formerly known as *Carcharodon*) teeth are mostly large, and the cutting edge is serrated. (Remains of the genus *Paratodus*—known only from the fossil record—are also classified in this group and are known from Miocene beds at Ipolytarnóc.) Overall, the great variety of teeth referred to this group is in accord with the fact that fish of different sizes and modes of life are classified in the group that contains the modern mackerel sharks by ichthyologists. For example, this group includes the porbeagle shark (*Lamna nasus*), widely known as "canned shark," and the much-feared white shark (*Carcharodon carcharias*). Indeed, the latter species belongs to those few that are known to attack people. In contrast, the harmless, filter-feeding basking shark (*Cetorhinus maximus*)—which may reach a length of 12 meters—is a member

Tertiary shark teeth from different localities (All original size, except #2). (1) *Notorhynchus* sp.—a fragmentary specimen from the Kiscell Clay, Újlak brick factory at Óbuda. (2) *Notorhynchus* sp.—a symphyseal tooth from the lower jaw from an unknown locality in Budapest. This kind of tooth grows in between where the right and left jaws meet and looks almost like two teeth joined together (2× magnification). (3) *Notorhynchus* sp.—a rounded tooth from the Eggenburgian at Felsőesztergály. (4) *Hemipristis serra* Agassiz—an incomplete specimen (tooth without root) from the Badenian at Mátraszőlős. (5) *Hemipristis* sp.—an exceptionally well preserved tooth from the Eocene at Porceşti. (6, 7) *Galeocerdo aduncus* Agassiz—a slightly worn tooth from a tiger shark, in external and internal views. This tooth was found at Danitz-puszta, near Pécs, where Badenian remains are found redeposited into younger (Pannonian) sediments. (8) *Sphyrna* cf. *zygaena* (Linné)—a tooth from a hammerhead shark from the sand pit near Hímesháza, where Badenian remains can be collected from Late Miocene (Pannonian) beds. (9) *Galeocerdo* sp.—fragmentary tooth of a tiger shark from the Eggenburgian shark tooth-bearing sandstone at Ipolytarnóc. (10) *Otodus obliquus* Agassiz—A relatively complete specimen with characteristic side cusplets from Eocene beds at Porceşti. (11) *Isurus hastalis* Agassiz—fragmentary remains from the Eggenburgian at Felsőesztergály. (12) *Isurus* sp.—fragmentary tooth from the Badenian at Mátraszőlős. (13) *Carcharocles* sp.—rootless tooth from the Badenian at Mátraszőlős. (14) *Carcharias acutissima* (Agassiz)—a well-preserved specimen with root and side cusplets, from Oligocene beds at the former Újlak brick factory at Óbuda. (15) *Carcharocles megalodon* (Agassiz)—a complete tooth from an unknown locality, with small striations on its enamel. (16) *Carcharocles megalodon* (Agassiz)—a slightly worn tooth from a large shark from redeposited Badenian sediments at the Danitz-puszta locality.

Exceptionally well preserved dentition of an Eocene ray (*Myliobatis* sp.) from the surroundings of Felsőgalla (0.8× magnification).

of another family (Cetorhinidae) and has more than 2,000 tiny appendages with claw-like endings in its mouth.

Genus *Cretolamna*—also known from Turnu Roşu—is one representative of an extinct family (Cretoxyrhinidae) that lived during the Cretaceous and Eocene. Denticles classified into this genus are generally small and have a main cusp that is usually flanked by side cusplets.

The teeth of *Galeorhinus*—which belongs with houndsharks (Triakidae)—are also strongly asymmetrical; these sharks have oblique teeth with single cusps. The teeth of the often mentioned genus *Hemipristis* are also very similar to these, but are classified into a distinct family (Hemigaleidae).

Finally, the family that includes the requiem sharks (Carcharhinidae) is represented by several genera from Turnu Roşu. This group includes—among others—the genera *Galeocerdo* and *Carcharhinus,* relatives of the living tiger shark (*Galeocerdo cuvieri*) and grey reef shark (*C. amblyrhynchos*). *Galeocerdo* teeth have one large, oblique crown and many small—also oblique—serrations. *Carcharhinus* teeth are similar, but the main cusp is either not, or is slightly, asymmetrical. This genus, *Carcharhinus,* thrives today and includes almost 40 species.

📖 Ciobanu 2002

Eocene Rays

The taxonomic separation between sharks and rays is artificial—there are several forms that can be considered to be transitional between the two groups. In general, however, it is true that in contrast to the usually rounded bodies of sharks, rays tend to be dorsoventrally flattened and the teeth of both forms are often found as isolated fossils. Rays are cartilaginous fish, as are sharks, but their remains are found less often as fossils.

Representatives of sting rays (in the widest sense), especially the fossil remains of eagle rays (Myliobatidae) are common in the Eocene. Recent representatives of Myliobatidae have a beautiful and dignified swimming mode that imitates flapping flight (hence their name). Eagle rays include the two most common Eocene ray genera, *Myliobatis* and *Aetobatis,* but there are also several other known forms. Eagle rays have a distinct flat, rhomboidal body with a strong, whip-like tail; under normal conditions only their "pavement-like" teeth—specialized for gnawing, crushing and grinding—fossilize. These flat, or slightly convex, grinding surfaces, found both in the lower and upper jaws, are made up of flat teeth which that close together. The laminae of these teeth often fall apart during fossilization and are buried separately. Their occlusal surfaces are smooth, but the on the other side they are striated; in lateral view these teeth resemble combs. Complete remains of teeth are scarce, yet the number of described species is large—probably without good reason.

📖 Ciobanu 2002; Zachár 2003

Sirenian Remains in the Transylvanian Basin

Mermaids, or sirens—who lured mariners to death—are common in old sea stories. One thing is certain: these legends are related to the occasional appearance of living—albeit less frequent—sea cows (manatees and dugongs). Extant sirenian families separated from one another during the Oligocene; however, the remains of earlier forms are frequent, even in Eocene beds.

Beautiful sirenian fossil remains can be found in large numbers in Paleogene beds across the Transylvanian Basin. Explosions carried out during the course of construction of the Someşul Mic Dam at Cluj-Monastir revealed canine teeth, alongside the remains of bony fishes, in an Upper Eocene (Priabonian) porous limestone which forms the banks of the river Someşul Mic. These finds, collected by Elek Pávay in 1866, were first falsely identified as crocodile teeth when in fact they belong to sirenians.

It was also Elek Pávay who, together with Count Kálmán Esterházy, collected a beautiful molar tooth from a deep ditch in the Middle Eocene nummulitic limestone between the villages of Gilău (located west of Cluj-Napoca) and Căpuşu Mare. Pávay classified this tooth into an extinct genus of sea cow, *Halitherium*. The fossil was later examined by Antal Koch, who remarked that "the whole dental enamel is ornamented with nice, delicate, wavy,

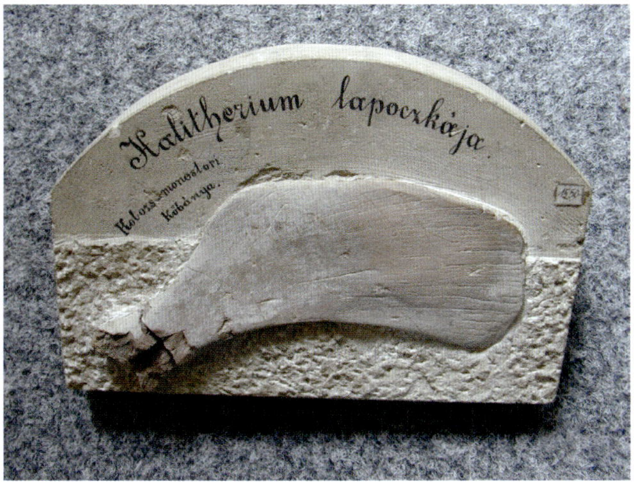

A shoulderblade from an Eocene sirenian in the old collections of Babeş-Bolyai University at Cluj-Napoca (approximately 0.3× magnification).

transversal wrinkles which can slightly be seen with the naked eye." It is noteworthy such a man as Count Esterházy, an aristocrat who played a crucial role in Hungarian life, also devoted his time and energy to finding and studying fossils.

In contrast to living sea cows (which have a greatly reduced dentition) the teeth of Eocene sirenians were much more abundant, forming complete dentition. However, with regard to Paleogene sirenians, the name *Halitherium* was rather generally applied in older paleontological literature: in subsequent publications the genus names *Eotherium, Prototherium* and *Protosiren* can also be found. Herman Brassói Fuchs (1915–1994) and János Tulogdy (1891–1979) studied Transylvanian remains in detail, and the most recent work has been done by Mátyás Vremir, Nicolae Şuraru, and Vlad Codrea.

📖 Brassói Fuchs 1959, 1990; Koch 1900; Şuraru & Codrea 1988; Tulogdy 1944

Sirenians in Hungary

Extant sirenians inhabit warm and shallow waters in near-shore marine environments around the world. Eocene remains of these animals from Hungary are known from limestones and marls which were also deposited in shallow and warm seas and are frequently accompanied by the fossils of larger foraminifera or mollusks. The Hungarian sirenian fossil remains have been studied in detail by Miklós Kretzoi (1907–2005) and by László Kordos. In one review paper, published in 2002, Kordos collected together all the available data on Hungarian sirenians and revised the fauna lists. His summary reveals that most of the fossil remains were in fact found in different drifts and open-pit mines in the Balinka, Dudar, Pusztavám, Tatabánya, and Oroszlány coal mining areas. Of these, the sirenian skeletons that are known from Keselő Hill at Tatabánya and the finds from Üröm are exceptionally beautiful. Indeed, one collection of Keselő Hill specimens comes from the mine itself while others were found during the course of railway cutting construction. The fossil from Felsőtárkány is also remarkable and it is named as the holotype of *Paralitherium tarkanyense*.

Because most of the sirenian fossil finds consist of fragmentary ribs and vertebrae, identifying them more precisely is not possible. However, some exceptional finds include teeth, skull remains, and more complete bones. While revising all these Hungarian fossils, Kordos realized that Hungarian sirenian remains should be classified into only four species and considered *Sirenavus hungaricus* (described by Kretzoi) and *Anisosiren pannonica* (which he described) to be valid. The other two forms were identified with the formerly described *Protosiren* and *Eotheroides*.

📖 Kretzoi 1941a; Kordos 1978a, 2002a

Land Vertebrates

Fossil Turtles in Transylvania

Turtles survived the end-Cretaceous mass extinction, and their body shapes have changed only slightly over the last 65 million years. The second evolutionary bloom for this group started in the Eocene and is still going on today with most remains, found in limestones across the Carpathian

The fragmentary skull of *Sirenavus hungaricus* Kretzoi from the Middle Eocene of Tatabánya (holotype, 0.4× magnification).

A sirenian fossil from the Middle Eocene at Tatabánya. This unique find came to light after an explosion in the mine. The hard limestone containing the fossil was broken apart as the explosion scattered bones over a great distance, leaving only cavities. Impressions of the vertebrae and ribs are the most prominent of the fossil remains that are left, although the impression of an elongate brainpan can also be seen. (The length of the piece of stone in the picture is 70 centimeters.)

Basin, classified as soft-shelled turtles (Trionychidae). Because these turtles are usually freshwater or brackish-water forms, they are considered here along with other "land vertebrates" in order to emphasize their differences from marine turtles. The moniker "soft" just refers to the fact that their shells are covered only with a strong and thick leathery skin, instead of scutes made up of keratin.

Trionyx is the most frequently mentioned soft-shelled turtle genus name that occurs in the literature. Exceptionally beautiful *Trionyx* remains are known from the Eocene at Cluj and the first description of turtles, found in Eocene beds across the Transylvanian Basin, was done by Lőrenthey and then later by Hermann Brassói Fuchs, followed by Mátyás Vremir (who at one time worked as a paleontologist at Babeş-Bolyai University at Cluj and is now affiliated to the Transylvanian Museum Society, also in Cluj-Napoca).

📖 Lőrenthey 1903; Brassói Fuchs 1994a, 1994b; Vremir 1995, 2004

Turtles in Overburden above Bauxite

Eocene remains of the bodies of turtles (other than shells) are known from Hungary, but they are rare. The impressions, or remains, of either the carapace or the plastron are most common, and less frequently limb bones are found. Descriptions of freshwater and brackish-water turtle remains from Middle Eocene sandstones and marl beds at Nyíres-puszta (east of Sümeg, in the area of the Southern Bakony Mountains) were published by Attila Ősi, who determined one of the ten fragments found in the Darvastó Formation as *Chinemys* cf. *strandi* in the family Emydidae.

Trionyx remains from the Eocene at Cluj. On the left, an almost complete body fossil of the upper shell of a turtle described by Imre Lőrenthey can be seen. In this fossil, the carapace itself has been preserved, and on the right is an impression of another carapace that represents the inner surface of the shell.

The bulk of the finds from this site were classified by Ősi into the family Trionychidae, which is not surprising since this Middle Eocene succession (at least the beds that certainly belong to it) overlies a bauxite that also contains *Nummulites* deposited in a marine environment.

The question is this: How were the remains of these soft-shelled turtles—which were freshwater, or at least brackish-water, dwelling animals—buried in marine deposits? According to Ősi, these animals may have lived in a river that flowed out onto the Lutetian floodplain, or even in the river mouth or delta itself. It even might be possible that they lived in the shallow-marine region close to the river mouth, where salinity was low. After death their bodies were transported by the river or, in the case of animals living in river mouth habitats, by currents to their actual place of burial.

📖 Szalai 1934; Młynarski 1966; Ősi 2001

The Pelican from Cluj-Monastir: The Only Known Eocene Bird Found from the Carpathian Basin

Fossil remains from the Carpathian Basin provide scant information for the bird fauna of this area. Just one bird, belonging to Pelecanidae, is known from the Eocene (*Eostega lebedynskyi*), and was described by Kálmán Lambrecht (1889–1936) from Cluj-Monastir. For many years, this fossil material represented the oldest fossil bird known from the Carpathian Basin. However, more recently described Cretaceous remains are now known.

📖 Lambrecht 1929

Brachydiastematherium transilvanicum: *The Fossil Ungulate from Rădaia*

In 1871 Elek Pávay, who was a military officer in the Revolution of 1848 and a geologist at the Royal Geological Institute of Hungary, had just returned from the United States. He read one of his dissertations to a session of the Hungarian Geological Society, titled "Kolozsvár és Bánfy-Hunyad közti vasútvonal ingatag talajának geologiai szerkezete" (The geological structure of the unstable ground of the railway between Kolozsvár and Bánfy-Hunyad). This work was subsequently published in the seventh volume of the *Bulletin of the Hungarian Geological Society*. Pávay mentioned a noteworthy fossil:

> Andrásháza [Rădaia], located to the West of Bács and next to it, on the left side of the Nádos brook, an approximately 40-feet-high, steep natural collapse can be found made up of russet sandstone. This is the best exposure of the bluish-gray clay stripes, or thin laminae, which are characteristic of the sandstone succession. As already mentioned, based on the exposures in the Kolozsvár [Cluj] Basin, this sandstone belong to the lowermost part of the Eocene sequence but has been considered as barren of fossils so far, even residual samples have not yielded discernible microscopic animals, yet (to say nothing of larger fossils such as mollusks etc.). However, things are different now: the fossil lower jaw of a remarkable animal proves that this rock belongs to the Lower Eocene. . . . The lower jaw I have found, together with the bone remains next to it . . . belong to the genus *Palaeotherium*. I will postpone publishing its illustration, detailed description and information about its position in the rock until my examination will be concluded with the determination of the species, which is probably a new one. At least—in consequence of munching—the double heart-shaped pattern (double coeur) on the molars indicates its affiliation to *Palaeotherium*.

Thus, Pávay found this fossil, realized its importance, and made a preliminary determination. He thought that his unique find was related to the genus *Paleotherium* (within the odd-toed ungulates), which was known at the time thanks to the work of Cuvier, who had described it from the gypsum at Montmartre. However, because of his sudden death, Pávay was unable to complete a detailed description of the new fossil. This work was left to János Böckh, who was director of the Institute at that time. In his dissertation, published in 1875, Böckh carefully described the jaw outlining the position and size of the teeth, the pattern on the enamel lobes, and the characteristics of the grinding surface. Finally, he stated that the Rădaia find is a peculiar fossil that shows different characteristics from other related forms and he described it as a new species within a new genus, *Brachydiastematherium transilvanicum*. The scientific importance of this Transylvanian fossil remains internationally outstanding: according to current, generally accepted opinion, *Brachydiastematherium* (which is to date known only from Rădaia) belongs to rhinoceroses in the strict sense within the odd-toed ungulates, and is classified in family Brontotheriidae (also called Titanotheriidae). Representatives of this group are well known in the North American Oligocene, and subsequently their remains have also been recovered from Asia. Their largest representatives were huge, rhinoceros-like animals that had variable horns above their noses and on their foreheads. Indeed, *Brachydiastematherium* is among the very rare Titanotheriidae finds in Europe and from a paleogeographic point of view it presents a clear link between the Eocene and Oligocene faunas of America and Asia.

📖 Pávay 1871; J. Böckh 1881b

The Most Ancient Rhinocerotids

Similarly to *Brachydiastematherium*, *Prohyracodon*—the other Paleogene fossil land vertebrate of outstanding importance in the Carpathian Basin—is also found at Andrásháza, (today Rădaia) in the Upper Eocene (Priabonian) terrigeneous Nádasvölgy beds. The largely incomplete skeleton, mentioned as the new species *Prohyracodon orientale*

Brachydiastematherium transilvanicum Böckh and Matyasovszky—the jaw of an odd-toed ungulate with teeth described from Rădaia, near Cluj-Napoca. The unique find is an exceptional treasure in the Museum of the Geological Institute of Hungary (0.6× magnification).

Pénzes Gyuri [Moneyed Georgie]—The Eocene Man

Any list of Hungarian vertebrates would be incomplete without noting the strange "find" that was reported in September 1987 by the journal *Mai Magazin*. Most people would find this short, one-page report funny if it did not contain so many errors and was not so full of nonsense. The article, headlined "Pénzes Gyuri lábnyoma" (The strange footprint of Moneyed Georgie), provides information about a find, which—according to its finder—is the footprint of a caveman: "I found this fossil in August 1984 at the border of Pénzesgyőr near Várpalota. This cave child—I call him Moneyed Georgie—had stopped in a muddy area. He stepped ahead with his left leg and stepped into a reed, which pricked his sole under the outer ankle. Then he snatched his foot away tilting his heel inwards and the mud lithified together with the reed. More than ten characteristics can be observed on it." A footprint-like stone is described in the article, which is derisive about the doubts of experts consulted in connection to it, implying that the stone is indeed a footprint, but the truth has been concealed by researchers. Alongside all this, several remarkable statements were also made about the age of the find: "The find is the living image of the footprint of a child, but how do you know that it is ten thousand years old?" "By means of my profession [the finder worked for a drilling company], I am interested in geology and I studied paleontology. There is no specialist book in Hungary on this subject that I have not studied." That is the exhaustive explanation. However, the stone is much older than 10,000 years: it is of Eocene age. This was pointed out by László Kordos, who was also interviewed for the report (although his name was misspelled "Kardos" in the article). Thus, the find could not have been a human footprint; the calcareous mud of which it is composed was deposited several tens of millions of years before humans appeared. And what is it if not a footprint? Nothing else but a joke of nature, which manifested itself in a footprint-shaped piece of rock.

by Antal Koch, was long considered to be the oldest rhinoceroid (in a geological point of view). Later, however, the fossil was reexamined and mentioned by several paleontologists (among others Othenio Abel, Henry Fairfield Osborn [1857–1935], Miklós Kretzoi, and Costin Rădulescu [1932–2002], as well as Vlad Codrea) in their works.

🕮 Koch 1897; Codrea 2000

Sparse Remains from the Hungarian Eocene: Hyrachyus *and Other Animals*

In the area of present Hungary, only a few Eocene land vertebrate remains have been found. The majority of these, mostly incomplete tooth remains, belong to early groups of Rhinocerotids. The latest find, the fifth, was made in overlying Eocene beds in the Csordakút II bauxite mine and consists of three well-preserved molar teeth that belong to an upper jaw. These fossils were described by the secondary school teacher and geologist László Kocsis, who determined that they represent a close relative of *Hyrachyus stehlini.*

The remarkable bones of *Prohyracodon orientale* (0.3× magnification).

This find is older than the Transylvanian *Prohyracodon;* its geological age is Lutetian or Bartonian, which makes it the oldest rhino fossil ever found in the Carpathian region. The area—in which the Csordakút locality is situated—may have been an island in the Eocene sea, somewhere between Africa and the more European part of the Adria plate. This island—probably part of a larger archipelago—was connected with European areas from time to time and so land animals were able to migrate to and fro. The Csordakút remains show relationships with Eocene faunas of the northern (European) region, although according to our present knowledge, the African connection can be excluded. Besides the above finds, some tooth remains classified into the family Amynodontidae are also known (one of them has been collected from the brown-coal exploratory borehole [which was deepened at Nagysáp], from a depth of 509 meters; the locality of the other is unknown). A fragmentary Palaeotheriidae tooth of the lower jaw and some Lophiodontidae limb bone remains have also been found in the Transdanubian Range.

🕮 Kretzoi 1940; Kordos 1978a; Kocsis 2002a

An unidentified sea urchin from the Early Oligocene Kiscell Clay. Unlike the Eocene and Miocene Series, the Oligocene has no true emblematic fossil. By that time, larger foraminifers so characteristic of the Eocene disappeared or became much smaller as a consequence of the end-Eocene cooling event. On the other hand, the living world still had to wait more than four million years for the warm period called the Miocene Climatic Optimum, when the diversity of animals reached another peak in the Carpathian region. The specimen is about 7 cm across.

6

The Oligocene

The last epoch of the Paleogene, which spanned about 11 million years, is called the Oligocene. By this time, in the area of the present-day Carpathian Basin, the central basin of the Paratethys (already isolated from the rest of the world's oceans) and continental terrane rising up above the sea can be found. However, from a geological standpoint the basin is not uniform: dislocation of huge masses from the basement along fractures (transform faults), which already took place in the Eocene, continued into the Oligocene. By the end of the Oligocene, the Magura Ocean, which was of considerable size in the Eocene and occupied approximately the area of the present-day Carpathians, was completely subducted. The folded mountains of the Outer Carpathians were later formed from the compressed and folded rocks that once formed this ocean.

Across present-day Hungary, both continental formation and the deposition of shallow and deeper marine sediments took place at this time. However, because these processes took place within the Carpathian Basin—by this time isolated from the rest of the world's oceans—Oligocene sediments in this area cannot accurately be correlated with those from other places. Because of this and extensive denudation that also took place in the Oligocene, fixing the upper and lower boundaries of the Oligocene has led to numerous debates.

THE PARATETHYS

Due to convergent movements of Africa and Europe the Tethys Ocean, which covered a huge area during the Mesozoic, with its axis along the equator, was gradually consumed and closed like a pair of scissors during the Paleogene. Simultaneous to the uplift of the Alps sometime in the Early Oligocene, the basin system between the two continents became an open sea. This water body, which still covered a large area, is called the Paratethys and within it geologists distinguish three regions: Western, Central, and Eastern Paratethys. The Carpathian region belongs to the Central Paratethys, and the history of the Neogene in this area is closely intertwined within the history of this sea. Connections between these smaller basins and the rest of the world's ocean changed from time to time as sometimes seawater flowed freely and sometimes certain areas of the sea became more restricted. Isolation, such as that of the Paratethys, always resulted in a decrease in salinity and the evolution of specific, endemic faunas. After a few million years, the Central Paratethys became totally isolated and evolved into a freshwater sea that covered an ever-decreasing area. Rivers running into this basin from the Carpathian belt—which was, at the same time, experiencing uplift and denudation—eventually filled up the lake, which was finally all that remained of the former sea.

📖 Báldi 1980

FOSSIL-RICH FORMATIONS

The earliest phase of the Oligocene, taking into account areas where sedimentation took place at all, can be recognized in northern Hungary and in the Carpathians in basinal sediments that contain fossil fish—fish shales. These beds, which were deposited in an oxygen-depleted environment, are practically free of benthic fossils. In the second phase of the Oligocene, when the basin became well oxygenated again, the famous successions of the Kiscell Clay and the basinal schlier (widespread in northern Hungary) were deposited. Marginal facies of the latter are represented by coarse-grained and more or less cemented sediments that surround basinal deposits on their southeastern and northwestern sides along the Hungarian Range.

Fish Shales in the Carpathian Basin

Although fish remains occur in most marine successions in the Carpathian Basin, there is a stratigraphic horizon in the Lower Oligocene in which osteichthyans in particular are especially abundant. This horizon is the Tard Clay (formerly "fish shale") in Hungary, while across the Carpathians this sequence is called Menilite. These

Series/Epoch	Regional Stage/Age	Millions of Years Ago
Oligocene	Egerian (partly)	30
	Kiscellian	34

Subdivision of the Oligocene into stages, with their approximate geological ages.

fish-rich sediments were formed in the deep basin and were cut off from other oceans. As the water column stratified (water masses separated into layers of different salinity and density) exchange between the surface and the bottom stopped and an oxygen-poor region formed close to the seafloor. Because of this, few organisms could live close to the bottom and the environment of the Tard Clay comprised free-swimming animals that lived in the upper regions of the water column (such as fish). Dead animals sank to the bottom and, in the absence of scavengers and mud dwellers, were preserved intact. Overall, however, fossils are rare in these fish shales; it is just that fish are the most common of the infrequently collected fossils—although the Tard Clay does contain plant fossils in large numbers. These are described in the section that describes Oligocene fossil plants.

Kiscell Clay and Hárshegy Sandstone

In the Early–Late Oligocene a new connection opened up between the enclosed sea basins of the Carpathians and the world's ocean. As a result, formation of fish shales was followed by deposition of new fine sand and clay sediments that are usually gray in color, the so-called Kiscell Clay. Because of this renewed oceanic connection, bottom anoxic conditions ended at this time and the deep sea was again inhabited by different groups of benthos. One reason the Kiscell Clay is well known is because is it suitable for brick making; from the clay pits dug for this purpose, paleontologists have collected often well-preserved fossils for more than 100 years. Most common from this unit are gastropods and bivalves, although nautiloids, echinoids, fish, and other fossils can also be found. Tamás Báldi was able to identify 169 species of mollusks; this alone suggests a diverse and variable fauna. Indeed, the microfauna of the Kiscell Clay is itself especially abundant and was first studied by Miksa Hantken. To date, more than 400 species can be distinguished from the foraminifera alone.

The so-called Hárshegy Sandstone, a near-shore facies, is almost coeval with the Kiscell Clay. This sandstone, made up of interbedded coarse-grained and fine-grained layers, contains few fossils because their remains were disintegrated by the grinding action of pebbles moved by waves in the surf zone.

Vöröskővár Hill at the Hármashatár-hegy glider port is composed of Hárshegy Sandstone, and this rock can also be studied in abandoned quarries on the Ezüst Hill, near Üröm. The fossil-bearing facies can also be seen at Várerdő Hill, near Solymár—which is also a recorded Eocene locality.

In northern Hungary, a gradual transition is seen from the Kiscell Clay into a characteristic sediment called schlier, which covers large areas. This rock type is also called "apoka," from the local "Palóc" word (the Palóc are an ethnic group with a particular folk culture). Schlier is a light-colored and well-bedded clayey silt which was deposited at more than 200-meter depths in marine basins. The most frequent fossil found in this usually fossil poor rock is the small, thin-shelled bivalve *Lentipecten* (*Amusium*), which lived in the calm bottom waters of the deep sea.

📖 Báldi et al. 1976; Báldi 1986

Mollusk-Rich Clays and Sands

The traditional name for the sediments that are found in the area of the Pilis Mountains, Esztergom, the Buda Hills, and in northern Hungary—and which are restricted to the late Oligocene Egerian stage—is based on the fossils they contain. This succession was often referred to in the past as the mollusk-bearing clay, and it is famous in professional circles. The most remarkable outcrops of this rock type are in the vicinity of Eger, including the type succession of the so-called Eger Formation and the regional stage "Egerian" of the Paratethys. (See also the description of the Wind Brick Factory clay pit, below.) These sediments, "*Pectunculus* (or *Glycymeris*) beds" and "*Cyrena* beds" in older literature, are nowadays known as the Törökbálint Sand Formation and the Mány Sand Formation. On the one hand, the *Pectunculus* sand is definitely marine in origin—probably a shelf facies; on the other hand, the *Cyrena* beds are brackish-water sediments deposited under fluctuating salinity conditions.

📖 Báldi 1963

REMARKABLE FOSSIL SITES

The most famous fossil localities from the Hungarian Oligocene are in the vicinities of Budapest and Eger. The Kiscell district in Óbuda is the type area for the Kiscellian stage (used in the area of the Central Paratethys), whereas the succession at the Wind Brick Factory in Eger is the type locality of the Egerian stage. Sadly, many classic exposures of the Kiscell Clay in the Budapest area cannot be studied anymore because clay pits have either been closed or built on.

The sand pit between Úny and Máriahalom is a rich fossil locality for mollusks, especially gastropods, and has been known since the beginning of the 1970s.

The Wind Brick Factory and Other Localities in the Vicinity of Eger

The former Wind Brick Factory boasts a remarkable exposure and is well known in Hungarian geological and paleontological circles. This easily accessible clay pit is located at the southeastern margin of Eger, and here fossil-rich beds of the Eger Formation can be seen as part of a continuous transition from the underlying Kiscell Clay. This site is also the type section of the Egerian stage in the Central Paratethys, and beds exposed here were deposited in shallow-bathyal and lagoonal environments. As a result, the Eger succession also preserves a diverse fauna and has been studied in detail. One of the beds, marked with a K by Ferenc Legányi and Károly Telegdi-Roth (1886–1955), for example, is especially remarkable because at one time its fossil contents were considered representative of the entire Egerian fauna. Revisionary work on this famous mollusk fauna was done by Tamás Báldi.

On many gastropods and bivalves, as well as on other invertebrate fossils, traces of the biological activities of carnivores, parasites, and others who used these calcareous shells as places to live can be seen. An investigation into this rich trace fossil assemblage was started by Árpád Dávid, a staff member at the Károly Eszterházy Teacher-Training College. Over the last two decades, Dávid and students have broadened our knowledge of this assemblage with a number of valuable publications within this theme.

Near Eger, in the direction of Kiseged, outcrops of both the fish shale and the Tard Clay can be found near the main road to Noszvaj. In this area, however, fossil collecting requires caution because of the curves of the road. In this shaly rock, bedding planes often reveal fish scales and sometimes even fish impressions; underlying layers contain the fossil nannoplankton species *Reticulofenestra* and *Orthozygus*. Remains of the endemic bivalve *Cardium lipoldi* can also be collected from this section. Nannoplankton research was first carried out by András Nagymarosy; the fish fauna of the area was first studied by Wilheim Weiler (1890–1972), and the fossil plants have been worked on by Gábor Andreánszky, Lilla Hably, and others (see sections that discuss fossil vertebrates and plant remains).

📖 Telegdi-Roth 1914; Weiler 1933, 1938; Báldi 1966, 1973; Hably 1979, 1993; Dávid 1993, 2009; Kvaček and Hably 1991

Fossil Localities in the Vicinity of Budapest

Near Budapest both the lower and upper boundaries of the Oligocene can be seen in successions that are close to one another. Indeed, the answer to debates over the boundaries, a question of great interest both within Hungary and internationally, was provided by Tamás Báldi and Mária Báldiné Beke based on work done close to Budapest. From the nannoplankton research of Mária Báldiné Beke we know that most of the Buda Marl was deposited in the Late Eocene, but that the top of this unit is Early Oligocene in age. Therefore, the base of the Lower Oligocene can be found in the classic Szép Valley area in Óbuda. The top of the Upper Oligocene was identified and studied by Tamás Báldi at the Pacsirta Hill in Budafok. Indeed, the name of the Kiscell Clay is taken from Kiscell in Óbuda. Although most clay pits have been filled in, or built on, some outcrops, including the clay pit at the Rozália Brick Factory in Solymár, do still exist.

There is another small sand pit near the road from Úny to Máriahalom. Some years ago beautiful Upper Oligocene mollusks were collected in large numbers from the wall of this pit. Here the most frequent fossil mollusks are shells of *Tympanotonos margaritaceus* that have lost their luster over millions of years and become fragile. Along with bivalves and gastropods, fossil vertebrate remains

The Kajmát Quarry at Bodajk. A rich Oligocene vertebrate fauna is preserved here in fissures in Triassic dolomite.

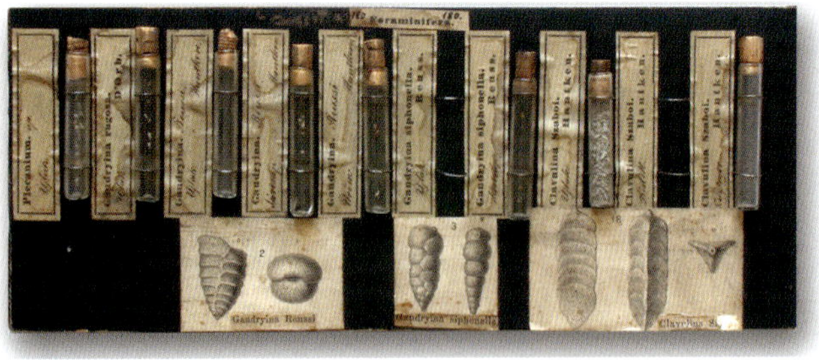

An old preparation of foraminifera from the Kiscell Clay that was made for educational purposes.

are also found—but much less frequently. In this collection—gradually enriched over decades—fossil sharks, rays, osteichthyan fish, turtles, lizards, crocodiles, birds, and marine and land mammals are represented, and it is interesting that the assemblage consists of remains that accumulated in a special environment, close to a tidal channel. Another significant feature of this locality is that before it was found large Egerian land mammals were almost unknown in Hungary.

📖 A. W. Janssen 1984; Botfalvai and Rabi 2007

Localities in Cluj and in the Vicinity of the Town

The beds that are exposed at Cluj, within the town on the side of the Someșul Mic River, have been known to paleontologists for many years because they contain Oligocene fossil vertebrates. This diverse fauna is briefly described in the section that deals with Oligocene fossils. The wider surroundings of Cluj (Mera, Aghireşu, and Gilău) have also yielded valuable Oligocene fossil vertebrates.

OLIGOCENE FOSSILS FROM THE CARPATHIAN REGION

It might seem strange, but the best-studied Oligocene fossils from the Carpathian Basin are probably the considerably variable foraminifera from the Kiscell Clay. These microfossils have been studied in detail ever since the beginnings of Hungarian paleontological investigations. From several successions of this age, a rich macrofauna (especially mollusks) is also known, but this reflects the diligence of collectors and researchers rather than the abundance of fossils in sediments of this age. Of the vertebrate fauna, Oligocene fish are most common; land mammal remains are extremely rare, and almost every fossil discovered has been cataloged and listed.

Microfossils

Because the Oligocene successions of the Carpathian Basin were deposited in the Paratethys Ocean, mollusks that are traditionally used for stratigraphic correlation and age determination are not suitable for international correlation based on Tethys faunas. However, one breakthrough in this area was made in nannoplankton stratigraphy, which uses the tiny fossils of these calcareous heterokonts. One of the best-known scientists to use this method is Mária Báldiné Beke, and as a result of her investigations, the Oligocene successions of the Carpathian Basin can now be correlated into zones marked with the letters "NP" (for Nannoplankton—Paleogene) and serial numbers.

Foraminifera of the Kiscell Clay

The foraminifera assemblage of the Kiscell Clay is the best known among the Oliogocene faunas of the Carpathian Basin. The rich and well-preserved fauna in bluish-gray clay and silt beds became known internationally thanks to Miksa Hantken's 1875 monograph. Later researchers—including László Majzon, Károly Sztrákos, Ágnes Nagyné Gellai, and Mária Horváth—worked in this area with the aim of finding subdivisions in this thick and monotonous succession. To date, all efforts to do this have failed—and are likely to continue to do so, as the fauna comprises mostly benthic elements. One such element is *Clavulina* (present name *Tritaxia*), which Hantken considered most characteristic to this fauna; he discusses it in his classic work "Die Fauna der Clavulina Szabói-Schichten" (The fauna of the Clavulina Szabói Beds). *Tritaxia* has a very slow evolutionary rate and its coverage in terms of area depends on local ecological conditions. Overall, however, it is planktonic elements that are more useful from a stratigraphic standpoint and are also more frequent. The major taxa that make up

Nowadays fossil foraminifera and ostracods are usually stored in so-called FRANKE cells. Micropaleontologists are lucky because there is no shortage of space for their fossils: there is room for almost their entire life's work in a cabinet!

Paleobotanists in the Department of Botany of the Hungarian Natural History Museum. Lilla Hably, the director (center), conducts research on Tertiary floras and the focus of her work is the description and comprehensive analysis of the Oligocene, Miocene, and Pliocene floras of Hungary. Based on work with thousands of fossils from the Mecsek Coal Formation, Mária Bókáné Barbacka (left) has contributed significant details to our knowledge of Early Jurassic plant life. Boglárka Erdei (right) has research interests in the Tertiary floras of Hungary, with a focus on climate reconstructions.

planktonic foraminiferal zones established on the basis of Tethyan successions (deposited under warmer climatic conditions) have not been found in the cold-water Kiscell Clay.

For the most part, the biostratigraphic zonation based on forams, which is used across Northwestern Europe, is suitable for the classification in Hungary. Although these assemblages are used primarily for stratigraphic classification, they are still reasonably accurate environmental indicators. Primarily, these elements play an important role in determining and proving the deep-water origin of the succession, especially based on the vertical distribution of extant forms, the ratio of planktonic and benthic representatives, and the relative abundance of hyaline and agglutinated taxa. Foraminifera indicate that the Kiscell Clay was deposited at a depth of several hundred—or even thousands—of meters, and although other Oligocene sediments are also rich in forams, these are not considered classical and are not well documented.

📖 Hantken 1868, 1881

Ostracods

A number of Oligocene rocks contain ostracod fossils and these are valuable, especially as facies indicators. The best-studied and -documented ostracod fauna is Lower Oligocene in age and was described in 2004 by Miklós Monostori, who noted the presence of 80 species. Of interest from a paleogeographic point of view, Oligocene faunas differ from their Eocene counterparts; the former are more similar to those in Italy, France, and Great Britain. Oligocene faunas more closely resemble those from Germany.

📖 Monostori 2004

A Paleontologist in Action: Let's Wash!

One of the classic methods for obtaining and studying microscopic fossils is washing sediment and then examining of the washed residue. This method can be used for all sediments, such as clay and marl, that fall apart after soaking. For example, sediments of the Kiscell Clay can be efficiently washed, and so the microfauna of this formation has been studied using this approach.

To do this, one kilogram of rock is crumbled into a bowl or smaller basin. The sediment is then thoroughly soaked with water and left to stand. Cloudy water is drained off and the remaining sediment repeatedly hand-crumbled into the water. This process should be repeated four or five times—the same method is used by gold prospectors—until the light clay and sand grains are poured out in the water—and the heavier mineral grains, fossil skeletons, tiny gastropods, fragments of corals and other colonial animals, fish teeth, and calcareous protozoans are left in the bowl.

The valuable fossils can then be sorted from the washed residue using a microscope. Rocks that have a darker color (and thus are rich in organic matter) can be much more easily soaked if a few drops of hydrogen peroxide (H_2O_2) are added to the material as it is crumbled and soaked. The organic matter will be oxidized and destroyed. (Be careful! Hydrogen peroxide is a dangerous chemical that requires caution when working with it!) An upgraded version of the abovementioned classical washing method is another process whereby the soaked material is washed through a series of metal sieves that are superimposed over each other. The advantage of this method is that one can wash a larger quantity of sediments at the same time, within a relatively short time, and the washed residue can be sorted according to the different (increasingly smaller) mesh sizes of the sieves. In this way, it is easy to sort the smaller and larger embryonic shells of gastropods and bivalves from the usually smaller fossil foraminifera.

Nannoplankton

In the Carpathian Basin there are two important stratigraphic horizons that are very rich in nannoplankton. The first of these so-called nannoplankton enrichments is found at the Jurassic-Cretaceous boundary, and resulted in massive formations of limestone called maiolica (see the discussion of microfossils in chapter 4). The second of these enrichments occurred at the time of deposition of the Tard Clay and certain beds—for example, those that make up hillsides at Kiseged and which resemble limestone—are almost entirely composed of nannoplankton. Work on nannoplankton in Hungary was initiated by Mária Báldiné Beke, followed by András Nagymarosy, who later became the head of the Department of Geology at Eötvös Loránd University. Nagymarosy was also a bass player in a folk band, a musicologist, and a oenophile as well as a nannoplankton specialist.

In the Carpathian Basin, nannoplankton occurs locally in rock-forming quantities in a rock called menilite, which is analogous to the Tard Clay.

📖 Báldiné Beke 1977; Nagymarosy 1985

Fossil Plants

Numerous fossil plant assemblages characterized by high diversity have come to light in the Oligocene sediments of the Carpathian Basin. Of these, the Early and Late Oligocene floras are clearly distinct from one another: the floras of the first half of this epoch indicate a warm climate, but investigations on younger assemblages suggest

Early Oligocene (Kiscellian) fossil plants from the Tard Clay, excavated from a former clay pit in the Budapest Óbuda–Nagybátony-Újlak Brickyard (original size). (1) *Osmunda* sp.—leaflet of a fern closely related to extant royal ferns. (2) *Ceratozamia floersheimensis* (Engelhardt) Kvaček. This parallel-veined leaflet is assigned to a group of cycads native today to Mexico, Belize, and Guatemala. (3) *Zizyphus zizyphoides* (Unger) Weyland. This leaf belongs to a plant in the buckthorn family. *Ziziphus* is a genus comprising numerous species of mainly shrubs, lianas, and trees. One of its well-known species is the jujube tree, also known as Chinese date. The plant commonly known as Jerusalem thorn or Christ's thorn is also among its close relatives. (4) *Engelhardia orsbergensis* (Wessel and Weber) Jähnichen, Mai, and Walther—leaf of a plant in the walnut family that was a frequent and characteristic element of the thermophilic paleofloras of Hungary. Its present-day relatives are native to Southeast Asia. (5) *Quercus lonchtis* Unger. This leaf is assigned to oaks. (6) *Pteris budensis* Andreánszky—the pinnate leaf of an extinct fern that is related to the recent *Pteridium*. The figured specimen is the holotype of the species. *Pteris* is a genus comprising numerous species native to the tropics and subtropics. (7) *Sloanea eocenica* (Rásky) Z. Kvaček, Hably, and Manchester—the fruit of a plant in the Elaeocarpaceae family native to the tropics. (8) *Engelhardia macroptera* (Brongniart) Unger—a winged fruit related to the walnut family. (9) *Raskya vetusta* (Ettingshausen) Manchester and Hably—This winged, propeller-like fruit belongs to an extinct plant and is of uncertain systematic affinities.

1
2
3
4
5
6
7
8
9

cooler conditions. A novel approach, combining evidence from plate tectonics and paleobotany, has led to the idea that the climatic cooling indicated by fossil plants is not due to abrupt global climatic shifts but, at least in part, is due instead to the fact that the microplate involving the Pannonian region was displaced northward into a colder climatic belt during the Oligocene.

📖 Hably & Kázmér 1996

Early Oligocene Flora of the Tard Clay

The flora that flourished in the Early Oligocene is best documented in the fine-grained shaly rocks of the Tard Clay. Persistent collecting and research activities have resulted in a huge collection comprising nearly 10,000 megafossil specimens—mostly leaves, but also occasional flowers and fruits. The Kiseged Hill near Eger and the Vécsey Valley are well-known localities, but leaf remains were also collected during historical excavations of similar sediments in Óbuda. Indeed, one of the most abundant assemblages known was found during the construction of the metro station in Batthyány Square (Budapest); in Transylvania, Mera and Ileanda are noteworthy localities. Forests at this time were composed of both deciduous and evergreen species—and in addition to leaves, some interesting fruit types (*Raskya, Tetrapterys*) possessing propeller-like wings have also come to light.

📖 Hably 2006

Late Oligocene Flora

One of the most diverse Late Oligocene assemblages is documented in the clay quarry of the Wind Brick Factory in Eger. Other remarkable occurrences are known at Andornaktálya, Csörög, Pomáz, and Kesztölc, as well as in the vicinity of Tatabánya (Baromállás).

The Late Oligocene flora is dominated by thermophilic elements. For example, an ancient species of plane tree, *Platanus neptuni,* has been found frequently and the leaves of *Daphnogene,* a laurel, occur in great numbers. The occurrence of elm leaves at this time is also significant; they provide the first indication of accompanying deciduous elements. Scientific study of these Hungarian Oligocene floras was pioneered by Gábor Andreánszky, former professor of paleobotany, and by Lilla Hably, at the Hungarian Natural History Museum. The richest Paleogene assemblages of fossil plants in the Transylvanian Basin are found near Cluj-Napoca and in the vicinity of Cornesti and Aghireşu. These fossils—including ferns, gymnosperms, and angiosperms—have been recovered from levels in the kaolin that is excavated in these regions. Work on them by paleobotanists at the University of Cluj-Napoca is in progress.

📖 Hably 1980, 1989, 1990, 2006

Invertebrates

Corals

Fossil corals occur in almost every Oligocene succession that was deposited in normal-salinity seawater, and were the subject of a 1962 monograph by Gyula Hegedűs. Of note, the Kiscell Clay has yielded the fan-shaped, solitary coral genus *Flabellum,* which is characteristic of deepwater environments. Shallow-marine sediments such as the Eger Formation generally contain many more fossil corals. From these sediments, among others, come the genera *Ceratotrochus, Odonthocyathus,* and *Caryophyllia.* According to work done by Rozália Fodor, who was a teacher and geology student, about half of the known Oligocene corals show some evidence of bioerosion, in particular the burrow traces from polychaete worms.

📖 Hegedűs 1962; Fodor 2001

Mollusks

Fossil gastropods and bivalves are the most abundant remains in most Carpathian Basin Oligocene successions, and often occur in large numbers. However, these faunas are quite different from those known from the Eocene—as even shallow marine sediments tend to lack large, thick-shelled forms. This is thought to be due to deterioration in climactic conditions and a decrease in water temperature. In the second half of the nineteenth century the fossil-rich sediments of the Upper Oligocene caught the attention of a number of paleontologists, including Miksa Hantken, Antal Koch, and Károly Hofmann. The names "Pectunculus obovatus sand" and "Cyrena beds" date from this time, and monographic description of the faunas was begun by Károly Telegdi-Roth, who described the Eger assemblage in 1914. Telegdi-Roth showed that characteristically Egerian, well-preserved mollusk remains are

Early Oligocene (Kiscellian) fossil plants from the Tard Clay Formation outcropped in the Kiseged hillside near Eger (original size). (1) *Eotrigonobalanus furcinervis* (Rossmassläler) Kvaček and Walther—leaf of an extinct genus in the beech family. This genus is referred to as *Dryophyllum* in earlier literature. (2) *Platanus neptuni* (Ettingshausen) Bůžek, Holý, and Kvaček—leaf of an extinct, ancient plane tree. This leaf shape is different from most of the plane or sycamore trees that are well known in modern-day parks, as the only similar living species is native to Southeast Asia. Indeed, the occurrence of this species in the Oligocene is indicative of a warmer climate than today. (3) *Callicoma egedensis* Andreánszky—impression of a leaf. This figured specimen is the holotype of this species. (4) *Sassafras tenuilobatum* Andreánszky—leaf of a genus in the laurel family. This figured specimen is the holotype of this species. (5) *Ceratopetalum articulatum* Andreánszky—impression of a leaf. This figured specimen is the holotype of this species.

1
2
3
4
5

1
2
3
4
5
6
7
8
9
10
M 60 727
11
12
13

found predominantly in a 2-meter-thick loose sand bed in the succession at the Wind Brick Factory in Eger, which has since been considered a classic locality.

Another rich fauna—the assemblage in the Kiscell Clay—was studied by Jenő Noszky, Sr. (1880–1951), director of the Department of Paleontology and Geology of the Hungarian Museum of Natural History, in papers published in 1939 and 1940. Documentation of this fauna can be attributed to a series of favorable circumstances and to the conscientious work of István Harmat, a mine inspector who worked at the Újlak Brick Factory and who had been collecting remains from the mollusk-poor deepwater sediments for decades. Harmat carefully prepared the fossils that he found and eventually donated his valuable collection to the National Museum in Budapest. Preparation work on these fossils deserves at least the same respect and appreciation as initial collecting work; the macrofossils in this clay, in contrast to the microfossils, are usually deformed and poorly preserved.

These Oligocene fossil remains were also examined by other researchers, including László Bogsch and Ilona Csepreghyné Meznerics. One research goal has been the determination of the age of these successions as either Oligocene or Miocene and their integration into the international chronostratigraphic framework, but this proved impossible on the basis of the fossil mollusks. As a result, the unit "Egerian stage" was named for this sequence in the area of the Central Paratethys.

There are detailed and up-to-date studies that deal with Hungarian Oligocene fossil mollusks—including the work of Tamás Báldi, professor in the Department of Physical and Historical Geology of Eötvös Loránd University. In addition to several shorter papers, Báldi published two monographs, one in 1973 that deals with the Egerian stage and another in 1986 that deals with the fauna of the Kiscell Clay. A taxonomic revision of these mollusks was also needed because Noszky, for example, had earlier distinguished no less than 764 forms in the Újlak fauna. According to Báldi (and without detracting from the scope of Noszky's work) the real numbers are as follows: 66 gastropods, 98 bivalves, 1 scaphopod, and at least 4 nautiloids. A similarly rich Upper Oligocene fossil assemblage from Slovakia that includes the fauna of the "*Pectunculus*-bearing" Kováčov Sand was described by Ján Seneš (1924–1992), a professor from Bratislava.

📖 Telegdi-Roth 1914; Seneš 1958; Báldi 1973, 1986

Gastropods

The fossil remains of gastropods are found predominantly in shallow-marine sediments like the Mány Formation, which was deposited in a marine environment of variable salinity. As a result it comprises rocks of variable grain sizes—from clay to pebbles, as well as strips of coal. Certain beds in the formation contain the high-spired shells of *Tympanotonos margaritaceus* in large numbers; occasionally these shells can reach a length of 6–7 centimeters. *Tympanotonos* (although redeposited), is common, for example, in the small sand pit between Úny and Máriahalom, along with the slender *Pirenella plicata, Turritella,* and *Ampullina.* Freshwater gastropods, including *Viviparus* and *Brotia,* that were buried at the bottoms of rivers and lakes also occur in these sediments.

Gastropods are also characteristic and common in the well-preserved Egerian fauna, including taxa like *Turritella, Pirenella, Drepanocheilus, Ficus, Athleta, Turris,* and *Turricula.* The deepwater assemblage of the Kiscell Clay also contains gastropods—*Aporrhais, Tibia, Cassidaria, Voluthilites,* and *Turricula* being the most frequent.

Gastropods and bivalves from the former exposures of the Kiscell Clay at Óbuda (all original size). (1, 2) *Gibbolucina (Eomiltha) rectangulata* (Hofmann). This shell, with a rounded-rectangular outline, is a frequently occurring fossil bivalve found in the Kiscell Clay. It was first described by Károly Hofmann. (3) *Angaria scobina apenninica* Sacco—a gastropod ornamented with spiral ribs that have knobs along them. (4) *Lithoconus ineditus* Michelotti. This relatively large form is an extinct, deep-sea relative of cone shells that inhabit predominantly shallow-marine environments. (5) *Turritella conofasciata* Sacco. Given the small number of specimens and the habitat of its extant relatives, it is most likely that the high-spired shells of these gastropods would have been swept from the shallow environments that encircled the Kiscell basin into deeper water by turbidity currents. (6) *Streptochetus elongatus* (Nyst). This gastropod has a strongly ribbed shell, belongs to the fossil fauna of the Kiscell Clay, and is characteristic to the Oligocene in the northern regions of Europe (boreal). (7) *Lyria decora* Beyrich. This high-spired form is a rare element in the Óbuda fauna. (8) *Perotrochus budensis* (Hofmann)—a large sea snail belonging to the Pleurotomaridae and considered a living fossil among gastropods. The shell of this taxon has a characteristic slit. (9) *Pecchiola argentea* Mariti. This genus is a characteristic element of deep-sea bivalve assemblages. (10) *Aporrhais mathiasi* Noszky. Similar to its extant relatives, this species—characterized by remarkable elongate projections around its aperture—may have been a semi-infaunal form: it lived partially buried in the sediment, in a horizontal position. (11) *Atrina* sp. Pen shells—including the specimen in the picture—live predominantly in the shallow marine zone, occasionally inhabiting deeper waters. (12) *Arctica islandica rotundata* (Braun in Agassiz). The recent species *A. islandica* is a large-sized, long-lived inhabitant of northern seas with a lifespan that often exceeds 100 years. (13) *Pelecyora* sp.

1
2
3
4
5
6
7
8
9
10
11
12
13
14
15
16
17

Bivalves

Bivalves are also among the most common elements in Carpathian Basin Oligocene fossil assemblages. Indeed, the old name for the Törökbálint Sandstone Formation comes from the characteristic large bivalves of the genus *"Pectunculus,"* now placed in the species *Glycymeris latiradiata*. The assemblage of the Mány Formation, which was deposited along the boundaries of marine and continental environments, is also characterized by bivalves, although these are less spectacular forms. This formation is locally rich in the shells of *Polymesoda* ("Cyrena beds") and this species is often found alongside the dark-gray and well-preserved shells of a small-sized oyster.

The shallow-marine but slightly deeper water Eger Formation contains frequent fossils of the non-ornamented, smooth-shelled *Pitar polytropa* and the beautiful, finely costate *Laevicardium*. The deep-marine Kiscell Clay is dominated by forms that inhabit similar environments today: *Nuculana, Cuspidaria, Pseudamussium,* and *Limopsis*. Bivalves are also common in the fauna of another Upper Oligocene deepwater succession, the Szécsény Schlier, where *Nucula, Yoldia, Lentipecten,* and *Corbula* frequently occur. However, because of their poor preservation these fossils were not studied by researchers for many years.

Costellamussiopecten pasinii (Meneghini)—A scallop shell from Dejtár (0.8× magnification).

Upper Oligocene (Egerian) mollusk remains (all original size). (1) *Tympanotonos margaritaceus* (Brocchi). This gastropod—ornamented with spirally arranged knobs that resemble a string of pearls—inhabited variably brackish Egerian lagoons in large numbers. Its species name also refers to its characteristic appearance (margaritaceus means "pearly") (Úny). (2, 3) *Athleta rarispina* (Lamarck). This common gastropod is characterized by a wide inner lip and few spines (rarispina) ornamenting its body whorl (Eger). (4) *Melanopsis impressa hantkeni* Hofmann. *Melanopsis* has inhabited freshwater environments since the Cretaceous. Nevertheless, the shell of this form can also be found in brackish-water marine sediments, such as the form shown in the picture (Úny). (5) *Turritella venus* d'Orbigny. This beautiful species of *Turritella* (worthy of its species name) is ornamented with several spirally arranged knobs. Its remains are frequent in the Eger fauna (Eger). (6) *Ficus condita* (Brongniart). The square, network-like pattern of the shell formed by the narrow spiral and axial protuberances gives a peculiar appearance to this relatively small gastropod (Eger). (7) *Nerita plutonis* Basterot. The shell grew rapidly during ontogeny. Due to the thick shell, the remains are usually well preserved (Úny). (8) *Turricula regularis* (Koninck). The beautiful gastropod shell is characterized by a long siphonal canal and a slit in the aperture margin, as well as the V-shaped growth lines on the body whorl, which indicate the former position of the slitted aperture (Eger). (9) *Typhis pungens* (Solander in Brandner). The shell—which is made up of about six whorls—is ornamented with spines in three rows. The upper spines—which are also seen on the older whorls—are cavernous and perforated on their ends (Eger). (10) *Ostrea cyathula* Lamarck. Oyster remains are especially frequent in the locality in Úny. Their dark, bluish-gray shells strongly differ from the white shells of other bivalves. Occasionally they are found cemented to gastropod shells (Úny). (11) *Lutraria oblonga* (Chemnitz). This thin-shelled bivalve that lived burrowed in the substrate occurs predominantly together with *Pitar polytropa* (Eger). (12) *Pitar polytropa* Anderson. This form is one of the most frequent bivalves of the Eger fauna (Eger). (13) *Polymesoda convexa* (Brongniart). Beds deposited in the Upper Oligocene lagoons of low or strongly changing salinity—and currently classified into the Mány Formation—were formerly distinguished as "*Cyrena* beds" after the former name of this frequent fossil bivalve (Úny). (14) *Arctica islandica rotundata* (Braun in Agassiz)—the largest form of the one-time bivalve assemblage, with a shell length of up to 9 centimeters. *A. islandica* (Linné)—in the strict sense—is an extant form; it inhabits the seas which encircle Europe on the North (Eger). (15) *Glycymeris latiradiata* (Sandberger in Gümbel). This thick-shelled bivalve that lived in shallow depths in the bottom, under the transitional zone of substrate and water, is the index fossil of a shallow-marine succession formerly known as "Pectunculus sand." In older, mainly nineteenth-century geological literature it was mentioned as *Pectunculus obovatus* Lamarck (Törökbálint). (16, 17) *Laevicardium tenuisulcatum* (Nyst). Worthy of the name, the ribs ornamenting the shell of this nice heart-shaped shell are divided by narrow grooves (Eger).

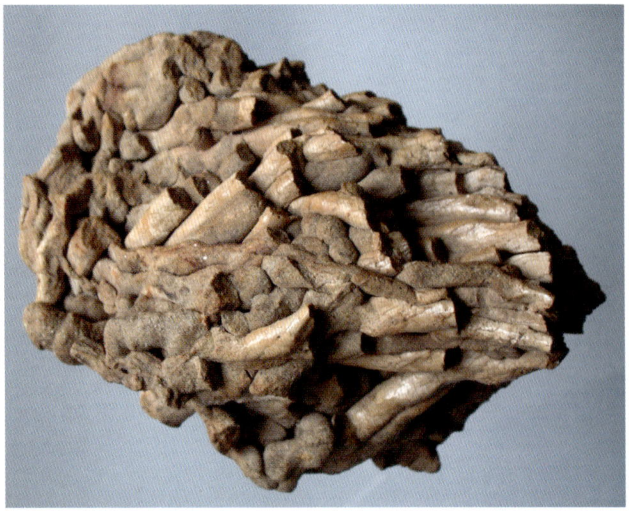

This photo shows petrified infillings of boring bivalve burrows (*Teredo*) from the sand pit at Úny. The predecessors of recent ship-worms bored into logs as they floated in the sea (0.5× magnification).

Sepia harmati—described by Erzsébet Szörényi from Oligocene sediments at Óbuda—is named in honor of István Harmat, the tireless fossil collector who worked at the Nagybátony-Újlak Brick Factory (original size).

Bivalves are almost the only invertebrate fossils to be found in one assemblage in the lowermost Oligocene, the Tard Clay Formation. These sediments were deposited in anoxic water and comprise beds characterized by three small forms—*Janschinella melitopolitana, Cardium lipoldi* and *Ergenica cimlanica*—which are of insignificant appearance but which occur in large numbers. However, this faunule is of great importance because it can be traced over huge distances, from Switzerland all the way to the Aral Sea. This is the oldest endemic assemblage known from the Paratethys, and it occurs exclusively in these beds. Poorly preserved bivalves are also characteristic fossils in another usually fossil poor rock, the shallow-marine Hárshegy Sandstone. In this assemblage scallops are the most common elements.

Cephalopods

In addition to gastropods and bivalves, fossil cephalopods are also found in the Oligocene beds of the Carpathian Basin. *Nautilus* remains, which have yet to be examined in detail, occur in relatively large numbers alongside other squid fossils and one interesting loliginid squid, unknown elsewhere. According to András Galácz, an expert on Jurassic ammonites and who is also familiar with Cenozoic cephalopods, the Oligocene coleoid assemblage of the Carpathian Basin is the richest fossil fauna known to date from anywhere in the world.

📖 Wagner 1938; Kretzoi 1942a

Crustaceans

The remains of the decapod crustacean *Ctenocheles* are frequent in the Kiscell Clay Formation, even though in recent faunas this genus lives in the deep sea. Less frequently encountered are plates of the barnacle *Scalpellum:* certainly, these crustaceans lived attached to free-swimming organisms and not in the deep basin.

📖 Lőrenthey & Beurlen 1929

Vertebrates

The most frequent fossils in the marine Oligocene beds of the Carpathian Basin are fish remains. In several sediments, shark teeth and well-preserved skeletons of bony fish can be collected in large numbers. The remains of land vertebrates are rare in Oligocene rocks, as they are in the Eocene.

Marine Vertebrates

Well-Preserved Fossils of Bony Fish

Historically, the fossils that are the reason for the naming of the fish shales in the Carpathian Basin, especially the Kiscell Clay Formation, were studied in detail by paleontologists. The first data about bony fish from menilites sediments in the Eastern Carpathians were published in 1849 by Johann Jakob Heckel (1790–1857) and a complete review of the fauna was done by M. Paucă. To the south of the Kiscell Clay, a similar kind of rock, but black shale, that is exposed close to Covasna and Comandău has also yielded fossil fish. These remains were described by Boleslaw Böhm (1900–1964), and the most beautiful collection of Oligocene fish remains from the Carpathian Basin can be seen in the museum at Piatra Neamţ.

Wilhelm Weiler (1890–1972) was among the first to publish a detailed paleontological treatment of the fish remains from the "silicified shaly marl at Eger" (the Tard Clay) and

A fossil crustacean (*Thaumastocheles*) from the Kiscell Clay at Óbuda. The extant representatives of this genus can be classified into two species: one of them lives in deep-sea habitats in the Indo-Pacific Sea, whereas the other lives in the Atlantic Ocean. A peculiar feature of *Thaumastocheles* is that its right claw is much larger than its left one.

from the Kiscell Clay. In his 1933 paper, Weiler described several species of fish that were already known from elsewhere in the Carpathian Basin, mentioning and illustrating the following taxa: *Alosa, Clupea* (which belongs to the group that includes herring), *Serranus* (white grouper), *Seriola, Scomber* (a relative of extant mackerels), *Aulostoma, Belone, Nemopteryx,* and *Ammodytes* (sand lance). He also noted that the Eger fossil fish fauna comprised mostly young specimens, with *Serranus simionescui* the most frequently encountered taxon. This fish was also described by Paucă.

An overview of the Eger fish fauna would, however, not be complete without mentioning the fossil ray egg capsule from this site that was first illustrated by András Tasnádi-Kubacska.

Weiler also described the remains of a beautiful fauna from the Kiscell Clay. Besides several types of sharks, he determined fossils of seven different bony fish (including, among others, herring, mackerel, and barracuda). Given what we know about the ecologies of the extant relatives of these fossil fish, it is reasonable to conclude that this fauna lived in a subtropical climate. Weiler noted that among the fossil fish from Eger there were forms that could tolerate brackish-water conditions, while his work on the fauna of the Kiscell Clay definitely indicated a more marine environment, without freshwater forms.

📖 Weiler 1933, 1938; Paucă 1934; Böhm 1941, 1942

The Shark from Csillaghegy

As discussed in connection with Eocene fossil chondrichthyans, teeth are usually the only parts of sharks and rays that are preserved as fossils. These elements are also usually found apart from one another, as they have fallen out of the jaw. Complete cartilaginous fish fossil remains are rare globally; this is why one Carpathian Basin fossil specimen—found by Ottó Csomor, an amateur fossil collector in the clay pit of the Csillaghegy Brick Factory—is so remarkable. The find included a remarkable 106 shark teeth and 37 vertebrae lying next to each other—probably from a single animal because some pieces of rock, found close to one another, also contained the deformed jaw. Some vertebrae were arranged in rows in the clay, overlapping half or two-thirds of one another, like dominoes, and on closer examination small particles around 1 millimeter

Predominantly coarse-clastic Oligocene successions are poor in fossil echinoids: The brittle shells of these invertebrates, which are otherwise frequent in the Tertiary, are well preserved only in fine-grained sediments deposited in calm environments like the Kiscell Clay. The rock in this photo containing several specimens is also derived from the Kiscell Clay (almost original size).

in diameter are seen to occur in large numbers. These particles are arranged in mosaic patterns, overlapping like fish scales, and are ornamented with slight central prominences. These represent the remains of the rough skin of the shark that was similar in appearance to sandpaper.

The collector, Ottó Csomor, donated his unique find to the Geological Institute of Hungary and the two meter long fossil was described by Péter Solt, a researcher at the institute, as the new species *Odontaspis* (*Synodontaspis*) *divergens*.

📖 Solt 1988

Otoliths and Scales

When we do not have skeletons and bones, we can still deduce the characteristics of fossil fish faunas because washed residues often comprise tiny scales and so-called otoliths. These otoliths are hard parts from the balance organs of fish, are very resistant to weathering, and can usually be seen using micropaleontological methods. Their significance is described in more detail in connection with Miocene marine vertebrates, and their taxonomic status is often uncertain.

Otoliths are small particles—usually around 0.5–15 millimeters—made up partly of organic matter and partly of calcium carbonate, and are found in the inner ears of fish beneath their three semicircular canals. In accordance with the structure of the labyrinth organ, otoliths are classified into three forms—saccular, utricular, and lagenal—with the most variation characteristic of species seen in saccular otoliths. Otoliths have a role mainly in the perception of static equilibrium for fish, and in some sedimentary rocks they are found in large numbers. Their identification is usually enough to establish the ecological demands of the fossil fauna. Recently, a beautiful otolith fauna was reported from Lower Oligocene Dîncu Tămaşa beds in the vicinity of Mera and from Egerian sediments in the sand pit at Máriahalom.

📖 Nolf & Brzobohaty 1994; Reichenbacher & Codrea 1999; Ráki 2001

Tamás Báldi (left) headed the Department of General and Historical Geology at Eötvös Loránd University in Budapest for decades. Generations of students studied from his course book and lecture notebooks. As a paleontologist and stratigrapher he worked on the Oligocene and Miocene of the Paratethys. Mária Báldiné Beke (right), is a researcher at the Geological Institute of Hungary and one of the leading Hungarian specialists working on Mesozoic and Cenozoic nannoplankton.

Marine Mammals

Taking the number of species and fossil specimens into account, Oligocene fossil marine mammal faunas known from the Carpathian Basin are far behind those known from the Miocene and Eocene. Only 1 sirenian (*Halitherium*)—found in the clay pit at the Újlak Brick Factory—and 30 whale vertebrae (which were found during grave digging in the Farkasrét Cemetery) from the Kiscell Clay are worth mentioning.

Land Vertebrates

Fossil Vertebrates under the Hotel Transylvania in Cluj

Oligocene beds—exposed on the side of the Fortress Hill (Cetăţuia) in Cluj, on the left side of the Someşul Mic—have long been famous because of their fossil vertebrate remains. Outcrops can be found on the hillside beneath the Hotel Transylvania and this locality has yielded, for example, a fossil of *Chinemys strandi*, a freshwater turtle, and several fossils that were originally described as tapirs. In fact, these fossils belong to Antracotherioids. In addition, the fossils of fish, lizards, snakes, crocodiles, and birds are also known from this site. Birds include one small wetland form that was a good swimmer—a kind of rail described as *Rallicrex kolozsvarensis* by Lambrecht—and a fragment of a leg bone recently published by Jenő Kessler and considered an anseriform, a "ducklike" bird.

📖 Lambrecht 1929; Szalai 1934; Kessler et al. 1998

Kochictis centennii: *The Ancient Predator of Uncertain Taxonomic Status*

In a 1928 obituary of Antal Koch, geology professor in Cluj, Mór Pálfy wrote, "When he retired in 1913 after turning 70 and finishing his 41-year-long activity as a professor, he left no unelaborated materials behind." Indeed, Koch's life work is magnificent and by and large complete, but (as noted by Miklós Kretzoi in 1943 on the occasion of Koch's centenary) a correction must be made to Pálfy's statement.

In 1891, in a session of the Section of Medical and Natural Sciences of the Transylvanian Museum Society in Cluj, Antal Koch showed off the flattened facial part of the skull of a fox-sized mammal, which had been collected from Upper Oligocene coaly beds at Aghireşu (Egeres). According to the minutes of this session, after his introduction, he stated that "we must consider this fossil to be a representative of the extinct order Creodonta,"

Vlad Codrea, professor at the University of Cluj-Napoca, in the field. His scientific activity is partly focused on the study of Cretaceous reptiles and partly on the mammalian fauna of the Cenozoic.

and added: "At present five families of this extinct order are known; however, for lack of proper comparative material and literature, I cannot take a stand on the question, which family does our fossil animal belong to. One thing is certain, namely that this is the first such find in our country; therefore it is especially noteworthy from a scientific point of view" (Kretzoi 1943).

Koch set this fossil aside and finally its formal description was carried out by Miklós Kretzoi. The following quotation is from Kretzoi's 1943 paper:

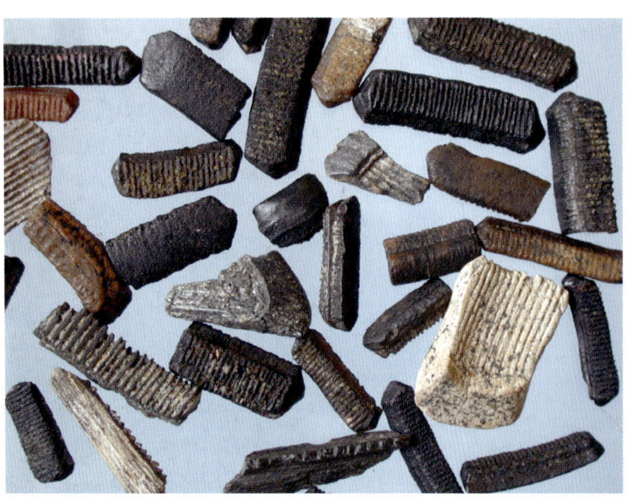

Ray teeth from the sand pit at Úny (approximately original size).

Fish remains (*Serranus budensis* [Hackel]) of Egerian age from the vicinity of Eger (approximately original size).

> The preliminary statements of Antal Koch are reliable in all respects, in spite of the half century which has passed. Nevertheless, he was wrong in judging his knowledge of the literature, as his notes enclosed with his description indicate that he knew the entire literature of his time well and did indeed put the find in its rightful place in the taxonomic framework. Koch knew that besides the known *Brachydiastematherium* and *Prohyracodon* remains from Andrásháza, the third Transylvanian curiosity of fossil mammalogy lay in front of him. He laid the fossil aside and in the meantime changed his place of work to the department in Budapest, and never dragged up the Aghireşu find again. In the last years of World War I, Tivadar Kormos (1881–1946) considered working on the fossil, as indicated by a photo of the find from this time and a drawing made by Teréz Dömök, which I found beside the fossil, when—a decade later—I started my own examination of it in the Geological Institute of Hungary. My manuscript from 1928 was put away, unpublished, because I myself was also afraid of a surprise that somewhere in the world someone else has described a similar fossil. However, my doubts were unfounded, and there is nothing to stop me from introducing the find in what follows.

Kretzoi went on to make a detailed description of this precious fossil and, in honor of the Koch centenary, he published on the predator and gave it the name *Kochictis centennii,* a new species within a new genus.

📖 M. Pálfy 1928; Kretzoi 1943

Coal-Animals in the Carpathian Basin

Anthracotherium (also called the "coal-beast" or "coal-animal") was a large artiodactyl ungulate, frequent in the Oligocene. *Anthracotherium* may have inhabited the large coal swamps that were prevalent at this time, as its remains are usually found in the beds that characterize these swamps. The smaller animals within this group

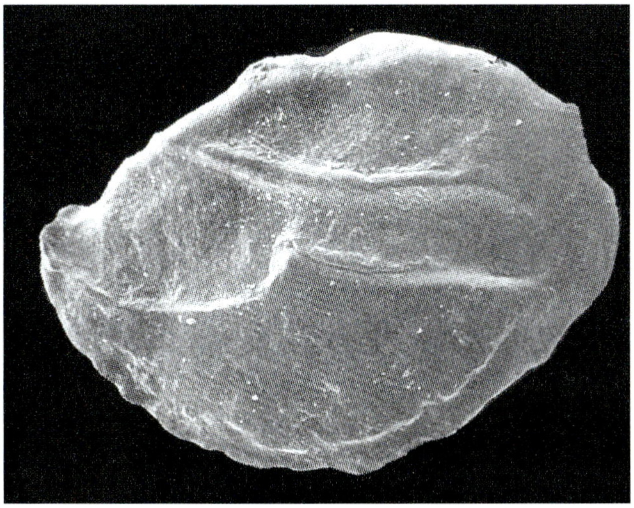

Otolith from Oligocene beds in the Transylvanian Basin. The size of the particle is 1–2 millimeters. After Reichenbacher and Codrea 1999.

were as big as swine; the largest, such as *Antracotherium magnum,* could be the size of rhinos.

Antracotherioids appeared in the Early Eocene and reached their acme in the Late Eocene and Oligocene. Their distant, extant relatives are the hippopotamids, and *Antracotherium* remains, especially their teeth, are known from several sites across the Carpathian Basin. The *Antracotherium* from Szápár was described by Gyula Éhik (1891–1964) and among other localities, *Antracotherium magnum* has been described from Oligocene beds at Jibău, Aghireşu, and Cristolţel, as well as from lignite seams in the Jiu Valley. *Elotherium hungaricum,* described from Huedin by Kretzoi, and *Elodon transsilvanicus,* a new species placed in a new genus from Merisor-Crivadia (Gilău Mendru), are both also hog-like artiodactyls from the Transylvanian Oligocene.

📖 Mártonfi 1890; Kretzoi 1941b

Rhinocerotids and Tapirs

Usually very incomplete remains (comprising mostly teeth) of the perissodactyl rhinocerotids and tapirs are also characteristic, but rare, fossils from Oligocene and Miocene sediments in the Carpathian region. At the time, the subtropical climate of the Carpathian Basin—which was warmer and much rainier than today—was probably suitable for these animals, and the bulk of their remains were described by Miklós Kretzoi. In this area the latest, comprehensive study is the 2000 work of Vlad Codrea, professor at the Babeş-Bolyai University, which describes the series of finds from the territories of present-day Romania.

The following species in this diverse fauna are worth mentioning: *Ronzotherium kochi, Cadurcodon zimborensis, Paraceratherium prohorov,* and *Benaratherium gabuniai.* Of these, the species *Ronzotherium kochi* was described by Kretzoi from Oligocene beds at Mera, while the type remains of the other three species derive from localities at Dobîrka, Zimbor, and Turea-Cornesti. Descriptions of these fossils have been done by Romanian colleagues, including Vlad Codrea, Nicolae Şuraru (1924–2009), Constantin Rădulescu, and Petre-Mihai Samson (1930–2001).

A jaw fragment of *Ronzotherium* (with two teeth) was also found in the Kiscell Clay at the Bohn Brick Factory; the fossil lophiodontid (relative of the tapir) that was collected from the Hárshegy Sandstone near Csobánka is the same age.

📖 Kretzoi 1940; Codrea 2000

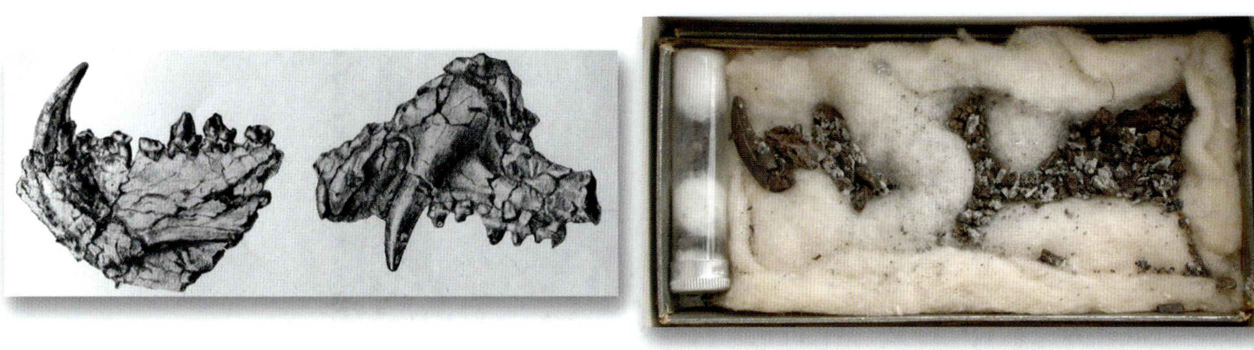

The famous fossil of *Kochictis centennii* Kretzoi from the Oligocene at Aghireşu shown in the illustration made for the initial description of the find. The pyrite content of the bone becomes sulphuric acid as it joins with water vapor in the air and this has damaged the bone so only small portions have remained of this valuable find (0.7× magnification).

The fragmentary jaw of an Oligocene sirenian, with two teeth, from the Újlak Brick Factory in Budapest. This find, together with several other fossils, was donated to the Collection of the Hungarian Natural History Museum by István Harmat, who worked in the mine (original size).

The tusk of *Anthracotherium magnum* Cuvier from the coal mine at Lupeni (above) and the jaw of the suid *Hyopotamus* sp. (related to *Anthracotherium*) from Cluj Napoca (bottom) (both almost original size).

Late Oligocene vertebrate remains from a sandpit dug near the road between Úny and Máriahalom. This locality allows us a unique insight into the animal life of the Late Oligocene because the sand exploited there, once deposited in tidal channels, contains bones and teeth left behind by both marine and dry-land vertebrates. Remains of near 30 species have been recorded from this small sand-pit so far. Original size unless otherwise indicated. (1, 2) Fragment of carapace of a turtle from the group Trionychoidea, genus and species not determined. The element displayed here is a right rear pleuralia seen in ventral (1) and dorsal (2) views. (3) *Crocodylia* sp.—Armor element of an undetermined crocodile. (4–6) *Potamotherium* sp.—Remains of an otter-sized, weasel-like animal well adapted to the aquatic mode of life. Tibia (4) and outer (5) and inner (6) view of a molar (1.8× magnification.) (7) Remains of a sirenian—rib fragment of an undetermined sea-cow. (8) Canine tooth of a carnivorous mammal (1.8× magnification). (9–11) Shark teeth. (9) *Carcharias cuspidata* (Agassiz); (10) *Carcharoides catticus* (Philippi); (11) *Odontaspis acutissima* Agassiz (1.25× magnification).

Remains of scallops can be found abundantly in most marine Early and Middle Miocene sediments in the Carpathian region and are rightly considered emblematic fossils of that interval. Where better index fossils are absent, some species can be used as stratigraphical markers. The specimen displayed in this photo, *Flabellipecten leythajanus*, was a scallop present in large numbers in the Paratethys Sea during the Badenian Age (original size).

7

The Early and Middle Miocene

The first long period of the Cenozoic, the Paleogene, was immediately followed by the Neogene. The first subdivision of this time period, which lasted about 18 million years, was the Miocene. Geological remnants of this long period are preserved in numerous sequences in the Carpathian region, and include a range of diverse fossil assemblages. There was also environmental change of great significance at the turn of the Middle and Late Miocene in the Carpathian Basin, as the connection between the Paratethys Sea and other global oceans ended.

Accordingly, the early period of the Miocene is represented mainly by marine successions. By the Late Miocene, the Paratethys was entirely separated—effectively a lake that was ultimately filled up with sediment. As a result, this period is characterized by deposits that are largely different from older ones: lake, river, and terrestrial sediments and fossils. In light of this, the Early and Middle Miocene geo- and biohistorical events and fossils of the Carpathian Basin are discussed separately from the Late Miocene.

The Central Paratethys seaway, which lay in the present-day Carpathian region, may have been periodically (during the Eggenburgian and Badenian) connected with the Western Paratethys Sea and, through it, even with the Atlantic Ocean during the first half of the Miocene. This is indicated, for example, by the varied occurrences of marine top predators in geological sequences of this age, including toothed whales and sharks. The impoverished fauna of the Ottnangian similarly indicates simultaneous sea level fall (regression) when the former connection with the world ocean discontinued. The consequence of the latter event was also the evolution of a peculiar, endemic mollusk fauna.

When a connection with the world ocean was restored—by a worldwide sea level rise during the Badenian—shark fossils are found again, as well as whales (by the middle Badenian). This is also the youngest age from which the last fully tropical marine deposits are found in Hungary and neighboring regions.

We also know that there was significant volcanic activity in the dynamically changing Alp-Carpathian region during the first half of the Miocene. The development and appearance of volcanoes is explained by tectonic changes in the region, including the subduction of pieces of crust. The presence of volcanoes is also important in the context of the evolution of life. They provide, for example, basement rock types that are well suited for corals to settle and thus for reefs to evolve. These are seen in the Dunazug and Börzsöny Hills in the Badenian archipelago; the famous fossil footprints near Ipolytarnóc are preserved in tuff.

The water body that covered the Carpathian-Pannonian region at this time again became separated and distinct by the end of the Middle Miocene. This explains the fact that the fossil diversity known from the Sarmatian Sea is not comparable with that from the Badenian.

📖 Báldi 1971

FOSSIL-RICH FORMATIONS

In the Early Miocene a variety of sediments (terrestrial, paludal, coastal, and deep water) were deposited because of diverse paleoenvironmental conditions. In earlier professional literature, these sandy, pebbly sequences are generally denominated by the names of the most common mollusks (usually shells) that are found in them. The most peculiar rock found in the Middle Miocene is the fairly fossil rich Lajta Limestone, named after the Austrian Leitha River (Lajta in Hungarian). In addition, the fossil-poor basin deposit—schlier—also known from the Oligocene, is spread across Northern Hungary at this time.

Early Miocene "Bivalve Facies"

The Early Miocene marine transgression left behind different sandy, pebbly sediments that can found in the environs of Budapest and in northern Hungary. The traditional names of these rocks refer to the (usually large) bivalves that occur in them—abundantly in places. Thus, in earlier geological literature one encounters the expressions "large pecten layers" or "anomia sand." These layers are now known to make up the Budafok Formation in reference

	Regional Stage/Age	Millions of Years Ago
Early/Lower and Middle Miocene	Sarmatian	13
	Badenian	17
	Carpathian	17.5
	Ottnangian	18
	Eggenburgian	21
	Egerian (partly)	23

Division of the Early and Middle Miocene into stages or ages.

to their typical area of occurrence. (See section on fossil sites, below.)

Lajta Limestone from Rákos, containing red algae, bivalve, crab, and gastropod fossils (original size).

The Extraordinarily Fossil Rich Lajta Limestone

This rock was formed in the tropical archipelago of the Badenian epoch. It is usually pale, and generally buttery yellow or white. This is, for example, the reason Fehér ("White") Street in Budapest is white: the loose limestone, mined in the neighborhood and carried by carts, covered the street leading into town with a white powder as dust fell from the wagons. The Lajta Limestone is composed of both reef-building and reef-dwelling organisms: forams, red algae (*Lithothamnium*), colonial corals (*Favia, Porites, Stylophora*), bryozoans, mollusks (gastropods, bivalves, scaphopods), and many others. The decapod crabs (crustaceans) that can be collected from this rock are famous. Organisms with carbonate skeletons have been fossilized in different ways: the originally calcitic ones have been preserved intact, whereas the aragonitic ones—like corals, most gastropods, and many bivalves—have been dissolved during diagenesis. The (most often incomplete) skeletons of some marine mammals, mainly whales, have also been recovered from this rock alongside plenty of fossil mollusks.

Siliceous Earth (alias Diatomite)

The volcanic activity that reached its peak in the Badenian led to huge amounts of dissolved silica entering into the waters surrounding volcanoes. This, in turn, allowed diatoms—small unicellular golden algae (chrysophytes)—to proliferate to an unprecedented degree. Siliceous earth, or diatomite, is the result of diatoms accumulating in large numbers. Pure diatomite is a porous, generally pale rock of very low density (0.45 grams per cubic centimeter) that is also characterized by fine, laminar stratification. This stratification indicates that, at one time, a lake bottom—or sea bottom—was free of life, so nothing disturbed the original structure of the deposit, although its composition

A Paleontologist in Action: Making Moldings

It is common in some rocks that fossils are preserved only as internal molds, or stone casts. The fossils of the Lajta Limestone are often preserved in this way. This is especially true of the mollusk fossils (see above): one often finds such a mold clinking within the rock, like the seed in a dried almond. The mold perfectly reflects the inner surface of the carbonate skeleton of a gastropod or bivalve; however, this mold—especially in the case of gastropods—is often inadequate for precise identification. The outer ornamentation of the skeleton is also important. In such cases it is necessary to make a mold of the imprint of the fossil, which can be done using cheap gypsum, two-component resin, or silicone rubber. To mold a small-sized, finely ornamented fossil, however, requires some practice: it is important that no bubbles get onto the surface, and it is also essential that the mold itself does not get stuck in the negative. One can use a release agent that suits the material of the mold if necessary.

Naturally, we can make molds not only of the mollusks of the Lajta Limestone but also of other imprints in other rocks. Similar materials are often used to make copies of rare or valuable fossils. A negative is made first; the molding of the real copy can follow only afterward. The preparation of three-dimensional copies of fossils, such as bones or whole skeletons, requires great patience and a good deal of practice.

The Lajta Limestone quarry near Zebegény. Remains of large species of *Pecten* and *Pholadomya* are common in the loose rock.

changed from season to season. The best-known use for diatomite rock was in the production of dynamite, but today it is used instead for the filtration of fluids and as abrasive filler for toothpastes. Diatomite was even once eaten mixed with flour in times of starvation—although, since it is an inorganic material, it is not really nourishing. The Carpathian Basin is one of the early centers for fossil diatom research: the stratigraphic value of these microfossils was first recognized by József Pantocsek (1846–1916), a physician and botanist from Bratislava. His collection is one of the treasures of the Department of Botany at the Hungarian Natural History Museum in Budapest.

Remarkable diatomite resources are found in Hungary at the foot of the Mátra, near Szurdokpüspöki, where this raw material was long mined. As long as quarrying continued at these mines, a large number of fish, insect, and mammal fossils (bones) were also recovered from the marine and freshwater layers deposited around the former landmass. Fine fossils were also found near Hasznos, and the remains of marine and terrestrial vertebrates were collected from tuffaceous diatomite layers outcropping on the slopes of the local Castle Hill. The bones of many fish (shark, rays), amphibians (*Pelobates, Bufo, Rana*), reptiles (*Testudo*) and large and small mammals (cervids, insect eaters, squirrels, dormice, and hamsters) were collected over the lifetime of the quarries. Findings were concentrated in a single 5-to-10-centimeter-thick bed, which also contained mainly tortoiseshell fragments.

The insect remains collected from near Tállya were also preserved in diatomaceous-tuffaceous laminated lake sediments.

REMARKABLE FOSSIL SITES

Numerous natural and artificial exposures of Lower and Middle Miocene rocks are known in Hungary and from

The railway cutting at Rákos, a rich Badenian fossil site.

the surrounding area. Many of these exposures are considered to be outstanding fossil sites: the Lajta Limestone and Badenian sands are particularly rich in fossils, and quarries that expose them are always good places to make finds. In particular, the famous Ipolytarnóc site is noteworthy because a volcanic pyroclastic flow has preserved not only well-known fossil mammal footprints but also the remains of a unique and rich Miocene fauna.

The Pacsirta Hill, the Hont Cleft, and the Ilona Valley

Traces of the earliest Miocene (Eggenburgian) transgression can be seen in the Buda Hills and to the northeast of them (Cserhát, Mátra). Fossil-poor schlier has been deposited in the deeper basins of this region since the Oligocene; sandy, coarse pebbly sediments are found nearer to the coastline. One beautiful exposure of schlier is located, for example, in the Hont Cleft in Nógrád, whereas better-studied and fossil-rich shallow-marine successions were once excellently exposed in sandpits on the slopes of the Kereszt Hill and Pacsirta Hill in Budafok. Unfortunately, these classical fossil sites have been filled up with rubbish in most places: not long ago, beds packed with large shells (*Ostrea, Chlamys, Glycymeris*) could be seen, constituting real "shell rocks" in places. Often all the preserved shells were lined up with their convex side facing upward in the shallow water because of the strong currents. The most characteristic fossils from these sequences are sea acorns, or *Balanus*.

Another fine outcrop of Eggenburgian age, known since the time of Pál Rozlozsnik, is located south of Parádfürdő, under the waterfall in the picturesque Ilona Valley. The rich fauna found here comprises more than 70 taxa (mainly mollusks), all collected from a conglomeratic layer. Corals, Bryozoa, and *Balanus*—as well as shark teeth—can also be collected here, the richest Eggenburgian fossil-assemblage known from Hungary.

📖 Rozlozsnik 1939; Báldi 1959, 1971, 2003; Főzy & Leél-Össy 1985

Badenian Fossil Sites

The Badenian Lajta Limestone occurs around Vienna, in western Hungary, in and around Budapest, in the Visegrád

Not a building or a cave, but an artificial wall cut into natural stone. Visitors are shown here in the Lajta Limestone Quarry at Fertőrákos.

The huge Lajta Limestone open-pit mine on the southern slope of the Fruška Gora Mountains (Mutalj Quarry). Here a fossil-rich reefal limestone is used by the Beočin Cement Factory.

Hills, and in the Börzsöny, as well as in the Mecsek of southern Hungary. This soft, friable, or consistent rock has been mined—for various purposes—for centuries, and these quarries generally are excellent fossil sites.

The Fertőrákos quarries were first mined in Roman times: this very loose, friable limestone is still being mined in a quarry only a few meters from the Austrian border, not far from the old ones that now serve as an open-air exhibition. If one manages to get in, fist-sized *Ostrea*, *Lithothamnium*, and other fossils can be collected at the foot of the quarry conveyor belt.

Biatorbágy (Nyakas Cliff), a number of sites on the Tétény Plateau and Örs vezér Square (around Fehér Street) should be mentioned as being among the best-known occurrences of this limestone in the vicinity of Budapest. Especially beautiful specimens (mollusks, sea urchins, crabs) were recovered from the foundation pit when the Sugár and Árkád shopping centers were built. In addition, if one approaches Budapest Keleti (east) Station by train, one will rumble through a railway bridge a few minutes before arrival: this is a classical outcrop from the railway cutting in Rákos. One can get back to this exposure on foot if one walks to the end of Tárna Street and goes down to the rails instead of going on through the bridge. To the left there is a narrow cutting; the excavated Badenian and Sarmatian succession on the right can be seen, shining white, from a distance.

The quarries and outcrops (quarries around Zebegény, Törökmező, Nagymaros, Szob, Szokolya, and Kemence) in the western and southern foothills of the Börzsöny Hills are also considered to be good fossil sites. In addition, the Badenian noncalcareous sand in the Szabó Quarry at Várpalota is extremely rich—primarily in gastropod and bivalve remains. This quarry is a nature reserve today, so the "eyes-only" rule must be followed here.

The fossils collected from two Transylvanian sites, Coşteiu de Sus and Lapugiu, are also very famous. The preservation of fossils from these localities is so exceptional they look like they were collected on a living beach. Although the main fossil site at Lapugiu is protected, collecting is possible outside the protected area.

📖 J. Pálfy & Pazonyi 2007

Devínska Nová Ves

Among the known Badenian fossil localities, this Slovakian site, also known by its Hungarian and German names (Dévényújfalu and Neudorf an der March) is of special importance. Fossils were collected here from a small mound

A reconstruction of the strange "horse-gorilla" (*Chalicotherium*) that inhabited the Carpathian Basin in the Middle Miocene. Their remains were also found at Devínska Nová Ves. This fearsome-looking, 2-meter-high animal was in fact a peaceful herbivore. AFTER GÉCZY 1993.

Mosaic of ossicles from *Psephophorus polygonus* v. Meyer, a large leatherback turtle (close to original size). *Psephophorus,* a close ally of the extant and cosmopolitan dermochelyid sea turtles, is the only Miocene leatherback turtle known from Europe. In the Carpathian Basin it was documented from Devínska Nová Ves.

(Devínska Kobyla) where the Danube and Morava rivers meet, and bones were found in the Stockerau Limestone Quarry and at a nearby locality called Sandberg ("Sand hill"). The quarry was active for more than a century, and during this time fossil-rich karstic crevices yielded numerous terrestrial fossils. The known assemblages comprise a wide range of species belonging to different groups, including amphibians, reptiles, flying mammals, insectivores, primates, rodents, predators, ungulates, and proboscidians—all from a site that is now protected.

The Sandberg succession is slightly younger and was deposited in the shallow sea that slowly covered this region. As a result, among the frequent marine invertebrate remains that are found, fossil fish—including sharks and turtles, seals, sirenians, and cetaceans—can also be collected and occasionally terrestrial vertebrates also occur. However, although bird remains were collected from this region (from Devínska Kobyla), the precise locality is unknown. Historically, the first specimens from this area were collected about 160 years ago; subsequently, dozens of papers have been published on the fossils from this site. Of the numerous researchers who worked here, the most important was Helmuth Zapfe (1913–1996), a professor at the University of Vienna, who published both thin papers and thick monographs on the rich terrestrial mammal fauna known from the region. This fauna includes more than 60 specimens of the bizarre "horse-gorilla," *Chalicotherium,* and one of the earliest hominid primates, *Griphopithecus suessi*—the latter collected from the Sandberg locality.

Giving proof to the old saying that there is always something new under the sun, in the middle of the 1980s a new fossil site was discovered at Devínska Nová Ves; it has so far yielded both marine and terrestrial fossil remains. Researchers have called this site a fossil bonanza, which suggests that there are more fossils to come.

🕮 Zapfe 1960, 1979; Švagrovský 1978, 1981

The Sarmatian fossil site in Tăşád, where—among remains of other animals—fossil whales have been recovered. Márton Venczel, research fellow at the Museum of Oradea and expert of the fossil herpetofauna of the Carpathian Basin, is climbing at the top of the outcrop.

Sarmatian Fossil Sites

The only really good—but, unfortunately, already destroyed—site for Sarmatian macrofossils was the Sőregi Quarry near Tinnye. There are basically two sorts of Sarmatian sediments in the Carpathian Basin from a petrological point of view. In the basin, clayey-marly fine-grained successions were deposited but are seldom exposed because they are so soft. These rocks are also almost useless commercially, so they were never quarried—in contrast to sequences of shallow-water limestone, which are usually porous and easily carvable. These sediments are, in fact, calcareous sandstones, because their grains are often foram or mollusk fragments coated with a calcareous crust rather than quartz. This Sarmatian limestone has been mined in many small and larger quarries across the Carpathians—the most famous being the Sóskút Quarry. This quarry, however, cannot be considered a fossil site. The only known location where a well-preserved and (considering the general nature of the fauna) rich fossil assemblage can be collected from a Sarmatian limestone is the Sőreg Quarry, near Tinnye. An abandoned small quarry on Sőreg Hill, about 500 meters west of the first bend in the road leading from Tinnye to Perbál, exposes these limestone beds. This site was already known to Miksa Hantken in the middle of the nineteenth century: the 6-meter-thick sequence was visible for quite a while, even after the mine was closed. Jenő Boda (1921–1996), an expert on Sarmatian fossils, distinguished 13 layers in this quarry during his research, the results of which were published in 1974 in the Sarmatian volume of a series introducing the successions and stages of the Central Paratethys. The approximately 20-centimer-thick, second from bottom bed in this quarry was particularly rich in fossils: a single cubic decimeter of rock contained 160 fairly large gastropod and bivalve specimens! These fossils, in contrast to those usually found in Sarmatian limestone, were preserved intact with their shells. Although the shells of bivalves are snow white and slightly soft like chalk, the shells of gastropods can be found relatively fresh at this site.

Boda might well have known that although humans create geological exposures, they can also finish them off, so he deposited a huge amount of rock from this quarry into the collection of the Department of Paleontology at Eötvös Loránd University for further study. He was proven right: during a study excursion at the end of the 1980s, a group of geology students found the fossil layer at this outcrop filled up with bones and dried and liquid pig manure.

📖 Boda 1959, 1970, 1971, 1974, 1979

"Shark Teeth–Rich" Layers

The most frequently collected Miocene vertebrate fossils from the Carpathians are fish, primarily sharks. In some places, where the once-rising sea spread its sandy sediments onto gently sloping beaches, so-called shark teeth–rich layers were deposited. One such classical site is Fehér

Participants of the Fifth Hungarian Paleontological Regional Meeting at the Ipolytarnóc shark teeth site in 2002. These once-rich layers yield far fewer fine fossils these days.

Hill, near Ipolytarnóc, where Antal Koch collected most of the fossils he described. Koch also collected shark teeth at other fossil sites—including Horné Strháre, 20 kilometers west of Ipolytarnóc and Pôtor. Miocene rocks containing abundant cartilaginous fish teeth are found near Kazár, to the side of the hollow leading to Aranyosipuszta, and at Mučín and Hímesháza, where fossils have been discovered in the granite sand laid down by the Miocene sea.

📖 Koch 1903; Solt 1992; Kocsis 2003, 2007

Whale Remains from the Carpathian Basin

Miocene fossil whales are rare from the Carpathian Basin. An almost complete list from Emese Kazár's PhD dissertation on the known fossil sites is presented below, divided by stages. Eggenburgian sites—Horné Strháre, Pôtor, Mučín, Ipolytarnóc, and Felső-Pálfalva (Salgótarján)—are located in present-day Slovakia and in northern Hungary. A significant proportion of the known Badenian sites are in present-day Austria: Walbersdorf, Rohrbach, Stotzing, Mannersdorf, St. Margarethen, and Kaisersteinbruch. Further Badenian fossil sites are known around Devínska Nová Ves, Komló, Danitz-puszta, and Mátraszőlős; Sarmatian whale remains are known from Austrian, Croatian, Hungarian, and Transylvanian localities. Several sites are located around Vienna, at the margin of the Leitha Mountains. The Podsused, Radoboj, Daruvar, and Pakrac sites are in Croatia; Holič is in Slovakia. Tăşád, Cluj-Napoca, and Tăureni are in Romania, and fine Sarmatian fossils are known from Kovácsszénája, Pécs-Vasas, Danitz-puszta, Hird, Pécsvárad, Hímesháza, Budafok, Budapest-Kőbánya, and Kozárd in Hungary.

📖 Kazár 2003

Ipolytarnóc: The Fossil Drinking Place

Ipolytarnóc is located at the northern boundary of Nógrád, close to the Slovakian border. East of this quiet village, in the Borókás Cut, one of the most remarkable fossil sites in Hungary, and famous all over Europe, was found. This is where a fossilized, silicified tree trunk over 40 meters long, the "Tarnócz hornbeam stone," used to lie: the stump of this massive tree is still on view at the spot. Bedding planes containing the fossilized footprints of Miocene vertebrates were discovered near this tree: they show evidence of a herd making for a drinking place that was covered and preserved by tuff from a volcanic eruption.

The already well-known giant fossilized tree trunk was visited in 1900 by Hugó Böckh (1874–1931) from Nagysúr, a professor in Selmecbánya, in the company of János Tuzson, a botanist. Tuzson discovered that the hard sandstone surface under the tree trunk hid the footprints of fossil mammals. As they suspected the significance of the finding,

A group of visitors at the dome hall erected above the Ipolytarnóc footprints, later named after András Tasnádi-Kubacska.

they broke up a particularly beautiful part of the bed with footprints and carried it to the Geological Institute. The rock bed, broken to pieces then carefully reassembled, is still on display in the lecture hall of the institute. The site, although not forgotten, was not studied in detail until 1928 when Baron Ferenc Nopcsa, then director of the institute, organized an international paleontological conference. The Ipolytarnóc site was revisited during the conference field trip—this led to the so-called rediscovery of the fossil footprints. Othenio Abel from Vienna—who, together with Nopcsa, is considered a father of paleobiology—also participated in the field trip. Encouraged by his colleagues, Abel elaborated the footprint of Miocene vertebrates of the Ipolytarnóc site with scientific accuracy. His 1935 book deals with the Hungarian footprints, along with others from all over the world.

In 1937 new excavations were made in the area, with the modest support of the Hungarian National Museum and the Hungarian Academy of Sciences. The works were coordinated by András Tasnádi-Kubacska, then a very young research fellow at the Natural History Museum. Tasnádi-Kubacska started working with the uncompromising and overwhelming enthusiasm and willingness that would characterize his entire life. In his popular 1977 book, he recollected those days:

> I spent about three weeks in the field in July 1937. I took along Viktor Haberl Jr., preparator at the museum. The local countrymen could not carve the silicified, hard sandstone, so I had stone-cutter specialists brought from the Salgótarján coal mines. The stone was so hard that their half-meter long, 2-inch thick steel chisels were worn off in a few minutes. . . . After the rock pieces were indented about 25–30 centimeters in depth, plates were broken open, sketches were drawn of the rock fragments, and they were numbered and packed up. We arrived at the field at five in the morning and finished work at dusk.

Stetten—A huge excavated fossil oyster reef turned into a popular vistor's center.

> We succeeded in packing the last piece for transport in two and one-half weeks only by working at such a pace; I then borrowed a truck and carried about 20 to 25 hundred kilograms of fossil material back to the museum.

In this way, piece by piece, the footprint-bearing bed was returned to the natural science collections at the National Museum. In all, 25 giant fossilized footprints (mostly rhino) are visible on an eight-square-meter slab. This fossil drinking place was finally placed under national protection after several decades of effort and a number of petitions. A domed hall suitable for visitors was built above the area containing the most beautiful footprints, 80 years after its discovery. These works were led by László Kordos, paleontologist at the Geological Institute of Hungary. Kordos has also done significant additional exploratory work at these old sites and in their vicinity. As a result, the surface of the footprint-bearing sandstone that can be studied in the field has been increased by tens of square meters and the number of discovered footprints by a few hundred pieces. More than 3,000 specimens are known today. Up-to-date scientific processing of the Ipolytarnóc footprints has now been finished, in line with the conservation work: the local display also exhibits the geology and valuable plant fossils found in the area. These fossils have been treated in detail by Klára Rásky and Lilla Hably.

For decades, only suppositions were made about the age of this footprint-bearing sandstone: reliable data were obtained for the first time in 2006. The radiometric age of small zircon crystals extracted from this rock turned out to be 17.5 million years, according to the painstaking work of József Pálfy and colleagues. These workers used a uranium-lead method that is considered to be highly accurate.

📖 Tasnádi Kubacska 1977; Hably 1985; Kordos 1985a; J. Pálfy, Mundil, et al. 2007

Fossil Sites of Proboscideans

The most spectacular Early and Middle Miocene continental vertebrate remains dated from after the Ipolytarnóc footprint-bearing sandstone are made up of proboscidean fossils newly arrived from Africa. Hungary's oldest proboscidean remains originate from the Lower Miocene carbonaceous succession of the Salgótarján Basin, from localities such as Salgótarján, Etes, Szuhakálló, Kotyháza, and Zagyvapálfalva.

📖 Gasparik 2004

Stetten in Austria: The Largest Known Fossil Oyster Reef in the World

About 16.5 million years ago, during the Middle Miocene, rocks of the Karpatian stage in the vicinity of Korneuburg, to the north of Vienna, formed in a small gulf in the Paratethys Sea. The southern edge of this gulf comprised an estuary with poor mangrove vegetation and extended

Ostrea reefs. Toward the mainland stretched brackish-water marshes, shallow lakes, and meandering river systems—as evidenced by the remains of freshwater mollusks and aquatic vertebrates. At the same time, the northern part of this gulf was occupied by a shallow marine environment where, at about 20–30 meters in depth, small coral patches lived on the sandy sea bottom. This assemblage contained many kinds of mollusks—including frequent, specialized predatory gastropods, sponges, and echinoderms. In the Karpatian beds at Korneuburg, the largest geological excavations conducted in Austria were carried out, and the results of both field and laboratory work were summarized by the researchers in two large volumes.

The local community and the Natural History Museum in Vienna decided to develop a fossil park at Teiritzberg, near Stetten, so that the public could get an introduction to life in the past and the fossils found there. Around 160 people spent about 12,000 hours working on this project, which was completed in 2010. The highlight of the park is the 400-square-meter pavilion, which is above the largest known fossil *Ostrea* reef in the world. It is not only the size of the reef that is impressive—many of the 20,000 individual *Ostrea* valves that make up the reef are more than 40 centimeters in length.

📖 Sovis & Schmid 1998, 2002

EARLY AND MIDDLE MIOCENE FOSSILS FROM THE CARPATHIAN REGION

Vertebrate and invertebrate remains give a true reflection of the large-scale and dynamic paleoenvironmental changes that took place in the Carpathian Basin during the Early and Middle Miocene. The Central Paratethys—according to one widely accepted explanation—was separated from the world ocean by the end of the Middle Miocene (Sarmatian), its salinity decreased considerably compared to a normal value, and there was a significant change in the composition of fossil assemblages. The sediments deposited in the basin of the Sarmatian Sea separated from the world ocean contain fairly specific, unique fossil assemblages—chiefly gastropods and bivalves. The number of species known from these sequences has been estimated to be in the dozens. The macrofauna is characterized by one crab genus and bryozoans, in addition to mollusks. Scallops, sea urchins, and corals—so familiar in the Badenian Sea at normal marine salinity—are all missing. The salinity of the water at that time is thought to have been about 10–12 percent—similar to that of the present-day Black Sea—but the number of species was much poorer. Lately, however, Austrian experts have suggested that instead the hypersalinity of the water might be responsible for the disappearance of so many species.

Microfossils

Diatoms

Although diatoms either cannot be found—or are only rarely found—in older sequences, they occur in rock-forming quantities in the Miocene.

The skeletons of diatoms are made of opal that lacks crystalline structure. The cytoplasm of individuals is closed into two box-like shells that touch; these can be prismatic, cylindrical, or elongate oval in shape. Diatoms are usually 10–100 microns in size, the extremes being 2 microns and 2 millimeters. They live solitarily or constitute bushy, ligamental, or irregularly spreading colonies. They are present on continents and in fresh- and marine waters: the radially symmetrical ones are always marine, whereas the bilaterally symmetrical ones are only partly marine. The characteristic sediment 4,000–6,000 meters deep in present-day oceans is diatomic mud, from which diatomaceous earth or diatomite can evolve as a result of digenesis.

The thickest Miocene diatomite succession in the Carpathian Basin is located west of Mátra—between Szurdokpüspöki and Gyöngyöspata—where it was mined in large quarries for decades. These exposures allowed the only Hungarian researcher to work on the famous ocean-bottom-exploring Glomar Challenger drilling ship, Márta Hajós (1916–2000), to examine diatom floras in detail. She published taxonomic descriptions in and a stratigraphic evaluation of the Miocene diatoms in Hungary; these monographs are still essential to the study of Miocene diatoms. The lower part of the diatomite sequence was deposited in fresh- and brackish water, according to Márta Hajós's examinations. This is supported by the presence of the dominant *Melosira, Stephanodiscus,* and *Navicula* genera—as well as that of others. The upper part of this succession contains forms—*Coscinodiscus, Liradiscus,* and *Actinoptychus* being the most common—that indicate a warm sea. Interestingly, a fairly large number of freshwater and marine gastropods and bivalves can also be found in the rock along with frequent fish remnants, which indicates that the environment may have become variably favorable for benthic animals.

📖 Hajós 1968, 1986

Coccoliths

Besides diatoms, the tiny calcareous skeletons of coccolithophorids, also a group of golden algae, occur in Lower and Middle Miocene marine sediments; Badenian successions are the richest of all. A detailed examination of this flora was conducted by nannoplankton expert András Nagymarosy. His determination of species within the genera *Reticulofenestra, Coccolithus, Cyclicargolithus,* and others has made it possible to fit the Carpathian sequences

A Paleontologist in Action: Studying Fossils with Computer Tomography

For years paleontologists have wished to be able to study the inside of fossils without them being destroyed. Making X-ray images is one way to make concealed things visible. It uses the principle that elements with larger atomic numbers absorb radiation better than those with smaller atomic number. On traditional X-ray images, however, the spatial arrangement of objects is not easily recognized. The development of computer tomography (CT) in the 1970s was revolutionary in this regard. Now the radiation source and the detectors that substitute the film turn themselves around the target object. The changes in the strength of radiation getting through the detectors is transformed into a digital signal that projects the three-dimensional image of the outer or inner surface—or optionally oriented section of the examined object—onto the screen, with the help of a suitable software. CTs of large resolution made for the examination of objects a few centimeters in size are especially appropriate to the study of microfossils. The first CT images were made of the skull of a Cretaceous crocodile, *Iharkutosuchus,* by Tamás Földes of the Hungarian Oil and Gas Company and Attila Ősi. The first paleontological micro-CT survey was executed by Balázs Szinger, Ágnes Görög, Emőke Tóth, and János Viszkok in 2006.

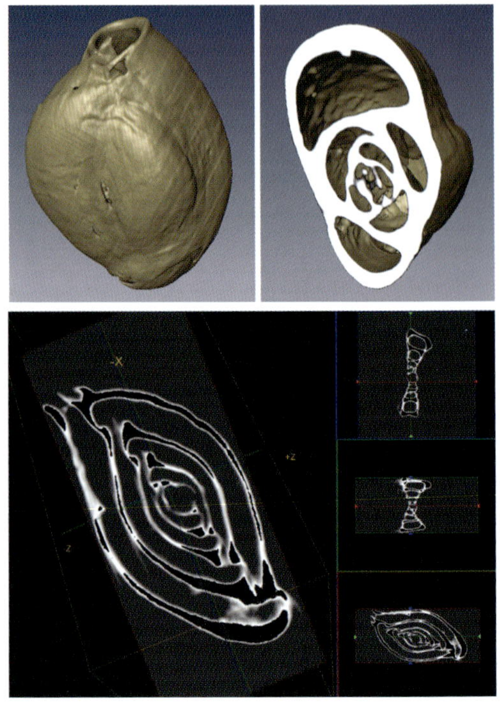

Micro-CT images of forams. The size of these specimens is about 0.5 millimeters.

into the worldwide framework of Neogene nannoplankton zones.

📖 Nagymarosy 1980

Spores and Pollen

Fossil spores and pollen have been extracted from almost every Lower and Middle Miocene sequence across the Carpathians. Eszter Nagy (1914–2012), head of the former Department of Paleontology of the Geological Institute in Budapest, spent decades studying them; her first study was published in the 1950s, and her most recent was published in 2005. Her 1985 systematic description of Neogene sporomorphs and her 1992 stratigraphic evaluation of the flora were both issued as part of the prestigious Geologica Hungarica series Palaeontologica. Her 2005 book, which discusses the development of Neogene climates, was based on detailed palynological data accumulated over the course of her career.

📖 E. Nagy 1985, 1992, 2005

Forams

Lower and Middle Miocene sequences in the Carpathian Basin also contain a hidden microfauna that is similarly diverse to floral microfossils. Forams occur in sediments from all stages, from the Eggenburgian to the Sarmatian. Although no comprehensive monographs have been published about these assemblages, it has been established that Badenian successions are the richest. Ágoston Franzenau (1856–1919) was the first to touch on these in an 1881 study, in which he introduced the fauna from the succession exposed in the Rákos railway cutting in Budapest. In this remarkably illustrated study, it is clear that the characteristic taxon in the Lajta Limestone is the sometimes even millimeter-sized *Borelis melo* of the group of miliolid forams. Globular tests of *B. melo* appear in this rock, as well as white spots, and its surface is ornamented by microscopic meridian-like grooves along the septa. This special appearance is also indicated by the name of this species (*melo* means "melon").

The investigation of Early and Middle Miocene forams long remained a neglected area. However, the works of Ilona Koreczné Laky (1930–1990) must be mentioned, alongside a number of shorter studies. Koreczné Laky described foram faunas from several regions, including Mecsek, Börzsöny, the East Borsod coal basin, and the Tokaj Hills.

Sarmatian sediments also contain a diverse foram assemblage as compared to their macrofauna. However, the planktonic forams that are so beneficial for the biostratigraphic division of Cenozoic marine sequences are missing at this age: these faunas contain only benthic forms. Nevertheless, zone division of the Sarmatian was possible even based on these forams. Ágnes Görög, who has since become a renowned foram expert, examined 26,000 specimens from three deep boreholes in the Zsámbék Basin as part of her graduate thesis in 1988. This work enabled Görög to correlate the successions at Zsámbék with the stratigraphic levels in the Vienna Basin. The most characteristic, age-indicating Sarmatian foram is the beautifully ornamented *Elphidium,* which has smooth and spiny variants. Another useful age indicator, *Porosononion,* also occurs in this assemblage and comprises about 60 species.

Finally, one very special foram can also be found in Carpathian Sarmatian sediments. According to Jenő Boda's 1979 study, an encrusting form named *Nubecularia* (now called *Sintzowella)* constitutes an enormous reef around Páty that is almost 10 meters thick and several hundred meters long.

📖 Franzenau 1881; Boda 1979; Görög 1992

Shell-Covered Crustaceans (Ostracods)

Although Early and Middle Miocene ostracod assemblages can be found in nearly every rock type, they have received less attention than their Eocene and Oligocene counterparts. The last monographs on these forms were published by Béla Zalányi (1887–1970). However, after a hiatus of a few decades, Emőke Tóth started a new study of Sarmatian ostracods in 2000. So far, we know from her work that this fauna is composed of about two dozen species, of which the most typical is *Aurila,* named for its peculiar earlike shape.

📖 Tóth 2004

Fossil Plants

Bridge over Time: A 50-Meter-Long Giant Fossil Tree Trunk

In all probability the largest fossil known to date in the Carpathian Basin is a huge silicified tree trunk, found lying in a field at Ipolytarnóc—or rather, what remains of the tree trunk. Once this giant trunk spanned a tributary of the Ipoly and local people used it as a bridge, believing it to be a petrified hornbeam. For this reason the fossil trunk was given the folk name Gyurtyánkő, which means "stone hornbeam." This trunk was first mentioned by Ferenc Kubinyi (1796–1874) in 1837, in a report that was illustrated by Károly Markó, Sr. (1791–1869). Kubinyi was guided by local shepherds to the giant trunk, the "stone bridge," which was still 42 meters long at that time. József Szabó, who also visited the site several times, mentioned in 1867 that it was 24.5 fathoms long. Szabó continued to estimate the total size of the tree, given that the apical part had long been destroyed, at nearly 30 fathoms—more than 56 meters—high. The story of this unusual fossil was summarized nicely by András Tasnádi-Kubacska in 1977, and the paleobotanist János Félix (1859–1941) was the first to make thin-sections for microscopic examination of splinters cracked off the giant trunk. Kubinyi had previously given the trunk the name *Petrefactum giganteum Humboldtii,* the giant fossil of Humboldt, but Félix considered it to be from an ancient pine; he gave it the name *Pityoxilon* sp.

Later this giant trunk was studied by Móricz Staub, and then by János Tuzson. Tuzson abandoned the earlier genus name established by Félix and instead related the trunk to modern pines. This is how the name *Pinus tarnocziensis* came into being. Tuzson visited the locality in Ipolytarnóc and sampled the ancient trunk himself; he was able to discover that the pieces taken earlier by Félix originated from parts of the trunk that had been destroyed before it was silicified. He wrote in 1901, "During my studies I came to realize that the trunk was attacked by fungi and woodborers because there are various cavities, cracks, and hollows in the trunk. The threads of woodborers, which are 1–2 millimeters in diameter, form regular cavities penetrating the trunk and are lined on their insides by tiny crystals. . . . Most of the thin-sections I made show well these preserved structures." The latest study of this trunk was completed by Pál Greguss: first he gave it the name *Pinuxylon lambertoides,* and later more closely followed the rules of nomenclature, referring it to *P. tarnocziense.* Sadly, however, the fate of this remarkable fossil was sealed from the moment of its discovery. First local people carried away small pieces and sold them as hard whetstones in markets; later, as news of the Gyurtyánkő spread, area landowners took away larger pieces for display in their gardens. These larger chunks were soon eroded by weather—mainly frost—and some were later recovered and removed to the permanent exhibitions of the National Museum and to the Hungarian Geological Institute. These rescued pieces have, fortunately, survived.

From time to time the Hungarian media have shown great concern about this fossil tree trunk, and politicians and influential members of the public—recognizing that it has long been on the brink of destruction—have often called for its protection. The pieces that remain at the original site were at one point buried for their own protection, subsequently re-excavated, and then in 1985 covered with a protective roof made of bricks. The sad state of this fossil was even mentioned by János Tuzson in one volume of the *Természetrajzi Füzetek* (Booklets in natural history), published by the Hungarian National Museum: "This fossil, which deserves careful preservation, is subjected again to decay. Around the trunk, crumbled pieces lie en mass, and hikers young and old from Losonc and elsewhere have been smashing the trunk to pieces so it is already damaged on all sides and is slowly losing its original form."

📖 Félix 1887; Tuzson 1901; Greguss 1954

Micrograph of a thin-section from the Ipolytarnóc silicified trunk that was published in János Tuzson's original work.

What remains of a giant tree trunk. Silicified wood surviving millions of years under the surface is degraded at the surface in just a few years by frost and—particularly—vandalism. Only a remnant stub a few meters in length can be visited inside the area of the nature preserve.

Leaves from Ipolytarnóc

Although the most notable fossil plant by far from Ipolytarnóc is the giant stone hornbeam, nearly 10,000 fossil leaves have also been collected from the surrounding area over a period of decades. Most of these fossil plants were recovered from the lower, 20- to 40-centimeter-thick section of the volcanic tuff that overlies the fossil track–bearing sandstone, but some have also come to light from deeper sandstone layers. The tuff immediately above the sandstone is comprised of fine-grained material attributed to a Miocene volcano, and in earlier geological studies it was referred to as the "lower rhyolitic tuff." This term refers to the fact that above the lower level, middle and upper tuff levels can also be found.

The abundant fossil leaves collected from this locality have so far been investigated in detail by Jenő Jablonszky (aka Eugene Jablonsky [1892–1975]), Klára Rásky, István Pálfalvy, and Lilla Hably. The first thorough study was published by Jablonszky—who, in addition to describing a large flora of nearly 30 species, determined the climate at this site (based on the flora) and the geological age of the Ipolytarnóc fossil plants. Jablonszky concluded that the plants in this area flourished on wet soils in a subtropical climate at an age younger than Oligocene but older than the Late Miocene. Later, Rásky described 92 already known and 11 newly recognized plant species from this area, and this long floral list was subsequently revised by Lilla Hably.

Early Miocene plant remains from Ipolytarnóc. These excellently fossilized leaf and twig remains were embedded and preserved in volcanic tuff (original size). (1) *Woodwardia muensteriana* (Presl in Sternberg) Kräusel—a pinnate and compound leaf of an ancient chain fern belonging to the family Blechnaceae. (2) *Laurus princeps* Heer. This entire-margined, beautiful leaf possesses a drip tip and belongs to the laurel family. (3) *Tetraclinis salicornioides* (Unger) Kvaček. This impression of a gymnosperm twig is from an ancient member of the cypress family. Remains of *Tetraclinis* occur quite often in the Hungarian Oligocene fossil record. (4) *Platanus neptuni* (Ettingshausen) Bůžek, Holý, and Kvaček—leaf of an extinct thermophilic plane species. Like surrounding elements, this tree was a characteristic feature of Oligocene floras and was dominant in the Ipolytarnóc flora. The impressions of leaves and twigs from this tree are often found as mass occurrences. (5) *Sabal major* (Unger) Heer—leaf of an extinct fan palm. These large, palmately compound leaves are extremely characteristic elements of the Ipolytarnóc flora. Species in the extant genus *Sabal* are native to the southeastern United States, the Caribbean, and Central and South America. (6) *Cyclocarya cyclocarpa* (Schlechtendal) Knobloch. This leaflet, with a fine serrated margin, belongs to the walnut family and was common at Ipolytarnóc. (7) *Daphnogene bilinica* (Unger) Kvaček and Knobloch—leaf of an extinct species that belongs to the laurel family. The venation of these strictly entire margined leaves is characteristic in that they have a strong basal pair of secondary veins. (8) *Engelhardia orsbergensis* (Wessel and Weber) Jähnichen, Mai, and Walther. These asymmetric, serrated leaflets belong to the walnut family. *Engelhardia* remains were first recorded from the Eocene and the genus is thought to have reappeared several times during the Paleogene and Neogene in phases of warm climate. *Engelhardia* was another dominant element in the Ipolytarnóc flora.

1
2
3
4
5
6
7
8

Next to Ipolytarnóc, on the other side of the Hungarian-Slovakian border is the locality of Lipovany, where volcanic ash preserves rich plant remains.

Over the course of these reinvestigations, some remains described under various older names have turned out to belong to the same species, and some species regarded as novel taxa by earlier workers are actually not supported. The detailed work of Hably has, for example, shown that ferns and lichens were present in the flora, gymnosperms (Cupressaceae and Pinaceae families) were also quite frequent, and dicotyledonous angiosperms were extremely abundant. From the latter group, *Magnolia* and lauraceous remains are noteworthy; numerous species belonging to various genera within the Lauraceae were also identified. From this latter group, the species *Daphnogene bilinica* is the most frequent at Ipolytarnóc; in addition, various groups of dicots have also come to light, including barberries, plane trees, species from the beech and rose families, maple, ivy, and walnut. *Engelhardia,* another of most characteristic elements of the Ipolytarnóc flora, belongs to the latter group. Among monocotyledonous plants, fossils of *Smilax* and palms have been found—and among the former both *Calamus noszkyi* and *Sabal major* are noteworthy; indeed, *Calamus noszkyi* is another of the dominant species in this flora. Jablonszky named this species in honor of geologist Jenő Noszky, Sr., who collected a vast number of fossil plants at Ipolytarnóc. Considering the ecology of the modern plants that are thought to be related to this diverse flora, the Ipolytarnóc site was likely a humid subtropical forest and, according to the work of both Rásky and Hably, there was no frost; the mean annual temperature must have been nearly 20°C. Indeed, some of the conclusions drawn by Lilla Hably are interesting in that they have significance beyond paleobotany: based on the kind of vegetation preserved, this Early Miocene flora cannot be related to older Egerian (Late Oligocene) ones, and thus it is possible that Ipolytarnóc had no close connection with the vegetation of the surrounding area. Thus, this flora might represent an outlier in both space and time—results that are in good agreement with regional tectonics that postulate northward movement of the Pannonian microplate during the Middle Miocene until it reached a latitude of 40°N. Ipolytarnóc is, thus, one of the most remarkable Miocene fossil plant localities in Hungary; the bulk of its fossil plants are stored in the Hungarian Natural History Museum and in the Hungarian Geological Institute.

📖 Jablonszky 1915; Rásky 1959; Pálfalvy 1976; Hably 1985

Silicified Wood and Coalified Remains

Pieces of silicified wood that have come to light—mostly from northern Hungary—are also noteworthy as Miocene plant specimens. The process of silicification is, of course, not age dependent and can happen at any time if the environmental conditions are correct; however, in Hungary most remains preserved in this manner are from the Miocene. In the silicification process original organic material is replaced by silica at the cellular level and so fine details of tissue structure are often excellently preserved. Investigation of these remains is, however, quite difficult because serial thin-sections have to be made from the rock-hard silicified plant pieces.

Investigation of this type of silicified (or petrified) wood in Hungary was initiated in the second half of the nineteenth century by János Félix, who then worked in Leipzig. Based on a "wood-stone" collection of mostly unknown origin, he described several new species with the thoroughness of a naturalist using the microscope. Later his work was continued by Pál Greguss, who was able to identify 99 silicified specimens in this collection from the lower part of the Middle Miocene. Of these fossils, 33 belonged to conifers, 63 to dicots, and 3 to palms; they had been collected mostly from the Cserhát and Mátra Mountains. The high silica content in fossil wood from this northern mountain range is a result of the silica-rich thermal waters that accompanied Miocene volcanism.

Particularly beautiful specimens have been collected from Megyaszó, where the internal parts of various silicified trunks and twigs is often composed of a colorful variety of quartz called opal. In Hungary, coalified wood remains are frequently found in Middle Miocene deposits (such as in Várpalota), where the original structures of plant tissues have been altered and become indiscernible. Thus, paradoxically, the formation of coal deposits from the accumulation of large amounts of plant material is usually not useful in paleobotanical studies.

📖 Greguss 1954, 1969

Erdőbénye and Tállya: Fossil Plant Localities in the Hungarian Sarmatian

Among the fossil plant occurrences that are known from the younger parts of the Miocene, the Middle Miocene localities in the Mecsek Mountains (i.e., Szászvár, Hidas) and those in the northern mountain range certainly deserve attention. In the second of these two areas, beautiful Sarmatian fossil plants are mainly preserved in volcanoclastic rocks. The first to work on these fossils was Gyula Kováts, author of the first Hungarian paleontological monograph: His paper on this fossil flora, "Fossile Flora von Erdőbénye," was published in 1856 in the journal *A Magyarhoni Földtani Társulat Munkálatai,* the first publication series of the Hungarian Geological Society. Kováts later described the beauty of the Tállya fossil plants he collected in the enthusiastic way that characterized both him and his age.

Portrait Gallery

Baron Gábor Andreánszky (1895–1967)—Botanist and Paleobotanist

Gábor Andreánszky was born in Alsópetény and served in World War I, thus interrupting his scientific studies at the University of Budapest for several years. Following his graduation in 1920 he worked in the university's Department of Systematic Botany, and in 1940 became the first-rate museum keeper of the Department of Botany of the Hungarian National Museum. In 1942 Andreánszky was appointed to a professorship at the university and between 1943 and November 1945 he was director of the Department of Botany. In 1945 the Baron became a corresponding member of the Hungarian Academy of Sciences; however, for political reasons he (along with many other excellent scientists), was relegated to the position of consulting member in 1949. In 1945 he returned to the university as a senior lecturer but in 1953 he had to leave because of his continued resistance against the autocratic regime in Hungary. During one ceremony he demonstrably did not take part in the rhythmic applause and cheering when the speaker mentioned the name of comrade Rákosi, the "wise teacher of the nation."Following his forced departure from the university, Andreánszky worked in the Department of Botany at the museum until nearly the end of his life. As a botanist he studied all the floras of the Mediterranean region, Alps, and Carpathians and he made significant contributions to paleobotany with his investigations on Tertiary floras. Andreánszky is also remembered because he wrote the first paleobotany textbook published in Hungarian.

📖 Pálfalvy 1969

1
2
3
4
5
6
7
8
9

This Sarmatian fossil flora from northern Hungary was later studied by Gábor Andreánszky, who described more than 500 species in a 1959 monograph. This large number of species is perhaps justified on the basis of the abundant fossil material and Andreánszky's factual knowledge, but his attitude—to describe each tiny difference as a specific, distinctive feature—no doubt has also contributed to the high species number. Sarmatian localities rich in fossil plants were also mentioned by Andreánszky from Balaton, Erdőbénye, Bánhorváti, Sály, Mikófalva, Füzérradvány, and Felsőtárkány—and the most recent studies that are focused on these assemblages—were published by Boglárka Erdei. Erdei described nearly 160 plant taxa based on an investigation of more than 5,000 specimens from six Hungarian fossil plant localities. She named one leaf, belonging to the barberries *Berberis andreanszkyi*, in honor of her renowned predecessor.

We know that the diversity of the fossil plant assemblages from the Sarmatian deposits of northern Hungary reflects various habitats and, as a consequence, they vary slightly from site to site. Among others, an extinct legume called *Podocarpium* and a swamp cypress called *Glyptostrobus*, are noteworthy taxa that occur in this rich flora.

📖 Kováts 1856a, 1856b; Andreánszky 1959; Erdei & Hír 2002

Invertebrates

Corals

Among Lower and Middle Miocene sediments, the Badenian Lajta Limestone and similar rocks include rich coral assemblages in some places. August Emmanuel Reuss (1811–1873), appreciated for his results on Cretaceous and Cenozoic faunas, was the first to publish a study about these fossils in a classical 1872 monograph on the fossil corals of the Austro-Hungarian Empire. The type locality for many new species he described is the Fekete Hill at Visegrád, the fauna of which was reexamined by Gábor Scholz at the end of the 1960s. Scholz also studied Triassic reefs and Cretaceous ammonites.

Middle Miocene (Sarmatian) Plants from Erdőbénye (original size). (1) *Carpinus neilreichii* Kováts—winged fruit of an extinct hornbeam. (2) *Carpinus betulus fossilis* Engelhardt and Kinkelin—winged fruit of a Miocene hornbeam species resembling the extant Common Hornbeam. (3) *Acer* sp.—winged fruit of a maple. (4) *Zelkova zelkovifolia* (Unger) Bůžek and Kotlaba—leaves described as a member of zelkovas. (5) *Quercus mediterranea* Unger—leaf of an extinct oak. (6) *Ulmus braunii* Heer—leaf of an extinct elm. (7) *Zelkova zelkovifolia* (Unger) Bůžek and Kotlaba—leaves belonging to the zelkovas. (8) *Quercus kubinyii* (Kováts ex Ett.) Czeczott—impression of an oak leaf called the Oak of Kubinyi. Originally the plant possessing these characteristic and impressive leaves was described as a *Castanea* species. It was assigned by the first Hungarian paleobotanist to an ancient chestnut species. (9) *Podocarpium podocarpum* (A. Braun) Herendeen—the pod and seed of a plant belonging to the legume family. Remains of this species (including beautiful impressions of leaflets) were mentioned in older paleobotanical literature as *Podogonium knorrii* Herr.

A colonial coral (*Montastraea*) from the Badenian at Lăpugiu de Sus (close to original size).

A number of colonial, reef-building corals occur in these Badenian assemblages. Their reappearance in assemblages indicates a warming in climate as coral reefs spread to the north during the Neogene. This indicates that the seawater temperature was generally above 16–17°C, based on what we know about the environmental demands of present-day corals. However, the size of this reef is also quite modest, so conditions were clearly not yet optimal for the corals.

In life, coral skeletons consist of aragonite, which is dissolved during diagenesis into the Lajta Limestone facies. These skeletons remain as internal molds. Since the stone that infills the space around the coral walls shows the so-called negative form of the real skeleton, usually silicone rubber—or resin—molds have to be made to determine which coral is which. Pál Müller, a recognized expert in the study of fossil decapod crabs, excelled in the study of Badenian corals. Since the walls were rather thin in several genera, the space formed during their dissolution is also very narrow. In such cases, the filling material can be forced into the fissures of the rock only by using a centrifuge!

The remains of fine corals can be found in the Mecsek around Pécsszabolcs, in the southern part of the Börzsöny in Visegrád, and in nearby Veszprém in Márkó. Some of the most characteristic forms are *Tarbellastraea, Porites,* and *Montastraea*.

📖 Reuss 1872; Kopek 1954; Scholz 1970; Hegedűs 1970; Hegedűs & Jankovich 1972

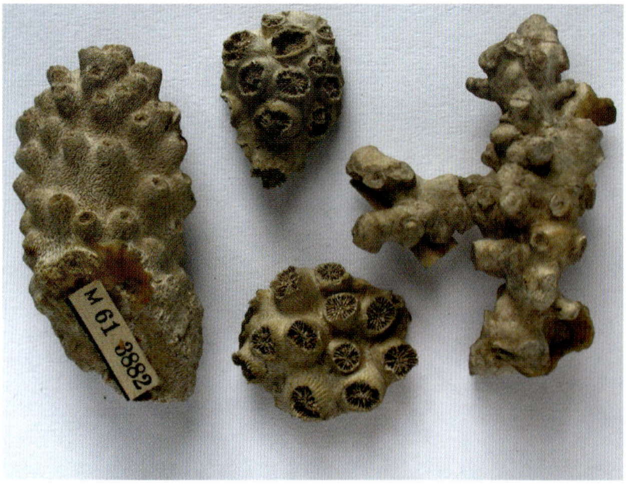

Colonial corals from Miocene sands around Sámsonháza (close to original size).

Gastropods and Bivalves

Bivalves in the Earliest Miocene

Mollusks are the most common macrofossils in rock sequences deposited during the Early and Middle Miocene. Some rocks were even named after their characteristic gastropod (for example, "pleurotoma clay") or fossil bivalve ("large pecten sand"). A great number of studies have been published about these massed and often well-preserved fossils over the past 150 years, but there are surprisingly few up-to-date works. Monographic studies are available only for Badenian gastropods and scallops.

The Eggenburgian age is generally indicated by the beginning of a new sedimentary cycle. The only exception to this is the Szécsény Schlier basin, where deposition of the deep-water "amusiumic schlier," named after the bivalve fossil *Lentipecten,* earlier *Amusium,* continued. A paleoenvironmental evaluation of this bivalve assemblage, mostly poorly preserved forms with generally thin shells, was published by Tamás Báldi and Szabolcs Leél-Őssy in 2003.

The coastal, shallow-marine Budafok Formation, earlier called the "large pecten" or "anomia" sand, contains a completely different mollusk assemblage. As implied by its old name, the most common fossils are the sometimes palm-sized specimens of *Pecten* and *Chlamys,* both genera of bivalves. The sandpit at Pacsirta Hill in Budafok offered good exposures of this succession until it was filled with waste. Nevertheless, thick oyster (*Crassostrea*) fragments can still be found in the neighborhood.

The Ottnangian is also represented, curiously enough, by marine pebble and sandstone layers that contain a fauna comprising mainly scallops. Rocks of this age can be seen in an isolated area near Várpalota, in the vicinity of Bántapuszta. At the same time, in northern Hungary, the Salgótarján Coal Formation was deposited: the barren beds of this unit contain a massive amount of freshwater (*Unio, Congeria*) and brackish-water (*Cardium*) fossil bivalves. The Kazár Sandstone, which caps this carbonaceous sequence, contains evidence for the second complete isolation of the Paratethys Sea; the small bivalve, *Rzehakia,* occurs in large numbers. Next in the succession in the Carpathian Basin is the "small pecten-rich sandstone," now called the Egyházasgerge Sandstone, which contains an assemblage of comb shells (*Chlamys*), leading into fresh–brackish water layers. The fauna of the Fót Formation ("*Balanus*-rich" sand) is similarly comprised mainly of species of *Chlamys.*

📖 Csepreghyné Meznerics 1954; Bohnné Havas 1971, 1985; Báldi & Leél-Őssy 2003

Miocene (Badenian) mollusks (original size). (1, 2) *Tudicla rusticula* (Basterot)—a beautiful gastropod with a long canal. The present-day representatives of this genus live in the Indian Ocean (Várpalota). (3) *Solecurtus candidus* (Renieri)—a burrowing bivalve with thin valves (Várpalota). (4) *Ficus condita* (Brongniart). This species also occurs in Hungarian Oligocene (Egerian) deposits. The genus is represented by numerous species even today. It is often called *Pyrula* in older literature (Várpalota). (5) *Cymatium affine* (Deshayes). This small predatory gastropod belongs to the group of tun shells and is closely related to one of the biggest gastropods living today, Tritonium (*Charonia tritonis*) (Borsodbóta). (6) *Conus antediluvianus* Bruguière. Small and large species of *Conus* belong to a group of gastropods with complicated color pattern, living in present-day shallow sandy-bottom seas and reefs. However, patterns and colors are preserved on fossil specimens only in exceptional cases. These forms belong to Toxoglossa; the word *toxoglossa* literally means "poison tongue" and the poison of some species is a danger even to humans (Borsodbóta). (7) *Fusus vindobonensis* Hoernes and Auinger—a gastropod with a long canal and a tall, slim shell (Borsodbóta). (8) *Murex brandaris* Linné. This gastropod is one of the most noteworthy of the rock shells or *Murex:* it was one of the most important Imperial Purple–producing species. The fabrics woven for royals used to be dyed with the precious extracts from recent Murex gastropods (Buituri). (9) *Aporrhais* sp. *Aporrhais,* or "pelican's foot shell," lives on muddy gravel in the subtidal zone (Várpalota). (10) *Phallium (Semicassis) miolaevigatum* Sacco. This gastropod and the closely related *Cassis,* or cameo shell (as well as *Cassidaria*) belongs to the group of tun shells (Tonnoidea). The name refers to the last bend of their shell because it bulges out (Borsodbóta). (11) *Cassidaria cingulifera* Hoernes and Auinger—a beautiful form ornamented with spiral ribs and nodes (Borsodbóta). (12) *Natica millepunctata* Lamarck—*Natica,* also called necklace shells, are predators. This fairly large gastropod creeps through the sandy bottom seeking prey. They bore a round hole through the shell of bivalves and gastropods and eat out their soft flesh (Borsodbóta).

1
2
3
4
5
6
7
8
9
10
11
12

1
2
3
4
5
6
7
8
9
10

Badenian Mollusks

The best-preserved and richest Miocene mollusk assemblages are known from Badenian sediments. In particular, the variety of gastropods found at this time is amazing. The rocks containing a great diversity of mollusks can be categorized into three groups. The best-known is the Lajta Limestone, which is well exposed around Sopron, in the western part of the Bakony Mountains, around Budapest, in Nógrád County, in the Mecsek Mountains, and in the classical name-bearing locality in the Lajta Mountains in Austria. Gastropods can be collected from the Lajta Limestone only as molds or imprints that preserve the outer shape of the shell and surface ornamentations—sometimes down to the tiniest details. The most common fossils are bivalves: scallops (*Flabellipecten, Pecten*) and oysters (*Cubitostrea*) have been preserved with their shells; others are known only from internal casts. These include frequently encountered benthic forms such as *Acanthocardia, Ringicardium, Linga,* and *Pelecyora*. There are also remains of specimens of rock borers, fossilized on the spot in these ancient coral reefs. In addition to the date seed–shaped *Lithophaga,* the intriguing, round *Jouannetia* is also common in these sediments.

Although the bivalve fauna of the Lajta Limestone Formation has interested a good many researchers, no one yet has written a monograph. The only exception to this lack of research is the scallops: the Hungarian Neogene forms were described in a book by Ilona Csepreghyné Meznerics. The sporadically outcropping clayey and sandy layers of this formation are mostly famous for well-preserved gastropod remains, although they do also contain a large number of bivalves. One such deposit is the deep-water Baden Clay, which used to be called the *pleurotoma*-rich clay, after the traditional name of one common gastropod fossil (*Clavatula*). In contrast, the Szilágy Claymarl was deposited in slightly shallower water: in this rock the gastropod *Turritella* and the bivalve *Corbula* are frequently found.

Our knowledge of the Badenian gastropods from the Carpathian Basin is mostly due to the work of László Strausz, who studied them for decades. He published his results in numerous papers of various lengths, as well as in a 1966 monograph in German. Strausz also wrote a 1962 identification book that included 79 photo tables—a truly rare contribution today. The richest gastropod site is the Badenian sand at Szabó Quarry in Várpalota, the fauna of which was also described in detail by László Strausz.

At one time Miocene brown coal was mined at Herend and in its vicinity. The rich mollusk fauna that came to light during the exploration and mining in this area was described by József Kókay, a prolific researcher, respected for being most familiar with the geology of the neighborhood of Várpalota. Kókay was also the first to reexamine and document the well-preserved assemblage that was unearthed during the digging of one of the Illés Street canals in Budapest at the end of the nineteenth century. His 2006 work is a monographic description of the Early and Middle Miocene non-marine gastropods and bivalves found in the Bakony Mountains.

The fauna of the sites around the Börzsöny Hills (Szob, Nagymaros, Letkés) contain about 280 gastropod and 80 bivalve species and were documented by Hugó Böckh, a short-lived director of the Geological Institute of Hungary and hydrocarbon expert who discovered reserves in Iran, and by Ilona Csepreghyné Meznerics. It was also

Miocene (Badenian) gastropods (original size). (1) *Hexaplex turonensis pontileviensis* (Tournouer)—*Murex*-type gastropod. Representatives of this group are predators or scavengers. A lot of species fight virtually hand to hand with bivalve prey. It will force the valves of the bivalve open and suck out the soft body using its proboscis (Várpalota). (2) *Nassarius rosthorni* (Partsch in Hörnes). Specimens of this genus are all small gastropods with squat shells. Present-day species eat organic debris, are scavengers, or are predators. They search for prey by reaching around with their proboscides (Hidas). (3) *Melongena cornuta* (Agassiz). Species of this still-living genus that have the most beautiful shells live in Caribbean waters. In Hungary the best-preserved and biggest specimens were found at Várpalota, and are similar to the one shown here. (4) *Babylonia brugadina* Grateloup. Present-day relatives of this plain, smooth-shelled gastropod live in Indo-Pacific waters (Borsodbóta). (5) *Pereiraea gervaisi* Vézian—A rare gastropod that has been found at just a few fossil sites (Szentgál). (6) *Fissurella italica* Defrance. The members of this genus, together with a number of relatives (such as *Patella*), live on rocky beaches. Since they are capable of tightly clinging onto stones, they are able to endure being out of water from time to time and will stay in one place at low tide until the water returns. When they are underwater, these gastropods browse the seaweed stuck to stones (Borsodbóta). (7) *Turritella aquitanica* Tournouer—*Turritella,* or tower shells, move slowly on the sea bottom and eat tiny organic debris from the sediment. These gastropods are found in practically all present-day seas, although some of the largest forms tend to live in tropical waters. Small, regular holes that mark the (mostly successful) attacks of gastropod predators on these shells are common in Hungarian Miocene specimens (Várpalota). (8, 9) *Strombus coronatus* Defrance. Present-day *Strombus* have heavy, strong shells ending in a fanlike ear (this has broken off the illustrated specimens). One interesting thing about these gastropods is that they do not creep along the sediment, but throw themselves forward (Lăpugiu de Sus). (10) *Xenophora testigera* Sacco. *Xenophora,* also called carrier shells, build smaller or larger pebbles, gastropod, bivalve, or other skeletal fragments into their growing shells. The strengthened, heavier shells of these gastropods are able to resist wave action (Borsodbóta).

1
2
3
4
5
6
7
8
9
10
11
12
13
14
15
16
17
18
19

Csepreghyné Meznerics, and later Margit Bohnné Havas, who studied the fossils from Hidas in the Eastern Mecsek Mountains. The original ornamentation of the cone shells (*Conus*) found at this site can be seen under ultraviolet light. The strangest form among these gastropods is *Pereiraea*, which is taller than 10 centimeters and has a last whorl which is ornamented with spikes perpendicular to the axis of the shell. However, these are not the reasons it is unusual; rather, it is known only from three distinct regions that are far from each other (Szentgál, Szurdokpüspöki, Slovenia). The biggest Badenian gastropod found in great numbers in the Szabó Pit at Várpalota is *Galeodes* (which earlier authors called "*Melongena*"): the 20-centimeter-tall, thick shell of this taxon is ornamented with large spikes, as is expected for a tropical form. Cone shells are extremely common in Badenian sediments and indicate tropical, warm seas. These shells belong to a group called "poison tongues" (Toxoglossa), whose pricking spike can even be dangerous to humans.

The sand layers of the Szabó Pit also comprise a rich bivalve fauna which includes *Anadara* (related to *Arca*), *Glycymeris*, *Linga*, and *Anomia*. The latter species has remarkable yellowish-brown valves.

The sequence at the Sámsonháza Formation also contains Badenian sandy-calcareous sediments. The exposures of these sediments and their faunas in the Mátra have been studied by numerous workers—including Endre Kutassy, László Bogsch, and Zita Korom. One striking characteristic of this fossil assemblage is the small size of the specimens: the evolutionary and paleoenvironmental interpretation of this so-called dwarf fauna—which includes large numbers of *Ancilla* and *Nassa* gastropods and *Chlamys*, *Variocorbula*, and *Linga* bivalves—has been the subject of several studies.

The deposition of deeper-water schlier successions in the Carpathian Basin continued into the Badenian. In these sediments, poorly preserved mollusk faunas that include bivalves such as *Amusium*, *Nucula* and *Variocorbula*, gastropods such as *Turritella* and *Ancilla*, and Pteropods ("winged feet" in Greek) predominate. The latter group, because they floated in the open sea, spread a great distance across the Paratethys Sea and is useful for stratigraphic correlation.

Research on pteropods has improved significantly over the last few years because of the work of Margit Bohnné Havas and Irene Zorn. The fauna of the Vienna Basin and the fossil assemblages that have been collected from Buituri, Hunyad County, and from around Lăpugiu are particularly noteworthy outside Hungary. This fossil site, a classic locality, is situated in a riverbank and is eroded continually by water. In spite of this, collecting is strictly prohibited as the site, and its vicinity, is under protection. However, this is not a problem for the locals, who offer beautiful fossil specimens for a considerable price to tourists.

🕮 Zilch 1934; Bogsch 1936, 1943; Strausz and Szalai 1943; Csepreghyné Meznerics 1960; Strausz 1966b; Kókay 1966, 1996, 2006; Bohnné Havas 1973

Middle Miocene (Badenian) mollusks (original size). (1, 3) *Subula fuscata* Brocchi. These gastropods, together with other similarly tall, slim forms, burrow themselves into loose, sandy sediment by moving their bodies in a way that resembles a screw penetrating wood. This species is usually mentioned as *Terebra* in earlier literature ([1] Lăpugiu de Sus, [3] Buituri). (2) *Dentalium* sp.—These so-called tooth shells are not real gastropods: they constitute a separate class (Scaphopoda, aka "elephant tusk shells") within mollusks. True to their names, they burrow into the sand, and only the tip of the approximately vertically positioned shell sticks out of the sediment (Lăpugiu de Sus). (4) *Cancellaria (Sveltia) varicosa* (Brocchi)—a gastropod ornamented with spiral ribs (Buituri). (5) *Conus extensus* Partsch—a cone shell with an extremely elongated shell (Kostej). (6) *Turritella riepeli* Partsch—a very large tower shell. Note that the shell illustrated here was damaged several times when the gastropod was still alive, and also that the animal patched these injuries up (Lăpugiu de Sus). (7) *Venus multilamella* Lamarck. The movement of this bivalve is helped by its finely-bowing concentric ribs when it burrows itself into the sand (Lăpugiu de Sus). (8) *Bufonaria marginata* (Martini)—a medium-tall, fairly thick-shelled gastropod (Kostej). (9) *Anadara diluvii* (Lamarck). This small bivalve is generally called *Arca diluvia*, that is, "ark of deluge" in older geological literature (Lăpugiu de Sus). (10, 11) *Neritopsis moniliformis* Grateloup—a small, thick-shelled gastropod ornamented with pearly nodes. Even the original color is preserved on the specimen illustrated (Lăpugiu de Sus). (12) *Ancilla glandiformis* (Lamarck). A characteristic feature of this species is the wide stripe on its smooth whorl, where stripes of different colors can be seen (Lăpugiu de Sus). (13) *Ranella* sp.—gastropod with a long canal, characterized by ribs, nodes, and a strengthened aperture (Kostej). (14) *Athleta rarispina* (Lamarck)—a gastropod also known from the Oligocene. The last whorl in this species almost completely covers the earlier ones (Kostej). (15) *Phallium (Semicassis) saburon* (Lamarck)—a round, plain gastropod that conforms to the Hungarian name for this group (the "tun shells") (Roşcani). (16, 17) *Solarium caracollatum* Lamarck. The earliest, microscopic whorls of this flat gastropod have sinistral coiling, while the rest of the shell is dextral (Lăpugiu de Sus). (18, 19) *Cypraea* sp. In general, *Cypraea* are also called kauri or cowrie shells. The peculiar fine shine and pattern seen in this specimen are rarely preserved on fossil specimens (Lăpugiu de Sus).

Portrait Gallery

Ilona Csepreghyné Meznerics (1906–1977)—Researcher of Miocene Mollusks

This prominent scientist, who was often regarding affectionately as Ilus by her personal acquaintances, was born in Subotica. She received a teaching degree in chemistry and physics from the Budapest University of Sciences in 1929. She earned a doctoral degree in geology in 1930 and then received a state scholarship from the Collegium Hungaricum, a school attended by many Hungarian paleontologists, in Vienna. Afterwards, with the help of a private scholarship, she worked for another two years at the Naturhistorisches Museum in Vienna. She returned to Hungary in 1934 and was employed by the National Committee for Unemployed Professionals as a trainee and later as a teacher. She then moved to the Department of Paleontology and Geology at the Hungarian Natural History Museum in 1940. Meznerics was head of this department from 1951 until she retired in 1970, and was responsible for the hard task of reorganizing this collection, which burned in 1956. She became a doctor of earth and mineral sciences in 1957.

After her retirement, Csepreghyné Meznerics worked as a research fellow at the Geological Institute of Hungary for three years. She was chair of the Paleontology Section of the Hungarian Geological Society for three years and was elected an honorary member of the Society in 1975.

As a paleontologist, Csepreghyné Meznerics is known mostly for her examination of Miocene mollusk faunas. She worked on Austrian fossil assemblages during her stay in Austria, but otherwise she dealt exclusively with Hungarian fossils—publishing more than 50 studies on her results, many of which were issued as monographs in the series Annals of the Hungarian Geological Institute. Her 1960 study, "Pectinidés du Neogène de la Hongrie et leur importance stratigraphique," was published in the *Mémoire de la Société géologique de France* and dealt with Hungarian Miocene scallops and their stratigraphic importance.

📖 Bogsch 1977

Sarmatian Mollusks

Jenő Boda published a description of about 160 invertebrate species and subspecies in a monograph on Sarmatian fossil mollusks from Hungary in 1959. The majority of the species he described are gastropods: the shells of these air-breathing snails were introduced into the sea by streams and rivers and are preserved in the rock beds.

The most common genera is *Cerithium,* usually found preserved as a mold without a shell. The remaining voids are characteristic to the Sarmatian Limestone. The fossils are usually from the genus *Pirenella,* which is basically marine but can tolerate an elevated salinity. This rock also conceals a huge number of specimens of a closely related species called *Pirenella* alongside other frequently found marine gastropods called *Gibbula.* Both these genera are alive today.

In lower Sarmatian clayey strata, small gastropod shells are found in great numbers. These include *Mohrensternia,* a lower Sarmatian "key fossil" for the Central Paratethys Sea region.

Although less frequent than gastropods, bivalves are also common in Sarmatian hard limestone. The thick valves of one oyster species of broad environmental tolerance, *Crassostrea gingensis,* occur in a few places—sometimes in large numbers. These oysters can be found, for example, in the white limestone cliffs of the Haviboldogasszony Hill in Pécs. The most frequent species among other bivalves are *Obsoletiforma vindobonensis* and *Sarmatimactra vitaliana,* both relatives of the edible cockle. The latter is an index fossil for the upper part of the Sarmatian, alongside small specimens of *Ervillia,* which also characterize this stratigraphic level. The narrow, bent valves of *Musculus,* a genus similar to *Modiolus,* are also eye-catching not for their size but because they are so common.

Worms with a calcareous shell also lived in the Sarmatian Sea alongside mollusks. The spiral, gastropod-shaped *Spirorbis,* which lived attached to plants, is one common worm—as is *Hydroides,* which initially had a spiral shell that became more undulated in a later stage of its ontogeny.

Oligocene and Early Miocene scallops (original size). (1) *Flabellipecten leythajanus* (Partsch). This regular scallop with two prominent auricles is one of the most characteristic fossils in Hungarian Badenian sediments (Budapest, Rákos). (2) *Oopecten gigas* (Schlotheim). This shell is an example of one of the most prominent fossils in the Early Miocene (Eggenburgian) large pecten sand and certainly deserves its name: some examples of this species reach 15 centimeters in diameter (Budafok).

1

2

Miocene scallops (original size). (1) *Flabellipecten kocheni* Fuchs—a small, round, wide, flat form with moderately sized ribs (Kemence). (2) *Flexopecten crestensis* Fontannes—a small form with large auricles. The ornament on the right-hand valve is extraordinary among pectens: radial grooves and very wide and flat ribs decorated with secondary ribs alternate. This species is commonly mentioned as *Chlamys palmata* in older literature (Budafok). (3) *Chlamys elegans* (Andrzejowski)—a small form with a rounded and quite tall beak (Pécsszabolcs). (4) *Aequipecten seniensis* (Lamarck). This beautiful, finely ornamented bivalve is one of the reasons that the small pecten sand was given this name. In older literature this form is known mainly as *Chlamys scabrella* (Lamarck) (Egyházasgerge). (5) *Pecten beudanti* Basterot—a small form with one valve that is very round, while the other is flat or slightly concave. Not long ago, very nice looking specimens of this scallop could be collected from the Pacsirta Hill in Budafok (Budafok). (6) *Chlamys biaense* Meznerics—a small form ornamented with very fine ribs (Bia). (7) *Pecten aduncus* Eichwald—a medium-sized form with a very round dextral valve and a flat or slightly concave left valve. This is one of the most common bivalves found in the Lajta Limestone (Sankt Margarethen). (8) *Flexopecten crestensis* Fontannes. The flat left valve of this species is seen here encrusted with barnacles (Balanuses) (Budafok). (9) *Gigantopecten nodosiformis* (De Serres in Pusch). This is a form that grows even bigger than palm size. Its ribs are strong and are oddly ornamented with nodes on one valve. This species is commonly referred to as *Chlamys latissima* in older literature (Budapest).

Bivalves feeding on rocks—these so-called rock borers actually eat themselves into coral reefs or limestone cliffs on seashores. Bivalves spend their lives in self-made narrow hollows and are safe from dangerous waves on the beach. The elongated form in the photo is a real rock eater (*Lithophaga*); the other, rounded form is *Jouannetia*. Representatives of both genera are abundant in the Lajta Limestone Formation (close to original size).

In addition to these fossils, only a few genera of bryozoans complete the Sarmatian invertebrate macrofossil assemblage in these limestone successions.

📖 Boda 1959; Harzhauser & Kowalke 2002

Arthropods

Fictitious Fossils: Digger Wasps from the Lignite at Herend

In the middle of the Bakony Mountains, in a small area bordered by fault lines, a preserved Middle Miocene sequence includes interbedded lignite banks. These banks were intensively mined in the 1950s and 1960s. As a result of this exploration, previous deep-drilling excavations, and the hard work of József Kókay, the geology of this region and its fossil assemblage are well known and documented. The most famous fossil from this site, however, was not mentioned in Kókay's 1966 monograph, but was described by none other than Endre Dudich (1895–1971), an eminent zoology professor and two-time member of the Hungarian Academy of Sciences—two-time because Dudich, together with a number of other excellent scientists who were unwilling to serve the new communist system, was downgraded to consulting member in 1949. After several years of being deprived of all rights as an academician, he was restored to an ordinary member of the academy in 1964.

The fossil in question was a real novelty. Dudich's study, published in the *Bulletin of the Hungarian Geological Society,* dealt with digger wasp remains that had been found in the woody lignite mined near Herend. Digger wasps place their eggs in protected places, sometimes in burrows scraped in trees, and drag paralyzed insects inside to feed their hatched larvae. Dudich described burrows in the lignite and the insect remains found inside them in his study. Although this excellent scientist was doubtful that the findings he reported were real fossils, Elemér Vadász—senior editor of the journal at the time and a geological authority—assured the author and his readers in a footnote that according to a "geologist's opinion" (referring surely to himself) Dudich was, indeed, describing a rare Miocene-aged fossil.

This paleontological sensation did not last long. Sándor Pordán (1940–1999), who graduated from the University of Budapest in 1964, did not respect the fossils shown to the students during a study excursion but pressed them with his fingers. Some disgusting material, the adipose body, gushed forth from the supposed Middle Miocene insect remains. Dudich's fossils turned out to be the eggs of present-day digger wasps, embedded in the Herend lignite, instead of the remains of 10-million-year-old insects.

📖 Dudich 1961

Decapods

Decapods, less important in Oligocene fossil assemblages, became more common in the Miocene, especially in the Badenian. In Hungary, a few species were first described by the French paleontologist Paul Brocchi (1838–1898) from the railway cutting in Rákos, Budapest (specimens were collected for him by Hébert and Munier-Chalmas). Later, Alexander Bittner, a specialist working on Triassic invertebrate fossils, published a series of studies about the faunas of other fossil sites. Imre Lőrenthey also worked on these remains at the end of the nineteenth century, but his enormous volume of work, including the descriptions of many new species, remained in manuscript form for a long time because of his sudden death. According to Pál Müller, who carried on and completed his work, he fell behind with publishing his study because the figures that he had were of very poor quality and he wanted to redraw them. (Apparently Lőrenthey had bad luck with his

Mollusks other than gastropods and bivalves sometimes also occur in Badenian sediments. In this photo, the rear section of the calcareous skeleton of a so-called sea cradle shell (chiton, Polyplacophora). This fossil comes from Bánd, in the foreland of the Bakony Mountains (SEM image). After Dulai 2005.

Gastropods and bivalves from the Sarmatian sequence around Tinnye (1–4, 7, 8: original size; 5, 6: 2× magnification). (1) *Sarmatimactra eichwaldi* (Laskarev). (2) *Obsoletiforma vindobonensis* (Partsch). (3) *Pirenella picta* (Defrance). (4) *Gibbula podolica* (Dubois). (5) *Modiolus incrassatus* (d'Orbigny). (6) *Musculus sarmaticus* (Gatuev). (7) *Plicatiforma latisulca* (Münster). (8) *Venerupis gregarius* Partsch.

illustrations. Some parts of the figures in his paper about Pannonian fossils from Nagymányok in the northern Mecsek Mountains do not even resemble the specimens, as certified by the apologetic note from the author in the accompanying table caption.) Since Beurlen, who edited the manuscript for publication, did not check the faithfulness of these drawings, they were published in rather imperfect form. (It is well known that Beurlen later left Europe and moved to South America, not out of shame for these poor figures but because of his Nazi past.)

In the 1960s, Pál Mihály Müller started to deal again with decapods from the Paratethys. We must pay tribute to his work in this field because it must be remembered that he studied fossils only in his free time: Müller was employed as a geologist, and worked in other areas. One result of his work was a huge 1984 monograph, published in English in the Palaeontologica series. Müller earned the title of doctor of the Hungarian Academy of Sciences for his prodigious scientific activity.

In addition to examining specimens collected by former specialists and kept in museums, Müller needed newly collected fossils to write this monograph, which included a description of the complete Badenian fauna. There are still heaps of small, centimeter-sized lumps of rock at many of the Badenian outcrops in the Carpathian Basin that were probably left behind by Pál Müller.

The shells of fossil decapods are very fragile, so they are preserved only under special conditions, such as in the protected holes and cavities of coral reefs on their lee side. The fossilized specimens often require strengthening; in many cases, silicone molding is the most suitable method.

As a result of this work, 102 species and subspecies (21 of them new), were documented, and the Badenian decapod assemblage of the Central Paratethys is one of the most diverse fossil faunas known from anywhere in the world. The railway cutting at Rákos provided the most forms from all of the fossil sites. Considering the number of specimens of determined genera, *Calappa, Matuta,* and *Callianassa,* a burrowing form that leaves strange traces in the deposit (*Ophiomorpha* trace fossil genus) are worthy of mention.

📖 Müller 1984, 2006

Cirriped Crustaceans

Remains of crustaceans other than decapods also occur in the Early and Middle Miocene, sometimes in such numbers that they determine the character of the rock. Such deposits, including successions of sand and gravel layers, are exposed across the Pest plain, especially around Fót. These are often called "*Balanus* sands" in the older professional literature, after the small but spectacular fossil *Balanus,* widespread genus of "curl-footed" crustaceans or barnacles. These are crustaceans that have given up moving freely, live attached to a hard substrate, and have a skeleton of six tightly set and two movable lamellae. Their frustum of cone-shaped shells often cover large numbers of stones and pebbles that lie on the bottom of shallow seas and along beaches.

Pál Mihály Müller (right)—geologist, paleontologist, and specialist on decapod crustaceans—and his former classmate László Trunkó—professor at the University of Karlsruhe—on Rysy Peak in the High Tatras Mountains.

Balanus larvae—without a steady sea bottom—stick, for example, to floating, drifting tree trunks or to slowly moving whales. Sediments containing the remains of curl-footed crustaceans are generally considered to be certain indicators of shallow water or former beaches.

The remains of these crustaceans are not too attractive to most paleontologists, but Charles Darwin (1809–1882) made a name for himself as a paleontologist by describing the known fossil *Balanus* from the United Kingdom. In Hungary, Gábor Kolosváry, also a coral expert, published a few short papers about *Balanus*.

📖 Kolosváry 1948, 1949b, 1950

Sarmatian Insect Remains from the Tokaj Hills

In the Tokaj Hills, a number of lakes of different sizes, both freshwater and salty, formed on the uneven surface of rhyolite tuffs during the Sarmatian. The silica content of these volcanic rocks led to massive deposition of siliceous algae and finally siliceous earth (diatomite) deposits. These finely laminated diatomites were slowly deposited in less than 10-meter depths and preserve the remains of insects and other arthropods, including extremely fragile forms. György Encsy from Tállya has collected a number of fine specimens, and his spectacular private collection can be viewed by appointment. The first report on the fossil insects from this area was published in 2002 by entomologist György Sziráki and geologist Alfréd Dulai, of the Hungarian Natural History Museum. Sziráki and Dulai noted the remains of eight insect genera alongside a false scorpion and a short-tailed scorpion among the arthropods they examined. Indeed, the fossil insect fauna of Tállya is the second most diverse such fauna known from the Carpathian Basin, after Radoboj in Croatia. Baetidae remains from Erdőbénye deserve special attention because previously fossils of mayflies were known only from the Pliocene and younger.

📖 Sziráki & Dulai 2002

Bryozoans

In the Early and Middle Miocene bryozoans became abundant in the Carpathian Basin. Remains of these fossils are found in older rocks, of course, but the only really rich assemblages are found in the Badenian. Work on them started just a few years ago, probably because scanning electron microscopes, necessary to examine and document these tiny fossils, only became routine recently.

Alfréd Dulai was the first to publish a study on two spectacular, freely living forms. He was later joined by Pierre Moissette, a professor at the University of Lyon and a well-known expert on bryozoans (and he speaks Hungarian perfectly). As a result of their joint research with Pál Müller, an outstandingly rich assemblage containing 238 species has

1
2
3
4
5
6
7
8
9
10
11
12
13
14

been described. Importantly, the work of these three has confirmed the widespread opinion that the Central Paratethys Sea was connected to the Indo-Pacific region in the Middle Miocene.

📖 Dulai 1995; Moissette et al. 2006

Brachiopods

Brachiopods, which are especially common in Paleozoic and Mesozoic marine deposits, are rather rare fossils from the Cenozoic. These taxa are often referred to as lamp shells in the older professional literature and in popular books because of their resemblance to nightlights. Nevertheless, the sporadic occurrences of brachiopod fossils, collected over many decades, were compiled and discussed in a short 1944 monograph by Ilona Meznerics.

The 60 years that passed since this work made a revision of these species necessary. This work was undertaken by Alfréd Dulai and Maria Alexandra Bitner, a research fellow at the famous Paleobiological Institute in Warsaw. They stated in their 2004 study published that Karpatian and Badenian sediments do contain brachiopods and that altogether around 600 specimens are known. From the Karpatian they identified 9 species and from the Badenian 13, as well as verifying that these faunas contain mostly species of *Terebratula*, all extremely alike. Many of the badly preserved specimens of larger-sized *Terebratula* specified by Meznerics cannot be more precisely assigned.

📖 Meznerics 1944; Bitner & Dulai 2004; Dulai 2007

Echinoderms

Sea Urchins (Echinoids), Sea Stars, Starfish (Asteroidea), and Sea Lilies (Crinoidea)

Sea urchins are also common fossils in some rocks in the Early and Middle Miocene, and research on these animals has followed a similar tradition to that done on the Crustacea. One pioneer in sea urchin research, Hardouin Michelin

Miocene (Badenian) fossil decapods (all natural size except [3]). The extraordinary abundance in forms and varieties of decapods is illustrated by the fact that the number of species known for this group approximately equals that for living birds! Although decapods were common during the Tertiary, complete remains of these shallow marine animals are rare because their fragile shells very easily broken by the sea. In rocks that represent reefs—such as the Miocene Lajta Limestone—calcified decapod shells, feet, and claws are common. Only one of the more than a hundred Badenian crustacean species so far described from Hungary can be placed into a living species. This contrasts with much higher similarity at the genus level: 85% are found in seas today, a good number in the Mediterranean. (1) *Rakosia carupoides* Müller—claw of a swimming crab (Portunidae family). The name of the genus commemorates one famous fossil site, the Rákos railway cutting in Budapest, which has provided many decapod remains. Its describer, Pál Mihály Müller, also introduced the *Kerepesia* genus, named after Kerepes Street in Budapest; however, this is not a cancer crab but a sponge crab (Budapest, Rákos). (2) *Callianassa munieri* Brocchi—partial claw. Species of the *Callianassa* genus are burrowing shrimps that spend most of their time deep in sediment, processing and reworking mud and sand. *Callianassa* mixes sand with mucus in the loose deposit and strengthens the walls of their burrows with the mixture. They are fairly common in the fossil record: their remains are preserved because of the burrowing mode of life (Budapest, Rákos). (3) *Matuta brocchii* Glaessner. This small animal is a member of the family of box crabs (Calappidae), so named because they hold out their claws "shyly," covering their fronts almost completely (Budapest, Gyakorló Street, 2× magnification). (4) *Portunus monspelliensis* Milne-Edwards—partial claw of a crab belonging to swimming crabs (Kishajmás). (5) *Portunus neogenicus* Müller—partial carapace of a crab belonging to swimming crabs (holotype) (Budapest, Gyakorló Street). (6) *Daira speciosa* (Reuss)—carapace of a small mud crab belonging to Xanthidae. This family is the largest group of living short-tailed crabs (Budapest, Rákos). (7) *Dromia eotvoesi* (Müller). This animal is a member of the sponge crabs (Dromiidae). This name refers to the dense "fur" that often covers the back shell and the feet of this crab; indeed, many sponge crabs cover themselves—as a disguise—with a small object (generally a bivalve shell) and walk around with it. Species of *Dromia*, which live in the Mediterranean Sea, place a disk cut from a sponge onto their backs. When the shell on their back grows these crabs must look for another suitable piece of sponge—often they simply steal from another crab (Budapest, Kerepes Street). (8) *Calappa heberti* (Brocchi)—strong and very wide claw of a box crab. These fossils are common in the Badenian crab fauna and are known from the Oligocene until recent times (Budapest, Rákos). (9) *Maja biaensis* Lőrenthey—back portion of crab shell belonging to a spider crab (Majidae). Young specimens of these crabs usually hang seaweed onto their "furry" backs—the seaweed does not die but grows on and conceals the animal. The rarity of spider crabs is striking in the Badenian crustacean fauna from Hungary (Budapest, Örs vezér Square). (10) *Matuta brocchii* Glaessner—fragile back part of shell from a box crab (Budapest, Gyakorló Street). (11) *Calappa heberti* (Brocchi)—wide claw of a box crab, with which it can almost completely cover itself (Budapest, Rákos). (12) *Petrochirus priscus* (Brocchi)—claw of a hermit crab. These crabs (Paguridae) often hide their soft, vulnerable abdomen inside empty gastropod shells. However, since the shell does not grow with the animal, they must seek larger and larger empty shells. One of their claws is often much bigger than the other (Budapest, Rákos). (13) *Charybdis mathiasi* Müller—back part of shell from a swimming crab (holotype) (Budapest, Rákos). (14) Ventral view of an indeterminate, complete specimen from Ófalu in the Mecsek Mountains. Because important taxonomic features are usually found on the dorsal surfaces of crabs, exact determination based just on the ventral surface is generally difficult.

Large specimen of a curl-footed crustacean (*Balanus*) from Szécsénke, in Nógrád County (original size).

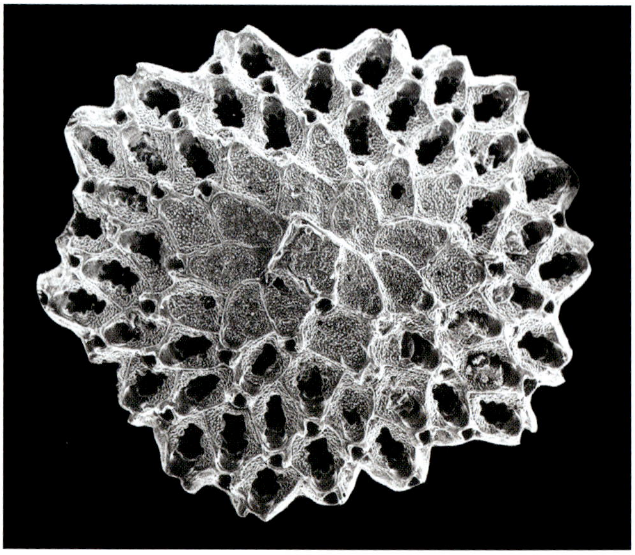

A bryozoan from the Badenian sequence at Szob. Colonies of this shape are called lunuliform. AFTER DULAI 1995, 15× magnification.

(1786–1867) mentioned three species from the Kemence fossil site in the Börzsöny Hills in a 1861 monograph that dealt with the genus *Clypeaster*. Gustav Laube (1839–1923), a paleontologist and Arctic explorer from Prague, also described two new species from close to Bia in 1871.

The first, and so far only, monograph to deal with echinoderms from what was at one time called the Mediterranean and is today called the Badenian of Hungary was published by Elemér Vadász. This study came out in 1915 as a volume in the Geologica Hungarica series. Vadász intended for it to be exhaustive, and so he discussed and illustrated the known remains of sea urchins, starfish, and sea lilies that had been found in the Carpathian Basin. Sea urchins are fairly common locally, whereas the other two taxa are much rarer. Of these, one very interesting starfish, *"Pentagonaster" muelleri,* described by Vadász from Lajta Limestone deposits close to Sankt Margarethen, Austria, is worth mentioning. The traces of brittle stars covering whole bedding planes (*Asteriacites*), at first incorrectly noted in the literature as body fossils, are more frequent in certain shallow marine sequences in northern Hungary.

Study of sea urchins was continued in the 1980s by Sándor Mihály, who also studied Paleozoic fossils. Construction work in Örs vezér Square (Sugár shopping center and IKEA store), Budapest, was the impetus behind renewed interest in these fossils; exceptionally fossil rich Badenian layers were exposed in foundation pits, and other good fossil sites were created in the neighborhood. These assemblages were found to differ from that seen at the nearby classical site, the Rákos railway cutting, as large scallops and sea urchin remains were abundant in the loose, crumbling calcareous sand layers. These latter taxa are mainly representatives of a still-living, extremely flat form that resembles a sand dollar, from the genus *Parascutella*. The peristome, or "Archimedes lantern," of sea urchins was also found often at these sites.

The sad and untimely death of Sándor Mihály brought an end to his enthusiastic research. He did not live to see these fossil-rich Badenian successions excavated again just a few years ago, when the foundations of the Árkád shopping center in Budapest were laid. The excellently preserved specimens that were unearthed as a result of this excavation have spread all over the mineral and fossil markets of Western Europe. Unfortunately, this will be the last fossil El Dorado to come from Örs vezér Square for a while, since there is no more available space for another big building. Nevertheless, thanks to the work of Sándor Mihály, about 3,000 carefully prepared, marked and cataloged specimens await study in the drawers of the National Geological Museum in Budapest. The latest milestone in the study of Badenian sea urchins from Hungary was the 1995 graduate thesis by Artúr Köhler, which

The 3- to 5-centimeter-long, excellently preserved spikes of *Cidaris*-type sea urchins from Mátyásföld.

A rare fossil: remains of a sea star from the Lajta Limestone around Sankt Margarethen. The specimen in this photo was identified as *Pentagonaster muelleri* by Elemér Vadász, but it probably belongs to the genus *Metopaster* (0.3× magnification).

dealt with the fauna of the Mecsek Hills. Today Köhler is a renowned environmental geologist.

Specimens of *Echinolampas, Cidaris,* and *Schizaster* are also common in Badenian deposits. However, a quite different story is seen in older deposits, in which *Spatangus* is most frequent in the less diverse faunas of the Ottnangian, and *Cidaris, Brissopsis,* and *Echinolampas* dominate in the Karpatian. Overall, the rocks of the Carpathian Basin contain the richest echinoderm assemblage known from the Central Paratethys Sea.

Fossil sea urchins can also be found in quite unusual places. One of the fossil sites around Kemence, exceptionally rich in a European context, is today a barred rock cavity called the "Holy Grave." When it was excavated about 100 years ago, vast numbers of fossils came to light and were built into the walls of the side-chapel of the church, where they are still on display.

📖 M. E. Vadász 1915; Mihály 1985, 1989

Vertebrates

The Early and Middle Miocene marine vertebrate fauna of Hungary, and indeed the whole Carpathian Basin, is much richer than known from older (Paleogene) successions: diverse fish and marine mammal (sirenian, cetacean) fossils are known from many sites. The most spectacular terrestrial vertebrate remains are the bones and teeth of proboscideans, and the Ipolytarnóc site in particular is worth mentioning—as the fossil footprints found here are unique.

📖 Kordos 1978b

Fish

Shark Teeth

Antal Koch worked on shark teeth from fossil sites in Nógrád County at the beginning of the 1900s. These well-preserved specimens were revised exactly100 years later by László Kocsis, who was then a graduate student. Based on observations of living sharks, Kocsis demonstrated that the shape of teeth can vary greatly depending on their position in the jaw; with this in mind he showed that the number of earlier described species must be much smaller than initially thought. Many of the species described by Koch were lost in this revision; at the same time, new specimens were shown to occur at Ipolytarnóc as a result of new collections of several hundred teeth.

An enormous number of chondropterygian teeth were described from around Kazár, in Heves County, by Péter Solt, a research fellow at the Geological Institute of Hungary. According to him, more than 80 percent of the sharks from this site belong to the family Odontaspididae, a group which was common in the Miocene, and the size of the Kazár specimens is generally smaller than is seen at other sites. Hammerhead sharks (*Sphyrna*), tiger sharks (*Galeocerdo*), cat sharks (*Scyliorhinus*), smooth-hounds (*Mustelus*), and angel sharks (*Squatina*) are important genera at this fossil site.

Although the Ipolytarnóc teeth are Eggenburgian and the Kazár ones are a little younger, from the Karpatian, teeth that have been described from the Lajta Limestone are younger still—Badenian in age. Generally, teeth from fossil sites of different ages are dominated by different genera, a feature that is apparently also connected with the fact that the sediments were deposited in different paleoenvironments as connections between branches of the Miocene Sea and other more distant basins changed continuously. The most spectacular specimens from Mátraszőllős are the 8-centimeter, serrated capital teeth of the giant shark *Carcharocles megalodon*. For a long time, this fish was thought to be a Miocene ancestor of the living great white shark (*Carcharodon carcharias*) but new research has shown that its extant counterpart is probably more closely related to the ancestors of *Isurus* than to *C. megalodon*.

📖 Koch 1903, 1904; Solt 1987, 1992; Kocsis 2003, 2007

Otoliths

Otoliths are 0.5-to-15-millimeter pebble-like objects that are found in the heads of fish. They are usually aragonite but can also sometimes consist of apatite or another (organic) material (such as protein). The word "otolith" literally means "ear stone," because these structures are located toward the bottom of the labyrinth, an organ below the three semicircular canals in the ear region. The exact function of otoliths in fish is not yet known, but they surely play a role in balance and in the coordination of movements.

Small, concentric otoliths are even sometimes found in fine-grained detrital sediments whereas other, more

1
3
5
2
6
4
7
8
9
11
10

Transported (allochtonous) remains (teeth, vertebrae) of Early and Middle Miocene marine and continental vertebrates (sharks, rays, other fish, crocodiles, and mammals) from Pannonian layers at Danitz-puszta (about original size).

fragile, fish remains are not commonly preserved. This is another reason they are important.

Otoliths can be separated from sediment by screening, washing, and then sorting. The fauna represented by these remains can then be determined even to the species level, since the shapes (including the outlines and ornamentation) of otoliths are so characteristic. Typical features of the paleoenvironment—such as water depth, salinity, and temperature—can also be deduced by comparison to living fish if the otoliths are identified correctly.

Although otoliths are found in practically all the Cenozoic aqueous sediments from the Carpathian Basin, they are not particularly striking fossils. Nevertheless, specialists have been studying them for over 100 years. The history of research on finds from Hungary was summarized in the 2006 graduate thesis by Mariann Bosnakoff, a student from Eötvös University, Budapest. She is the most recent worker to examine otoliths, having starting her work at the Károly Eszterházy College in Eger. This institution is well known for research in this area: many students have participated in otolith and trace fossil research under the direction of Árpád Dávid.

This team achieved very good results from a comparative analysis of Badenian otoliths collected from around Ipolydamásd and Bükkmogyorósd. More than 2,000 otoliths were separated from about 100 kilograms of sediment comprised of numerous genera and species. Although this diversity is not surprising, this work at least proves, for the first time, that the fish fauna of this subtropical sea was quite varied. Various kinds of gobies (*Gobius*), sea bream (*Sparidae*), picarels (*Spicara*), cod (*Gadidae*), and lantern fish (*Diaphus*) made up the more than 31 recognized species. However, the faunas from the two sites were also found to be quite dissimilar despite the fact that they are about the same age. This can be explained by the different environments: although Ipolydamásd was part of a sublitoral shallow marine system, the sediments at Bükkmogyorósd come from a deeper environment. It is also interesting that the relative dominance of Atlantic and cosmopolitan species is opposite in the two fossil sites, which might even mean that the two areas were gulfs of two different seas. Although the Ipolydamásd area was open toward the west, Bükkmogyorósd is thought to have been connected to the Indo-Pacific region.

📖 Bosnakoff 2001

Miocene (Badenian) sea urchins (0.5× magnification). (1, 2) *Echinolampas* sp.—a round or ovoid frequent form. The petals (petalidiums) are straight; the mouth is located in the center, and it is wide rather than long. The anus is in edge position. About 300 species belong to this genus and are found from from the Eocene up until recent times (Budapest, Örs vezér Square). (3, 4) *Parascutella vindobonensis* (Laube). This genus and related forms are characterized by a flat lower (oral) side, a slightly bulging upper (aboral) side and a sharp edge. *Parascutella* is round or pentagonal, with wide petals that are also quite short. The mouth is located in the center, and the anus has moved to the lower (oral) surface and is run through with branching food grooves. This genus used to be referred to as *Scutella* and is known from the Miocene of Southern Europe (Budapest, Örs vezér Square). (5, 6) *Amphiope bioculata* (Desmoulins). The most significant feature of this *Scutella*-like sea urchin is the presence of two oval holes located toward the rear of its skeleton. Although this genus is rather rare in Hungary, it is known from outside Europe in Oligocene–Miocene rocks (Budafok). (7, 8) *Schizobrissus cruciatus* (Agassiz)—ovoid, large-statured form. The lower side is flat and wide, and the upper side is slightly bulging. The large anus is located at the edge, on an oblique surface. It is known from the Mediterranean Sea and the eastern coast of the Atlantic Ocean from Eocene–Miocene sequences. The Eocene sea urchin specified as *Deakia* genus by Pávay may belong to this genus (Ófalu). (9) *Schizaster* sp. Many species classified within this genus are known worldwide since the Eocene. The remains of *Schizaster* that live under the bottom surface are common in Tertiary sediments in Hungary (Ófalu). (10) *Trachypatagus* sp.—very bulging, rare form. It is known from Eocene and Miocene layers (Ófalu). (11) *Clypeaster* sp. The often large, bulging, rounded or pentagonal *Clypeaster* belong to the most common group of Tertiary sea urchins, known in the fossil record since the Eocene. Their petals are wide, whereas the lower side is flat and the edge is rounded. These urchins are frequent worldwide, especially in tropical–subtropical seas. A number of fossil species have been described, but it is often difficult to distinguish individual forms because of the commonness of transitional forms (Hetvehely).

A tooth of *Carcharocles megalodon* from Sankt Margarethen (original size). This giant Miocene shark was much bigger than the largest living predatory shark, the great white shark. A sitting person could have easily fit in its open mouth.

Marine Mammals

Seals

Sarmatian vertebrate faunas are less well known and less diverse that their Badenian counterparts. Even so, they are still fairly diverse. The most interesting elements in the Sarmatian fauna are the seals, which were first described from Érd by Miklós Kretzoi. Miocene seal fossils are also known from Danitz-puszta near Pécs and from Holíč in Slovakia.

Seals have long been thought to have been dwellers of cold, polar seas, but few people know that one species lives in the Mediterranean Sea—hardly cold water. A monograph was recently published about Miocene fossil seals from the Paratethys Sea.

📖 Kretzoi 1941c; Koretsky 2001

Sirenians

Interestingly, far fewer Miocene sirenian remains have been found than are known from the Eocene. Antal Koch was the first to mention fossils of this age; one of his specimens, (*Halitherium* cf. *christoli*), is from Horné Strháre in Slovakia, and a second taxon, *Thalattosiren petersi,* is from the Badenian Clay in Marz, Austria. *Thalattosiren* was originally named "*Metaxytherium*" by Zoltán Schréter. The third fossil described by Koch was found in an andesite tuffaceous quarry in the Lajta Limestone, near Mátraszőlős in Hungary. This specimen was actually initially described in the professional paleontological literature by Miklós Kretzoi as *Haplosiren leganyii,* a new species in a new genus he named in honor of Ferenc Legányi. Further Miocene sirenian fossils are stored in a number of museums in Budapest and around the country, and they need to be worked on.

Remains of a Sarmatian bony fish (*Blennius*) from Transylvania (about original size).

The sirenian remains found at the type locality of the Eggenburgian in the Vienna Basin are also famous. The most beautiful specimens (*Metaxytherium krahuletzi*) are on display at the Krahuletz Museum in Eggenburg.

📖 Koch 1900; Schréter 1917; Kretzoi 1951a, Domning & Pervesler 2001

Cetaceans from the Carpathian Basin

An important time in the evolution of cetaceans was the Miocene, when they adapted themselves to all sorts of new environments. It was also at this time the first representatives of the living families appear in the fossil record.

Thus, it is fortunate—if one is interested in doing research on these animals—that during the Miocene most of the Carpathian Basin was covered by sea; cetacean bones, along with other fossils, are commonly preserved. The finest fossil representatives of toothed whales that turned up during the nineteenth century and at the beginning of the twentieth were described by the experts of the time (Hugó Böckh, Dragutin Gorjanović-Kramberger [1856–1936], Antal Koch, and Károly Papp—among others).

A very detailed, carefully compiled PhD dissertation about the suborder of toothed whales (Odontoceti) was written by Emese Kazár, who reviewed all the known finds from the Carpathian Basin. She described all the remains held in both museums and private collections, altogether representing 41 toothed whale fossil sites. Kazár also revised the taxonomic statements of earlier experts and performed taphonomic analysis on bones where possible, using methods from both classical comparative anatomy and cladistic character analysis. As a consequence of the nature of the material, she was able to extend her character analyses to include arms as well as simply skull bones (as was standard in earlier works).

📖 Kazár 2003

Early and Middle Miocene Toothed Whales

Early and Middle Miocene toothed whales from the Carpathian Basin belong, in part, to the family that includes living sperm whales (Physeteriadae) as well as to other extinct families. One well-known taxon from Felső-Pálfalva, *Orca semseyi,* described in 1899 by Hugó Böckh, has been mentioned several times in the professional literature; however, it is in reality a sperm whale of uncertain taxonomic position according to Emese Kazár. In addition, remains from Mučín, which—according to Emese Kazár—belong to the genus *Squalodon,* are also of Early Miocene age.

The Badenian contains more fossils, and its whale fauna differs somewhat from those of earlier ages. The finest remains were found in Burgenland (Walbersdorf, Stotzing, Sankt Margarethen) and most Badenian findings belong in the genera *"Heterodelphis"* and *Delphinodon* (?), which are in extinct families. The diversity of toothed whales in the region reached its peak in the Sarmatian, with most fossils known as allochtonous clasts from Late Miocene Pannonian sediments, primarily at Danitz-puszta. The articulated whale remains that have been collected from the Western Mecsek Hills, around Kovácsszénája, are also peculiarly fine.

📖 H. Böckh 1899, Kazár 2003, 2005

The Sankt Margarethen Dolphin

The dolphin remains from the Lajta Limestone, described as *Heterodelphis leiodontus* by Károly Papp, comprise some of the finest Miocene whale fossils from the whole Carpathian region. This precious fossil was found just a few kilometers from the western margin of Lake Fertő (Neusiedler See), near Sankt Margarethen, in 1880. The stone containing it was split exactly along the body of the dolphin, so that either half of the skeleton can be seen in separate limestone plates. These remains were prepared first by Károly Papp. This "paleodolphin," which is kept in the museum of the

Limb bones of a seal preserved together from Holíč (0.3× magnification).

1

2

3

Geological Institute of Hungary, was "re-operated" on 120 years later: László Kordos and Emese Kazár opened its skull and removed the ear bones, the right perioticum, and the right tympanicum. This enabled a more precise taxonomic determination for this fossil, as these bones are crucial in whale systematics. The remains turned out to represent an independent genus within Kentriodontidae—a dolphin family that went extinct at the end of the Miocene—rather than belonging to the genus *Heterodelphis*. Kordos and Kazár were unable to name the specimen as a new genus because of the absence of other characteristic features in the skull.

Nevertheless, this fossil is especially important because articulated vertebrate remains—which preserve the original position of bones—are extremely rare from the Lajta Limestone. Besides its skull, a number of vertebrae and ribs of the Sankt Margarethen dolphin were also preserved, together with articulated scapula and limb bones. As Károly Papp had already noted, the surface of the rock between the fifth and ninth vertebrae is full of tiny fragments of fish vertebrae, fin remains, and fish teeth—these must represent fossil stomach contents. The dead body of this animal must have been rapidly covered with sediment in stormy, shallow, near-shore water.

📖 Papp 1905; Kazár & Lantos 2001; Kazár 2003

The Whale of Walbersdorf

The story goes that a few connected animal vertebrae turned up during clay digging in the brick factory of János Prost in Walbersdorf, Sopron County in January 1899. Among these finds the most noticeable were remains of a giant animal: these came to the attention of the press, and the January 13 issue of the newspaper *Pester Lloyd* reported the sensational discovery.

Miocene whale fossils. (1) Physeteridae sp.—a jaw fragment with teeth from an unknown sperm whale. This rare fossil was described more than 100 years ago by Hugó Böckh; it comes from Felső-Pálfalva. Böckh was an excellent geologist who also became well known for discovering hydrocarbon occurrences in Iran. The Eggenburgian finding was first mentioned as a fossil orca (*Orca semseyi*) in the Hungarian paleontological professional literature (0.47× magnification). (2) *Sophianaecetus commenticius* (Kazár)—skull of a toothed whale originally described as representing a new genus *Mediocris* from Sarmatian porous limestones around Kovácsszénája, in south Transdanubia. Later, however, the name *Mediocris* proved to be pre-occupied, thus introduction of a new name *Sophianaecetus* was necessary. A number of vertebrae and limb bones were preserved alongside this dolphin skull (0.35× magnification). (3) *Heterodelphis leiodontus* Papp—half of the Sankt Margarethen fossil dolphin, reprepared 120 years after its discovery. This fossil belongs to a family of extinct toothed whales (Kentriodontidae), and was described by Károly Papp from the Badenian Lajta Limestone (0.23× magnification).

This was how the history of one famous fossil from the Carpathian Basin began. Several published studies provided descriptions of and geological information about the remains, and these are summarized here. It turns out from reports published at the time that although the scientists at the Geological Institute immediately (that very January) got hold of the fossil for their collection, they did not do much to preserve it. In fact, the whale was only saved from decay by the interest of the factory owner where it was found: "Mr. Prost had the excavation works done under his own supervision as soon as the first eight vertebrae turned up in the clay. The work was stopped by unfavorable weather, but the owner had the intriguing fossil site guarded around the clock for fear of unauthorized hands doing harm to the bones that possibly remained there" (Szontagh 1904).

Excavations at the site were resumed in July. Lajos Bella, the "head science teacher and noted archaeologist in Sopron," assisted Prost during the works as "a shaft was cautiously dug into the about 4-meter-high, steep clay wall, and further skeleton pieces of the animal were searched for with costly, slow work. As luck would have it, at this very time, the Imperial and Royal professors Eduard Suess from Vienna and Rudolf Hörnes from Graz were on vacation at nearby Márczafalva and were able to give instructions to the local diggers as to the best ways to preserve the skeleton as much as possible. The professors, especially Suess, also dug in the clay and cleaned the bones as they appeared. Suess was the first to note that the skeleton was from a balaenopterid whale" (Szontagh 1904).

Prost asked the Royal Geological Institute of Hungary for further help in August, since he probably thought the fossil would end up there anyway and Tamás Szontagh (1851–1937), soon to be the deputy director, traveled to the spot himself. "Since none of my colleagues were in Budapest at the time, our worthy director sent me immediately to Walbersdorf. Here I found 17 vertebrae already under a roof, but the rest and the trunk still lay in the field. . . . The bones were very flattened and easily crumbled." Szontagh "pondered the fossil for a minute with sad disappointment and resignation at what remained to be done" but finally "he started to get the skull loose with great expectation." However, Szontagh, although indispensable to the institute, had the attitude of an official and soon lost his enthusiasm for the task, entrusting it instead to the newly arrived István Sedlyár, a laboratory assistant "whose patient and lucky hands were already well known." Sedlyár completed the excavation while Szontagh returned to Budapest.

The lonely assistant Sedlyár embedded several thousand whale bone fragments in paraffin so that they would not fall apart during transportation. As the *Budapest Journal* reported on July 14, 1900, the paraffined balaenopterid

Photo from the excavation of the Whale of Walbersdorf.

skeleton was later viewed by Franz Joseph the First (1830–1916) who "awarded János Prost, whose brick factory had hidden the precious fossil, with the Gold Cross of Honor for his selfless gift."

Embedding in paraffin was a good idea, but the bones should not have been left in it for one and one-half years: the paraffin melted in the heat of the summer and the bones sank. The clay parts dried out and "the bones were fragmented so much that most of them had completely fallen apart by the time the systematic preparation of the skeleton resumed" (Kadić 1907). This raised difficulties, especially during the later reconstruction of the skull, which had become too flat.

Although Károly Papp gave a short osteological evaluation of the fossil whale in his preliminary study, Ottokár Kadić, once a biologist and always an enthusiastic vertebrate paleontologist, was appointed to work on the remainder. The whale was finally prepared by Kadić, with rather dubious results, and the professional value of the fossil was doubted even by Lajos Lóczy, Sr., director of the institute. Kadić certainly tried everything; as he noted, "I carefully took every connected fragment out of the clay and paraffin, then cleaned them, wrapped them in wire gauze and cooked them in a dilute glue solution for a few hours. At the end of the process the bone fragments were free of any clay or paraffin. . . . The bones were not only cleaned but also impregnated by glue. . . . Then I had to try my best to sort the dried bones and glue together the matching pieces."

Kadić then made up the missing parts with plaster, providing his critical colleagues with a new avenue of attack. They nicknamed the result of his preparation "plastocetus" or "plastotherium." despite the fact that Kadić painted the plaster a gray color, different from that of the bones, during preparation so that replacements could be distinguished from the original. The puzzle gluing lasted for two years, but when the preparation and comparative anatomical analysis was complete, Kadić could proudly say, "The fossil Whale of Walbersdorf is a new species among *Mesocetus* in my opinion, and I introduce it into the paleontological literature as *Mesocetus hungaricus,* the finest fossil known from Hungary."

The completely finished skeleton was fitted into an iron frame and was put up in a ship-shaped case at the museum of the Geological Institute in February 1904. The complex history of this fossil, however, was not over: it was later moved to the Natural History Museum and was very badly damaged during the fire of 1956.

The Whale of Walbersdorf was found in sediments that belong to the Baden Clay Formation. The age of this formation is the same as that of the Middle Miocene Lajta Limestone, but it was deposited in deeper water. This young whale was 6.5 meters long, with a 1.8-meter head, and it belongs to the group that includes the baleen whales (suborder Mystacoceti). Although this fossil is much smaller than some living representatives of this group (for example, the blue whale can reach 20 or even 30 meters in length), the Whale of Walbersdorf is still the longest fossil animal so far discovered from the Carpathian Basin.

📖 Szontagh 1904; Kadić 1907; Kazár 2003

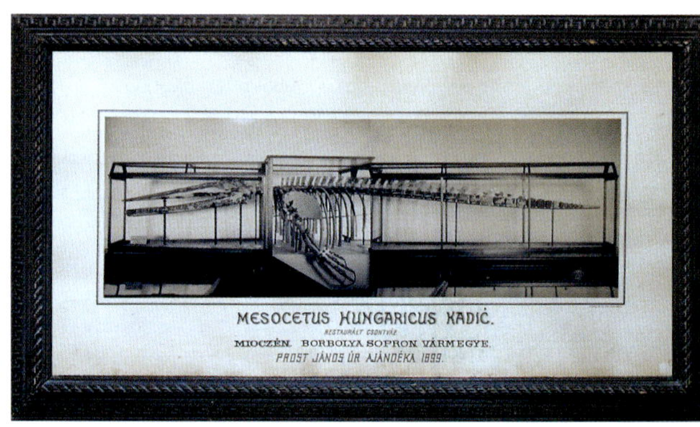

Prepared skeleton of *Mesocetus hungaricus,* aka the Whale of Walbersdorf, as it was displayed at the Royal Geological Institute of Hungary in 1904. The transport, preparation, and exhibition cost exceeded 3,000 crowns—which was covered by Andor Semsey, member of the Upper House and the honorary director of the institute. Parts of the fossils from disannexed areas were withdrawn from the exhibition after World War II because, according to notions of the time, they did not fit the profile of the collection. This is how the Whale of Walbersdorf—and presumably also this large-sized, nicely framed photo—got into the natural science collection of the Hungarian National Museum. In 1956, when the mineralogical and paleontological collection of the museum was bombed, the bulk of the collection and numerous bones from the skeleton of the Whale of Walbersdorf burned. It is lucky that the skull, which was covered in soot, and some other bones have remained reasonably intact.

Portrait Gallery

Ottokár Kadić (1876–1957)—Paleontologist, Paleoarcheologist, and Cave Scientist

This zealous, excellent, and often unrecognized expert in Hungarian paleontology and cave research was born in Opazova, Slavonia. Kadić went to secondary school in Bjelovar, and then in Zagreb, and he graduated in Munich under the supervision of R. Herwig and K. A. Zittel, two of the most outstanding zoologists and paleontologists of the time.

Originally Kadić planned to be a zoologist; he even wrote his dissertation on a comparative analysis of the mouthparts of Coleoptera (beetles), but later he became a geologist and finally a paleoarchaeologist. He was urged by his Hungarian mother to find a job in Budapest, but there was no vacancy in the Department of Zoology at the University of Pest. Because of this he was eventually appointed to the Geological Institute, at that time directed by János Böckh, where his task was to care for the vertebrate paleontology collection, which had remained without a curator following the early death of Gyula Pethő. Kadić was also asked to take part in geological mapping, which must have been a real challenge for the young entomologist. He then worked with the Whale of Walbersdorf and described Plio-Pleistocene vertebrates from around Lake Balaton.

Gorjanović-Kramberger's discovery of the world-famous Neanderthal site around Krapina also coincided with the start of Kadić's scientific career. Kadić responded to the discoveries of his one-time mentor with great zeal, and published several scientific and popular studies connected to them. Because of this work he was asked to study the controversial tools of early man that were being unearthed during new excavations at Miskolc-Avas.

These excavations, started in the Szeleta Cave in 1906, mark the beginning of regular paleoarcheological research in Hungary. However, the authenticity of the beautiful bay leaf–shaped stone tools found in the cultural layers in the cave was met with skepticism by the scientific world of the time, and Kadić himself fell under suspicion. That he was unfairly accused was recognized only years later. "Only those who knew Ottokár Kadić really recognized what the ordeal the four long years of paleolith-debate centering on Szeleta meant to this honest, immeasurably precise, peace-loving, humble and quiet researcher," remembered Miklós Kretzoi, a friend, student, and colleague of Kadić.

It was also Kadić who—among other things—discovered the *Hipparion* fauna at Csákvár, excavated the caves at Felsőtárkány, and systematically and scientifically explored the Buda Castle caves. Indeed, although it was Kadić who carried out, managed, and organized these cave explorations he worked only on the paleoarcheological findings himself. He always delegated the study of fossils the younger scientists working around him.

📖 Kretzoi 1958; Kadić 2010

Continental Vertebrates

We have quite limited knowledge of Paleogene and Early Miocene continental vertebrate (mainly large mammal) faunas from the Carpathian Basin. This can be explained in part by the small number of fossils that have been found; it is only in the Middle Miocene (Badenian) when faunal patterns become much clearer as the number of fossils increases. One reason for this increase is the speed of continental mammal evolution and their increased role in stratigraphy is revealed by the fact that 17 successive mammal zones are determined in the European Neogene.

Middle Miocene Small Vertebrates

For years, collectors and scientists have been interested in spectacular large bones, such as the remains of proboscideans. No one studied the small vertebrate faunas known from Early and Middle Miocene continental and coastal deposits for many years. The first results on the faunas of Hasznos and Szentendre were published at the beginning of the 1980s by László Kordos.

In the last few years, large-scale excavations at several localities in northern Hungary have begun under the supervision of János Hír, director of the museum at Pásztó. As a result, a much more precise picture has begun to take shape regarding Badenian, and especially Sarmatian, small vertebrate faunas. Fossils of this age have been collected mainly from localities at Sámsonháza, Mátraszőlős, Felsőtárkány, and Tăşad, close to the Pădurea Craiului Mountains. Most of the Hungarian finds, however, have been from various strata in the Sajóvölgy Formation. Bones are rather rare in these non-marine (lacustrine, continental) layers and so the long faunal lists that are known are the result of collectors washing huge amounts of sediment before carrying out microscopic separation and then identification of tiny

János Hir screening tons of Miocene terrestrial sediment. Thanks to patient study of the tiny bones found in these strainers, a hitherto unknown small vertebrate fauna has been revealed.

bones. The fragile remains of the amphibians (newts and frogs) and reptiles (lizards and snakes, including chiefly colubrids and vipers but sometimes also blind snakes) were studied by Márton Venczel, for example. On the basis of screened fossils from four Hungarian Middle Miocene fossil site (Hasznos, Szentendre, Mátraszőlős és Felsőtárkány) the presence of at least seven distinct species of frog have been documented, two of them (*Palaeobatrachus hiri* and *Pelobates sanchizi*) entirely new to science.

The rare, usually very fragmentary (and thus hard to identify) bird bones from these sites have been studied by Erika Gál. A large pheasant—alongside remains of small herons, teal, flamingo and bones of a few songbirds—have thus far been reported. The rodent fauna from this area is peculiarly diverse; bones of a variety of squirrel (including flying squirrels), dormice, Eomyidae (extinct rodents), hamsters, whistling hares, and beavers—have all been described by János Hír.

The insectivores known from these small vertebrate faunas have been worked on by Lukács Mészáros, and so far the dozen or so species that have been described have been placed into the families that comprise the shrew-mice, hedgehogs, and moles—some of the oldest insectivores described to date from anywhere in Hungary. These faunas can also be subdivided according to Mammalian Neogene (MN) zones based on the small mammals—with the oldest fauna known from Hungary, at Litke in Nógrád County, belonging to MN 5 (early Badenian in age). The faunas from Mátraszőlős and Felsőtárkány are younger and belong to Badenian and Sarmatian biozones MN 6 and MN 7–8, respectively.

Interestingly, the small mammal fossils found at sites close to one another in time and space are often very

As far as it can be determined on the basis of its well-preserved, but slightly incomplete, skeleton this fossil is an ancient representative of the green toad (*Bufo viridis*). This specimen was donated by geologist Péter Vincze to the Eötvös Museum of Natural History, but the name of its collector is unknown. The specimen was probably found in the diatomite quarry at Szurdokpüspöki in the Mátra Mountains—as suggested by Zoltán Szentesi on the basis of the diatoms separated from the surrounding rock. The toad may have come from a freshwater level within this basically marine succession (close to original size).

dissimilar. These differences could be explained either by the preservation and method of collection of the remains, or by the fact that different animals (amphibians, reptilians, birds, and small mammals) can have very diverse faunal patterns depending on their microenvironments. Evaluation of the Middle Miocene small mammals from the Carpathian Basin is still in an initial phase, but new results are expected in the near future.

📖 Gál et al. 1999, 2000; L. Gy. Mészáros 2000a; Hír 2001, 2003; Hír et al. 2001; Venczel 2004; Szentesi 2008

A Fossil Deer Found beside a Whale

To everybody's surprise, after the discovery of the Whale of Walbersdorf, an almost complete skeleton of a large deer (Cervidae) was also found in the clay at the Prost brick

Complete skull of the deer (*Palaeomeryx*) from Walbersdorf (about 0.2× magnification).

factory. This new fossil was first identified as *Palaeomeryx*, but following a revision by Miklós Kretzoi, the specimen was referred to a new species (*aquensis*) within the genus *Spatuloceras*. Sadly, this nicely prepared and mounted skeleton was badly damaged during World War II. However, several parts of it (including antlers) remain, and they are preserved in the collections of the Geological Institute. A closely related form (*Palaeomeryx eminens*) is now known from the Badenian diatomite beds at Hasznos.

🕮 Kretzoi 1976a; Kordos 1981

Fossil Footprints at Ipolytarnóc

The remains that are found in the "footprint-bearing" sandstone at Ipolytarnóc in Nógrád County are of considerable importance. Although numerous experts have studied these footprints, which were discovered in 1900, they were most thoroughly evaluated by László Kordos in 1985. Altogether, Kordos identified more than 1,600 footprints and the major part of the slab (about 80 percent) is exhibited under a dome built in 1980. The traces belong to 11 distinguishable species and in total about 3,000 footprints have now been found at Ipolytarnóc, along with the ones discovered and mapped most recently in 1993.

The most common of the traces are the peculiar, palm-sized footprints of rhinoceroses. These consist of three oval-shaped hoof imprints at the front of each mark; the two to the side are about the same size as one another, while the middle one is a little larger. Second in frequency are footprints, formerly called "small deer" traces. These footmarks, from obviously even-toed ungulates, are thought to belong to the genus *Pecoripeda*. The tracks that are about 4 centimeters long and 2 centimeters wide are varied in shape, probably because of the strength and dynamics of the step as well as the conditions of the soil. Because of this, it is possible that marks we suppose to be from different species actually belong to the same one.

Another common kind of footprint is hooves consisting of two elliptical, round, and symmetrical similarly sized imprints that close an 8–15° angle. Another frequent, but slightly larger, footprint type of a different shape was described by Kordos as *Megapecoripeda miocenica,* a new ichnospecies. Bird traces, which are also fairly common in the sandstones at Ipolytarnóc, were first studied by Kálmán Lambrecht, a famous paleoornithologist. His cautious work indicates that the imprints at Ipolytarnóc most closely resemble footprints of the common snipe (*Gallinago gallinago*). Subsequent researchers have distinguished more bird traces: Tasnádi-Kubacska recognized five, for example, but Kordos disputed their distinctness and found just four types. Of these, one was named *Ornithotarnocia lambrechti* in honor of the site and Lambrecht, and another was called *Tetraornithopedia tasnadii* after Tasnádi-Kubacska. Footprints of predators are rare but diverse: there are some large traces (*Bestiopeda maxima*) that were possibly left behind by a relative of living bears (*Amphicyon*), smaller ones (*Carnivoripeda nogradensis*), and tiny ones (*Mustelipeda punctata*) that resemble the traces of weasels. Even traces of claws can be sometimes seen.

🕮 Kordos 1985a, 1987a

Were There Any Proboscideans at Ipolytarnóc?

The question of whether proboscidean (mastodon) tracks can be found among the fossil footprints from Ipolytarnóc has long intrigued scientists. This idea was floated in the

Rhinoceros footprints in sandstone. Each of these footmarks is about the size of a dessert plate. These footmarks at Ipolytarnóc are thought to be traces left by a species called *Rhinoceripeda tasnadyi.* Altogether 753 imprints have been counted; they probably belonged to males of different sizes as well as females and youngsters. Most traces are also parts of longer tracks. This photograph was taken in the natural conservation area at Ipolytarnóc, where visitors can see the preserved footprints in situ. One sandstone slab with especially superb rhino footprints on it is displayed in the footprint room, toward the rear of the ceremonial hall, at the Geological Institute of Hungary.

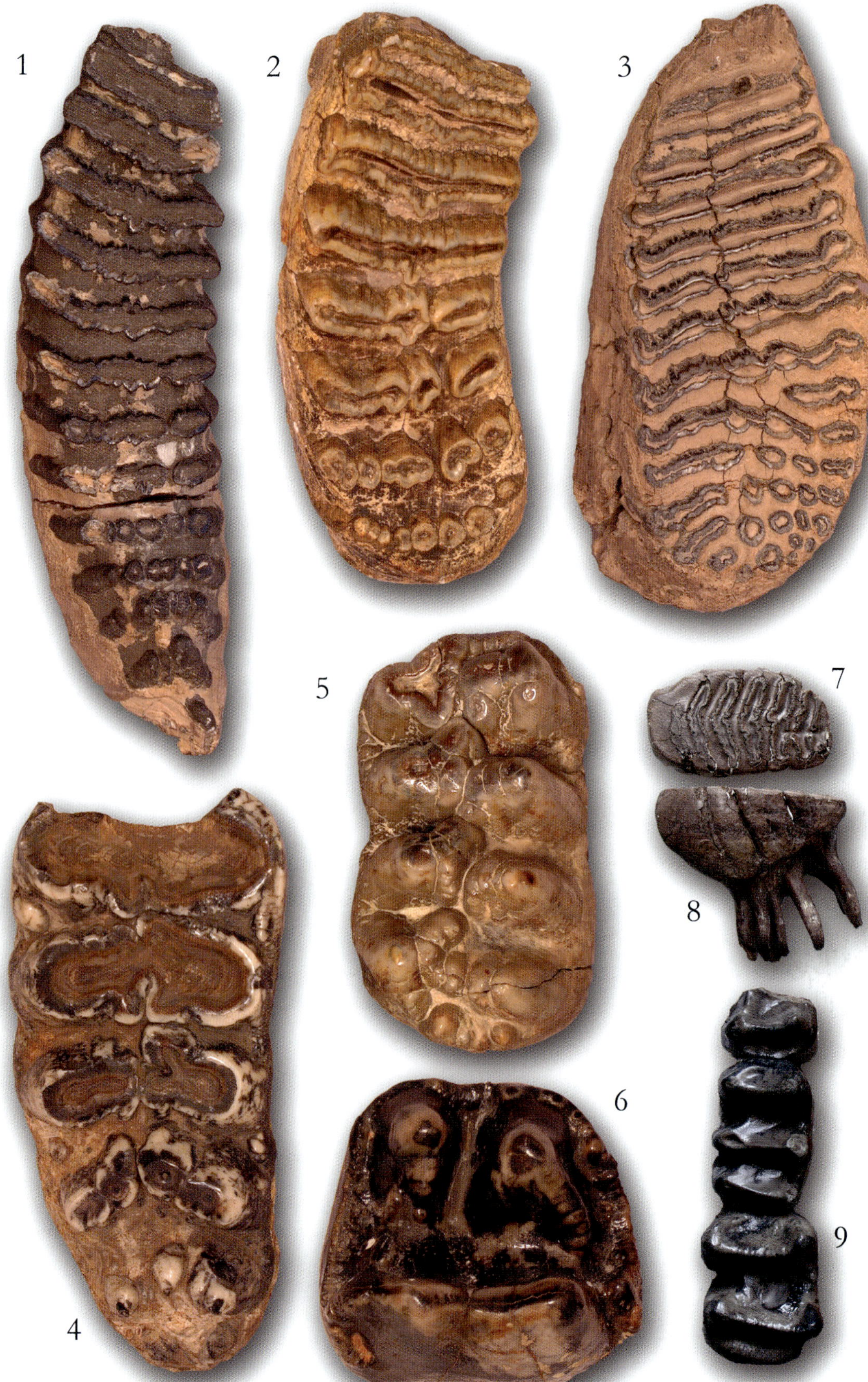
1
2
3
4
5
6
7
8
9

professional literature by O. Abel, who even published a picture of a four-toed proboscidean in his 1935 book. Tasnádi-Kubacska was convinced that proboscideans were present at the site, and wrote about them in a number of popular publications. However, neither Kretzoi nor Kordos were able to confirm the presence of this animal, suggesting instead that rhinoceros footprints superimposed on top of one other had confused earlier experts. According to this hypothesis, one other special find that was identified as a mastodon dropping by Tasnádi-Kubacska must belong to something else. Tasnádi-Kubacska, who unwaveringly believed in the Ipolytarnóc mastodon, may have thought that accumulated pine needles, washed together then covered with tuff, were fossil proboscidean droppings.

The possible presence of mastodons is, however, important not only in terms of the fauna: as far as we know, proboscideans appear in Europe only in the Ottnangian, thus their footprints are not expected to have been present in the Ipolytarnóc sandstone, which is thought to be older (Eggenburgian). According to the current view of the site, mastodons did not walk across the ground at Ipolytarnóc.

📖 Abel 1935; Kretzoi 1950; Kordos 1985a

Beach or Wallow?

What was the paleoenvironment of Ipolytarnóc like? This question, understandably, has also intrigued scientists for years. As Kordos writes, the idea that Ipolytarnóc was either a fossil beach or a sandy coastline cannot be supported because there is no evidence to support this hypothesis. Marine fossils are completely absent, for example, and so it is more probable that these footprint-bearing beds were formed around a spring where water broke to the surface through a pebble layer. So, the fossil tracksite is thought to represent some kind of a wallow from the early stages of the Miocene. None of the vertebrate trace types derive from water- or marsh-dwelling animals.

Undulating ground is also traceable along the footprint-bearing sandstone surface excavated from under volcanic tuff and the spatial distribution of the footmarks is also very irregular. Bird traces, for example, are frequent at the valley floor, although they are almost completely absent from the hillsides that were exposed in 1993. Here rhinoceros traces are most common, but at some other locations the animals stamped the ground flat by treading on their own footmarks, and at others—especially steep parts of the site—traces are rarely found at all. The animals either never frequented these areas or their loose-edged footprints were deformed and destroyed by the weight of the overlying tuff.

📖 Kordos 1985a, 2002b

The Appearance of the Horse Family: Immigrants from the New World?

The evolution of animals in Europe in general, and so also in the Carpathian Basin, was affected a great deal by faunal waves arriving from different directions at different times. One of the first such newcomers from afar (in this case from North America) in the Tertiary was the Transylvanian *Brachydiastematherium*. This rhino-like animal probably did not arrive alone, but so little is known about of the evolution of Paleogene land vertebrates in the Carpathians that nothing can be stated for certain.

The first similar Neogene event took place in the Early Miocene, when *Anchitherium* arrived from North America. These leaf eaters (folivores) belong to the horse family (Equinae) within the odd-toed ungulates and, as noted by Barnabás Géczy in his vertebrate paleontology textbook,

Teeth of Tertiary and Quaternary proboscideans (0.4× magnification). These initially small, short-crowned, and bulbous molars develop into bigger, long-crowned teeth that have more and more peaks and crests rather than thin plates throughout evolution. As a result of these changes, larger and larger chewing surfaces were formed, so proboscideans that originally ate leaves and soft vegetation could switch to feeding on harder grasses. These changes were also forced by the continuously changing environment, as the closed forests that prevailed under the warm climates of the Miocene gradually gave way to grass lands in the cooler Pliocene and Pleistocene. (1) *Elephas antiquus* (Falconer and Cautley)—a molar from Early Pleistocene layers at Győrújfalu. (2) *Mammuthus meridionalis* (Nesti)—a molar from unknown site. Early Pleistocene *meridionalis* remains are generally referred to as *Archidiskodon* in earlier literature. (3) *Mammuthus trogontherii* Pohling—a molar. These proboscideans, also known as steppe mammoths, developed thinner coats than their counterparts, the woolly mammoths. This fossil was found at Bugyi, in Middle Pleistocene strata. (4) *"Mastodon" grandincisivus* (Schlesinger)—giant proboscideans, referred to in earlier literature as *Stegoterabelodon*. This genus is characterized by two huge and strong (lower and upper) pairs of tusks; the lower tusks of "M." *grandincisivus* are much larger than those of its relatives. The upper-right molar in this photo derives from Late Miocene (Pannonian) sediments at Pestszentlőrinc. (5) *Anancus arvernensis* (Croziet and Jobert)—a very fresh and hardly worn molar from Pliocene layers at Százhalombatta. (6) *Deinotherium proavum* Eichwald—front premolar from a "hoe tusker" (a prehistoric elephant) from Late Miocene (Pannonian) layers at Pannonhalma. Several giant proboscideans similar to *D. proavum* are mentioned in the literature, and so the exact taxonomic placement of individual teeth is problematic. (7, 8) *Mammuthus primigenius* (Blumenbach)—top and side view of a milk tooth, with long roots clearly visible. This fossil comes from an unknown Late Pleistocene site in Hungary. (9) *Prodeinotherium hungaricum* Éhik—top view of the teeth from the lower left jaw. This fossil was found in Early Miocene beds at Putnok.

1
2
3
4
5

they were usually not much larger than a rough collie. *Anchitherium* is thought to have immigrated via the Bering Way (then a land bridge, not a strait!) and their arrival shows that the two continents of Eurasia and North America were connected at this time. Faunal relationships between the two landmasses make it possible to correlate historical and geological events between the two distant continents. *Anchitherium* disappears from the Carpathian region at the beginning of the Late Miocene.

The next significant event to take place in the Carpathian region was the arrival of proboscideans, much larger than prehistoric horses. These animals arrived from the south via the "*Gomphotherium* land bridge"—from sunnier Africa rather than America.

The *Gomphotherium* Land Bridge and the Proboscidean Datum

The original homeland of proboscideans was probably Africa, as the oldest (Eocene) fossils of these mammals are found there. This is thought to be the center of their evolution, from which they migrated and came to inhabit Eurasia. At the start of the Miocene, a land bridge formed between Africa and Eurasia as a result of the rotation of Arabia and a considerable drop in sea level. This is called the *Gomphotherium* land bridge because the first proboscideans to get across it were members of the genus *Gomphotherium*. This group populated the Carpathian region rapidly after crossing the bridge—appearing in Europe in the Early Miocene (Ottnangian), during the MN 4 mammal zone. The time at which proboscideans arrived in Europe is also called the Proboscidean Datum.

This *Gomphotherium* land bridge existed permanently from the Ottnangian up until the very beginning of the Badenian. However, land connection between Africa and Eurasia broke off in the Early Badenian because of another rise in sea level that affected the whole Mediterranean. As a result, the evolution of the proboscidean lineages that got stuck in Europe continued independent from the evolution of African forms, and because of this new and distinct forms were seen in the Carpathian Basin.

📖 Kordos 1992; Gasparik 2001b

"Hoe Tuskers" that Never Hoed

Deinotheriums (prehistoric elephants) also appeared in Africa in the earliest Miocene, although their anatomy was much less uniform by this time. The most characteristic feature of this group, unique among proboscideans, was the presence of curiously downward-curved tusks attached only to the lower jaw. These features account for the origin of the old, but still used, term for this group, the "hoe tuskers." This name suggests that these animals scraped and dug in the ground for food using their tusks. However, one look at the animal shows that this name is inaccurate. To make this movement, deinotheriums

Mihály Gasparik, a research fellow of the Hungarian Natural History Museum, is the leading expert on Hungarian fossil proboscideans. In his PhD dissertation he reviewed all the known Hungarian Neogene and Early Pleistocene proboscidean finds and made a list of fossils, especially teeth, that are present in different collections around the country. Gasparik also clarified the stratigraphic position and taxonomy of all these fossils and refined our knowledge of the evolution of proboscideans based on teeth. This photo was taken in 1994.

Fragmented jaws of Tertiary and Quaternary proboscideans with molars (0.25× magnification). (1) *Prodeinotherium hungaricum* Éhik—a lower-left jaw from Early Miocene layers at Putnok. (2) *Mamut borsoni* Hays—a lower-left jaw from Pliocene layers at Százhalombatta. (3) *Mammuthus primigenius* (Blumenbach)—a lower-left jaw from the Late Pleistocene pebble bed of the Danube. A new molar ready to emerge can be seen as it pushes the full-grown and worn old tooth forward. (4) *Gomphotherium* (?) *angustidens* (Cuvier)—a lower-left jaw from Early Miocene layers at Salgótarján (Zagyvapálfalva). (5) *Deinotherium proavum* Eichwald—a lower-left jaw from Late Miocene (Pannonian) layers at Budapest (Kőbánya).

Proboscideans appeared in the Eocene and flourished into the relatively recent past, up until the Pliocene and Pleistocene. Their extinction was caused partly by changes in climate, environmental changes, and damage caused by human activity. H. F. Osborn, a famous American paleontologist, distinguished 352 proboscidean species, 350 of which have already died out, in his 1942 volume. The only surviving representatives of this order are elephants, favorites in zoos, which still live in Africa and Asia in their natural environments. In the past, however, proboscideans also lived in the area of present-day Hungary. In this figure the evolution of the hoe tuskers (family Deinotheridae) is shown based on Hungarian fossils. A gradual increase in size characterizes the evolution of the group.

would have had to stand beneath the ground that it wished to hoe, or at least would have had to kneel down. Mihály Gasparik remarked on this question in his PhD dissertation, suggesting that "both possibilities seem to be rather unlikely." Hoe tuskers never hoed, in other words, but their peculiar, downward curving tusks may have helped them peel tree trunks or pull down branches.

🕮 Gasparik 2004

The First Hoe Tusker: A Genuine Hungaricum

The appearance of proboscideans in Hungary is marked by the Proboscidean Datum. One of the first of the known Hungarian fossils, *Prodeinotherium hungaricum,* is a small-bodied early deinotherid known from the Ottnangian. Fossils of this proboscidean are known from coal sequences at nine sites in Nógrád and North Borsod Counties, and on the basis of its occurrences and teeth, this animal is thought to have lived in areas with dense vegetation (closed forests, fen woods). It was probably a leaf eater that fed upon leaves, crop, soft seeds, and other tender plant parts. Some workers, however, do not consider *Prodeinotherium hungaricum* to be a distinct species but regard it as an early, small form of the later diverging and much larger "Bavarian hoe tusker," *Prodeinotherium bavaricum.* Another proboscidean, *Prodeinotherium petényii,* presumably also belongs to the same species as *Prodeinotherium hungaricum* as there are many similarities, including size and habitat, between the two.

🕮 Éhik 1930; I. Vörös 1989; Gasparik 2001a, 2004

Other Early and Middle Miocene Proboscideans

Two other proboscideans (*Gomphotherium angustidens, Zygolophodon turicensis*) are known from Hungarian Early Miocene strata, in addition to *Prodeinotherium hungaricum.* These animals are mastodons (the families Gomphotheriidae and Mammutidae), elephant groups that are well known in the popular and professional literature. All three species appear in the fossil record around the Proboscidean Datum; yet another Early and Middle Miocene proboscidean, *Prodeinotherium bavaricum,* turns up in the fossil record a little later, in the Badenian.

🕮 Gasparik 2004

Fossil droppings (a coprolith) left behind by a large-sized predator in Miocene rocks at Salgótarján. On closer examination, both plant and animal remains (bones) can be recognized (original size).

Far from the shore in the Pannonian Lake, fine-grained sediments (mainly clay, silt, and sand) were deposited. The subsequent layers give a spectacular cake-like aspect to the succession. This image was taken in the clay pit at Mályi.

8

The Late Miocene

The last 6 million years of the Miocene correspond with the Pannonian and Pontian ages across the area of the regressing Central Paratethys seaway. The likely already brackish Sarmatian Sea became completely separated from neighboring basins, and by the beginning of the Late Miocene a deep lake evolved in the area of the Carpathian Basin. Mountains began to stand out as islands and an endemic mollusk fauna evolved in the lake, making the correlation of these sedimentary sequences with those in distant areas difficult. This correlation is easier using fossil vertebrates from the coast of the lake or marshes, including the three-toed fossil horse, *Hipparion*. Huge amounts of debris from the rising mountains rapidly filled up the lake, which eventually became a shallow freshwater basin by the beginning of the Pliocene and then dried out entirely.

Some attempts to reconstruct the evolution and alteration of Lake Pannon by combining results from different stratigraphic methods (dinoflagellate, mollusk, mammal, and magneto-stratigraphy) and radiometric age determinations have proved successful in recent years. For example, it has been shown that the water level in the lake was quite low about 12 million years ago when the basin formed, and then later it rose slowly and the lake became deeper and deeper. The lake reached its largest extent about 9.5 million years ago, and from then on it tapered off due to the progression of rivers and increase in delta areas. The bulk of the debris arrived from the north, so the coastline of Lake Pannon advanced southward. By the start of the Pliocene, most of the lake had disappeared, and river plains took its place.

In order to understand the thick sedimentary sequences deposited in Lake Pannon and the animals and plants that lived within it, we need to know how the evolution of the lake is connected with climate changes and with the development of neighboring areas (mountains, distant lakes, seas). Research in this field is ongoing and many questions remain unanswered. We do know that smaller lakes are short lived in a geological sense, and even a vast lake like Lake Pannon is extremely sensitive to the balance between rainfall, evaporation, and drainage. If this balance is lost, then the lake can begin to dry up, and as a threshold is crossed much water can be lost. Even the salt content of a sea that has turned into a lake can be considerably diminished due to a quick flushing—greatly affecting biological diversity: only some animals and plants are able to adapt to such rapidly changing environmental conditions.

The surroundings of Lake Pannon comprised the continuously, but not necessarily evenly, rising Carpathian Mountains, which had other similar water bodies beyond them. According to one new theory, the water of Lake Pannon was naturally drained at least once, or even several times, toward the present-day Iron Gate in Romania.

📖 Kázmér 1990; Geary et al. 2000; Uhrin 2011

FOSSIL-RICH FORMATIONS

Lake Pannon was very deep in places, its basement sunk unevenly but constantly. These deep basins were filled with rock debris—sand, silt, and clay—derived from the nearby rising mountains as a large amount of sediment was transported into the lake by rivers and swept along the coastline. These thick, sandy successions show characteristic features of delta, lagoon, and river environments and have been studied in detail because many of these Pannonian depressions contained (or still contain) important hydrocarbon reserves. Shallow lacustrine, mainly coarse detrital deposits accumulated along these dissected coastlines, and around islands and peninsulas at lesser thicknesses. Terrestrial sediments include cave deposits and freshwater limestones, some of which contain fossil vertebrates.

Six Thousand Meters of Pannonian

The amount of sediments that filled up the continuously sinking Pannonian basin is amazing. Borehole Hódmezővásárhely-I, drilled between 1969 and 1971, used to be the deepest in Hungary, at a depth of 5,842.5 meters.

	Regional Stage/Age	Millions of Years Ago
Late/Upper Miocene	Pontian	7
	Pannonian	12

Division of the Late Miocene into stages or ages.

But this was exceeded in 2006, near Makó, to a depth of 6,000 meters.

Originally, based on data from 42 samples, these drill cores were thought to stop in Badenian sediments beneath a thick Pannonian succession, but Andrea Szurominé Korecz, a micropaleontologist, and her colleagues from the Hungarian Oil and Gas Company reexamined the sequence and found that even the lowermost layers are Pannonian. This means that even the deepest borehole in Hungary does not reach the base of the Pannonian—even at 6,000 meters! Evidence for this is based principally on dinoflagellates and ostracods studied by the micropaleontologists. Older, Badenian fossils (mainly forams and nannoplankton) in the deeper part of the sequence turn out to have been redeposited into sandy and silty layers during the Pannonian, and had confused earlier workers who had examined the borehole.

📖 Szuromi-Korecz et al. 2004

From a Lake to a Sea (of Stones): Sandstone Boulders in the Káli Basin

Most Pannonian sediments—at least those on the surface—are loose and soft, since the clay and sand layers deposited in Lake Pannon only partly lithified as they lost

Congeria czjzeki Hörnes, the most frequently occurring fossil bivalve in the clay pit at Mályi.

their water content. In some exceptional cases, however, hard stones do occur. One such area is in the picturesque Káli Basin around Kővágóörs, Salföld, and Mindszentkálla—where upwelling siliceous solutions have hardened the covering white quartz sand. In this area, sandstone lenses and beds have been tilted from their original positions and now form spectacular "seas of stones"—protected natural areas. This rock type, a popular building stone, is also exposed in a small quarry at Mindszentkálla, and the miners have collected occasional fossils. Although this site has been known since the 1970s, a detailed examination of the fossil assemblage was completed only in 1986 by Imre Magyar, who was then a geology student. Now Magyar is a researcher with the Hungarian Oil and Gas Company and is a recognized expert on the fossils

What is "Pannonian"?

The layers overlying Sarmatian deposits were named "Pannonian"—or "Pannon" in more popular parlance—by Lajos Telegdi-Róth (1841–1928), after the former Roman province of Pannonia. The introduction of a new name was necessary because the sediments covering the Sarmatian successions showed locally characteristic features. This is in accord with present international opinion about the "Pannonian": it is one of the last stages in the history of the shrinking Central Paratethys Sea, and is thus a regional stage. On the basis of radiometric measurements, Pannonian layers formed during the Pannonian age, which lasted from about 11 million years ago until about 5 million years ago.

Despite this apparently clear situation, use of the concept of "Pannonian" is far from uniform. Until the 1960s, the upper part of the Pannonian was usually identified using layers described from around the Black Sea—and, thus, with another regional stage called the Pontian. In other words, in older work the Pannonian was defined and used in a wider sense, and this stage was thought to have been longer: today, in accordance with international agreement, the stage is thought to be shorter in duration. The background to this is that the Pannonian is not viewed uniformly in Hungary: many experts see it as the last, long regional period of the Central Paratethys Sea, containing the whole Pliocene. However, this view assumes that the Pliocene was short and that the Pannonian lasted up until the Quaternary—almost twice the internationally accepted length. This is not a merely theoretical issue: it is necessary to know what is meant by the expression when reading old, or modern, literature. Following international practice, we consider the Pannonian to be sandwiched between the Sarmatian and the Pontian stages and we retain usage of the Pliocene.

Sea of stones at Kővágóőrs.

of Lake Pannon. He has documented the occurrence of seven mollusks and eight formerly unknown plants on the basis of specimens found earlier or that he has collected.

The preservation mode of Lake Pannon fossils is also very interesting. Often the calcareous shells of bivalves and gastropods have dissolved away during lithification, so only casts or imprints remain. Bivalve casts often have the marks of both the external and internal shell surfaces, which indicate that the shells had already dissolved before the complete consolidation of sediment. Plants are also generally preserved as imprints and cones that left voids in the sand. These can be reconstructed and identified accurately by molding the voids with silicon rubber.

Of the extinct dwellers of Lake Pannon, *Lymnocardium* remains are the most common of the animal fossils; frequent fossil plants include members of the living species *Rhamnus alaternus,* typical today of macchia vegetation around the Mediterranean Sea, and of the Aleppo pine (*Pinus halepensis*). Both these kinds of plants indicate the Mediterranean nature of the Pannonian climate, which was much milder than it is today.

🕮 Magyar 1988

REMARKABLE FOSSIL SITES

Different Pannonian deposits have been used by the building trade as raw materials for many years and, since they are often situated near the surface, Pannonian clay and sand pits can be found throughout Hungary. In general, lake fossils are collected from deep-water sediments but in coastal layers both terrestrial and aquatic forms are found. Small mammal (rodent) remains are known from the Fehér-part ("White Coast") at Tihany, which is also a classical site for Pannonian mollusks. Indeed, vertebrate fossils can be used in these areas for sequence correlation.

Hungarian Pannonian fossil sites that have yielded small and large mammal remains are also of outstanding significance internationally. Some of the important sites that are known (in time sequence) include Rudabánya, Sümeg, Tihany, Csákvár, Baltavár, Hatvan, Tardos, and Polgárdi. Of these localities, Rudabánya, which is famous for hominid fossils, is probably the best known.

Fossil Plant Sites

All along the extensive lake margins of the Late Miocene Lake Pannon, marsh forests grew in the warm, temperate climate. As these plants died and accumulated over time, mineable lignite deposits formed. Although well-preserved plant fossils are not found in the lignite itself, they can often be collected from the overlying layers. Pieces of stems and leaves occur often, and occasionally entire tree trunks can be found in life position. The most important Pannonian plant localities are in the vicinity of Gyöngyösvisonta, Bükkábrány, and Dozmat.

As part of the complex evaluation of the famous Rudabánya fossil hominid locality, a study of the fossil plants was also completed. These fossil remains give an insight into the ecology and living environment of the early

The open cast lignite mine at Bükkábrány with *Glyptostrobus* trees preserved in situ.

hominids; therefore, these results are of special interest. It is definitely more than a coincidence that this area had both marsh vegetation and mesophyll forests—good environments for the early hominids.

The Fehér-part (White Coast) at Tihany and Mollusk Sites of Transdanubia

Pannonian deposits providing rich, or at least promising, faunas can be found on or near the surface across large areas of Transdanubia. One of the longest known of these exposures is at Tihany, where late Pannonian strata can be studied at a more or less vertical sequence called the Fehér-part.

In these lagoonal, coastal, river, and shallow-lake sediments, shells of the *Congeria balatonica* are known alongside famous "goat's hooves"—shells of *Congeria ungulacaprae,* which can be collected along the path toward the so-called Hermits' place. Other fossil sites among numerous Transdanubian localities are Várpalota (Bántapuszta), Balatonfűzfő (Papvásár Hill), Kötcse (Csillagó), Tab, Nagyárpád, Hird, Pécsvárad, Sopron, Kisbér, Dáka, Danitz-puszta, Bátaszék, Somberek, and Tata. Some of these exposures are well-known classical sites, whereas the fauna from others is only just beginning to be understood. Indeed, many of these sites are clay pits—some still providing brick factories with raw material.

Vertebrate Fossil Sites

The rich fossil vertebrate fauna of the dissected and often marshy coast of Lake Pannon has been preserved in many places. Plenty of fossil sites are known, in addition to the most famous ones mentioned above, where small and large Pannonian vertebrate faunas have been discovered. Indeed, Miklós Kretzoi mentioned 200 fossil sites known from the Carpathian Basin in his 1982 thesis—in which he described the so-called *Hipparion* fauna, comprising Pannonian vertebrate remains. He also provided a list of around 170 Hungarian fossil sites, together with fauna lists. Most of these sites, understandably, are located around the margins of mountains, especially the Hungarian Range, although a few are known from the foothills of the Alps and in the Mecsek Mountains.

Part of these terrestrial vertebrate remains accumulated in caves and rock cracks. The famous fossils from Csákvár, for example, were found in the Esterházy Cave, where bones were piled up partly by water and partly by predators. In another example, in the valuable fossil sites at Polgárdi, rock cracks acted as traps, and bone breccias with hardly any matrix were formed.

Indeed, to date almost the whole Pannonian is now well known to paleontologists thanks to around a dozen rich faunas and specimens from about 200 other smaller collections of sporadic fossils. The continental stratigraphic classification that has resulted from these fossils was internationally recognized and Miklós Kretzoi, who is also very successful in other fields of vertebrate paleontology, has gained a good deal of fame in this area. Kretzoi is known for work determining and defining the mammalian stratigraphic units Bódvaium, Csákvárium, Sümegium, Hatvanium, Bérbaltavárium, and Csarnótánum based on the fossil faunas of Hungary. Of these units, the last belongs to the Pliocene.

Rudabánya: A Primate Cemetery

The village of Rudabánya in Borsod-Abaúj-Zemplén County is one of the oldest miner's settlements in Hungary. From earliest times, clay pigments from the earth, and then copper and iron ore, were mined here. Although the Slavic word *ruda* means "ore," and this is how the village got its name, Rudabánya is now famous internationally for Late Miocene (Pannonian) fossil vertebrates, particularly primates (hominids).

The Rudabánya fossils were found in a brown, coal-bearing, clayey succession that overlies the iron ore–bearing metamorphic Triassic limestone, deposited on the marshy coast of Lake Pannon. Although collectors had already known bones (chiefly spectacular proboscidean remains) from the area, Rudabánya found its place as a globally important fossil site after the discovery of the first hominid remains. The scientific reputation of Rudabánya was first established by geologist Gábor Hernyák, who found most of the first significant fossils, including the first hominid remains. Later, Miklós Kretzoi, who was quick to recognize the significance of these findings, and László Kordos, who led many successful excavations at the site, further enhanced the picture of the Rudabánya Pannonian fossil vertebrate fauna.

About 40 paleontologists and geologists have so far taken part in systematic research at Rudabánya, and several thousand promising fossils have been found. This ongoing research first started seriously at the beginning of the 1990s, when the site was re-explored and new fossils were collected by vertebrate paleontologists from a range of countries working as part of a Hungarian-American team. Results from this work were finally published in 2005 in a special volume of *Palaeontographia Italica;* they provide a uniquely complete picture of a very rich fauna comprising about 100 vertebrate species. All in all, a range of experts contributed papers to this issue, which dealt with the Rudabánya amphibians, reptiles, insectivores (hedgehogs, shrew-mice, and moles), rodents (squirrels, dormice, and hamsters), predators, proboscideans, horses, rhinos, pigs, and ruminants. Now we have a much more precise and detailed picture about the diverse living world around Lake Pannon and about the ecology of early hominids.

📖 Kretzoi 2002; Kordos 1985b, 2001; Bernor, Kordos, & Rook 2005

Sümeg: Backbones on the Backbone Hill

On a ridge called the "Backbone," north of Sümeg, bones representing the last known faunal period of the Pannonian were found in a crack. For a long time, these bones were thought to represent fossils from the period between the Csákvár and the Hatvan faunas, but more recent work has shown that they are older. Miklós Kretzoi was the first to study these remains; the site is about as fossiliferous as the Csákvár locality, so far yielding 3 amphibians (frogs), 5 birds, and 47 mammals—including insectivores, bats, rodents, small and large predators, rabbits, and even- and odd-toed ungulates. The large proportion of insectivores within the variable small mammal fauna at this site is particularly striking and can only be explained by accumulation in the vicinity of sleeping and resting places of owls. The pellets dropped by these birds fell apart, and the tiny bones within them (such as the teeth and bones of shrew-mice) formed a layer. A further curiosity of the Sümeg fauna is relatively high diversity of cats and hyenas, as well as the presence of a sheep-like ungulate.

📖 Kretzoi 1985b

Csákvár: Bones in a Cave

The Esterházy (aka Báraczháza) Cave is located in the foothills of the Vértes Hills, 2 kilometers southwest of Csákvár. In 1924, a tourist group from Székesfehérvár made a series of test excavations in this 10-meter case and found the first fossil vertebrate bones. Two years later, most of the cave infilling was removed and a very rich *Hipparion* fauna was collected from Holocene and Pleistocene deposits. Finally, after further fieldwork had almost completely exhausted the cave deposit, Miklós Kretzoi described the known fauna and identified 2 fish, 3 amphibians, 6 reptiles, 6 birds, and a remarkable 66 mammal species. As a result of this rich fauna, the Esterházy Cave is well known to paleontologists.

📖 Bogsch 1928; Kretzoi 1951b

Baltavár: The "Wayside" Fossil Site

The village of Baltavár (now called Bérbaltavár because of the contraction of two villages) in Vas County has been a well-known site for Hungarian vertebrate paleontology for about 150 years. In the 1850s, fossil bones were found on Szőlő Hill during the construction of a post road to connect Buda with Graz; curiously, Eduard Suess, a famous professor from the University of Vienna, was the first to study them. He traveled to the spot and collected a large number of fossils that he later published on, mentioning saber-toothed cats, hyenas, hoe tuskers, rhinos, ancient giraffes, antelopes, pigs, and hipparions. Although Suess's faunal list was later revised in several aspects, he correctly noted that this fauna can be correlated with a similar one from Pikermi, near Athens. From this point on, the term "Pikermi-type" is often seen in the Hungarian paleontological literature. Later, as a result of the work of Antal Brunner, a road master who built a house near the exposure, and subsequent geologists, the number of collected bones and identified ancient animals has gradually increased. Ferdinand Stoliczka (1838–1874), Gyula Pethő, Tivadar Kormos, and László Benda (1904–1977) also enriched the paleovertebrate collection known from Baltavár, and most of the fossil bones found their way into collections in Vienna, Budapest, Szombathely, and Vasvár.

However, while the fauna from Baltavár was becoming more and more known to experts, the condition of the fossil site itself was getting worse and worse: it became weedy, then covered with garbage, and then the exact location of the bone-bearing layer was lost until new excavations were started in 1998 under the guidance of László Kordos and American Raymond L. Bernor. The premise of this joint work was the former Hungarian-American research on hominids of similar age at Rudabánya, and it aimed to enrich the faunal list and describe the paleoenvironmental conditions of the fossil site. Kordos has preserved the Baltavár site by buying the area and enclosing it with a fence and a sign that reads, "no trespassing, excavation area." By law, all further research is subject to licensing, and the discovered fossils have to be handed over to museums. Those resulting from the last series of excavations are now held in the collections of the Geological Institute of Hungary.

At the site, the clayey lower layers of the Baltavár sequence were deposited in a floodplain or lake environment. Later the clay was cut repeatedly by a river of changing size that also carried a considerable number of bones and bone fragments; these got stuck on shoals and were cemented with minerals into the sediment as they filtered upward. All of this activity happened 7.5–8 million years ago, according to the latest paleomagnetic results.

📖 Kordos 2005

Hatvan: The Clay Pit on the Strázsa Hill

The clay pit of the First Steam Tilery of Hatvan operated on Strázsa Hill during the first half of the twentieth century, and local workers put the bones that turned up during mining away for years—until István Gaál (1877–1957) discovered the collection and described the fossils. Gaál was first a professor at the University of Cluj-Napoca, then a research fellow, and later the head of the Department of Mineralogy-Geology in Pest and then in Szeged.

Middle Miocene whale vertebrae from Pannonian deposits in the sandpit at Danitz-puszta. The average size of these vertebrae is 3–4 centimeters.

Most of the fossil mammals from Hatvan were deer, but horses, rhinos, bears, and even giraffes were also found. Most of the bones were found in a bluish layer within a succession of river and lake deposits; and a number of other sporadic fossils also from the neighborhood of Hatvan (Ecséd, Szücsi, Rózsaszentmárton) were also mentioned by Gaál.

📖 I. Gaál 1943; Bernor et al. 2003

Polgárdi, Kőszár Hill

Mostly old Devonian limestone was mined in a huge limestone quarry near the town of Polgárdi. The stone has been mined there since Roman times. As a result of excavations, a series of small caves and karst caverns have been found, and many contain a famous Pannonian vertebrate fauna. The first excavations in this area were conducted by Tivadar Kormos, although the site he examined, along with a few others, does not exist anymore. The bones from these cavities, destroyed by later mining, are held in museum collections. Nevertheless, the infilling of the cavern that Kormos explored contained five layers characterized by tortoise, *Ictitherium* (hyenas), *Hipparion, Gazella,* and *"Capreolus"* remains, as well as those of other small mammals. Indeed, the scientific significance of Polgárdi is that it was the first site to be found to contain a rich microfauna, in addition to the already well-known *Hipparion* fauna.

Another large cavity yielding very rich bone material was discovered in 1985; the small mammals, including the hamsters *Cricetus polgardiensis* and *C. kormosi* named after the fossil site and its first researcher, were examined by László Kordos. The shrew-mice found at Polgárdi were described by Lukács Mészáros, the unique bird fauna by Dénes Jánossy (1926–2005), and the remarkable herpetofauna (reptile and amphibian remains) by Márton Venczel from Oradea.

The fauna preserved in the karstic caverns of Kőszár Hill near Polgárdi is biostratigraphically uniform and all belongs to the earliest stage of the Miocene, the MN 13 mammal zone. This means that the fauna is about 5–6 million years old, although the fauna from the different sites varies somewhat within this zone.

📖 Kormos 1911a; Venczel 1998a; L. Gy. Mészáros 1999a, 1999c

Danitz-puszta: Everything in One Place

At Danitz-puszta, near Pécs, redeposited marine and terrestrial (mainly vertebrate) remains can be found preserved in Pannonian sand. The fossils here represent at least two stratigraphic units, the Badenian and Sarmatian.

The Pannonian Beočin Marl in the Filijala open-pit mine southeast of Beočin (Serbia). Certain levels of this 200-meter-thick succession are rich in fossil mollusk remains, and the well-preserved fossil fish from this site were first described by Antal Koch.

Fish remains (gullet teeth, fin spikes, bone fragments), soft-shelled and land tortoise shells (*Trionyx, Testudo*), as well as mammal fossils (even- and odd-toed ungulates, proboscideans) are common among fossils from the younger of the two stratigraphic units.

Seal, sirenian, and whale remains (vertebrae, limb bones, and rare ear bones—of crucial importance in taxonomic classification) are characteristic elements of the older marine vertebrate fauna. Indeed, the fossil site at Danitz-puszta has long been a favorite, not just for professionals but also for amateur fossil collectors. As Emese Kazár remarks, whale bones have been found in the thousands, and they are not the commonest fossil remains. Large numbers of fine, mostly small vertebrate fossils (primarily fish teeth) have also been recovered from the sandpit at Hímesháza, in south Hungary, a geological environment similar to Danitz-puszta.

📖 Kazár et al. 2001; Kocsis 2002b; Kazár 2003

Fossil Sites in Transylvania

One of the large bays of Pannonian Lake system was the Transylvanian Basin. In this basin, Pannonian sediments can be found as far afield as the foot of the Eastern Carpathians, in the region of present-day Odorheiu Secuiesc. Bituminous sand occurrences around Brusturi in Bihor County contain diverse plant fossils, including giant leaves, and valuable fossil vertebrate assemblages, including rhino bones.

LATE MIOCENE FOSSILS FROM THE CARPATHIAN REGION

As if it were a giant evolutionary laboratory, the distinct and separate basin that became Lake Pannon played host to a peculiar endemic mollusk fauna. The fossil remains of this fauna aroused the curiosity of scientists very early on. The first paleontological debate known in the Hungarian language concerned the most characteristic of the Pannonian fossils, the goat's hoof, and took place about 200 years ago.

The Pannonian fossil vertebrates that are mostly preserved in the marsh and coastal environments of the lake are also worthy of note. Of these, the most famous ancient mammals of this period are the *Hipparion* (three-toed horses)—but a large range of other fossil mammals including ungulates, predators, and proboscideans are also known from Pannonian deposits. The hominid remains that have also been found around the margins of the lake at Rudabánya are, of course, world famous.

Microfossils

The microscopic living world of Lake Pannon differed considerably from earlier Sarmatian assemblages as salinity levels in the lake dropped. Forams, for example, that generally require salt water disappeared. Single-celled animals are represented by the silica-shelled amoeba *Silicoplacentina,* which was first described by micropaleontologist József Kőváry in 1956 from Pannonian deposits exposed in deep drill cores for hydrocarbon exploration. In contrast, organic-shelled microplankton were much more diverse at this time; unicellular flagellates—dinoflagellates—usually classified as algae along with seaweed, are represented by a number of endemic species. This diversity of dinoflagellates has made it possible to recognize seven successive biozones in Pannonian sequences, as these microfossils are the only fossils that are found in both shallow and deeper waters. As a result, even changes in the shape of the outline of Lake Pannon can be determined on the basis of dinoflagellate distributions. Of the stratigraphically important forms, a number of species and subspecies belong to the genus *Spiniferites,* which had a long duration geologically; *Pontiadinium pecsvaradensis* and *Mecsekia ultima,* both named after Hungarian localities, are also critical zone indicators. Mária Sütőné Szentai, a researcher at the Komló Laboratory of the Hungarian Geological Research and Drilling Company, first recognized the importance of these microfossils and described them in detail when she was working at the Komló Natural History Collection.

The most diverse microfossils from the Pannonian are probably ostracods, known from both fresh- and saltwater sequences. Some forms, like the smooth-shelled *Candona* and the richly ornamented and nodal *Leptocythere,* are markers of freshwater and slightly saltwater environments. For years—after the early works of Béla Zalányi, Gyula Méhes, and Imre Héjjas (1869–1957)—little attention was paid in Hungary to these important environmental marker fossils, although they were studied from the southern part of the Carpathian Basin by Nadežda Krstić, who worked in Beograd, Serbia. Hungarian research was renewed in the early 1990s when Andrea Szurominé Korecz, a specialist from the Hungarian Oil and Gas Company, turned her attention to microfossils.

Fossil Plants

Vegetation was lush around the Pannonian Lake, and huge areas in the foreland of the northern mountain range and in Transdanubia were also covered with swamps. In these regions, swamp forests covered huge areas comprising just a few arboreal taxa and herbaceous aquatic plants. In shallow water, water chestnuts (*Trapa*) were abundant and the lake was surrounded by bulrush-like plants. In the marshes, ferns, swamp cypress, and another extremely flood tolerant extinct alder (*Alnus cecropiifolia*) were common, while shoots of *Byttneriophyllum* would have made the vegetation

Excavated tree trunks on top of the lignite preserved in grey sand. The darker, oxidized, yellow-to-brown sands above contain no traces of any fossil tree remains.

impenetrable. Riparian forests along the rivers had more diverse floras and were characterized by *Liquidambar europea*—an extinct relative of the American sweet gum—by fossil plane trees (*Platanus leucophylla*), and by another species of alder (*Alnus ducalis*), different from those living in the marshes.

The Buried Forest at Bükkábrány

A fossil forest was discovered and excavated in the Hungarian Bükkábrány open-pit lignite mine in the summer of 2007. Pieces of fossil wood had long been known from this lignite, from the overlying beds at Bükkábrány, and from the nearby locality of Visonta; occasionally even stumps in standing position had been found. Indeed, earlier researchers had described shoots, leaves, and fruits belonging to more than 20 different fossil plant species from Bükkábrány. Sandy deposits above the lignite level were also known to be especially rich in fossil leaves and the remains of water chestnuts ("*Trapa* level").

In 2007 the remains of huge trees were found at this site, not just isolated plant remains, indicating the presence of a real forest. Miners working on the removal of overburden sand 60 meters down the mine excavated tree trunks from atop of the coal seam. All in all 16 stumps were excavated from the mine; they were 1.8 to 3.6 meters in diameter at their bases and up to 6 meters in height. The distance between the trees varied from 5 to 15 meters. Directed to preserve these unusual fossils, the miners removed the sand layers with care and were able to expose the trees and leave them in situ. Such preservation, an in

Some of the Bükkábrány stumps were transported away from the mine and erected again in front of the visitor center in the Ipolytarnóc Geological Conservation Area. However, the woody material of these trees became dry in air and started to shrink, presenting a difficult task for scientists.

situ fossil forest with trees preserved in growth positions and with the original structure of the forest intact, is extremely rare; examples are nevertheless known on all continents and from most geological ages. In these other fossil forest examples the tree wood is mostly mineralized—turned to silica, carbonate, or (rarely) to pyrite. But in the case of the Bükkábrány trees, preservation is different: in these trees the wood remains are preserved intact with an appearance just like that of real wood. This type of preservation is extremely unusual.

Initial anatomical studies of this fresh fossil woody material allowed taxonomic identifications to be made for some of the trunks. *Taxodioxylon germanicum* (a tree related to modern *Sequoia*) and *Glyptostroboxylon* sp. were initially determined. From the topmost layer of the coal abundant foliage and cones of *Glyptostrobus* were also collected.

This unique mode of preservation was possible because of a sudden and dramatic rise in the water level of Lake Pannon that drowned the forest and buried it in delta sand. As a result, trees were completely saturated for millions of years and mineral precipitation was minimal. Both the biological and geological ages of these trees then became an issue of special interest; when the first samples were cut for tree ring analysis, narrow rings less than 1 millimeter in width were observed, and up to 400 rings were counted in a tree with a trunk just 80 centimeters in diameter. Unfortunately, however, the freshly cut surfaces of the wood were oxidized within a few hours, and the ring boundaries can no longer be observed. Much older trees, however, would have been expected in a mature forest of this kind. The famous living cypress in Santa Maria del Tule (Oaxaca, Mexico), for example, is about 2,000 years old. The age of the sand preserved above the lignite, which indicates the minimum geological age of these trees, is between 7 and 8 million years old (as revealed initially by geophysical reflection sections). These reflection profiles were laid down through Bükkábrány toward the deeper part of Lake Pannon, where the age of sediments was already known. Following these reflection boundaries along the section meant that correlation of the age of the Bükkábrány beds, deposited along the shore of the basin, was simple.

This ancient *Taxodium* swamp was later covered with a huge amount of sand, which also buried the forest. Maybe a river entering the lake changed its direction, or maybe sand dunes along the shoreline suddenly collapsed over the area because of an intense storm. In any case, the flow of this sand stopped the formation of lignite and finally preserved the fossil forest. These giant trees stood along the edge of Lake Pannon for a long time but now only their

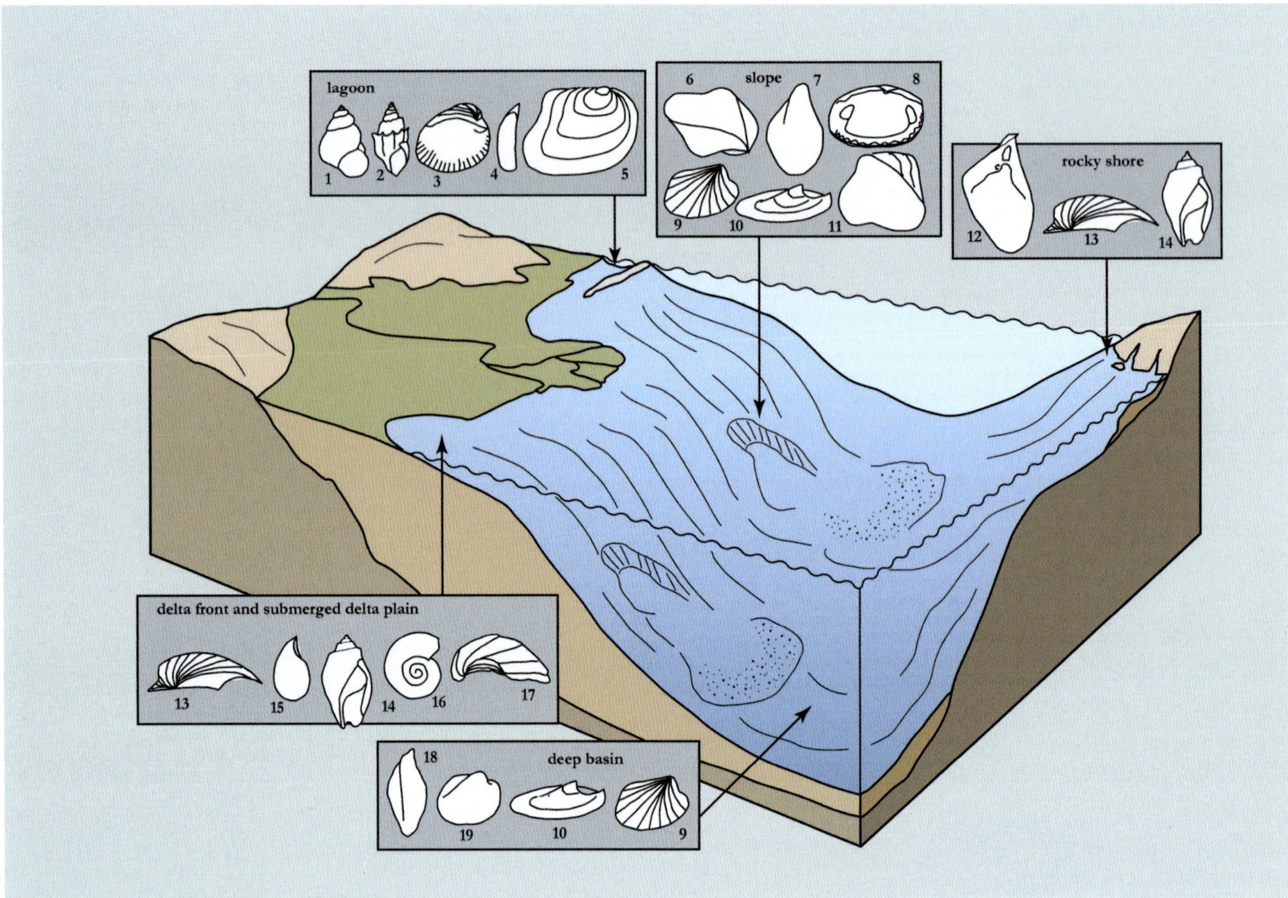

Different mollusk biocenoses in Lake Pannon. (1) *Viviparus.* (2) Small, ornamented *Melanopsis.* (3) *Prosodacnomya.* (4) *Dreissena serbica* Brusina. (5) *Anodonta.* (6) *Congeria rhomboidea* Hörnes. (7) *C. zagrabiensis* Brusina. (8) *Lymnocardium majeri* (Hörnes). (9) *Paradacna abichi* (Hörnes). (10) *Valenciennius* sp. (11) *Pteradacna pterophora* (Brusina). (12) *C. ungulacaprae* (Münster). (13) Large *Lymnocardium* species. (14) Large *Melanopsis* species. (15) *Dreissena auricularis* (Fuchs). (16) *Gyraulus.* (17) *C. balatonica* Partsch. (18) *Dreissenomya digitifera* (Andrusov). (19) *"Pontalmyra" otiophora* (Brusina). After Geary et al. 2000.

lower portions remain. The upper parts of these trees were eroded away over time.

📖 László 1992; Kázmér 2008; Erdei et al. 2009; Hámor-Vidó et al. 2010

Invertebrates

Gastropods and Bivalves

Lake Pannon, a Natural Evolutionary Laboratory

Remains of organisms that lived at one point in the Carpathian region, especially during the Paleozoic and Mesozoic eras, can sometimes be found at fossil sites thousands of kilometers away. This is never the case with Pannonian fossils, which comprise some of the most characteristic faunal assemblages in the region.

During the Late Miocene the whole Carpathian Basin (Pannonian) and part of the area to the south was covered for a long period by a vast lake that was sometimes deeper than 1,000 meters. As the lake formed, water from incoming rivers would have rapidly decreased salinity and caused the animals and plants that lived in the salt-water Sarmatian Sea to disappear almost completely. A very characteristic Pannonian fauna of relic marine and freshwater mollusks evolved rapidly and in isolation in the lake: an endemic fauna of species, genera, and even families that are not known from anywhere else. Indeed, several hundred papers, including vast monographs, have been written about the remarkable and often very well preserved fossils of Lake Pannon; over the last decade it has become possible to determine even the relative geological ages of these fossil assemblages using accurate geophysical methods. For all these reasons, Lake Pannon is regarded as one striking global example of a natural evolutionary "laboratory," as recent research has characterized the environments inhabited by its special fauna.

Scientists come from all over the world to work on the fossils of Lake Pannon. For example, Dana Geary, presently a professor at the University of Wisconsin, arrived in Hungary as a student on the recommendation of her professor, Stephen Jay Gould (1941–2002), a well-known evolutionary biologist and one of the most popular and famous American scientists of the 1980s and 1990s. Geary worked on the evolution of the famous *Melanopsis* gastropods from Lake Pannon.

Indeed, paleontologists have labored for around 150 years to determine the precise geological ages of the Pannonian sequences using fossil mollusks. At first they endeavored to divide the not very thick successions at individual exposures using the gastropods and bivalves, but were not successful because it was soon recognized that differences between assemblages reflect local changes in depositional environments rather than large-scale processes affecting the whole basin. It was a great leap forward, for example, when it was noticed that different forms could live in coastal and deep-basin environments at the same time. The real breakthrough in the recognition of evolutionary processes in Lake Pannon, however, was made when the relative ages of the fossil sites were determined for the first time using geophysical methods. These data were brought together in a 2004 publication by Imre Magyar, in which he recounts how promising ideas and facts are overturned by the presence of age indicator fossils occurring unexpectedly in the "wrong" places—that is, at incorrect stratigraphic levels. Magyar's paper is a gripping read, even for non-geologists.

📖 Magyar 2004, 2010

Margaritifera flabellatiformis Grigorowich—a bivalve Methuselah in Lake Pannon? The ancestors of living freshwater pearl mussels were common in rivers flowing into the lake. The characteristic nacreous, flaky shells of these mollusks are mentioned as *Unio wetzleri* in former literature. The representatives of living *Margaritifera* species sometimes live as long as 120–140 years. It is not known whether their Late Miocene ancestors were such long-lived creatures as well, since the Pannonian specimens have not yet been studied from this point of view (about original size).

Endemic Gastropod Fauna

Gastropods are the most widespread mollusks, so their remains can be found in most Pannonian sediments, including coastal marshes or the deepest parts of the lake. However, these mollusks have turned out to be of little use for stratigraphy and are hardly mentioned in the different stratigraphic classifications that are based on Pannonian mollusks.

The most frequently occurring fossil gastropod shells are *Melanopsis,* which lived in coastal waters; *Viviparus,* which still lives in the Carpathian Basin, and *Theodoxus;* which often has its color pattern preserved. However, there are some fossil sites where much more extraordinary gastropods can be found, such as *Valenciennius,* which used to live in the deeper parts of Lake Pannon, and specimens of the unusually large *Tinnyea,* described by Miksa Hantken.

The most often cited gastropods from the Pannonian belong to the genus *Melanopsis,* which is known from rocks dating back to the Cretaceous. Traditionally, at least 3 species (*impressa, fossilis,* and *vindobonensis*) within this genus are generally distinguished among the larger-sized forms—with the first, earliest described form found in the lower parts of the classical Pannonian succession in the Vienna Basin. At first it was thought to be a good age indicator, but the species are in fact hard to tell apart because of innumerable transitional forms. A group of Austrian and German experts—Mathias Harzhauser, Thorsten Kowalke, and Oleg Mandic—concluded in a 2002 study that in addition to the 3 species mentioned above there are at least 10 other "species" that represent extreme morphological variations of *Melanopsis fossilis.* A similar confusing situation is seen in *Viviparus,* but with less diverse and characteristic forms.

Tinnyea, which is taller than 7 centimeters and has a diameter of 3–4 centimeters, is a strikingly large genus compared to the other gastropods from Lake Pannon. The early whorls of this form are ornamented with oblique ribs that run parallel to the axis of the shell but that later change into rare, oblique ribs that bend toward the shell tip. This genus was named after the village of Tinnye, where specimens of the type species *Tinnyea vasarhelyii* were found in a long-lost, then later rediscovered, fossil site. The living relatives of this genus live in rivers, so the occurrence of their remains in the otherwise undisturbed environment of Lake Pannon suggests that fossils of autochthonous organisms were occasionally covered by sediment together with shells carried into the lake by rivers.

Peculiar fossils from often light-colored calcareous rocks were also deposited in the basin of Lake Pannon, far from coastal areas. Examples of these rocks include the Beočin calcareous marl, of long debated age, which contains the characteristically dome shaped gastropod *Valenciennius,*

Imre Magyar, geologist at the Hungarian Oil and Gas Company, has made a lasting contribution to studies on Pannonian fossil mollusks as well as working as a hydrocarbon expert. He has always viewed the endemic Pannonian fauna as a form of Hungarian national heritage, even though all countries that border modern Hungary, including the Czech Republic, have a share in it. This is because the distribution of Pannonian deposits tends to follow the boundaries of Hungarian settlements in most places, likely because Hungarian people have always preferred to live on plains. Pannonian sediments are never found in mountain areas: either they were not deposited at all or have subsequently been eroded

which can reach 10–15 centimeters in diameter. In the early stage of ontogenesis a pointed, bent shell of several whorls is formed out of center by this gastropod, while the last whorl is spread out and plate like. The line of the siphon in *Valenciennius* is marked by a wrinkle that runs between the margin close to the peak and the peak itself: The siphon allows fresh water into the pallial cavity and drains away waste water. The last whorl in this genus is ornamented with strong, concentric grooves and ribs.

📖 Hantken 1887; Harzhauser et al. 2002

Goat's Hooves from Lake Balaton and other Pannonian Bivalves

Pannonian bivalve fossils were known long before the seventeenth century and include *Congeria ungulacaprae,* popularly known as the goat's hoof, from Lake Balaton. This bivalve is characterized by a shell that is thickened below the umbo, the area that is most often preserved. According to the well-known legend, these goat's hooves are indicative of a herd of goats that drowned in the lake long ago. However, the fact that these hooves are really fossil bivalves was also recognized early on: János Korabinszky (1740–1811) described them as bivalves in 1777, and the first person to illustrate them was C. D. Bartsch, who published some fine woodcuts of three specimens in the *Ungarisches Magazin* in 1782. Bartsch thought the goat's hooves were oyster remains, as did the adventurous Ignác Born (1742–1791), one of the most versatile characters of the Central European enlightenment. Born was equally expert in mining, metallurgy, chemistry, and the study of living and ancient mollusks; according to some sources, he was the model for Sarastro in Mozart's *Magic Flute.* He published the scientific description of the goat's hooves in his last work, the printed catalog of the Raab collection.

The first paleontological debate in the Hungarian language was also about these goat's hooves, and it can be traced in contemporary documents. In the journal *Scientific Collection,* Paul Partsch (1791–1856), then the director of the Cabinet of Natural Sciences in Vienna, suggested that the fossils belonged to the genus *Mytilus,* a blue mussel. He thought that these originally marine forms got on land because of some kind of continental revolution and afterward were eroded out by waves. The last of these notions is appropriate but the others are not: erosion had already been noted by János Ries, a physician from Balatonfüred, who observed and described these white bivalve fossils washed into the water of Lake Balaton by wave action from the rock layers of the Tihany Peninsula. However, this first outcrop—the original "source" of goat's hooves—was lost for many years to geologists after the publication of Ries's writings. It was rediscovered much later in 1902 by Gyula Halaváts (1853–1926), an expert on Pannonian sediments and fossils and supporter of the library of the Geological Institute in Budapest.

Final clarification on the taxonomy of goat's hooves can also be credited to Partsch, who identified them as representatives of a new genus, *Congeria,* in a study with fine lithographs published in 1835 in the annals of the Natural History Museum of Vienna. Many subsequent works were published on these fossils, the most notable

A massive accumulation of *Congeria* shells. This boulder containing fossils comes from a clay pit at Doba. In some of the Pannonian layers only the thick central part of these shells, known as goat's hooves, is preserved.

Real goat's hooves from Lake Balaton (Tihany) (2× magnification).

of which include Imre Lőrenthey's work that appeared in 1902 in the prestigious journal *Palaeontographica;* István Vitális's (1871–1947) paper, which was published as an appendix to the Balaton monograph, and Melchior Neumayr and C. M. Paul's fine 1875 monograph about Slavonian faunas. Neumayr, who later became famous for his paleogeographical interpretations of European Jurassic faunas, was among the first to recognize the evolutionary significance of Pannonian mollusks in light of Charles Darwin's revolutionary work published not long before (in 1859).

Many species within *Congeria* that also evolved rapidly can be counted, besides the species *ungulacaprae,* and are useful for stratigraphic purposes. Different species of *Congeria* lived coevally in shallower regions and in the deeper basins of the lake; the largest form, *C. rhomboidea,* is particularly "stratigraphically appropriate." This species was described by Moritz Hörnes (1815–1868) in his voluminous study on the Tertiary fossils of the Vienna Basin, originally from an exposure near Nagyárpád, a village belonging to the town of Pécs, which has even today preserved its rustic character. The specific, almond-shaped *C. czjzeki* is also an age-indicator taxon alongside its contemporary, the deep-basin species *C. banatica,* known in Hungary only from deep boreholes in the Great Hungarian Plain. Fossils of an additional endemic form, the deep burrowing, peculiar, and flat *Dreissenomya,* are found only in the sediments of Lake Pannon.

Another common Pannonian fossil bivalve is *Lymnocardium,* which is thought to have had marine ancestors and is a relative of living cockles. These mollusks adapted to the decreasing salinity conditions in Lake Pannon well and occur in large numbers in great diversity. Indeed, the various *Cardium* species known from the lake used to be thought to belong to the generally misspelled genus *"Limnocardium,"* but later numerous genera (including *Phyllocardium, Pteradacna,* and *Caladacna*) were distinguished based on increasingly obvious differences: some experts even think that these forms comprise distinct families. Another frequent and characteristic species, *Lymnocardium arpadense,* was also described from Nagyárpád by Hörnes.

Getting bogged down in details, scientists generally forgot to consider the Pannonian cockles with suitable perspective. Indeed, in 1991 Pál Müller and Imre Magyar showed, on the basis of specimens derived from fossil sites of known ages, that two species considered by former workers to represent distinct genera (or even distinct families) are actually connected by a series of transitional forms. This means that

1
2
3
4
5
6
7
8
9
10
11
12
13
14
15
16
17

the species genera and even families in question were not real biological units. Familiar forms also occur among the fossil bivalves, including the remains of ancestors of the living painter's mussel, *Unio,* which are common in sediments deposited in shallow water environments.

📖 Neumayr & Paul 1875; Lőrenthey 1911; Halaváts 1911; F. Bartha et al. 1971; Korpásné Hódi 1983; Müller & Magyar 1991

Vertebrates

Fish

Fish remains are frequent in Pannonian deposits as scales, tiny bones, and otoliths. The latter have not been studied in detail but can be common in washed residual sediments. Whole fish fossils are rare, but the best-preserved fishes from the Beočin calcareous marl were reviewed by Antal Koch. There are a number of fossil fish remains from this site, mostly valued pieces in old collections. According to Koch, the rich fish fauna at this site included cods (*Gadus*), burbots (*Brosmius*), barracuda-like predators, and others—for example, fish with pavement-like teeth. He also mentions a cod specimen that was probably more than a meter in length. Although the fish fauna from the Beočin calcareous marl was essentially marine, fossils from younger Pannonian layers lived in freshwater. Older deposits were often redeposited on the coast of Lake Pannon, so older marine and younger freshwater fish fossils can often be collected from the same layer at some sites.

📖 Koch 1902

Amphibians and Reptiles

A fairly diverse amphibian and reptile fauna is known to have lived in the Carpathian Basin during the Pannonian. One of the most recently and thoroughly studied faunas is known from Rudabánya. Five tailed and 9 tailless amphibians, along with 11 snakes, have been identified from this site. Of these records, *Bombina* and tree frogs (*Hyla*) are worth mentioning because, among frogs, these fossils are the earliest occurrences of their respective genera. Although colubrids (Colubridae) dominate the fossil snake fauna at Rudabánya, venomous snakes such as adders (*Vipera*) and cobras (*Naja*) have also been described. This snake fauna can even be correlated with extant ones at the genus level, although it also includes a number of extinct species. At least the Rudabánya snake fauna is known to differ significantly from other known European Middle Miocene sites.

📖 Szyndlar 2005

Birds

Late Miocene bird fossils from the Carpathian Basin are relatively rare, at least compared to those of similarly aged regions. Only a few remains have been found at the relevant fossil sites, including 12 different species (teals, gallinaceous birds, quails, earless owls, and songbirds—including corvids) described from Rudabánya by Dénes Jánossy.

Just a falcon, a partridge, and a few songbird fossils have so far been described from Sümeg, but much data has been collected from the similarly aged Csákvár site. This is a rich ancient vertebrate locality, at one time the richest *Hipparion* fauna known anywhere in the world. Kálmán Lambrecht described a swan, *Cygnus csakvarensis,* alongside several other fragments of unclear affinities. Lambrecht's list was later updated and completed by Kretzoi, who also described crane and owl fossils from Csákvár. Lambrecht also described a Miocene fossil snakebird, *Plotus pannonicus,* from the bituminous sands at Brusturi.

The Polgárdi site is unique among Late Miocene fossil bird localities. Dénes Jánossy examined hundreds of

Pannonian mollusks (all original size, except [6] and [7]). (1, 2) *Viviparus cyrtomaphorus* Brusina. Shells of *Viviparus* are frequent in deposits of the former Lake Pannon. Some shells are pointed, with rounded whorls, and some are cylindrical. Their specification is greatly hampered by the numerous transitional forms between the distinct species. According to a supposition gaining more and more ground, the variations of only one species, very similar to that in the photo, *V. sadleri* Partsch, can be collected. If this is so, then the bulk of species names enrich the list of synonyms (Fonyód). (3) *Planorbarius* sp. This planorbid gastropod of Lake Pannon was a dweller of calm waters in marshy areas. (4) *Tropidomphalus doderleini* (Brusina). Representatives of land genera also occur in the Pannonian gastropod assemblages dominated by aquatic forms. The species in the photo lived in slightly wet parts of the dry land around the lake. It is also known in former literature as *Mesodontopsis* and *Tacheocampylaea.* (5, 8–12) *Melanopsis* spp. Pannonian *Melanopsis* are quite large with smooth or spirally ornamented as well as small, spiky, ribbed forms. The distinction of the formerly described numerous species is—like *Viviparus*—not always possible because of the transitional forms ([5] Sopron, [8–10, 12] Tinnye, [11] Dáka). (6, 7) *Theodoxus radmanesti* Fuchs. Pannonian *Theodoxus,* like living ones, had variously colored shells (Várpalota) (1.5× magnification). (13, 14) *Tinnyea escheri vasarhelyii* Hantken. The largest gastropod among the Pannonian fossil assemblage, which probably lived in the water of rivers flowing into the lake instead of in Lake Pannon itself. This specimen derives from the type locality (Tinnye) of the genus. (15) *Valenciennius reussi* Neumayr—a peculiar gastropod from the deepest, most peaceful waters of Lake Pannon (Unknown fossil site). (16, 17) *Lymnocardium schmidti* Hörnes. A large bivalve compared to other *Lymnocardium,* its length can reach 8 centimeters. In Hungary it is exposed only south of Lake Balaton (Hidas).

1
2
3
4
5
6
7
M 63 7112
8
9
10
11
12
M 62 9944

bones from this site and described about 20 species—including peacock, brook ouzel, great bustard, snipe, tern, tattler, owl, warbler, swift, wagtail, shrike, and thrush, among other bird remains. Of the fauna, however, the most interesting element is probably *Palaeocryptonyx hungaricus*, a Miocene galliform bird that was described by Jánossy on the basis of more than 100 bones. Thanks to this work we know that by Late Pannonian times the fauna of Polgárdi was diverse, a forerunner of the bird assemblages of the Plio-Pleistocene.

📖 Lambrecht 1929; Kretzoi 1957a; Jánossy 1991, 1993

Mammals

Small Mammals

It has always been well known, since the beginning of research on Pannonian vertebrates, that the mammal fauna at this time was diverse. However, only more recently has the stratigraphical and paleoecological significance of the different mammal groups been recognized. Rich small mammal faunas are known, for example, from Alsótelekes, Rudabánya, Sümeg, Csákvár, Tardos, and Egyházasdengeleg, as well as from several different cracks in the Polgárdi Quarry. It is also generally held that small mammals include the remains of shrew, hedgehogs, bats, squirrels of different sizes, dormice, mice, hamsters, lagomorphs (rabbits), and small predators such as weasels.

Sometimes the bodies of these dead animals were washed away by water and their remains were finally caught in a crack or a cave but it is also common that because larger predators (for example, birds) eat small animals, the bones of small mammals accumulate around nests and resting places.

Small mammal research in Hungary dates back to the dawn of Hungarian vertebrate paleontology and to the work of János Petényi Salamon (1799–1855). Kormos was also among the first workers to study small fossil mammals, followed by Jánossy, Kretzoi, and Kordos, who all dealt with Late Miocene (the Plio-Pleistocene in particular) small mammals—primarily rodent and insectivore fossils. Lukács Mészáros has also recently studied the Hungarian Late Miocene shrew fauna in detail in several publications. His examinations have revealed that the often-large numbers of remains of these tiny animals (teeth and jaws in the first place) can be used for paleogeographical and paleoenvironmental interpretations. Small mammals, such as shrews, are very sensitive to changes in their local environments, and according to the most recent work, white-toothed and red-toothed shrews appear to be the most suitable for detecting changes in climate.

As is common in paleontology and biology, contemporary scientists often name a newly discovered species after a former expert, perhaps a professor, or after a well-known fossil site. The genus names *Mimomys petenyii* and *Petenyia,* the former a vole and the latter a geographically widespread shrew, preserve the name of one pioneer in Hungarian vertebrate paleontology. The first of these names was created by Lajos Méhelÿ (1862–1953) and the second by Tivadar Kormos. Later the shrew name *Crusafontina kormosi* was introduced to commemorate this Hungarian scientist by his foreign colleagues. The name *Kordosia topali* recognizes the scientific achievements of László Kordos, former director of the Geological Institute of Hungary and the Geological Museum of Hungary and György Topál, a bat specialist working at the Hungarian Natural History Museum.

📖 Kubinyi 1864; Kormos 1911a, 1913; Kretzoi 1956, 1985b; Kordos 1987b; L. Gy. Mészáros, 1999a, 1999b, 2000a, 2000b

Predators

Early Pannonian predators were relatively little known until the detailed examinations of fossils from Rudabánya. At least 17 species belonging to bears, raccoons, weasels, cats, viverrids, false saber-toothed cats, hyenas, and an extinct group (Amphycionidae) have been discovered at this famous site. Indeed, today Late Pannonian predators are known from a series of fossil sites. Hyena (*Ictitherium*) remains have been found at Csákvár, Sümeg, and Polgárdi; *Atticofelis* remains from Sümeg are probably the size of the extant wildcat. The size of *Parapseudailurus* from the same site (also known from Csákvár) falls between that of a lynx and a panther. Representatives of saber-toothed cats (*Machairodus*) are now known from several sites (including Csákvár and Baltavár).

The fossil bear that is known from Baltavár, identified as *Ursus ponticus* by Kormos, has subsequently been placed within the genus *Indarctos* by Kretzoi. Some teeth from Rózsaszentmárton and Hatvan, which were described by Miklós Kretzoi and thought to belong to an additional new genus (*Agriarctos*) contain two species and are also

Pannonian bivalves (original size). (1, 2, 4–7, 9, 10) *Congeria* spp.—(1) *C. rhomboidea* Hörnes, the largest *Congeria,* its length can reach 10 centimeters (Szekszárd). (2) *C. hoernesi* Brusina (Sopron). (4, 5) *C. partschi* Czjzek (Sopron). (6, 7) *C. triangularis* Partsch (Radmanest). (9) *C. vutkitsi* Brusina (Radmanest). (10) *C. suemeghyi* Strausz (Nyárád). (3) *Dreissenomya schroeckingeri* Fuchs—an endemic burrower bivalve from Lake Pannon (Radmanest). (8) *Anodonta horvathi* Brusina—the Pannonian ancestor of a thin-shelled lake bivalve (*A. cygnea* Linné) still living in Hungary (Radmanest). (11, 12) *Unio atavus* Partsch. Ancestors of the living painter's mussel (*U. pictorum* Linné) lived in the shallow parts of Lake Pannon and in the inflowing rivers in large numbers (Szentlőrinc).

1
2
3
4
5
6
7
10
8
9
11
12
13

Relatively large (almost 1 centimeter in length) otoliths from the Pannonian Lenke clay pit, near Sopron.

of Pannonian age. These species, *A. gaali* and *A. vighi,* are extinct representatives of a subfamily within the family of bears that includes the living giant panda.

🕮 Kormos 1914; Kretzoi 1942b, 1952, 1985a; Werdelin 2005

Even-Toed Ungulates

Incomplete fossils representing a range of different pigs, stag, deer, and antelope are known from a number of sites and occasionally deer antler remains are also collected (luckily for paleontologists, these ungulates shed their antlers once a year, so more can be found in the fossil record!). One of the most interesting of these specimens, a real riddle from the even-toed ungulate fauna collected at Hatvan, is an extraordinarily well-preserved and intact kneecap. As István Gaál, who published on this fossil, wrote that based on the size of the bone it could have come only from the largest mammals known from this time. Proboscideans and rhinos, however, can be excluded because of their shape; this left only giraffes. Gaál was finally able to get access to a giraffe skeleton in the Animal Anatomy Institute of the József Nádor Technical University and this allowed him to be certain about the identification of the enigmatic kneecap.

🕮 I. Gaál 1943

*Odd-Toed Ungulates: The "*Hipparion *Event"*

At the beginning of the Late Miocene North American three-toed prehistoric horses, the hipparions, invaded the Old World and made their way into the Carpathian Basin. This wave of fauna invasion that took place about 11 million years ago is reflected in the fossil assemblages. They changed significantly compared to older assemblages—and anchitheriums, advanced proboscideans, also appeared. *Hipparion* remains can be collected from Pannonian layers across the Carpathian Basin in large numbers, and characterize the whole of this age. For this reason, the Pannonian fauna is often also referred to as the *Hipparion* fauna.

Indeed, for a while, the huge numbers of generally disarticulated hipparion remains and the apparent diversity of forms baffled scientists, until it became clear that the proper method to evaluate the fossils was using statistics. Based on these approaches, it now appears that the *Hipparion* fauna of the Pannonian Basin was in fact less diverse than similar faunas known in Southeastern Europe and Southwestern Asia. Several, presumably four, different prehistoric horse species lived in the region that is now Hungary; the earliest forms, including *Hippotherium intrans* from Rudabánya and *H. primigenium* and *H. sumegense* from Sümeg, are small-statured animals similar to Late Pannonian forms (*H. microdon*) known from Baltavár.

Hipparions even lived in the Late Pannonian dry, maybe even semiarid, climate in great numbers together with their relatives, gazelles. Their remains are easy to recognize (they have characteristic teeth and limb bones) and occur in several places; they persisted until the very Early Pleistocene. This long-surviving lineage was supplanted by the real horses (genus *Equus*) in only the Late Pliocene, about 2.5 million years ago, having also arrived from North America, just like their ancestors, *Hipparion.*

One intriguing historical paleontological fact that relates to this story is that the genus *Equus* later became

Pannonian bivalves (original size). (1–7, 10, 11) Cockles—(1) *Lymnocardium majeri* (Hörnes) (Nagymányok). (2) *L. apertum* (Münster) (Tihany). (3) *L. riegeli* (Hörnes) (Hidas). (4) *L. penslii* (Fuchs), early form (Dáka). (5) *L. schmidti* (Hörnes) (Szekszárd). (6) *L. ponticum* (Halaváts) (Fonyód). (7) *Pontalmyra chyzeri* (Brusina) (Radmanest). (10) *L. penslii* (Fuchs), late form (Radmanest). (11) *L. rogenhoferi* (Brusina) (Szekszárd). (8, 9) *Congeria hoernesi* Brusina (Sopron). (12, 13) *Budmania semseyi* (Halaváts)—a bizarre-shaped endemic bivalve from Lake Pannon. The peculiar extinct form belongs to the diverse group of cockles; however, the very high hollow ribs on the outer surface of the shells are unique among them. *Budmania* used to be a dweller of deeper, tranquil waters of the lake. Its striking ornamentation probably helped the animal anchor itself safely in the sediment.

Shrew-mouse (*Blarinella*) jaws from Polgárdi. After L. Gy. Mészáros 1999a.

1
3
4
5
6
2
7
8
9
10
11

A *Hipparion* skull from Isle Naxos. This exceptionally well preserved specimen was collected by Tivadar Kormos. These fossil ungulates arrived in Europe from America and related forms are now known from five continents (0.2× magnification).

extinct in America. When the Spanish conquerors took horses back to where they had originated, Natives were taken aback by these giant domesticated animals. Another strange twist is that Natives quickly learned how to treat horses and soon kept them as if they had domesticated them themselves. *Chalicotherium* is one representative of an extinct odd-toed ungulate group that was strange in appearance, similar in form to a gorilla. Fossils of this so-called horse-gorilla are known from Baltavár and Rózsaszentmárton, among other places.

📖 Kretzoi 1965, 1982, 1983

Rhinos and Tapirs

Spectacular remains of rhinos have long been known from a range of different Pannonian fossil sites. *Aceratherium incisivum* is the most frequently mentioned species in the literature. The exact translation of the term *hornyolt szarutlanóc* sounds very strange even to Hungarian ears, but it was introduced into the paleontological literature by the pioneer Salamon János Petényi. He was full of enthusiasm following the modernization of the Hungarian language in the nineteenth century, but this

Late Miocene (Pannonian) predator and herbivore fossils (original size). (1) *Allohyaena kadici* Kretzoi—canine tooth of a large extinct hyena from Csákvár. (2) *Eomellivora orlovi* Kretzoi—jaw fragment with two molars of an ancient predator from the quarry of the brickyard in Győrszentmárton (holotype). (3, 4) *Ictitherium* sp.—canine teeth of a small early hyena from the Gerinc Quarry in Sümeg. (5, 6) *Allohyaena kadici* Kretzoi—molars of a large extinct hyena from Csákvár. (7) *Adcrocuta eximia* (Roth and Wagner)—jaw fragment with teeth of a large hyena from Bérbaltavár. (8, 9) *Varanus* cf. *hofmanni* Roger—vertebra of a varanid reptile from Polgárdi, front and top view, fossil site 5. (10, 11) *Propotamochoerus palaeochoerus* (Kaup)—jaw fragment of a pig with molars from Rudabánya, top and side view.

initiative was of limited success. This animal had grooved teeth and short limbs, and lacked horns; its remains are known from Rudabánya, Sümeg, and a range of locations in Transylvania—including Brusturi and Ungurei. *A. incisivum,* together with *Hoploaceratherium belvederense*—also from Rudabánya—and a third species all belong to an extinct group of rhino.

Rhinoceros schleiermacheri is also found in literature that deals with Late Miocene rhinos. In Hungary, this taxon was first described from Baltavár by Tivadar Kormos, although later authors tend to mention it using the genus *Dicerorhinus*. *D. hungaricus* was described by István Gaál from Pannonian layers in Hatvan.

Finally, tapir remains are less common than fossil rhino in the Carpathian Basin. *Tapiriscus pannonicus* was described by M. Kretzoi from a site near Şărmăşag; later Vlad Codrea from Cluj Napoca discussed all the known Romanian rhino and tapir fossils, including the Pannonian ones, in a detailed monograph.

📖 Kormos 1914; Codrea 2000; Heissig 2005

Proboscideans in the Late Miocene

Three Middle Miocene proboscidean species were already present in Late Miocene strata and because another seven species (*Deinotherium giganteum, Deinotherium proavum, Tetralophodon longirostris, "Tetralophodon" gigantorostris, "Mastodon" grandincisivus, Anancus arvernensis,* and *Mamut borsoni*) first appear at this time, we know that proboscideans flourished toward the end of the Miocene—not only in terms of numbers of species and, presumably, individuals. It is also the case that originally small- and medium-sized forms were replaced by some real giants. The huge size of some of these taxa is reflected in the Latin names listed above, by which sporadic fossils of Late

Three-toed prehistoric horses. According to the most recent taxonomic work the name *Hippotherium* should be used instead of *Hipparion*. Pencil drawing by András Szunyoghy.

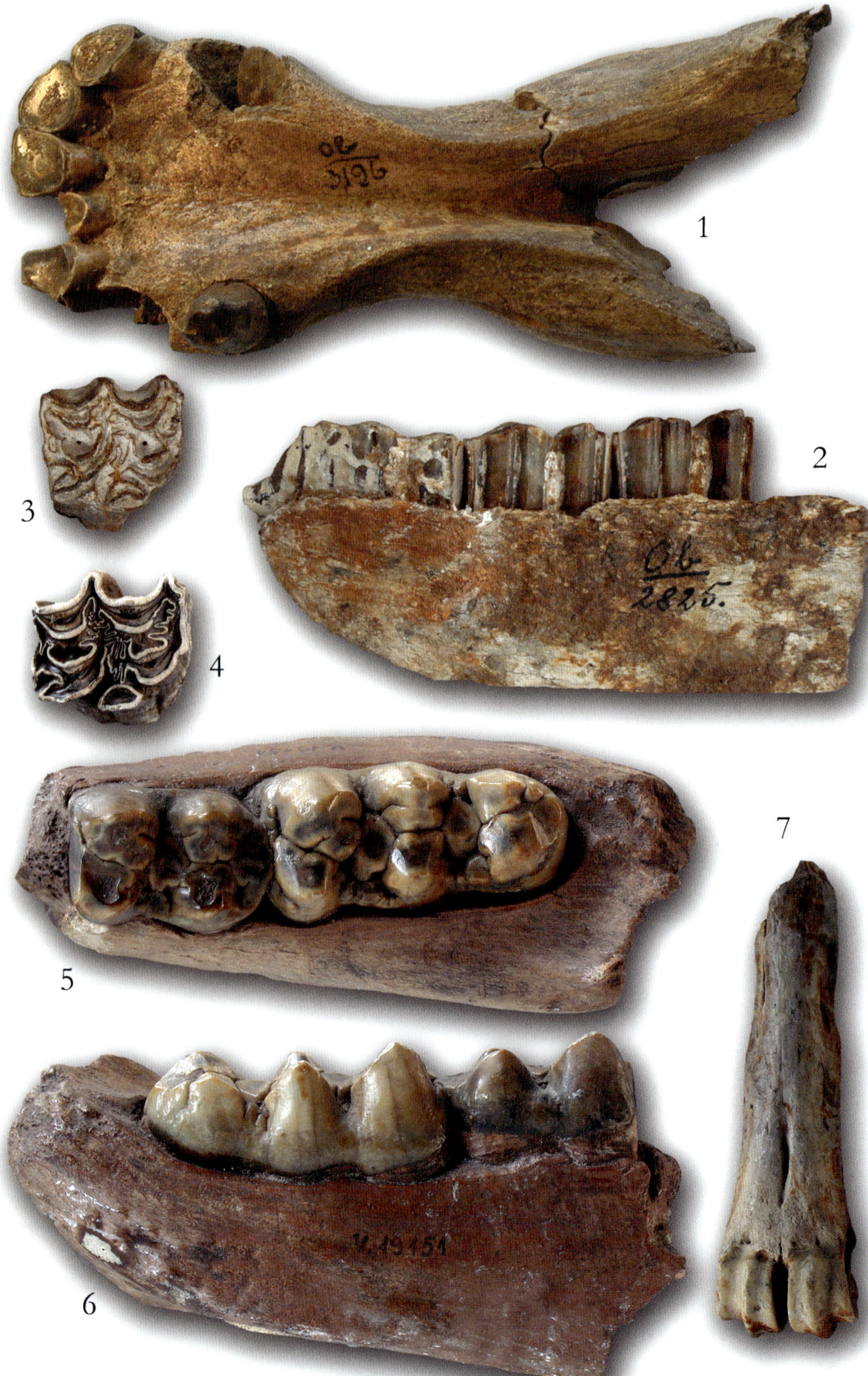
1
2
3
4
5
6
7
Ob.
2825.
V.19151

Miocene proboscideans from a number of Hungarian fossil sites (including Szurdokpüspöki, Tinnye, Sopron, Pannonhalma, Kemendollár, Polgárdi, Egercsehi, Budapest-Kőbánya, Rudabánya, Csákvár, and Baltavár) are known.

📖 Gasparik 2005

Primates

Rudi, Gabi, and Friends: Ancient Apes from Rudabánya

In 1965 Gábor Hernyák, the sharp-eyed head geologist who worked in the iron mine at Rudabánya, discovered a fragmentary jaw in overlying coal layers deposited in a tropical marsh environment above the ore in the Vilmos Quarry. Hernyák recognized that his find did not belong to any of the fossil animals so far found at the site and so, thinking that it required safe custody, he wrapped it in a candy bag and stowed it away in his desk. The fossil was still stored there two years later, when András Tasnádi-Kubacska reached Rudabánya on his tour of Hungary. Tasnádi-Kubacska was looking for minerals and fossils for an upcoming centenary exhibition to be held at the Geological Institute of Hungary.

Tasnádi-Kubacska took these fossil remains straight to the institute and put them on the desk of Miklós Kretzoi, one of the resident paleontologists. Kretzoi immediately recognized the significance of the find and went on to describe this jaw fragment of an ancient hominoid as *Rudapithecus hungaricus*. Two years later, a second fossil of *Rudapithecus* was found; later numerous bones of several dozen additional ancient hominoids were found during excavation works into the mid-1970s, all led by Miklós Kretzoi. As time passed, more and more information was discovered about this ancient hominoid, now popularly called Rudi, and Kretzoi described parts of other, newly collected remains as three new ancient genera: *Bodvapithecus altipalatus, Anapithecus hernyaki,* and *Ataxopithecus serus.*

Excavations at Rudabánya were later led by the paleontologist László Kordos and one fossil, RUD-77—a shattered skull, but nonetheless more complete than any others previously collected—was unearthed in 1985. However, when Kordos put the pieces of this broken skull back together, he found that the jaw and parts of the bones from around the orbit were missing. Therefore an exact reconstruction of the facial skeleton of this small hominoid could not be completed.

As time passed, Gábor Hernyák, discoverer of the first fossil at Rudabánya, found more and more remains. In July 2000 he found one remarkable specimen, named Gabi in his honor, that is even more complete than the famous RUD-77. This specimen includes (finally) a jawbone, an almost entire upper set of teeth, the cheekbone that surrounds the nasal cavity, and the orbit and part of the cerebral cranium that all belong together. So now nothing has kept paleontologists from a complete and realistic reconstruction of this ancient hominoid from Rudabánya.

📖 Kretzoi 1976b, 1984; Kordos 1985b, 1990, 1991b; Kordos & Begun 1997, 2001a, 2001b

From a Shoe Box to a Safe

Miklós Kretzoi was an excellent paleontologist, but also very much his own man. His first report about the new ancient ape from Rudabánya was, for example, published in a Hungarian daily newspaper, the *Magyar Nemzet,* rather than in a professional journal as would have been expected by other scientists. He did publish scientific papers: Kretzoi was the first Hungarian paleontologist to publish in the renowned British scientific journal *Nature.* (Naturally, this paper is also about the hominoids from Rudabánya.)

Kretzoi, later amiably known as "Uncle Miklós" by many people, however, showed these important Rudabánya fossils to no one while he completed their examination. He stored them, perfectly illegally, in a shoebox in his quiet flat on Lövőház Street. One day János Szentágothai (1912–1994), another well-known academic, visited the Geological Institute and had asked to see these fossils. Kretzoi was notified in advance and was told to bring the fossils for examination and to place them in the safe at the institute. However, on the day of Szentágothai's visit, Kretzoi took a plaster model of one of the fossils out of the safe and started explaining its anatomy. The other academic, however, interrupted this mini-lecture and, according to those present, merely asked, "Is this all you want to show me?" The event grew into a scandal, as it was difficult for Kretzoi's colleagues at the institute to explain where the genuine fossils were being stored and why they could not be examined.

In the end these fossil remains did end up in a collection, but in the Hungarian National Museum rather than the Geological Institute. Uncle Miklós still insisted that others could study the fossils only in his presence and with his special permission until he had completed his publication. This finally happened in 2003, when his

Remains of Late Miocene (Pannonian) herbivores (original size). (1) *Hipparion microdon* Kormos—jaw fragment of a three-toed prehistoric horse from the famous roadside fossil site in Bérbaltavár, top view. (2) *Hipparion primigenius* (Meyer)—jaw fragment from the Beszakadt Cave, on the Szár Hill near Polgárdi. (3) *Hipparion primigenius* (Meyer)—grinding tooth from the Beszakadt Cave on the Szár Hill near Polgárdi, top view. The looped enamel crest that is characteristic of Hipparions is clearly visible in the lower middle part of this tooth. (4) *Hipparion* sp.—a grinding tooth from Rudabánya, top view. (5, 6) *Microstonyx* sp.—jaw fragment of a pig from Bérbaltavár with molars, in top and side views. (7) *Gazella* cf. *baltavarensis* Benda—an accreted metatarsus from a now infilled cave on the Szár Hill near Polgárdi.

Remains of Late Miocene (Pannonian) herbivores (original size). (1) *Gazella* cf. *baltavarensis* Benda—antler pedicle from a now infilled cave on the Szár Hill near Polgárdi. (2, 3) *Tapiriscus pannonicus* Kretzoi—a grinding tooth of a tapir from the Esterházy Cave at Csákvár, top and side view. (4, 5) *Lartetotherium* aff. *sansaniensis* (Lartet)—premolars of an ancient rhino from Rudabánya. (6, 7) *Propotamochoerus palaeochoerus* (Kaup)—left and right lower tusks (canines) of a pig from Rudabánya. The remains of ancient pigs are the most frequent even-toed ungulate fossils at this site. (8, 12) *Procapreolus loczyi* Pohlig—antler fragment and a grinding tooth. The remains of this extinct deer were discovered in a cave on the Szár Hill near Polgárdi. (9, 10) *Miohystrix parvae* Kretzoi—incisor in side and front view. Characteristically, only the front part of this tooth is covered with colored enamel. The remains of this rodent were found in the Esterházy Cave at Csákvár. (11) *Lucentia* aff. *pierensis* Thomas—an antler fragment from Rudabánya. These are the most frequent finds among fossils of ancient ruminants at this site. (13) *Tapirus priscus* Kaup—jaw fragment of an ancient tapir with teeth from lower Pannonian strata in the sandpit at Diósd.

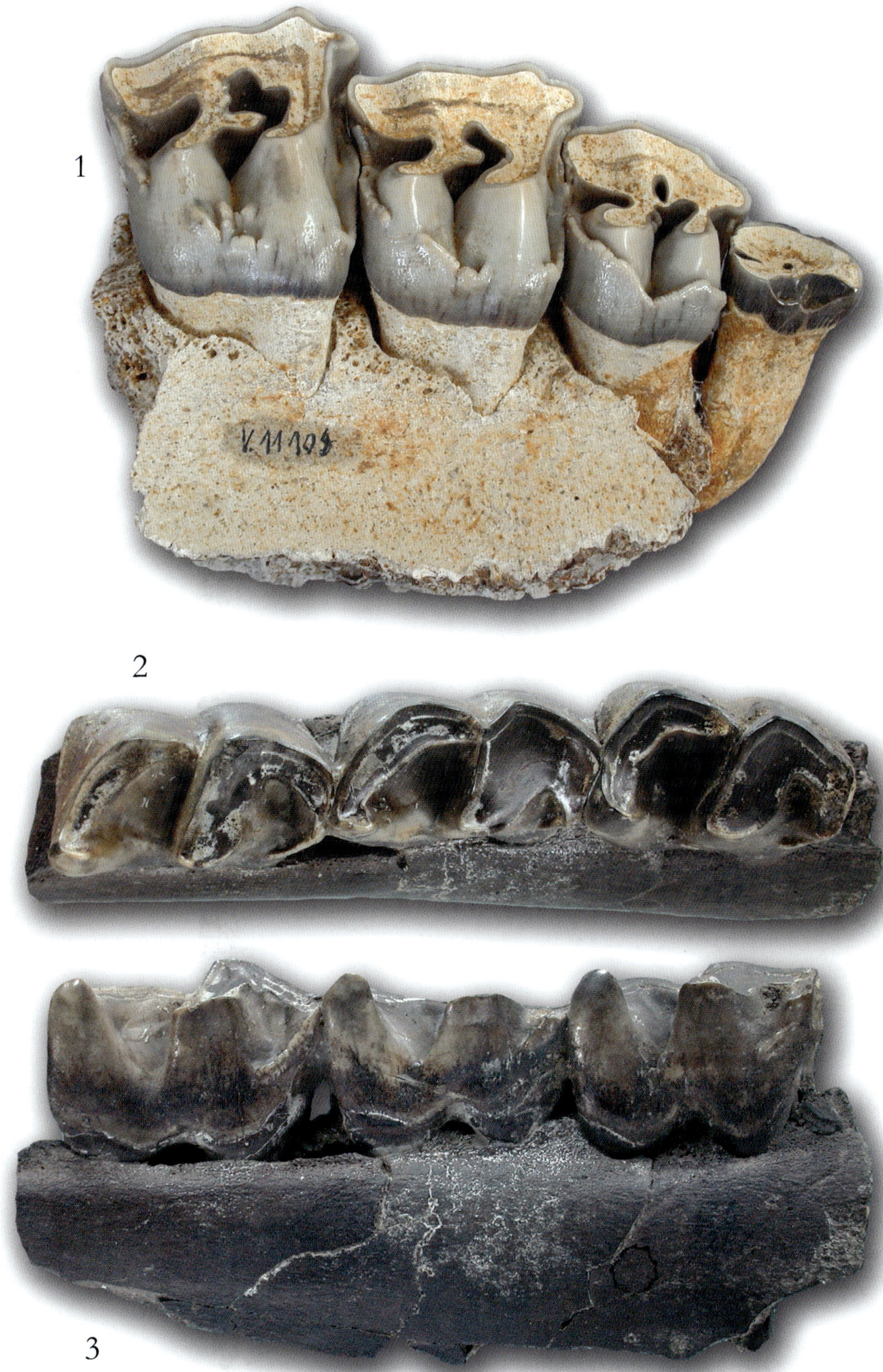

Remains of Late Miocene (Pannonian) rhinos (original size). (1) *Aceratherium incisivum* Kaup—jaw fragment of an ancient hornless rhino with grinding teeth from lower Pannonian beds in the Diósd sand pit. (2, 3) *Dicerorhinus schleiermacheri* (Kaup)—fragmentary jaw of a rhino with teeth from Brusturi.

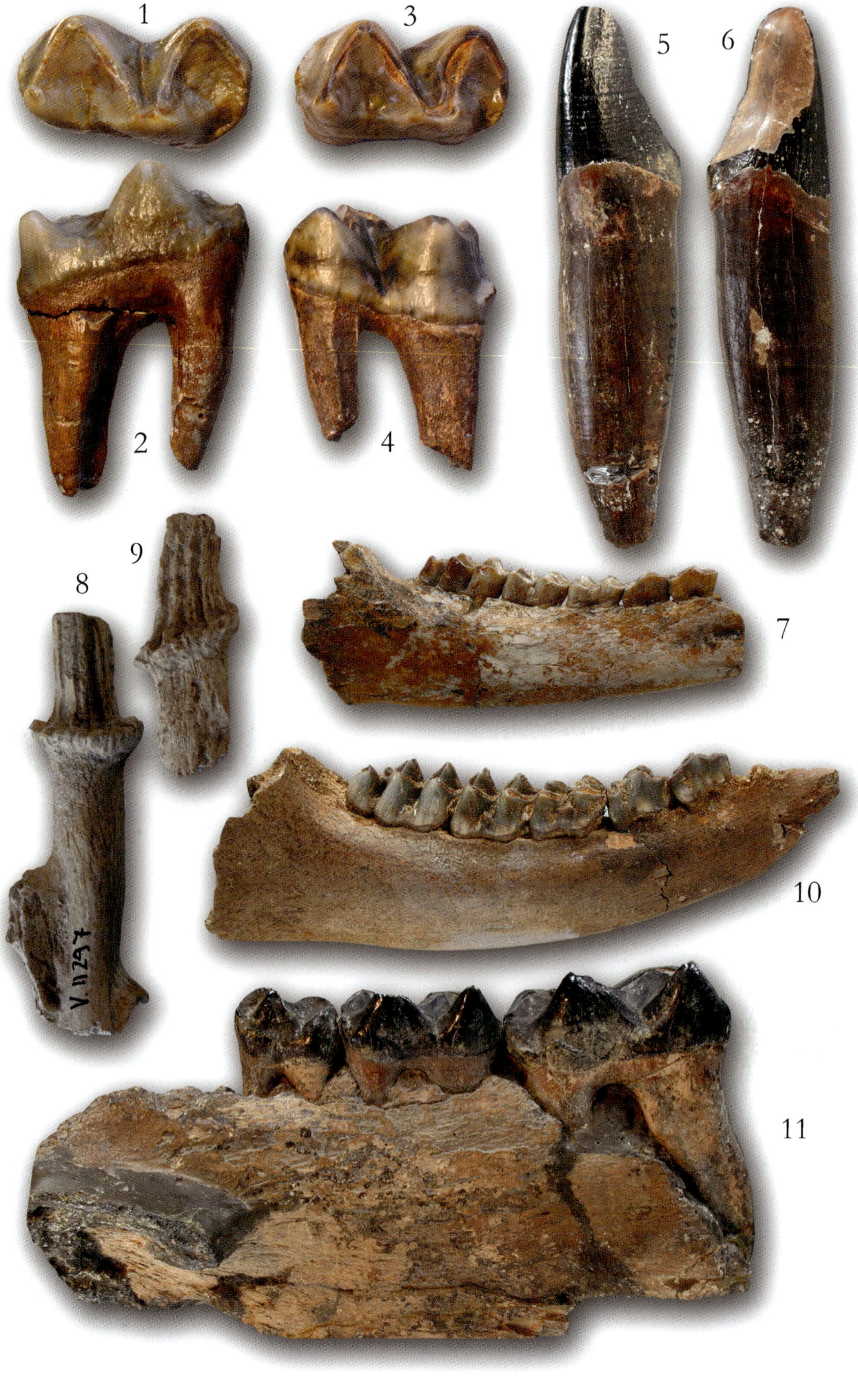
1
3
5
6
2
4
9
8
7
10
11

monograph about the hominoids of Rudabánya was finally published in English. The fossils were taken out of their safe the following year and were displayed in a nicely organized exhibition, timed to coincide with the 97th birthday (on February 9) of Miklós Kretzoi, who was still working and completely mentally alert.

Rudi with Fresh Eyes

The Rudabánya fossils are all about 10 million years old and the approximately 120-to-130-centimeter *Rudapithecus* demonstrates an early stage in human evolution—when chimpanzees (*Pan*), gorillas (*Gorilla*) and humans (*Homo*) had not yet separated from one another. Characteristics of skull morphology at this stage in evolution are considered most important for the interpretation of evolutionary relationships and include the width of the root of the nose, the distance between the eyes, the elevation of the nose, the ratio of the facial and cerebral cranium, and the size of the cerebral cranium. Postcranial elements are also important as they can help to reveal such things as posture.

At the time of the discovery of *Rudapithecus,* most known pre-hominid and hominid remains were from Africa, and it was widely supposed that the so-called dark continent was the early cradle of human evolution. However, as new fossils came to light, including the 8-million-year-old *Ramapithecus* from India, some workers began to speculate that the early homeland of humanity might have been in Eurasia instead. According to another theory, both African and Eurasian evolutionary lines existed in parallel. However, this has turned out to be incorrect: a complete skull from Pakistan that was found in the 1980s showed that *Ramapithecus* is in fact an ancestor of orangutans (*Pongo*), and has nothing to do with the immediate ancestors of humans. Current opinion holds that the remains of Rudi do not prove the presence of early humans in Eurasia; rather, *Rudapithecus* is considered as a European counterpart of the African *Dryopithecus.* Similar to the living anthropoid apes, these hominoids did not have tails—and, in contrast to earlier reconstructions, *Rudapithecus* did not walk upright, but used all four of its limbs for moving among branches. Based on the morphology of its teeth we know that this animal likely fed on soft forest fruits and lived alongside another ancient ape, *Anapithecus,* many even- and odd-toed ungulates, the three-toed prehistoric horse *Hipparion,* and mastodons.

Professor László Kordos, former director of the Geological Institute of Hungary, with a jaw fragment of the early hominoid *Rudapithecus hungaricus.*

Remains of Late Miocene (Pannonian) herbivores (original size). (1–4) *Chalicotherium baltavarense* Pethő—grinding teeth from a large, clawed, and marsh-dwelling odd-toed ungulate, the so-called horse-gorilla from the classical site of Bérbaltavár, in lateral and top views. (5, 6) *Lartetotherium* aff. *sansaniensis* (Lartet)—a worn canine of an ancient rhino from Rudabánya. (7) *Gazella* sp.—a jaw fragment from Tivadar Kormos's collection from Polgárdi. (8, 9) *Cervavitulus mimus* Kretzoi—antler fragments from the Esterházy Cave at Csákvár. (10) *Procapreolus loczyi* Pohlig—A jaw fragment with teeth, from Gyula Pethő's collection from Bérbaltavár. (11) *Chalicotherium goldfussi* Kaup—a jaw fragment with teeth from Late Pannonian brown coal layers at Rózsaszentmárton.

The ancient apes that lived at Rudabánya are thought to have arrived from Africa, just like the first Miocene elephants. The dense subtropical vegetation around Lake Pannon would have provided an ideal environment for them. However, as time passed, the shrinking surface of the lake meant that these forests were replaced by ever more extensive grasslands, which led to the disappearance of *Dryopithecus* and related genera at around the same time. At the end of the Pannonian, new mammal faunas dominated by macaque-like monkeys are known from the Pliocene onward. The distant relatives of these monkeys can still be found in Europe; their last refuge is on the famous cliffs of Gibraltar. The dark continent where the ancestors of these monkeys came from can even be seen from Gibraltar on a clear day: hindered by the ocean, however, these monkeys are bound to stay on the old continent to entertain themselves with tourists.

📖 Kordos 2001

The “Alginite Pit” in Pula, one of the few Hungarian sites for Pliocene fossils.

9

The Pliocene

The last 3.5 million years of the Tertiary is called the Pliocene. By then, the area of the former Lake Pannon had been filled in completely and the Carpathian Basin became largely an alluvial plain. It is not easy to study terrestrial sediments since they contain only a few fossils, and the continuity of deposition sequences is difficult to demonstrate. Consequently, the available data are imperfect, and Lower Pliocene deposits are particularly little known in Hungary. Due to the redefinition of the Pliocene-Pleistocene boundary, many fossil sites, successions and faunas (Kisláng, Villány) that had earlier been thought to belong to the Pleistocene have been reassigned to the Pliocene.

CHANGES IN THE WIDER ENVIRONMENT OF THE CARPATHIAN BASIN

The end of the Pannonian and the beginning of the Pliocene resulted in an extraordinary transformation in the wider environment of the Carpathian region. Boreholes drilled into the basin of the Mediterranean Sea show that the whole sea dried out several times during this period and left deposits of precipitated salt. Eventually the Atlantic Ocean streamed in through the Strait of Gibraltar and flooded the basin. Although these events affected the climate and environment of the Carpathian Basin as well, their exact influence on sedimentation and the evolution of local ecosystems is unknown.

The so-called Csarnóta fauna level of the Late Pleistocene, described from the Villány Mountains, indicates the presence of warm, humid monsoon forests, whereas a younger fauna assemblage from Beremend suggests a drier climate, somewhat warmer than today. More recent faunas from Villány and Kisláng, earlier assigned to the Pleistocene, also suggest a warm and dry climate.

Although the Pliocene depositions of peculiar red clayey successions are probably related to some sort of climate change, more dramatic climate changes are usually associated with the start of the Ice Age at the end of the Pliocene.

FOSSIL-RICH FORMATIONS

The basically terrestrial to fluvial deposits of the Pliocene are easily distinguished from the Pannonian sequences deposited in deep lacustrine, coastal, marshy, lagoonal, or fluvial to deltaic habitats. Red clayey beds that are known from deep boreholes along the Körös Rivers in the Great Hungarian Plain and outcrops of sandy to clayey sediments bearing vertebrate remains in the Gödöllő Hills are typical.

Bones of land animals accumulated in crevices and caves in karst formations, just as they did earlier and continue to do today. The sporadic paleontological data indicate that the older freshwater limestones in Hungary, such as those in Széchenyi Hill (earlier Szabadság Hill), are from the Late Miocene instead of Pliocene as previously thought. Alginite, which consists basically of fossil algae, is a characteristic deposit associated with crater lakes in basalt volcanoes from western Hungary.

Alginite, the Deposit of Crater Lakes

Young (Plio-Pleistocene) volcanic activity produced both the lava flows in the Balaton Highlands, and rings of tuff formed by pyroclastic flows. With the cessation of volcanic activity, craters in the tuff filled with water, creating tranquil lakes. Weathering of the volcanic rocks gradually increased the nutrient level in those lakes, supporting blooms of algae whose mass death and accumulation on the lake bottoms led to the formation of alginite. Still waters contributed to the formation of very regular laminae, occasionally interspersed with the remains of other plants or animals. One species, *Botryococcus braunii* (Chlorophycaceae), played a particularly important role in the formation of alginite and can be said to be literally rock forming. Because *Botryococcus* forms sheets held together by a lipid biofilm, its alginite deposits have a particularly high organic content and can be considered to be a type of oil shale.

The alginite can be best studied in the Balaton Highlands near Pula, where the loose rock is mined from time to time. (The alginite is rich in trace elements and mainly used for soil amelioration.) Fish fossils are the

Series/Epoch	Regional Stage/Age	Millions of Years Ago
Pliocene	Romanian	4.3
	Dacian	5.3

Division of the Pliocene into stages or ages.

most common animal remains from the lake. Even 20-to-30-centimeter long specimens may occur frequently in certain stratigraphical levels. The fossils are difficult to prepare and study in the loose rock, but Andrea Pászti managed to identify the remains of tittlebats (*Perca fluviatilis*) and round chubs (*Leuciscus cephalus*). The latter species dwells in fast-flowing brooks, so its appearance in a calm crater lake is surprising.

The hydrology of the quiet waters in pot-like crater lakes 1–2 kilometers in diameter produces periodic bouts of anoxic conditions, possibly promoted by algal blooms. Such conditions may have led to mass die-offs among the fish. In the absence of current, their fragile bodies have often been laid gently on the bottom, where additional layers of algae contributed to their intact preservation.

📖 Hably 1997; Pászti 2004

REMARKABLE FOSSIL SITES

A classical site for study and collecting in the continental clayey sequences was created in a road cutting constructed during work on the Gödöllő-Máriabesnyő Railway. The famous small mammal faunas from Villány (Csarnóta, Beremend, Villány) and the fauna of Esztramos in northern Hungary were preserved in crevices in the Mesozoic limestones and have been collected in quarries. Some of these are still in operation, but others have long been abandoned. The remains from Kisláng in Fejér County occur in gravel pits. The best site for fossils that have been preserved in alginite is the pit in Pula.

📖 Mottl 1939; Jánossy 1986

The laminae of the dried alginite resemble pages in an old, charred book.

Fossil Mammals of Hajnáčka

The Csontos (literally, "Bone") Ravine at Hajnáčka in Slovakia (known mostly by its older Hungarian name, Ajnácskő, in old paleontological literature) is a famous fossil site. Peasants recognized large white bones in the ravine—especially after heavy showers—in the nineteenth century and some of these finds were shown to the landowner, Emil Ebeczky. Finally some of the bones were sent to Pest and Vienna, where specialists were able to examine them and recognized them "as tapir remains, regarding their beauty and completeness representing the best European specimens" (Krenner 1867). Besides these fossil tapirs, a beaver species was also found, which later turned out to be new to science. This fossil beaver was named *Castor ebeczkyi* in honor of the landlord and was described in a beautifully illustrated paper published in 1867 by József Krenner, who was otherwise known as a distinguished mineralogist.

Many of the fossils from Hajnáčka were collected from volcanic deposits or from sediments that are mixed with volcanic lapillis. Although excellent geologists, including József Szabó and Gyula Pethő, and paleontologists (like Tivadar Kormos and, much later, Oldřich Fejfar) made collections and did fieldwork at this site, the age and geological position of the bones long remained unresolved.

The list of species recognized from this region is long, described over time in a series of separate papers and monographs. Of particular interest are fossil mastodon, rhino, deer, goats, hyenas and even leaf monkeys (Colobinae sp.). One rare find was even described by Kormos as a panda (*Parailurus hungaricus*). The known smaller mammals are less spectacular but are also very interesting from a scientific standpoint. Among others, a new hamster species (*Baranomys loczyi* Kormos) and a range of voles (*Mimomys* spp.) have been documented.

Geologists excavated artificial trenches and made drillings in the vicinity of the ravine in order to discover the precise geological position of the fossil finds. It was demonstrated, for example, that bones had accumulated in a crater lake of a Pliocene low-relief volcanic crater. The glimmering water had perhaps lured the unsuspecting animals to their deaths in the lake (due to the lethal content of evaporation from the volcano), or they were simply unable to get out from inside the deep and loose-walled crater and then their cadavers were buried by subsequent eruptions. All this is thought to have happened about 2.8–3.3 million years ago, as estimated on the basis of radiometric age determination. One interesting additional fact is that the position of the former-crater lake remained wet for a long time and, as the region slowly emerged, the rim of the crater eroded and the lake sediments containing the fossil bones were redeposited. Today, the Csontos Ravine is a protected area, and fossil collecting is strictly prohibited. In any case, collecting would also be extremely difficult because the once treeless ravine is now covered by lush vegetation and a young forest.

📖 Krenner 1867; Kormos 1917; Fejfar 1961; Sabol 2004; L. Gaál 2010

Beremend, Szőlő Hill

The vast quarry in Beremend lies at the northeastern margin of the village. Mining has been going on since the nineteenth century and Szőlő Hill has been almost entirely mined away. Rich Pliocene and some Pleistocene vertebrate faunas were found in the clayey infillings of crevices in karst that were exposed and usually destroyed by the mining. Nonetheless, the recovered material has been abundant and provided work for generations of vertebrate paleontologists.

The fauna of the Szőlő Hill site were first studied by Salamon János Petényi (in 1864). The study represents the beginning not only of research on the ancient vertebrates of Beremend but also of Hungarian vertebrate paleontology. The sites where Petényi collected his fossils are no longer identifiable and were probably destroyed during mining. As mining has progressed over the last 150 years, some 25 crevices and caves have been uncovered. Some have been subsequently destroyed, but others still exist.

Many outstanding experts of the Hungarian vertebrate paleontology have followed in Petényi's footsteps, including Tivadar Kormos and the notorious racist and war criminal convicted by the People's Tribunal Lajos Méhelÿ, as well as Miklós Kretzoi, Dénes Jánossy, and László Kordos.

Thanks to their efforts, Beremend can be considered as the type locality of a number of ancient vertebrates. A shrew-mouse (*Beremendia*), a possible ancestor of the mole rats (*Prospalax*), and a hare (*Lepus terraerubrae*) were described.

Many recently discovered fossils were on display in Harkány between 2004 and 2009, thanks to the work of environmentalists and researchers who keep an eye on activity in the quarries. László Pongrácz, a collector and researcher, has distinguished himself by organizing and attending to this collection. The most precious and spectacular pieces include the nicely preserved skull and jaws of a macaque monkey (*Macaca prisca*); some baboon-like monkey (*Dolichopithecus ruscinensis*) remains; and fossils of hyenas (*Pachycrocuta*, discussed below), bears (*Indarctos*), and proboscideans. An ancient wolverine (*Gulo*), a northern interloper, is unusual by its presence in the fauna; the other known constituents all point to a warmer climate.

A view of the limestone quarry in Beremend in 2004. A very rich fossil assemblage, consisting mainly of remains of small mammals, has been found in the dykes of the mined Cretaceous limestone. The fauna was first studied by Salamon János Petényi (1864).

However, site no. 22, where this unusual fossil was found, may have been deposited during a cooler period.

📖 Kubinyi 1864; Kretzoi 1956; Jánossy 1986, 1996; Pongrácz 1999

Villány and Kisláng

The Templom Hill near the railway station in Villány, already mentioned in connection with Jurassic fossils, is also a famous Pliocene-Pleistocene fossil site. The bones were found in red clayey deposits ("terra rossa") that have been preserved in crevices in Mesozoic limestones. The history of the fossil site in Villány offers a good example of the changing interpretation of the Pliocene-Pleistocene boundary as well as prehistoric climatic processes. Although the bones in Villány were discovered by Lajos Lóczy Jr., the first detailed collections were done by Tivadar Kormos. He visited the site first in the winter of 1910, but was interrupted by World War I and only managed to return in 1916. He then returned regularly for eight years and collected a large amount of material from various crevices. Kormos held a conservative, monoglacialist opinion about the Ice Age and also the fauna in Villány, which he published in a long series of papers. According to him, the Ice Age was one single climatic event and the fauna in Villány represented a preglacial period immediately at its start.

After the death of Kormos in 1946, Miklós Kretzoi set about studying the rich fauna of Villány and Betfia. In contrast to Kormos, he took a polyglacialist view, which took into consideration both climatic and faunal changes and led to the recognition of large, relatively stable, faunal complexes known as the Villányium (and Biharium). The type locality for the Villányium is fossil site no. 3, which is still exposed at the eastern end of the quarry in Templom Hill.

Kretzoi subdivided the faunal units of Villány into older Beremendium and a younger Kislángium. The type locality of the latter is an area dissected by gravel pits near Kisláng, in Fejér County. He saw the fauna of Villányium as representing the early (lower) part of the Pleistocene. Owing to the reinterpretation of the Plio-Pleistocene boundary, and the fact that it is about 1.8 million years older than previously thought, these faunas are today thought to be Pliocene. The fauna collected near Kisláng is one of the richest paleovertebrate assemblages of the period. Miklós Kretzoi identified about 100 species, mostly mammals, from the area.

Portrait Gallery

Salamon János Petényi (1799–1855)—The First Hungarian Vertebrate Paleontologist

Petényi was a versatile, pioneer Hungarian natural scientist from Ábelfalva (present-day Ábelová, Slovakia), in Nógrád County. He pursued his studies in Lučenec, Banská Bystrica, and Banská Štiavnica. Initially, he was going to follow in his father's footsteps and become an evangelical pastor, and studied theology in Bratislava and Vienna. In Vienna he became acquainted with the natural historical collection of the local museum and met various Viennese zoologists and scientists. Later, he even met Christian Ludwig Brehm himself, with whom he kept up an active correspondence. He learned how to prepare a bird skins and made several collecting trips. He spent all of his free time studying nature, especially birds, and briefly worked as a forest guard. At one point, he stayed at the farm of Ágoston Kubinyi (1799–1873), who was later the director of the National Museum. He completed his religious studies and became a pastor in Cinkota in 1826, but resigned in 1833 and moved to Pest. There he got a job as a preparator at the Department of Natural History of the Hungarian National Museum, which later became the collection of the Natural History Museum.

He is rightly recognized among zoologists and paleontologists for his work on living and fossil vertebrates, and he is considered to be the founder of Hungarian ornithology and vertebrate paleontology. Much of his paleontological research was based on material from Beremend. He also produced important studies of local fish and was the first to carry out thorough studies of bats in Hungary.

In the first half of the nineteenth century, rising national consciousness led scientists to create Hungarian names for the plants and animals of the region. Petényi also participated, introducing apt phrases for several animals, including mole rats (*Spalax*) and a rhinoceros (*Aceratherium*). Upon his death, he left behind innumerable unpublished manuscripts on a wide diversity of topics. Some of them were edited and published by Ferenc Kubinyi and Titusz Csörgei (1875–1961). Many others, however, have been lost forever. Only a few pages have been found, for example, from his vast ornithological work, which he had left to the Hungarian Academy of Sciences. Ottó Herman discovered them being used as wrapping paper.

📖 F. Mészáros & Gazda 2000

Dénes Jánossy, former director of the Department of Geology and Paleontology of the Hungarian Natural History Museum, gathered teams of specialists and young enthusiasts to excavate the site in Villány for several years. Detailed study of the huge collection of washed and partly sorted small vertebrate fossils commenced recently.

📖 Jánossy 1986; Kormos 1934; Kretzoi 1956, 1969; Venczel 1998b

PLIOCENE FOSSILS FROM THE CARPATHIAN REGION

Our understanding of the Pliocene fauna of Hungary is based almost entirely on the remains of terrestrial animals. The large mammals from Gödöllő, Beremend, Villány, and Kisláng have been exhaustively studied, but a growing amount of new data on small mammals is helping us to reconstruct a world that has long since disappeared. The Pliocene fauna consists of the established, usually larger forms of the Pannonian *Hipparion* fauna and newcomers, probably arriving from Southeast Asia.

Fossil Plants

Lilla Hably was the first to describe plant macrofossils from fossil localities that can be dated with certainty to the Pliocene using radiometric methods. As she pointed out, the fossil flora that comes from these oil-shale successions was deposited in maar-lakes inside the Pula and Gérce paleovolcanoes and indicates that the zonal vegetation of this landscape was made up from mesophile forests that included *Quercus kubinyii, Zelkova zelkovifolia,* and *Ulmus brauni* as the dominant species. Indeed, the Pliocene flora of Transdanubia is known to also include some exotic elements such as *Ginkgo* (the maidenhair tree), which is now restricted to Asia and regarded as a living fossil. It is also worth mentioning that this flora was well adapted to the moderately warm climate of the

Portrait Gallery

Miklós Kretzoi (1907–2005)—An Outstanding Figure of Hungarian and International Paleontology, Geology, and Paleoarchaeology

Miklós Kretzoi was born and studied in Budapest. He got his doctoral degree from the University of Pécs in 1930. He worked at the Royal Geological Institute of Hungary beginning in 1926—for some years as an unpaid volunteer research fellow, then as an official member of the staff. He became a mapping geologist and geophysicist for Eurogasco (later Hungarian-American Oil [MAORT]), in 1933. Then he worked as an assistant museum guard in the Hungarian Natural History Museum, which at that time had tight bonds with the National Museum; later he became a custodian and then director of the collection. In 1950 he returned to the Geological Institute and took over the management of the most significant prehistoric vertebrate collection of Hungary. By that time, the institute had lost its crown, as it were, and become a state institute; the "Royal" was deleted from its name.

His unusual talent had an effect on everyone. This bright intellectual radiation probably played a role in his being chosen director by his colleagues in 1956—at a time when politics made everyday life very complicated in Hungary. He resigned from the director's post in December 1957, commenting characteristically, "My most important action was to get rid of the spittoons in the corridors of the institute" (personal communication). He stayed on as a research fellow for another twenty years, and was head of the Department of Evolutionary Zoology and Anthropology of the University of Debrecen for four years.

He had a wide range of paleontological interests and a talent for recognizing unsuspected significance and value in the tiniest new discovery or some ancient dust-covered specimen that had languished on the shelves of the museum. At first he dealt with the description of smaller, then larger taxonomical groups. His interpretations of Plio-Pleistocene waves of faunal migrations and associations are enduring. He also led excavations in the Ördöglyuk Cave in Solymár, in Gombasek, in Betfia, and at other sites. He is most widely recognized for his work on hominids from Rudabánya and was awarded the Széchenyi Prize for his results in Hungarian paleovertebrate research in 1992.

The last years of Miklós Kretzoi's long and busy life were very active as well. At well over 90 he got acquainted with a new tool, a computer, and, with the support of his legendary home library, he set himself newer and newer tasks. He published a vast volume with a comprehensive survey of the genera of extinct and living mammals in Leiden. "My catalog cards have been published," he said, and waved his hand modestly. He passed away at the age of 99.

📖 Kretzoi & Kretzoi 2000; I. Vörös 2005

Pliocene and has been shown to contain some relict elements previously thought to be restricted to the Middle and Late Miocene. Such warmth-dependent forms include, for example, *Sassafras ferretianum* and *Engelhardia orsbergensis,* a laurel-like and a nut-like plant, respectively.

The composition of the Pula and Gérce paleovolcano flora, as well as morphological features of their leaves (size and shape), indicate that a warm-to-moderate climate prevailed during the Pliocene in the Carpathian Basin.

📖 Hably 1997

Vertebrates

Amphibians and Reptiles

The Hungarian Pliocene amphibian and reptile fauna is rather poorly known. One of the most diverse fossil herpetofaunal associations occurs in a bone breccia that was found in a crevice in the Triassic limestone of the Esztramos Hill, between Bódvarákó and Tornaszentandrás. Remains of at least 7 different frogs (Anura) and 14 scaled reptiles (Squamata) have been discovered there. Both living and extinct genera and species accumulated together in the herpetofauna of the Esztramos Hill. The fauna includes fire-bellied toads (*Bombina bombina*) and other frogs (Ranidae), geckos (Gekkonidae), lizards (Lacertilia), slow worms (Anguinidae), colubrid snakes (assigned to the paraphyletic family Colubridae), and vipers (Viperidae). The composition of the fauna reflects mild and wet climatic conditions and also the fact that in this area various environments occurred together in the Pliocene, just as they do today.

📖 Venczel 2001; Venczel & Gardner 2005

Mammals

The Early Pliocene Mammal Association

The major part of the Pliocene vertebrate fossils is familiar from the *Hipparion* faunas of the Pannonian period. The genus *Hipparion* itself can be found—as well as modern horses represented by the *Equus* genus, which had

Miklós Kretzoi at home in Budapest in 2004.

"Catalog cards" amounting to two vast volumes.

emigrated from North America and appeared only in the early part of the Pleistocene, during the Villány faunal period.

Proboscideans are the largest of the Pliocene fossils and include all four species known from the Late Miocene (*Deinotherium proavum, "Mastodon" grandincisivus, Anancus avernensis, Mamut borsoni*). Their sporadic remains appear as rarities in many Hungarian museums. Rhinoceros remains occur alongside the proboscideans. (Specimens from Pula are dealt with below.) Among the ungulates *Parabos,* which may have originated in India, is also worth mentioning.

The small mammals also form a characteristic assemblage of presumed forest dwellers that includes mice (*Mus*), dormice (Gliridae), flying squirrels (*Petromys*), shrewmice (*Petenyia*), and an ancient type of rooted-tooth vole (Arvicolinae).

Small pandas (*Parailurus*) and many other predator species from South Asia also lived here, the latter being the relatives of civet cats and sun bears. Large felines included saber-toothed cats (*Epimachairodus hungaricus*) and lynx (*Lynx*).

Most animal associations from the first half of the Pliocene indicate a warmer climate than today, namely monsoonal forests. The so-called ancient wolverine (*Gulo*) from Beremend seems to be the odd one out, since its living relatives live in cold climates.

📖 Gasparik 1997; Jánossy 1986

The Late Pliocene–Early Pleistocene Animal Association: The Camel of Kisláng, the Ostrich, and the Others

In contrast to faunas of the warm, wet, monsoon forests during the first half of the Pliocene, the subsequent animal associations are positively steppe like. One of the richest fossil sites for this diverse fauna occurs near Kisláng, but others of similar age are known from Villány and the quarry at Esztramos Hill.

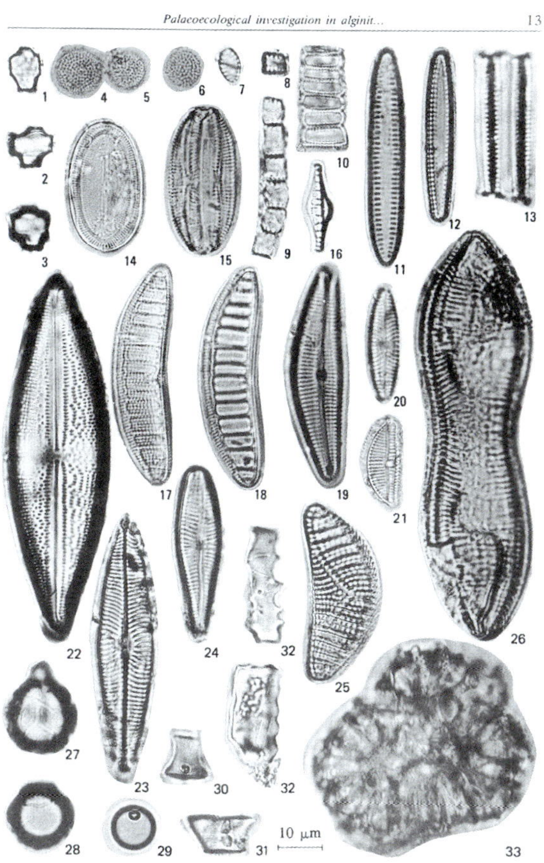

Microscopic plant remains, mainly diatoms, from the Pula alginite. The high percentage of organic material characteristic of this sediment is due to the presence of the planktonic green alga *Botryococcus braunii,* which occurs here in rock-forming quantities, and is figured in the lower right corner of this photo.

Fossil insects have also been preserved in alginite. In the picture you can see a cricket-like insect from Pula (1.2× magnification).

It is striking that while in the northern areas (for example, Esztramos Hill) remains of ancient deer (Cervidae) are common among proboscideans; in the south (primarily in the Villány Mountains) antelopes (tentatively assigned to Bovidae) were more common. These animals lived together with the last *Hipparion* and with horses (*Allohippus, Macrohippus*) that had appeared earlier in the Pliocene. A small-statured bear, *Ursus etruscus*—which is probably identical to *U. arvernensis*—is known from Kisláng.

The first modern elephants (Elephantidae) appear among other proboscideans from the Early Pliocene (*Anancus arvernensis, Mamut borsoni*). The earliest representative is the so-called southern elephant, (*Archidiskodon meridionalis*). For many years geologists remained uncertain about the age of the gravel layers (the so-called meridionalis gravel) that contained its scattered remains.

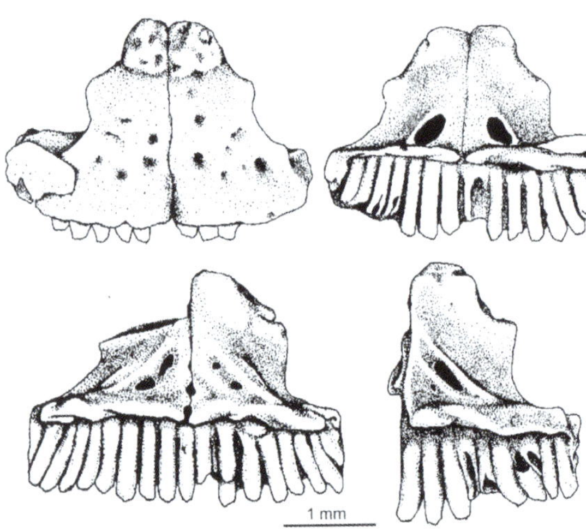

Another "Last of the Mohicans"—a premaxillary bone of a salamander-like tailed amphibian, *Albanerpeton pannonicus*, from the Pliocene in Csarnóta. The interesting thing about the species, identified in 2005, is that it is the last representative of an ancient group. Remains of the Albanerpetonidae family are known from as early as the Early Jurassic. On a family level, the Plio-Pleistocene herpetofauna shows more similarities with living amphibians and reptiles than with earlier Neogene or even older Mesozoic faunas. After Venczel & Gardner 2005.

Tapirs (Tapiridae) were also present among large mammals. These coastal, marsh-dwelling animals were present in the Carpathian Basin since the Early Miocene. A characteristic form, the "Etruscan rhinos" (*Dicerorhinus etruscus* and *D. megarhinus*) are relatives.

Artiodactyl mammals were also fairly varied. Deer with branchy antlers (*Eucladoceus*), red deer (*Cervus*), and giant prehistoric deer (*Megaloceros*) all lived there. A camel (*Camelus*) described from Kisláng was first thought to be Early Pleistocene, but is now believed to be Late Pliocene. Its presence indicates a dry and warm climate. The giant ostrich (*Pachystruthio pannonicus*), described from the same site by Miklós Kretzoi, suggests a similar climate.

The fauna from Kisláng also included giant beavers (*Trogontherium*), modern beavers (*Castor*), a relative of the mole rat (*Prospalax*), and otters (*Lutra*).

Among the small mammals, mice declined, the rooted-toothed voles (*Dolomys, Mimomys*) thrived, and the first rootless-toothed voles (*Lagurus*) appeared. Modern shrews (*Sorex*) and musk-shrews (*Crocidura*) appeared beside Early Pliocene shrew-mice (see *Beremendia* and *Petenyia*).

Small predators, such as weasels (Mustelidae), are also present in the faunas from Villány. Large predators also are surprisingly varied. Bones of prehistoric hyenas (*Pachycrocuta*), panthers and lions (*Panthera*), cats (*Felis*) and saber-toothed cats (*Epimachairodus*) have been found in large numbers in Kisláng. Foxes (*Vulpes*), wolves (*Canis*), and bears (*Ursus*) are also present.

📖 Kretzoi 1964; Gasparik 1997

Rhinoceros in Pula

Since the Pliocene alginite pit at Pula opened in the 1980s, the remains of at least six rhinoceros (Rhinocerotidae) have been found among plants and fish. Most of the skeletons were once complete, but broke up when the oil shale was excavated. The rhinoceroses generally lay on their side, with the ribs pressed against each other, and the skull bones strongly flattened. The presumably warm water of the former crater lakes, fed by thermal springs and overgrown with plants, probably provided a pleasant living and resting place for the animals. But why did they perish here and become buried in one piece? Perhaps poisonous gases generated by decomposing sediments or by volcanic activity poisoned the wallowing animals. Recently people and animals have died around some crater lakes in Africa due to catastrophic eruptions of poisonous gas.

László Kordos has tentatively identified the Pula rhinoceros as *Dicerorhinus megarhinus*. Smaller and larger, presumably male and female, animals have been found among the six specimens. A few remains, together with an almost complete animal have been taken to the collection of the Natural History Museum of Bakony Mountains in Zirc. The successful rescue operation was largely due to the energetic leadership of geologist János Futó, director of the museum. A more comprehensive scientific survey of the material is still to come; for the moment, only sporadic data are available.

📖 Futó 2001

A rhino from the alginite in Pula in the corridor of the museum in Zirc. The skeleton has been flattened by geological pressures but is essentially complete.

A team of paleobotanists, led by Enikő Magyari (Research Group of the Hungarian Academy of Sciences), is preparing to take undisturbed core sample from the sediments of Lake Bucura (Retezat Mountains, Southern Carpathians). The mud sample obtained here is about 6 meters long and represents about 15,000 years—the entire Holocene and the top portion of the Pleistocene. Age data in years were based on a radiocarbon study of plant macrofossils; pollen, plant macrofossils, siliceous algae, and trace elements were used to reconstruct environmental and water-depth changes through time. Together with plants, the remains of animals are also embedded in the mud, including specimens of Chironimididae (insects belonging to Diptera) that are tiny and common enough to be found in mass occurrences. Because these insects are very sensitive to environmental changes, fluctuations in chironimid assemblages provide a useful additional tool allowing reconstruction of average summer paleotemperatures. Data from these approaches are comparable with the stable isotope temperature record from similarly aged Greenland Ice.

10

The Pleistocene

The last 2.6 million years of Earth history following the Tertiary period is called the Quaternary. This period includes the time of the last big glaciation (the Pleistocene) as well as the geological present (the Holocene). The study of the wildlife of the Holocene forms the subject matter of botany, zoology, and anthropology.

During the Pleistocene the large amount of snow that had gradually accumulated in the Carpathian Mountains turned into firn (compacted ice) and subsequently into glacial ice. However, the European continental ice shield, constantly moving from north to south, never reached the Carpathian Basin, which remained a refuge for plants and animals. Deposits of the Ice Age (loess, gravel, sand, and travertine) can nevertheless be found in many places—on the surface or just below it—all over the Carpathian Basin. These sediments can be rich in vertebrate fossil remains, especially mammal bones, and this is one reason why Hungarian Pleistocene research is so well known internationally.

CLIMATE CHANGES THROUGHOUT THE ICE AGE

Within the Pleistocene, which was basically cold, individual pulses of extra-cold climate are called glacial periods, and indicate times when the polar ice sheet was expanding. Between these intervals, mild periods are also distinguishable; these are called interglacials. These intermittent periods had climates that were often warmer then today. Across the Alp-Carpathian region, the most pronounced glacial periods are termed (from oldest to youngest) the Günz, Mindel, Riss, and Würm—all names that refer to rivers originating in the Alps. The glaciers carved deep valleys—the primary physical effect of glaciation. The valleys of these rivers, where glacial phenomena were first studied, were formed by glaciers. Over the last 2 million years, the climate of the Carpathian Basin has changed a great deal, but significant tectonic changes also have occurred that have altered environmental conditions considerably. All the time, the Carpathian Mountains have continued to emerge while lowlands have rapidly sunk: in some places on the Great Hungarian Plain (Alföld) subsidence has created accommodation place for Quaternary sediments exceeding 600 meters in thickness. In this context the hydrogeography of the Carpathian Basin also changed a great deal during the glacial cycles, forming the rivers that we see today.

FOSSIL-RICH FORMATIONS

During the Pleistocene, the meandering rivers on the Carpathian plain left behind thick pebbly, sandy, muddy, and clayey deposits. From the warm springs around the foothills of the mountains bog lime and travertine (see below) were formed; in the fissures and caves in the hills clayey sediments were deposited; and constant winds from Asia created airborne, fine-grained loess deposits. All these sediments may contain fossils.

Cave Deposits

Karstic cavities, fissures, and caves are extremely important for fossilization. Bones that reach the bottom of caves are not damaged by daily and seasonal temperature oscillations or by sunshine and frost. Deep in the corridors of the cave, carcasses are hidden and protected from scavengers, so bones that accumulate can easily be preserved. Because of the absence of vegetation, humic acids cannot attack the bones, so microbial decay is also less an issue.

Bone accumulations in caves may have a range of different origins. For cave dwellers (such as bats) subterranean cavities are natural habitats—these animals live, die, and are often embedded in the sediments of caves. Bones can also be dragged into caves by predators that use them as temporary or permanent shelters. It is also common for a large amount of small vertebrate remains to accumulate beneath bird (mainly owl) nests in the form of pellets. These pellets comprise the indigestible parts of prey, such as their bones. However, overall the most common accumulation method for bones into caves is simply gravity: animals fall into sinkholes, or a carcass may drift

System/Period	Series/Epoch	Millions of Years Ago
Quaternary	Holocene	0.01
	Pleistocene	2.6

Subdivision of the Quaternary into epochs.

underground in creeks supplied by thunderstorms or by the rapidly melting snow.

Fossils covered by sediment in caves can survive for a long time. In the Carpathians—like everywhere in Europe—most of the fossil bones found in caves are from the Ice Age, but there are other, much older, fossils that are also preserved in this way. The Cretaceous fossil reptiles that are known from the Bihor Mountains, for example, were fossilized in karstic holes partly filled with bauxite. At Bodajk, in the Bakony Mountains, an Oligocene terrestrial mammal fauna was preserved in karstic fissures in an Upper Triassic dolomite, while the extremely rich Pliocene vertebrate assemblage known from Beremend was found in karstic caverns in a Cretaceous limestone.

Travertine and Bog Lime

Travertine is a type of limestone formed by the precipitation of carbonate minerals deposited in warm, or even hot, springs. It is generally white and porous but can be also relatively hard and compact. In contrast, so-called bog limestone, or bog lime, is always soft and not yet cemented. This lime is still sediment, and not really rock.

The high porosity of travertine is due to the rich vegetation that lived in the vicinity of hot springs and in the pools supplied by rising water. In warmer water, bacteria, algae, and other primitive plants—such as mosses—lived in masses; precipitating carbonate formed a thick crust around them and sometimes preserved them as well. During photosynthesis plants extract carbon dioxide from water, and this process can also support the precipitation of carbonates. Vegetation also acts as a sedimentary trap, and the limestone as it forms hardens immediately. Bogs are formed in the inner, deeper part of these pools, where vegetation has no effect and the limey mud remains soft. Precipitating and crusting carbonate minerals cover everything in the pond—including plants and animals such as mollusks, and the bones of vertebrates.

Travertine and bog deposits are found all across the Carpathian Basin, but they mainly occur where Pleistocene thermal springs flow to the surface—many of these are still active today. The hard varieties of travertine limestones have been quarried for building purposes since earliest times and have yielded many important fossils over the years. Fossils from these unique limestones range from tiny gastropods to the internal molds of freshwater turtle shells, elephant skulls, and even archaeological artifacts.

Pebbly, Sandy, and Clayey Deposits

The eroding rocks of the emerging Carpathian Mountains were transported by rivers—the predecessors of today's rivers—toward the basin surrounded by the Carpathian Chain. Waters arriving onto these plains became slower, and as they lost energy they left behind a complex system of terraces that can be traced even today. In the thick pebbly, sandy, or clayey deposits of these terraces disarticulated fossil bones—especially of the large mammals of the Ice Age—collected over the centuries. The so-called blue clay of the ancient riverbed of the River Tisza, an exceptionally fossil rich formation, is discussed in the section on remarkable fossil sites.

Sediment from the Air: The Loess

During the drier periods of the Pleistocene, winds from the East carried with them a huge amount of fine dust. This fine powder settled down and was trapped by the grassy vegetation of the steppe and forest and was turned into a rock typical of the Quaternary—loess. This soft and porous rock type can be thick and form stable vertical walls if not saturated with water—unlike unconsolidated sand, which collapses. Loess deposits can be found over large areas on plains and also on the leeward side of hills. This powder had little effect on life in the Ice Age, but it sometimes preserved the fossil remains of animals and plants. In loess the most frequent fossils are the shells of terrestrial gastropods, known as loess gastropods. Bones are rare, but thanks to patient collecting over the decades, a small and pretty fauna has been gathered that includes large and small vertebrate remains. Among the known fossil sites, two localities at Tokod are of particular interest—bones from these localities were studied and published first by Dénes Jánossy and later by Mihály Gasparik.

Loess preserves remains of ancient vegetation as well and frequently carbonized pieces of plants are found along with common and well-preserved pollen. From the youngest loess successions, the remains of prehistoric people have also been collected.

One of the thickest and best studied of the Hungarian loess outcrops can be found along the Danube, at Paks. Nearly 50 meters of thick loess wall have been carefully investigated bed by bed. Studies of fossil soil levels and paleontological results on the plant and animal remains from this site have allowed a high-resolution climatic reconstruction for the past 2 million years to be built. Pál Kriván (1927–1985), professor at Eötvös Loránd University

During winters Carpathian lakes are covered with thick ice, but paleobotanists and paleoclimatologists do not rest. Samples can now be taken by standing on the ice, allowing penetration into deep lake sediments that extend back to the Late Glacial. This provides an exceptional opportunity to study past environmental changes, as lake core samples can be studied centimeter by centimeter, at high resolution. Today, as climate change is one focus of public interest, these kinds of studies attract attention from specialists and non-specialists alike. Researchers believe that by better understanding the past we can be prepared for future climate change and develop strategies to survive.

and highly memorable personality in Hungarian geology, published on the loess wall succession at Paks in a now-classic study.

📖 Kriván 1955; Jánossy 1986; Gasparik 1993

REMARKABLE FOSSIL SITES

The plains and riverbeds of the Carpathian Basin are mostly filled with Pleistocene-aged deposits, and most caves contain Ice Age sediments. As a consequence, the number of Pleistocene fossil sites (most of them vertebrate bearing) within the Carpathian Basin is very high: over 500 Hungarian mammoth localities alone are known! This high number naturally includes a lot of sporadically collected data, including isolated single bones, which have proven hard to verify geographically (river terraces, for example). Of the known Carpathian fossil sites, caves and fissures often preserve extremely samples of bones and are important for this reason. A number of famous caves are found across Hungary and in the Romanian Bihor Mountains. Among the known karstic fissures, special importance is afforded to the Betfia locality, the type locality of the Biharian stage.

Famous Caves and Crevices

With few exceptions, most of the famous karst topography of the limestone hills in the Carpathian region was formed during the Pleistocene. In caves, sinkholes, and fissures, different-sized fossils from the Ice Age are commonly preserved as a consequence of weathered carbonate rocks on the surface. In Hungary, most caves and crevices are found in the Villány, Mecsek, Bakony, Gerecse, Pilis, Bükk, Uppony, and Aggtelek-Rudabánya Mountains; in Romania the most important sites are in the Bihor Mountains.

A group of paleontologist visiting the Esztramos Plio-Pleistocene fossil locality near Aggtelek, Hungary.

Because of space constraints, we will not give a complete list of fossil-bearing limestone cavities, but some of the more important and famous sites include these: Nagyharsány Hill and Somssich Hill in the Villány Mountains; Bajót Rock Shelter no. 3, Jankovich Cave, Pilisszántó Rock Shelter, Ördöglyuk at Solymár, Bivak Cave, Remete Hill Rock Shelter, and the Castle Hill Cave in the Gerecse, Pilis, and Buda mountains and hills; Tarkő Rock Shelter, Istállóskő Cave, Subalyuk, Peskő Cave, Petényi Cave, Pongor-lyuk, Hórvölgy Cave, Lambrecht Kálmán Cave, Kiskőháti Sinkhole, Kőris Cave, Diósgyőr-Tapolca Cave, and Szeleta Cave in the Bükk Mountains; Baradla Cave, Nagy oldali Sinkhole, Porlyuk, and Esztramos in the Aggtelek-Rudabánya Mountains; and Igric Cave and Oncsásza in the Bihor Mountains.

Bones in the Net: Fossil Vertebrate Remains of the Large Rivers

Occasionally, during the early part of nineteenth century, huge fossil bones—including the skulls of woolly rhino, antlers of *Megaloceros,* and mammoth remains—were caught and brought ashore by fishermen working the Tisza River. These mostly Late Pleistocene fossils had been washed out from the bluish clay deposits of the riverbed; fishermen were not happy to find these remains because the heavy bones destroyed their fine nets. Bones were lifted into boats or collected from the riverside and were often taken to out-of-the-way fishing lodges that have since proved to be treasure troves for researchers. According to contemporary reports, scientists traveling

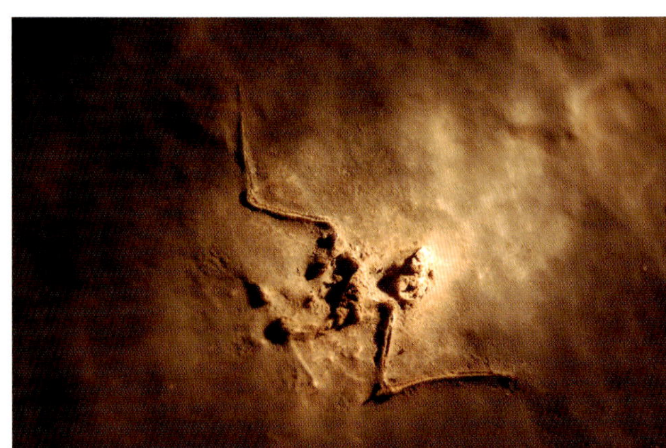

A subfossil vertebrate. Here a recently deceased bat starts on its long road toward fossilization, with its arms expanded in the famous Baradla Cave at Aggtelek. On the wet walls of this cave, precipitating calcium carbonate forms a thin crust over bones preserved in this way. It is also common that mass occurrences of bones can be cemented by a similar calcareous crust—this is how bone breccias form.

Portrait Gallery

Tivadar Kormos (1881–1946)—A Colorful Personality in Hungarian Vertebrate Paleontology

Tivadar Kormos, geologist and paleontologist, was born in Győr and first studied law, then biology and earth sciences, at the University of Budapest. After receiving his doctorate he became an assistant in the Department of Geology at the university, and between 1908 and 1919 he worked as a geologist at the Royal Geological Institute. Since he openly supported the Hungarian Soviet Republic in 1919, he was forced to leave his workplace and his position was suspended, just like those of a number of his other colleagues, including Kálmán Lambrecht and Gyula Leidenfrost. Finally, after he retired in 1922, Kormos was involved in raw material exploration. At first his paleontological carrier focused on malacology, but after discovery of the Late Miocene vertebrate fauna in karstic crevices in the Paleozoic limestone at Szár Hill in Polgárdi, Kormos focused his work on vertebrates. Two years after the discovery of the site, Kormos published a short, but important, contribution on the fossil assemblage, in which he pointed out the relevance of the fauna. Later he performed regular examinations of the fossil vertebrate being excavated by the campaigns at Baltavár.

This well-known site continues to yield a rich fauna, containing *Hipparion,* and Kormos was the first to realize that the layer yielding terrestrial vertebrate remains is found above a mollusk-bearing clay horizon. This realization was especially important for the correlation of these terrestrial and lacustrine Pannonian deposits. Subsequently, Kormos carried out research at Püspökfürdő in Betfia and at Beremend, publishing one of his most important monographs on the fossil vertebrate fauna of the Pilisszántó Rock Shelter. In this paper he gave an up-to-date summary of the known fossil vertebrates and the paleoarchaeological results; another of his important works, *A tatai őskőkori telep* (The Paleolithic settlement at Tata), deals exclusively with archaeological subjects.

Kormos continued to be a strong proponent of his own monoglacial theory (that is, that there was only a single glaciation within the Ice Age) until his death but this, as it turns out, erroneous point of view is, in part, due to the fact that Middle Pleistocene vertebrate faunas from the Carpathian Basin were very little known at this time. In spite of this, mostly based on the Villány Hills fossil assemblage, he worked painstakingly to understand the evolution of mammal faunas in the Ice Age, pioneering work that was later carried on by Miklós Kretzoi.

📖 Kormos 1912; Kretzoi 1969; Jánossy 1981

along the Tisza banks in carts were able to buy enormous amounts of Pleistocene fossil bones from local people. Later on, during the wars and in the course of collection moves, many of these bones were lost.

In particular, the Tisza yielded bones between Tiszafüred and Szolnok. Because the meandering Tisza was straightened in the middle of the nineteenth century, erosion of the riverbanks decreased considerably—and even stopped in some places—and as a result few additional fossil bones were exhumed. Finds are rather rare nowadays.

Similar to those of the Tisza, fossil terraces of smaller rivers in northern Hungary—including the Zagyva, Tarna, Sajó, Hernád, and Galga—have also yielded rich mammal bone assemblages. Fine specimens have even been excavated from the ancient riverbed of the Danube. In the vicinity of Győrújfalu, for example, operational gravel pits are found even today and often fossils come from them—however, these bones are found in the water that fills the pit and are not collected with nets. The gravel is pumped from the bottom of ponds and so the bones also sometimes come to light. As gravel is sorted using sieves, bones also collect on strainers.

Travertine Limestone Localities

Travertine limestone is found mostly close to the Danube, in the vicinity of the Buda and Gerecse Hills, where hot springs running along tectonic faults supplied warm ponds during the Ice Age. Most of the known fossil sites are quarries at localities such as Tata, Dunaalmás, Dunaszentmiklós, Süttő, Lábatlan, Üröm, Budakalász, and Budapest. These pools, once filled with warm water, have yielded important fossils; notable among them is the mandible of an extinct tapir from Süttő and hippopotamus bones collected from Üröm. From Tata, a beautiful collection of Late Pleistocene plant remains were collected, and the hominids from Vértesszőlős are also famous.

The famous fossil site at Betfia, near Oradea, is also often referred to as Püspökfürdő in older paleontological literature. Early Pleistocene bones are preserved here in fractures in the flat, rolling limestone surface of the Somlyó Hill. Over a relatively small area there are lots of fissures that have yielded a rich fauna of slightly different ages; the different crevices are distinguished using a numbering system by researchers. Fossil site (fissure) no. 2 is the type section of the so-called Biharium—a stage introduced by Miklós Kretzoi for the Middle Pleistocene. The Betfia locality was discovered by Mihály Tóth; later Tivadar Kormos, Miklós Kretzoi, Tibor Jurcsák, and Elene Terzea all made excavations here. Recently the fossil birds, the herpetofauna, and the microvertebrates of the site were studied by Erika Gál and Jenő Kessler, Márton Venczel, and János Hír. 🕮 Kormos 1930; Kretzoi 1941d; Terzea 1996; Hír & Venczel 1998; Gál 2002

PLEISTOCENE FOSSILS FROM THE CARPATHIAN REGION

All the common groups of Quaternary fossils are rightly considered to be renowned and important paleontological resources from the area. These fossils include both the spectacular remains of large mammals that had stimulated the imaginations of early collectors as well as less conspicuous skeletal elements of smaller vertebrates, whose scientific importance well exceeds their size. Invertebrate fossils, although often less eye-catching in appearance, are of considerable scientific value particularly for geological age determination and for paleoenvironmental reconstructions.

Fossil Plants

Victims of Pleistocene Climate Change

It was a challenge for all members of the biota, both plants and animals, to adapt to climate change in the Pleistocene, which was fast on a geological time scale. Expansion of areas with ice cover suppressed the vegetation, although during later warming cycles, many plants gradually returned to their earlier habitats. Yet many species went extinct as a result of these climate cycles, and European flora gradually decreased in diversity throughout the Pleistocene. This decline is especially evident in the arboreal flora of northwestern Europe: compared to the diversity of trees at the end of the Tertiary, nearly two-thirds of arboreal genera disappeared during glacial periods. This overall decrease in diversity was, however, less drastic in the Carpathian Basin, where at least some of the biota could find refuge even during the most severe glaciations.

🕮 Medzihradszky 2002

The Early Pleistocene: Incomplete Data

Although detailed data are available on the pattern of floral development over the last few hundred thousand years, at the same time—as is normal in paleontology—these data are patchy. However, some of the earliest data that document the so-called Günz-Mindel interglacial in the Pleistocene is from the gravel pit at Győrújfalu. Along with bones of vertebrates, a great number of giant tree trunks have been found in this mine, and as result of detailed paleobotanical investigations a number of floral elements have been recorded that today are extinct in Hungary and across the rest of Europe. The hemlock (*Tsuga*), for example, is considered to be a Tertiary relic element.

The Flora of the Mindel and Riss Glacials

In the Pleistocene, vegetation occurred at higher diversity close to thermal waters, springs, and lakes, and the calcareous tufa precipitated from freshwater, all of which contributed to the preservation certain plant organs—such as leaves, twigs, and pollen. One example of this kind of preservation can be found in the calcareous tufa at Vértesszőlős that formed during the Mindel glacial.

Bones of large mammals in the old store of the Department of Paleontology of the Hungarian Natural History Museum. These specimens are too big to put into drawers and have therefore been placed on the tops of cupboards.

A drawer full of fossil bones from the Pleistocene site at Gombasek. Among the most spectacular faunal elements described from here are remains of the "lion of Gombasek" (*Leo gombaszoegensis*) and a fossil hyena (*Crocuta crocuta spelaea*).

István Skoflek (1934–1981), a paleontologist who, sadly, died young, started to investigate the fossil plants from this locality as well as others from the famous Tata locality. The huge collections that he made are stored in the Kuny Domokos Museum in Tata.

Another archaeologist and paleobotanist who worked in this area, Zsófia Medzihradszky, gives a graphic description of the vegetation that can be reconstructed based on the plant remains: "Alder, willow, and poplar trees inhabited the damp, wet soils, and alder buckthorn, spindle, currants, and common buckthorn were flowering in the oak-elm gallery forests. Hops and grapevines were climbing up the trees and the meadows were crowded with the colorful flowers of common hepatica, yellow buttercup, and white meadowsweet. Higher elevations were covered with mixed oak, birch, and Scots pine forests, with hazel and lilac in the lower shrub layer." This idyllic-sounding Pleistocene landscape is explained by the fact that mean-temperatures in January are estimated to have been warmer than 3°C during the warmer phases of the Mindel glacial. In the colder phases of this glacial the area covered by these hornbeam–European spruce forests expanded, and as the climate became colder needle-leaved vegetation displaced the deciduous forests. Severe conditions of the glacial period were followed by the Mindel-Riss interglacial, and during this drier phase steppe vegetation became widespread across the Carpathian Basin. During wet phases the deciduous forests reappeared. This interglacial was followed by the next glacial period (Riss) that brought with it a cold phase that must have been even more severe than the previous one: the Hungarian Plain was covered with steppe vegetation and taiga forests prevailed in Transdanubia.

📖 Medzihradszky 2002

The Last 135,000 Years

It is well known that the Würm, the last significant period of glaciation, included numerous short phases of cooling and warming. The response of vegetation to these climatic fluctuations was definitely reflected both in lowlands and in mountainous areas, as the ice cover reached its greatest areal extent about 20,000 years ago. However, needle-leaved trees (probably in forests) occurred across Hungary even in the coldest phases of the Würm, and as ice cover retreated, broad-leaved forests reappeared. This cycle is thought to have happened several times during the Cenozoic glaciations. Thanks to high-resolution spore and pollen analyses, the paleobotanical and climatic evolution of this last glacial phase is known in more detail than any of the earlier time slices.

Invertebrates

Among the known Pleistocene invertebrates from the Carpathian Basin, gastropod and bivalve shells and ostracod valves are the most important. Because the Carpathian

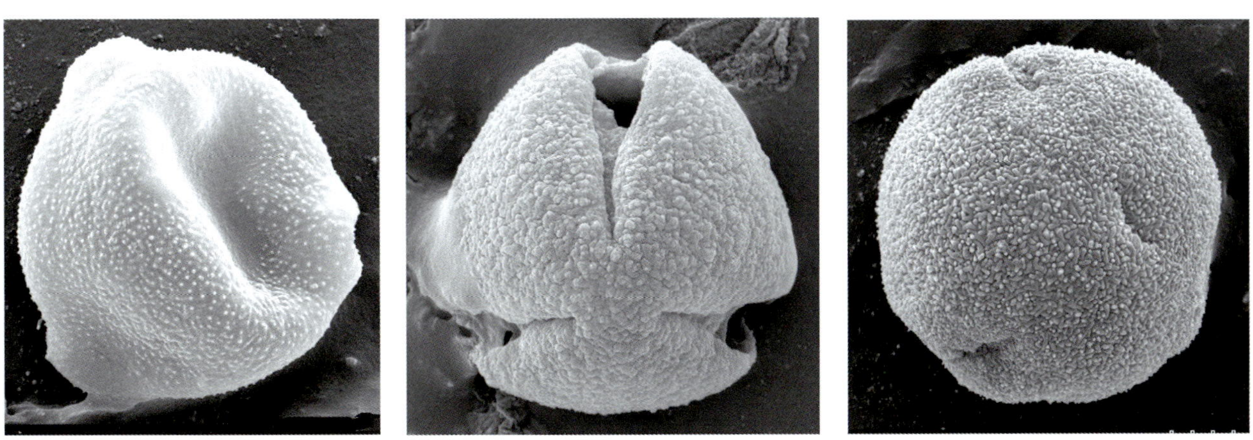

Subfossil remains of pollen (SEM images). From left to right: birch (*Betula*), beech (*Fagus*), and elm (*Ulmus*).

Old Trees, Young Scientists: Dendrochronology

It is well known that the growth speeds of trees in seasonal climates are not even. These changes in rates are conspicuously indicated by growth rings, the number of which refer to the age of the tree (growth is terminated by the death of the tree); patterns of growth rings can be used to determine the systematic affinity of the tree, its habitat, and the climatic conditions under which it was growing. Among the environmental factors that show seasonal variation and consequently affect tree growth, temperature, amount of precipitation, and the impact of pests are important examples. Even the thickness of the annual rings of a tree can vary, as the factors affecting growth speed can vary from year to year. This is the science of dendrochronology, based on the metric analysis and comparison of growth rings of tree trunks or wooden objects that were preserved in sedimentary rocks or excavated from archaeological sites. Nowadays a range of software has been developed for the purpose of aiding dendrochronological research. In general, the older the tree the more suitable it is for dendrochronological analysis.

Overall there is a high chance of some at least partly contemporaneous specimens being found among Methuselah trees, so at the time one of them was several hundred years old another would have just started to grow. Regional chronologies can be developed based on series of growth ring data that overlap in time, and with the help of these chronologies (cross dating) it is often possible to estimate with extreme accuracy (to a few months in some cases) when a wooden object was made or when a fossilized piece of wood was alive. In certain exceptional regions, fully anchored chronologies have been achieved that date back several thousand years. Indeed, dendrochronology can, in some cases, provide estimates with a higher degree of accuracy than can radiocarbon-based measurements. This method is also much cheaper.

Besides archaeological age determination, analysis of growth rings can also contribute essential data to our knowledge of past climate change. In Hungary around the year 2000, this field attracted the attention of Miklós Kázmér, professor in the Department of Paleontology at Eötvös Loránd University. "Kazi," a multifaceted researcher and an experienced geologist—along with András Grynaeus, a pioneer of this research area, and geographer and mathematician Zoltán Kern—established a dendrochronological lab. Research in the lab is focused on surface-forming events (slope movements and floods) in the recent past and on the processes and chronology of phenomena that document climate changes.

Basin was only temporarily waterlogged during the Ice Age, the remains of the former are extremely abundant. Most of the Pleistocene invertebrate genera found in this area are representatives of extant taxa.

Mollusk Remains

Gastropod shells are found in nearly all Pleistocene deposits and, in particular, representatives of the genera *Fagotia, Radix, Succinea, Granaria, Vallonia, Gyraulus, Bithynia,* and *Planorbis* are abundant in travertine and in cave deposits as well.

To find fossils in Carpathian loess deposits it is generally necessary to screen large volumes of sediment to find gastropod remains, including *Pupilla, Succinea,* and *Vallonia*. Indeed, these genera are often referred to as loess gastropods. In fine-grained river deposits, gastropod remains—including *Viviparus, Lymnaea,* and *Theodoxus*—are often found alongside bivalves such as *Unio* and *Pisidium*. Travertine gastropods were described by Endre Krolopp (1935–2010), a geologist at the Hungarian Geological Institute; loess gastropod assemblages were studied by Mihály Rotarides (1893–1950). Rotarides published a particularly important paper on this topic in 1931.

In older research, fossil gastropods were only used for biostratigraphical subdivision of Pleistocene successions, but more recently they have also been used widely for paleoenvironmental reconstruction. Today the Pleistocene is subdivided into five units based on the stratigraphic distribution of gastropod genera, and the index species of the zones are the following genera: *Viviparus, Perforatella, Helicigona,* and *Bythinia*. Krolopp and also Pál Sümegi, who works in Szeged as Rotarides did before him, have published important papers that have advanced understanding of environmental changes during the Ice Age. Since the environmental needs of the different living *Vertigo* species are well known, this genus is particularly important for paleoenvironmental reconstruction. In contrast, use of bivalves as a tool for reconstructing paleoenvironments seems to be rather limited.

📖 Rotarides 1931; Krolopp 1995

Ostracods

Relatively significant ostracod assemblages are also found commonly in travertine deposits. One such assemblage was described from the famous Vértesszőlős site by Kurt Diebel and Erika Pietrzeniuk from Germany, who found a number of extant freshwater genera with large geographical distributions—such as *Darwinula, Ilyocypris,* and *Candona*—to be characteristic.

📖 Diebel & Pietrzeniuk 1990

Vertebrates

During the Ice Age, the distribution of animals—like that of plants—was mainly controlled by the location of the advancing and subsequently retreating ice sheets. Animals that were unable to adapt to rapid periods of either cooling or warming became extinct or migrated out of the Carpathian Basin. Empty niches were taken by incoming migrants.

Because of very rich fossil collections from across the basin we are now able to predict which animals followed one another in and out of the Carpathian area. From many fossil sites hundreds of bones of large vertebrates and thousands of bones of smaller vertebrates, such as small mammal teeth and tiny bones that can only be studied under the microscope, have been collected. Again, the number of fossil localities is large and the number of known Pleistocene vertebrate species is also very high (hundreds of species).

The Higher Part of the Lower Pleistocene: The Fauna of Betfia

As discussed earlier, the fauna of Villány, although originally considered Lower Pleistocene, is now thought to be Pliocene in age. Only the uppermost part of the Villányian "stage" is thought to extend into the Lower Pleistocene.

The Bihar faunal phase, also known as the Biharian or the "Biharian stage," was first introduced by Miklós Kretzoi and represents the top of the Lower Pleistocene. The type locality of this stage is at Betfia (also known as Püspökfürdő or by its rarely used Romanian name, Episcopia) in the vicinity of the city of Oradea. Many Hungarian localities from this time period are also known, including Villány Somssich Hill and sites within the Uppony and Rudabánya Hills. In general, the Biharian fauna is characteristic in that it already lacks the last representatives of the earlier *Hipparion* fauna and contains no more antelope and mastodon, as these older forms have by this time been replaced by their modern counterparts.

Another characteristic of this fauna is the mass appearance of snake vertebrae—generally found disarticulated—as well as a relatively rich bird fauna compared to those of earlier periods. Among the small mammals, voles and hamsters are most common—although these differ at the species level from related forms known from an earlier fauna found at Kisláng. As increasingly novel faunal elements appear, the Bihar assemblage becomes increasingly similar to recent faunas. Among newcomer species, many are characteristic to extant northern European faunas; this is definite evidence for a cooling climate.

Miklós Kázmér was one of the founders of the modern science of dendrocronology in Hungary. He is shown here taking samples from a stem of a dwarf juniper (*Juniperus*) on Chichi-jima (Bonin Islands, Japan).

The fossils from the travertine at Üröm Hill in Budapest are also representative of the Biharian stage. Of special interest, however, is that in contrast to the assemblages discussed above, which comprise of grassland animals, the Üröm fauna is made up of animals that lived around a lake and also includes numerous fish remains. Among interesting elements are remains of giant beavers (*Trogontherium*) that are also known from earlier deposits, and hippos (*Hippopotamus*). The latter is rare in the Hungarian Pleistocene because it lived in very special environmental conditions, just as it does today.

📖 Kretzoi 1941d; Gasparik 1997

The Early Phases of the Middle Pleistocene: First Northern Newcomers

The youngest part of the Biharian stage, the first period of real glaciation, is traditionally referred to as Middle Pleistocene in age. Fossils from this time period are known, for example, from the Tarkő Rock Shelter in the Bükk Mountains, from Vértesszőllős, and from the Uppony Rock Shelter no. 1. These faunas were studied first by Dénes Jánossy and Miklós Krtezoi, who listed localities in stratigraphic order within the Middle Pleistocene and found that the rich fossil material allowed them to generate a precise biostratigraphy. Fossil assemblages of a similar age are also known from the travertine at the Buda Castle Hill and from the gravel below it. Indeed, the

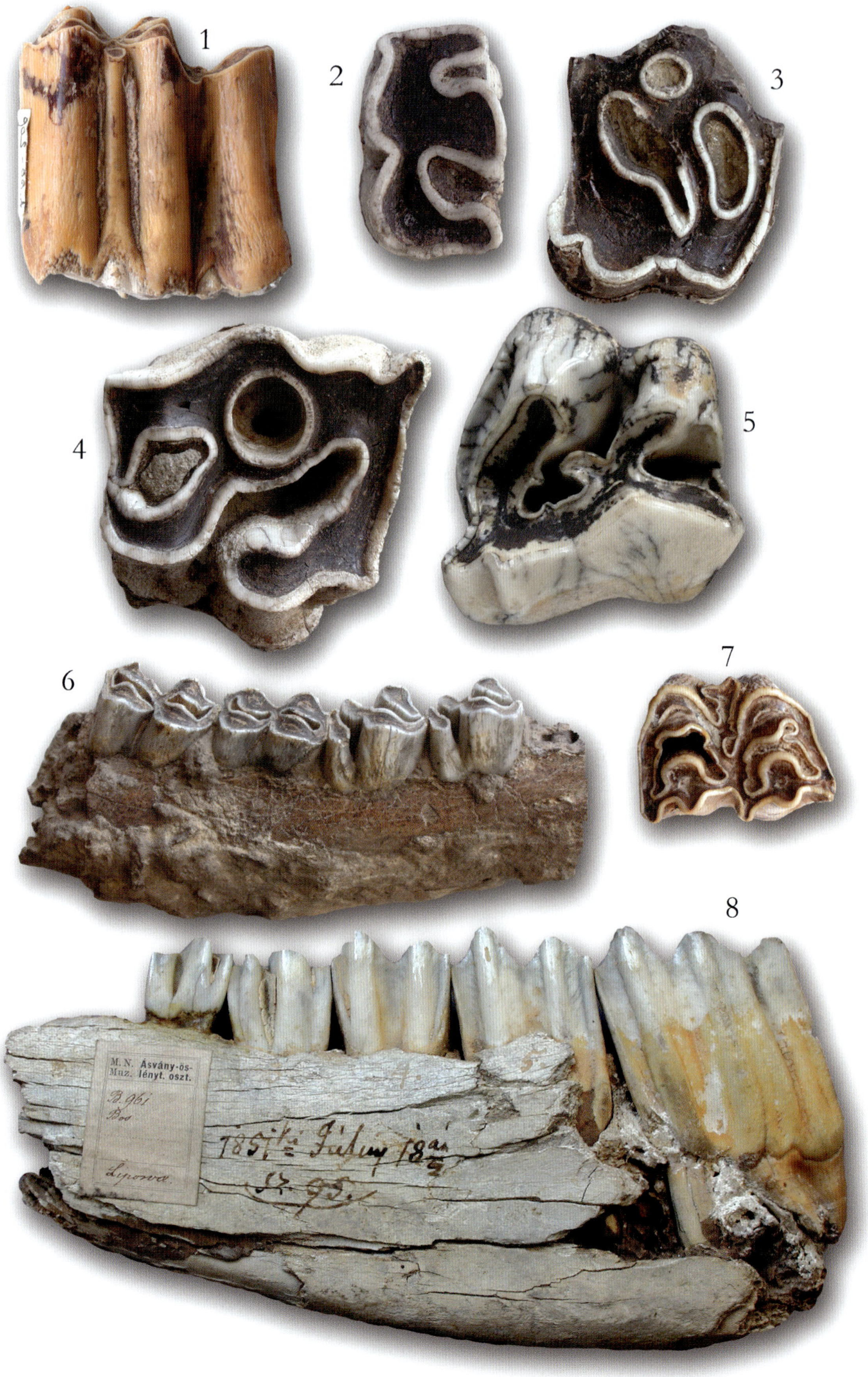
1
2
3
4
5
6
7
8
M. N. Ásvány-ős-
Muz. lényt. oszt.

pretty fauna from the castle district (including remains of elephantids, horse, rhino, bear, hyena, lion, bison, and many others) found in basements of houses on Országház, Szentháromság, Úri, and Fortuna Streets were described by Mária Mottl and Dénes Jánossy. The mollusks from the same localities were revised by Endre Krolopp.

The exceptionally well-preserved fossil bones that are known from Gombasek are also Middle Pleistocene in age. This locality, which is in Slovakia today, is the type locality for the "Lion of Gombasek," *Leo gombaszoegensis*.

At the beginning of the Middle Pleistocene animals like wolverine (*Gulo*), reindeer (*Rangifer*), and musk ox (*Ovibos*) were already living north of the Carpathians. All of them came from Asia and gradually adapted to the colder climate of the region. Significant changes also occurred in the Carpathian region as, for example, the proboscidean *Archidiscodon* was replaced by another elephantid (*Parelephas trogontherii*); the bulk of the larger mammal fauna in this area became essentially modern.

In general, faunas of this age are rich in fossil deer (*Capreolus, Cervus*) and bison (*Bison*). Predators—including many new forms—are also well represented, and among these are characteristic doglike and small steppe wolves (*Canis mosbachensis*). Cats are larger and include *Felis magna, Panthera pardus sickenbergi, Leo gombaszoegensis*, and the cave lion, *Leo spelaeus*. All these animals lived alongside a number of species already known from the Early Pleistocene, including the ancestor of the cave bear (*Ursus deningeri*), an ancient hyena (*Pachycrocuta*), a saber-toothed tiger (*Epimachairodus*), the Etruscan rhino (*Dicerorhinus etruscus*), and a giant fossil beaver (*Trogontherium*).

In terms of smaller mammals the most conspicuous change in fauna at this time is again seen among voles: many of the older forms (such as *Mimomys*) had become extinct by this time, and the survivors were living together with newcomers. Different species of voles can be distinguished by statistic evaluation of large numbers of fossil remains, especially teeth. The group of red-toothed shrews (including *Sorex subaraneus*) and *Lagurus transiens*, a rodent thought to be part of a lineage that led to the living steppe lemming, and water voles (including *Arvicola cantina*) all appeared in the Middle Pleistocene. The mouse genus (*Mus*), which is thought to have come from Asia, also evolved in this region; they are also common elements of the fauna, alongside different dormice (*Glis*), squirrels (*Sciurus*), and hedgehogs (*Erinaceus*). Hamsters (*Cricetus*) are also present in the Carpathian Basin, but the Early Pleistocene mole rat (*Prospalax*) was replaced by modern *Spalax*. Indeed, the latter is a rare, but still living, element of the Carpathian Basin mammal fauna. Excavations done at the Uppony Rock Shelter have allowed for bed-by-bed study of these faunas and have revealed that forest species gradually became dominate over the steppe species.

Fossil teeth of Pliocene and Pleistocene herbivores (all original size except [8]). (1) *Bison priscus* Bojanus—lateral view of a molar of a steppe bison from Budapest Castle Hill. This specimen was collected by Ottokár Kadić and was identified and determined by Mária Mottl. This information is on the label below the fossil. (2, 3, 4) *Coelodonta antiquitatis* (Blumenbach)—molars of the woolly rhino, all from the same specimen. These pretty fossils were discovered in clayey marl during construction work at Budakeszi Street 51, Budapest. (5) *Dicerorhinus etruscus* (Falconer)—a moderately worn molar of a so-called Etruscan rhino from Gombasec. (6) *Rangifer tarandus* Linné—fragmentary jaw of a reindeer with molars in place, from the old Nagy Hont County. (7) *Bison priscus* Bojanus—an upper view of a molar from a steppe bison collected from Budapest Castle Hill. (8) *Bison schoetensacki* Freudenberg—a fragmentary jaw bone of an extinct bison from the Lipova Cave near Hronec (Slovakia) (0.7× magnification).

Reindeer incisors from Pleistocene deposits in Jankovich Cave, in the Gerecse Mountains (almost original size).

📖 Jánossy 1986; Gasparik 1997

The Younger Part of the Middle Pleistocene

This part of the Quaternary in the Carpathian Basin is best known thanks to the fauna from the Ördöglyuk Cave at Solymár, a very important Hungarian locality. Fossil bones of a similar age are also known from Nagyharsány Hill (Villány Mountains) and from the Hór Valley Cave in the Bükk Mountains.

Large-scale excavations in the Ördöglyuk Cave were led by Dénes Jánossy, paleoarchaeologist László Vértes (1914–1968), and bat specialist György Topál. Fossil bones were collected from a red clay deposit inside a sinkhole, close to the entrance of cave no. 1. Among the fossils collected, bats were very common because they lived in the cave—but other finds included rodents, insectivores,

1
3
5
7
2
4
9
8
11
12
10
6
13
14
15

Portrait Gallery

Dénes Jánossy (1926–2005)—Tireless Researcher of Small Mammals and Birds

Dénes Jánossy completed his studies at Pázmány Péter University in Budapest, studying first geography and chemistry and then later geology. He then joined the staff of the Department of Geology and Paleontology at the Hungarian Natural History Museum in 1946, where he worked for more than 40 years—for some time as head of the department. He earned a doctor of science in paleontology, received a number of scholarly awards,, and later become an honorary university professor. Above all, he was a passionate field ornithologist and was proudest of his title of "founding chairman" of the Hungarian Ornithological and Nature Conservation Society.

Jánossy's main scientific interest was focused on the Hungarian Plio-Pleistocene small mammal and bird fauna. He published dozens of articles and catalogs dealing with archaeological and paleontological assemblages of bird bones; the new fossil species and subspecies that he described will stand as his major contribution to this field. Notably, he recognized the presence of the Late Pleistocene faunal horizon (Varbó Phase) in the Bükk Mountains and subsequently worked on older and older Pleistocene and Pliocene fossil faunas, conducting regular field collecting campaigns at the northeastern Hungarian Esztramos locality in the 1970s and at the classic southern Somssich Hill fossil site in the 1980s. After publishing numerous smaller papers, Jánossy summarized his paleontological and stratigraphical results in a book entitled *Pleistocene Vertebrate Faunas of Hungary,* published in English in 1986.

and other kinds of small mammals that must have come into the cave from outside. Shells of snails were collected and documented by Mihály Rotarides, alongside charcoal of unknown origin that might have been a result of forest fires. Among the larger mammals collected one rhino (*Dicerorhinus kirchbergensis*) is regarded as a biostratigraphic marker: this species is probably the descendent of the older Etruscan rhino. From the long list of mammals collected from the Ördöglyuk Cave, a fossil elk (*Alces brevirostris*) and a close ally of the extant mountain hare (*Lepus praetimidus*) are of special interest.

🕮 Gasparik 1997

Late Pleistocene Animal Assemblages

The Late Pleistocene is the last period of the Ice Age and the best documented part of the Pleistocene in Hungary. Famously, this was the time when woolly rhinos, mammoths, and cave bears lived alongside many other animals in the Carpathian Basin. The older part of this time interval is called the Riss-Würm interglacial, and the upper part is considered equivalent to the Würm glacial. This latter period (Würm glacial) was both the harshest and coldest time in the Ice Age and is often called the reindeer period, after its most characteristic animal.

What we know about life in the Carpathian Basin in the Late Pleistocene is mostly thanks to the work of Miklós Kretzoi and Dénes Jánossy. Subsequent assemblages, also known as phases, are all named after the fossil sites (Süttő, Varbó, Subalyuk, Istállóskő, Piliszántó, and Palánk) that yielded the characteristic faunas.

The Süttő Phase of the Late Pleistocene, introduced by Kretzoi, is based on the fauna found in crevices in travertine at Süttő. This succession starts with an Arctic assemblage and is followed by a temperate sub-Mediterranean one, as the fossils clearly reflect fluctuations in climate. Among the characteristic cold-adapted taxa are the rock

Pliocene and Pleistocene remains of herbivores (all original size). (1, 2) *Alces alces* (Linné)—lateral and upper views of a tooth with broken roots from a Eurasian Elk collected from Herman Ottó Cave in the Bükk Mountains. (3, 4) *Megaloceros giganteus* Blumenbach—lateral and upper views of a tooth with broken roots from an Irish elk collected from Szeleta Cave in the Bükk Mountains. (5, 6) *Equus* sp.—upper and lateral views of a horse molar with characteristic long root, collected from the Lóczy Cave in Veszprém County. (7, 8) *Rangifer tarandus* Linné—lateral and upper views of a reindeer molar from the Pálffy Cave near Bratislava. (9, 10) *Trogontherium* sp.—incisor of a giant beaver from the quarry near Tarpa. (11) *Hystrix vinogradovi atavus* Jánossy—lower jaw of a large rodent, a close ally of the modern crested porcupine, with molars and long incisor intact. This figured specimen is the holotype of this subspecies. (12) *Equus mosbachensis* Reichenau—Upper view of a molar from a fossil horse. This tooth was found at Úri Street 72–73, in the Budapest castle district. (13) Incisors of different voles in a glass vial. (14, 15) *Lepus timidus* Linné—skull of the "timid" mountain hare, seen from above and below. This specimen was collected from Ördöglyuk ("Devil Pit") Cave at Solymár.

1
2
3
4

ptarmigan (*Lagopus mutus*) and collared lemming (*Dicrostonyx*), while the Greek tortoise (*Testudo graeca*), which lives in the Mediterranean region today, is a marker of a milder climate. Among larger mammals besides predators (*Crocuta, Leo, Canis*), fallow deer (*Dama*), which also inhabit the Mediterranean today, is also characteristic.

The type material of the Varbó Phase is from deposits in the Lambrecht Kálmán Cave at Varbó, in the Bükk Mountains. This extremely rich fossil assemblage contains reptiles, amphibians, birds, and, of course, many mammals—large and small. Based on charcoal results and evaluation of the ecological needs of the animals, this assemblage is thought to comprise animals that lived on the edge of steppe and woods. The bird fauna is dominated by the western capercaillie (*Tetrao urogallus*)—a species that was once abundant but today is very rare in the Carpathians. Among the predators, small-sized bears (*Ursus arctos*), wolves (*Canis lupus*), and lions (*Leo spelaeus*) have been collected; the hamster fauna is dominated by a large variety of the European hamster (*Cricetus cricetus major*), known in the fossil record from the Early Pleistocene onward, and a small-sized form called *Allocricetus*. These two hamster species lived alongside one another and coexisted with a porcupine (*Hystrix vinogradovi*) that is also typical and is regarded as an index fossil all across Europe. We know this animal on the basis of not only its own bones but also bite marks that it left behind on horse bones, part of a faunal association that was first noted by Jánossy in 1986: "The mild and wood/steppe characters of this assemblage is proven by the presence of forest species (i.e., deer, wild boar, wildcat, European badger) and also by the steppe and semi-desert faunal elements (i.e., *Allactaga*, a kind of jumping mouse) that are known. Typical Alpine elements (i.e., chamois and Alpine ibex) are missing from the fauna, but cooling is nevertheless marked by the presence of some northern European species, like rock ptarmigan, Arctic fox, and reindeer. Of special interest is the presence of European ass (*Asinus hydruntinus*), an index fossil of this period."

The next faunal phase represented in the Carpathian is named for Subalyuk Cave, which is on the western side of the Hór Valley in the Bükk Mountains. Fossils of this phase, in contrast to those of the Varbó Phase (in which the quantity of fossil remains from forests and steppe habitats are balanced), are overwhelmingly semi-desert and include jumping mice and steppe lemmings. However, some Alpine species, such as chamois and the Alpine ibex, are also found. The Alpine wolf (*Cuon alpinus*), widespread across Europe, is regarded as a biostratigraphical marker found at this site, and in the bird fauna abundance of black grouse (*Lyrurus tetrix*) is conspicuous. The hazel grouse (*Tetrastes bonasia*) had disappeared by this time. It is thought to have returned to the Carpathians much later, during the Holocene.

The subsequent faunal phase of the Carpathian Pleistocene can be characterized by the fossil bone assemblage that is found in Istállóskő Cave in the Bükk Mountains. Some small mammals, so abundant in preceding faunas, are missing from this site, whereas remains of cave bear are quite abundant. Indeed, this bear is represented by about 15,000 fossils just from this one cave! Rock ptarmigan (*Lagopus mutus*) again dominates in numbers when compared to its Arctic relative (*Lagopus lagopus*) and, in addition to these elements, the fauna here comprises many other birds and large and small mammal species. Many of these elements (such as fox, bear, European badger, European polecat, deer, wild pig, and rabbit) are thought to be from the same genus, or even species, found in the Carpathians today; some (such as reindeer, mammoth, Alpine ibex, and cave lion) either subsequently became extinct or disappeared later from the Carpathian fauna. The Istállóskő Cave is also famous, not just for its rich so-called Middle Würm fossil assemblage, but also for the polished bone artifacts that were left behind by earlier humans living in the area. A series of similar faunas and archaeological finds have also been documented from Herman Ottó and Szeleta Caves in the Bükk Mountains.

So-called Late Würm faunas are represented in the Carpathian Basin by the Pilisszántó and Palánk phases. The type locality of the first of these phases is Pilisszántó Rock Shelter no. 1, close to Budapest. The extremely rich paleontological and archaeological material from here has been studied by Kadić, Kormos, Lambrecht, Vértes, and Jánossy. The type locality of the second phase, Palánk, is at Szekszárd-Palánk in southern Hungary.

However, none of the known equivalent fossil assemblages of this age can be compared with the Pilisszántó fauna in terms of sheer species richness. The larger mammal fauna from this site is dominated by reindeer (*Rangifer tarandus*); because 1,500 fossils alone have been collected from the Pilisszántó Rock Shelter, this phase is also sometimes called the reindeer age. In general, the composition of the mammal fauna found here resembles that seen on the Asian tundra today. Dénes Jánossy has provided a graphic description of the Pilisszántó wildlife:

> The willow grouse and the ptarmigan (*L. lagopus* and *mutus*) are numerically predominant in the avifauna, with

Skulls of Pleistocene predators (0.25× magnification). (1, 2) *Panthera leo spelaea* Goldfuss—skull of a cave lion from above and from below; it was collected from the Igric Cave in Bihor County. (3, 4) *Ursus spelaeus* Rosenmüller and Heinroth—skull of a cave bear from below and from above; it was collected from the Igric Cave in Bihor County.

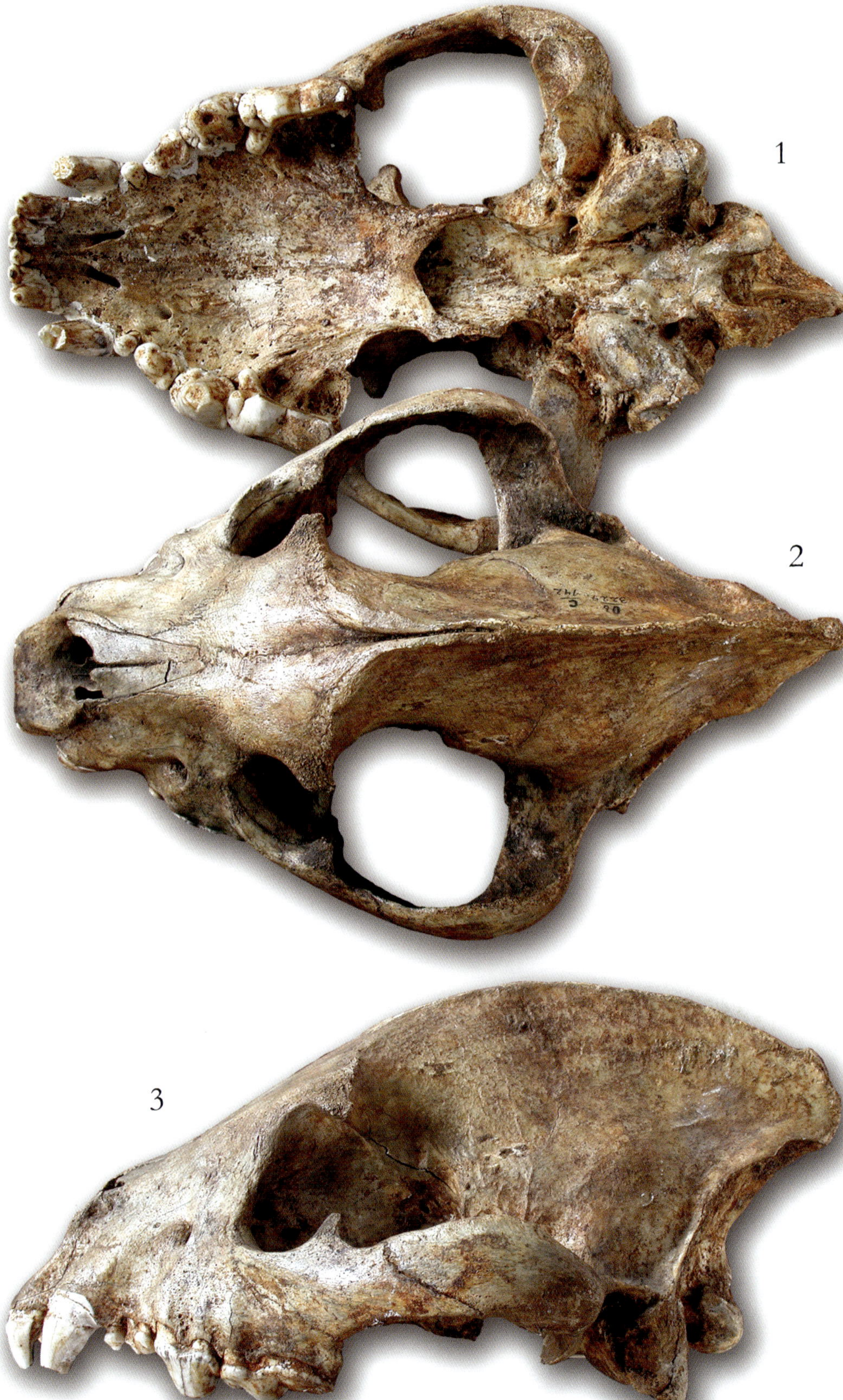
1
2
3

A rare fossil—a snakeskin impression. In the so-called Aragonite Quarry in the vicinity of Corund (Transylvania), very young (Pleistocene, or maybe even Holocene) carbonate minerals are mined. This snake died close to a spring and its skin was preserved as a mold.

lesser numbers of black grouse and capercaillie. Among mammals, the sporadic occurrence of the desman (*Desmana*), the Arctic fox (*Vulpes lagopus*) and the wolverine (*Gulo*), and the absolute dominance of the pika (*Ochotona*) and the Arctic lemming (*Dicrostonyx*) are characteristic. The narrow-skulled vole (*Microtus gregalis*) and the snow vole (*Microtus nivalis*) occurred in appreciable numbers, especially in the upper layers. Except for the reindeer, all the Upper Pleistocene large mammals were disappearing rapidly in the geological sense. These included, in addition to the cave bear (*Ursus spelaeus*), hyena (*Crocuta spelaea*), cave lion (*Leo spelaeus*), Alpine ibex (*Capra ibex*), chamois (*Rupicapra*), woolly rhinoceros (*Coelodonta*) and mammoth (*Mammuthus primigenius*).

A characteristic form of the gallery forests was the beaver (*Castor*), which was also present in upper layers.

Animals of the Palánk Phase lived about 10,000 years ago. The composition of this fauna is similar to that of the Pilisszántó Phase, but the number of tundra species is less and forest species are more abundant. This fauna marks the end of the Pleistocene.

📖 Jánossy 1986

Amphibians and Reptiles

The Pleistocene herpetofauna of the Carpathian Basin is rather diverse and has considerable similarities to living amphibian and reptile faunas. Most species found in the Ice Age today live in Hungary or in neighboring countries, but many Early Pleistocene species are now found only in the Mediterranean. At some localities, fossil reptiles are so abundant that they furnish most of the small vertebrate fossils extracted by screening. Most are disarticulated bones, predominantly snake vertebrae. This is not surprising because snakes have hundreds of vertebrae: in these reptiles, the vertebrae as well as the disintegrating cranial elements are characteristic and used for systematic studies.

Cave hyena skull (0.4× magnification). Skull of the cave hyena (*Crocuta crocuta spelaea*) from the Igric Cave in Bihor County from below (1), above (2), and in lateral view (3).

Freshwater turtle remains are known from a number of different travertine limestone localities. This specimen of a European pond turtle (*Emys orbicularis*) was collected from the Pleistocene travertine at Kiscell, near Budapest (0.2× magnification).

The first herpetological studies of fossils from the Carpathian Basin were done by Baron Géza Fejérváry (1894–1932), István Bolkay (1887–1930), Lajos Méhelÿ, and János Szunyoghy (1908–1969). Their work was then continued by Márton Venczel, a worker from the Ţării Crişurilor Museum in Oradea. In his paper, the "Quaternary Snake from Bihor (Romania)," Venczel focuses on the mainly Early and Middle Pleistocene herpetofauna of Betfia.

📖 Venczel 1997, 2000a, 2000b

Turtles from Süttő

Freshwater turtle remains are not very abundant but are rather typical fossils known from Carpathian Basin Pleistocene travertine deposits. Especially beautiful specimens have been found near Budapest (Kiscell) and Süttő, although sometimes just the cast of the carapace is preserved as an internal mold with occasional bones. Many remains are from the European pond turtle (*Emys orbicularis*); *Clemmys mehelyi* was described by Kormos as a new species, believed to be a relic that had apparently survived in ponds supplied by warm springs along tectonic lines. According to Kormos, the relic characteristics of this animal are supported by the accompanying mollusk fauna (*Theodoxus* gastropods) and decapod crabs (*Telphusa fluviatilis*). The latter—which is otherwise known only in the northern hemisphere from Dalmatia (Croatia)—is based on the determination of Lőrenthey.

Subsequently, *Clemmys mehelyi* was replaced by usage of the genus *Clemmydopsis*.

📖 Kormos 1911b, Szalai 1934

Birds

The Plio-Pleistocene Bird Fauna

The Pliocene and Early Pleistocene bird fauna of the Carpathian Basin differs considerably from recent avifauna. This means that some birds seen in earlier time periods have already disappeared from Hungary and neighboring countries; it also means that although some species are extinct, their distant relatives are nevertheless living in the Carpathian Region today. One example is the ostrich known from the Pliocene of Kisláng. Remains of this bird were unearthed in local sand and gravel pits in the southern part of the village and it was described as a new species in the ostrich genus *Struthio (Pachystruthio) pannonicus* by Miklós Kretzoi on the basis of a single phalange and about 100 eggshells and coproliths. Indeed, this represents the first Hungarian and also the first European report of a Pliocene ostrich. This bird, regarded when first described as being Pleistocene in age, was likely one of the largest ostriches of all time.

The famous Early Pleistocene avifauna from Betfia was first studied and described by the Czech specialist Vaclav Čapek (1862–1926). At the time of Čapek's publications, Betfia was known to yield the most abundant early Ice Age bird material known not only from the Carpathian Basin, but also from across Europe. The work of Čapek has since been

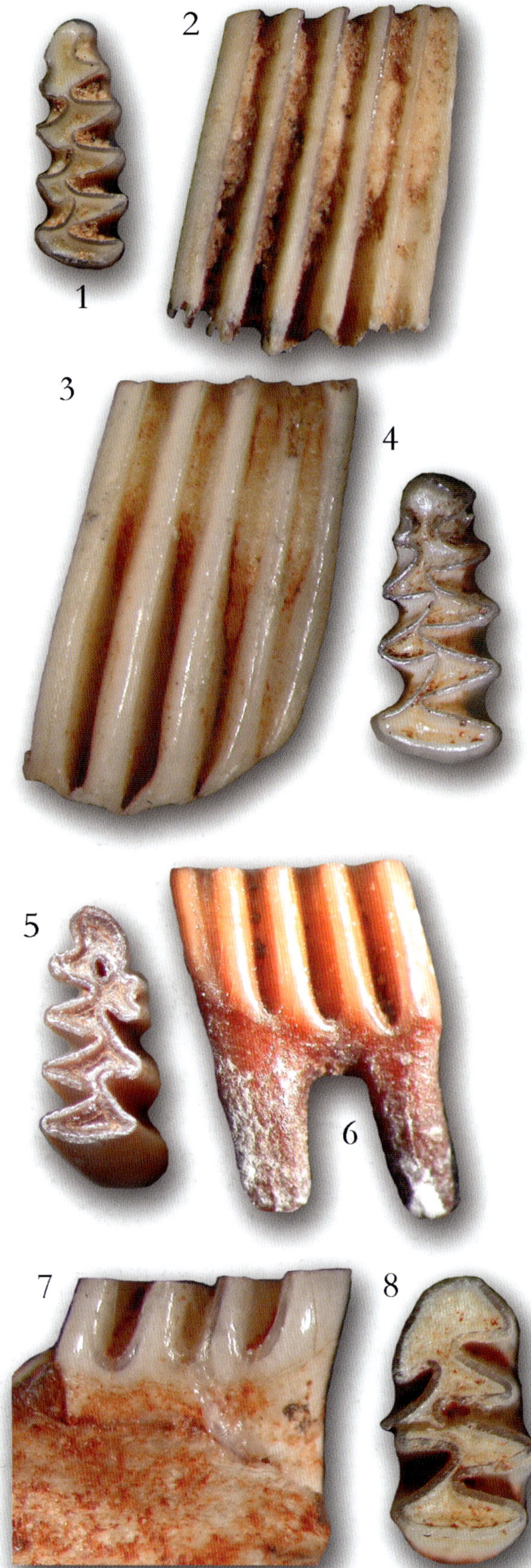

Pliocene and Pleistocene voles. Evolutionary trends in teeth are illustrated by fossils from some representative species. Toward the top of this figure, teeth get younger in age and similar evolutionary changes in the morphology of vole teeth can be recognized in several different parallel evolutionary lines (8× magnification). (1, 2) *Microtus* sp.—upper and lateral views of a tooth from Somssich Hill in the Villány Mountains that comes from Late Pleistocene deposits. These teeth are typically high crowned and grew continuously throughout the life of the animal. They have no roots and there is a considerable amount of cement in the wrinkle of the enamel. (3, 4) *Mimomys ostramosensis* Jánossy and Meulen—lateral and upper views of a tooth from the Esztramos Quarry, from the Plio-Pleistocene boundary beds. These teeth are highly crowned, rootless, and have only a small amount of cement in the wrinkle of the enamel. (5, 6) *Cseria gracilis* Kretzoi—upper and lateral views of a tooth from Pliocene beds at Csarnóta, in the Villány Mountains. The crown is moderately high, roots exist, and there is no cement at all. (7, 8) *Promimomys cor* Kretzoi—lateral and upper views of a tooth from Pliocene beds at Csarnóta, in the Villány Mountains. This is a typical form of an early vole tooth because the crown is very low, roots exist, and there is no cement at all.

Vole teeth from Beremend placed in plasticine and standing side by side.

continued by Miklós Kretzoi, and more recently by Jenő Kessler and Erika Gál. Among other taxa, fossil storks, falcons, and great bustards have been documented from Betfia.

Today, sites called Beremend 16 and 17, in southern Hungary, are thought to be among the richest earliest Pleistocene bird sites. Dénes Jánossy documented 45 species from these localities; among the taxa that make up this diversified avifauna, fossil storks are common (*Ciconia stehlini*) along with small-sized teal (*Anas*), francolin (*Francolinus*), and grouse (*Lyrurus*). These sites have also yielded the first Hungarian records of Pleistocene crakes (*Porzana*), Bohemian waxwings (*Bombicilla garrulus*), and Calandra larks (*Melanocorypha calandra*). In contrast, the fossil bird assemblages reported by Dénes Jánossy from the Bükk Mountains (Puskaporos, literally "Gunpowder" Cave) and from the vicinity of Budapest (Remete Hill and Pilisszántó) are much younger. Another Bükk Mountain locality (Istállóskő Cave) has yielded around 30 species of birds that are slightly older. A notable difference among these different Late Pleistocene bird assemblages is in the composition of species of grouse. These birds are less mobile than are many others, and remain in the same place for the whole year. As a result, the ratio between the willow ptarmigan (*Lagopus lagopus*) and the rock ptarmigan (*Lagopus mutus*) can be used to characterize minute changes in climate.

🕮 Jánossy 1977, 1986

Mammals

Voles: The Forams of the Pleistocene

Extant voles are the most common small mammal fossils from the Carpathian Basin and are also frequent in Pliocene and Pleistocene sediments. Because of their prominent role in terrestrial biostratigraphy, their importance is often compared to Tertiary forams and is the reason for the title of this section.

The earliest, hamster-like, representative of this group is *Pannonicola brevidens,* first described by Miklós Kretzoi from Late Miocene (Pannonian) beds. From the Pliocene locality of Beremend, genuine voles are already known—including *Mimomys silasensis* and *Promimomys microdon,* described by Dénes Jánossy. These taxa are among the oldest voles known from Europe, and are supposedly Asian invaders.

An explosion in vole evolution took place during the Pliocene when the number of specimens and species increased dramatically. As part of this radiation, the first voles with rootless molars (*Lagurus*) are known from the Kisláng fauna; another fauna of similar age, from Esztramos, has also yielded numerous *Mimomys* species. This diverse Pleistocene vole fauna contains several parallel evolutionary lines, all containing subsequent genera and species.

The evolution of voles can most easily be traced by studying the changes in their teeth, as these are resistant elements and are found frequently in screened sediments. These characteristics also make teeth very suitable for detailed stratigraphic research. One of the most important features of vole teeth is their size. The initially low crown of their teeth becomes higher during evolution, as the

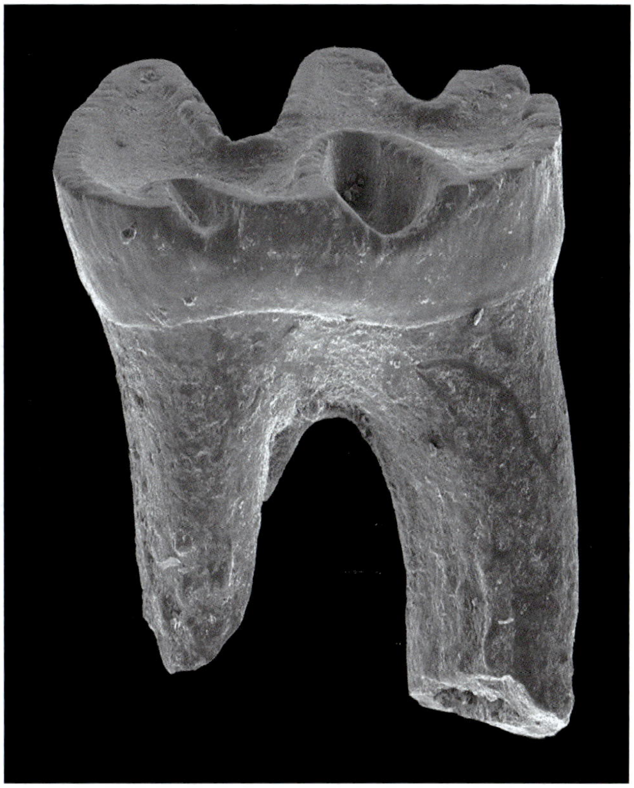

SEM photograph of a tooth of a very early vole (*Pannonicola brevidens*) collected from Pannonian deposits.

roots also get longer. The amount of cement, which fills the corners of the grinding surfaces, has also changed during evolution—but these trends differ considerably among parallel vole lineages. Lemmings, for example, are characterized by molars that lack roots and cement.

Kretzoi 1957; Kordos 1991a

The "Vole Thermometer"

Voles are not only important tools for biostratigraphy, but they also play an important role in paleoenvironmental reconstruction. This is, in part, because their remains are frequent and accumulated in owl pellets—and so can be statistically evaluated. Indeed, it is now well known that the geographical distribution of vole species correlates well with the distribution of certain plants and with climate. This is one reason voles, frequent in paleontological and archeological collections, are regarded as excellent paleoclimate indicators. This method, dubbed the "Vole thermometer," is based on the fact that each species can be characterized by existence at a specific temperature, optimal for the animal in a given period of the year. For example, the common vole, short-tailed vole, Siberian vole, and Arctic lemming can be characterized by July temperature optima at 21°C, 19°C, 10°C, and 7.5°C, respectively. The scale can be refined by adding further species, and in this way the Vole thermometer can be made even more precise. By separating and determining different vole species in micropaleontological samples and by counting numbers of specimens that make up each species, it is possible to calculate an average temperature that was characteristic for a given time. This Vole thermometer was used across the Early Holocene, and good results were also obtained when tracing Late Pleistocene climate change.

Kordos 1977

Bears and Bear Studies

Alongside the woolly mammoth and rhino, the cave bear (*Ursus spelaeus*) is a characteristic animal from the Ice Age. Cave bears were carnivores, like extant brown bears (*Ursus arctos*), but also fed on plants. The largest specimens were bigger than the fearsome North American grizzly.

Spectacular bones of cave bears, especially skulls, attracted the imagination of early naturalists. The first discoveries of these fossils were reported by János Kovács in 1863 in a journal that was the forerunner of the *Bulletin of the Hungarian Geological Society*. Later, György Primics (1849–1893) dedicated an entire paper to this topic, entitled "Spuren des Höhlenbären (Ursus speleus Blumenb.) in Ungarn" (The cave bear in our homeland) in 1890.

Most of the fossils of cave bear known from the Carpathians are from the second part of the Late Pleistocene, from a time internal between 70,000 and 15,000 years before present. In museum collections, especially rich material is known from the Oncsásza and Igric Caves (Bihor County, Romania). From the Oncsásza Cave, for example, Primics discovered a nearly complete cave bear skeleton, which was subsequently mounted and exhibited in the Museum of the Royal Geological Institute. From Igric Cave, more than 300 nearly complete bear skulls were collected and then studied by István Majer (1887–1953). However, Majer later became a chemist at the Geological Institute, and so published just a few papers on the evolution of bears; most of his results remain in manuscripts. Subsequently, Mária Mottl (1906–1980) worked on this topic; her MA thesis, entitled "Zur Morphologie der Höhlenbärenschädel aus der Igric-Höhle" (Morphological studies on cave bear skulls from the Igric Cave), was published in 1932. The fossil bear material from this cave is thought to be especially rich because the cave acted as a shelter for these animals for long periods of time. Skeletons therefore accumulated on the spot, in the same place in which they were also living. Bear skeletons drifted into the deepest parts of the cave and accumulated in the so-called bear whirl, also known as the bone hall, as a result of heavy showers or floods.

The huge number of fossil bones from these sites that slowly accumulated in collections over time also allowed the unique possibility for paleopathological studies. In this field, the book published by András Tasnádi-Kubacska, *Paläopathologie (Pathologie der vorzeitlichen Tiere)* (Pathology of prehistoric animals), offers the most comprehensive review. Damage and many kinds of illness can be traced accurately on teeth, joints, and bones and Tasnádi-Kubacska illustrated many examples of deformed fossil bear bones. Among them and of special interest is a broken penis bone from a bear, which was subsequently healed with a callus. According to Tasnádi-Kubacska, this type of damage can be a result of "extreme impatience of the animals, especially of the females."

Cave bears were likely fearsome enemies of ancient people and, according to fossil and archaeological evidence, there were many fight incidents as well as clear evidence for an ancient bear cult. The latter, for example, is evidenced by the three carefully arranged skulls discovered in Kőlyuk Cave in the Bükk Mountains and by many other finds in which bear skulls have been found under huge flat stones in the halls of caves. Drilled bear teeth were also worn as jewelry by prehistoric people.

The predecessor of the cave bear was probably *Ursus deningeri,* known along with a number of other species names in the paleontological literature. Miklós Kretzoi, for example, first described other Middle Pleistocene species, *U. gombaszoegensis* and the even older *U. stehlini.* The first

Penis bones (baculum) of the cave bear. On the right is a broken and subsequently healed specimen (close to original size).

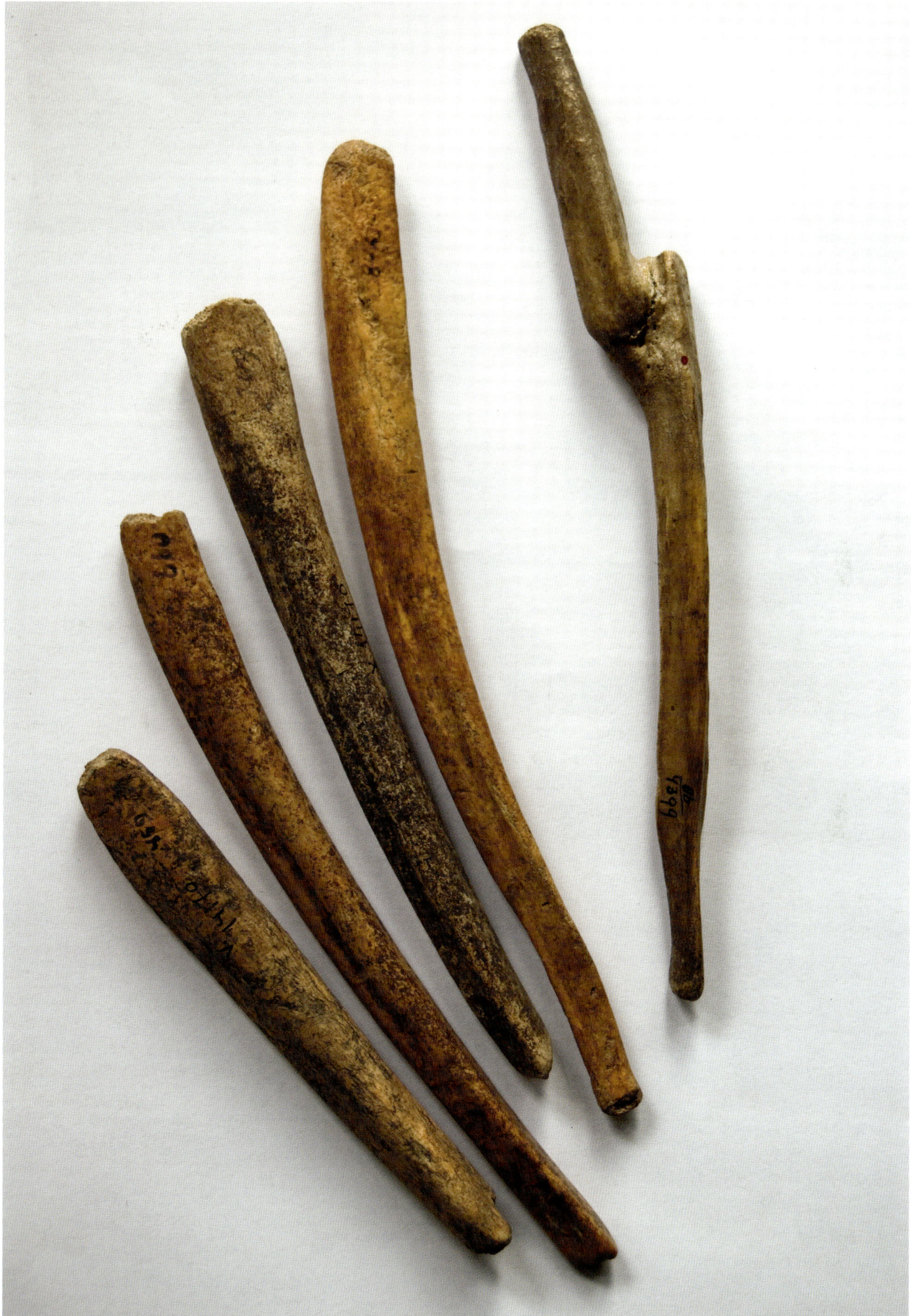

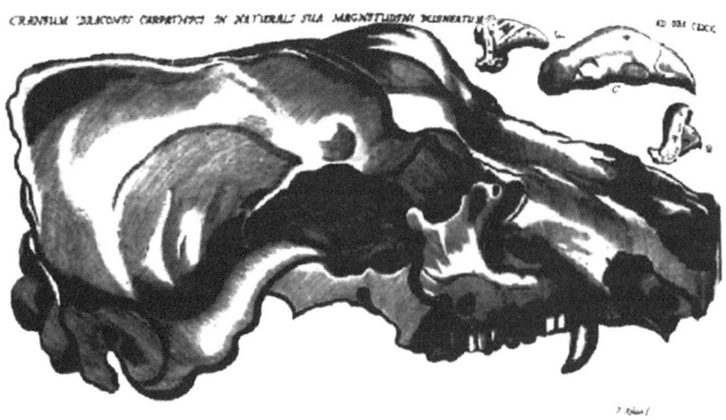

Dragon bones—skull of a cave bear as shown by J. Patersonius Hain in 1673. His woodcarvings of fossil bones appeared as illustrations in the first Hungarian paleontological paper focused on the so-called dragons of the Carpathians.

of these species was found in Gombasek, in Slovakia, and the latter comes from Villány, in Hungary. The brown bear, still extant in the Carpathians (although decreased in numbers), is not a member of the *deningeri-spelaeus* lineage but represents another branch of the bear evolutionary tree.

📖 J. Kovács 1863; Primics 1890; Majer 1928; Mottl 1932; Kretzoi 1942b; Tasnádi-Kubacska 1962

Large Cats and Dogs

Felinae is the group that includes the cats and their relatives of all sizes, adapted to carnivorous lifestyles. In the diminishing forests of the Carpathian region the wildcat (*Felis silvestris*) and the Eurasian lynx (*Lynx lynx*) still occur, but their fossil remains are abundant in Pleistocene deposits. Additionally, the occasional remains of larger cats, including leopard (*Panthera pardus*), saber-toothed cat (*Epimachairodus* and related forms), and lion, are also sometimes found. Together with the so-called common lion (*Panthera leo*), the extinct cave lion (*Leo spelaeus*) is also a characteristic predator of the Ice Age. Interestingly, we know from cave paintings that neither of these species had a mane or a tasseled tail, but that their fur was longer and thicker and their ears were smaller than is seen in their present-day African relative. All these features are clear adaptations to colder climates. These Carpathian cats were massive animals—about one-third again as large as the extant African lion. These species were contemporary with ancient man and went extinct—like many other larger mammals—toward the end of the Würm glacial period, about 10,000 years ago.

The "lion of Gombasek" (*Leo gombasszoegensis*) is another extinct large cat known from the Carpathians. German paleontologist Helmut Hemmer first suggested that this animal might be more closely related to the recent Jaguar (*Panthera onca*) than to lion, and therefore should correctly be referred to as *Panthera onca gombaszoegensis*. Hemmer's conclusions were based on anatomical study of the skull and research on the supposed habitat of this

Cave bears were fearsome enemies of prehistoric people. PENCIL DRAWING BY ANDRÁS SZUNYOGHY.

animal. Eszter Hankó has also supported this idea on the basis of cladistic analysis of the relatively poor fossil record of cats from the Carpathian Basin.

At certain fossils sites, the remains of the cave hyena (*Crocuta crocuta spelaea*) are also abundant. These animals used their stocky and extremely strong jaws to crunch thick animal bones to get to the marrow.

The early invasion of the *Hipparion* into the Carpathians was followed by the arrival of the first real horses from North America during the Pleistocene. This migration event also involved species of dog: because these animals hunt in packs and are known from relatively large numbers of fossils, they are more useful stratigraphically than predators that hunted alone or in small groups.

📖 Kretzoi 1929; Hemmer 2003; Hankó 2007

The Wolverine, Marbled Polecat, and Others: Pleistocene Small Predators

The wolverine, which appeared in the fossil record in the Pleistocene, is just 1 meter long; nevertheless, it is a fearless predator, weighing about 20–30 kilograms. It is the largest member of the family Mustelidae (which also includes stoats and weasels) and looks rather like a small

Jaws of Pliocene and Pleistocene predators. (1) *Parailurus anglicus* Dawkins—lower jaw of a lesser panda with teeth in place. This unique and beautiful specimen was found in lignite deposits at Căpeni, in Covasna County (original size). (2) *Ursus deningeri* Reichenau—lower jaw of "Deninger's bear" with teeth in place, from Tarkő Cave in the Bükk Mountains (0.5× magnification). (3) *Ursus boeckhi* Schlosser—fragmentary jaw of an extinct bear from lignite deposits at Căpeni, in Covasna County (original size). (4) *Ursus spelaeus* Rosenmüller and Heinroth—lower jaw of a large cave bear. The huge specimen was collected in 1881 from the Haligóczi Cave (Mnichová Dolina) (0.4× magnification).

1
2
3
4

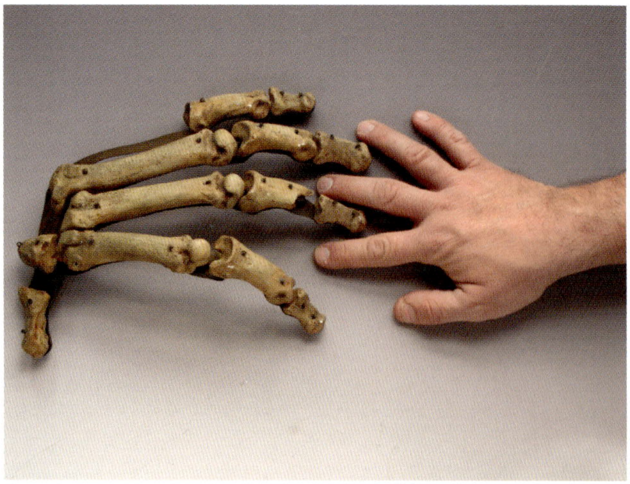

Huge hand bones of a cave lion from Ördöglyuk ("Devil Pit") Cave at Solymár.

Washed and sorted bones of different small vertebrates (close to original size).

bear. Wolverines are stocky in appearance, with thick fur and a lethal bite. They often feed on carrion left behind by wolves but can also hunt on their own, killing caribou and young deer. These mustelids can also defend themselves well against larger predators, and there are reports of them fighting off bears. Indeed, this frightening creature is well known in the folk beliefs of Nordic people; it disappeared from the Carpathian Basin at the end of the Ice Age and became restricted to Scandinavia.

Smaller predators than wolverine, such as the European polecat and the beech marten, common in the vicinity of people, are frequently encountered even today. However, despite a good fossil record the paleontology of small Ice Age predators has been little studied. The only comprehensive, but short, paper so far published is Miklós Kretzoi's "Tigeriltis, Iltis und Nerz im Ungarischen Pleistozän" (The marbled and the European polecat and the European mink in the Hungarian Pleistocene).

📖 Kretzoi 1942d

The Giant Elk

The giant elk (*Megaloceros giganteus*) is often refereed as the "Irish elk" because many of the best-preserved known specimens come from the peat bogs of Ireland. However, this animal was neither exclusively Irish nor an elk. *Megaloceros* is an extinct giant deer, the largest species ever, that stood up to 2.1 meters tall at the shoulder, had antlers spanning close to 3.6 meters, and weighted about 1,400 kilograms. Recent DNA-based studies have shown that the closest allies of *Megaloceros* are among the fallow deer (*Dama dama*), and that they appeared in Eurasia about 400,000 years ago.

Usually just the scattered bones of *Megaloceros* transported and washed by rivers are found in the Carpathian Basin—but some excellent specimens, including antlers, have been collected from the Tisza riverbed in the days before the river was straightened. Although the antlers of *Megaloceros* are the largest known from any species of deer, what they were used for remains debated. It is very likely that *Megaloceros* used their huge antlers to fight each other, as most deer do today, during the mating season; it is also thought that these were shed each year again, just as they are by extant deer. It is also clear that to grow such a large antler set would have required a huge amount of calcium and phosphorus, and so it is probable that during their growth the deer would have suffered from an osteoporosis-like state, since it would have had to extract, at least partially, the required minerals from its bones. Some people have even argued that could be one reason these animals went extinct, but it seems more reasonable to conclude that *Megaloceras,* like many other animals from the Ice Age, was simply unable to adapt to dramatic changes in climate, and that this is really the main reason for their disappearance. Maybe the spacious meadows, where plants which would have been

Pleistocene predator remains (original size). (1) *Ursus etruscus* Cuvier—canine tooth from a small-sized extinct bear. The specimen was collected by Tivadar Kormos from the "Kalkberg" (today Mész) Hill at Villány. (2) *Canis falconeri* Major—canine tooth of a fossil dog from Küspád. This specimen was donated to the Royal Geological Institute of Hungary by Viktor Kaertner, an evangelical pastor, in 1913. (3, 4) *Ursus spelaeus* Rosenmüller and Heinroth—incisors of a cave bear from Herman Ottó Cave, at Miskolc. (5) *Ursus spelaeus* Rosenmüller and Heinroth—a slightly worn large canine tooth of a cave bear from an unknown locality. (6, 7) *Gulo schlosseri* Kormos—lower jaw fragments of an extinct wolverine from the Betfia locality near Oradea. (8) *Ursus etruscus* Cuvier—worn molar of a small extinct bear, from Mész Hill in the Villány Mountains. (9) *Canis spelaeus* Goldfuss—lower jaw of a cave wolf—a close relative of extant wolves—from Igric Cave at Aleşd.

1
2
3
4
5
6
7
8
9

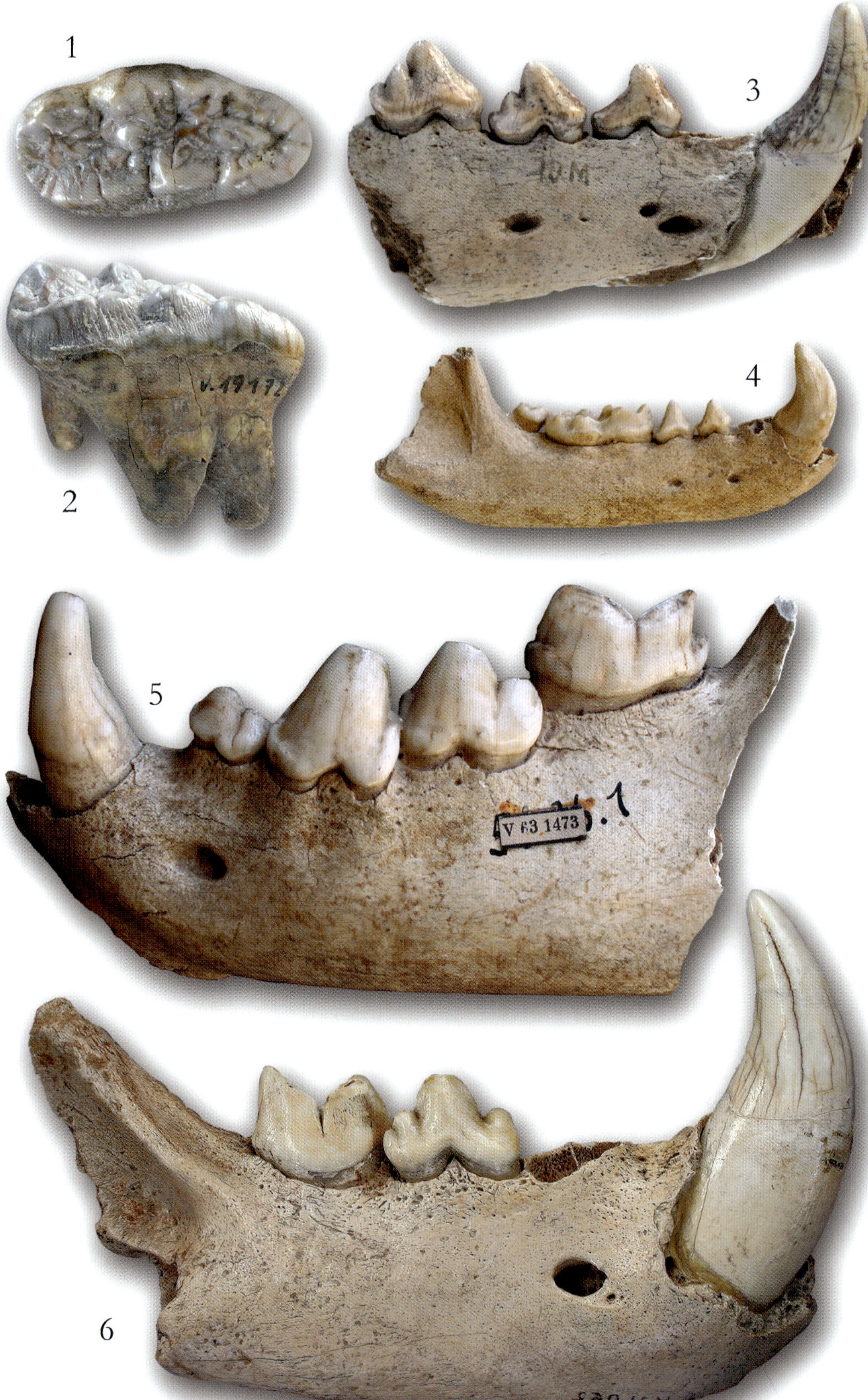
1
2
3
4
5
6
V 63 1473

particularly essential to the deer for their antler growth, also disappeared or changed; it is also possible that human hunting decimated *Megaloceros,* which were already slowly decreasing in numbers. We know that the last Irish elk died out about 7,700 years ago.

The Musk Ox

In 1853, three years after the publication of the first report on the Ice Age vertebrate fauna of Hungary, by János Salamon Petényi, the first musk ox fossils (*Ovibos moschatus*) were found in the Carpathian Basin. Because this unusual herbivore is found today only in northern circumpolar regions, including Canada and Greenland, its presence in the Carpathian Basin indicates a much cooler climate at the time. Indeed, musk oxen are an excellent example of adaptation to extremely severe climates. Their Hungarian name, Pézsmatulok, or "musk bullock," is misleading because these animals have no musk gland and are not bullocks; instead, they belong to the family Bovidae. They are known for the strong smell emitted by the males, and it is from this that the name "musk" derives. These oxen are not large, standing on average just 1.2 meters high at the shoulder, although the males are larger than the females. Their fur is very long and insulates them perfectly as they live in herds and feed on grasses. When musk ox herds are threatened, these animals face outward to form a stationary ring formation, with their horns pointing toward the enemy. It is interesting that these oxen, unlike many other Ice Age animals, are not extinct but have survived in Nordic regions.

📖 Kretzoi 1942e

Fossil Bison

One of the most characteristic creatures of the Eurasian Pleistocene was the steppe bison, also known as the steppe wisent (*Bison priscus*). This bison was somewhat larger than its extant relatives, the American and European bison (*B. bison* and *B. bonasus*), and very likely it grazed in vast herds on the Carpathian steppe. The steppe wisent was over 2 meters tall and had angular and often broken horns that were projected upward and forward. The tips of these horn cores were 1.2 meters apart, and the horns themselves were over half a meter long. Steppe wisent lived mostly on low-growing herbs and grasses—but as the vegetation changed toward the end of the last glaciation and grasslands were replaced by arboreal forest and tundra, these animals are thought to have been unable to adapt; they went extinct about 40,000 years ago in Eurasia. Human hunters and other animals are also thought to have accelerated the extinction of the steppe bison.

However, long after the steppe wisent became extinct its close ally, the European bison, still populated the forests of the Carpathians. This animal was a royal prize and was eagerly hunted by nobleman and kings. One story about them, published in the *Bulletin of the Hungarian Geological Society* by vertebrate paleontologist Mária Mottl, is of special interest. One day in 1911, a group of treasure hunting Transylvanian gypsies discovered a brown bear and a bison skeleton in a cave in Krassó-Szörény County (in present-day Romania). Lying next to the bear was an iron arrowhead from the twelfth or thirteenth century, but the bison was much older, as proven by a thick dripstone crust that covered many of the bones. Mottl rediscovered these bones in the collection of the Geological Institute, was able to assemble them, and put the skeletons on display—where they remained for many years. However, the bison turned out not to be a steppe wisent but the skeleton of a female European bison (*B. bonasus*). The last representatives of these magnificent animals, relics of the Ice Age, disappeared from the territory of Transylvania in the eighteenth century.

📖 Mottl 1935

Happy Hippos of the Ice Age

The so-called Üröm hippopotamus lived about 750,000 years ago in lukewarm ponds at Üröm Hill, near Budapest. At this time, the Carpathian region was deep in the Ice Age, so the discovery of hippos, heat-loving creatures, was a surprise for paleontologists. These bones were discovered by Tamás Báldi, at the time an enthusiastic young secondary school student, and later a professor of geology at Eötvös Loránd University.

The details of his find were noted by Professor Báldi in an excellent and readable book in Hungarian. He recollects that as a young man he was wandering around Csúcs Hill, where his family had a small fruit garden, and walked down the nearby Üröm Hill, where the remains of old travertine quarries were still accessible. In one of the quarries he found an old military bunker cut into the soft

Pleistocene predator remains (all original size). (1, 2) *Ursus spelaeus* Rosenmüller and Heinroth—upper and lateral views of a bulbous molar of a cave bear from Pálffy Cave at Plaveckỳ Mikuláš. (3) *Canis lupus* Linné—a fragmentary lower jaw of a wolf with the canine tooth and three molars intact from Pálffy Cave at Plaveckỳ Mikuláš. (4) *Meles meles* Linné—nearly complete right lower jaw of European badger from Bajót in the Gerecse Mountains. (5) *Crocuta spelaea* (Goldfuss)—fragmentary left lower jaw of a cave hyena, with canine tooth and sharp molars intact, from the vicinity of Miskolc—possibly the ancient riverbed of the Sajó. (6) *Panthera onca gombaszoegensis* (Kretzoi)—fragmentary right jaw, with the canine tooth and two molars intact, from Gombasec. According to the latest results, this large cat, known as the "lion of Gombasec" is more closely related to recent jaguar (*P. onca*) than it is to lion (*P. leo*).

Oblique view of a woolly rhino skull that was literally trawled from the riverbed of the Tisza. The rhino horn lacks the bony core seen in bovine horns—and because of its smooth nose, the skull is somewhat funny in appearance.

Natural cast of the braincase of the fossil equivalent of the modern horse (*Equus caballus fossilis*) from a Pleistocene travertine.

lime bog and roofed by the harder, upsequence travertine. He found numerous small gastropods shells alongside "brown objects of stone appearance." He showed these finds to the vertebrate paleontologist Dénes Jánossy, who immediately recognized them as fossil bones, specifically, the "pretty and well-determinable" limb bones and teeth of an ancient hippo. Similar bones from similar geological strata were later also documented from the nearby Budakalász locality.

This hippo from Üröm (*Hippopotamus antiquus*) may have been a direct predecessor of the extant African hippopotamus (*Hippopotamus amphibius*), as their bones are nearly indistinguishable and they likely had similar ecologies. But the African hippo is a semiaquatic creature that is adapted to warm climates. What was a hippo doing in the Carpathians during the last Ice Age? This is not really a contradiction, because the hippos at Üröm lived about 700,000 years ago, during the Günz-Mindel interglacial period, and thus experienced a mild Mediterranean climate, albeit sandwiched between two colder periods. Hippos would at least have been able to tolerate such conditions—and, as their ponds were likely supplied with warm spring water, they may have also been able to survive more severe glacial periods as well. However, their food supplies—plants that lived in and around the ponds—were unlikely to have survived the colder glacial periods, and so it is possible that the decline of the hippo population at Üröm was caused by a decline in their food supply.

Hippos lived in the Carpathians during the last Ice Age and supplied by warm spring water, just as hippos are able to live in the Budapest Zoo today. These mammals are kept in basins that are supplied with hot water from the famous nearby Széchenyi Thermal Bath, which is only a few hundred meters away. These hippos enjoy warm water from one of the largest spa complexes in Europe, just like tourists from every part of the world. Indeed, the hippos at Budapest Zoo breed well in captivity—a clear sign that the animals are content with their water temperature.

🕮 Jánossy 1961; Báldi 2003

The Woolly Rhinoceros

The woolly rhinoceros (*Coelodonta antiquitatis*) was 3–4 meters in lengths, up to 2 meters tall, and weighed around 2–3 tons on average. These animals also had thick, long fur and two curved horns on their heads, the anterior of these reaching up to 1 meter in length. Its stocky head was generally held lower than those of its African allies, but it likely had a similar biology to extant African rhino: bulls would have roamed alone for most of the year, while females lived in groups with their calves. Woolly rhinoceros appeared about 350,000 years ago on the Eurasian steppe, and went extinct about 10,000 years ago, again probably due to sudden climate change—although some relic populations may

Pliocene and Pleistocene predator remains (original size). (1, 2) *Pannonictis pilgrimi* Kormos—a fragmentary skull with preserved teeth, from Mész Hill in the Villány Mountains. (3) *Pannonictis pilgrimi* Kormos—a fragmentary lower jaw with preserved teeth, from Mész Hill in the Villány Mountains. (4) *Canis mosbachensis* Soergel—molars of a small wolf from Mész Hill in the Villány Mountains. (5) *Vulpes vulpes* Linné—fragmentary jaw of a fox from Szelata Cave, in the Bükk Mountains. (6) *Canis lupus* Linné—an incomplete jaw of a wolf from Igric Cave at Aleşd. (7) *Epimachairodus* sp.—canine of a saber-tooth cat from Gombasec. Large Pliocene and Pleistocene saber-tooth cats are all referred to using the generic name *Homotherium* in recent paleontological literature. (8, 9) *Epimachairodus hungaricus* Kretzoi—copy of the canine of a saber-tooth tiger. The original fossil was collected from the Betfia site, near Oradea.

1
2
4
3
5
6
8
9
7

Pleistocene herbivore remains from the Tisza riverbed. (1) *Bos primigenius* Bojanus—skull with curved horncore. Horn shape was very characteristic in aurochs and was pointed forward and curved inward. This animal also had a flat brow. The largest distance between these horncores is 102 centimeters. (2) *Bison priscus* Bojanus—fragmentary skull from a steppe bison, also known as a wisent, with horncores in place. Wisents had short, horizontal horns that pointed upward, and this animal had a higher brow. The largest distance between these horncores is 92 centimeters.

Pleistocene herbivore remains. (1) *Alces alces* (Linné)—a large, nearly complete antler of a European elk from Pleistocene deposits in the Tisza River. This antler is 85 centimeters wide. (2, 3) *Rangifer tarandus* (Linné)—antlers of a reindeer found in loess at a prehistoric reindeer hunters' settlement near Ságvár. These fragmentary fossils are about 30–40 centimeters long. (4) *Cervus elaphus* Linné—antler of a fossil red deer from Pleistocene deposits in the Tisza. The antler is 90 centimeters long.

Megaloceros was the largest deer of all time. Its common name (Irish elk) is misleading; it is not an elk, but a deer, and it was found all across Europe, Asia, and into North Africa.

have survived in Siberia until about 8,000 years ago. The best fossil specimens from the Carpathian Basin come from the sediments of the Tisza and its tributaries.

The Rise and Fall of Proboscideans

Toward the end of the Pliocene the once-diverse Carpathian Basin proboscidean fauna was diminishing and only two species—*Anancus arvernensis,* with numerous cusps on its molars, and *Mammut borsoni,* with crests on its molars—crossed the Pliocene-Pleistocene boundary. The most famous of all Ice Age creatures, the so-called woolly mammoth (*Mammuthus primigenius*) is known only from the second half of the Pleistocene. This animal was successfully adapted to extreme cold but was unable to survive when climate started to warm; it was also decimated by hunting by humans. The last known populations of mammoth disappeared a few thousand years ago in Siberia, but were extinct much earlier in the Carpathian Basin. The very last specimens, a dwarfed race, were still living on the Wrangel Island until roughly 1,700 years ago.

The woolly rhino was much larger than its close extant relatives, who live in much warmer climatic conditions.

Mammoths in the Carpathian Basin

The woolly mammoth is the most famous and most characteristic animal from the Ice Age, and many superb specimens are known, for example, from the permafrost of Siberia. Here, in the ice, entire animals—including flesh and fur—have been preserved and frozen carcasses reveal all anatomical features. We even know, for example, that there was a small fleshy cap over the anus that served to prevent the opening from freezing. From the Carpathian Basin, just the bones of mammoths are known, and most finds are incomplete—usually disarticulated bones and teeth. Across the whole of Hungary, however, more than 500 mammoth fossil sites are listed.

Very rarely, however, complete or near complete specimens of mammoth are found in the Carpathians. Some finds are known from Hungary—including from sites near Zalaegerszeg, Pécsbányatelep, Dunaföldvár, and Mátraderecske—as well as close to the city of Oradea in present-day Romania. One of these specimens, the mammoth from Pécsbányatelep, was even considered to be a

Woolly mammoth—emblematic creature of the Ice Age.
Pencil drawings by András Szunyoghy.

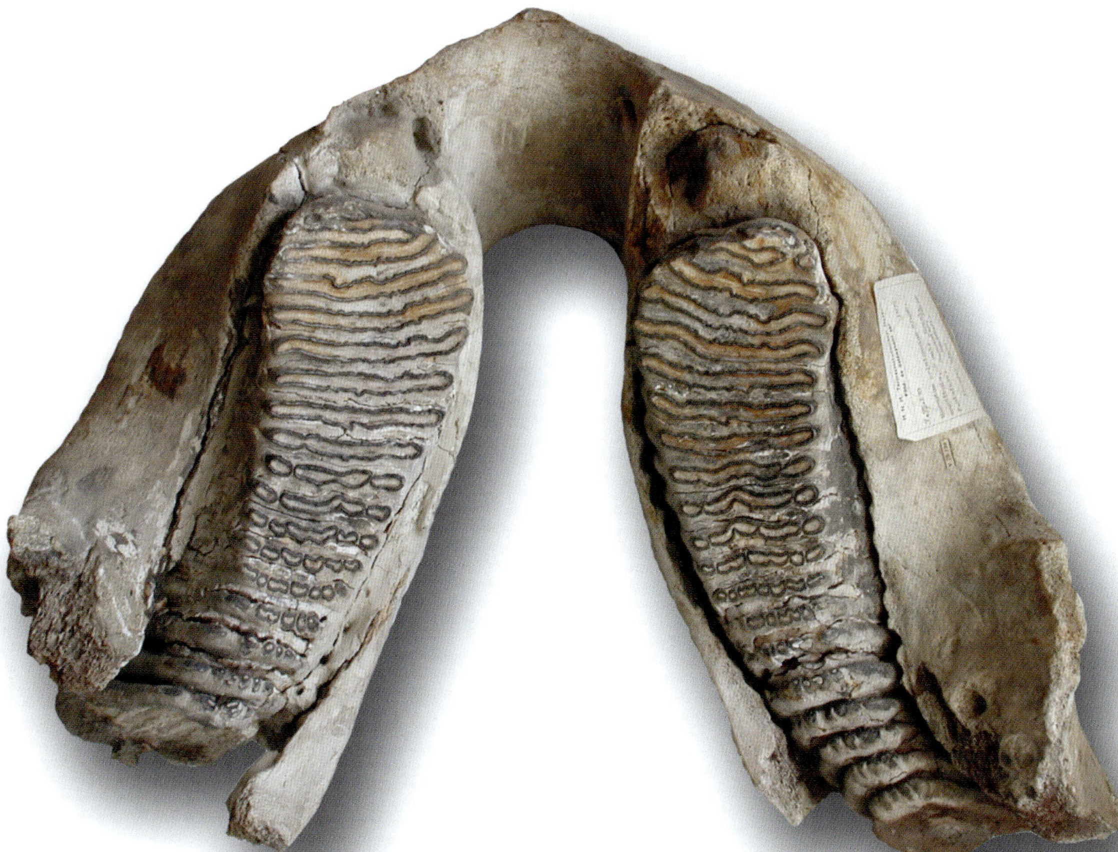

Lower jaw of the woolly mammoth, with two huge molars in place (0.2× magnification).

new species by Miklós Kretzoi; he described it as *Mammonteus hungaricus*. Indeed, according to Kretzoi, this proboscidean differs from other known woolly mammoths because it had shorter legs. However, later researchers have not followed this argument, and today the so-called Hungarian mammoth is not treated as a distinct species.

Another of these relatively complete mammoth specimens, from Mátraderecske, was found in a small sandpit at the end of a garden and it was donated to the Paleontological Collection at the István Dobó Castle Museum in Eger. Subsequently, other missing parts of this skeleton were found, and now the complete mounted skeleton is a highlight at the Mátra Museum in Gyöngyös, the only nearly complete woolly mammoth skeleton on exhibit in Hungary.

Even more recently, further mammoth fossil remains were discovered at Zók in southern Hungary and close to Veszprém, where a calf and an adult were unearthed during construction of a road along the shore of Lake Balaton, a popular summer resort in Hungary. The latter of these two discoveries received significant news coverage: a "baby mammoth crossed the river at Lake Balaton," reported one tabloid. To be honest and correct, at the time, of course, Lake Balaton had not formed: the lake is young geologically and only formed about 8,000–10,000 years ago, several thousand years *after* the last mammoths had disappeared from the area. Even more controversial was that since the new bones were found between two villages, both claimed them; the local mayors even got into a fight in the Hungarian media. Finally, the specimens were donated to the nearest natural history museum, in Zirc.

📖 Halaváts 1899; Kretzoi 1941e; I. Vörös 1975, 1980, 1989; Konrád et al. 2010

The Mammoth Herd from Jobbágyi

Mammoth remains from the Carpathian Basin are generally found disarticulated and as single elements; articulated remains are rare. So one find of special interest is from Jobbágyi, in northern Hungary, where a bone bed was uncovered more than 100 years ago during railway construction work between Hatvan and Salgótarján. Fossil bones were found in a Pleistocene gravel terrace formed by the Zagyva River. The site was visited by geologist Gyula Halaváts, who showed that this bone bed, more than half a meter thick,

Skull of a woolly mammoth trawled from the riverbed of the Tisza. The tusks in this animal originate from the elongate part of the skull; the nostril, where the proboscis is positioned in these animals, is located in the upper middle part of the skull.

contained almost exclusively mammoth bones. An excavation more than "25 footsteps wide" and "50–60 steps in length" was carried out at this site, and a huge number of fossil bones, presumably the remains of a mammoth herd, were unearthed. Halaváts examined this bone bed in detail and concluded that the bones had not been transported and redeposited by the river, but that the animals had died on the spot. He supposed that a startled herd had run into a swamp and that the mammoths had all died quickly. Unfortunately, however, because the bones from Jobbágyi were in a poor state of preservation, most of them were eventually lost to science.

📖 Halaváts 1899

Where Large Mammals Grazed: The Mammoth Steppe

The mammoth steppe was a typical Late Pleistocene biotope, ranging from Western Europe to Alaska, and named for its most famous inhabitant, the woolly mammoth. Under extremely cold and dry climatic conditions a vast belt of grassland formed; it served as the base for an entire fragile ecosystem. The soil on the steppe was firm, with a very thin humus layer—but because of the continuous sedimentation of loess, it was very productive. This environment was rich in nutrients, especially calcium, and the mammoth steppe was likely characterized by a very diverse flora and fauna with a large variety of species.

According to current thinking, the grazing and trampling activities of large mammals would have been essential to the formation and maintenance of this environment. As animals herded about, the appearance of moss and tundra vegetation was likely prevented, and without it the surface of the steppe would not be insulated. During the short summer months, then, the frozen soil melted deeper than is the case in extant tundra, allowing nutrients and the roots of plants to penetrate deeper into the soil.

In other words, the mammoth steppe has no recent analog. The bulk of the megafauna at the time were herbivorous, and all the animals would have fed only on grasses and other types of low-lying plants. There were no trees—or herbivores feeding on the foliage. Besides the eponymous mammoth (*Mammuthus primigenius*), characteristic grass eaters also included bison (*Bison priscus*), woolly rhino (*Coelodonta antiquitatis*), horses (*Equus*), and reindeer (*Rangifer tarandus*). The habitats of these mammals overlapped with the habitats of other species, including Irish elk (*Megaloceros giganteus*), Eurasian elk (*Alces alces*), Saiga antelope (*Saiga tatarica*), and musk ox (*Ovibos moschatus*). The most important predators on the steppe were the wolf (*Canis lupus*), Arctic fox (*Alopex lagopus*), wolverine (*Gulo gulo*), cave hyena (*Crocuta crocuta spelaea*), and cave lion (*Panthera leo spelaea*); on the European part of the grassland, which included the Carpathian Basin, cave bear (*Ursus spelaeus*) and lynx (*Lynx lynx*) were

An mammoth tooth encased in iron; it once hung over the city gate at Győr.

Cyclops Skulls, War Elephants, Ironed Bones, and Wall Decorations of the Apothecary

In antiquity, huge mammoth skulls were thought to be the remains of giants, and it is also likely that this is the origin of the legend of the Cyclops. These mythical creatures, with just a single eye, were likely based on the skulls of proboscideans, which have a large hole in their middle upper part, where the trunk attaches.

Luigi Ferdinando Marsigli (1658–1730)—opened-minded traveler, military engineer, politician, and scientist—gave a more scholarly explanation. He believed instead that the huge skulls found in the Tisza riverbed were the remains of war elephants from Roman times.

It was also the case that some bones, just because of their extraordinary size, were highly valued by ancient peoples and thought to afford magic powers. Some mammoth bones were even encased in iron and hung over the gates of cities to protect the walls. Invaders may also have been scared by these large bones. Another mammoth bone, this time not used to scare people away, was noted by the geologist Ferenc Schafarzik in an apothecary shop in Medgyes (present-day Mediaş, in Romania). Writing in the *Bulletin of the Hungarian Geological Society,* he noted that "visiting the hydrocarbon deposits of the Transylvanian Basin in 1907, I recognized a well-preserved *Elephas primigenius* lower jawbone, used as a wall decoration in the drugstore of Frigyes Schuster at Medgyes. There were two milk teeth still sitting in the jaw, but the piercing new molars had partly pushed them away from the alveoli. This particular specimen had become the possession of the mother of the apothecary during the last century, when, as I was told by her, it was washed out by a heavy shower from the ancient terrace of the River Maros, just above the city." Indeed, mammoth remains are relatively frequent even today and many fossil bones and teeth are stored in family display cabinets.

📖 Schafarzik 1927

predominant. This unique mammoth steppe ecosystem is thought to have collapsed toward the end of the Ice Age—to be replaced by the modern tundra, taiga, and steppe belts of Eurasia.

📖 Pazonyi 2006

Early Man

Sámuel: The Early Man of Vértesszőlős

One of the most remarkable Hungarian fossil finds that bears on the evolution of humans comes from Vértesszőlős, in the southern foothills of the Gerecse Mountains. These excavations were supervised by paleoarchaeologist László Vértes (the similarity of his family name to that of the village is pure coincidence) and yielded a number of sensational

A life-size reconstruction of a woolly mammoth in the Ludovika Square Exhibition Hall of the Hungarian Natural History Museum. The long hair of this model is made of fibrous plant material. All around the model mammoth bones are scattered. When we think about mammoths today, we tend to think about a time when our forebears lived close to the natural environment. With this in mind it is easy to forget current opinion, which holds that one reason for the disappearance of large Ice Age mammals might have to do with the hunting activities of ancient people. The extinction of mammoths and other Ice Age mammals is the most recent and certainly one of the largest mass extinctions in Earth history, and is often referred to as the sixth extinction. It is also hard for us to give an open-minded evaluation of the dynamics and effects of this mass extinction—because we play an active role in it, because it has just started, and because its eventual outcome is unpredictable.

The "*tjurunga*" from Tata. This image shows a carefully polished plate of a woolly mammoth molar that was once painted, as proven by earth-dye remains that are still preserved faintly on its lower part. The strange object is about 100,000 years old and was found at the famous prehistoric settlement at Tata. It is not clear what kind of artifact this plate represents but László Vértes, leader of the excavations at Tata, supposed that it was used by Stone Age man for some form of ritual practice—perhaps it was used as a *tjurunga* (sometimes spelled *Churinga*), or "bullroarer." If this interpretation is correct, then the plate would have hung on the end of a long cord and would have been spun around fast in the air to produce a monotonous roaring vibrato sound to accompany ritual ceremonies. Similar bullroarers made of stone, wood, or even bone are found all over the world, and are still in use by aboriginal Australian people today. This simple piece of a fossil indicates that the history of prehistoric animals overlaps with human prehistory. The science of archeology starts approximately where paleontological research ends (close to 0.5× magnification).

finds thanks to systematic fieldwork in the 1960s. As well as the fossil remains of early humans, tools, a fireplace, and a place for chopping prey were also uncovered. Among these finds is a famous fragmentary occipital bone, found August 21, 1965, and given the name Samuel (Sámuel in Hungarian). This bone has become widely known by its nickname, "Samu," and was described as *Homo erectus* seu *sapiens palaeohungaricus* by anthropologist Andor Thoma (1928–2003). This rather complicated name is intended to express uncertainty about the affiliations of the fossils—specifically whether it belongs to *Homo erectus* or *H. sapiens*. At the time they were found, these remains were believed to be about 500,000 years old, but later investigations have revealed that they are probably much younger, about 250,000–300,000 years old—which means that the early man of Vértesszőlős lived during the Mindel glacial period. Now the Vértesszőlős occipital bone is thought to represent the species *Homo heidelbergensis*.

Today, the Vérteszőlős site is open to the public, and visitors can see fossil footprints preserved in once soft calcareous mud. Among fossilized bear and hoof prints there are some other tracks that were long thought to be human, but subsequent studies have not confirmed this interpretation.

📖 Vértes 1964; Kretzoi & Dobosi 1990; Kordos 1994

Early Man in Subalyuk Cave

Subalyuk Cave is in the Hór Valley, near Cserépfalu in the Bükk Mountains. Its first fossils were collected by the enthusiastic fossil hunter from Eger, Ferenc Legányi, who was followed by János Dancza (1899–1985)—a naturalist and speleologist, also from Eger, who unearthed Neanderthal remains from clay deposits inside the cave. These early discoveries were made in 1932, during hard times in Hungary, when the cave was temporarily named for Italian dictator Benito Mussolini. Later, Ottokár Kadić and Dancza led a joint expedition to the cave and excavated a large collection of fossil bones, including ibex and cave bear. Around this time, it also became clear that the cave had been used as a shelter by *Homo neanderthalensis,* who used carefully elaborated stone tools to hunt for ibex and other animals. The exact age of the Subalyuk assemblage remained unclear for a long time because the ^{14}C method, widely used for radiometric age determination, cannot be applied reliably to this earliest phase of the last glaciation.

Further small, but important, new discoveries were made in 1990—when along with small mammal remains (*Lagurus lagurus*), which were largely unstudied at the time of Kadić, additional teeth of *neanderthalensis* were also collected. *Lagurus* is a small vole and a little Asian newcomer, which arrived in Europe about 60,000 years ago. These little rodents rapidly spread over the cold steppe, all the way across Europe to France. Their presence in Subalyuk Cave provides an age constraint on the previously collected fossils, including the human remains.

As we humans have become more and more intelligent and have probed deeper into our own fossil history; discovering the age of these Hungarian fossil humans was finally only possible thanks to evidence from a small fossil vole.

📖 Bartucz et al. 1939

Faces from the Past

Looking at a skull one might wonder what the person's face was like when they were alive. It is a mystery whether she or he was beautiful or ugly, young or old. However with the help of modern science the original face can be reconstructed from any complete skull. To do this, wire pins are cut and fixed to the skull to show the thickness of the soft tissue at forty-five anthropological measuring points. The more uneven and developed the surface of the bone, the thicker the muscles placed on and adhering to that site will be. The most important morphological features are marked with long needles (see the reconstruction in progress [right]). The muscles and soft tissues are then rebuilt from plasticine.

The final phase of the reconstruction is to cover the muscles with "skin" and harmonize the lines of the face. This step requires sculptural skill or the resulting face can appear lifeless and strange. Each person's skull is unique so the reconstruction will naturally show the unique features of every face.

Making Faces. Ágnes Kustár, anthropologist of the Hungarian Natural History Museum, prepares scientifically accurate facial reconstructions based on human skulls.

Facial reconstruction of a Neanderthal child. In 1932, the bones of a twenty-five- to thirty-year-old woman and a three- to four-year-old child were found in the Subalyuk Cave (Bükk Mountains, N Hungary). Both were Neanderthals who lived about 60,000 years ago. It is unclear whether they had been buried or had died in the cave that preserved their bones. The image shows a facial reconstruction of the child found in the cave done by the late Gyula Skultéty.

Facial reconstruction of an adult male *Homo neanderthalensis* by the late Gyula Skultéty.

Paleontological exhibition of the Hungarian National Museum around 1900.

PART FOUR

MUSEUMS AND COLLECTIONS

The newly constructed building of the Hungarian National Museum, around the mid-nineteenth century. The first significant collection of fossils from the Carpathian Basin was housed here.

11

Museums and Collections

There have always been people, even in very early times, who picked up and collected the eye-catching fossils that they found in the rocks. Prehistoric tombs all across the Carpathian Basin have yielded fossils, especially mollusks, worn as jewels by Early Man. Special finds, such as huge bones from Ice Age mammals, ended up in the collections of noblemen, pharmacists, and naturalists. However, the rapid development of earth sciences from the middle of the nineteenth century onward resulted in the systematic quest for fossils and the enlarging of private and state collections. These turned into the independent collections of state and national natural history museums and special collections in university departments; they served as teaching aids for paleontology and Earth history. Systematic geological mapping of the Carpathian region began after World War I, and this resulted in the further rapid accumulation of fossils in the geological institutions of different countries.

THE DEPARTMENT OF PALEONTOLOGY AND GEOLOGY OF THE HUNGARIAN NATURAL HISTORY MUSEUM

In 1802 the patriotic aristocrat Count Ferenc Széchényi (1754–1820) offered his precious collection of books, manuscripts, carvings, and coins to establish a nationwide museum. This donation gave rise to the Hungarian National Museum, and as the collection got larger additional museums, including the Hungarian Natural History Museum, were founded. Széchényi's first offer was followed by subsequent donations of minerals, fossils, and plant preparations, and finally the Department of Paleontology at the museum was created. The valuable fossil cabinet of Count Pál Szápár (1753–1825), and the Pleistocene large-mammal bones that belonged to Archduke József Rainer (1783–1853) and that literally arrived by carriage, were among the earliest elements of the paleontological collections. The Brunswick Collection, containing precious minerals and fossils, and the collection of the Lobkowitz Dukes also arrived in the middle of the nineteenth century. The latter collection arrived in Pest (at that time Buda and Pest were separate cities) in 300 chests in 11 railway carriages directly from Bílina, a historic city in Bohemia, where the family had an estate. Around the turn of the century, this collection was augmented by the regular purchases of Andor Semsey, the greatest benefactor of Hungarian natural sciences. Some of the finest, most spectacular specimens in today's museum were bought and donated by Semsey.

Some of the stars of early of Hungarian earth sciences, especially paleontology—including figures such as János Salamon Petényi, Gyula Kováts, Miksa Hantken, Lajos Lóczy Sr., and Ágoston Franzenau—worked at the museum in its early years.

However, the work of the museum was hampered by wars and, perhaps saddest of all, the Hungarian Revolution in 1956—a spontaneous nationwide revolt against the government and the Soviet Union. During street fighting a bomb struck the National Museum (where the natural history collection was housed), just opposite the Radio Budapest building, an obvious target. The museum caught fire and the collection was heavily damaged. The fire reached the paleontological collection from the direction of the Department of Mineralogy and after a large, plaster-covered door in the wall burned through, the fire became uncontrollable. All the minerals and nearly the entire paleontological collection were destroyed.

Later the collection was rebuilt and reorganized, and today the number of cataloged items is over 100,000. Of the collection of invertebrate fossils, Triassic ammonites and brachiopods from the Balaton Highlands, Jurassic ammonites and belemnites from the Bakony Mountains, Cretaceous ammonites from the Bakony and Gerecse Mountains, and Tertiary mollusks are of special importance. Vertebrates are treated separately, and they include the small fossil mammal collections of Tivadar Kormos and Dénes Jánossy, which were gathered from different Hungarian Pliocene and Pleistocene localities. Among the highlights of this collection are the Cretaceous reptiles from Iharkút, all of which were collected over the last 10 years.

Flying reptile model display in the hall of the Hungarian Natural History Museum in Budapest. Collections of rarities from older times that contained prehistoric bones and fossil remains gave birth to modern museum collections. In older museum collections thousands of specimens were often put out on display all together, whereas in modern museums hundreds or thousands or millions of specimens are often held in storage, and more models and reconstructions of fossils are put on display for the public.

The Hungarian Natural History Museum originated as an outgrowth of the Hungarian National Museum. Although it reached maturity, at least legally, a long time ago, it entered true adulthood only when it moved to its own building in 1996. Previously the exhibitions were organized in various locations, and the museum had no clear public image. The new Natural History Museum is an old, but partly renovated, building designed by the celebrated Hungarian architect Mihály Pollack (1773–1855); it was a military high school during the Austro-Hungarian Empire. The building later served as the home of different institutions before the 200-year-old Natural History Museum finally moved there.

📖 Kecskeméti & Nagy 1991

THE GEOLOGICAL MUSEUM OF HUNGARY

This museum is located within the Hungarian Geological Institute in Budapest, which itself is housed in a gorgeous palace built in 1899 and designed by the famous Hungarian architect Ödön Lechner, a pioneer of the Secessionist style. The forerunner of the Geological Museum was established in 1868, well before the foundation and construction of this institute.

The royal decree founding the Geological Institute was issued by the emperor and king Franz Joseph the First in 1869. It was founded to map the country and to perform geological research for the Hungarian state, its economy, and its science—with the help of the Hungarian Geological Society. (This was the third such society that existed at the time.)

The museum collection thus become larger and larger as prospecting geologists collected samples all over the country; and it was again Andor Semsey, the great benefactor, who financially supported this enlargement. Semsey purchased the famous Coquand Collection, one of the most important fossil collections at the time, and among the early keepers of the museum were such talented researchers as Antal Koch, Ottokár Kadić, Mária Mottl, Miklós Kretzoi, and András Tasnádi-Kubacska.

After World War I, Hungary lost two-thirds of its territory and the profile of the collection also changed as the country shrunk and geology developed as a science. After World War II, as a result of raw material–based research mandated by the socialist government, the number of incoming geological and paleontological samples increased dramatically. Most of these samples were sent to the recently opened stores of the institute in the countryside, and only the main body of the collection remained on Stefánia Street (at the time known first as Vorosilov Street, after a Soviet marshal, and subsequently as Népstadion [People's Stadium] Street as time passed and Hungarian politics changed).

Today, the collections of the museum constitute more than 150,000 cataloged mineral and fossil specimens. The paleontological collection is mainly organized according to geological stages—but separate paleobotanical, echinoderm, and vertebrate collections were also established. Among the most precious fossils in the museum are the skeleton of the famous pebble-toothed placodont (*Placochelys placodonta*), large slabs with footprints from Ipolytarnóc, and the Late Miocene hominid remains from Rudabánya.

It was clearly expressed in the founding documents of the institute that the museum contained within it had to have a public exhibition. This was not always the case, but the 1900 dedication of its gorgeous art nouveau–style palace marked the beginning of a new era. A freshly erected permanent exhibition occupied the entire second floor,

Part of the collections of the Hungarian Natural History Museum in a crowded cellar of the Hungarian National Museum, before it was moved.

Huge compact store in the recently completed subterranean stores of the Hungarian Natural History Museum. On the left, part of the Zoological Collection, a forest of antlers—deer trophies, many of which were shot by counts and dukes—has social and and natural historical value.

1,470 square meters in footprint. Thumbing through the little *Vezető a Magyar Királyi Földtani Intézet Múzeumában* (Guide to the Museum of the Hungarian Royal Geological Institute), which was published in 1909 to celebrate the 40th anniversary of the institute, it is possible to form an idea about the size and beauty of the exhibited collection. It is possible to estimate that the exhibition contained about 100,000 specimens—of which 30,000 belonged to the Coquand Collection.

However, as a consequence of the World War II, this beautiful exhibition was taken apart in 1939 and was never reassembled. After the war, the whole idea of the museum was reconsidered, and (unbelievably) specimens that came from regions that were at that point detached, including Transylvania, were removed from the collection. Many of these were donated to the Hungarian Natural History Museum, only to be destroyed when the museum burned down in 1956. In any case the institute lost tens of thousands of specimens, perhaps in part because the Communist Party, which had frequent meetings in the building, also needed the upstairs ceremonial hall that had formed a central part of the former exhibition. This, at least, was the prevailing rumor at the time.

For a long time afterward there was no public museum in the Geological Institute, although some superb minerals and fossils were always on show in the freestanding cabinets in the corridors. The museum was reorganized and reopened a few years ago, and today visitors to the Geological Museum of Hungary can admire the exhibited objects and learn about the past and present of Hungarian geological research. One can also visit most parts of the beautiful institute building, which is roofed by famous

From riding school to cinema to museum: the new exhibition halls of the Hungarian Natural History Museum in Budapest. The main building was a military academy, and this part served as the riding school for cadets. Later the academy was closed and the riding school was turned into a movie theater. Finally the movie theater was closed, and the roof of the abandoned building burned down completely—another mysterious story. After renovation work, the building complex serves as a home for the more then 200-year-old Natural History Museum—forever, or at least we hope.

Allosaurus versus *Stegosaurus*—an old-fashioned plaster of Paris reconstruction, once the highlight of the paleontogical exhibition of the Hungarian Natural History Museum. The sculptures, made by Margit Szilágyi, stood at the rear of the exhibition hall in the Hungarian National Museum and captured audience imaginations for decades.

bluish Zsolnay ceramics and which attracts the attention of visitors to Budapest.

📖 Anonymous 1909; Kordos 1991c

THE TRANSYLVANIAN MUSEUM SOCIETY

The Transylvanian Museum Society (TMS) is the biggest scientific society of an academic leaning in Transylvania. Although the administrative foundations for this society were laid in 1841–1842 by the Transylvanian parliament,

Splendid palace on Stefánia Street in Budapest. The interior of the more than 100-year-old building of the Hungarian (once, Royal) Geological Institute. The first directors of this institute included Miksa Hantken, János Böckh, and Baron Ferenc Nopcsa, who were also prominent personalities in Hungarian geology and paleontology.

Up-to-date reconstruction of Hungarian dinosaurs in the new the Hungarian Natural History Museum exhibition, inaugurated in 2011. The authentic mounted skeletons and body reconstructions are side by side.

The collection of the Geological Institute was long stored in hundreds of large oak cupboards, but as the number of specimens increased the quest for modern solutions became paramount. In the early 1980s, a loft was built to accommodate the large number of workers, and a very novel, compact series of cupboards was built to store the enlarging collection. The interior design for this enlargement was done by the studio of Ernő Rubik, maker of the world-famous Rubik's Cube. These days, as a result of the global recession in geological research, this loft is rather empty, but metal cabinets still hold the documents of the early researchers.

Portrait Gallery

Andor Semsey (1833–1923)—The Great Benefactor of Hungarian Science

Andor Semsey was educated both in Hungary and abroad so that he would learn how to manage his 40,000-acre estates in northern Hungary. However, Semsey did not enjoy farming at all, so he leased his estates in 1866 and moved to Budapest. Once there he dedicated special attention to earth sciences and spent nearly all of his fortune to support them; he bought minerals, fossils, and books for the National Museum and the Royal Geological Institute of Hungary, and he financially supported scientific research and the foreign trips of geologist and geographers. One generous donation, for example, allowed the purchase of the famous Coquand Collection and the publication of the Balaton monograph series. Semsey became a member of the Hungarian Academy of Sciences, was made an honorary doctor at the University of Pest, and became an honorary director of the Hungarian Geological Institute. He also supported the building of the fabulously beautiful palace that houses the Geological Institute in Budapest by paying half the costs of construction. Semsey long supported Hungarian earth sciences, but was a legendary miser when it came to his own needs and those of his employees: he never bought new clothes and paid his own people extremely poorly. Nevertheless, his name is preserved on a street in Budapest.

Hallway of the geopaleontological department at Babeş-Bolyai University in Cluj. This collection comprises a pretty selection of invertebrate and vertebrate fossils, mostly from Transylvanian fossil localities.

the TMS finally came together for the first time in 1859 due to the efforts of Count Imre Mikó. Mikó donated both his house and garden to the society and was elected its first president. Indeed, the foundation of this society was a result of strong public will; Hungarians living in Transylvania needed their own association and museum collections that were considered independent from the scholarly circles of Pest (Budapest) and the Transylvanian Saxons, who already had their own organization. Principally the TMS acted as a group of collectors: in the first decades of its existence, members of the society created a valuable collection of books, handwritten documents, coins, and antiquities, as well as natural science pieces and a letter archive—which are now parts of the historical and natural treasures of Transylvania. However, the society also served to provide a solid base for scientific work. This is why, when the University of Kolozsvár (later the Hungarian Royal Science University, then Ferenc József University, and today Babeş-Bolyai University) was created and the two academic institutions began to collaborate, the Transylvanian Museum Society even let its book collection be used by the university for a sum allocated by the Hungarian state. Later the society offered the Mikó Garden in order that the clinic of the university could be built.

The paleontological collections of the society were deposited and kept in the Department of Geology and Paleontology at the university. This collection was reorganized and enhanced by subsequent professors at the university, and today it contains some important (mainly Mesozoic) invertebrate specimens that were collected by Ferenc Herbich, Melchior Neumayr, and Elemér Vadász. The historical part of this informal vertebrate collection

Repository within a Repository: The Coquand Collection

Henri Coquand (1813–1881) was a French geologist who performed pioneering work in Morocco and visited most of the Mediterranean countries, up to and including Albania. He became a professor in Besançon in 1852, and later moved to Marseille, where he retired in 1863. During his field trips and journeys he collected a huge number of fossils: his collection later comprised more than 30,000 specimens. Most of these fossils were invertebrates that came from southern France, Spain, and North Africa. Some of his specimens are now considered types—unique fossils that served as the basis for description of new species. It was Coquand's intention to sell his large collection intact to an institution, where it could be preserved in his name, for a reasonable price.

In the last year of his life, during a congress of French naturalists that was held in Algeria, Coquand made a deal with the Hungarian geologist József Szabó to sell his collection to a Hungarian institution. However, the price he was asking, about 30,000 francs, was very high and neither state institutions nor private individuals were able to pay; the deal almost came to nothing. However, when Henri Coquand died his brother again offered the collection for sale, but at a much lower price. Andor Semsey again footed the bill; the Coquand fossils were intended to be a comparative collection for ongoing research.

However, when the Hungarians unpacked the samples it turned out that many specimens were missing, and Paul Coquand was forced to give an additional discount. At the end of the day, this collection—which arrived in 24 chests—cost the institute 8,000 francs and was exhibited entirely as Henri Coquand had wished. Cataloging the specimens proceeded slowly, however, and this process was hampered by the two world wars. After World War II, a large part of the collection was finally passed to the Hungarian Natural History Museum, where the specimens were cataloged in stages. By chance it turned out that the bulk of Coquand's Jurassic invertebrates went to the Hungarian Natural History Museum, and the majority of his Cretaceous fossils remained in the collections of the Geological Institute—today the Geological Museum of Hungary.

📖 Bácskay 1991

was realized by Antal Koch, and some of the most recently collected specimens (mainly reptiles) have come from Late Cretaceous sediments across Transylvania.

Although the historical parts of the Transylvanian Museum Society's collection are now scattered and incorporated into different collections at different institutions (the fossils, for example, are housed in Babeş-Bolyai University in Cluj Napoca), the society itself was recently reorganized and revitalized after more than 40 years, during which individual members had kept small study groups up and running. After the Romanian Revolution in 1989 the Transylvanian Museum Society resumed its activities as a legal entity, based on former rules, and there is a clear drive to reestablish the collections to their former glory. This endeavor received an excellent impetus with the 2009 donation of the superb theropod dinosaur skeleton *Balaur bondoc*.

📖 Csíky 1991

An old etiquette label on a fossil in the collection of Babeş-Bolyai University, that denotes the former owner (Erdélyi Múzeum Egyesület, the Transylvanian Museum Society).

THE COLLECTION OF EÖTVÖS MUSEUM OF NATURAL HISTORY (EMNH)

Of course Babeş-Bolyai is not the only university that possesses a fine collection of Carpathian Basin fossils for research and teaching purposes. The collection of the Department of Paleontology at Eötvös Loránd University, which now forms a part of the collection of EMNH, also has a long history. This department was created in 1882 and was the second of its kind in Europe. Its first professor was Miksa Hantken, a multifacted geologist and the former head of the Hungarian Royal Geological Institute. Hantken brought a collection of about 4,000 specimens, representing all major fossil groups, with him to the university.

The biological and paleontological exhibition of the Natural History Museum of Eötvös Loránd University, Budapest. The specimens on show here are intended to demonstrate the uniform and integrated nature of the living and extinct animal and plant world.

Among these fossils, forams were particularly common, since Hantken had a special interest in micropaleontology. This is why some of the famous so-called green cassettes arrived at the department; these carefully prepared and mounted examples of *Nummulites* had won a gold medal at the International Exhibition held in Vienna in 1873.

After Hantken's death, Antal Koch took over administration of the department in 1895. Koch, a former professor at the University of Kolozsvár (Cluj) had ample experience in organizing a collection. He maintained the invertebrate section but focused on fossil vertebrates, and systematically developed the collection with the regular collection and purchase of fossils. One of his most valuable procurements was a specimen of *Ichthyosaurus* from Holzmaden. Later, when the earth sciences departments of the university were reorganized, the entire collection, including the minerals, was uniformly exhibited in beautiful new showcases in the Múzeum Boulevard building of the university. These cases were known as "coffin-cabinets" because of their shape.

After the wars and other troubles, the Department of Paleontology celebrated its centennial in yet another building—this time in Ludovika Square, the same place that was later offered to the Hungarian Natural History Museum. Today the department is situated in a brand new campus along the Buda side of the Danube, and some of the best fossil specimens are on display at the recently opened Biological and Paleontological Exhibition of the Natural History Museum of Eötvös Loránd University, which occupies a central place in the southern university building.

📖 Kázmér 1987

THE COLLECTION OF THE UNIVERSITY OF SZEGED

The history of the University of Szeged (a picturesque city in Southern Hungary), and the problems of its scientific collection, clearly reflect the dramatic political changes that have taken place in Central Europe over the last century. The university owes its existence to the foundation of the

Eastward-Straying Sparkle of Western European Enlightenment

One of the most precious fossils in the collection of the Eötvös Museum of Natural History is a palm-sized flying reptile, the holotype of *Pterodactylus micronyx,* which was first described by Hermann von Meyer (1801–1869). The story of the nearly lost and practically forgotten specimen was revealed and published by Barnabás Géczy, then head of the Department of Paleontology at Eötvös University. *Pterodactylus* is a rare fossil and this specimen, also known as the Pester Exemplar, was probably one of the first ever found. The fossil belonged to the collection of Archduchess Mária Anna of Austria (1738–1789), one of the daughters of Maria Theresa (1718–1780), queen of Hungary and Bohemia. Ignaz Edler von Born (aka Ignatius von Born—or Ignác Born, to give him his Hungarian name), the private teacher of the archduchess, rearranged this collection; in the process, he misidentified and cataloged the fossil pterosaur as an invertebrate (a decapod crab). Born was a leading scientist of his age, but the Pester Exemplar has no skull, and the tiny elongated phalanges of the fourth finger resemble the antenna of decapods. Probably he was misled by the rock that surrounds the fossil, the Solhofen lithographic limestone, which often contains decapod remains. Vertebrates, apart from fish, are rare.

The recently rediscovered Budapest specimen of *Pterodactylus micronyx,* one of the first flying fossil reptiles to be described.
🕮 Ősi, Prondvai, & Géczy 2010

This precious fossil, part of the Mária Anna Collection, then passed into the possession of the University of Pest as a result of the intervention of Count Ferenc Eszterházy (1715–1785). It then languished at the university after Mária Anna—having been made the abbess of the Imperial and Royal Convent for Noble Ladies in Prague—left the city.

According to Barnabás Géczy, this forgotten and then rediscovered holotype of *Pterodactylus micronyx* is the most important fossil in Hungary from the standpoint of cultural history, and can be regarded as an eastward-straying sparkle of Western European Enlightenment, which after two centuries, still glows.

🕮 Géczy 1991

University of Kolozsvár because—after World War I, when Hungary lost a good deal of its territory—this Transylvanian university was moved to Szeged. Later, in 1940, when Hungary temporarily received Northern Transylvania back as a result of the agreement known as the Second Vienna Award, a part of the University of Szeged was repatriated to Kolozsvár. The setbacks of this institute, its moving, and renaming can be seen clearly in the books in its Geological and Paleontological Library; they have up to five different stamps, but have always belonged to the same institution.

After the Treaty of Trianon, most of the fossil specimens remained in Kolozsvár, and a new collection was badly needed. Of the present collection of the department, the quaternary gastropod fauna that was collected by Mihály Rotarides is of special importance. This collection is not large, but it is representative and fits well with the main scientific interests, namely Holocene and Quaternary climate change, of the department's researchers today.

🕮 Molnár 1991

The Natural History Museum building at Sibiu. Among other fossils, a large collection of Eocene fossil cartilaginous fish is housed here.

OTHER MUSEUMS

Many other institutions across the Carpathian region have important paleontological collections. The Natural History Museum of Vienna (Naturhistorisches Museum) must be mentioned first. The historical part of this collection originated during the Austro-Hungarian Empire, when Viennese geologists and paleontologists did mapping and fieldwork across the Carpathians. Smaller, but rich and beautiful, local museums can also be found in Lower Austria; of these, the Krahuletz Museum in Eggenburg has a superb collection of Early Miocene Eggenburgian fossils. The highlights of this collection are gavial-like marine crocodiles (*Gavialosuchus eggenburgensis*) and sea cow skeletons (*Metaxytherium krahuletzi*)—all found nearby.

The vineyards near the village of Gainfarn in the southwestern Vienna Basin contain one of the most important Neogene molluscan Lagerstätte of Europe. A carefully arranged exhibition in the city museum at Bad Vöslau gives an introduction to the general geology of the region and shows a great selection of Miocene invertebrate fossils from Gainfarn. The main attraction of this exhibition is the in situ presentation of a disarticulated sirenian—and, in a reconstruction of a cave near the ruins of Castle Merkenstein, famous for its cave bear findings, the occurrence of *Panthera* and dense rodent remains from the youngest Pleistocene.

In Hungary, some collections of local importance can be found in the towns of Zirc, Sümeg, Tata, Pásztó, Gyöngyös, Szombathely, and Komló. Additional important collections are in Romania, at Oradea, Deva, Sibiu, and Piatra Neamţ.

The collection of the Natural History Museum of Zirc is housed in an old abbey. The basis of this collection was the fossil material, mainly Triassic and Jurassic invertebrates, gathered by Dezső Laczkó. What is less known in the international community is that some additional parts, including teeth and bone, of the famous "pebble-toothed" placodontid (*Placochelys placodonta*) are housed here. It is also worth mentioning that there is an important collection of Tertiary mollusks from the Bakony Mountains in this museum, alongside an as yet unstudied Miocene fish collection that contains about 40 specimens from Várpalota. A complete Pliocene rhino skeleton, the main attraction of the exhibition, is a more recent acquisition.

The large collection of several thousand fossils, amassed by former quarryman Lajos Kocsis (1909–2002), is housed in the Kisfaludy Memorial House at Sümeg. Most of these specimens (mainly invertebrates), gathered with dedication and expertise over the course of many decades, originate from the Upper Cretaceous formations found around the city.

The most important collection in the natural history section of the Kuny Domokos Museum at Tata is Pleistocene fossil plant material that was gathered by excellent local teacher and paleobotanist István Skoflek.

The exhibition in the local museum at Pásztó in northern Hungary focuses on Tertiary plant and vertebrate remains from the neighboring countryside. Some of the earliest European proboscidean remains and some fine Miocene plants from Ipolytarnóc are on display here.

The basis of the fossil collection at the Mátra Museum in Gyöngyös is the Legányi Collection, which moved from Eger in 1987. Most of these specimens originate from Oligocene beds near Eger and from the Paleozoic in the Bükk Mountains. A collection of Pleistocene and Holocene mollusks was also recently established by Levente Fűköh, the museum's director.

Many of the Pannonian (Late Miocene) vertebrate remains from Baltavár, collected by László Benda (later his named changed to Bendeffy) are deposited in the Savaria Museum at Szombathely, in western Hungary.

The Komló Museum in Southern Hungary has some fossils from the Jurassic of the nearby Mecsek Mountains. This collection also includes microfossils and some footprints of the dinosaur *Komlosaurus carbonis*.

The paleontological collection at the Muzeul Ţării Crişurilor in Oradea contains an important collection of Middle Triassic marine vertebrates from nearby localities, Jurassic plants from the Pădurea Craiului Mountains, Early Cretaceous terrestrial reptiles from the Bihor Mountains, and Pleistocene mammals from Betfia. Nowadays, the collection is situated in the baroque buildings of the former Episcopal Palace and is waiting to move into a new edifice.

The local museum at Deva (Muzeul Civilizaţiei Dacice şi Romane [the Deva Museum of Dacian and Roman Civilization]), in accordance with its somewhat strange name, contains numerous objects of historical interest as well

Ferenc Legányi, the Legendary Fossil Collector

Probably the most effective, and definitely the best known, Hungarian amateur fossil hunter lived as a farmer on a small estate in the vicinity of Eger, Northern Hungary. Legányi became acquainted with fossils and fossil collecting on trips into the nearby Mátra and Bükk Mountains, and became a passionate collector. Because he paid special attention and wanted to always learn more and more about the nature of his finds, Legányi soon also became a noted paleontologist. His farming career ended early when a road was constructed across his land from Eger to the nearby Eged Hill. Although the city authorities had promised to refund half his expenses, the money never arrived; Legány was forced to sell his farm. His enthusiasm toward fossils remained: although his life was simple and he made his living transporting farm products, he regularly sent fossils that he found to both the Hungarian Geological Institute and the Hungarian Natural History Museum. Between 1951 and 1963, he was finally able to continue with some activity of his own as an employee of the Museum of Eger. Besides fossils, Legány also collected archaeological dates, folk songs, and folklore; the image of this eccentric gentleman returning in the evenings from his collecting trips carrying a wooden butt packed with fossils and minerals, is still remembered by elderly people in and around Eger. Legány contributed immensely to understanding of the geology and paleontology of Hungary; for example, he was the discoverer of a strange looking brachiopod (*Leptodus*) in the Bükk Mountains that turned out to be the earliest evidence of Permian deposits in that area. A street in Eger bears his name, and the largest part of his mineral and fossil collection is in the Mátra Museum in Gyöngyös.

Ferenc Legányi (1884–1964) in the courtyard of the Eger Museum. The figure of Legányi, burdened with heavy books, appears to symbolize the eternal endeavor to investigate the secrets of nature as well as one's own past. Let the efforts of the tireless old fossil collector set an example for future generations!

as some Cretaceous ammonites and fossil vertebrates, including dinosaurs from the nearby Haţeg Basin.

The Museum of Natural History at Sibiu forms part of the Bruckenthal National Museum, pride of the Transylvanian Saxon community, and it has a long history. The permanent exhibition has a paleontological section that includes some well-preserved mounted skeletons of Pleistocene large mammals. One of the highlights is a complete steppe wisent, which was discovered in 1901. Of special interest here is the large collection of foraminifera and Eocene shark teeth that were gathered by Johann Ludwig Neugeboren, former curator of Baron Samuel Bruckenthal's collection.

The local museum at Piatra Neamţ is known in particular for its collection of Oligocene fish, which were gathered from the Eastern Carpathians and studied by Mihai Ciobanu in the 1970s.

OPEN-AIR GEOLOGICAL MUSEUMS

Open-air geological sites have special appeal because they combine the feel of a traditional museum visit with outdoor activity, often as part of a geopark. A brand new institution of this kind was recently inaugurated at Stetten, in the Korneuburg Basin near Vienna. Here, the highlight is a Miocene oyster reef of more than 400 square meters and exhibited in situ. Within this structure the largest known fossil pearl was found: a blister pearl 5 centimeters in diameter from a mytilid bivalve. The paleoenvironment represented by this reef was a tropical estuary some 16.5 million years ago, and oyster reefs extended along the muddy shores. Even these oysters would have been giants for their time, some reaching up to 40 centimeters in length. Stetten is a unique place and definitely worth a visit.

In Hungary, the Tata open-air geological museum and nature conservation area managed by Eötvös Loránd

University is situated inside the town in an abandoned limestone quarry complex that exposes most of the Triassic to Cretaceous carbonate facies found in the Transdanubian Range. The geology and paleontology of this site has been documented by numerous papers and monographs, but of special interest here is the flint mine that was used by prehistoric people. Ancient man used Jurassic cherty rocks (radiolarites) to make tools.

A similar prehistoric chert mine was unearthed in the vicinity of Sümeg, where a Late Jurassic–Early Cretaceous succession is exposed in nearly vertical position. Flint was excavated from Early Cretaceous carbonates, which contain siliceous nodules—and the site, which opened to the public a few decades ago, is little visited these days.

Very young, Pleistocene-age formations can be visited at Vértesszőlős, where the surroundings of a prehistoric thermal spring are protected by a glass pavilion as a fossil wallow preserves the fossilized footprints of a range of animals.

One of the most visited open-air geological museums in Hungary is at Ipolytarnóc, in the northern part of the country. Here, Lower Miocene strata contain both marine and terrestrial fossils—including shark teeth, petrified trees, leaf impressions, and tracks. Footprints of eleven different vertebrate species have been described from this footprint-bearing sandstone which was conserved by a covering of volcanic tuff. A detailed description of the site and its fossils was provided in the Miocene chapter.

PRIVATE COLLECTIONS

Collecting fossils is a noble passion, but it takes time, energy, and even some money. Perhaps this is one reason why it is not so popular nowadays in Central Europe, and the number of collectors and non-professional paleontologists is rather low. However, this was not always the case. In the past, among aristocrats and leading politicians in the Carpathian Basin there were quite a number of people interested in Earth history and fossils. It is not well known, but Lajos Kossuth (1802–1894) Hungarian freedom fighter and later regent-president of Hungary, while a celebrated politician in England and in the United States, also collected fossils during his later years of exile in Italy. Unfortunately, most of his carefully labeled fossil specimens were lost.

Archbishop Councilor Rezső Streda (?–1960), perhaps the most important Hungarian collector of the twentieth century, always returned from his trips heavily laden with fossils. After World War II, he sold his large collection to the Hungarian National Museum; today these specimens can be found in the Hungarian Natural History Museum.

György Encsy, a resident of the little Northern Hungarian village of Tállya, has also built up an important private collection, mainly consisting of Miocene fossils from the neighboring countryside. His collection can be visited upon request.

The quest for fossils has always been exciting for many people and it will surely remain so. Fossils hunters, both professional and amateur, have contributed enormously toward our understanding of the nature of fossils and life on Earth. As decades and centuries have passed, new techniques have developed that have given further insights into our science. However, traditional fieldwork-based paleontology, the collecting, cleaning, and organizing of fossils into collections, has somehow become old fashioned. This tendency has resulted in a slow decrease in the number of traditional paleontologists; at the same time, non-professionals have achieved more space in the field and more recognition for their invaluable contributions. The traditional boundaries between scientists and amateurs are fading and private collectors are sure to play a more and more important role in the future of paleontology.

Epilogue

Although the Earth is more than 4 billion years old, humans are the first creatures to evolve the ability to recognize the past. Because of this unique ability, it seems important to us to search and to understand our environment and its past history.

While working on this comprehensive book, we realized again and again that some of the unique geological successions in the Carpathian Basin document the past history of the Earth and life extremely richly—and that the intellectual achievements of earlier researchers, who discovered Carpathian Basin fossils and understood their true nature, provide good examples for later generations of paleontologists. Although many important periods of Earth history and environments are not documented by rocks in the Carpathian Basin, a huge number of important fossils are known and have been described. These fossils can tell us about both the history of the Carpathian Basin and of our planet, and can also teach us humility. Anyone who is able to glimpse the passage of time evidenced by fossils during their short lifespan is also able to catch a glimpse of eternity.

GLOSSARY

This section contains short explanations, in alphabetical order, of some selected terms and phrases which are not otherwise defined in the main text or in the figure captions of the book.

Ammonite. Ammonites are extinct marine invertebrates. From a geological point of view, ammonites are the most important of the lineages of cephalopods—the clade that also include octopus, squid, cuttlefish, and nautiloids. The name derives from the Egyptian god Ammon, who was often depicted ornamented with two coiled ram's horns behind each of his ears; these resemble the generally spirally coiled shells of ammonites. The smallest adult-sized ammonites were about 1–2 centimeters across the spiral, with the largest on record exceeding 1.5 meters. Most had a planispiral flat coil but others had straight shells or loosely coiled spirals, the most bizarre of the shapes to evolve. The most spectacular, so-called heteromorph (mostly uncoiled) forms were especially frequent in the Cretaceous. Ammonites had an outer shell that began with a minute, bubble-like chamber (called the protoconch) and then continued into a chambered region (the phragmocone). The successive chambers that made up the shell became progressively larger and were separated by walls known as septa. The last, and largest, chamber was the body chamber, where the soft body of the animal was situated; a thin tube (called the "siphuncle") passed through the septa, connecting it with the chambers of the phragmocone. Using this tube the ammonite was able to control the relative proportions of liquid and gas inside its shell and, thus, its buoyancy. With this mechanism, ammonites were able to rise and fall in the water column. The intersection of the septum with the outer shell, a structure called the suture line, was often very complicated in these cephalopods, and had many folded, curved saddles and pointed lobes. Indeed, the suture line is characteristic of certain groups of ammonites; their shells were either smooth or, in most cases, slightly or heavily ornamented with simple or branching ribs—which could be either weak or strong—marked by spines or rows of tubercles. The ventral portion of their shells had one or more grooves or was keeled.

Although we know a lot about what ammonites looked like, less is known about how they lived. Their closest living analog is the *Nautilus,* which has a very simple suture line and a gently curved chamber wall, but it is hard to imagine that all ammonites—so different from one another and from the *Nautilus*—had the same mode of life. The shell of these cephalopods suggests that they could swim, but not very efficiently, and surely they were masters of flotation in the water column. The first ammonites appear in the fossil record in the Early Devonian; they diminished in number around the Permo-Triassic boundary, and nearly went extinct toward the end of the Triassic. Ammonites rebounded in number during the Jurassic and Cretaceous but finally went extinct, along with the dinosaurs, 65 million years ago—at the end of the Mesozoic era.

Ammonites are excellent index fossils—in some sequences almost each bed is characterized by a unique ammonite, or set of ammonites. As a group they enjoyed a long time span and their rapid evolution was reflected in their changing ornamentation. They also inhabited the open sea, and so their fossils are found over vast areas. All these features, in combination, make them excellent fossils for sedimentary correlation over space and time.

Bedding surface, or bedding plane. In most sedimentary rocks the single beds that make up a succession (a packet of rocks) are bounded by flat, or slightly undulating, surfaces; these are the bedding surfaces, or bedding planes. In most cases, these surfaces have an erosional origin. Unconsolidated sediment was swept away and resulted in a surface that represents a small stratigraphic gap in the succession. Later on this eroded surface might be cemented, bored into, or bioturbated: all these later processes can make observation of the bedding surface problematic.

Belemnite. An extinct group of marine invertebrates closely related to modern squid. Within cephalopods, belemnites are the most important group in subclass Coleoidea. Belemnites had an inner shell that had a heavy calcite guard on its rear. The main role of this structure appears to have been to counterbalance the front of the organism—although, in most cases, this guard is all that remains of the animal as a fossil. Belemnites were active swimmers and had an important role in ecological communities during the Jurassic and Cretaceous. They went extinct along with ammonites and many other creatures at the end of the Mesozoic era.

Benthic. This term refers to the community of organisms which live on, in, or near the seabed, riverbed, or on the bottom of a lake. The term also refers to a bottom-dwelling way of life.

Brachiopod. Brachiopods—also known as "lamp shells" because some species are shaped like oil lamps—are definitely one animal group that is little known among non-scientists. However, these invertebrates are very common in the fossil record and are traditionally classified in the same phylum as Bryozoans. Recent work, however, has suggested that the two lineages may be quite different, and now both groups are classified into a separate phylum (indeed, they are very different). Brachiopods resemble bivalves, since both animals have two basically calcareous valves that protect the soft body and can be moved by muscles, but the former have valves that are not laterally located but rather dorsal and ventral to the body axis. Modern species of brachiopods are rare and are often just a few centimeters in length, but some extinct forms exceeded 30 centimeters and were highly variable in shape. Valves can be concave or convex, smooth or ornamented, and have a hinge. Again, in contrast to bivalves, active muscle control is used to open the valves (in bivalves, this is passive). As a result, after the death of the animals, shells commonly remain jointed; so fossil brachiopods are often preserved with double valves. In their soft anatomy, brachiopods are also very distinct to bivalves and characteristically possess a lophophore, a crown of tentacles that filters food particles out of the water. The lophophore sits on a skeleton that can be permineralized and thus preserved; the skeleton is easily studied by making serial sections of fossils.

Brachiopods are generally marine and benthic animals, although a small group—*Lingula* and relatives—live within the sediment, and others are cemented or attached to the seafloor by a stem. In the past there were free-living brachiopods as well; these forms lived like scallops do today. As a group, brachiopods first appeared in the Early Cambrian and remained as characteristic benthic faunal elements throughout the Paleozoic. Indeed, this group is often referred to as "the bivalves of the Paleozoic; the fauna in the Devonian is especially diverse. After the extinction of many groups, brachiopods had another acme in the Carboniferous, but after the Permian mass extinction they never again achieved the same role in ecosystems. Overall, brachiopods can be used as relatively good biostratigraphic markers in some Paleozoic and Mesozoic formations. Since their larvae had a short time span, they were unable to cross certain paleogeographic barriers, such as wide oceans. Many species thus had a restricted geographic distribution and can be useful for paleobiogeographical studies.

Breccia. A kind of rock composed of angular to subangular, randomly oriented clasts that are cemented by a generally much finer matrix. Breccias can have very different origins; one can find references to sedimentary, tectonic, volcanic, and impact breccias.

Bryozoa. Also sometimes known as "moss animals," broyozoa are extant, aquatic, and mainly marine invertebrates. Traditionally they are classified, along with brachiopods, in the phylum Tentaculata, although some taxonomists classify them in their own phylum. All bryozoans are colonial, except for one genus, and their remains are found attached to both recent and fossil mollusk shells. Each colony is built up from individuals called zooids, which live in closely packed separate capsules. Zooids are generally smaller than 1 millimeter in length, have simplified bodies, and perform specialized tasks. Most common among them are autozooids, which are the colony members responsible for feeding; there are also zooids that clean and defend the colony using tiny whips, which they can move around. Some zooids also play a special role during reproduction. Overall the hard skeleton of the colony is calcareous and can have a range of shapes—including brushes, fans, and sheets. Bryozoans filter the food out of surrounding water; this is done using small tentacles arranged into the shape of a crown, a tiny organ called a lophophore. Interestingly, brachiopods—different looking but closely related animals—also possess this organ.

Bryozoans first appear in the fossil record in the Ordovician and during the Paleozoic experienced several acmes in their diversity. Modern bryozoan groups are known from the Jurassic onward, and it is believed that their latest bloom, which is still flourishing, began in the Cretaceous. In general these animals are not very good age indicators but can provide information about paleoenvironments. Fragile branching forms would have needed calm waters, for example, whereas compact colonies indicate more agitated water conditions. Roughly speaking, the number of recent species is around 4,000, while the number of known fossil species is around 16,000.

Carbonate platform. This term is used to describe bodies of carbonate rocks originated from autochtonous deposition of lime mud. The lime mud is produced by living organisms, especially calcareous algae. The body of accumulated calcareous sediments can be very thick, in some cases reaching thousands of meters. In these cases the sedimentation kept pace with the subsidence for a long time. Carbonate platform grow mostly in the tropical belt. The surface of carbonate platforms is characteristically uneven, occasionally dry areas and deeper lagoons can be encountered. The best-known example of present-day carbonate platforms is the Bahama Bank.

Carbonized fossils. If an organism or sediment is buried, a process known as carbonization begins. Because of high pressure and temperature volatile components disappear from the sediment and the relative carbon content of the material increases. Finally, coal—graphite in some cases—is all that remains. Plants are often carbonized as fossils—this is how coal deposits form—but animal remains are often preserved in this way as well.

Cladistics. Cladistics, developed in the 1960s, is a widely used method of building hypotheses about phylogeny by grouping organisms, according to their most recent common ancestor, into clades. Thus, a clade includes an ancestor and all of its descendants—living and fossil. All features, the characters, of an organism considered in this way are hypothesized to either be ancient or subsequently derived (and thus used for grouping); all organisms that share common derived characters are thought to descend from a common ancestor, and thus form a clade. The result of a cladistic analysis is a cladogram that shows relationships between species; the more features that are included in the analysis, the more robust is the result.

Coleoidea. A large group of marine invertebrates that belong to Cephalopoda but that generally lack an outer shell. Most coleoids have a more, or less, reduced inner shell (such as *Sepia*) that is used for buoyancy or support. Some species (such as *Octopus*) have lost this inner bone completely; in others it has been replaced with a new cartilaginous or calcareous support structure (as in some squid and *Argonauta*). Coleoidea is a subclass that comprises three extinct and four extant orders and includes all living cephalopods apart from *Nautilus* and *Allonautilus,* which has an outer shell. From a geological point of view belemnites (order Belemnitoidea) are the most important of the groups of coleoids; the first representatives of this lineage appear in the Devonian, and by the Carboniferous the group was already diverse. Coleoids played an important role in ecosystems throughout the Mesozoic. It is hard to prove because they do not fossilize well, but it is generally believed that coleoids without shells bloomed after the end of the Cretaceous. The colossal squid—the prototype of Kraken, depicted as a fearsome predator in literature and folklore, and one of the largest extant invertebrates—is also a coleoid.

Conglomerate. This term refers to a type of rock in which rounded clasts (pebbles) are cemented by generally finer rock material such as carbonates, silica, or limonite.

Conodont. Traditionally this term was used to describe minute (0.1–6 millimeter) phosphatic tooth-like structures of unknown origin. Now, however, it is clear that conodonts were extinct eel-like chordates; to avoid confusion these teeth are now referred to as conodont elements. Not much is known about the body of the conodont animal, but some fossil imprints suggest that they possessed bilateral symmetry and used their teeth to filter their food. Conodonts first appear in the Cambrian and went extinct at the end of the Triassic. Within this time interval, their remains are found in marine sediments, usually deposited in moderately deep waters in the tropical and subtropical belts. Due to their resistant material, conodont elements are common even in slightly metamorphosed rocks, where no other fossils are preserved. Conodont teeth also occur worldwide and can be extracted from host rocks with relative ease.

Coral. Corals are extant marine invertebrates that belong to a large group of colorful animals called Anthozoa, within the even larger phylum Cnidaria. There are several known kinds of solitary corals, but many live in colonies, in which each individual coral animal is called a polyp. Many polyps have their own outer calcareous skeleton, which can easily be preserved as a fossil. Indeed, in contrast to hard, reef building corals the so-called soft corals—which include sea fans, sea feathers, and other similar forms—have no hard skeleton at all and thus are not well represented in the fossil record.

Corals have radial symmetrical structures and biologists often classify the recent forms into a group called Hexacorallia. However, in the past many corals had a different form of symmetry, markedly different from anything seen today. Corals are sedentary marine animals that feed on plankton and use the tentacles of their polyps to catch food. Many, especially the reef-building species, live in symbiotic relationships with a kind of algae called zooxanthellae. The first Anthozoa are known from Precambrian rocks, but these soft-bodied creatures are only distant relatives of modern corals. Many groups, like the rugose and tabulate corals, appeared, bloomed, and went extinct during the Paleozoic; the first representatives of modern corals (Scleractinia) are not known until the Triassic. Large-scale diversification of the modern reef-building corals commenced after the end-Cretaceous mass extinction event. This group filled the niche vacated by the extinct Rugosa and Tabulata corals. Indeed, the golden age of scleractinian corals is now.

Corals may occur in rock-forming quantities and some of them can be used as index fossils or indicators of paleoenvironment. Porous fossil reefs built by corals and other associated organisms often contain hydrocarbons of great commercial value.

Crinoid. Crinoids are extant marine invertebrates, and the most important group of echinoderms from a geological point of view. The common name for crinoids, sea lilies, implies that many are attached to the seafloor during adulthood and resemble flowers (lilies). However, the fragile skeletons of these invertebrates are composed of tiny particles and are made up from three basic sections: the stem, calyx, and arms. The stem is attached to the ground by a rootlike radix and can be very long—up to 30 or 40 meters in some cases.

In some fossil species of crinoids, the stem or arms can also be reduced. These bizarre forms look like aliens from sci-fi movies rather than some of the earliest representatives of life on Earth. Most crinoids live attached to the seabed, but there are some exceptional swimming or floating forms, as well as a few that slowly crawl across the sediment. Most Paleozoic and Mesozoic forms lived in shallow marine environments, whereas almost all living crinoids live in deep seas. These invertebrates feed on small plankton particles, filtered by minute featherlike appendages—pinnules—on their arms. Fossils of crinoids can be abundant—and so crinoidal limestones are often thick, suggesting that these echinoderms formed forests on the seafloor. Crinoids, gently moving together deep in tranquil waters, would have been spectacular. The group first appears in the fossil record in the Early Ordovician and flourished during the Paleozoic, but after the end-Permian mass extinction event their numbers diminished dramatically. The forms that are still seen today experienced their greatest diversity during the Mesozoic and are especially important because of their rock-forming potential. After the death of the animal and the disintegration of the calcareous skeleton, small particles form a crinoidal limestone, or encrinite. Because these particles are generally transported and washed around before final burial, however, the use of crinoidal limestones as indicative of paleoenvironments is quite limited. However, there are some good index fossils among the Paleozoic species because their systematics is based primarily on the position and shape of the plates in the calyx. Indeed, the majority of the thousands of known species is Paleozoic in age and is classified into four subclasses. Among these, only the Articulta, which has members that date back to Triassic, boasts extant species.

Diagenesis. This term describes the chemical, physical, and biological processes that turn soft sediments into hard rocks. Because at first sediments must be buried, pressure and temperature play a significant role during diagenesis.

Endemic. This term indicates that an organism is native to a particular place or biota. For example, armadillos are said to be endemic to the New World. The phrase is occasionally also used in an ecological sense, as in the phrase "brackish endemic," which indicates that the organism in question lived exclusively in brackish water (a mixture of fresh and saline water).

Facies. This is a widely used geological term that turns out to be quite hard to define precisely. It refers in general to the so-called face of the rock—in other words, the specific characteristics of the sediment that reflect a particular process or environment. The term "lithofacies" refers to the lithological features of a rock, while "biofacies" can be used to describe the fossils contained. In the case of sedimentary rocks, facies refers to the environment in which the rock formed (for example, shallow marine facies).

Foraminifera. Foraminifera (or forams for short) are single-celled organisms that typically live in seas and have a mineralized test, or outer shell. The name literally means "hole bearers," and refers to the numerous tiny holes that perforate their shells and from which their pseudopods (bulging ectoplasm) emerge. The foram cell overall is rather complex, as it has evolved to perform all the necessary functions of the organism—including locomotion, capturing and digesting food, and reproduction. Indeed, some foram shells can be simple and single chambered, while others can be multi-chambered and elaborately structured. Some can be entirely smooth and some can be beautifully ornamented, and the size of tests can vary considerably. Extant forms are normally small, less than 1 millimeter, but some of the known fossil forams achieved much larger sizes, up to 10 centimeters in diameter.

Foram tests are generally composed of secreted calcium carbonate ($CaCO_3$), but their internal structure varies, being composed of organic materials or cemented particles (including sand grains or other shell fragments) that they have picked from the seafloor. Forams are all aquatic, mostly marine—and most prefer normal salinity. Some species, however, are able to tolerate changes in salinity. Most belong to benthic communities and move around slowly on the seafloor, but there are a few species that either live in the sediment or are sessile (that is, they live attached to the seabed). Still other lineages live in the water column as part of the plankton. The earliest forms are known from Cambrian deposits.

The first boom in foram diversity happened in the Carboniferous and Permian, when spindle-shaped larger forminifera called fusulinids flourished. The next boom in diversity occurred during the Jurassic; the Cretaceous can be characterized by the appearance of a new larger foram group called the orbitolinids and high diversity of the planktonic *Globotruncana* and related forms. The Early Cenozoic is characterized again by a new group of larger forms called *Nummulites*.

Foraminifera provide excellent age and environmental indicators. The larger forams may indicate shallow marine environments, whereas planktonic forms are especially useful when analyzing deeper marine successions. The ratio of benthic/planktonic specimens, for example, can characterize the water depth at which the foraminifera-bearing sediment was deposited, and the stable isotope composition of the test may reveal information about the temperature of ancient waters. Indeed, this is a burgeoning research area given the current concern about climate change. Mass occurrences of larger forams always indicate high carbonate productivity, and thus tropical paleoenvironments; other groups can provide information about levels of salinity, oxygen, and nutrient supply.

Formation. In geological terms, a formation is a formally named unit of rock that can be identified easily in the field. Names, such as Dachstein Limestone Formation, typically combine a geographical specifier of where the rock body occurs (or was identified) as well as the rock type. However, use of the term "formation" also refers to lithology, and presents an age and environmental implication as well. The Dachstein Limestone, for example, is a typical Late Triassic shallow marine sediment known from the Alpine-Carpathian region.

Gastropod oolith. This is a special kind of limestone composed of rounded grains (ooids) and a matrix which cements the grains together. The ooids themselves are tiny gastropod shells covered with a thin calcareous crust that formed in shallow marine environments, similar to other oolites. In these environments, the grains are agitated by intertidal water, covered with a spherical crust, and finally cemented together.

Global Stratotype Section and Point (GSSP). Because the same periods and stages of geology are represented by different facies and formations in different parts of the world, conceptions on the limits and durations of time intervals also varies among geologists around the world. The GSSP is thus an internationally agreed upon geological section that serves as the reference section for a particular boundary in the geologic time scale, in order to prevent confusion and disagreement between researches. The GSSP program, launched in 1977, is conducted by the International Commission on Stratigraphy, under the aegis of the International Union of Geological Sciences. An ideal GSSP section must be accessible for further research, is well documented, contains a radiometrically dated bed at the boundary in question and includes well-defined markers (such as fossils) that can be applied globally. Most, but not all, GSSPs are defined based on changes in the fossils that they comprise along the section. Selecting and analyzing GSSP candidate sections is scientific work, but the formal decision is made via a bureaucratic process of voting by members of commissions. At the time of writing, about half the required 96 GSSPs have been approved. (See also Stratigraphy.)

Graptolite. This extinct group comprises small- to medium-sized marine, colonial animals, which are commonly preserved in Paleozoic formations. The name "graptolite" literally means written stone, since so many of these fossils are preserved as black carbonized films that resemble hieroglyphics written on rocks. The tests of these animals was organic, so their fossils are preserved only under special environmental conditions—the poor bottom circulation, low oxygen content, and an absence of scavengers. Graptolites appeared in the Cambrian and went extinct in the Carboniferous. They enjoyed a large geographic distribution, and because they evolved rapidly they serve as excellent index fossils—especially for Ordovician and Silurian formations.

Hemipelagic. This geological term refers to sedimentary rocks deposited in seas close to the margins of continents and that thus contain particles of both marine and terrestrial origin. In contrast to pelagic rocks, which often are formed much slower in oceans far from the continents and which thus contain almost no terrestrial input, hemipelagic sediments accumulate at a higher rate. In oceanography, this term is also used to refer to the belt of sediments that spans from the outer shelf down to the start of the abyssal plain, including the entire continental slope.

Internal mold. Also known as "mold," "cast," "internal cast," or "steinkern" (the latter after a German term), this is one of the most common ways an organism can be preserved as a fossil. After death, soft tissues decay—but harder elements, such as bones and shells, can be buried by sediments. Thus, an impression of the negative surfaces of the organism can remain in the surrounding sediments, which later become lithified. This eventually comprises a mold, or external mold, as the external surface is preserved. If this negative is filled with sediment and hardens into rock, then it forms an internal mold. Such molds may or may not be informative, depending on the impression that is formed: for example, gastropods that have smooth, corkscrew-shaped shells often cannot be determined from their molds, whereas ammonites often lead to molds that preserve important taxonomic features. The internal molds of ammonites, then, are often highly informative and species can be determined.

Nautiloid. Nautiloids are an extant group of cephalopods that have an external shell. Similar to their extinct allies, the ammonites, nautiloids have a protoconch (an internal chamber), a chambered phragmocone, and a body chamber. In contrast, their shells are usually smooth or just moderately ornamented, with simple chamber walls that usually bend backward. In the Paleozoic there were some nautiloids that were straight or loosely coiled, and some of these achieved giant size. During the Mesozoic, on the other hand, flat or inflated planispiral forms predominated. Today, this once very diverse lineage is represented by just a few species, most classified in the genus *Nautilus.* These amazing animals have 90 arms, inhabit the Indo-Pacific region and were well known to ancient peoples. Indeed, the shell of *Nautilus* follows a logarithmic spiral and has inspired artists and craftsmen for thousands of years.

The cephalopods are able to swim, but not very well. They have, however, perfected the art of buoyancy. They are scavengers: modern *Nautilus,* apparently, likes very much to eat chicken meat (used as bait by some Native peoples)! Nautiloids first appear in the fossil record in the Cambrian, and the group as a whole remained diverse throughout the Paleozoic; it is generally accepted that the large tube-like forms that flourished at that time lived near the seafloor. Nautiloids almost went extinct at the Triassic-Jurassic boundary, and from that time on have played a more subordinate role in marine ecosystems; they also are of minor geologic importance. Indeed, the evolutionary rate of nautiloids is thought to be much less than that of ammonites; nevertheless, some Paleozoic species do have index value. The extant *Nautilus* and *Allonautilus* are regarded as living fossils because they have changed little over the course of geological time.

Oolith. These are small, generally calcareous, cemented grains composed of concentric layers that form in constantly agitated shallow marine waters.

Ostracod. Ostracods are tiny aquatic—and mainly marine—arthropod crustaceans. The bodies of these small invertebrates are reduced in comparison with other crustaceans, and they have only 5 to 7 pairs of appendages. Of particular interest is their miniature, paired, and bivalve-like shell that is made up of chitinous and calcareous material and protects their flattened soft body. Ostracod shells are typically small, around 1 millimeter in diameter, but can reach up to 3 centimeters; generally only their valves are preserved in the fossil record. Ostracods live in the water column, on the seafloor, or in the upper part of soft sediments. The group contains representatives with a wide variety of diets, but most are scavengers. Ostracods first appear in the fossil record in the Cambrian and first move into fresh water in the Carboniferous. Today these invertebrates are found in almost all aquatic environments, including places that are wet only temporarily. Fossil ostracods are excellent indicators of ancient environments, as different lineages do not appear to have changed their habitats throughout geological time.

Palynology. Palynology is a branch of science devoted to the study of recent and fossil palynomorphs, spores, and pollen. Today this branch also includes the study of other kinds of microfossils that possess an organic skeleton and includes forms like dinoflagellate cysts. Diatoms, radiolarians, and forams are, however, excluded from palynology because they possess inorganic, calcareous, or siliceous skeletons.

Plankton. This general term is used for any drifting, floating, and typically small sized organisms (including bacteria, plants, and animals) that are found in the water column, typically in the ocean. These are all creatures that have a planktonic mode of life—in contrast to the nekton, which are animals that can actively swim and are in control of their position in the water column.

Platform. An extended, flat region in a shallow marine environment, often including reefs, where intensive in situ carbonate formation can occur.

Radiolaria. Radiolarians are extant, marine, and pelagic unicellular microscopic organisms that have siliceous or strontium sulfate skeletons. For a long time it was thought that this group is characterized by a low evolutionary rate, but as scanning electron microscopy became more widely used, it became clear that very tiny elements of the radiolarian skeleton are constantly changing. These differences imply different species and have allowed a precise biostratigraphy to be developed based on their variations. Radiolarian skeletons range in size from a few tenths of a millimeter to a few millimeters and have very diverse shapes. Some are built up from concentric spheres ornamented with spines that resemble bones carved into weird shapes (often called Chinese devil's work). These bizarre forms generally belong to the suborder Spumellaria, whereas the other big radiolarian suborder (Nassellaria) is represented mainly by monoaxial and bilateral forms. Some of these forms look like tiny spacecraft and still others have even stranger shapes; some resemble eagles. Radiolarians first appear in the fossil record in the earliest Cambrian and even today comprise a considerable fraction of the plankton. It is estimated, however, that up to 90 percent of all radiolarian species are extinct.

Siltstone (Aleurolite). This is a fine-grained siliciclastic sedimentary rock. The grain size of aleurolite is between $^1/_{16}$ and $^1/_{256}$ millimeters. Siltstone is deposited underwater, from mud, and most grains are composed of either quartz or feldspar.

Storm Deposit (Tempestit). This term refers to sediments deposited by a storm, often on the continental slope. Waves generated by strong winds often agitate and redeposit these unconsolidated sediments and such storm deposits are often characterized by massive accumulations of bivalve, gastropod, or brachiopod shells.

Stratigraphy. This is the study of layered or stratified rocks, with special focus on their duration and position within the geological time scale. Subdisciplines of stratigraphy include lithostratigraphy, which is based on the lithological features of rocks; biostratigraphy, which can be based on fossil contents; and chronostratigraphy, which refers explicitly to the age of rocks. The cross correlation of distant sedimentary successions is also an important stratigraphic issue. (See also Global Stratotype Section and Point [GSSP].)

Stratotype. This term refers to a section that has been carefully selected by an authorized group of scientists and that is supposed to provide the best representation of the boundary between two chronostratigraphic units, such as stages. Before a stratotype can be selected, it is very important that a detailed documentation of the section in question is provided. Any chosen stratotype will serve as a basis for future study and as a reference.

Stromatolite. A stromatolite is a fossil of microbial origin, generally formed as a result of the activity of blue-green algae. Algae or cyanobacteria can trap fine-grained sediment and form a subtly laminated sedimentary structure. As sediment is trapped and cemented, layers or domes are formed that have a very characteristic shape. Indeed, stromatolites are among the first organic fossils to be found in the rock record, and were much more abundant during the Precambrian than they are today. For a long time it was believed that stromatolites exclusively form in shallow marine environments, but it is now clear that they can also form deep in the ocean.

Taxon. This refers to any systematic unit, named or not, that is part of a hierarchy based on the conception of Carl Linnaeus (1707–1778). A taxon can be assigned a rank (species, genus, family, order, class, or phylum) and can refer to higher and higher groups of organisms, depending on usage. A correctly assigned taxon is necessarily a group of organisms (plants or animals) that have the same origin (in other words, they are monophyletic: they all belong to the same lineage and share a single common ancestor). However, many named taxa are not very well known and can comprise organisms of differing origins (in other words, polyphyletic). A researcher who is handling taxa is called a taxonomist, and the science itself is known as taxonomy.

Trilobite. A very important but extinct marine group of arthropods. Generally, only the external skeletons—composed of calcite, calcium phosphate minerals, and chitin—of trilobites are preserved in the fossil record. Trilobites are so named because of their three longitudinal portions, or lobes: left, right, and middle. Otherwise the bodies of these arthropods are divided into three major sections: the cephalon or head, the thorax, and the pygidium (the rear end of the animal). Trilobites generally ranged from between 5 and 30 centimeters in length; however, the smallest species were about 5 millimeters and the largest were around 80 centimeters. These arthropods were all marine and had very different lifestyles. Some had a very complicated eye, whereas others were blind. Many were likely active predators or scavengers and moved freely on the seafloor; others lived inside the sediment and were filter feeders. Some trilobites even swam around and fed on plankton. Trilobites appear in the fossil record at the beginning of the Cambrian (this was also the time of their highest diversity), and while their diversity decreased during the Ordovician, their end-Permian extinction was preceded by a long decay.

MAPS

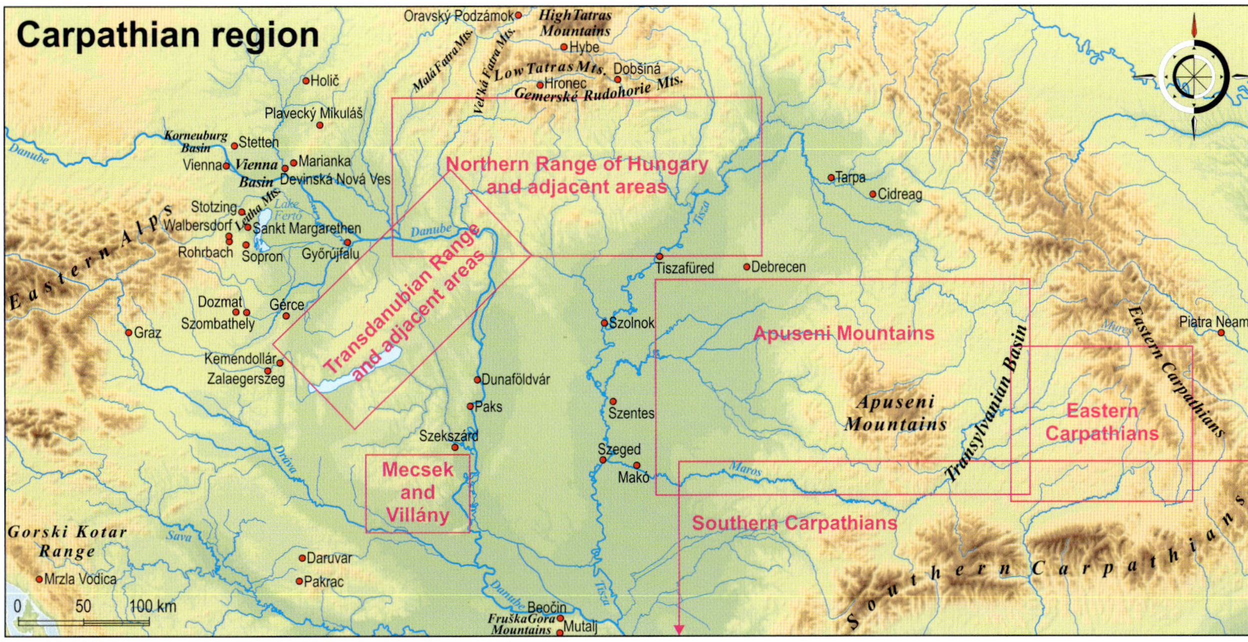

Map 1. Fossil localities from the Carpathian Region that are mentioned in this volume. (This map does not include localities that are illustrated on the more detailed maps.)

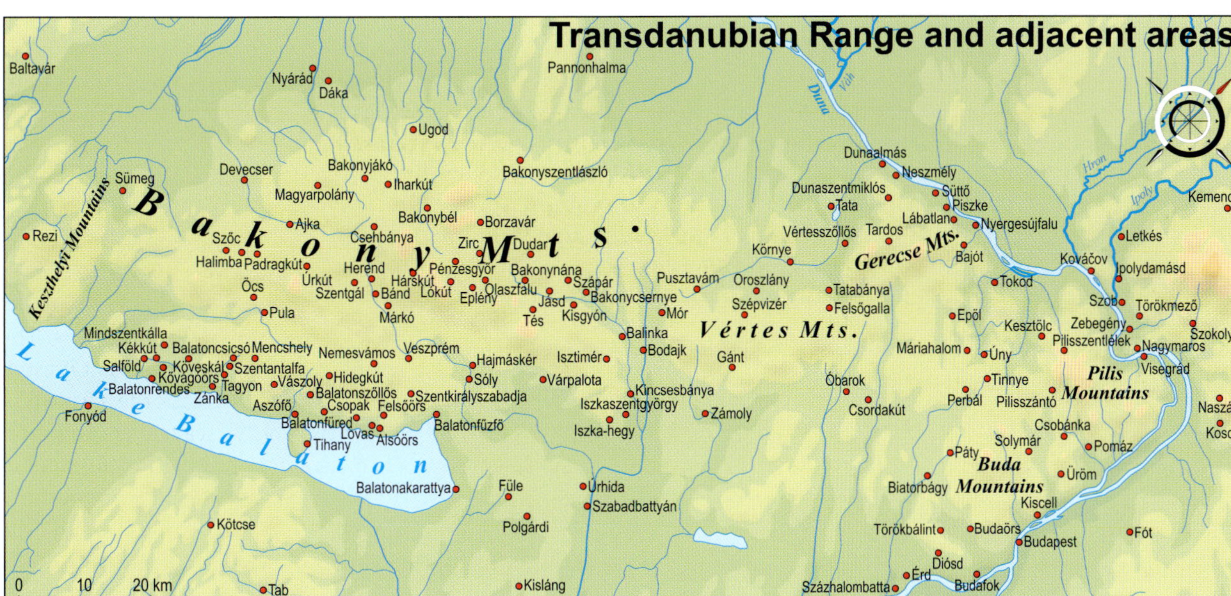

Map 2. Fossil localities from the Transdanubian Range and surroundings that are mentioned in this volume.

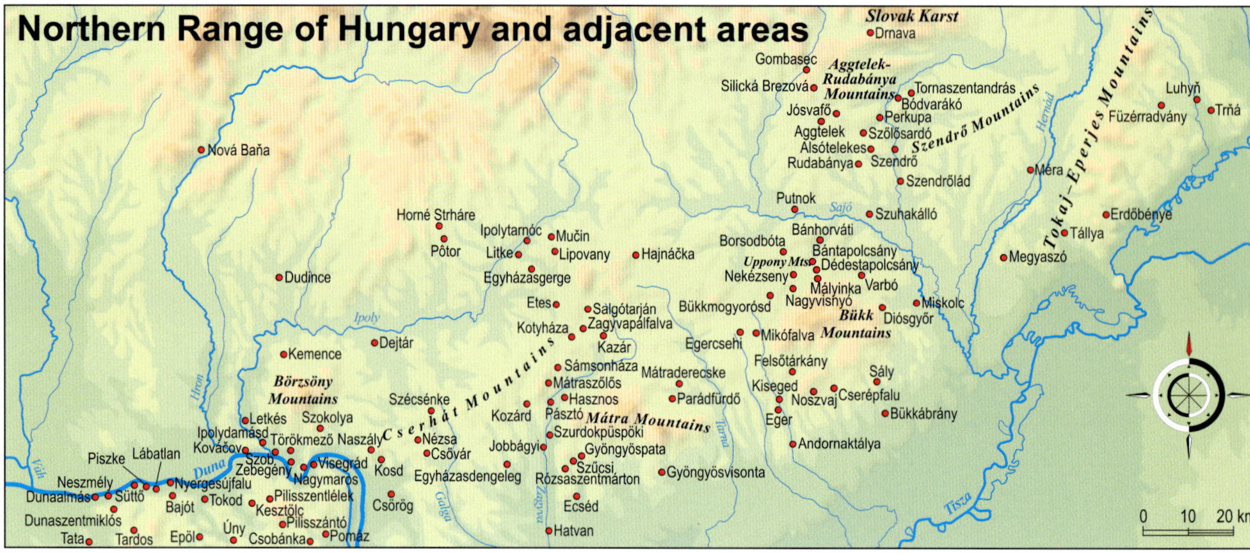

Map 3. Fossil localities from the northern Hungarian Range and surroundings that are mentioned in this volume.

Map 4. Fossil localities from the Mecsek and Villány Mountains and surroundings that are mentioned in this volume.

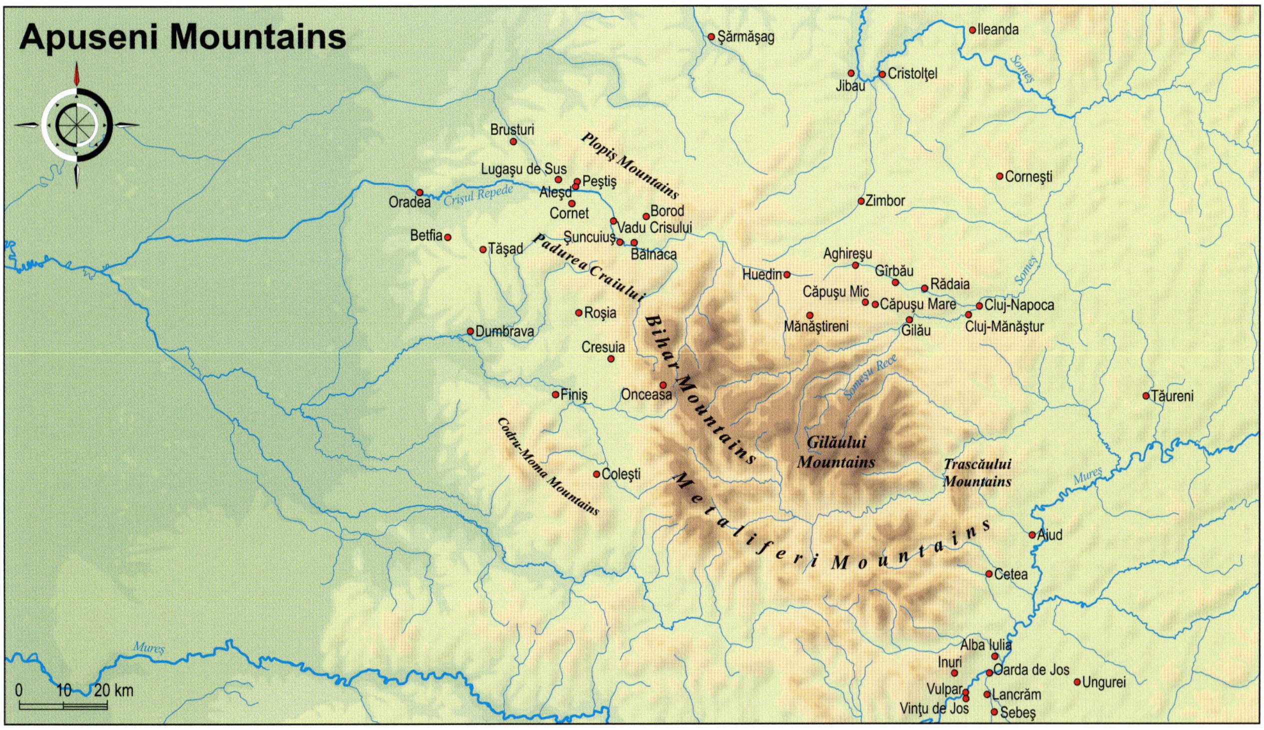

Map 5. Fossil localities from the Apuseni Mountains and surroundings that are mentioned in this volume.

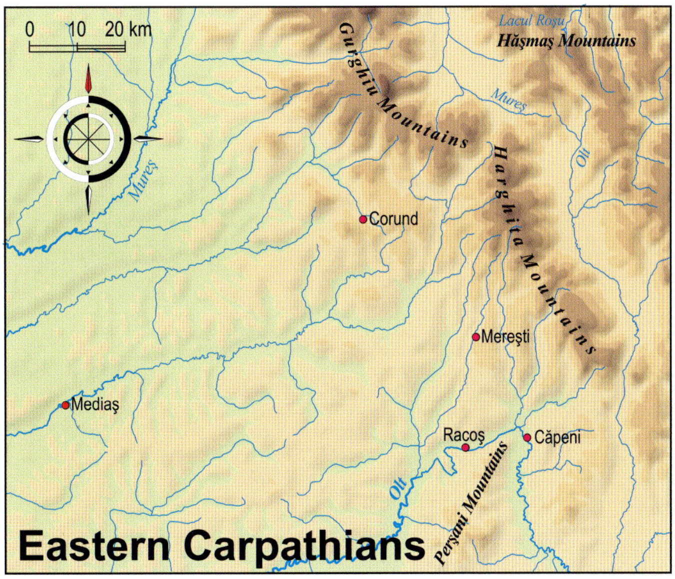

Map 6. Fossil localities from the Eastern Carpathians and surroundings that are mentioned in this volume.

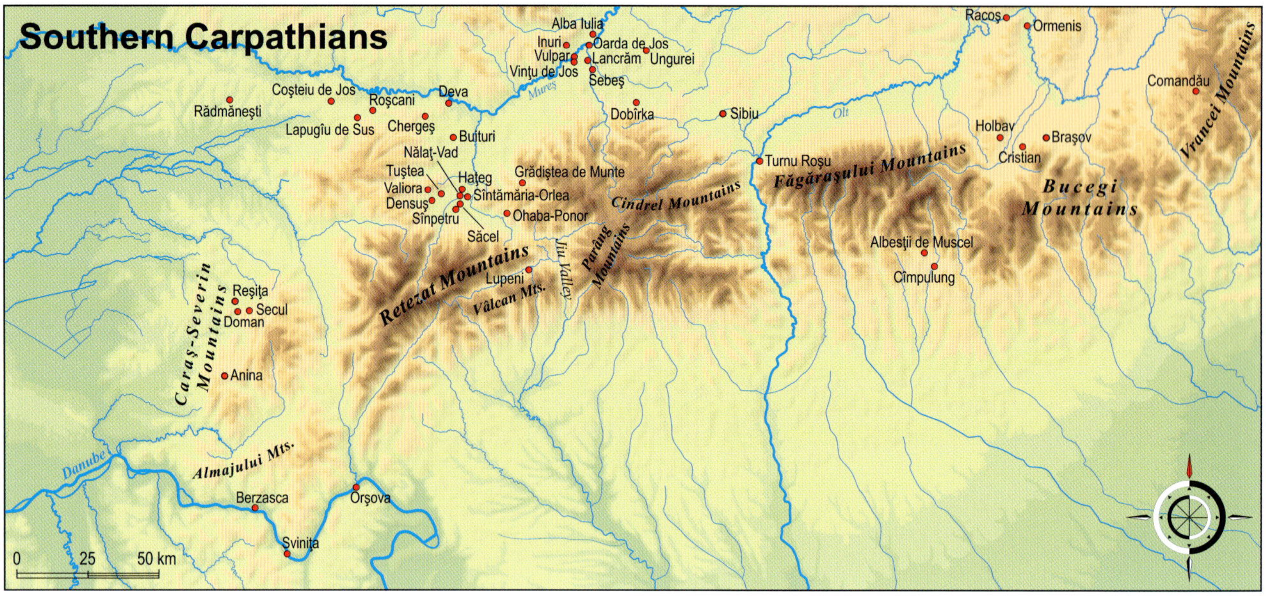

Map 7. Fossil localities from the Southern Carpathians and surroundings that are mentioned in this volume.

REFERENCES

A detailed study on the history of paleontological research carried out in the Carpathian region up to the end of First World War I, as well as on the history of Hungarian paleontology up the 1980s, can be found in Géczy (1994). Lists of fossils recorded from the whole territory of present-day Hungary up to the time of publication as well as references to the relevant literature are given in the explanatory booklets issued in the 1960s and 1970s, associated with the geological map series of scale 1:200,000, by the Hungarian Geological Institute. Paleontological papers considered to be the most up-to-date and comprehensive in the mid-1980s are listed, according to the respective lithostratigraphic units, in the bibliographic guide compiled by Kázmér (1986). Descriptions of several fossil localities in Hungary and Transylvania are given in the field guidebook edited by J. Pálfy and Pazonyi (2007). The most frequent Cenozoic fossil animal remains found in Hungary are reviewed in Monostori (2006). Boda (1964) published the catalog of type specimens of fossil animal species described from Hungary. A catalog of invertebrate and vertebrate paleontological type specimens, predominantly from the Carpathian region and housed in the Hungarian Natural History Museum, was compiled by J. Pálfy, Dulai, et al. (2008). A large amount of data on the fossils recorded from the Carpathian region can be found in the textbooks of Galácz and Monostori (1992) and Géczy (1972, 1984, 1993a, 1993b).

Abel, O. (1935). Vorzeitliche Lebensspuren.—Gustav Fischer, Jena, 644 pp.

Ager, D. V., & Callomon, J. H. (1971). On the Liassic age of the "Bathonian" of Villány (Baranya).—Annales Univ. Sci. Budapest., Sect. Geol., 14: 5–16.

Albani, R., Lelkesné Felvári, Gy., & Tongiorgi, M. (1985). First record of Ordovician (Upper Arenigian, Acritarchs) beds in Bakony Mts., Hungary.—Neues Jb. Geol. Paläont., Abh., 170(1): 45–65.

Andreánszky, G. (1948). Baumstämme aus der Unteren Kreidezeit.—Földtani Közlöny, 79(5–8): 243–252.

Andreánszky, G. (1959). Die Flora der sarmatischen Stufe in Ungarn.—Akadémiai Kiadó, Budapest, 360 pp.

Andreánszky, G. (1960). Calamites-Rest vom Bánya-hegy bei Füle (Westungarn).—Acta Universitatis Szegediensis: Sectio Scientiarum Naturalium: Acta Biologica Szeged, Nova Series, 6(1–4): 7–8.

Anonymous. (1909). Vezető a Magyar Királyi Földtani Intézet múzeumában [Guide to the Museum of the Royal Hungarian Geological Institute].—A Magyar Királyi Földtani Intézet Népszerű Kiadványai 1: 316 pp. (In Hungarian)

Aradi, V. (1905). Liász és dogger a budai hegységben. [Liassic and Dogger in the Buda Hills].—Földtani Közlöny, 25(2–3): 79–83. (In Hungarian)

Arthaber, G. von. (1911). Neue Funde in den Werfener Schichten und im Muschelkalke des südlichen Bakony und Revision der Cephalopodenfauna des Muschelkalkes.—Resultate Wiss. Erforsch. Balatonsees, Paläont. Anhang, 3(2): 26 pp.

Bácskay, E. (1991). The Coquand Collection. In Gy. Vitális & T. Kecskeméti (Eds.), Museums and Collections in the History of Mineralogy, Geology and Paleontology in Hungary. Annals of the History of Hungarian Geology Special Issue, 3 (pp. 197–204).—Hungarian Geological Society, Budapest.

Báldi, T. (1959). Paläoökologische Fazies-Analyse der burdigal-helvetischen Schichtreiche von Budafok in der Umgebung Budapest.—Annales Univ. Sci. Budapest., Sect. Geol., 2: 21–38.

Báldi, T. (1963). Das Alter der "Pectuculussande" von Törökbálint und die Frage der Oligozän-Miozän-Grenze).—Földtani Közlöny, 93(3): 204–216. (In Hungarian with German abstract)

Báldi, T. (1966). Die Oberoligozäne Molluskenfauna von Eger und die Neuuntersuchung der Schichtenfolge.—Annales Hist-Nat. Mus. Nat. Hung., 58: 99–101.

Báldi, T. (1971). Das untere Miozän in Ungarn.—Földtani Közlöny, 101: 85–90. (In Hungarian with German abstract)

Báldi, T. (1973). Mollusc fauna of the Hungarian Upper Oligocene (Egerian). Studies in Stratigraphy, Palaeoecology, Palaeogeography and Systematics.—Akadémiai Kiadó, Budapest, 512 pp.

Báldi, T. (1980). The Early History of the Paratethys.—Földtani Közlöny, 110(3–4): 456–472.

Báldi, T. (1986). Mid-tertiary stratigraphy and paleogeographic evolution of Hungary.—Akadémiai Kiadó, Budapest, 201 pp.

Báldi, T. (2003). Egy geológus barangolásai Magyarországon. [A geologist's walks in Hungary].—Vince Kiadó, Budapest, 198 pp. (In Hungarian)

Báldi, T., Báldiné Beke, M., Horváth, M., Kecskeméti, T., Monostori, M., & Nagymarosi, A. (1976). Alter und Bildungsverhältnisse des Hárshegyer Sandsteins.—Földtani Közlöny, 106(4): 353–386. (In Hungarian with German abstract)

Báldi, T., & Leél-Őssy, Sz. (2003). The palaeoecology of the Eggenburgian biofacies of Hungary: Silty facies: Szécsény Schlier.—Földtani Közlöny, 133(4): 501–514. (In Hungarian with English abstract)

Báldiné Beke, M. (1965). The genus Nannoconus (Protozoa, inc. sedis) in Hungary.—Geol. Hung. Ser. Pal., 30(2): 107–179.

Báldiné Beke, M. (1972). The nannoplankton of the Upper Eocene Bryozoan and Buda Marls.—Acta Geologica Hungarica, 16: 211–229.

Báldiné Beke, M. (1977). Stratigraphical and faciological subdivisions of the Oligocene as based on nannoplankton.—Földtani Közlöny, 107(1): 59–89. (In Hungarian with English abstract)

Báldiné Beke, M. (1984). The nannoplankton of the Transdanubian Paleogene formations (Hungary).—Geol. Hung. Ser. Pal., 43: 307 pp.

Báldiné Beke, M., & Nagymarosy, A. (1986). A nannoplankton: Előnyei, hátrányai, alkalmazási lehetőségei a biosztratigráfiában [The nannoplankton: Its advantages, disadvantages, as well as potentialities of its use in biostratigraphy].—Őslénytani Viták, 32: 59–76. (In Hungarian with English abstract)

Balogh, K. (1940). Daten zum Geologischen Kenntnis der Umgebung von Pelsöcardó.—Tisia, 4: 149–200.

Balogh, K. (1964). Die geologischen Bildungen des Bükk-Gebirges.—A Magyar Állami Földtani Intézet Évkönyve, 48(2): 557–713.

Balogh, K. (1970). Dr. Schréter Zoltán emlékezete. [In memoriam Zoltán Schréter].—Földtani Közlöny, 100(3): 238–242. (In Hungarian)

Balogh, K. (1976). Pelecypods from the Late Triassic of the South Gemerikum I.—Acta Min. Petr., 22(2): 285–296.

Balogh, K., & Kovács, S. (1976). Sphinctozoa from the Wetterstein reef limestone of Alsóhegy Mount (South Gemericum, West Carpathians, Northern Hungary).—Acta Min. Petr., 22(2): 297–310.

Bandel, K., & Riedel, F. (1994). The Late Cretaceous gastropod fauna from Ajka (Hungary, Bakony Mountains): A revision.—Ann. Naturhist. Mus. Wien A, 96: 1–65.

Barabás, A. (1995). Bathonian Radiolaria from the Mecsek Mts (South Hungary).—Annales Univ. Sci. Budapest., Sect. Geol., 30: 83–92, 219–220.

Barabásné Stuhl, Á. (1981). Microflora of the Permian and Lower Triassic sediments of the Mecsek Mts (South Hungary).—Acta Geologica Hungarica, 24(1): 49–97.

Barbacka, M. (1994a). Pachypteris banatica (Humml) Doludenko from Liassic locality in Mecsek Mountains, Hungary.—Acta Palaeobotanica, 34(1): 5–19.

Barbacka, M. (1994b). Komlopteris Barbacka, gen. nov., a segregate from Pachypteris Brongniart.—Review of Palaeobotany and Palynology, 83: 339–349.

Barbacka, M. (2000). Bennettitales from the Mecsek Mountains Liassic, Hungary.—Acta Palaeobotanica, 40(2): 111–127.

Barbacka, M., & Konijnenburg-van Cittert, J. H. A. (1998). Sun and shade leaves in two Jurassic species of Pteridosperms.—Review of Palaeobotany and Palynology, 103: 209–221.

Bárdossy, Gy., Dercourt J., & Durand-Delga, M. (2000). Role of French geologists in the development of geology during the 19th century in Central Europe, particularly in Hungary.—Comptes Rendus de l'Academie des Sciences, Series II, A, Earth and Planetary Science, 330(8): 585–594.

Bartha, A. (1992). Upper Oligocene Echinoidea from Buda Hills, Hungary.—Annales Univ. Sci. Budapest., Sect. Geol., 29: 190–216.

Bartha, F. (1962). Examen biostratigraphique du complexe houillier du Crétacé supérieur de la partie meridionale de la montagne Bakony.—Acta Geol. Acad. Sci. Hung., 7(3–4): 359–398.

Bartha, F., Kleb, B., Kőrössy, L., Szabóné Kilényi, É., Szatmári, P., Széles, M., Szénás, Gy., & Tóth, K. (1971). A magyarországi pannonkori képződmények kutatásai [Investigations of Pannonian formations in Hungary].—Akadémiai Kiadó, Budapest, 361 pp. (In Hungarian)

Bartucz, L., Dancza, J., Hollendonner, F., Kadić, O., Mottl, M., Pataki, V., Pálosi, E., Szabó, J., & Vendl, A. (1939). Die Mussolini-Höhle (Subalyuk) bei Cserépfalu.—Geol. Hung. Ser. Pal., 14: 348 pp.

Bather, F. (1911). Triassic echinoderms of Bakony.—Resultate Wiss. Erforsch. Balatonsees, Paläont. Anhang, 1(6): 288 pp.

Benkőné Czabalay, L. (1964a). Examen malacologique du Crétacé supérieur du Sümeg (Montagne Bakony).—A Magyar Állami Földtani Intézet Évi Jelentése 1961-ről: 263–289. (In Hungarian with French abstract)

Benkőné Czabalay, L. (1964b). Die obersenone Gastropodenfauna von Sümeg im südlichen Bakony.—Sitzungsberichte der Österreichischen Akademie der Wissenschaften, Mathematischnaturwissenschaftliche Klasse, Abteilung 1, 173(3–4): 155–188.

Benkőné Czabalay, L. (1965). Les Gastéropodes de l'Aptien et du Cénomanien de la Montagne Bakony (Massif Central hongrois).—Geol. Hung. Ser. Pal. 31: 181–291.

Benton, M. J., Cook, E., Grigorescu, D., Popa, E., & Tallódi, E. (1997). Dinosaurs and other tetrapods in an Early Cretaceous bauxite-filled fissure, northwestern Romania.—Palaeogeography, Palaeoclimatology, Palaeoecology, 130: 275–292.

Bércziné Makk, A. (1969). Die obertriadischen Ammonoiden des Fazekas Berges.—Földtani Közlöny, 99(4): 351–367. (In Hungarian with German abstract)

Bércziné Makk, A. (1970). Kampiler Ablagerungen des Iszka Berges.—Földtani Közlöny, 100: 160–172. (In Hungarian with German abstract)

Bércziné Makk, A. (1999). Foraminiferal stratigraphy of Jurassic beds in Bükkzsérc, North Hungary.—Földtani Közlöny, 129(3): 363–392. (In Hungarian with English abstract)

Bernor, R. L., Feibel, C., & Viranta, S. (2003). The vertebrata locality of Hatvan, Late Miocene (Middle Turolian, MN12), Hungary. In A. Petculescu, C. Rădulescu, E. Ştiucă, & P. M. Samson (Eds.), Advances in Vertebrate Paleontology: Hen to Panta (pp. 105–112).—Romanian Academy "Emil Racovita," Institute of Speleology, Bucharest.

Bernor, R. L., Kordos, L., & Rook, L. (Eds.). (2005). Multidisciplinary research at Rudabánya.—Palaeontographia Italica, 90: 1–313.

Beudant, F. S. (1822). Voyage minéralogique et géologique en Hongrie, pendant l'année 1818.—Librarie de Verdière, Paris, 3 vols.

Bitner, M. A., & Dulai, A. (2004). Revision of Miocene brachiopods of the Hungarian Natural History Museum, with special reference to the Meznerics collection.—Fragmenta Palaeontologica Hungarica, 22: 69–82.

Bitner, M. A., & Dulai, A. (2008). Eocene micromorphic brachiopods from north-western Hungary.—Geologica Carpathica, 59(1): 31–43.

Bitner, M. A., Dulai, A., & Galácz, A. (2010). Middle Eocene brachiopods from the Szőc Limestone Formation (Bakony Mountains, Hungary), with a description of a new genus.—N. Jb. Geol. Paläont. Abh., 259(1): 113–128.

Bitoianu, C. (1988). Macroflore Stephanienne de Lupac.—Studia Univ. Babeş-Bolyai, 33(2): 51–63.

Bittner, A. (1912a). Lamellibranchiaten aus dem Trias der Bakonyer Waldes.—Resultate Wiss. Erforsch. Balatonsees, Paläont. Anhang, 2(3): 107 pp.

Bittner, A. (1912b). Brachiopoden aus der Trias des Bakonyer Waldes.—Resultate Wiss. Erforsch. Balatonsees, Paläont. Anhang, 2(1): 60 pp.

Böckh, H. (1899). Orca Semseyi, eine neue Orca-Art aus dem unteren Miozän von Salgó-Tarján.—Mitt. Jb. Kön. Ung. Geol. Anst., 13(3): 106–107.

Böckh, J. (1873). Die geologischen Verhältnisse des südlichen Theiles des Bakony I.—Mitt. Jb. Kön. Ung. Geol. Anst., 2(1): 28–180.

Böckh, J. (1874). Die geologischen Verhältnisse des südlichen Theiles des Bakony II.—Mitt. Jb. Kön. Ung. Geol. Anst., 3(1): 1–180.

Böckh, J. (1881a). Geologische und Wasser-Verhältnisse der Umgebung der Stadt Fünfkirchen.—Mitt. Jb. Kön. Ung. Geol. Anst., 4(4): 152–328.

Böckh, J. (1881b). Brachydiastematherium transilvanicum Bkh. et Maty.—Mitt. Jb. Kön. Ung. Geol. Anst., 4(3): 126–150.

Böckh, J. (1881c). Adatok a Mecsekhegység és dombvidéke jurakorabeli lerakódásainak ismeretéhez II. Paleontológiai rész [Contributions to the knowledge of the Jurassic in the Mecsek Mountains and the surrounding hilly region. II. Paleontological part].—Értekezések a Természettudományok Köréből, 11(9): 1–106. (In Hungarian)

Boda, J. (1959). Das Sarmat in Ungarn und seine Invertebraten-Fauna.—A Magyar Állami Földtani Intézet Évkönyve, 47(3): 569–862.

Boda, J. (1964). Catalogus Originalium Fossilium Hungariae.—Magyar Állami Földtani Intézet, Budapest, 229 pp.

Boda, J. (1970). A magyarországi szarmata ritka ősmaradványai [Rare fossils from the Sarmatian of Hungary].—Földtani Közlöny, 100: 397–398. (In Hungarian)

Boda, J. (1971). Gliederung des Sarmats von Ungarn auf Grund der Invertebraten-Faunen).—Földtani Közlöny, 101: 107–113. (In Hungarian with German abstract)

Boda, J. (1974). Faziostratotypus Sőreg. In A. Papp, F. Marinescu, & J. Seneš (Eds.), Chronostratigraphie und Neostratotypen, Miozän der Zentralen Paratethys M_5 Sarmatien (Vol. 4, pp. 199–202)—Slowakische Akademie de Wissenschaften, Bratislava.

Boda, J. (1979). Nubecularia-félék (Foraminifera) kőzetalkotó mennyiségben a hazai szarmatában [Sediment-forming Nubecularia (Foraminifera) in the Sarmatian of Hungary].—Földtani Közlöny, 109: 288–293. (In Hungarian)

Bodó, K. (1992). Study of Late Eocene bivalves from Buda Hills.—Annales Univ. Sci. Budapest., Sect. Geol., 29: 217–235.

Bodrogi, I. (1989). Planktonic Foraminifera stratigraphy of the Pénzeskút Marl Formation. (Middle Cretaceous).—A MÁFI Évkönyve, 63(5): 127 pp. (In Hungarian, with English abstract)

Bódy, M. (2000). A Mecsek-hegység dogger krinoideái [Middle Jurassic crinoids of the Mecsek Mountains]. In Program, Előadáskivonatok, Kirándulásvezető, 3: Magyar Őslénytani Vándorgyűlés 2000, május 5–6, Tihany (p. 5). (In Hungarian)

Bogsch, L. (1928). Die Hipparionen der Báraczháza-Höhlung bei Csákvár (Kom. Fejér).—Földtani Közlöny, 58: 223–229.

Bogsch, L. (1936). Tortonische Fauna von Nógrádszakál.—A Magyar Királyi Földtani Intézet Évkönyve, 31(1): 144 pp.

Bogsch, L. (1939). Nachruf auf Dr. Andreas Kutassy.—Földtani Közlöny, 69(1–3): 1–10. (In Hungarian with German abstract)

Bogsch, L. (1943). Tortonische Fauna von sandiger Fazies aus der Umgebung des Szentkuter-Klosters bei Mátraverebély (Komitat Nógrad).—A Magyar Királyi Földtani Intézet Évkönyve, 35(4): 194–297.

Bogsch, L. (1950). Triászbeli daonellás rétegek az Alföld medencealjzatában [Triassic Daonella-bearing beds in the basement of the Great Hungarian Plain].—Földtani Közlöny, 80(2): 189–191. (In Hungarian)

Bogsch, L. (1977). Csepreghyné Meznerics Ilona emlékezete, 1906–1977 [In memoriam Ilona Csepreghyné Meznerics, 1906–1977].—Földtani Közlöny, 107(3–4): 275–281. (In Hungarian)

Böhm, B. (1941). Die fossilen Fische von Kovászna und Kommandó in Siebenbürgen.—A Magyar Királyi Földtani Intézet Évkönyve, 35(5): 179–203.

Böhm, B. (1942). Beiträge zur tertiären Fischfauna Ungarns.—Geol. Hung. Ser. Pal., 19: 6–42.

Bohn, P. (1966). Senonemys sümegensis nov. gen., nov. sp.: Ein neuer Schildkröten fund aus Ungarn.—Földtani Közlöny, 96(1): 111–118.

Bohnné Havas, M. (1971). Oncophorák (Rzehakiák) a kazár-gyulakeszi (Nógrád m.) alapszelvényben [Oncophora (Rzehakia) in the Kazár-Gyulakeszi section (Nógrád County)].—A Magyar Állami Földtani Intézet Évi Jelentése 1968-ról: 131–144. (In Hungarian)

Bohnné Havas, M. (1973). Tortonische Molluskenfauna des östlichen Mecsek-Gebirges.—A Magyar Állami Földtani Intézet Évkönyve, 53(4): 948–1161.

Bohnné Havas, M. (1985). A study of Ottnangian molluscs from the Eastern Borsod Basin (N Hungary).—Geol. Hung. Ser. Pal., 48: 97–177.

Bóna, J. (1976). Triadische Conodonten aus dem Villányer Gebirge.—Geol. Hung. Ser. Geol., 17: 229–252.

Bóna, J. (1995). Palynostratigraphy of the Upper Triassic formations in the Mecsek Mts (Southern Hungary).—Acta Geologica Hungarica, 38(4): 319–354.

Bosnakoff, M. (2001). Investigations of the Badenian otoliths of two localities (Ipolydamásd, Börzsöny Mountains and Bükkmogyorósd, Bükk Mountains): A comparison.—Földtani Közlöny, 131: 197–220. (In Hungarian with English abstract)

Botfalvai, G. (2004). Késő-kréta teknősök az iharkúti gerinces lelőhelyről [Late Cretaceous tortoises from the vertebrate locality Iharkút]. In Program, Előadáskivonatok,

Kirándulásvezető, 7: Magyar Őslénytani Vándorgyűlés, 2004, május 6–8, Beremend (pp. 5–6). (In Hungarian)

Botfalvai, G., & Rabi, M. (2006). A korai Bothremydidaek (Chelonia) és Alligatoroideak (Crocodylia) paleobiogeográfiája az új késő-kréta bakonyi fosszíliák ismeretében [Paleobiogeography of early Bothremydids (Chelonia) in the light of Upper Cretaceous specimens found recently in the Bakony Mountains]. In Program, Előadáskivonatok, Kirándulásvezető, 9: Magyar Őslénytani Vándorgyűlés, 2006, május 18–20, Ajka (pp. 7–8). (In Hungarian)

Botfalvai, G., & Rabi, M. (2007). Előzetes taxonómiai és paleoökológiai vizsgálatok a máriahalmi felső-oligocén (egri) feltárás (Mány-Zsámbéki-medence) gerinces faunáján [Preliminary studies on the vertebrate fauna of the Upper Oligocene (Egerian) exposure at Máriahalom (Mány-Zsámbék Basin]. In Program, Előadáskivonatok, Kirándulásvezető, 10: Magyar Őslénytani Vándorgyűlés, 2007, május 24–26, Budapest (pp. 12–13). (In Hungarian)

Brassói Fuchs, H. (1959). Ein Sirenenfund aus Siebenbürgen.—Földtani Közlöny, 89(3): 326–328. (In Hungarian with German abstract)

Brassói Fuchs, H. (1990). Einige Angaben zur Kenntsnis von fossilen Sirenen (VII.) Sirenenzähne aus der Umgebung von Cluj (Klausenburg, Romänien).—Földtani Közlöny, 120(1): 89–92.

Brassói Fuchs, H. (1994a). Contributions to the knowledge of the species "Euclastes" kochi Lőrenthey, 1903 (Testudinata, Cheloniidae?) described from the Upper Eocene beds of Cluj, Romania.—Földtani Közlöny, 124(4): 479–482. (In Hungarian)

Brassói Fuchs, H. (1994b). The carapace fragment of superordo Trionychoidea sensu Gray 1873 from Eocene sediments of Cluj, Romania.—Földtani Közlöny, 124(4): 483–488. (In Hungarian)

Broglio Loriga, C., & Berti Cavicchi, A. (1972). Praeaplocoma hessi n. gen., n. sp., un'ofiura del Werfeniano (Trias Inferiore) del Gruppo della Costabella, Dolomiti. Memorie Geopaleontologiche Dell'Università di Ferrara, 2(1): 185–197.

Broglio Loriga, C., Góczán, F., Haas, J., Lenner, K., Neri, C., Oraveczné Scheffer, A., Posenato, R., Szabó, I., & Tóthné Makk, A. (1990). The Lower Triassic sequences of the Dolomites (Italy) and the Hungarian Mid-Mountains (Hungary) and their correlation.—Memoire di Scienze Geologiche, 42: 41–103.

Buffetaut, E., Grigorescu, D., & Csiki, Z. (2002). A new giant pterosaur with a robust skull from the latest Cretaceous of Romania.—Naturwissenschaften, 89: 180–184.

Bystrický, J. (1964). Slovenský kras: Stratigrafia a Dasycladaceae mezozoika Slovenského krasu [Slovak Karst: Stratigraphy and Mesozoic dasycladaceans of the Slovak Karst]. 204 pp. (In Slovakian)

Ciobanu, R. (2002). Selacienii paleogeni din România [Paleogene sharks of Romania].—Editura Universitati "Lucian Blaga," Sibiu, 216 pp. (In Romanian)

Codrea, V. (2000). Rinoceri şi tapiri tertiari din România [Rhinoceros and tapirs from the Tertiary of Romania].—Editura Presa Universitarâ Clujeanâ, 174 pp. (In Romanian)

Codrea, V., Vremir, M., Jipa, C., Godefroit, P., Csiki, Z., Smith, T., & Farkas, C. (2010). More than just Nopcsa's Transylvanian dinosaurs: A look outside the Haţeg Basin.—Palaeogeography, Palaeoclimatology, Palaeoecology, 293(3–4): 391–405.

Czabalay, L. (1971). Pachyodontes crétacés de la Montagne de Mecsek.—A Magyar Állami Földtani Intézet Évi Jelentése 1968-ról: 185–211.

Czabalay, L. (1982). La faune des rudistes des environs de Sümeg.—Geol. Hung. Ser. Pal., 41: 221 pp.

Czabalay, L. (1994). Korrelation der Molluskenfaunen des Urgon von Ungarn (Villány- und Mecsek-Gebirge) und Österreich (Vorarlberg). In H. Lobitzer, G. Császár, & A. Daurer (Eds.), Jubiläumsschrift 20 Jahre Geologische Zusammenarbeit Österreich-Ungarn (pp. 209–224).—Geologische Bundesanstalt, Wien.

Császár, G. (2002). Urgon formations in Hungary: With special reference to the Eastern Alps, the Western Carpathians and the Apuseni Mountains.—Geol. Hung. Ser. Geol., 25: 209 pp.

Császár, G., & Turnšek, D. (1996). Vestiges of atoll-like formations in the Lower Cretaceous of the Mecsek Mountains, Hungary.—Cretaceous Research, 17: 419–442.

Csepreghyné Meznerics, I. (1954) Helvetische und tortonische Fauna aus dem östlichen Cserhátgebirge.—A Magyar Állami Földtani Intézet Évkönyve, 41(4): 1–185.

Csepreghyné Meznerics, I. (1960). Pectinidés du neogène de la Hongrie et leur importance stratigraphique.—Mém. Soc. Geol. France (Nouvelle Série), 39: 1–56.

Csiki, Z. (1999). New evidence of armoured titanosaurids in the Late Cretaceous: Magyarosaurus dacus from the Haţeg Basin (Romania).—Oryctos, 2: 93–99.

Csiki, Z., & Grigorescu, D. (1998). Small theropods from the Late Cretaceous of the Haţeg Basin (Western Romania): An unexpected diversity at the top of the food chain—Oryctos, 1: 87–104.

Csiki, Z., & Grigorescu, D. (2001). Paleobiogeographic implications of the fossil mammals from the Maastrichtian of the Haţeg Basin, Romania.—Acta Palaeontologica Romaniae, 3: 89–95.

Csiki, Z., Vremir, M., Brusatte, S. L., & Norell, M. A. (2010). An aberrant island-dwelling theropod dinosaur from the Late Cretaceous of Romania—Proceedings of the National Academy of Sciences, USA, 107(35): 15357–15361.

Csíky, G. (1991). The Foundation and Progress of the Transylvanian Museum Association. In Gy. Vitális & T. Kecskeméti (Eds.), Museums and Collections in the History of Mineralogy, Geology and Paleontology in Hungary. Annals of the History of Hungarian Geology Special Issue, 3 (pp. 413–422).—Hungarian Geological Society, Budapest.

Dávid, Á. (1993). Trace fossils on molluscs from the Molluscan Clay (Late Oligocene, Egerian)—a comparison between two localities (Wind Brickyard, Eger and Nyárjas Hill, Novaj, NE Hungary).—Scripta Geol., Spec. Issue, 2: 75–82.

Dávid, Á. (2009). Bioeróziós és patológiás elváltozások az egerien Mollusca faunáján [Bioerosion and pathological phenomena in the tests of Egerian Age molluscs]. Disszertációk az Eszterházy Károly Főiskola Földrajz Tanszékéről, 3.—EKF Földrajz Tanszék, Eger, 230 pp. (In Hungarian)

De Wever, P. (1984). Triassic radiolarians from the Darno area (Hungary).—Acta Geologica Hungarica, 27(3–4): 295–307.

Deák, M. (1965). Recherches palynologiques des dépôts aptiens de la Montagne Centrale de Transdanubie.—Geol. Hung. Ser. Pal., 29: 105 pp.

Dénes, I., Zólya, L., Both, J., & Papucs, A. (2005). Protected geological values from Székelyland.—Földtani Közlöny, 135(2): 263–292. (In Hungarian with English abstract)

Detre, Cs. (1971a). Revision of the brachiopods from the Carboniferous of Szabadbattyán.—Őslénytani Viták, 18: 77–88. (In Hungarian with English abstract)

Detre, Cs. (1971b). A Villányi-hegységi triász makrofauna újravizsgálata. [Revision of the Triassic macrofauna of the Villány Mountains].—Unpublished doctoral dissertation, Eötvös Loránd University, Budapest, 192 pp. (In Hungarian)

Detre, Cs. (1973). Über den ersten, stratigraphisch auswertbaren Ammonoiden-Fund von bester Erhaltung der Mecseker Trias.—A Magyar Állami Földtani Intézet Évi Jelentése 1971-ről: 277–282. (In Hungarian with German abstract)

Detre, Cs. (1989). Upper Triassic Daonella limestone from the S margin of the Bükk plateau, NE Hungary.—A Magyar Állami Földtani Intézet Évi jelentése 1987-ről: 259–266 (In Hungarian with English abstract)

Detre, Cs. (1993). Carnian brachiopods of Hungary. In J. Pálfy & A. Vörös (Eds.), Mesozoic brachiopods of Alpine Europe (pp. 27–30).—Hungarian Geological Society, Budapest.

Detre, Cs., Dosztály, L., & Hermann, V. (1988). The Upper Norian (Sevatian) fauna of Csővár.—A Magyar Állami Földtani Intézet Évi Jelentése 1986-ról: 53–67. (In Hungarian with English abstract)

Detre, Cs., Lantos, M., & Ó-Kovács, L. (1992). Biofaciological, biochronological and biometrical investigations based on the Middle Triassic Coenothyris vulgaris (Schlotheim) found in Hungary.—A Magyar Állami Földtani Intézet Évi Jelentése 1990-ről: 395–461. (In Hungarian with English summary)

Detre, Cs., & Mihály, S. (1987). Two new ophiuroid finds from the Triassic of the Balaton Highland.—A Magyar Állami Földtani Intézet Évi Jelentése 1985-ről: 449–452. (In Hungarian with English abstract)

Detre, Cs., Szentes, I., & Szente, E. (1986). Coenothyris vulgaris (Schlotheim) coenoses from Hungary: a biometric and quantitative taxonomic evaluation.—A Magyar Állami Földtani Intézet Évi jelentése 1983-ról: 219–233. (In Hungarian with English abstract)

Di Bari, D., & Rettori, R. (1996). Morphological features of Triasina hantkeni Majzon, 1954 (Foraminiferida, Aulotortidae) and remarks on the test wall structure.—Revue de Micropaléontologie, 39(4): 305–313.

Diebel, K., & Pietrzeniuk, E. (1990). Pleistocene ostracods from Vértesszőlős. In M. Kretzoi & V. T. Dobosi (Eds.), Vértesszőlős: Site, Man and Culture (pp. 145–162).—Akadémiai Kiadó, Budapest.

Domning, D. P., & Pervesler, P. (2001). The osteology and relationships of Metaxytherium krahuletzi Depéret, 1895 (Mammalia: Sirenia).—Abhandlungen der Senckenbergischen Naturforschenden Gesellschaft, 553: 1–91.

Dosztály, L. (1989). Triassic radiolarians from Dallapuszta (Mount Darnó, N. Hungary).—A Magyar Állami Földtani Intézet Évi Jelentése az 1988. évről, 2: 193–201.

Dosztály, L. (1993). The Anisian/Ladinian and Ladinian/Carnian boundaries in the Balaton Highland based on radiolarians.—Acta Geologica Hungarica, 36(1): 59–72.

Dosztály, L. (1998). Jurassic radiolarites in the Transdanubian Central Range.—Földtani Közlöny, 128(2–3): 273–279.

Dudich, E. (1961). Ein Insektenfund aus der Lignite von Szentgál (Bakonygebirge).—Földtani Közlöny, 91: 20–31. (In Hungarian with German abstract)

Dulai, A. (1995). Middle Miocene (Badenian) lunulitiform Bryozoa from Szob (Börzsöny Mts., Hungary).—Annales Historico-Naturales Musei Nationalis Hungarici, 87: 35–54.

Dulai, A. (2002). Hettangian and Early Sinemurian (Early Jurassic) brachiopods of the Transdanubian Central Range (Hungary) I: Diversity, stratigraphic distribution, paleoecology, paleobiogeography, evolution.—A Bakony természettudományi kutatásának eredményei, 26: 1–112. (In Hungarian with English abstract)

Dulai, A. (2003). Hettangian and Early Sinemurian (Early Jurassic) brachiopods of the Transdanubian Central Range (Hungary) II: Systematic descriptions.—A Bakony Természettudományi Kutatásának Eredményei, 27: 5–144. (In Hungarian with English abstract)

Dulai, A. (2005). Badenian (Middle Miocene) Polyplacophora from the Central Paratethys (Bánd and Devecser, Bakony Mountains, Hungary).—Fragmenta Palaeontologica Hungarica, 23: 29–50.

Dulai, A. (2007). Badenian (Middle Miocene) brachiopods from Bánd and Deevecser (Bakony Mountains, Hungary).—Fragmenta Mineralogica Palaeontologica Hungarica, 24–25: 1–13.

Dunai, M. (1999). Egy különleges korallfaj a nagyvisnyói karbon agyagpalából [A peculiar coral species from the Carboniferous shale of Nagyvisnyó]. In Program, Előadáskivonatok, Kirándulásvezető, 2: Magyar Őslénytani Vándorgyűlés 1999, május 7–8, Noszvaj (p. 8). (In Hungarian)

Dyke, G. J., Benton, M. J., Posmoşanu, E., & Naish, D. W. (2011). Cretaceous (Berriasian) birds and pterosaurs from the Cornet bauxite mine, Romania.—Palaeontology, 54: 79–95.

Éhik, J. (1930). Prodinotherium hungaricum n. g. n. sp. Appendix by Tibor Szalai: On the geological occurrence of Prodinotherium hungaricum Éhik.—Geol. Hung. Ser. Pal., 6: 21 pp.

Erdei, B., Dolzych, L., & Hably, L. (2009). The buried Miocene forest at Bükkábrány, Hungary.—Review of Palaeobotany and Palynology, 155: 69–79.

Erdei, B., & Hír, J. (2002). Vegetation and climate reconstruction of Sarmatian (Middle Miocene) sites from NE and W

Hungary.—Acta Universitatis Carolinae Geologica, Praha, 46(4): 75–84.

Erős, Z. (1987). Pénzes Gyuri különös lábnyoma [The strange footprint of Monied Georgie].—Mai Magazin, 1987(9): 52. (In Hungarian)

Fejfar, O. (1961). Die plio-pleistozänen Wirbeltierfaunen von Hajnáčka und Ivanovce (Slowakei), ČSR, I. Die Fundumstände und Stratigraphie.—Neues Jb. Geol. Paläont., Abh., 111(3): 257–273.

Félix, J. (1887). Beiträge zur Kenntniss der fossilen Hölzer Ungarns.—Mitt. Jb. Kön. Ung. Geol. Anst., 8(5): 146–162.

Fenninger, A., & Nievoll, J. (1983). Der erste Nachweis einer phyllodonten Zahnplatte aus dem oberen Perm des Bükkgebirges (Ungarn).—A Magyar Állami Földtani Intézet Évi Jelentése 1981-ről: 201–207.

Flügel, E., Velledits, F., Senovbary-Daryan, B., & Riedel, P. (1992). Rifforganismen aus "Wettersteinkalken" (karn?) des Bükk-Gebirges, Ungarn.—Geologisch-Paläontologische Mitteilungen Innsbruck, 18: 35–62.

Fodor, R. (2001). Traces of bioerosion from test on Upper-Oligocene solitary corals (Wind Brickyard, Eger, NE Hungary).—Földtani Közlöny, 131(1–2): 179–196. (In Hungarian with English abstract)

Földvári, A. (1952). Lead ores and fossiliferous Dinantian (Lower Carboniferous) at Szabadbattyán.—Acta Geologica Hungarica, 1(1): 11–36.

Főzy, I. (1993a). Upper Jurassic ammonite biostratigraphy in the Gerecse and Pilis Mts. (Transdanubian Central Range, Hungary).—Földtani Közlöny, 123(4): 441–464.

Főzy, I. (1993b). Upper Jurassic ammonite biostratigraphy of the Mecsek Mts. southern Hungary.—Földtani Közlöny, 123(2): 195–205.

Főzy, I. (2001). Campanian (Late Cretaceous) cephalopods from Sümeg.—Fragmenta Palaeontologica Hungarica, 19: 25–37.

Főzy, I. (2003). What happened with the ammonites on the Jurassic/Cretaceous boundary?—Földtani Közlöny, 133(4): 383–397.

Főzy, I. (2004). A Kárpát-medence dinoszauruszai [Dinosaurs of the Carpathian Basin].—Focus Kiadó, Budapest, 176 pp. (In Hungarian)

Főzy, I., & Janssen, N. M. M. (2006). The stratigraphic position of the ammonite bearing limestone bank of the Márványbánya quarry (Zirc, Bakony Mts, Hungary) and the age of the Borzavár Limestone Formation.—Neues Jb. Geol. Paläont., Mh. 2006(2): 41–46.

Főzy, I., & Janssen, N. M. M. (2009). Integrated Lower Cretaceous biostratigraphy of the Bersek Quarry, Gerecse Mountains, Transdanubian Range, Hungary.—Cretaceous Research, 30: 78–92.

Főzy, I., Janssen, N. M. M., & Price, G. (2011). High-resolution ammonite, belemnite and stable isotope record from the most complete Upper Jurassic section of the Bakony Mts. (Transdanubian Range, Hungary).—Geologica Carpathica, 62(5): 413–433.

Főzy, I., Kázmér, M., & Szente, I. (1994). A unique Lower Tithonian fauna in the Gerecse Mts, Hungary.—Paleopelagos Spec. Publ., 1: 155–165.

Főzy, I., & Leél-Őssy, Sz. (1985). Comparative study on the mollusc fauna of two Lower-Miocene conglomerates in the Eastern-Mátra (N-Hungary).—Annales Univ. Sci. Budapest., Sect. Geol., 25: 117–130.

Franzenau, Á. (1881). Beitrag zur Foraminifer Fauna des Rákoser (Budapest) Ober-Mediterran Stufe.—Földtani Közlöny, 11(1): 83–17.

Frech, F. (1911a). Nachträge zu den Cephalopoden und Zweischalern der Bakonyer Trias (Werfener und Cassianer Estherienschichten).—Resultate Wiss. Erforsch. Balatonsees, Paläont. Anhang, 3(5): 30 pp.

Frech, F. (1911b). Neue Cephalopoden aus der Buchensteiner, Wengener und Raibler Schichten des südlichen Bakony.—Resultate Wiss. Erforsch. Balatonsees, Paläont. Anhang, 3(4): 70 pp.

Frech, F. (1912). Neue Zweischaler und Brachiopoden aus der Bakonyer Trias.—Resultate Wiss. Erforsch. Balatonsees, Paläont. Anhang, 2(2): 140 pp.

Fülöp, J. (1958). Die kretazeischen Bildungen des Gerecse-Gebirges.—Geol. Hung. Ser. Geol., 11: 122 pp.

Fülöp, J. (1964). Unterkreide-Bildungen (Berrias-Apt) des Bakony-Gebirges.—Geol. Hung. Ser. Geol., 13: 193 pp.

Fülöp, J. (1966). Les formations crétacées de la Montagne de Villány.—Geol. Hung. Ser. Geol., 15: 131 pp.

Fülöp, J. (1976). The Mesozoic basement horst blocs of Tata.—Geol. Hung. Ser. Geol., 16: 228 pp.

Fülöp, J. (1990). Magyarország geológiája: Paleozoikum I. [Geology of Hungary: Paleozoic I.].—Magyar Állami Földtani Intézet, Budapest, 326 pp. (In Hungarian)

Fülöp, J. (1994). Magyarország geológiája: Paleozoikum II. [Geology of Hungary: Paleozoic II.]—Akadémiai Kiadó, Budapest, 445 pp. (In Hungarian)

Futó, J. (2001). Orrszarvú leletek Puláról [Rhinoceros remains from Pula]. In Program, Előadáskivonatok, Kirándulásvezető, 4: Magyar Őslénytani Vándorgyűlés, 2001, május 4–5, Pécsvárad, (pp. 13–14). (In Hungarian)

Gaál, I. (1943). Unterpliozäne Säugetierreste aus Hatvan in Ungarn.—Geol. Hung. Ser. Pal., 20: 119 pp.

Gaál, L. (2010). Gömörország természeti öröksége [Natural Heritage of Gömör County].—Gömör-kishonti Múzeum Egyesület, Rimaszombat, 207 pp. (In Hungarian)

Gál, E. (2002). Betfia (Bihar-megye, Románia) alsó-pleisztocén madárvilága: paleoökológiai, paleoklimatológiai és taxonómiai eredmények [Early Pleistocene bird fauna of Betfia (Bihor County, Romania): Results of paleoecological, paleoclimatological and taxonomical studies].In Program, Előadáskivonatok, Kirándulásvezető, 5: Magyar Őslénytani Vándorgyűlés, 2002, május 3–4, Pásztó (p. 13). (In Hungarian)

Gál, E., Hír, J., Kessler, E., Kókay, J., Mészáros, L., & Venczel, M. (1999). Middle Miocene fossils from the sections at the Rákóczi Chapel at Mátraszőlős.—Folia Historico Naturalia

Musei Matraensis, 23: 33–78. (In Hungarian with English abstract)

Gál, E., Hír, J., Kessler, E., Kókay, J., Mészáros, L., & Venczel, M. (2000). Middle Miocene fossils from the section of the road at the Rákóczi Chapel, Mátraszőlős, II: Locality Mátraszőlős 2.—Folia Historico Naturalia Musei Matraensis, 24: 39–75. (In Hungarian with English abstract)

Galácz, A. (1980). Bajocian and Bathonian Ammonites of Gyenespuszta Bakony Mts., Hungary.—Geol. Hung. Ser. Pal., 39: 227 pp.

Galácz, A. (1987). Important fossil finds, famius localities in Hungary. In J. Hála (Ed.), Rocks, Fossils and History: Italian-Hungarian Relations in the Field of Geology. Annals of the History of Hungarian Geology Special Issue, 1 (pp. 133–144).—Hungarian Geological Society, Budapest.

Galácz, A. (1995). Ammonites and stratigraphy of the Bathonian red limestone of the Mecsek Mts, south Hungary.—Annales Univ. Sci. Budapest., Sect. Geol., 30: 111–150, 225–230.

Galácz, A. (2004). Nautiloid cephalopods from the Middle Eocene of Iszkaszentgyörgy, Transdanubian Hungary.—Annales Univ. Sci. Budapest., Sect. Geol., 34: 1–7.

Galácz, A., Géczy, B., & Monostori, M. (2008). Csernye revisited: New ammonite finds and ostracods from the Lower Jurassic Plienbachian/Toarcian boundary beds in Bakonycsernye, Transdanubian Hungary—Geologica Pannonica, 36: 187–225.

Galácz, A., & Monostori, M. (1992). Ősállattani praktikum [Paleozoological handbook].—Tankönyvkiadó, Budapest, 664 pp.

Galácz, A., & Vörös, A. (1969). Belemite fauna of the ammonite-rich Callovian bed at Villány, South Hungary.—Annales Univ. Sci. Budapest., Sect. Geol., 12: 117–138.

Galton, P. M. (2009). Notes on Neocomian (Lower Cretaceous) ornithopod dinosaurs from England—Hypsilophodon, Valdosaurus, "Camptosaurus," "Iguanodon"—and referred specimens from Romania and elsewhere.—Revue de Paléobiologie, 28: 211–273.

Gasparik, M. (1993). Late Pleistocene gastropod and vertebrate fauna from Tokod (NE Transdanubia, Hungary)—Fragmenta Mineralogica et Palaeontologica, 16: 89–116.

Gasparik, M. (1997). Az állatvilág jégkorszaki és jelenkori fejlődése Magyarország területén. Az utolsó 2,5 millió év. [Development of the animals in the area of Hungary during the Pleistocene and Holocene]. Lacertina Füzetek, 6.—Magyar Természettudományi Múzeum, Budapest, 24 pp. (In Hungarian)

Gasparik, M. (2001a). Neogene proboscidean remains from Hungary: An overview.—Fragmenta Palaeontologica Hungarica, 19: 61–77.

Gasparik, M. (2001b). A "Proboscidean datum" Magyarországon [The "Proboscidean Datum" in Hungary]. In Program, Előadáskivonatok, Kirándulásvezető, 4: Magyar Őslénytani Vándorgyűlés, May 3–4, Pécsvárad (pp. 15–16). (In Hungarian)

Gasparik, M. (2004). Magyarországi neogén és alsó-pleisztocén Proboscidea maradványok [Neogene and Lower Pleistocene Proboscidean remains from Hungary].—Unpublished doctoral dissertation, Eötvös Loránd University, 129 pp. (In Hungarian)

Gasparik, M. (2005). Proboscidean remains from the Pannonian of Rudabánya.—Palaeontographia Italica, 90: 181–193.

Geary, D. H., Magyar, I., & Müller, P. (2000). Ancient Lake Pannon and its Endemic Molluscan Fauna (Central Europe; Mio-Pliocene). In A. Rossiter & H. Kawanabe (Eds.), Ancient Lakes: Biodiversity, Ecology and Evolution. Advances in Ecological Research, 31 (pp. 463–482).—Academic Press, San Diego, CA.

Géczy, B. (1954). Studien über Cycloliten (Anth.).—Geol. Hung. Ser. Pal., 24: 158 pp.

Géczy, B. (1961a). Die jurassische Schichtenreihe des Tűzköves Grabens von Bakonycsernye.—A Magyar Állami Földtani Intézet Évkönyve 49(2): 507–563.

Géczy, B. (1961b). Cenoceras truncatus vadaszi n. ssp. from the Middle Liassic complex of Bakony Mts.—Földtani Közlöny, 91: 325–337.

Géczy, B. (1966). Ammonoides Jurassiques de Csernye, Montagne Bakony, Hongrie: Part 1 (Hammatoceratidae).—Geol. Hung. Ser. Pal., 34: 275.

Géczy, B. (1967). Ammonoides Jurassiques de Csernye, Montagne Bakony, Hongrie: Part 2 (excl. Hammatoceratidae).—Geol. Hung. Ser. Pal., 35: 412.

Géczy, B. (1972). Ősnövénytan [Paleobotany].—Tankönyvkiadó, Budapest, 365 pp.

Géczy, B. (1984). The Jurassic ammonites of Villány.—Annales Univ. Sci. Budapest., Sect. Geol., 24: 189–198.

Géczy, B. (1991). History of the "Pester Exemplar" of Pterodactylus.—In Gy. Vitális & T. Kecskeméti (Eds.), Museums and Collections in the History of Mineralogy, Geology and Paleontology in Hungary. Annals of the History of Hungarian Geology Special Issue, 3 (pp. 159–167).—Hungarian Geological Society, Budapest.

Géczy, B. (1993a). Ősállattan: Invertebrata Paleontologia [Paleozoology: Invertebrate paleontology].—Nemzeti Tankönyvkiadó, Budapest, 595 pp.

Géczy, B. (1993b). Ősállattan: Vertebrata Paleontologia [Paleozoology: Vertebrate paleontology].—Nemzeti Tankönyvkiadó, Budapest, 502 pp.

Géczy, B. (1994). Brief History of the Hungarian Palaeontology.—Annals of the History of Hungarian Geology, Special Issue, 6.—Hungarian Geological Society, Budapest, 68 pp.

Géczy, B. (1998). Lower Pliensbachian ammonites of Villány (Hungary).—Hantkeniana, 2: 5–47.

Géczy, B., & Galácz, A. (1998). Bathonian ammonites from the classic Middle Jurassic locality of Villány, South Hungary.—Revue de Paléobiologie, 17 (2): 479–511.

Géczy, B., & Meister, C. (1998). Les ammonites du Domérien de la montagne du Bakony (Hongrie).—Revue de Paléobiologie, 17(1): 69–161.

Géczy, B., & Szente, I. (2006). Middle Toarcian Ammonitina from the Gerecse Mts, Hungary.—Acta Geol. Hungarica, 49(3): 223–252.

Gheerbrant, E., Codrea, V., Hosu, A., Sen, S., Guernet, C., Broin, F., & Riveline, J. (1999). Découverte de vertébrés dans les Calcaires de Rona (Thanétien ou Sparnacien), Transylvanie, Roumanie: les plus anciens mammifères cénozoiques d'Europe Orientale.—Eclogae Geol. Helv., 92: 517–535.

Gnoli, M., & Kovács, S. (1992). The oldest megafossils of Hungary: Silurian orthocone nautiloids from Strázsa Hill, Uppony Mts., NE Hungary.—A Magyar Állami Földtani Intézet Évi Jelentése 1990-ről: 375–393.

Góczán, F. (1961). Stratigraphische Auswertung der Gastropodenfauna der transdanubischen und alpinen Trias.—A Magyar Állami Földtani Intézet Évkönyve, 49(2): 59–70.

Góczán, F. (1971). Contributions to the study of the microplankton of the Silurian siliceous shales at Alsóörs, Szalatnak and Uppony.—Őslénytani Viták, 18: 13–20. (In Hungarian with English abstract)

Góczán, F. (1997). New organic framework plant microfossils in the Lower Rhaetian beds of the Csővár Limestone Formation.—Acta Geologica Hungarica, 40(2): 197–239.

Góczán, F., & Siegel-Farkas, Á. (1989). Palynostatigraphy of the Rendek Member of the Pollány Marl Formation.—A Magyar Állami Földtani Intézet Évi Jelentése 1988-ról, 2: 47–87.

Goetel, W. (1917). Die rhätische Stufe und der unterste Lias der subtatrischen Zone in der Tatra.—Bulletin de l'Académie des Sciences de Cracovie, Classe de Sciences Mathématiques et Naturelles, Série A: 1–222.

Golej, M. (2005). Rhaetian bivalves of Hybe Formation: paleoecology and paleobiogeography. In J. Pálfy & P. Ozsvárt (Eds.), Program, Abstracts and Field Guide, 5th Field Workshop of IGCP 458 Project (Tata and Hallein, September 2005) (pp. 14–15).

Görög, Á. (1992). Sarmatian foraminifera of the Zsámbék Basin, Hungary.—Annales Univ. Sci. Budapest., Sect. Geol., 29: 31–153.

Görög, Á. (1993). Orbitolina-félék (Nagyforaminiferák) megjelenése magyarországi alsó-és középső-kréta képződményekben [Occurrence of Orbitolina (larger foraminifer) in the Lower and Middle Cretaceous of Hungary].—Őslénytani Viták, 39: 51–72. (In Hungarian)

Görög, Á. (1994). Early Jurassic planktonic foraminifera from Hungary.—Micropaleontology, 40(3): 255–260.

Görög, Á. (1995). Bathonian foraminifera form the Mecsek Mts (South Hungary).—Annales Univ. Sci. Budapest., Sect. Geol., 30: 7–82, 209–218.

Götz, A. E., Török, Á., Feist-Burkhardt, S., & Konrád, Gy. (2003). Palynofacies patterns of Middle Triassic ramp deposits (Mecsek Mts., S Hungary). A powerful tool for high-resolution sequence stratigraphy.—Mitt. Ges. Geol. Bergbaustud. Österr., 46: 77–90.

Götz, A. E., Ruckwied, K., Pálfy, J., & Haas, J. (2009). Palynological evidence of synchronous changes within the terrestrial and marine realm at the Triassic/Jurassic boundary (Csővár section, Hungary).—Review of Palaeobotany and Palynology, 156: 401–409.

Greguss, P. (1954). Les vestiges de bois silificié du Miocène inférieur d'Ipolytarnóc.—Földtani Közlöny, 84(1): 91–108. (In Hungarian with French abstract)

Greguss, P. (1961). Permische fossile Hölzer aus Ungarn.—Palaeontographica Abt. B, 109(5–6): 131–146.

Greguss, P. (1967). Fossil Gymnosperm Woods in Hungary, from the Permian to the Pliocene.—Akadémiai Kiadó, Budapest, 136 pp.

Greguss, P. (1969). Tertiary angiosperm woods in Hungary. Akadémiai Kiadó, Budapest, 151 pp.

Grigorescu, D. (1981). A stratigraphic, taphonomic and paleoecologic approach to a "forgotten land": The dinosaur-bearing deposits from the Haţeg Basin (Transylvania—Romania)—Acta Palaeontologica Polonica, 28(1–2): 103–121.

Grigorescu, D. (1984). New paleontological data on the dinosaur beds from the Haţeg basin. In Seventy-five Years of the Laboratory of Paleontology: Special Volume (pp. 111–118).—University of Bucharest, Laboratory of Paleontology, Bucharest.

Grigorescu, D. (2010). The Latest Cretaceous fauna with dinosaurs and mammals from Haţeg Basin: A historical rerview.—Palaeogeography, Palaeoclimatology, Palaeoecology, 293: 271–282.

Grigorescu, D., Garcia, G., Csiki, Z., Codrea, V., & Bojar, A.-V. (2010). Uppermost Cretaceous megaoolithid eggs from Haţeg Basin, Romania, associated with hadrosaur hetchlings: Search for explanation—Palaeogeography, Palaeoclimatology, Palaeoecology, 293: 230–374.

Grigorescu, D., Seclamen, M., Norman, D. B., & Weishampel, D. B. (1990). Dinosaur eggs from Romania.—Nature, 346: 417.

Grigorescu, D., Venczel, M., Csiki, Z., & Limberea, R. (1999). New latest Cretaceous microvertebrate fossil assemblages from the Haţeg Basin (Romania).—Geologie en Mijnbouw, 78: 301–314.

Gulyás-Kis, Cs. (2001). A nagyvisnyói felső-karbon agyagpala (Mályinkai Formáció) brachiopodái [Brachiopods of the Upper Carboniferous shale at Nagyvisnyó (Mályinka Formation)].—Unpublished MSc thesis, Eötvös Loránd University, 120 pp. (In Hungarian)

Gulyás-Kis, C. (2003). Upper Carboniferous flora from the Mecsek Mts (Southern Hungary)-summarized results.—Acta Geol. Hungarica, 46(1): 115–126.

Gulyás-Kis, Cs. (2004). Palaeobiogeographic analysis of Moscovian (Pennsylvanian) brachiopods Productida, Orthotetida, Orthida, Rhynchonellida from Bükk Mts, Hungary.—Hantkeniana 4: 17–35.

Haas, J. (Ed.). (1993). Magyarország litosztratigráfiai alapegységei: Triász [Basic lithostratigraphic units of Hungary: Triassic].—Hungarian Geological Institute, 278 pp. (In Hungarian)

Haas, J. (Ed.). (2004). Magyarország geológiája: Triász [Geology of Hungary: Triassic].—Eötvös Kiadó, Budapest, 384 pp. (In Hungarian)

Haas, J. (2007). Geological Garden in Tata, Transdanubian Range, Hungary.—Nova Acta Leopoldina, 94(349): 237–251.

Haas, J., Budai, T., Szente, I., Piros, O., & Tardiné Filácz, E. (2005). Upper Triassic slope and basin sequences in the Pilis Mts and the Tatabánya Basin).—Földtani Közlöny, 135(4): 513–543. (In Hungarian with English abstract)

Haas J., Góczán, F., Oravecznè Scheffer, A., Barabásné Stuhl, Á., Majoros, Gy., & Bérczinė Makk, A. (1986). Permian-Triassic boundary in Hungary.—Mem. Soc. Geol. It. 34: 221–241.

Haas, J., Jocha-Edelényi, E., Gidai, L., Kaiser, M., Kretzoi, M., & Oravecz, J. (1985). Geology of the Sümeg area.—Geol. Hung. Ser. Geol., 20: 365.

Hably, L. (1979). Some Data to the Oligocene Flora of the Kiscellian Tard Clay, Hungary.—Annales Hist.-Nat. Mus. Nat. Hung., 71: 33–53.

Hably, L. (1980). Platanus neptuni (Ett.) Bůžek, Holý et Kvaček in the Hungarian Oligocene.—Acta Botanica Academiae Scientiarum Hungaricae, 26(3–4): 299–316.

Hably, L. (1985). Early Miocene plant fossils from Ipolytarnóc, N Hungary.—Geol. Hung. Ser. Pal., 44–46: 73–255.

Hably, L. (1989). The Oligocene flora of Nagysáp.—Fragmenta Mineralogica et Palaeontologica, 14: 83–99.

Hably, L. (1990). Egerian plant fossils from Vértesszőlős, NW Hungary.—Studia Botanica Hungarica, 22: 3–78.

Hably, L. (1993). Egerian macroflora from the Andornaktálya sandpit (Hungary).—Studia Botanica Hungarica, 24: 5–22.

Hably, L. (Ed.). (1997). Early Pliocene volcanic environment, flora and fauna from Transdanubia, West Hungary.—Studia Naturalia, 10: 191 pp.

Hably, L. (2006). Catalogue of the Hungarian Cenozoic leaf, fruit and seed floras from 1856 to 2005.—Studia Botanica Hungarica, 37: 41–129.

Hably, L., & Kázmér, M. (1996). Short-term floristic changes due to terrane displacement in the Miocene of Hungary. In F. Corsetti & B. H. Tiffney (Eds.), Abstract Volume: Fifth Quadrennial Conference of the International Organisation of Palaeobotany (p. 39).—Department of Geological Sciences, University of California, Santa Barbara, Santa Barbara, CA.

Hagdorn, H., Török, Á., & Konrád, Gy. (1997). Crinoids from the Muschelkalk of the Mecsek Mountains and their stratigraphical significance.—Acta Geologica Hungarica, 40(4): 391–410.

Hagdorn, H., & Velledits, F. (2006). Middle Triassic crinoid remains from the Aggtelek platform (NE Hungary).—N. Jb. Geol. Paläont. Abh., 240(3): 373–404.

Hajós, M. (1968). Diatomeen der miozänen Ablagerungen des Mátravorlandes.—Geol. Hung. Ser. Pal., 37: 401 pp.

Hajós, M. (1986). Stratigraphy of Hungary's Miocene diatomaceous earth deposits.—Geol. Hung. Ser. Pal. 49: 339 pp.

Hála, J., Kecskeméti, T., & Voigt, V. (2004). Kőpénzek. Mondák, költemények és énekek [Coin-stones. Legends, poems, and songs].—Mentor Kiadó, Marosvásárhely, 330 pp. (In Hungarian)

Halaváts, Gy. (1899). Der Mammuth-Befund von Jobbágyi (Nógrader Com.).—Földtani Közlöny, 29(1–4):128–129.

Halaváts, Gy. (1911). Die Faune der pontischen Schichten in der Umgebung des Balatonsees.—Resultate Wiss. Erforsch. Balatonsees, Paläont. Anhang, 4(2): 80 pp.

Halupka, G. (1996). Miliolidák (Foraminifera) taxonómiai vizsgálata gánti gastropodák héjkitöltéséből [Taxonomic study of Miliolids (Foraminifera) from infillings of gastropod shells found at Gánt]. Unpublished MSc theses, Eötvös Loránd University, 108 pp. (In Hungarian)

Hámor-Vidó, M., Hofmann, T., & Albert, L. (2010). In situ preservation and paleoenvironmental assessment of Taxodiacea fossil trees in the Bükkalja Lignite Formation, Bükkábrány open cast mine, Hungary.—International Journal of Coal Geology, 81(4): 203–210.

Hankó, E. (2007). A revision of three Pleistocene subspecies of Panthera, based on mandible and teeth remains stored in Hungarian collections.—Fragmenta Palaeontologica Hungarica, 24: 25–44.

Hantken, M. (1868). A kiscelli tályag foraminiferái [Foraminifers of the Kiscell Clay].—Magyar Földtani Társulat Munkálatai, 4: 75–96. (In Hungarian)

Hantken, M. (1871). Az esztergomi burányrétegek és a kiscelli tályag földtani kora [On the geological age of the coral-bearing beds of Esztergom and of the Kiscell clay].—Értekezések a természettudományok köréből, 13: 3–18. (In Hungarian)

Hantken, M. (1881). Die Fauna der Clavulina Szabói-Schichten.—Mitt. Jb. Kön. Ung. Geol. Anst., 4(1): 1–93.

Hantken, M. (1884). A magyarországi mész- és szarúkövek górcsövi alkatáról. [On microscopic construction of limestones and cherts from Hungary].—Mathematikai és természettudományi értesítő, 2(8–9): 373–377. (In Hungarian)

Hantken, M. (1887). Tinnyea Vásárhelyii nov. gen. et nov. spec.—Földtani Közlöny, 17(4): 345–348. (In German)

Harzhauser, M., & Kowalke, T. (2002). Sarmatian (Late Middle Miocene) Gastropod Assemblages of the Central Paratethys.—Facies, 46: 57–82.

Harzhauser, M., Kowalke, T., & Mandic, O. (2002). Late Miocene (Pannonian) Gastropods of Lake Pannon with Special Emphasis on Early Ontogenetic Development.—Annales Naturhist. Mus. Wien, 103 A: 75–141.

Heer, O. (1878). Ueber permische Pflanzen von Fünfkirchen in Ungarn.—Mitt. Jb. Kön. Ung. Geol. Anst., 5(1): 1–18.

Hegedűs, G. (1962). Oligocene corals of Hungary.—A Magyar Állami Földtani Intézet Évi Jelentése 1959-ről: 231–260. (In Hungarian with English abstract)

Hegedűs, G. (1970). Coralliaires tortoniens de Herend.—Földtani Közlöny, 100: 185–191. (In Hungarian with French abstract)

Hegedűs, G., & Jankovich, I. (1972). Récif coralline du Badénien à Márkháza.—A Magyar Állami Földtani Intézet Évi Jelentése 1970-ről: 39–54. (In Hungarian with French abstract)

Heissig, K. (2005). The fossil Rhinoceros of Rudabánya.—Palaeontographia Italica, 90: 217–258.

Hemmer, H. (2003). Pleistozäne Katzen Europas: Eine Übersicht.—Cranium, 20(2): 6–22.

Herak, M., & Kochansky, V. (1963). Jungpaläozoische Kalkalgen aus dem Bükk-Gebirge (Nordungarn).—Geol. Hung. Ser. Pal., 28: 45–77.

Herbich, F. (1878). Das Széklerland mit Berücksichtigung der angrenzenden Landestheile, geologisch und paleontologisch beschrieben.—Mitt. Jb. Kön. Ung. Geol. Anst., 5(2): 23–363.

Herbich, F. (1886). Paläontologische Studien über die Kalkklippen des siebenbürgischen Erzgebirges.—Mitt. Jb. Kön. Ung. Geol. Anst., 8(1): 54 pp.

Hips, K. (1996) Stratigraphic and facies evaluation of the Lower Triassic formations in the Aggtelek-Rudabánya Mountains, NE Hungary.—Acta Geologica Hungarica, 39(4): 396–411.

Hips, K., & Pelikán, P. (2002). Lower Triassic shallow marine successions in the Bükk Mountains, NE Hungary.—Geologica Carpathica, 53(6): 1–17.

Hír, J. (2001). New Middle Miocene rodent faunas from Northern Hungary.—Lynx, N. S., 32: 107–122.

Hír, J. (2003). The Middle Miocene (Late Astaracian, MN8) rodent fauna of Felsotarkany 3/2 (Hungary).—Acta Palaeontologica Romaniae, 4: 125–136.

Hír, J., Kókay, J., & Venczel, M. (2001). Middle Miocene molluscs and microvertebrata from Tăşad (Bihor County, Romania).—Acta Paleontologica Romaniae, 5: 161–172.

Hír, J., & Venczel, M. (1998). New Allophaiomys material from Betfia IX/b and IX/c, Bihor County, Romania.—Mededelingen Nederlands Instituut voor Toegepaste Geowatenschappen TNO, 60: 308–312.

Hofmann, K. (1884). Bericht über die auf der rechten Seite der Donau zwischen Ó-Szőny und Piszke im Sommer 1883 ausgeführten geologischen Spezialaufnahmen.—Földtani Közlöny, 14: 323–342.

Horváthné Kollányi, K. (1988). Eocene benthonic Smaller Foraminifera fauna from Dudar.—A Magyar Állami Földtani Intézet Évkönyve, 63(4): 1–173.

Huene, F. von. (1931). Ergänzungen zur Kenntnis des Schädels von Placochelys und seiner Bedeutung.—Geol. Hung. Ser. Pal., 9: 16 pp.

Jablonszky, J. (1915). Die mediterrane Flora von Tarnóc.—Mitt. Jb. Kön. Ung. Geol. Anst., 22(4): 252–293.

Jaekel, O. (1911). Placochelys placodonta aus der Obertrias des Bakony.—Resultate Wiss. Erforsch. Balatonsees, Paläont. Anhang, 3(8): 91 pp.

Jámborné Kness, M. (1981). Alvéolines éocènes de la Hongrie.—Geol. Hung. Ser. Pal., 40: 143 pp.

Jámborné Kness, M. (1988). Les grands Foraminifères éocènes de la Hongrie.—Geol. Hung. Ser. Pal., 52: 629 pp.

Jánossy, D. (1961). Der erste Nachweis von Hippopotamus antiquus Desmarest, 1822 im ungarischen Altpleistozän.—Állattani Közlemények, 49: 63–74. (In Hungarian with German abstract)

Jánossy, D. (1977). Plio-Pleistocene bird remains from the Carpathian Basin. III. Galliformes. 2, Phasianidae.—Aquila, 83: 29–42.

Jánossy, D. (1981). Tivadar Kormos was born one hundred years ago (1881–1946).—Fragmenta Mineralogica et Palaeontologica, 10: 7–8.

Jánossy, D. (1986). Pleistocene Vertebrate Faunas of Hungary.—Developments in Palaeontology and Stratigraphy, 8.—Elsevier, Amsterdam and New York, 208 pp.

Jánossy, D. (1991). Late Miocene bird remains from Polgárdi (W-Hungary)—Aquila, (89): 13–35.

Jánossy, D. (1993). Bird remains from the Upper Miocene (MN9) of Rudabánya (N-Hungary).—Aquila, 100: 53–70.

Jánossy, D. (1996). Lower Pleistocene Vertebrate Faunas from the Localities 16 and 17 of Beremend (Southern Hungary).—Fragmenta Mineralogica et Palaeontologica, 18: 91–102.

Janssen, A. W. (1984). Late Oligocene molluscs from a sandpit near Máriahalom (Hungary): a preliminary study.—Annales Univ. Sci. Budapest., Sect. Geol., 24: 109–150.

Janssen, N. M. M., & Főzy, I. (2003). Neocomian belemnites from the Bersek Hill (Gerecse Mountains, Hungary).—Földtani Közlöny, 133(2): 291–294.

Janssen, N. M. M., & Főzy, I. (2004). Neocomian belemnites from the Bersek-hegy (Gerecse Mountains, Hungary), part I: Late Valanginian to earliest Barremian.—Fragmenta Palaeontologica Hungarica, 22: 27–49.

Janssen, N. M. M., & Főzy, I. (2005). Neocomian belemnites from the Bersek Hill (Gerecse Mountains, Hungary), part II: Earliest Barremian–Late Barremian.—Fragmenta Paleontologica Hungarica, 23: 59–86.

Jekelius, E. (1915). Die mesozoischen Faunen der Berge von Brassó. 1. Die Liasfauna von Keresztényfalva. 2. Die Neokomfauna von Brassó.—Mitt. Jb. Kön. Ung. Geol. Anst., 23(2): 30–135.

Jekelius, E. (1925). Die mesozoischen Faunen der Berge von Brassó.—Mitt. Jb. Kön. Ung. Geol. Anst., 24(2): 30–107.

Jekelius, E. (1935). Der weisse Triaskalk von Braşov und seine Fauna.—Anuarul Institutului Geologic al Romanei, 17: 107 pp.

Jurcsák, T. (1975). Tanystropheus biharicus n. sp. (Reptilia, Squamata) o noua specie Pentru fauna Triasica a Romaniei.—Nymphaea, 3: 45–52.

Jurcsák, T. (1976). Noi descoperiri de reptile fosile in Triascul de la Aleşd.—Nymphaea, 4: 67–105.

Jurcsák, T. (1978). Rezultate noi in Studiul Saureinilor fosili de la Aleşd.—Nymphaea, 6: 15–60.

Jurcsák, T. (1982). Occurrences nouvelles des sauriens mesozoiques de Roumanie.—Vertebrata Hungarica, 21: 175–184.

Jurcsák, T., & Kessler, E. (1986). New contributions to the Knowledge of the Lower-Cretaceous bird remains from Cornet.—Travaux du Muséum d'Histoire naturelle Grigore Antipa, 28: 289–295.

Jurcsák, T., & Kessler, E. (1991). The Lower Cretaceous paleofauna from Cornet, Bihor County, Romania.—Nymphaea, 21: 5–32.

Jurcsák T., & Popa, E. (1978). Resturi de dinosaurieni in bauxitele de la Cornet (Bihor).—Nymphaea, 6: 61–64.

Jurcsák T., & Popa, E. (1979). Dinosaurieni ornitopozi din bauxitele de la Cornet (Munţii Pădurea Craiului).—Nymphaea, 7, 37–75.

Jurcsák, T., & Popa, E. (1982). Reptile Zburatoare in Bauxitele de la Cornet.—Nymphaea, 10: 7–15.

Jurcsák, T., & Popa, E. (1984). Pterosaurians from the Cretaceous of Cornet, Roumania. In W. E. Reif & F. Westphal (Eds.), Third Symposium on Mesozoic Terrestrial

Ecosystems, Short Papers (pp. 117–118).—Attempto, Tübingen, Germany.

Kadić, O. (1907). Mesocetus Hungaricus Kadić, eine neue Balaenopteridenart aus dem Miozän von Borbolya in Ungarn. Mitt. Jb. Kön. Ung. Geol. Anst., 16(2): 24–91.

Kadić, O. (2010). Kadić Ottokár: Önéletrajz. [Ottokár Kadić: An autobiography]. Magyar Állami Földtani Intézet, 191 pp. (In Hungarian with extended English abstract)

Kaszap, A. (1961). Die Bath-Kallov-Schichten in dem Villányer Gebirge.—A Magyar Állami Földtani Intézet Évkönyve, 49(2): 659–664.

Kaszap, A. (1968). Korynichnium sphaerodactylum (Pabst) Einzelfährte im Perm von Balatonrendes (Transdanubia).—Földtani Közlöny, 98(3–4): 429–433.

Kazár, E. (2003). Miocén fogascet-leletek (Cetacea: Odontoceti) a Kárpát-medencében [Miocene toothed whales (Cetacea: Odontoceti) in the Carpathian Basin].—Unpublished doctoral dissertation, Eötvös Loránd University, 344 pp. (In Hungarian)

Kazár, E. (2005). A new kentriodontid (Cetacea: Delphinoidea) from the Middle Miocene of Hungary.—Mitt. Mus. Nat. Berlin, Geowiss. Reihe, 8: 53–73.

Kazár, E., Kordos, L., & Szónoky, M. (2001). Danitz-pusztai homokbánya. Pannon homok áthalmozott gerinces maradványokkal [Danitz-puszta, sand-pit. Pannonian sand with reworked vertebrate remains]. In Program, Előadáskivonatok, Kirándulásvezető, 4: Magyar Őslénytani Vándorgyűlés, 2001, május 4 N>5, Pécsvárad (pp. 42–43). (In Hungarian)

Kazár, E., & Lantos, Z. (2001). Taphonomy and palaeoecology of the holotype of Heterodelphis leiodontus Papp, 1905 (Cetacea, Mammalia) from Szentmargita (St. Margarethen, Austria).—Fragmenta Palaeontologica Hungarica, 19: 51–60.

Kázmér, M. (1986). Tectonic units of Hungary: Their boundaries and stratigraphy (A bibliographic guide).—Annales Univ. Sci. Budapest., Sect. Geol., 26: 45–120.

Kázmér, M. (1987). A brief history of the collections of the Department of Palaeontology of the University of Budapest.—In J. Hála (Ed.), Rocks, Fossils and History. Italian-Hungarian Realtions in the Field of Geology. Annals of the History of Hungarian Geology Special Issue (pp. 171–173).—Hungarian Geological Society, Budapest.

Kázmér, M. (1990). Birth, life and death of the Pannonian Lake.—Palaeogeography, Palaeoclimatology, Palaeoecology, 79: 171–188.

Kázmér, M. (1998). Pygopid brachiopods and Late Jurassic paleorelief in the Gerecse Mts., Hungary.—Földtani Közlöny, 128(2–3): 265–272.

Kázmér, M. (2008). The Miocene Bükkábrány Fossil Forest in Hungary: Field observations and project outline. In A. Galácz, (Ed.), 125th Anniversary of the Department of Palaeontology at Budapest University: A Jubilee Volume.—Hantkeniana, 6: 229–244.

Kecskeméti, T., & Nagy, I. Z. (1991). The Paleozoological Collection of the Hungarian Natural History Museum. In Gy. Vitális & T. Kecskeméti (Eds.), Museums and Collections in the History of Mineralogy, Geology and Paleontology in Hungary. Annals of the History of Hungarian Geology Special Issue, 3 (pp. 67–86).—Hungarian Geological Society, Budapest.

Kecskemétiné Körmendy, A. (1972). Die Eozäne Molluskenfauna der Doroger Beckens.—A Magyar Állami Földtani Intézet Évkönyve, 55(2): 144–377.

Kecskemétiné Körmendy, A. (1980). La faune des Mollusques du faciés de bassin éocène du Bakony nord-oriental (Transdanubie).—A Magyar Állami Földtani Intézet Évkönyve, 63(3): 227 pp.

Kecskemétiné Körmendy, A. (1990). La faune de Mollusques éocènes du bassin Nagyegyháza-Csordakút-Mány.—A Magyar Állami Földtani Intézet Évkönyve, 71(1): 269 pp.

Kecskemétiné Körmendy, A., & Mészáros, M. (1980). Mollusques éocènes du secteur oriental de la Montagne du Bakony (facies archipélagique).—A Magyar Állami Földtani Intézet Évkönyve, 63(2): 143 pp.

Kessler, E., & Jurcsák, T. (1984a). Lower Cretaceous birds from Cornet (Roumania). In In W. E. Reif & F. Westphal (Eds.), Third Symposium on Mesozoic Terrestrial Ecosystems, Short Papers (pp. 119–121).—Attempto, Tübingen, Germany.

Kessler, E., & Jurcsák, T. (1984b). Fossil bird remains in the bauxite from Cornet (Romania, Bihor County).—Travaux du Muséum d'Histoire Naturelle Grigore Antipa, 25: 393–401.

Kessler, E., & Jurcsák, T. (1984c). Fossil bird remains in the bauxite from Cornet (Pădurea Craului Mountains—Romania): Preliminary note. In Seventy-five Years of the Laboratory of Paleontology: Special Volume (pp. 129–134).—University of Bucharest, Laboratory of Paleontology, Bucharest.

Kessler, E., Codrea, V., & Vremir, M. (1998). A fossil bird from the Lower Oligocene of Cluj-Napoca (Western Romania).—Studia Univ. Babeş-Bolyai Geologia, 43(2): 7–12.

Kiss, J. (1951). Les conditions géologiques et metallogénétiques du mont Szár de Szabadbattyán.—Földtani Közlöny, 81(7–9): 264–274. (In Hungarian with French abstract)

Kittl, E. (1912a). Trias Gastropoden des Bakonyer-Waldes.—Resultate Wiss. Erforsch. Balatonsees, Paläont. Anhang, 2(5): 58 pp.

Kittl, E. (1912b). Materialen zu einer Monographie der Halobiidae und Monotidae der Trias.—Resultate Wiss. Erforsch. Balatonsees, Paläont. Anhang, 2(4): 230 pp.

Koch, A. (1884). Die alttertiären Echiniden Siebenbürgens.—Mitt. Jb. Kön. Ung. Geol. Anst., 7(2): 48–132.

Koch, A. (1896). Über das vorkommen und die Verbreitung des Gryphaea esterházyi Pávay.—Földtani Közlöny, 26(11–12): 360–366.

Koch, A. (1897). Prohyracodon orientalis, ein neues Ursaugethier aus den mitteleocenen Schichten Siebenbürgens.—Természetrajzi Füzetek, 20: 481–500.

Koch, A. (1900). A Magyar Korona Országai kövült gerinczesállat maradványainak rendszeres átnézete [A systematic overview of fossil vertebrates of the lands of the

Hungarian Crown].—Magyar Orvosok és Természetvizsgálók Vándorgyűléseinek Munkálatai, 30: 526–560. (In Hungarian)

Koch, A. (1902). Neuere Beiträge zu den geo-paläontologischen Verhältnissen des Beočiner Cementmergels.—Földtani Közlöny, 32(7–9): 311–322.

Koch, A. (1903). Tarnócz im Komitat Nógrád, als neuer, reicher Fundort fossiler Haifischzäne. Földtani Közlöny, 33: 139–164.

Koch, A. (1904). Fossile Haifischzähne und Säugetierreste von Felső-Esztergály im Komitate Nógrád.—Földtani Közlöny, 34: 260–273.

Kochanová, M. (1985). Middle Triassic Bivalvia from the area of Gemerská Hôrka in Slovak Karst (West Carpathians).—Západné Karpaty, Séria Paleontológia, 10: 55–73.

Kochanová, M. (1987). Obertriassische Muscheln von Ostré Vŕšky (Slowakischer Karst). Západné Karpaty, Séria Paleontológia, 12: 31–46.

Kocsis, L. (2002a). Middle Eocene Hyrachyus cf. stehlini (Mammalia, Perissodactyla) from the Gerecse Hills, Hungary.—Neues Jb. Geol. Paläont. Monatshefte, 2002(11): 649–658.

Kocsis, L. (2002b). Mecsek környéki miocén porcoshalmaradványok [Remains of cartilaginous fish from the environs of the Mecsek Mountains]. In Program, Előadáskivonatok, Kirándulásvezető, 5: Magyar Őslénytani Vándorgyűlés, 2002, május 3–4, Pásztó (p. 21). (In Hungarian)

Kocsis, L. (2003). Az ipolytarnóci cápafogak revíziója [Revision of the shark teeth from Ipolytarnóc]. In Program, Előadáskivonatok, Kirándulásvezető, 6: Magyar Őslénytani Vándorgyűlés, 2003, május 8–10, Zirc (p. 18). (In Hungarian)

Kocsis, L. (2007). Central Paratethyan shark fauna (Ipolytarnóc, Hungary).—Geologica Carpathica, 58(1): 27–40.

Kókay, J. (1966). Geologische und paläontologische Untersuchung des Braunkohlengebietes von Herend-Márkó (Bakony-Gebirge, Ungarn). Geol. Hung. Ser. Pal., 36: 147 pp.

Kókay, J. (1996). Palaeontological and geological revision of the Badenian mollusc fauna from Illés street, Budapest.—Földtani Közlöny, 126(4): 447–484. (In Hungarian with English abstract)

Kókay, J. (2006). Nonmarine mollusc fauna from the Lower and Middle Miocene, Bakony Mts, W Hungary.—Geol. Hung. Ser. Pal., 56: 196 pp.

Kollárová-Andrusovová, V., & Kochanová, M. (1973). Molluskenfauna des Bleskový prameň bei Drnava (Nor, Westkarpaten).—Slowakische Akademie der Wissenschaften, Bratislava, 234 pp.

Kolosváry, G. (1948). New Balanids from the Middle-Miocene of Várpalota in Hungary.—Földtani Közlöny, 78: 102–112. (In Hungarian with English abstract)

Kolosváry, G. (1949a). The Eocene corals of the Hungarian Transdanubian province.—Földtani Közlöny, 79(5–8): 141–242. (In Hungarian with English abstract)

Kolosváry, G. (1949b). New Balanids from the Hungarian Tertiary age.—Földtani Közlöny, 79: 111–118.

Kolosváry, G. (1950). Descriptions of four fossil Tertiary barnacles from Hungary.—Földtani Közlöny, 80: 271–274.

Kolosváry, G. (1954). Les coralliares du Crétacé de la Hongrie.—A Magyar Állami Földtani Intézet Évkönyve, 42(2): 67–163.

Kolosváry, G. (1957). Neue Korallen aus den unterkretazischen Schichten von Lábatlan.—Földtani Közlöny, 87: 81–84.

Kolosváry, G. (1958). Corals from the Upper Anisian of Hungary.—Journ. of Paleontology 32(3): 636–637.

Konrád, Gy., Kovács, J., Halász, A., Sebe, K., & Pálffy, H. (2010). Late Quaternary woolly mammoth (Mammuthus primigenius Blum) remains from southern Transdanubia, Hungary.—Comptes Rendus Palévol, 9(1): 47–54.

Kopek, G. (1954). Les coralliaries miocènes de la Hongrie septentrionale.—A Magyar Állami Földtani Intézet Évkönyve, 42(1): 1–63.

Kordos, L. (1977). Change in the Holocene climate of Hungary reflected by the "vole thermometer" method.—Földrajzi Közlemények 25(1–3): 222–229.

Kordos, L. (1978a). Major finds of scattered fossils in the Palaeovertebrate Collections of the Hungarian Geological Institute (Communication No. 3).—A Magyar Állami Földtani Intézet Évi jelentése 1976-ról: 281–290.

Kordos, L. (1978b). Eocene, Oligocene and Miocene Palaeovertebrate Localities in Hungary.—A Magyar Állami Földtani Intézet Évi jelentése 1976-ról: 291–295.

Kordos, L. (1981). The Age of the Upper Miocene Vertebrate Locality of Hasznos in Terms of Mammal Zonation.—A Magyar Állami Földtani Intézet Évi Jelentése 1979-ről: 459–464. (In Hungarian with English abstract)

Kordos, L. (1983). Major Finds of Scattered Fossils in the Palaeovertebrate Collection of the Hungarian Geological Institute (Communication No. 8). Dinosaur footprints (Komlosaurus carbonis n. g. n. sp.) from the Liassic of the Mecsek Mountains.—A Magyar Állami Földtani Intézet Évi jelentése 1981-ről: 503–511. (In Hungarian with English abstract)

Kordos, L. (1985a). Footprints in Lower-Miocene sandstone at Ipolytarnóc.—Geol. Hung. Ser. Pal., 46: 257–415.

Kordos, L. (1985b). Az első ötvemillió év: Ősök Rudabányán [The first fifty million years: Ancestors from Rudabánya].—Gondolat Könyvkiadó, Budapest, 206 pp. (In Hungarian)

Kordos, L. (1987a). A contribution to the footprint record of the Lower Miocene sandstone of Ipolytarnóc.—A Magyar Állami Földtani Intézet Évi Jelentése 1985-ről: 453–466. (In Hungarian with English abstract)

Kordos, L. (1987b). Neogene vertebrate biostratigraphy in Hungary.—A Magyar Állami Földtani Intézet Évkönyve, 70: 393–396.

Kordos, L. (1990). The new Rudabánya finds and the multiplying theories of becoming man.—Magyar Tudomány, 1990(1): 9–14. (In Hungarian with English abstract)

Kordos, L. (1991a). Evolution and biostratigraphic ranging of the Late Pleistocene vole fauna in Central Europe.—A Magyar Állami Földtani Intézet Évi Jelentése 1989-ről: 495–522.

Kordos, L. (1991b). Le Rudapithecus hungaricus de Rudabánya (Hongrie).—L'Anthropologie (Paris), 95(2–3): 343–362.

Kordos, L. (1991c). The Museum of the Hungarian Geological Institute. In Gy. Vitális & T. Kecskeméti (Eds.), Museums and Collections in the History of Mineralogy, Geology and Paleontology in Hungary. Annals of the History of Hungarian Geology Special Issue, 3 (pp. 169–180).—Hungarian Geological Society, Budapest.

Kordos, L. (1992). The Proboscidean and Hipparion Datum in Hungary.—A Magyar Állami Földtani Intézet Évi Jelentése 1990-ről: 535–539.

Kordos, L. (1994). Revised biostratigraphy of the early man site at Vértesszőlős, Hungary.—Courier Forschungs-Institute Senckenberg, 171: 225–236.

Kordos, L. (2001). Rudabánya: Az emberré válás világhírű lelőhelye [Rudabánya: A famous locality of becoming man]. In Á. Ringer (Ed.), Emberelődök nyomában: Az őskőkor emlékei Északkelet-Magyarországon (pp. 43–49).—BAZ Megyei Levéltár, Miskolci Egyetem Ős- és Ókortörténeti Tanszék, Herman Ottó Múzeum, Miskolc. (In Hungarian)

Kordos, L. (2002a). Eocene sea cows (Sirenia, Mammalia) from Hungary.—Fragmenta Palaeontologica Hungarica, 20: 43–48.

Kordos, L. (2002b). Ipolytarnóc: Az alsó miocén feltárások újdonságai [Ipolytarnóc: New results of the studies on Lower Miocene exposures]. In Program, Előadáskivonatok, Kirándulásvezető, 5: Magyar Őslénytani Vándorgyűlés, 2002, május 3–4, Pásztó (pp. 35–36). (In Hungarian)

Kordos, L. (2005). Diggings at a Palaeovertebrata Site of Baltavár.—Vasi Szemle, 59(1): 76–78, 134. (In Hungarian with English abstract)

Kordos, L., & Begun, D. R. (1997). A New Reconstruction of RUD 77, a Partial Cranium of Dryopithecus brancoi from Rudabánya, Hungary.—Am. J. of Physical Anthropology, 103: 277–294.

Kordos, L., & Begun, D. R. (2001a). Primates from Rudabánya: allocation of specimens to individaulas, sex and age categories.—Journal of Human Evolution, 40: 17–39.

Kordos, L., & Begun, D. R. (2001b). A new cranium of Dryopithecus from Rudabánya, Hungary.—Journal of Human Evolution, 41(6): 689–700.

Koretsky, I. (2001). Morphology and Systematics of Miocene Phocinae (Mammalia: Carnivora) from Paratethys and the North Atlantic Region.—Geol. Hung. Ser. Pal., 54: 109 pp.

Kormos, T. (1911a). Der pliozäne Knochenfund bei Polgárdi (Vorläufiger Bericht).—Földtani Közlöny, 41(1–2): 48–64.

Kormos, T. (1911b). Une nouvelle espèce de tortue (Clemmys Méhelyi n. sp.) du Pleistocène hongrois.—Földtani Közlöny, 41(6): 506–512.

Kormos, T. (1912). A tatai őskőkori telep [The Paleolithic settlement at Tata].—N.p., Budapest. (In Hungarian)

Kormos, T. (1913). Sciurus gibberosus Hofm. im Miozän Ungarns.—Földtani Közlöny, 43: 36–38.

Kormos, T. (1914). Über die Resultate meiner Ausgrabungen im jahr 1913.—A MKFI Évi Jelentése 1913-ról: 559–604.

Kormos, T. (1917). Az ajnácskői pliocén rétegek és faunájuk [Pliocene beds at Ajnácskő and their fauna].—A Magyar Királyi Földtani Intézet Évi jelentése 1915-ről: 582–654. (In Hungarian)

Kormos, T. (1930). Beiträge zur Präglazialfauna des Somlyóberges bei Püspökfürdő.—Állattani Közlemények, 27: 40–56.

Kormos, T. (1934). Neue Insektenfresser, Fledermäuse und Nager aus dem Oberpliozän der Villányer Gegend.—Földtani Közlöny, 64(10–12): 296–321.

Korpásné Hódi, M. (1983). Palaeoecology and biostratigraphy of the Pannonian Mollusca fauna in the northern foreland of the Transdanubian Central Range.—A Magyar Állami Földtani Intézet Évkönyve, 66(1): 1–163.

Kovács, J. (1863). Földtani kirándulások Biharmegyében a Sebes és Fekete Kőrös vidékén [Geological excursions in the country of the Sebes and Fekete Körös Rivers, in Bihar County].—A Magyarhoni Földtani Társulat Munkálatai, II: 54–64. (In Hungarian)

Kovács, S. (1977). New conodonts from the North Hungarian Triassic.—Acta Min. Petr., 23(1): 77–90.

Kovács, S., (1978a). New sphinctozoan sponges from the North Hungarian Triassic.—N. Jb. Geol. Paläont., Mh., 1978(11): 685–697.

Kovács, S., (1978b). Newer calcareous sponges from the Wetterstein reef limestone of Alsóhegy Karstplateau (Silica nappe, Western Carpathians, North Hungary).—Acta Min. Petr., 23(2): 299–317.

Kovács, S. (1986). Conodont-biostratigraphical and microfacies investigations in the Hungarian part of the Northeastern Rudabánya Mts.—A Magyar Állami Földtani Intézet Évi Jelentése 1984-ről: 193–244. (In Hungarian with English abstract)

Kovács, S., Papšová, J., (1986). Conodonts from the Paraceratites binodosus Zone (Middle Triassic) from the Mecsek Mts., Southern Hungary, from the Choč nappe of the Low Tatra Mts., Czechoslovakia.—Geol. Zborn. Geol. Carpath., 37(1): 59–74.

Kovács, S., & Rálischné Felgenhauer, E. (2005). Middle Anisian (Pelsonian) conodonts from the Triassic of the Mecsek Mts. (South Hungary), their taxonomy, stratigraphic significance.—Acta. Geol. Hung., 48(1): 69–105.

Kovács, S., Rálischné Felgenhauer, E., & Bóna, J. (2006). Conodont colour alteration related to a half-graben structure, its controversion with vitrinite reflectance data: an example from the Mesozoic of the Mecsek, Villány Hills area, Tisza Megaunit, Southern Hungary.—Geol. Carpath, 57(4): 257–268.

Kováts, J. (1856a). Fossile Flora von Erdőbénye.—A Magyarhoni Földtani Társulat Munkálatai, 1: 1–37.

Kováts, J. (1856b). Fossile Flora von Tállya.—A Magyarhoni Földtani Társulat Munkálatai, 1: 39–52).

Kozur, H. (1970). Neue Ostracoden-Arten aus dem obersten Anis des Balaton-Hochlandes (Ungarn).—Ber. Nat.-Med. Ver. Innsbruck, 58, 1–40.

Kozur, H. (1984a). Muellerisphaeridae eine neue Ordnung von Mikrofossilien unbekannter systematischer Stellung aus dem Silur und Unterdevon von Ungarn.—Geol. Paläont. Mitt. Innsbruck, 13(6): 125–148.

Kozur, H. (1984b). Fossilien aus dem Silur von Ungarn (vorläufige Mitteilung).—Radovi Geoinsituta, 17: 163–174.

Kozur, H. (1985). Neue Ostracoden-Arten aus dem oberen Mittelkarbon (höhere Moskovian), Mittel und Oberperm des Bükk-gebirges (N-Ungarn).—Geol. Paläont. Mitt. Innsbruck, Sb. 2: 1–145.

Kozur, H. (1993). First Evidence of Liassic in the Vicinity of Csővár (Hungary), and its Paleogeographic and Paleotectonic Significance.—Jahrbuch der Geologischen Bundesanstalt, 136: 89–98.

Kozur, H., & Mock, R. (1977). On the age of the Paleozoic of the Uppony Mountains (North Hungary).—Acta Min. Petr., 23(1): 91–107.

Kozur, H., & Mostler, H. (1973). Mikrofaunistische Untersuchungen der Triasschollen im Raume Csővár, Ungarn.—Verh. Geol. B.-A., 1973(2): 291–325.

Krenner, J. (1867). Ajnácskő ősemlősei [Fossil mammals of Ajnácskő]—A Magyarhoni Földtani Társulat Munkálatai, 3: 115–132. (In Hungarian)

Kretzoi, M. (1929). Feliden-Studien.—Budapest, 22 pp.

Kretzoi, M. (1940). Alttertiäre Perissodactylen aus Ungarn.—Annales Mus. Nat. Hung., 33: 87–99.

Kretzoi, M. (1941a). Sirenavus hungaricus n. g. n. sp., ein neuer Prorastomidae aus dem Mitteleozän (Lutetium) von Felsőgalla in Ungarn.—Annales Mus. Nat. Hung., 34: 146–156.

Kretzoi, M. (1941b). Siebenbürgische Elotheriiden.—Földtani Közlöny, 71(7–12): 345–348.

Kretzoi, M. (1941c). Seehund-Reste aus dem Sarmat von Érd bei Budapest.—Földtani Közlöny, 71: 350–356.

Kretzoi, M. (1941d). Die unterpleistozäne Säugetierfauna von Betfia bei Nagyvárad.—Földtani Közlöny, 71: 384–417.

Kretzoi, M. (1941e). Mamonteus hungaricus, ein neues Waldmammut aus Ungarn.—Földtani Közlöny, 71(7–12): 343–345.

Kretzoi, M. (1942a). Necroteuthis n. g. (Ceph. Dibr., Necroteuthidae n. f.) aus dem Oligozän von Budapest und das System der Dibranchiata.—Földtani Közlöny, 72(1): 124–138.

Kretzoi, M. (1942b). Zwei neue Agriotheriiden aus dem ungarischen Pannon.—Földtani Közlöny, 72: 350–353.

Kretzoi, M. (1942c). Eomellivora von Polgárdi und Csákvár.—Földt. Közl. 72: 318–323.

Kretzoi, M. (1942d). Tigeriltis, Iltis und Nerz im Ungarischen Pleistozän.—Földtani Közlöny, 72: 323–344.

Kretzoi, M. (1942e). Der Moschusochs in Ungarn.—Földtani Közlöny, 72: 357–363.

Kretzoi, M. (1943). Kochictis centennii n. g. n. sp., ein altertümlicher Creodonte aus dem Oberoligozän Siebenbürgens.—Földtani Közlöny, 73: 190–195.

Kretzoi, M. (1950). Le grès à empreintes de pieds de Ipolytarnóc et la question de l'Aquitanien.—Földtani Közlöny, 80: 259–261.

Kretzoi, M. (1951a). Neuer Sirenen-Typus aus dem ungarischen Miozän.—Földtani Közlöny, 81: 438–439.

Kretzoi, M. (1951b). The Hipparion-fauna from Csákvár.—Földtani Közlöny, 81(10–12): 402–417.

Kretzoi, M. (1952). Die Raubtiere der Hipparionfauna von Polgárdi.—A Magyar Állami Földtani Intézet Évkönyve, 40(3): 1–38.

Kretzoi, M. (1956). Die altpleistozänen Wirbeltierfaunen des Villányer Gebirges.—Geol. Hung. Ser. Pal., 27: 264 pp.

Kretzoi, M. (1957a). Bird remains from the Hipparion-fauna of Csákvár.—Aquila, 63–64: 239–248.

Kretzoi, M. (1957b). Wirbeltierfaunistische Aufgaben zur Quarterchronologie der Jankovich-Hohle.—Folia Archeologica, 9: 16–21.

Kretzoi, M. (1958). En mémoire de O. Kadić.—Föld. Közl. 88: 13–21. (In Hungarian with French abstract)

Kretzoi, M. (1964). A gerinces állatok fejlődése [Evolution of the vertebrates]. In A. Tasnádi Kubacska (Ed.), Az élővilág fejlődéstörténete (pp. 351–545).—Gondolat Kiadó, Budapest. (In Hungarian)

Kretzoi, M. (1965). Die Hipparion-Fauna von Győrszentmárton in NW-Ungarn.—Annales Hist.-Nat. Mus. Nat. Hung., 57: 127–142.

Kretzoi, M. (1969). Sketch of the Late Cenozoic (Pliocene and Quaternary) terrestrial stratigraphy of Hungary.—Földrajzi Közlemények, 3: 179–204. (In Hungarian with English abstract)

Kretzoi, M. (1976a). Wichtigere Streufunde in der Wirbeltiersammlung der Ungarischen Geologischen Anstalt. (1. Mitteilung).—A Magyar Állami Földtani Intézet Évi Jelentése 1974-ről: 415–429. (In Hungarian with German abstract)

Kretzoi, M. (1976b). Die Hominisation und die Australopithecinen.—Anthrop. Közl., 20: 3–11.

Kretzoi, M. (1982). Wichtigere Streufunde in der Wirbeltiersammlung der Ungarischen Geologischen Anstalt. (7. Mitteilung).—A Magyar Állami Földtani Intézet Évi Jel. 1980-ról: 385–394.

Kretzoi, M. (1983). Wirbeltier-Indexformen in ungarischen Jungneozoikum: Hipparion.—A Magyar Állami Földtani Intézet Évi Jelentése 1981-ről: 513–522.

Kretzoi, M. (1984). New hominoid form from Rudabánya.—Anthrop. Közl., 28: 1–96.

Kretzoi, M. (1985a). Neuer Amphicyonide aus dem Altpannon von Pécs (Südungarn).—Annales Hist.-Nat. Mus. Nat. Hung., 77: 65–68.

Kretzoi, M. (1985b). Upper Pannonian. The fauna and faunal age of Sümeg-Gerinc.—Geol. Hung. Ser. Pal., 20: 225–234.

Kretzoi, M. (2002). The Fossil Hominoids of Rudabánya (Northeastern Hungary) and the Early Hominization.—Hungarian National Museum, Budapest, 287 pp.

Kretzoi, M., & Dobosi, V. T. (Eds.) (1990). Vértesszőlős, Man, Site and Culture.—Akadémiai Kiadó, Budapest, 555 pp.

Kretzoi, M., & Kretzoi, M. (2000). Index generum et subgenerum Mammalium, I–II. In W. Riegraf (Ed.), Fossilium Catalogus Animalia, Pars 137. Section 1–2 (pp. 1–433, 434–726).—Backhuys, Leiden.

Kretzoi, M., & Noszky, J. (1951). Une dent de Saurien de la formation bauxitique du Bakony.—Földtani Közlöny, 81:333.

Kriván, P. (1955). La division climatologique du Pléistocène en Europe Centrale et le profil de loess de Paks.—A Magyar Áállami Földtani Intézet Évkönyve, 43(3): 364–512.

Krolopp, E. (1995). Biostratigraphic division of Pleistocene formations in Hungary according to their mollusk fauna. In L. Fűköh, E. Krolopp, & P. Sümegi (Eds.), Quaternary Malacostratigraphy in Hungary (pp. 17–78).—Mátra Múzeum, Gyöngyös.

Kubinyi, F. (1864) (Ed.). Petényi Salamon János hátrahagyott munkái [Unpublished works of János Salamon Petényi].—Magyar Tudományos Akadémia, Budapest, xii + 130 pp. (In Hungarian)

Kutassy, A. (1927). Beiträge zur Stratigraphie und Paläontologie die Alpinen Triasschichten in der Umgebung von Budapest.—Mitt. Jb. Kön. Ung. Geol. Anst., 27(2): 108–175.

Kutassy, A. (1930). Das Vorkommen des Heterastridien in der ungarischen Trias.—Mathematisch und Naturwissenschaftliche Berichte aus Ungarn, 37: 111–126.

Kutassy, A. (1933). Weitere Beiträge zur Kentniss der Fauna des Dachsteinkalkes in der Umgebung von Budapest.—Matematikai és Természettudományi Értesítő, 49: 222–250.

Kutassy, A. (1934a). Pachyodonta mesozoica (Rudistis exclusis). Fossilium catalogus 1. Animalia 68.—Backhuys, Leiden.

Kutassy, A. (1934b). Die Fauna des norischen Dachsteinkalkes von St. Anna bei Neumarktl (Oberkrain).—Földtani Közlöny, 64(4–6): 68–80.

Kutassy, A. (1936). Faunen aus dem Hauptdolomit und Dachsteinkalk des Budaer Gebirges.—Matematikai és Természettudományi Értesítő, 54: 1006–1054.

Kutassy, A. (1937). Triadische Faunen aus dem Bihar-Gebirge, 1: Gastropoden.—Geol. Hung. Ser. Pal., 13: 78 pp.

Kvaček, Z., & Hably, L. (1991). Notes on the Egerian stratotype flora at Eger (Wind brickyard), Hungary, Upper Oligocene.—Annales Hist.-Nat. Mus. Nat. Hung., 83: 49–82.

Laczkó, D. (1911). Die geologischen Verhältnisse von Veszprém und seiner Umgebung.—Resultate Wiss. Erforsch. Balatonsees, 1, 1 (Geol., Petrograph., Miner. und Mineralchem. Anhang): 1–207.

Lambrecht, K. (1929). Mesozoische und tertiäre Vogelreste aus Siebenbürgen. In E. Csiki (Ed.), Xe Congrès International de zoologie tenu à Budapest du 4 au 10 septembre 1927 (pp. 1262–1275).—Imprimerie Stephaneum, Budapest.

László, J. (1992). Fossil plant remains from the lignite open mining at Bükkábrány.—A Magyar Állami Földtani Intézet évi jelentése 1990-ről: 321–337. (In Hungarian with English abstract)

Lelkesné Felvári, Gy. (1978). Petrographische Untersuchung einiger Präpermische Bildungen der Balaton-Linie.—Geol. Hung. Ser. Geol., 18: 193–295.

Lelkesné Felvári, Gy., Kovács, S., & Majoros, Gy. (1984). Lower Devonian pelagic limestone in borehole Kékkút 4, Bakony Mts.—A Magyar Állami Földtani Intézet Évi Jelentése 1982-ről: 289–315. (In Hungarian with English abstract)

Lenner, K. (1989). Do Triassic Bellerophontacea exist in Hungary?—A Magyar Állami Földtani Intézet évi jelentése 1987-ről: 267–279. (In Hungarian with English abstract)

Less, G. (1987). Paleontology and stratigraphy of the European Orthophragminae.—Geol. Hung. Ser. Pal., 51: 373 pp.

Lóczy, L., Jr. (1915). Monographie der Villányer Callovien-Ammoniten.—Geol. Hung. Ser. Geol., 1(3–4): 255–502.

Lőrenthey, I. (1903). Zwei neue Schildkrötenarten aus dem Eozän von Kolozsvár.—Földtani Közlöny, 33(5–6): 249–266.

Lőrenthey, I. (1907). Vannak-e juraidőszaki rétegek Budapesten? [Are there any Jurassic rocks in Budapest?].—Földtani Közlöny, 37(9–11): 359–368. (In Hungarian)

Lőrenthey, I. (1911). Beiträge zur Fauna und stratigraphischen Lage der pannonischen Schichten in der Umgebung des Balatonsees.—Resultate Wiss. Erforsch. Balatonsees, Paläont. Anhang, 4(2): 216 pp.

Lőrenthey, I., & Beurlen, K. (1929). Die fossilen Dekapoden der Länder der Ungarischen Krone.—Geol. Hung. Ser. Pal., 3: 420 pp.

Lupu, D. (1976). Contributions à l'étude des rudistes sénoniens des monts Apuseni.—Institut de Géologie et Géophysique Mémoires, 24: 83–152.

Lupu, D. (1992). Les faciès à rudistes du Cénomanien de Haţeg (Carpathes meridionales, Roumanie).—Geologica Romana, 28: 351–358.

Magyar, I. (1988). Mollusc fauna and flora of the Pannonian quartz sandstone at Mindszentkálla, Hungary.—Annales Univ. Sci. Budapest., Sect. Geol., 28: 209–222.

Magyar, I. (2004). What lesson can we learn from the hundred-year history of Lake Pannon mollusc biostratigraphy in Hungary?—Földtani Közlöny, 134(3): 369–390. (In Hungarian with English abstract)

Magyar, I. (2010). A Pannon-medence ősföldrajza és környezeti viszonyai a késő miocénben őslénytani és szeizmikus rétegtani adatok alapján [Paleobiogeography and paleoenvironments of the Pannonian Basin during the Late Miocene as inferred from paleontological and seismic data].—GeoLitera, Szeged, 140 pp. (In Hungarian)

Majer, I. (1928). Die Stammesgeschichtliche Stellung vom Ursus böckhi Schl.—Földtani Közlöny, 56: 197–200.

Majoros, G. (1964). Reptilian footprint from the Permian of Balatonrendes.—Földtani Közlöny, 94(2): 243–245. (In Hungarian with English abstract)

Majzon, L. (1966). Foraminifera-vizsgálatok [Studies on Foraminifers].—Akadémiai Kiadó, Budapest, 939 pp. (In Hungarian)

Makádi, L. (2005). Moszaszauruszok a Bakonyból [Mosasaurs from the Bakony Mountains].—Élet és Tudomány, 39: 1236–1238. (In Hungarian)

Makádi, L. (2006). Bicuspidon aff. hatzegiensis (Squamata: Scincomorpha: Teiidae) from the Upper Cretaceous Csehbánya Formation (Hungary, Bakony Mts).—Acta Geologica Hungarica, 49(4): 373–385.

Makádi, L., Botfalvai, G., & Ősi, A. (2006). The Late Cretaceous continental vertebrate fauna from the Bakony Mountains I:

Fishes, amphibians, turtles, squamates.—Földtani Közlöny, 136(4): 487–502. (In Hungarian with English abstract)

Manni, R., Nicosia, U., & Szabó, J. (1992). Late Jurassic crinoids from the Eperkés-hegy (Bakony Mts., Hungary).—Fragmenta Mineralogica et Palaeontologica, 15: 115–137.

Mártonfi, L. (1890). Anthracotherium magnum Cuv. Kis-Krisztolczról [Anthracotherium magnum Cuv. from Kis-Krisztolcz].—Értesítő az Erdélyi Múzeum-Egylet Orvos-Természettudományi Szakosztályából II: Természettudományi Szak 15, 12(3): 317–324. (In Hungarian)

Mazin, J.-M., & Pinna, G. (1993). Palaeoecology of the armoured placodonts.—Paleontologia Lombarda, n.s., 2: 83–91.

Medzihradszky, Z. (2002). A Pannon-medence növénytakarójának múltja [The past of the vegetation of the Pannonian Basin].—História, 24(5–6): 15–21. (In Hungarian)

Méhes, Gy. (1911) Über Trias-Ostrakoden aus dem Bakony.—Resultate Wiss. Erforsch. Balatonsees, Paläont. Anhang, 3(6): 38 pp.

Méhes, Gy. (1936). Die eozänen Ostracoden der Umgebung von Budapest.—Geol. Hung. Ser. Pal., 12: 56 pp.

Méhes, K. (1964). The foraminiferal genus Orbitolina from Hungary.—Acta Geologica Hungarica, 8(1–4): 265–281.

Mészáros, F., & Gazda, I. (2000). Petényi S. János emlékezete születésének 200. évfordulóján [Memory of János S. Petényi, on occasion of the bicentenary of his birthday].—Magyar Tudománytörténeti Intézet, Magyar Természettudományi Múzeum, Piliscsaba-Budapest, 179 pp. (In Hungarian)

Mészáros, L. Gy. (1999a). An exceptionally rich Soricidae (Mammalia) fauna from the Upper Miocene localities of Polgárdi (Hungary).—Annales Univ. Sci. Budapest., Sect. Geol., 32: 5–34.

Mészáros, L. Gy. (1999b). Some insectivore (Mammalia) remains frim the Late Miocene locality of Alsótelekes (Hungary).—Annales Univ. Sci. Budapest., Sect. Geol., 32: 35–47.

Mészáros, L. Gy. (1999c). Taphonomical observations on Late Miocene Soricids (Mammalia).—Földtani Közlöny, 129: 159–178. (In Hungarian with English abstract)

Mészáros, L. Gy. (2000a). New results for the Late Miocene Soricidae stratigraphy in the Pannonian Basin.—Newsl. Stratigr., 38: 1–11.

Mészáros, L. Gy. (2000b). Palaeogeography and environment of the Late Miocene Soricidae (Mammalia) faunae of the Pannonian Basin.—Annales Univ. Sci. Budapest., Sect. Geol., 33: 107–120.

Mészáros, N. (1957). Fauna de moluste a depozitelor paleogene din Nord-Vestul Transilvaniei.—Acad. R. P. R., Bucharest, 174 pp.

Meznerics, I. (1944). Die Brachiopoden des ungarischen Tertiärs.—Annales Hist.-Nat. Mus. Nat. Hung., 36: 10–60.

Mihály, S. (1973). Revision der unterkarbonischen Korallen des Kőszár-Berg von Szabadbattyán.—A Magyar Állami Földtani Intézet Évi Jel. 1971-ről: 249–276. (In Hungarian with German abstract)

Mihály, S. (1975). Palaeoecological observation in the Middle Eocene of Gánt, Hungary.—Földtani Közlöny, 105(1): 75–81. (In Hungarian)

Mihály, S. (1978). Die Mitteldevonischen Tabulaten des Szendrőer Gebirges.—Geologica Hungarica Series Geologica, 18: 117–191.

Mihály, S. (1980). Felsőkarbon növénymaradványok a fülei Kő-hegyről [Upper Carboniferous plant remains from the Kő-hegy of Füle].—Veszprém Megyei Múz. Közl., 15: 21–28. (In Hungarian)

Mihály, S. (1985). Late Badenian Echinoidea from new exposures in Budapest.—A Magyar Állami Földtani Intézet Évi Jelentése 1983-ról: 235–262. (In Hungarian with English abstract)

Mihály, S. (1989). Two new Echinoidea species from the Miocen of Hungary.—A Magyar Állami Földtani Intézet Évi Jelentése 1987-ről: 237–249. (In Hungarian with English abstract)

Mihály, S., & Solt, P. (1983). Acrodus tooth from the Upper Permian of the Bükk Mountains (NE-Hungary).—A Magyar Állami Földtani Intézet Évi Jelentése 1981-ről: 209–212. (In Hungarian with English abstract)

Mihály, S., & Vincze, P. (1984). New paleoecological remarks concerning the Middle Eocene beds of the Bagoly-hegy at Gánt, Transdanubia, Hungary.—Földtani Közlöny, 114(3): 263–283. (In Hungarian)

Młynarski, M. (1966). Die fossilen Schildkröten in den ungarischen Sammlungen.—Acta Zoologica Cracoviensia, (11): 223–303.

Moissette, P., Dulai, A., & Müller, P. (2006). Bryozoan faunas in the Middle Miocene of Hungary: biodiversity and biogeography.—Palaeoegeography, Palaeoclimatology, Palaeoecology, 233: 300–314.

Molnár, B. (1991). The Paleontological Collection of the József Attila University, Szeged. In Gy. Vitális & T. Kecskeméti (Eds.), Museums and Collections in the History of Mineralogy, Geology and Paleontology in Hungary.—Annals of the History of Hungarian Geology Special Issue, 3 (pp. 255–262).—Hungarian Geological Society, Budapest.

Monostori, M. (1978). A szabadbattyáni karbon mészkő mikrofaunája [Microfauna of the carboniferous limestone at Szabadbattyán].—Geol. Hung. Ser. Geol., 18: 297–313. (In Hungarian)

Monostori, M. (1985). Eocene ostracods from the Dorog Basin (Northern Transdanubia, Hungary).—Akadémiai Kiadó, Budapest, 214 pp.

Monostori, M. (1991): *Triadogigantocypris balatonica* n. g. n. sp.: a giant ostracode from the Hungarian Triassic.—Neues Jb. Geol. Paläont., Monatshefte, 1991(2): 91–96.

Monostori, M. (1994). Ostracod evidence of the Carnian Salinity Crisis in the Balaton Highland, Hungary.—Neues Jb. Geol. Paläont., Abh., 193(3): 311–331.

Monostori, M. (1995a). Environmental significance of the Anisian Ostracoda fauna from the Forrás Hill near Felsöörs (Balaton Highland, Transdanubia, Hungary).—Acta Geologica Hungarica, 39(1): 37–56.

Monostori, M. (1995b). Bathonian Ostracods from the Mecsek Mts (South Hungary).—Annales Univ. Sci. Budapest., Sect. Geol., 30: 151–176, 231–234.

Monostori, M. (1996a). Ostracods and charophytes from the Triassic Kantavár Formation, Mecsek Mts., Hungary.—Acta Geol. Hung., 39(3): 311–317.

Monostori, M. (1996b). Eocene ostracods of Hungary Systematical part 1 (Cytheracea 1).—Annales Univ. Sci. Budapest, Sect. Geol., 31: 27–74.

Monostori, M. (1998). Eocene ostracods of Hungary Systematical part 2 (Cytheracea 2).—Hantkeniana, 2: 49–101.

Monostori, M. (2000a). Ostracoda fauna of the Pénzeskút Marl Formation (Albian-Cenomanian) of Bakony Mountains (Hungary).—Annales Univ. Sci. Budapest., Sect. Geol., 33: 5–61.

Monostori, M. (2000b). Eocene ostracods of Hungary part 3 (Cytheracea 3).—Annales Univ. Sci. Budapest., Sect. Geol., 33: 63–103.

Monostori, M. (2004). Lower Oligocene (Kiscellian) ostracods in Hungary. Systematic description.—Annales Univ. Sci. Budapest., Sect. Geol., 34: 27–141.

Monostori, M. (2006). Magyarország ősmaradványai—kainozoós ősállatok [Fossils of Hungary—Animal remains from the Cenozoic].—Hantken Press, Budapest, 139 pp. (In Hungarian)

Mottl, M. (1932). Zur Morphologie der Höhlenbärenschädel aus der Igric-Höhle.—Mitt. Jb. Kön. Ung. Geol. Anst., 29(4): 194–246.

Mottl, M. (1935). Wisentskelett in Museum der Kgl. Ung. Geologischen Anstalt.—Földtani Közlöny, 65(10–12): 363–364.

Mottl, M. (1939). Die mittelpliozäne Säugetierfaune von Gödöllő bei Budapest.—A Magyar Királyi Földtani Intézet Évkönyve, 32(3): 257–350.

Müller, P. (1984). Decapod crustacea of the Badenien.—Geol. Hung. Ser. Pal., 42: 317 pp.

Müller, P. (2006). New Decapods from the Miocene of Hungary.—Földtani Közlöny, 136(1): 25–37.

Müller, P., & Collins, J. S. H. (1991). Late Eocene coral-associated decapods (Crustacea) from Hungary.—Contr. Tert. Quatern. Geol., 28(2–3): 47–92.

Müller, P., & Magyar, I. (1991). Continuous record of the evolution of lacustrine cardiid bivalves in the late Miocene Pannonian Lake.—Acta Palaeontologica Polonica, 36(4): 353–372.

Nagy, E. (1968). Triasbildungen des Mecsek-Gebirges.—A MÁFI Évkönyve, 61(1): 198 pp.

Nagy, E. (1985). Sporomorphs of the Neogene in Hungary.—Geol. Hung. Ser. Pal., 47: 471 pp.

Nagy, E. (1992). A Comprehensive Study of Neogene Sporomorphs in Hungary. Geol. Hung. Ser. Pal., 53: 379 pp.

Nagy, E. (2005). Palynological evidence for Neogene climatic change in Hungary.—Magyar Állami Földtani Intézet, Budapest, 120 pp.

Nagy, I. (1986). Investigation on Calpionellids from the Mecsek Moutains—Acta Geologica Hungarica, 29(1–2): 45–64.

Nagy, I. Z. (1958). Teudopsis subacuta n. sp. aus dem Mecseker Lias.—Földtani Közlöny, 88(2): 240–242. (In Hungarian with English abstract)

Nagy, I. Z. (1961). Liassic plant remains of the Mecsek Mountains.—A Magyar Állami Földtani Intézet Évkönyve, 49(2): 609–658.

Nagy, I. Z. (1963). Nautiloides Crétacés de la Hongrie.—Földtani Közlöny, 93(1): 203–221. (In Hungarian with French abstract)

Nagy, I. Z. (1967). Unterkretazische Cephalopoden aus dem Gerecse-Gebirge I.—Annales Hist.-Nat. Mus. Nat. Hung., Pars. Min. et Palaeont., 59: 53–79.

Nagy, I. Z. (1968). Unterkretazische Cephalopoden aus dem Gerecse-Gebirge II.—Annales Hist.-Nat. Mus. Nat. Hung., Pars Min. et Palaeont., 60: 41–60.

Nagymarosy, A. (1980). The correlation of the Badenian in Hungary based on nannofloras.—Annales Univ. Sci. Budapest., Ser. Geol., 25: 33–86.

Nagymarosy, A. (1985). Mono- and duospecific nannofloras in the Lower Oligocene sediments of Hungary.—Őslénytani Viták, 31: 29–31. (In Hungarian with English abstract)

Naish, D., & Dyke, G. J. (2004). Heptasteornis was no ornithomimid, troodontid, dromaeosaurid or owl: The first alvarezsaurid (Diposauria: Theropoda) from Europe.—N. Jb. Geol. Palaont., Mh., 2004: 385–401.

Németh, G. (1989). Dr. Strausz László emlékezete [In memoriam Dr. László Strausz].—Földtani Közlöny, 119: 325–330. (In Hungarian)

Neugeboren, J. L. (1851). Zur vorweltlichen Conchyliologie Siebenburgens.—Verh. u. Mitt. d. Sieben. Ver. f. Naturw. in Hermannstadt, 2: 4–9.

Neumayr, M. (1873). Die Fauna der Schichten mit Aspidoceras acanthicum—Abh. k. k. Geol. Reichsanst., 5(6): 141–257.

Neumayr, M., & Paul, C. M. (1875). Die Congerien- und Paludinenschichten Slavoniens und deren Faunen. Ein Beitrag zur Descendenz-Theorie.—Abhandlungen der k.k. Geologischen Reichsanstalt, 7: 1–111.

Nolf, D., & Brzobohaty, R. (1994). Fish Otoliths from the Late Oligocene (Eger and Kiscell Formations) in the Eger Area (Northeastern Hungary).—Bulletin de L'Institut Royal des Sciences Naturelles de Belgique: Sciences de la Terre, 64: 225–252.

Nopcsa, F. (1899). Dinosaurierreste aus Siebenbürgen (Schadel von Limnosaurus transsylvanicus nov. gen. et spec.).—Denkschriften d. Kais. Akad. d. Wiss. Math. Naturw. Classe, 68: 555–591.

Nopcsa, F. (1902). Dinosaurierreste aus Siebenbürgen II (Schadelreste von Mochlodon). Mit einem Anhange: Zur Phylogenie der Ornithopodiden.—Denkschriften d. Kais. Akad. d. Wiss. Math. Naturw. Classe, 72:149–175.

Nopcsa, F. (1904). Dinosaurierreste aus Siebenbürgen III: Weitere Schadelreste von Mochlodon.—Denkschriften d. Kais. Akad. d. Wiss. Math. Naturw. Classe, 74: 229–264.

Nopcsa, F. (1915). Die Dinosaurier der Siebenbürgischen Landesteile Ungarns.—Mitt. Jb. Kön. Ung. Geol. Anst., 23(1): 1–24.

Nopcsa, F. (1923). Kallokibotium, a primitive amphichelydean tortoise from the Uppermost Cretaceous of Hungary.—Palaeontologia Hungarica, 1: 1–33.

Nopcsa, F. (1928). Palaeontological notes on Reptiles.—Geol. Hung. Ser. Pal., 1(1): 75–84.

Nopcsa, F. (1929). Dinosaurierreste aus Siebenbürgen V: Struthiosaurus transsylvanicus.—Geol. Hung. Ser. Pal., 4: 1–76.

Noszky, J. (1934). Adatok az északi Bakony krétaképződményeinek ismeretéhez.[Contributions to the knowledge of the Cretaceous of the northern Bakony].—Földtani Közlöny, 64(1): 99–136. (In Hungarian with German abstract)

Oravecz, J. (1964). Silurbildungen in Ungarn und ihre regionalen Beziehungen.—Földtani Közlöny, 94(1): 3–9. (In Hungarian with German abstract)

Oraveczné Scheffer, A. (1987). Triassic foraminifers of the Transdanubian Central Range.—Geol. Hung. Ser. Pal., 50: 331 pp.

Ősi, A. (2001). Middle Eocene turtles from Nyires-puszta (Bakony Hills, Hungary).—Földtani Közlöny, 131(3–4): 353–360.

Ősi, A. (2003). Dinoszauruszok Magyarországról [Dinosaurs from Hungary].—Élet és Tudomány 58(1): 6–8. (In Hungarian)

Ősi, A. (2004). Halászó repülő hüllők a magyarországi felsőkrétából (Bakony) [Fishing reptiles from the Upper Cretaceous of Hungary]. In Program, Előadáskivonatok, Kirándulásvezető, 7: Magyar Őslénytani Vándorgyűlés, 2004, május 6–8, Beremend (p. 20). (In Hungarian)

Ősi, A. (2005a). Krokodilok Magyarországról [Crocodiles from Hungary].—Élet és Tudomány 60(8): 230–232. (In Hungarian) Ősi, A. (2005b). Hungarosaurus tormai, a new ankylosaur (Dinosauria) from the Upper Cretaceous of Hungary.—Journal of Vertebrate Paleontology, 25(2): 370–383.

Ősi, A. (2008a). Cranial osteology of Iharkutosuchus makadii, a Late Cretaceous basal eusuchian crocodiyliform from Hungary.—Neues Jb. Geol. Paläont., Abh., 243(2): 169–177.

Ősi, A. (2008b). Enantiornithine bird remains from the Late Cretaceous of Hungary.—Oryctos, 7: 55–67.

Ősi, A., Apesteguía, S. M., & Kowalewski, M. (2010). Non-avian theropod dinosaurs from the early Late Cretaceous of Central Europe.—Cretaceous Research, 31: 304–320.

Ősi, A., Barbacka, M., & Szente, I. (2005a). Dino-ösvény [Dinopath].—Hantken Press, Budapest, 32 pp. (In Hungarian)

Ősi, A., Butler, R. J., & Weishampel, D. B. (2010). A Late Cretaceous ceratopsian dinosaur from Europe with Asian affinities.—Nature, 465: 466–468.

Ősi, A., Clark, J. M., & Weishampel, D. B. (2007). First report on a new basal eusuchian crocodyliform with multi-cusped teeth from the Upper Cretaceous (Santonian) of Hungary. Neues Jb. Geol. Paleont. Abh., 243(2): 169–177.

Ősi, A., Jianu, C. M., & Weishampel, D. B.(2003). Dinosaurs from the Upper Cretaceous of Hungary. In A. Petculescu, C. Rădulescu, E. Ştiucă, & P. M. Samson (Eds.), Advances in Vertebrate Paleontology: Hen to Panta.—Romanian Academy "Emil Racovita," Institute of Speleology, Bucharest, pp. 117–120.

Ősi, A., & Makádi, L. (2009). New remains of Iharkutosuchus tormai (Ankylosauria, Dinosauria) from the Upper Cretaceous of Hungary: Skeletal reconstruction and body mass estimation.—Paläont. Zeitschrift, 83: 227–245.

Ősi, A., Pálfy, J., Makádi, L., Szentesi, Z., Gulyás, P., Rabi, M., Botfalvai, G., & Hips, K. (2011). Hettangian (Early Jurassic) dinosaur tracksites from the Mecsek Mountains, Hungary.—Ichnos, 18(2): 79–94.

Ősi., A, Prondvai, E., & Géczy, B. (2010). The history of Late Jurassic pterosaurs housed in Hungarian collections and the revision of the holotype of Pterodactylus micronyx Meyer 1856 (the "Pester Exemplar"). In R. T. J. Moody, E. Buffetaut, D. Naish, and D. M. Martill (Eds.), Dinosaurs and Other Extinct Saurians: A Historical Perspective. Geological Society Special Publications, 343 (pp. 277–286).—Geological Society, London.

Ősi, A., & Rabi, M. (2006). The Late Cretaceous continental vertebrate fauna from the Bakony Mountains II: Crocodiles, dinosaurs (Theropoda, Aves, Ornitischia), pterosaurs.—Földtani Közlöny, 136(4): 503–526. (In Hungarian with English abstract)

Ősi, A., & Weishampel, D. B. (2009). Jaw mechanism and dental function in the Late Cretaceous basal Eosuchian Iharkutosuchus.—Journal of Morphology, 270(8): 903–920.

Ősi, A., Weishampel, D. B., & Jianu, C. M. (2005). First evidence of azhdarchid pterosaurs from the Late Cretaceous of Hungary.—Acta Paleontologica Polonica, 50(4): 777–787.

Ozsvárt, P. (2005). Oriented thin-sections of isolated Paleogene benthic foraminifera: A catalogue from the micropalaeontological collection of the Hungarian Natural History Museum, Budapest.—Fragmenta Palaeontologica Hungarica, 23: 51–58.

Ozsvárt, P. (2007). Middle and Late Eocene benthic foraminiferal fauna from the Hungarian Paleogene Basin: Systematics and paleoecology.—Geologica Pannonica, Special Papers: 1–129.

Pálfalvy, I. (1969). Dr. Andreánszky Gábor emlékezete [In memoriam Dr. Gábor Andreánszky].—Földtani Közlöny, 99(1): 13–17. (In Hungarian)

Pálfalvy, I. (1976). Fossil plants in the "Sandstone with footprints" at Ipolytarnóc Village (N Hungary).—A Magyar Állami Földtani Intézet Évi Jelentése 1974-ről: 95–96.

Pálfy, J. (1988). Middle Triassic rhynchonellids from the Balaton Highland (Transdanubian Central Range, Hungary).—Annales Hist. Nat. Mus. Nat. Hung., 80: 25–46.

Pálfy, J. (2003). The Pelsonian brachiopod fauna of the Balaton Highland.—Geol. Hung. Ser. Pal., 55: 139–158.

Pálfy, J., Demény, A., Haas, J., Carter, E. S., Görög, Á., Halász, D., Oravecz-Scheffer, A., Hetényi, M., Márton, E., Orchard, M. J., Ozsvárt, P., Vető, I., & Zajzon, N. (2007). Triassic-Jurassic boundary events inferred from integrated stratigraphy of the Csővár section, Hungary.—Palaeogeography, Palaeoclimatology, Palaeoecology, 244: 11–33.

Pálfy, J., Dulai, A., Gasparik, M., Ozsvárt, P., Pazonyi, P., & Szives, O. (2008). Catalogue of Invertebrate and Vertebrate Paleontological Type Specimens oft he Hungarian Natural History Museum.—Hungarian Natural History Museum, Budapest, 209 pp.

Pálfy, J., Mundil, R., Renne, P. R., Bernor, R. L., Kordos, L., & Gasparik, M. (2007). U-Pb and 40Ar/39Ar dating of the

Miocene fossil track site at Ipolytarnóc (Hungary) and its implications.—Earth and Planetary Science Letters, 258: 160–174.

Pálfy, J., & Pazonyi, P. (Eds.). (2007). Ősléntani kirándulások Magyarországon és Erdélyben [Paleontological excursions in Hungary and Transylvania].—Hantken Press, Budapest, 260 pp. (In Hungarian)

Pálfy, J., & Török, Á. (1992). Comparison of Alpine and Germano-type Middle Triassic brachiopod faunas from Hungary, with remarks on Coenothyris vulgaris (Schlotheim 1820).—Annales Univ. Sci. Budapest., Sect. Geol., 29: 303–323.

Pálfy, M. (1902). Die oberen Kreideschichten in der Umgebung von Alvincz.—Mitt. Jb. Kön. Ung. Geol. Anst., 17(3): 244–348.

Pálfy, M. (1928). Dr. Bodrogi Koch Antal emlékezete [In memoriam Antal Koch].—Földtani Közlöny, 58(1): 7–14. (In Hungarian)

Papp, K. (1905). Heterodelphis leiodontus nova forma aus dem miozänen Schichten des Comitates Sopron in Ungarn.—Mitt. Jb. Kön. Ung. Geol. Anst., 14(2): 26–60.

Papp, K. (1911). Trias-Korallen aus dem Bakony.—Resultate Wiss. Erforsch. Balatonsees, Paläont. Anhang, 1(5): 23 pp.

Pászti, A. (2004). A pulai alginitbánya halfosszíliáinak őslénytani és paleoökológiai vizsgálata [Paleontological and paleoecological tudy of fish fossils from the alginite quarry at Pula].—A Nógrád Megyei Múzeumok Évkönyve, 27–28: 291–297. (In Hungarian)

Patrulius, D. (1974). Duranddelgaia et Miseia, deux nouveaux genres de rudistes du Sénonien de Pădurea Craiului (Monts Apuseni).—Dări de seamă ale şedinţelor (3: Paleontologie), 60: 169–180.

Paucă, M. (1934). Die fossile Fauna und Flora aus dem Oligozän von Suslanesti-Muscel in Rumanien. Eine systematische und paläobiologische Studie.—Ann. Inst. Geol. Rumanie, 16: 577–663.

Pávay, E. (1871). A Kolozsvár és Bánfy-Hunyad közötti vasútvonal ingadozó talajának geológiai szerkezete [Geological structure of the unstable soil exposed along the railway between Kolozsvár and Bánfy-Hunyad].—Földtani Közlöny, 1(7): 130–145. (In Hungarian)

Pávay, E. (1873). Die geologischen Verhältnisse der Umgebung von Klausenburg.—Mitt. Jb. Kön. Ung. Geol. Anst., 1(5): 352–441.

Pávay, E. (1879). Die fossilen Seeigel des Ofner Mergels.—Mitt. Jb. Kön. Ung. Geol. Anst., 3(2): 1–179.

Pazonyi, P. (2006). A Kárpát-medence kvarter emlősfauna-közösségeinek paleoökológiai és rétegtani vizsgálata [Paleoecological and stratigraphical study of Quaternary mammal assemblages of the Carpathian Basin]. Unpublished doctoral dissertation, Eötvös Loránd University, 113 pp. (In Hungarian)

Peters, K. F. (1862): Über den Lias von Fünfkirchen.—Sitzungsberichte kais. Akad. Wiss., Mat.-Naturwiss. Kl., 46(6): 241–293.

Pethő, G. (1906). Die Kreide (Hypersenon) Fauna des Peterwardeiner (Pétervarader) Gebirges (Fruška Gora).—Palaeontographica, 52: 57–331.

Pienkowski, G., Popa, M. E., & Kedzior, A. (2009). Early Jurassic sauropod footprints of the Southern Carpathians, Romania: Palaeobiological and palaeogeographical significance.—Quart. Geol., 53(4): 461–470.

Piros, O. (2002). Anisian to Carnian carbonate platform facies and dasycladacean biostratigraphy of the Aggtelek Mts, Northeastern Hungary.—Acta Geologica Hungarica, 45(2): 119–151.

Pocsai, T., & Sasvári, Á. (2005). Úrkút, Csárda-hegy természetvédelmi terület: tanösvény-vezetőfüzet [Úrkút, nature conservation area: A guide to the study path].—Pangea Egyesület, Pénzesgyőr, 50 pp. (In Hungarian)

Počta, Ph. (1886). Über einige Spongien aus dem Dogger der Fünfkirchner Gebirges.—Mitt. Jb. Kön. Ung. Geol. Anst., 8(3): 110–121.

Pongrácz, L. (1999). A beremendi Szőlő-hegy természettudományi kutatásának 150 éve [150 years of research at the Szőlő Hill of Beremend].—Petényi-emlékkönyv, Beremend, 149 pp. (In Hungarian)

Popa, M. E. (2009). Late Palaeozoic and Early Mesozoic Continental Formations of the Resiţa Basin.—Editura Universitatii din Bucuresti, Bucharest, 197 pp.

Posmoşanu, E. (2003). Iguanodontian dinosaurs from the Lower Cretaceous bauxite site from Romania.—Acta Palaeontologica Romaniae, 4: 431–439.

Posmoşanu, E., & Cook, E. (2000). Vertebrate taphonomy and dinosaur palaeopathology from a lower Cretaceous bauxite lens, northwest Romania.—Oryctos, 3: 39–51.

Primics, Gy. (1890). Spuren des Höhlenbären (Ursus spelеus Blumenb.) in Ungarn.—Földtani Közlöny, 20(5–7): 213–224.

Prinz, Gy. (1904). Die Fauna der älteren Jurabildungen im nordöstlichen Bakony.—Mitt. Jb. Kön. Ung. Geol. Anst., 15(1): 1–142.

Prinz, J. (1906). Die Nautiliden in der Unteren Jura-Periode.—Annales Musei Nationalis Hungarici, 4: 201–243.

Rabi, M. (2004). Az első magyarországi késő-kréta krokodilus leletek [First finds of crocodyles from the Upper Cretaceous of Hungary]. In Program, Előadáskivonatok, Kirándulásvezető, 7: Magyar Őslénytani Vándorgyűlés, 2004, május 6–8, Beremend (p. 23). (In Hungarian)

Ráki, M. (2001). Csontoshalfogak a máriahalmi homokbányából [Teeth of bony fish from the sand-pit at Máriahalom]. In Program, előadáskivonatok, kirándulásvezető, 4: Magyar Őslénytani Vándorgyűlés, 2001, május 4–5, Pécsvárad (p. 32). (In Hungarian)

Rákosi, L. (1989). Some new Thallophyta remains from the Hungarian Upper Cretaceous and Tertiary.—A Magyar Állami Földtani Intézet Évi Jel. 1988-ról, 2: 15–47.

Rákosi, L. (1991). Az együttes-zónák korrelációs kísérlete a palynológiai vizsgálatok alapján [An essay of the correlation of assemblage zones on the basis of palynological studies].—Unpublished manuscript (cited in T. Kecskeméti [1998], A magyarországi epikontinentális eocén képződmények rétegtana. In I. Bérczi & Á. Jámbor [Eds.], A magyarországi geológiai képződmények rétegtana [pp. 403–417].—MOL, MÁFI, Budapest). (In Hungarian)

Rakós, M. (1994). Revision of ammonites from Marianka shales (Little Carpathians near Bratislava).—Mineralia Slovaca, 26(2): 118–125.

Rakusz, G. (1932). Die oberkarbonischen Fossilien von Dobšiná und Nagyvisnyó.—Geol. Hung. Ser. Pal., 8: 223 pp.

Rásky, K. (1945). Fossile Charophyten-Früchte aus Ungarn.—Magyar Nemzeti Múzeum, Budapest, 74 pp.

Rásky, K. (1954). Lower Cretaceous fossil plants from Hungary.—Földtani Közlöny, 84(1–2): 83–90.

Rásky, K. (1959). The fossil flora of Ipolytarnóc.—Journal of Palaeontology, 33(3): 453–461.

Reichenbacher, B., & Codrea, V. (1999). Fresh- to brackish water fish faunas from continental Early Oligocene deposits in the Transylvanian Basin (Romania).—Bulletin de l'Intitut Royal des Sciences Naturelles de Belgique Science de la Terre, 69: 197–207.

Reuss, A. (1872). Die fossilen Korallen des österreich-ungarischen Miozän.—Denkschriften d. Kais. Akad. d. Wiss. Math. Naturw. Classe, 31: 74 pp.

Rieppel, O. (2001). The cranial anatomy of Placochelys placodonta Jaekel, 1902, and a review of the Cyamodontoidea (Reptilia, Placodonta).—Fieldiana (Geology), 45: 1–10.

Rómer, F. (1860). A Bakony [The Bakony]. 101 pp. (In Hungarian)

Rosovskaya, S. E. (1963). Bükkhegységi Fusulinidák [Fusulinids from the Bükk Mountains].—Geol. Hung. Ser. Pal., 28: 43 pp. (In Russian)

Rotarides, M. (1931). A lösz csigafaunája, összevetve a mai faunával, különös tekintettel a szegedvidéki löszökre [Gastropod fauna of the loess, compared to the recent fauna, with special refernce to the loess at the surroundings of Szeged].—A Szegedi Alföldkutató Bizottság Könyvtára, VI: Szakosztály: Állattani Közlemények, 8: 1–180. (In Hungarian)

Rozlozsnik, P. (1924). Nummulinák Magyarország óharmadkori rétegeiből [Nummulites from the older Tertiary beds of Hungary].—Földtani Szemle, 1(4): 159–189. (In Hungarian)

Rozlozsnik, P. (1939). Geologische Studien am Nordfusse des Mátra-Gebirges in der Umgebung der Gemeinden Parád, Recsk und Mátraderecske]—A Magyar Királyi Földtani Intézet Évi Jelentése 1933–35-ről, 2: 545–601. (In Hungarian with German summary).

Sabol, M. (Ed.). (2004). Early Villanyian Site of Hajnáčka (Southern Slovakia): Paleontological Research 1996–2000.—Gemer-Malohont Museum, Rimavská Sobota, 143 pp.

Schafarzik, F. (1914). Böckh János emlékezete [In memoriam János Böckh].—Az MTA Elhunyt Tagjai fölött Tartott Emlékbeszédek, 16(12): 389–426. (In Hungarian)

Schafarzik, F. (1927). Ein alter Fund von Elephas primigenius Bl. aus Medgyes.—Földtani Közlöny, 57(1–9): 53.

Schafarzik, F., & Vendl, A. (1929). Geológiai kirándulások Budapest környékén [Geological excursions in the neighborhood of Budapest]—Magyar Királyi Földtani Intézet, Budapest, 341 pp. (In Hungarian)

Scholz, G. (1970). The Tortonian coral fauna of Fekete Hill at Visegrád.—Földtani Közlöny, 100(2): 102–106. (In Hungarian with English abstract)

Scholz, G. (1972). An Anisian Wetterstein limestone reef in North Hungary.—Acta Min. Petr., 20(2): 337–362.

Scholz, G. (1979). Die Ammoniten des Vracon (Oberalb, Dispar-Zone) des Bakony-Gebirges (Westungarn) und eine revision der Wichtigstein Vracon-Arten der West-Mediterranen Faunenprovinz. Teil 2.—Palaeontographica Abt. A, 165(3–6): 81–136.

Schréter, Z. (1917). Mediterranes Metaxytherium-Skelett von Márczfalva—Földtani Közlöny, 47: 176–177.

Schréter, Z. (1936). Lyttonia aus dem Bükk-Gebirge.—Földtani Közlöny, 66(3–4): 113–121.

Schréter, Z. (1948). Trilobites aus dem Bükk-Gebirge.—Földtani Közlöny, 78: 33–39.

Schréter, Z. (1963). Die Brachiopoden aus dem oberen Perm des Bükk-Gebirges in Nordungarn.—Geol. Hung. Ser. Pal., 28: 79–182.

Schréter, Z. (1974). Die Nautiloiden aus dem oberen Perm des Bükkgebirges. In M. Sidó, B. Zalányi, & Z. Schréter (Eds.), Neue Palaontologische Ergebnisse aus dem Oberpaläozoikum des Bükkgebirges (pp. 253–311).—Akadémiai Kiadó, Budapest.

Seneš, J. (1958). Pectunculus Sande und Egerer Faunentypus im Tertiär bei Kovacov im Karpatenbecken.—Geol. Práce Monogr. Ser., 1: 1–232.

Senowbari-Darian, B., & Velledits, F. (2007). Aggtecella, a new genus of Corallines (Rhodophyta) from the Anisian of the Aggtelek-Rudabánya Mountains, NE Hungary.—Facies, 53(3): 401–407.

Siblík, M. (1967). The brachiopods of the Norian locality Drnava (Southern Slovakia).—Geol. Práce, Zprávy, 43: 81–97. (In Czech with English abstract)

Siblík, M. (1986). Carnian Rhynchonellid Brachiopods from the Slovak karst area.—Západné Karpaty, Séria Paleontológia, 11: 7–33.

Sieverts-Doreck, H. (1961). Crinoiden aus dem Neokom des Bakony Gebirges, Ungarn.—A Magyar Állami Földtani Intézet Évkönyve, 49(3): 917–919.

Simony, F. (1850). Bericht über die Arbeiten der Section V.—Jahrbuch der k. k. Geologischen Reichsanstalt, 1(4): 654–657.

Smith, T., Codrea, V. A., Sasaran, E., Van Itterbeeck, J., Bultnyck, P., Csiki, Z., Dica, P., Farcas, C., Folie, A., Garcia, G., & Godefroit, P. (2002). A new exceptional vertebrate site from the Late Cretaceous of the Haţeg Basin (Romania).—Studia Univ. Babeş-Bolyai, Geologia, Special Issue, 1: 321–330.

Solt, P. (1987). Legányi Ferenc nyomában Mátraszőlősön, a Procarcharodonok lelőhelyén [Following in Ferenc Legányi's track to the locality of Procarcharodons at

Mátraszőlős].—Fol. Hist.-Nat. Mus. Matr., 12: 15–18. (In Hungarian)

Solt, P. (1988). Odontaspis (Synodontaspis) divergens n. sp. from the Oligocene of Csillaghegy.—A Magyar Állami Földtani Intézet Évi Jelentése az 1986. évről: 524–533.

Solt, P. (1992). Fish fossils of the shark-tooth-bearing bed at Kazár.—A Magyar Állami Földtani Intézet Évi Jelentése 1990-ről: 495–500. (In Hungarian with English abstract)

Soós, I., & Jámbor, Á. (1960). Oberkarbonische Pflanzenreste aus den Helvetschottern des Mecsekgebirges (Südungarn).—Földtani Közlöny, 90(4): 456–458. (In Hungarian with German abstract)

Sovis, W., & Schmid, B. (Eds.). (1998). Das Karpat des Korneuburger Beckens, Teil 1.—Beiträge zur Paläontologie, 23: 413 pp.

Sovis, W., & Schmid, B. (Eds.). (2002). Das Karpat des Korneuburger Beckens, Teil 2.—Beiträge zur Paläontologie, 27: 467 pp.

Strausz, L. (1966a). Die Eozängastropoden von Dudar in Ungarn.—Geol. Hung. Ser. Pal., 33: 199 pp.

Strausz, L. (1966b). Die miozän-mediterran Gastropoden Ungarns.—Akadémiai Kiadó, Budapest, 694 pp.

Strausz, L. (1974). Die Eozänmollusken von Neszmély.—Geol. Hung. Ser. Pal., 38: 158 pp.

Strausz, L., & Szalai, T. (1943). A várpalotai felső-mediterrán kagylók [Upper Mediterranean bivalves from Várpalota].—Beszámoló a Magyar Királyi Földtani Intézet Vitaüléseinek Munkálatairól, 5(1): 112–156. (In Hungarian)

Şuraru, N., & Codrea, V. A. (1988). Un dinte premolar (P4 dext) de sirenid (Mammalia, Sirenia) in orizontul calcarului de Cluj, de la Cluj-Napoca.—Crisia, 18: 689–695.

Švagrovský, J. (1978). Faziostratotypus: Devínska Nová Ves-Sandberg bei Bratislava. In E. Brestenska (Ed.), M_4 Badenien. Chronostratigraphie und Neostratotypen: Miozän der zentralen Paratethys, 6 (pp. 188–194).—Vyd. Slov. Akad. vied, Bratislava.

Švagrovský, J. (1981). Lithofazielle Entwicklung und Molluskenfauna des oberen Badeniens (Miozän M4d) in dem Gebiet Bratislava—Devínská Nová Ves, Západné Karpaty, Séria Paleontológia, 7: 5–204.

Szabó, I. (1972). Triász [Triassic]. In M. Deák (Ed.), Magyarázó Magyarország 200 000-es földtani térképsorozatához, L-33-XII, Veszprém [Explanatory notes to the geological map series of Hungary of 1:200 000 scale, Sheet L-33-XII, Veszprém] (pp. 35–72).—Magyar Állami Földtani Intézet, Budapest.

Szabó, J. (1979). Lower and Middle Jurassic Gastropods from the Bakony Mountains (Hungary), part I: Euomphalidae (Archaeogastropoda).—Annales Hist.-Nat. Mus. Nat. Hung., 71: 15–31.

Szabó, J. (1980). Lower and Middle Jurassic Gastropods from the Bakony Mountains (Hungary), part II: Pleurotomariacea and Fissurellacea (Archaeogastropoda).—Annales Hist.-Nat. Mus. Nat. Hung., 72: 49–71.

Szabó, J. (1981). Lower and Middle Jurassic Gastropods from the Bakony Mountains (Hungary), part III: Patellacea and Trochacea (Archaeogastropoda).—Annales Hist.-Nat. Mus. Nat. Hung., 73: 55–67.

Szabó, J. (1982). Lower and Middle Jurassic Gastropods from the Bakony Mountains (Hungary), part IV: Neritacea, Craspedostomatacea, Amberleyacea (Archaeogastropoda).—Annales Hist.-Nat. Mus. Nat. Hung., 74: 17–33

Szabó, J. (1983). Lower and Middle Jurassic Gastropods from the Bakony Mountains (Hungary), part V: Supplement to Archaeogastropoda; Caenogastropoda.—Annales Hist.-Nat. Mus. Nat. Hung., 75: 27–46.

Szabó, J. (2011). A revision of the nomenclature of well-known gastropod fauna from the Upper Triassic Dachstein Limestone of Budapest (Buda Hills) and thoughts on fate of the type collection.—Földtani Közlöny, 141(3): 217–232.

Szalai, T. (1934). Die fossilen Schildkröten Ungarns.—Folia Zool. Hydrobiol., 6(2): 97–142.

Szász, L. (1981). Contribution à l'étude des ammonites Coniaciens de la Roumanie.—Institut de Géologie et de Geophysique Mémoires, 30: 91–114.

Szente, I. (1995). Bivalvia from the Bathonian (Middle Jurassic) of the Mecsek Mts, Hungary.—Annales Univ. Sci. Budapest Sect. Geol., 30: 93–109, 221–224.

Szente, I. (1996). Bivalve assamblages from the Austrian and Hungarian Hierlatzkalk (Lower Jurassic): A Comparison. In E. Dudich & H. Lobitzer (Eds.), Advances in Austrian-Hungarian Joint Geological Research (pp. 137–145).—Geological Institute of Hungary, Budapest.

Szente, I. (1997). Bivalve assemblages from the Middle Triassic Muschelkalk of the Mecsek Mts, South Hungary: An overview.—Acta Geologica Hungarica, 40(4): 411–424.

Szente, I. (2003). Late Jurassic and Early Cretaceous bivalve assemblages from Transdanubia (Hungary).—Földtani Közlöny, 133(4): 477–499.

Szente, I., & Vörös, A. (1992). A Pliensbachian (Early Jurassic) bivalve faunula from the Harsány-hegy: First record of the Domerian Substage from the Villány Hills (southern Hungary).—Fragmenta Mineralogica et Palaeontologica, 15: 95–106.

Szente, I., & Vörös, A. (2003). The Pelsonian bivalve fauna of the Balaton Highland.—Geol. Hung. Ser. Pal., 55: 123–138.

Szentesi, Z. (2008). *Bufo aff. viridis* (Anura: Bufonidae) from the Middle Miocene diatomite of Szurdokpüspöki, northeastern Hungary.—Central European Geology, 51(4): 315–324

Szentesi, Z,., & Venczel, M. (2010). An advanced anuran from the Late Cretaceous (Santonian) of Hungary.—Neues Jb. Geol. Paläont., Abh., 256(3): 291–302.

Szentpétery, I., & Less, Gy. (Eds.). (2006). Az Aggtelek-Rudabányai-hegység földtana: Magyarország tájegységi térképsorozata [Geology of the Aggtelek-Rudabánya Mountains: Regional map series of Hungary].—Magyar Állami Földtani Intézet, Budapest, 92 pp. (In Hungarian)

Sziráki, G., & Dulai, A. (2002). Sarmatian (Late Miocene) arthropods from Tállya and neighbouring localities (Tokaj Mts, Hungary): Preliminary report.—Annales Hist.-Nat. Mu. Nat. Hung., 94: 31–44.

Szives, O. (1999). Ammonite biostratigraphy of the Tata Limestone Formation (Aptian–Lower Albian), Hungary.—Acta Geologica Hungarica, 42(4): 401–411.

Szives, O. (Ed.). (2007). Aptian-Campanian ammonites of Hungary.—Geol. Hung. Ser. Pal., 57: 187 pp.

Szontagh, T. (1904). Die Geschichte der Ausgrabung des Urwals von Borbolya.—Földtani Közlöny, 34: 279–284.

Szörényi, E. (1933). Neue Tertiäre Sepiinae aus Ungarn nebst Bemerkungen zu zeitlichen Auftreten und zu Entwicklung der Gattung Sepia.—Földtani Közlöny, 63(7–12): 183–189.

Szörényi, E. (1955). Échinides crétacés de la Bakony.—Geol. Hung. Ser. Pal., 26: 286 pp.

Szörényi, E. (1965). Échinides du Crétacé inférieur de la Hongrie.—Geol. Hung. Ser. Pal., 32: 295–370.

Szörényi, E. (1966). Laticlypus giganteus n. gen., n. sp. (Echinoidea) des asisses Jurassiques de la Montagne Bakony.—Acta Geologica Hungarica, 10: 445–452.

Szörényi, E. (1973). Magyarországi eocén echinoideák [Eocene echinoderms from Hungary].—Akadémiai Kiadó, Budapest, 102 pp. (In Hungarian)

Szőts, E. (1953). Mollusques éocènes de la Hongrie, 1: Les mollusques éocènes des environs de Gánt.—Geol. Hung. Ser. Pal., 22: 241 pp.

Szőts, E. (1956). L'Éocène (Paléogène) de la Hongrie.—Geol. Hung. Ser. Geol., 9: 318 pp.

Sztrákos, K. (1979). Le stratigraphie, paléogéologie, paléogéographie et les Foraminifères de l'oligocène du nord-est de la Hongrie.—Cahiers de Micropaléontologie, 3: 1–95, pls. 1–35.

Sztrákos, K. (1982). Les Foraminfères de la Marne de Buda et la limite éocène–oligocène en Hongrie.—Cahiers de Micropaléontologie, 4: 1–48, pls. 1–39.

Szuromi-Korecz, A., Sütőné Szentai, M., & Magyar, I. (2004). Biostratigraphic revision of the Hód-I well: Hungary's deepest borehole failed to reach the base of the Upper Miocene Pannonian stage.—Geologica Carpathica, 55(6): 475–485.

Szyndlar, Z. (2005). Snake fauna from the Late Miocene of Rudabánya.—Palaentographia Italica, 90: 31–52.

Tasnádi-Kubacska, A. (1962). Paläopathologie (Pathologie der vorzeitlichen Tiere).—Gustav Fischer, Jena, 269 pp.

Tasnádi Kubacska, A. (1967). Dinoszaurusz lábnyomok hazánkban [Dinosaur footprints in Hungary].—Élet és Tudomány, 24: 1118–1121. (In Hungarian)

Tasnádi Kubacska, A. (1969). Hantken Miksa [Miksa Hantken]. In J. Fülöp & A. Tasnádi Kubacska (Ed.), 100 éves a Magyar Állami Földtani Intézet (pp. 38–44).—Magyar Állami Földtani Intézet, Budapest. (In Hungarian)

Tasnádi Kubacska, A. (1977). Expedíció az időben [Expedition in time].—Gondolat Kiadó, Budapest, 125 pp. (In Hungarian)

Tausch, L. (1886). Über die Fauna der nicht-marinen ablagerungen der oberen Kreide des Csingerthales bei Ajka im Bakony: (Veszprimer comitat, Ungarn) und über einige conchylien der Gosaumergel von Aigen bei Salzburg.—Abhandlungen der k.k. Geologischen Reichsanstalt, 12: 1–32.

Telegdi-Roth, K. (1914). Eine oberoligozäne Fauna aus Ungarn.—Geol. Hung. Ser. Geol., 1(1): 1–77.

Terzea, E. (1996). Biochronology of the Pleistocene deposits at Betfia (Bihor, Romania).—Acta Zoologica Cracoviensia, 39: 531–540.

Till, A. (1910–1911). Die Ammonitenfauna des Kelloway von Villány (Ungarn).—Beiträge zur Paläontologie und Geologie Österreichs-Ungarns und des Orients, 23 (1910): 175–199, 251–272; 24 (1911): 1–49.

Tomas, R., & Pálfy, J. (2007). Revision of Early Jurassic ammonoid types from the Perşani Mts. (East Carpathians, Romania).—Neues Jb. Geol. Palänt. Abh., 243(2): 231–254.

Tomor-Thirring, J. (1934). A Bakony dudar oszlopi "Sűrű" hegycsoportjának földtani és őslénytani viszonyai [Geology and paleontology of the "Sűrű" Hills at Dudar-Oszlop in the Bakony Mountains].—A Földtani Szemle Melléklete, 48 pp. (In Hungarian)

Tóth, E. (2004). Sarmatian ostracods from Budapest.—Hantkeniana, 4: 161–170.

Tozer, E. T. (1984). The Trias and Its Ammonoids: The Evolution of a Time Scale. Geological Survey of Canada Miscellaneous Report, 35.—Geological Survey of Canada, Ottawa, 171 pp.

Tulogdy, J. (1944). Szirénafog a Bácsi-torok eocén felső durvamészkövéből [A Sirenian tooth from the Eocene upper coarse limestone of Bácsi-torok].—Múzeumi Füzetek Kolozsvár, 2: 56–59. (In Hungarian)

Tuzson, J. (1901). A Tarnóczi kövült fa (Pinus tarnócziensis n. sp) [The petrified wood from Tarnócz (Pinus tarnócziensis n. sp)].—Természetrajzi Füzetek, 24: 273–316. (In Hungarian)

Uhrin, A. (2011). A Pannon-tó só- és vízháztartásának számítása [Salt and water budget calculations for the Late Miocene Lake Pannon].—Földtani Közlöny, 141(4): 383–392. (In Hungarian)

Vadász, E. (1943). Emlékezzünk Koch Antalra [Let's remember Antal Koch]!—Földtani Közlöny, 73(1–3): 1–9. (In Hungarian).

Vadász, M. E. (1908). Über eine oberliassische Lytocerasart mit aufgelöster Wohnkammer.—Földtani Közlöny, 38: 131–136.

Vadász, M. E. (1909). Die paläontologischen und geologischen Verhältnisse der älteren Schollen am linken Donauufer.—Mitt. Jb. Kön. Ung. Geol. Anst., 18(2): 116–193.

Vadász, M. E. (1915). Die mediterranen Echinodermen Ungarns.—Geol. Hung. Ser. Geol., 1(2): 79–253.

Vaňová, M. (1987). Paleontological characteristics of some species from the Jeruzalemský Vrch mountain near Dobšiná.—Západné Karpaty, Séria Paleontológia, 12: 47–68.

Vašiček, Z., Michalik, J., & Reháková, D. (1994). Early Cretaceous stratigraphy, palaeogeography and life in Western Carpathians.—Beringeria, 10: 5–170.

Végh, S. (1964). A Déli Bakony raeti képződményeinek földtana.—Geol. Hung. Ser. Geol., 14: 109.

Végh-Neubrandt, E. (1982). Triassische Megalodontaceae: Entwicklung, Stratigraphie und Paläontologie.—Akadémiai Kiadó, Budapest, 526 pp.

Velledits, F., Péró, Cs., Blau, J., Senowbari-Daryan, B., Kovács, S., Piros, O., Pocsai, T., Szügyi-Simon, H., Dumitrică, P., & Pálfy, J. (2011). The oldest Triassic platform margin reef from the Alpine–Carpathian region (Aggtelek, NE Hungary). Platform evolution, reefal biota and biostratigraphic framework.—Rivista Italiana di Paleontologia e Stratigrafia, 117(2): 221–268.

Venczel, M. (1997). Amphibians and reptiles from the Lower Pleistocene of Osztramos (Hungary).—Nymphaea, 23–25: 77–88.

Venczel, M. (1998a). Late Miocene snakes (Reptilia, Serpentes) from Polgárdi (Hungary): A second contribution.—Acta Zool. Cracov., 41: 1–22.

Venczel, M. (1998b). Fossil amphibians and reptiles from Villány 8 (Hungary).—Nymphaea, 26: 147–162.

Venczel, M. (2000a). A kipusztult Palaeobatrachidae farkatlan kétéltűcsalád neogén és negyedkori nemzetségei a Kárpát-medencében [Neogene and Quaternary genera of the extinct tailless amphibian family Palaeobatrachidae in the Carpathian Basin].—Erdélyi Múzeumi Füzetek (Új folyam), Kolozsvár, 9: 209–213. (In Hungarian)

Venczel, M. (2000b). Quaternary snakes from Bihor (Romania).—Muzeului Ţării Crişurilor, Oradea, Romania, 144 pp.

Venczel, M. (2001). Anurans and squamates from the Lower Pliocene (MN 14) Osztramos 1 locality (Northern Hungary).—Fragmenta Palaeontologica Hungarica, 19: 79–90.

Venczel, M. (2004). Middle Miocene anurans from the Carpathian Basin.—Palaeontographica Abt. A, 271(5–6): 151–174.

Venczel, M., & Gardner, J. D. (2005). The geologically youngest albanerpetontid Amphibian, from the Lower Pliocene of Hungary.—Palaeontology, 48(6): 1273–1300.

Venkatachala, B. S., & Góczán, F. (1964). The spore-pollen flora of the Hungarian "Kössen facies."—Acta Geologica Hungarica, 8(1–4): 203–228.

Vértes, L. (1964). Az őskőkor és az átmeneti kor emlékei Magyarországon [Remains of the Paleolithic in Hungary].—Akadémiai Kiadó, Budapest, 385 pp. (In Hungarian)

Vinassa de Regny, P. (1911). Trias-Spongien aus dem Bakony.—Resultate Wiss. Erforsch. Balatonsees, Paläont. Anhang, 1(2): 22 pp.

Vincze, P. (1987). Hungarian implications of the oeuvre of Paolo Emilio Vinassa de Regny. In J. Hála (Ed.), Rocks, Fossils and History: Italian-Hungarian Relations in the Field of Geology. Annals of the History of Hungarian Geology Special Issue, 1 (pp. 269–274).—Hungarian Geological Society, Budapest.

Vogl, V. (1908). Über eozäne Nautiliden.—Földtani Közlöny, 38(7–10): 535–549.

Vogl, V. (1913). Die Paläodyas von Mrzla-Vodica in Kroatien.—Mitt. Jb. Kön. Ung. Geol. Anst., 21(5): 156–168.

Vörös, A. (1971). Lower and Middle Jurassic bivalves of the Villány Mountains.—Annales Univ. Sci. Budapest, Sec. Geol., 14: 167–208.

Vörös, A. (1983). Some new genera of Brachiopoda from the Mediterranean Jurassic.—Annales Historico-Naturales Musei Nationalis Hungarici, 75: 5–25.

Vörös, A. (1987). Preliminary results from the Aszófő section (Middle Triassic, Balaton area, Hungary): A proposal for a new Anisian ammonoid subzonal scheme.—Fragm. Min. et Pal., 13: 53–64.

Vörös, A. (1991). Hierlatzkalk: A peculiar Austro-Hungarian Jurassic facies. In H. Lobitzer & G. Császár (Eds.), Jubiläumsschrift 20 Jahre Geologische Zusammenarbeit Österreich-Ungarn (Vol. 1, pp. 145–154).—Geologische Bundesanstalt, Vienna.

Vörös, A. (1997). Magyarország júra brachiopodái. Faunafejlődés és paleobiogeográfia a Tethys nyugati részén [Jurassic brachiopods of Hungary. Faunal changes and paleobiogeography in the western Tethys].—Studia Naturalia, 11: 1–110. (In Hungarian with English abstract)

Vörös, A. (1998). A Balaton-felvidék triász ammonoideái és biosztratigráfiája [Triassic ammonoids and biostratigraphy of the Balaton Highland].—Studia Naturalia, 12: 1–104. (In Hungarian with English abstract)

Vörös, A. (2000). The Triassic of the Alps and Carpathians and its interregional correlation. In H. Yin, J. M. Dickins, G. R. Shi, & J. Tong (Eds.), Permian-Triassic Evolution of Tethys and Western Circum-Pacific (pp. 173–196).—Elsevier, New York and Amsterdam.

Vörös, A. (2001). Middle Triassic (Anisian) nautilid cephalopods from Aszófő (Balaton Highland, Hungary).—Fragmenta Palaeontologica Hungarica, 19: 1–14.

Vörös, A. (2003). The Pelsonian Ammonoid Fauna of the Balaton Highland.—Geol. Hung. Ser. Pal., 55: 71–121.

Vörös, A. (2009). The Pliensbachian brachiopods of the Bakony Mountains (Hungary).—Geol. Hung. Ser. Pal., 58: 1–300.

Vörös, A. (2010). Late Anisian Ammonoidea from Szár-hegy (Rudabánya Mts.)—a Dinaric-type fauna from North Hungary.—Fragmenta Palaeontologica Hungarica, 28: 1–20.

Vörös, A. (2011). Episodic sedimentation on a peri-Tethyan ridge through the Middle–Late Jurassic transition (Villány Mountains, southern Hungary).—Facies, 58(3): 415–443.

Vörös, A., Budai, T., Haas, J., Kovács, S., Kozur, H., & Pálfy, J. (2003). GSSP (Global Boundary Stratotype Section and Point) proposal for the base of Ladinian (Triassic).—Albertiana, 28: 35–47.

Vörös, A., & Dulai, A. (2007). Jurassic brachiopods of the Transdanubian Range (Hungary): Stratigraphical distribution and diversity changes.—Fragmenta Palaeontologica Hungarica, 24–25: 51–68.

Vörös, A., Szabó, I., Kovács, S., Dosztály, L., & Budai, T. (1996). The Felsőörs section: A possible stratotype for the base of the Ladinian Stage.—Albertiana, 17: 25–40.

Vörös, I. (1975). The Mammoth skeleton of Mátraderecske.—Fol. Hist.-Nat. Matr., 3: 151–157.

Vörös, I. (1980). Fossile Elephantidae of Hungary, I: The finds of Elephantidae of North Hungary.—Fol. Hist.-Nat. Mus. Matr., 6: 13–49.

Vörös, I. (1989). Prodeinotherium petenyii sp. n. from the Lower Miocene at Putnok (North Hungary).—Fragmenta Mineralogica and Palaeontologica, 14: 101–110.

Vörös, I. (2005). In memoriam Kretzoi Miklós (1907–2005).—Archeometriai Műhely, 2: 57–58. (In Hungarian)

Vremir, M. (1995). On the upper Eocene (Bartonian) sea turtle paleofauna (Cheloniidae: Testudines) from Transylvanian depression (Romania).—Studia Univ. Babeş-Bolyai, Geologia, 40(2): 55–62.

Vremir, M. (2004). Fossil turtle found in Romania: Overview.—A Magyar Állami Földtani Intézet Évi Jelentése 2002-ről: 143–152.

Vremir, M. (2010). New faunal elements from the Late Cretaceous (Maastrichitan) continental deposits of Sebeş area (Transylvania).—Terra Sebus, Acta Musei Sabesienis, 2: 635–683.

Vremir, M., & Codrea, V. (1996). Palaeochelys sp. (Testudines; Emydidae) from the Paleocene of the Transylvanian depression: Outcrops from Rona and Jibou (Salaj County, Romania).—Studii şi Cercetari, 2: 75–81.

Vremir, M., & Codrea, V. A. (2002). The first Late Cretaceous (Maastrichtian) dinosaur footprints from Transylvania (Romania).—Studia Univ. Babes-Bolyai, Geologia, 47(2): 93–104.

Wagner, H. (1938). Die dibranchiaten Cephalopoden der Mitteloligozanen (Rupelien) Tonschichten von Kiscell und neue Sepiinae aus dem ungarischen Eozän: Annales Historico-Naturales Musei Nationalis Hungarici, 31: 192–199.

Wéber, B. (1964). Neue oberkarbonische Schotter mit Pflanzenresten aus den Helvetschichten des Westlichen Mecsek.—Földtani Közlöny, 94(3): 379–381. (In Hungarian with German abstract)

Weiler, W. (1933). Zwei oligozäne Fischfaunen aus dem Königreich Ungarn.—Geol. Hung. Ser. Pal., 11: 54 pp.

Weiler, W. (1938). Neue Untersuchungen an mitteloligozänen Fischen Ungarns.—Geol. Hung. Ser. Pal., 15: 30 pp.

Weishampel, D. B., Grigorescu, D., & Norman, D. B. (1991). The Dinosaurs of Transylvania.—National Geographic Research and Exploration, 7(2): 196–215.

Weishampel, D. B., & Jianu, C.-M. (1996). New Theropod Dinosaur Material from the Haţeg Basin (Late Cretaceous, Western Romania).—N. Jb. Geol. Palaont., Abh., 200(3): 387–404.

Weishampel, D. B., Jianu, C.-M., Csiki, Z., & Norman, D. B. (2003). Osteology and phylogeny of *Zalmoxes* (n. g.), an unusual Euornithopod dinosaur from the latest Cretaceous of Romania.—Journal of Systematic Palaeontology, 1(2): 65–123.

Weishampel, D. B., Norman, D. B., & Grigorescu, D. (1993). Telmatosaurus transsylvanicus from the Late Cretaceous of Romania: The most basal hadrosaurid dinosaur.—Palaeontology, 36(2): 361–385.

Werdelin, L. (2005). The Carnivora of Rudabánya (MN9, Late Miocene, Hungary).—Palaeontographia Italica, 90: 163–180.

Zachár, I. (2003). Harmadidőszaki myliobatoideák elkülönítése [Distinguishing Tertiary mytilobatoideans]. In Program, Előadáskivonatok, Kirándulásvezető, 6: Magyar Őslénytani Vándorgyűlés, 2003, május 8–10, Zirc. (p. 33). (In Hungarian)

Zágoršek, K. (1993). New Carboniferous Bryozoa from Nagyvisnyó (Bükk Mts., Hungary).—Földtani Közlöny, 123(4): 417–440.

Zágoršek, K. (1995). First find of Bryozoan zoaria in the Jurassic of Hungary.—Annales Univ. Sci. Budapest., Sect. Geol., 30: 177–180, 235–236.

Zágoršek, K., & Kázmér, M. (2001). Eocene Bryozoa from Hungary.—Cour. Forsch.-Inst. Senckenberg, 231: 1–159.

Zalányi, B. (1959). Ostracoden-Faunen aus der Apstufe des Nördlichen Bakony-Gebirges.—A Magyar Állami Földtani Intézet Évkönyve, 47(2): 359–565.

Zalányi, B. (1974). Die oberpermischen Ostracoden des Bükkgebirges. In M. Sidó, B. Zalányi, & Z. Schréter (Eds.), Neue Palaontologische Ergebnisse aus dem Oberpalaozoikum des Bükkgebirges (pp. 95–251).—Akadémiai Kiadó, Budapest.

Zapfe, H. (1960). Die Primatenfunde aus der miozänen Spaltenfüllung von Neudorf a. d. March (Devínská Nová Ves), Tschechoslowakei.—Schweizerische Paläontologische Abhandlungen, 78: 1–293.

Zapfe, H. (1979). Chalicotherium grande (Blainv.) aus der miozänen Spaltenfüllung von Neudorf an der March (Devínská Nová Ves), Tschechoslowakei.—Neue Denkschriften des Naturhistorischen Museums in Wien, 2, 1–282.

Zilch, A. (1934). Zur Fauna des Mittel-Miozäns von Kostej (Banat). Typus-Bestimmung und Tafeln zu O. Boettgers's Bearbeitungen.—Senckenbergiana, 16(3–4): 193–302.

INDEX

Items referring to figures are followed by an f. *Items discussed in captions are followed by a* c.

István Főzy is Senior Curator and Researcher in the Department of Geology and Paleontology at the Hungarian Natural History Museum. Previously he was in charge of museum exhibitions and public education. He received his PhD from Eötvös Loránd University, Budapest. His main scientific interest is Jurassic and Cretaceous marine invertebrates, and his research focuses on the systematics of Late Jurassic and Early Cretaceous ammonites. Főzy has played a major role in numerous successful museum exhibits. His two previously published books concern the "Dinosaur Baron," Ferenc Nopcsa, and ancient reptiles of the Carpathian Basin.

István Szente, geologist and paleontologist, is Curator of the Paleontology Collections of the Museum of Natural History at Eötvös Loránd University, Budapest. After receiving his PhD from Eötvös Loránd University, he worked as a department librarian at the university's Paleontological Institute. His research interests are Mesozoic gastropods and bivalves. University courses taught by Szente include general paleontology, paleobiology of bivalves, fossil collecting and preparation techniques, and methods in paleontology.